T0320985

Fundamentals of 5G Communications

Connectivity for Enhanced Mobile Broadband and Beyond

Wanshi Chen, Ph.D.

Peter Gaal, Ph.D.

Juan Montojo, Ph.D.

Haris Zisimopoulos, M.Sc.

New York Chicago San Francisco
Athens London Madrid
Mexico City Milan New Delhi
Singapore Sydney Toronto

Library of Congress Control Number: 2021938432

Fundamentals of 5G Communications:
Connectivity for Enhanced Mobile Broadband and Beyond

3 4 5 6 7 8 9 CCD 26 25 24 23 22 21

ISBN 978-1-260-45999-9
MHID 1-260-45999-3

This book is printed on acid-free paper.

Sponsoring Editor
Lara Zoble

Editing Supervisor
Stephen M. Smith

Production Supervisor
Lynn M. Messina

Acquisitions Coordinator
Elizabeth M. Houde

Project Manager
Rishabh Gupta, MPS Limited

Copy Editor
Yashoda Rawat, MPS Limited

Proofreader
A. Nayyer Shamsi, MPS Limited

Indexer
Melissa Stearns Hyde

Art Director, Cover
Jeff Weeks

Composition
MPS Limited

Contents

Foreword

Several years ago, around the time that 3GPP was starting its work item to develop a 5G standard and almost 2 years before the first claimed deployment of new radio (NR), I started to wear the 3GPP 5G[1] lapel pin. To my amazement, many people outside of the cellular industry — I'm talking about artists and teachers — recognized 5G. This is a testament to the ubiquity and importance of cellular technology in society, and the visibility of previous generations of technology, epitomized by "4G LTE."

In 2010, when 4G was launched, the goal was better connectivity to the smartphone. 5G is so much more. As a result, I have often termed 5G as both a technology and a movement. In this book, the authors are addressing the technology; however, we cannot ignore the movement, as it is the movement which is driving technology. Central to the movement is the emerging ubiquity of connectivity. Today, almost everyone has a smartphone which has replaced the wired telephone, the TV, the camera, and the watch. The cellular subscriber's personal platform has evolved to include not only the smartphone, but also their smart watch, ear buds and microphone, augmented or virtual reality glasses, and pointer device. The connectivity is expanding to many other objects around us. Transforma Insights has estimated that by 2030, there will be over 24 billion internet of things (IoT) devices—about three connected devices for every human in the world, and many times this number when one considers just the developed world. This connectedness is leading to many businesses examining 5G to determine the benefits that it can provide, such as greater operational efficiencies. As a result, this has led to a shift in the vision of cellular communications to beyond the cell phone, to a world where almost everything around us is connected.

The Gs in cellular technologies seemingly magically appear about every 10 years. The first G, what we call analog cellular, appeared in the early 1980s. The most successful of these analog systems was advanced mobile phone system (AMPS) with the first service launched in the United States in October 1983. It was essentially an FM mobile radio, with a connection to a telephone central office, a bit of control to send dialed digits and to manage the call, and the ability to manage a radio channel and to perform handoff. It was primitive by today's standards, but it worked. Mobile telephone systems actually date from 1949 with the commercial introduction of mobile telephone system (MTS). However, its capacity was so low, and its cost so high, even for the enhanced improved mobile telephone system (IMTS), that the total number of customers in the

[1] 3GPP approved the logo in February 2017.

entire United States was limited to 40,000. AMPS, however, was a major success. By 1990—just 7 years after the commercialization of AMPS—there were 5 million customers in the United States. AMPS was deployed in several countries in addition to the United States. A derivative, with essentially different channel spacing, total access communications system (TACS) was deployed in the United Kingdom and many other countries, in both Europe and the rest of the world. JTACS (Japan TACS) was a variant deployed in Japan. In Europe, several other analog systems were developed including Nordic mobile telephone (NMT) operating at 450 and 900 MHz (launched in 1981), Radiocom2000 (400 MHz, 200 MHz, 160 MHz), C450 (having 800,000 subscribers) in Germany, TMA in Spain, and Radio Telefono Mobile Integrato (RTMI) in Italy.

With the rapid growth of AMPS, operators in the United States were demanding more capacity. The United States took two directions. The first standard that was developed, IS-54, called DAMPS (Digital AMPS), used time-division multiple access techniques to get three calls in the same 30 kHz channel bandwidth of AMPS. It became known as TDMA and was deployed by several major operators, which essentially coalesced to become what we now know as AT&T. Qualcomm, at the time a small startup company, claimed that it could obtain 10 times the capacity of AMPS by using a wideband code division multiple access system. At the time, it was called wideband, due to its 1.23 MHz bandwidth, as compared to the 30 kHz bandwidth analog and DAMPS (TDMA) currently in use. It was also standardized in the United States and became known as IS-95 or CDMA, and was deployed by several major operators, which today are known as Verizon and Sprint. It was also used in Korea by all the operators, in Japan, and in several other countries. The development of CDMA was a particularly important step, as it brought into the cellular industry advanced communications concepts and set the stage for the high-performance cellular systems that we have today.

While the United States was searching for additional capacity, Europe undertook a standardization program to create a pan-European system, which became known as global system for mobile communications (GSM). The GSM physical layer used TDMA techniques and was not particularly pioneering, but the overall system architecture and separation of functions was very logical and clean. Perhaps most important was that the development of GSM brought together many countries and created a single standard for Europe. An often-overlooked aspect was that the standardization was conducted in English and the standard was written in English. This avoided the complexities of translating between multiple languages, speeded-up the standardization process, and created a single definitive version of the standard, reducing ambiguities. Japan developed personal digital cellular (PDC) which was a derivative of North American TDMA.

During this era, there was considerable debate between proponents of cellular systems and proponents of short-range cordless systems. The Japanese-designed personal handyphone system (PHS) and the European DECT (Digital European Cordless Telephone/Digital European Cordless Telecommunications/Digital Enhanced Cordless Telecommunications) were the most deployed. However, by the end of the 1990s, it had become clear that cellular technology had become the communications mode chosen by the consumer.

In 1985, the International Telecommunications Union (ITU) started to study mobile telephony, 6 years before the first commercial 2G systems. Recommendation M.687, issued in 1990, put forth a vision for future public land mobile telecommunication systems (FPLMTS) which included many of the attributes of modern systems, including support for many services, international operation, and automated roaming. In hindsight, they got two things wrong. The first was that they viewed mobile operation as an extension of the public switched telephone network (PSTN) and integrated services digital network (ISDN)—a vision that was to hinder the development of good data communications through the early days of 3GPP. The second was their estimates of the required spectrum were from 110 to 160 MHz for voice services and 65 MHz for data services for a total of between 175 MHz and 225 MHz. This is an almost laughable number given the more than 1150 MHz and 700 MHz of sub-6-GHz licensed spectrum available today in China and the United States, respectively. In addition, the United States has now allocated more than 3 GHz of millimeter wave spectrum which is being used for 5G deployments; other countries have made or are making a large amount of millimeter wave spectrum available.

By 1992, when the ITU issued Recommendation M.816, "Framework for Services Supported on Future Public Land Mobile Telecommunication Systems (FPLMTS)," the system was being called third generation with service starting around 2000. In 1994, the ITU began using the term IMT-2000 instead of third generation (3G); however, the popular term has remained 3G. The vision was for one air interface (radio interface), though more than one could be acceptable, according to M.1035, "Framework for the Radio Interface(s) and Radio Sub-System Functionality for International Mobile Telecommunications—2000 (IMT-2000)": "IMT-2000 may need to use more than one radio interface in order to meet various operating environment or application needs." When issued in 2000, the IMT-2000 specification, M.1457, "Detailed Specifications of the Terrestrial Radio Interfaces of International Mobile Telecommunications—2000 (IMT-2000)," contained not a single common air interface, but five air interfaces, none of which had been developed by the ITU. The ITU had succeeded in setting the stage for 3G, but the air interfaces were developed by various other groups. An essentially similar stage-setting approach has been used for successive generations: IMT-Advanced (4G) and IMT-2020 (5G).

During the late 1990s, there were numerous public and behind-the-scene discussions on creating a common 3G air interface. While the goal of creating a single 3G air interface was not achieved, this era took a huge step forward in bringing together the world's cellular communications industry. The result was the creation of two partnership projects, 3GPP and 3GPP2, along with two CDMA-based systems: universal mobile telecommunications system (UMTS), often called WCDMA, and cdma2000. One common system was not attainable for both technical and business reasons. From a technical perspective, cdma2000 offered a smooth evolution path from IS-95 CDMA (called CDMAone) and significantly higher capacity, which was particularly a concern due to the more limited amount of spectrum that was available at the time. From a business perspective, beginning in the mid-1990s, GSM and cdma2000 had been in a very heated competition trying to secure operators for their technology, particularly in South America, Africa, and Asia. Over time, the larger Europe-based ecosystem of GSM secured more operators, but it took a few more years for that to fully play out. Three of the initial five air interfaces that were part of IMT-2000 came from 3GPP and 3GPP2. They were IMT-2000 CDMA direct spread (WCDMA), IMT-2000 CDMA multi-carrier (cdma2000), and IMT-2000 CDMA TDD. IMT-2000 CDMA TDD was a combination of a TDD air interface originally developed by 3GPP, and TD-SCDMA, developed by CWTS (China Wireless Telecommunications Standards), which was, eventually, incorporated into the 3GPP specifications. cdma2000 had both what was called a 1× mode which had a single 1.23 MHz bandwidth carrier, and what was called a 3× multi-carrier mode which had three carriers spaced by 1.25 MHz, hence the name IMT-2000 CDMA multi-carrier. This was the precursor for 3GPP's carrier aggregation, though the 3× mode was never commercialized. The other two IMT-2000 air interfaces were an evolution of TDMA and DECT.

Around this time, AT&T concluded that continuing the TDMA path was not practical and began deployment of GSM.

Thus, by the early 2000s the situation had clarified: 3G would be based upon the two major CDMA-based air interfaces, WCDMA and cdma2000, though there was deployment by China Mobile of TD-SCDMA. Standardization had consolidated into the two partnership projects, 3GPP and 3GPP2. For much of the 2000s 3GPP and 3GPP2 were rivals, each coming up with new technologies, which in some form were adopted by the other partnership project.

One of the weaknesses of the initial 3G standards was that they did not fully rethink handling of packet traffic to support the rapidly expanding internet. WCDMA could send on the Downlink about 2 Mbps using its 5 MHz carrier. Nevertheless, this was an enhancement over general packet radio service (GPRS) which supported up to 114 kbps, and its enhancement EDGE (enhanced data rates for GSM evolution) which supported up to 473.6 kbps. The CDMA community had focused on internet support from its early days and had incorporated significantly higher data rates in IS-95-B; the cdma2000 specification could support about 628 kbps on a 1.23 MHz bandwidth carrier.

A major step forward in providing better internet traffic came in late 2000, with 3GPP2's completion of the cdma2000 high-rate packet data (HRPD) specification. The common name for 3GPP2's HRPD was 1× EVolution Data Only (1×EV-DO), which was then changed to 1× Evolution Data Optimized. This air interface addressed Downlink transmissions through a single transmission stream, very short transmissions, high-rate feedback of the best transmission mode, the recognition that throughput is higher when the best link to a mobile is selected for transmission (proportional fair scheduling), the use of 8PSK and 16QAM modulation when the channel quality is good, and incremental redundancy. While maintaining the 1.23 MHz channel bandwidth of IS-95 and cdma2000 1×, it was able to attain peak transmission rates of 2.4 Mbps, though it used a separate channel. This was the source of considerable controversy at the time, as 1×EV-DO was not able to support circuit voice services and SMS—which were the primary services at the time—though later there were some trials of packet voice services. As a result, by the end of 2000, 3GPP2 started to develop 1× EVolution Data and Voice (1×EV-DV). While 1×EV-DV pioneered many new concepts, it was never commercially deployed, and high-rate operation on 3GPP2 systems was provided by 1×EV-DO.

3GPP soon started to develop its equivalent of 1×EV-DV which appeared as high-speed Downlink packet access (HSDPA) in Release 5 in 2002 with Downlink peak data rates of 14.4 Mbps. Two years later high-speed Uplink packet access (HSUPA) appeared in Release 6.

Due to the straightforwardness of the upgrade from IS-95 to cdma2000, the first commercial cdma2000 systems appeared in 2000. It wasn't until early 2004 that WCDMA Release 99 was performing well, resulting in many operators starting commercial service, though NTT Docomo in Japan had launched a simplified variant in 2001, called freedom of mobile multimedia access (FOMA). However, it was not until the launch of HSDPA at the end of 2005 with good internet packet performance that WCDMA showed its real potential. Further enhancements beyond Release 6 added even higher order modulation, dual-carrier, four carriers, simultaneous operation in multiple bands, and MIMO.

In the early 2000s after the completion of the first generation of IMT-2000, the ITU turned its attention to two phases for the development of mobile systems, as outlined in M.1645, "Framework and Overall Objectives of the Future Development of IMT-2000 and Systems Beyond IMT-2000," which was published in 2003. The first phase was the evolutionary enhancement of IMT-2000, which had been occurring through the various

3GPP releases and 3GPP2 specification revisions. The second was that "there may be a need for a new wireless access technology to be developed around the year 2010, capable of supporting high data rates with high mobility, which could be widely deployed around the year 2015 in some countries."[2]

The ITU's timeline would soon be eclipsed, with the first specifications of what might be called 4G emerging in 2007. A number of important factors contributed to this. First was the rapid increase in the users of cellular communications with 1.7 billion cellular subscribers in mid-2004 (ITU M.1645 had predicted that same number would be reached in 2010). A second was the rise of the data usage, spurred by several developments. At the end of the 1990s, the camera began to appear in phones and was in almost every high-end phone just a few years later. Users wanted access to their email on cell phones. Mobile location using global positioning system (GPS) was starting to appear. And there were frustrations with the time that it took to send and receive data, particularly photos. All this led to increasing demands for higher capacity and higher data rates. A third was the previously described frustration with the performance of WCDMA in the early 2000s. Orthogonal frequency division multiple access (OFDMA) as a wireless technology was causing a considerable amount of interest, due to its inherent resilience to multipath as compared to CDMA which tried to identify and process multiple paths. This became more difficult as the bandwidth increased due to the greater number of resolvable paths. Orthogonal frequency division multiplexing (OFDM) as a technology went back many years and had been used in the European digital audio broadcasting (DAB) and digital video broadcasting (DVB) systems, and the IEEE 802.11a and 802.11g standards for Wi-Fi. The push for OFDMA as a technology for cellular usage came from a small startup, Flarion Technologies, which developed the 1.25 MHz bandwidth flash OFDMA system in the early 2000s. But perhaps the most important incentive came from the development of WiMax using the IEEE 802.16 standard. The early versions of the 802.16 standard were designed for backhaul or point-to-multipoint data distribution, but the 802.16e amendment was meant for mobile operation.

This spurred 3GPP to have a future evolution workshop in November 2004 to study the evolution of 3GPP radio technologies. This led to the development of the 3GPP standard for long term evolution (LTE) based upon OFDMA which was completed as part of Release 8 in December 2008. The workshop also led to further enhancements of WCDMA, notably carrier aggregation.

Not to be outdone, 3GPP2 developed ultra-mobile broadband (UMB), a rather advanced OFDMA design, which was completed in mid-2007. In spite of its technical advantages, some key cellular operators which had been major backers of cdma2000, notably Verizon, decided that it was time to have a common worldwide cellular standard based upon LTE. As a result, UMB was never commercially deployed.

While 802.16e saw commercial deployment by a number of operators, particularly as the WiBro version in Korea, it took several years after the completion of the standard to adequately develop the ecosystem sufficiently for deployment. This delay, coupled with some design flaws which hindered performance, prevented 802.16e from becoming a success. The design flaws were mostly fixed in the later 802.16m amendment, but by then it was too late as operators had focused on LTE.

---

[2]ITU-R M.1645, "Framework and Overall Objectives of the Future Development of IMT-2000 and Systems Beyond IMT-2000," 2003, p. 3.

LTE turned out to be a great success. While there were a few small deployments beginning in December 2009, the massive 38-market launch in December 2010 by Verizon in the United States showed that LTE was commercially ready. Since then, almost every operator in the world has deployed LTE. The term "4G-LTE" has become known by the populace at large.

Yet the first release of LTE, Release 8, was not really "4G" in some countries as it preceded the IMT-Advanced specification of the ITU. In Japan, NTT Docomo called it Super 3G. The ITU IMT-Advanced designation came two releases later with Release 10. M.2012, "Detailed Specifications of the Terrestrial Radio Interfaces of International Mobile Telecommunications Advanced (IMT-Advanced)," was published at the beginning of 2012 and contained both LTE and 802.16m. By then, it was clear that there was going to be one cellular air-interface technology worldwide and that it was LTE. A number of people were concerned that the loss of competition between standards bodies might slow down innovation and lead to the stagnation of air interface design. The past ten years have shown these concerns to be unfounded. The desire of engineers to come up with good ideas, the pride that engineers take in creating the next G, and the rivalry between engineers and companies have kept innovation alive.

One of the trends of 4G, but really beginning with 3G, was the expansion of the technologies beyond just cell-to-device unicast connectivity. The first came with multimedia broadcast multicast service (MBMS) in Release 6 which provides 3G broadcast and multicast services for mobile operators. With LTE came other capabilities including proximity services (ProSe) communications, which enables direct communications without going through the network for devices that are close to each other. Release 13 specified mission-critical push to talk (MCPTT), which supports public safety and other users over unicast, MBMS, or ProSe transports. Release 14 introduced cellular vehicle-to-everything (C-V2X) which enables communications between automobiles and with road side units (RSUs). Of particular note has been all the activity on machine type communications (MTC) to support IoT, notably creating enhanced MTC (eMTC), a simplified version of LTE using a 1.4 MHz bandwidth, and creating a special air interface named narrow-band IoT (NB-IoT), using 200 kHz bandwidth carriers. These highlight just a few of the hundreds of work areas of 3GPP over the years.

This foreword has covered the first four generations of cellular communications comprising the past 40 years. It came from simple analog mobile radios to what is arguably the most sophisticated and complex man-made system on the planet with over two-thirds of the world population being subscribers and the number of devices exceeding the world's population.[3] People can communicate almost anywhere and to almost anyone in the world with a few keystrokes. Things, or IoT-capable devices, can likewise communicate to servers or to someone anywhere in the world.

These systems have been brought to you by a dedicated group of engineers with a vision to better communicate, who do research, design, standardization, product development, testing, and whatever else is needed to bring these systems to fruition.

I now turn the story of 5G over to the four authors of this book: Wanshi Chen, Peter Gaal, Juan Montojo, and Haris Zisimopoulos. They are eminently qualified. All have been involved with 3GPP for many years and together have close to 70 years of experience designing cellular communications systems.

Ed Tiedemann
Concord, Massachusetts

[3] GSM Association, "The Mobile Economy 2020."

Acknowledgments

This book would not have been possible without the motivation and help from many colleagues and our families.

As a former 3GPP TSG RAN1 (https://www.3gpp.org/specifications-groups/ran-plenary/ran1-radio-layer-1/home) Chair (from August 2017 to May 2021), Wanshi Chen is, in particular, grateful for the experiences of managing and driving one of the key 3GPP working groups—the physical layer group—for the standardization of 5G from its first release. Newly appointed to the post in August 2017, he had the challenging task to execute on the 5G standardization acceleration which had been agreed upon in March 2017 by delivering the early drop (also known as the non-standalone version of 5G, or architecture option 3) of the first 5G release, due in December 2017. The pressure was overwhelming—enormous workload, tight deadlines, hundreds of delegates and thousands of contributions per meeting, controversial discussions going until midnight almost daily, and difficult but necessary compromises offline and online. Even now when looking back, we are still amazed at the progress that 3GPP, as a group and as a community of experts, made in the past several years for 5G standardization. All the progress would not have been possible without the dedication and the constructiveness of the participants in 3GPP. Therefore, we are deeply grateful to all of them, our colleagues inside and outside our own company. In March 2021, Wanshi was elected 3GPP TSG RAN Plenary (https://www.3gpp.org/specifications-groups/ran-plenary/ran-plenary) Chair and is looking forward to continuing working with all 3GPP colleagues to further enhance 5G and to build a solid foundation for future 6G standardization.

When approached about writing a 5G book, we were a bit hesitant, primarily due to the workload we have every day—3GPP standardization continues to evolve without any gaps. It was very hard to make room for any additional time to write a book. We were, thus, very thankful for the support from Baaziz Achour, Lorenzo Casaccia, Durga Malladi, John Smee, and Ed Tiedemann, who encouraged us to embark on this project. We are also very thankful to Lara Zoble of McGraw Hill for her support and patience throughout the publication process.

We deeply appreciate the reviews and comments from our colleagues, who spared their precious time during their daily busy activities, particularly their 3GPP standardization–related work. This greatly helped improve the quality of the book. In particular, we would like to acknowledge the help from Albert Rico Alvarino, Sudhir Baghel, Luca Blessent, Yiqing Cao, Lena Chaponniere, Kausik Ray Chaudhuri, Aleks Damnjanovic, Adrian Escott, Stefano Faccin, Valentin Gheorghiu, Miguel Griot,

Georg Hampel, Chenxi Hao, Gavin Horn, Kianoush Hosseini, Tingfang Ji, Masato Kitazoe, Eddy Kwon, SooBum Lee, Jing Lei, Le Liu, Tao Luo, Alexandros Manolakos, Ozcan Ozturk, Shailesh Patil, Gabi Sarkis, Joseph Soriaga, Sebastian Speicher, Jing Sun, Fred Takeda, Xiaofeng Wang, Yongbin Wei, Huilin Xu, Wei Yang, Juan Zhang, Xiaoxia Zhang, Yu Zhang, and Yan Zhou. Special thanks to Ed Tiedemann for, in addition, writing the Foreword for this book and to Luis Lopes, who contributed material for the 5G RAN architecture.

We would also like to thank Qualcomm executives Steve Mollenkopf, Cristiano Amon, Jim Thompson, Alex Rogers, and Don Rosenberg for their support and encouragement in getting this work completed. Once the manuscript was ready for publication, the Qualcomm legal team, especially Melissa DeVita and Chris Smith, provided invaluable support to get us through final production.

Last but not least, this book would not have been possible without the support of our families. Many times, we had to sacrifice late evenings, early mornings, weekend hours, and vacation time to gradually write the book piece by piece. The understanding and the encouragement we received from our families were crucial in making the book happen.

Wanshi Chen, Ph.D.
Peter Gaal, Ph.D.
Juan Montojo, Ph.D.
Haris Zisimopoulos, M.Sc.

Introduction

It is not a straightforward decision to write a 5G book. All of us have been actively involved in standardization for wireless communications, particularly in 3GPP (3rd generation partnership project; https://www.3gpp.org/), for more than 10 years. Our involvement is full-scale and systematic, starting from the initial design often requiring extensive analysis and evaluations, to forming detailed proposals in 3GPP for a potential study item (a necessary phase for the eventual standardization) and subsequently for a potential work item leading to the final 3GPP specifications, and to maintaining the specifications and participating in the implementation efforts resulting in commercial deployments. It is relatively straightforward for us to introduce what has been standardized in 3GPP particularly for 5G, also known as new radio or NR and 5G Core (5GC). However, such description, at least in our views, only brings limited benefits for the readers. Therefore, this book is NOT intended to merely provide an introduction to 5G in terms of 3GPP specification details.

Given our firsthand and rich experiences in the standardization process, we would rather prefer to focus on *how* and *why* 5G is standardized, i.e., the stories behind the decisions. Standardization of a set of features in 3GPP is generally driven by commercial needs. Each feature is often associated with a certain set of use cases (e.g., smartphones, vehicles, industrial internet of things (IIoT), augmented reality, cloud gaming) and under a set of deployment scenarios (e.g., outdoor, indoor, macro cells, small cells). For each potential enabling technique, there are usually multiple possible design options, each with its pros and cons from performance and complexity perspectives. Extensive analysis and evaluations are necessary in order to converge to a particular design providing a best tradeoff for commercial deployments. This is particularly necessary since in 3GPP all decisions are consensus driven. In other words, any particular design requires consensus from the entire group (i.e., no sustained objections), which consists of hundreds of members participating in the standardization process. To achieve such consensus, it often takes a long time, sometimes even more than one year, for the discussion via various means, e.g., face-to-face meetings, emails, and to a much less extent, conference calls. It is typical that the final design may not be the same as any of the originally proposed design options by each individual proponent, but a solution that is a result of compromise based on two or more originally proposed design options. It is hard to imagine the difficulty in achieving consensus when in a same meeting room, there are hundreds of delegates with diverging views and sometimes diverging incentives involved in the decision process. This is possible largely because of a common goal shared by all 3GPP participants: being constructive and dedicated to in-time and quality delivery of 3GPP specifications to enable successful commercial deployments.

Although it is quite involved, we decided to write this book with the goal of not only describing the new 5G standards, but also telling the reasons and stories behind them. One difference of 5G compared to other wireless communication standards is that it offers an end-to-end possibly turnkey system that covers all the aspects of the network deployment, from physical layer to core network, from the radio network architecture and protocol to regulatory services support. We aim therefore to provide an end-to-end picture of how 5G works as a complete system, with focus on the architecture and physical layer related design and specifications.

Mobile communications have made a leap approximately every 10 years (as shown in Fig. 1), starting from the 1st generation (1G) in the years of the 1980s enabling mobile voice communications, followed by the 2nd generation (2G) in the years of the 1990s improving voice communication efficiency and reachability. The focus started to shift to mobile data communications in the 3rd generation (3G) in the years of the 2000s, when wireless internet became a reality. Mobile broadband took off in the 4th generation (4G), also known as the long-term evolution (LTE), achieving unprecedented wireless communications speeds and driving the proliferation of smartphones. 4G LTE also witnessed the emerging expansion of wireless communications into other fields. 5G is built on top of these previous generations, driven by continuous research and innovations.

5G is designed to be a common communication fabric connecting everything to everything from the very beginning, supporting diverse services, diverse spectrum, and diverse deployments, as illustrated in Fig. 2.

5G was conceived from the beginning to be a unified future-proof platform. Figure 3 illustrates the timeline for 5G NR in 3GPP. The pioneering event for 5G is the first 5G workshop in September 2015 [1]. Similar to IMT-2020 [2], three emerging high-level use cases were identified [1]:

- Enhanced Mobile Broadband (eMBB),
- Massive Machine Type Communications (mMTC), and
- Ultra-Reliable and Low Latency Communications (URLLC).

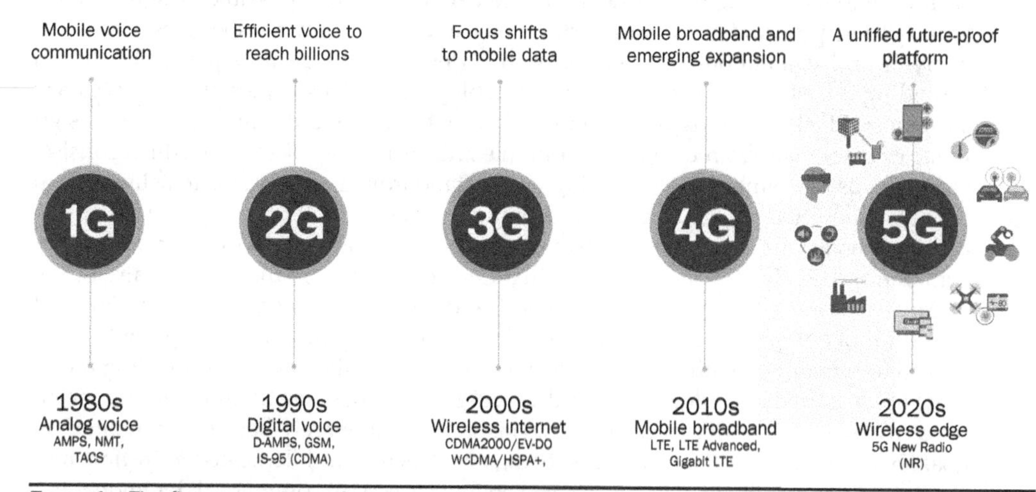

Figure 1 The five generations of wireless communications.

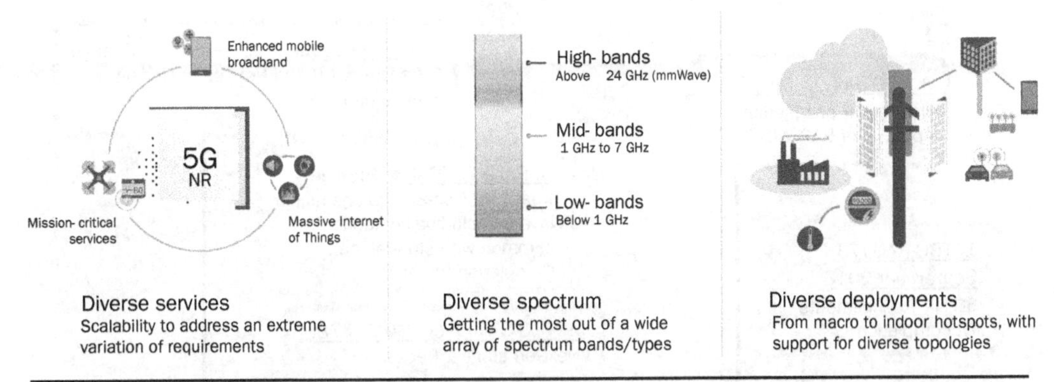

FIGURE 2 5G: a unified connectivity fabric for existing, emerging, and unforeseen connected services.

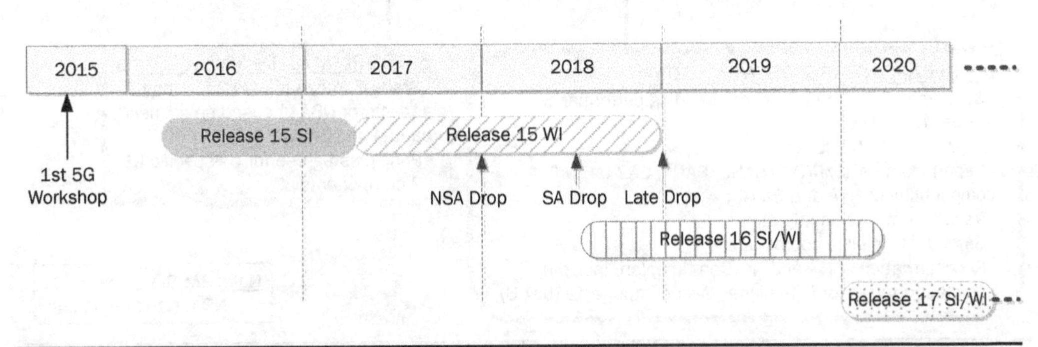

FIGURE 3 The timeline for 5G NR in 3GPP.

It was generally agreed during the workshop that 5G NR should support a variety of new services, such as automotive, health, energy, manufacturing, etc. Compared with 4G LTE, 5G NR would be a new and a non-backward-compatible radio access technology (RAT). It was envisioned at the very first workshop that the design of 5G NR should be forward compatible, so that additional use cases and requirements can be gradually introduced later in a compatible and optimal manner.

One important difference of 5G compared to previous generations of wireless standards is that it provides the option of a *staggered* evolution in radio and core network deployments. This effectively allows not only the deployment of NR with an *old* core network (Evolved Packet Core as in LTE), but also a later migration to a *new* core network (i.e., 5GC). The staggered evolution would accelerate availability of 5G products into the market. This concept was first agreed in a joint RAN (Radio Access Network) and SA (System Architecture) plenary session in June 2016 [3]. Figure 4 shows the original RAN-SA workplan agreement [4]. After a few twists and 9 months later after the joint session, 3GPP agreed to a final workplan proposal [5] for 3GPP 5G NR specifications as the global 5G standard. This way forward was the result of months of online (i.e., official decision-making sessions) and offline discussion and involved a coalition of 47 companies! The final timeline that had emerged from these discussions is shown in Fig. 3.

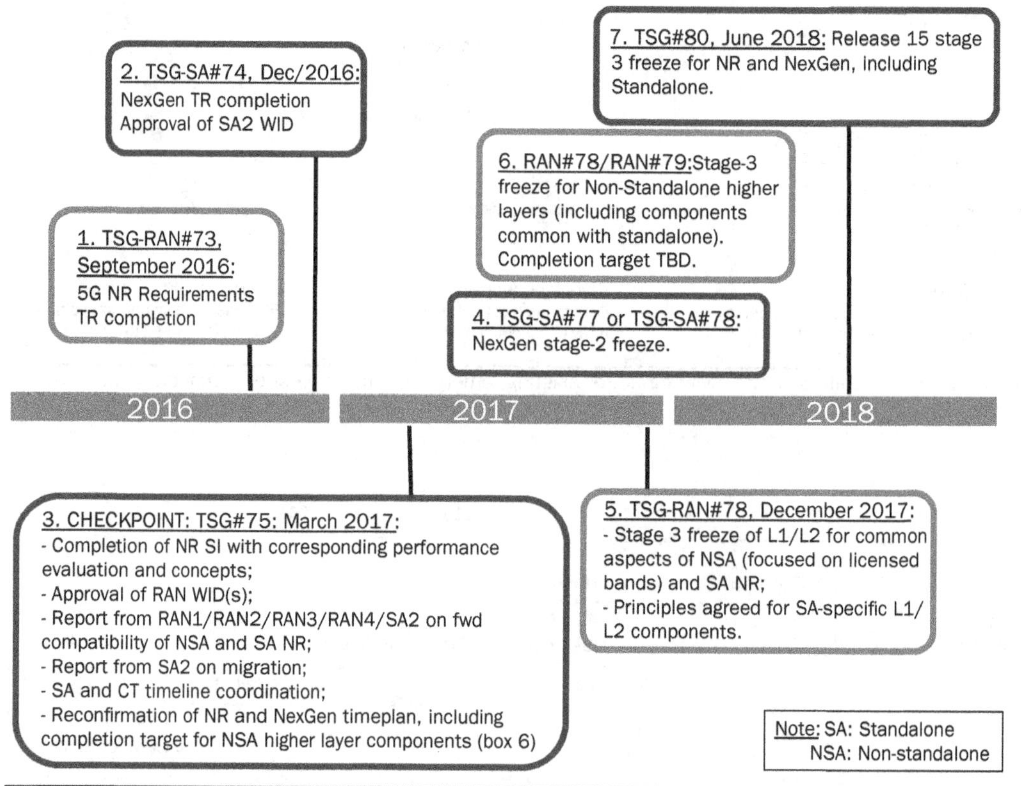

FIGURE 4 The original RAN-SA workplan agreement (as in [4]).

It is obvious that any standardization work has to be done in a phased approach. The standardization work requires extensive evaluations, careful analysis, meticulous end-to-end system design including handling possible error cases, practical performance characterization and testability, and reconciliation of various views from all participating parties. In addition, standardization work in 3GPP is increasingly witnessing close interaction and cooperation among different standardization organizations, most notably due to the expansion of wireless communications into other vertical domains, such as automotive industry, smart factory, satellites, usage of unlicensed or shared spectrum, etc. As one example, liaisons are often being used by 3GPP to communicate to and from standard bodies such as 5GAA (http://5gaa.org) for V2X (vehicle to everything) related topics, and IEEE (http://standards.ieee.org) particularly for the coexistence between 3GPP specified operation and Wi-Fi in unlicensed or shared spectrum. Moreover, new information such as new use cases, additional deployment scenarios, new commercial or regulatory requirements, etc., may be identified later, which would drive continuing evolution of standardization.

In 3GPP, the phased approach is managed using a terminology called a *release.* Each release usually lasts 1 to 2 years. The releases in 3GPP are indexed contiguously. After GSM Phase 1 and Phase 2 and the first 3GPP release called Release 99 (which was named by the year of the anticipated release), starting from Release 4 in the year 2000, the releases in 3GPP were indexed with a specific number, incremented after each release. For 5G NR, there have been three releases so far, namely:

- Release 15: from 1Q 2016 to 4Q 2018, first 5G release,
- Release 16: from 3Q 2018 to 1Q 2020, second 5G release, and
- Release 17: From 2Q 2020 to 2Q 2022, third 5G release.

This is illustrated in Fig. 3. The prolonged duration of Release 17 is due to inevitable impact from COVID-19. Note that it is typical that a release may consist of a study phase and a specification phase (also known as work items in 3GPP). After the end of each release, the resulting specifications may undergo a series of corrections and fixes, often lasting several months or even longer.

As the first release of 5G, Release 15 was done in a very short time—a 1-year study item from 1Q 2016 to 1Q 2017, followed by a 9-month work item for a first drop of the release focusing on non-standalone (NSA) deployments using the EPC (what is commonly known as architecture option 3; see Chap. 2 for more details). Then, this was followed by another 6 months for standalone (SA) deployments using the new core network (5GC) (what is commonly known as architecture option 2; see Fig. 5), and yet another 6 months for a late drop delivering additional use cases and additional architecture options including connecting evolved LTE to the new 5G core (what is commonly known as architecture options 5 and 7). Yet, the first 5G release delivers an overwhelmingly large set of features—it not only covers almost all the features offered by 4G LTE, which were developed and standardized in a 10-year period, but also provides support of additional use cases, additional deployment scenarios, additional spectrum ranges, and a bouquet of different architecture options as mentioned above. This was deemed by many as a "mission impossible" task, but was successfully completed only after the constructive and goal-oriented cooperation among thousands of delegates spending extra hours with creative ways of working (e.g., to use a projector for offline discussion projecting onto a wall, the back of a sofa, or a ceiling). For both the second and third 5G releases, while there are continued evolutions for the traditional eMBB, expansion of 5G into vertical domains is addressed in Release 16 (e.g., V2X, Non-Public/Private Networks, URLLC and IIoT, support of 5G in unlicensed and shared spectrum), and is further accelerated in Release 17 (e.g., additional V2X/URLLC/IIoT enhancements, support of non-terrestrial networks, support for proximity services, multicast-broadcast).

Figure 5 shows the system architecture of 5G NR revised based on the ones in [6], which consists of:

- NG interface, connecting 5GC (e.g., functions such as access and mobility management and user plane, with more details in Chap. 3) and gNB (next generation node B) or ng-eNB (next generation evolved node B);
- Xn interface, connecting among gNBs and/or ng-eNBs (more details can be found in Chap. 3);
- Uu interface, the over-the-air connection between eNBs and user equipments (UEs), comprising Downlink (DL) and Uplink (UL); and
- PC5 interface, connecting among UEs, also known as sidelink (SL).

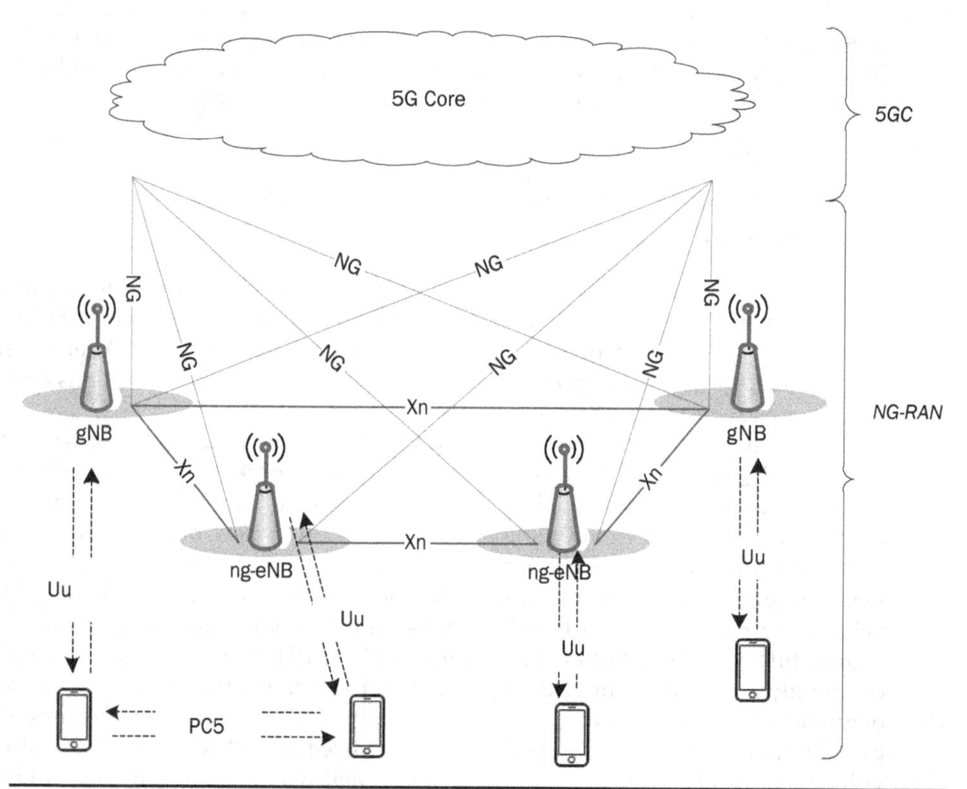

Figure 5 System architecture for 5G NR (using option 2 as an example).

A UE may be configured with a single carrier, or multiple carriers in the form of carrier aggregation (CA) or dual-connectivity (DC) [6], where the former assumes a tight coordination among the carriers and the latter assumes a non-ideal link between the multiple carriers (e.g., limited capacity, non-negligible latency). When multiple carriers are configured for a UE, a UE is configured with a primary cell (Pcell), and one or more secondary cells (Scells). The UE may be additionally configured with a master cell group (MCG) and a secondary cell group (SCG), where the management of the two cell groups is generally separate especially from Downlink control and Uplink control perspectives. More details can be found in Chap. 3.

In comparison, Fig. 6 shows the system architecture of 4G LTE revised based on the ones in [7], which consists of:

- S1 interface, connecting EPC (e.g., functions such as mobility management entity and serving gateway, with more details in Chap. 3) and eNB (evolved node B);
- X2 interface, connecting among eNBs (more details can be found in Chap. 3);
- Uu interface; and
- PC5 interface.

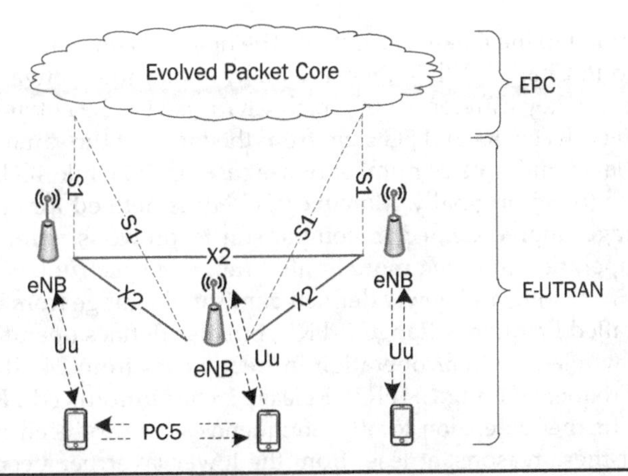

Figure 6 System architecture for E-UTRAN (4G LTE).

Throughout the book, we will use 4G, LTE, and 4G LTE interchangeably. Similarly, we will also use 5G, NR, and 5G NR interchangeably.

The development of 5G is not coming from scratch. Instead, it is a natural evolution from 4G LTE. In 2008, 4G LTE completed its first release (Release 8), enabling mobile broadband services in an unprecedented manner. The offered speed by 4G LTE started with tens of Mbps, and gradually upgraded to hundreds of Mbps and even up to several Gbps covering both indoor and outdoor scenarios. 4G LTE enjoys enormous commercial success, particularly in terms of driving the proliferation of smartphones.

One natural question would then be:

With 4G LTE being extremely successful, why do we still need 5G?

In other words, is it possible to do a small surgery of 4G LTE specifications to accommodate the 3 pillars of 5G high-level use cases (eMBB, URLLC, and mMTC)? It is clear that when 4G LTE was first standardized, there were certain assumptions and techniques not friendly to introducing new features, e.g.:

- An always-on cell-specific reference (CRS),
- A fixed frame structure,
- A fixed transmit time interval (TTI) for scheduling packets, and
- A fully-frequency-distributed control channel over the entire cell bandwidth.

Therefore, even if additional evolutions for 4G LTE are possible to accommodate some use cases identified for 5G NR, the design is subject to the existing restrictions in order to minimize any impact to legacy operations and to maintain the existing structure. It is worth noting that some of the restrictions in LTE were recognized toward the end of the first LTE release, although not addressed primarily due to the late stage of the release, as detailed in Chaps. 1 and 5.

On the other hand, the expertise in 4G LTE in the last decade was proved to be invaluable for the design and standardization of 5G NR. This is particularly possible because both 4G LTE and 5G NR are OFDMA-based. The accelerated pace to standardize the first release of 5G NR would not have been possible if 4G LTE design had not been leveraged to the maximum extent. The detailed comparison of 4G LTE and 5G NR is provided in Chap. 1. While there are many similarities between 4G LTE and 5G NR, there are some key differences in which Chap. 1 will present and discuss. Most of the differences between 4G and 5G stem from the fact that the same radio interface will be used for a large and diverse number of use cases in 5G while 4G LTE primarily targeted mobile broadband originally. Moreover, 5G NR is defined for operation in a large frequency range unprecedented in comparison to previous wireless generations which aimed at operation in much more limited range of spectrum. 5G NR not only defines operation in existing and newly defined bands in the range from 410 MHz to 7.125 GHz or the so-called Frequency Range 1 (FR1), but also defines operation in completely new bands for wireless cellular operation in frequencies from 24 GHz to 52.6 GHz or the so-called Frequency Range 2 (FR2). Release 15 has introduced FR1 and FR2 as defined above but further extension to other frequencies is introduced in subsequent releases of NR. For these reasons, at least from the lower-layer perspective, the OFDM-based access introduced in LTE gets generalized in 5G NR to multiple numerologies and multi-beam operation. Special emphasis to forward compatibility also modulates some aspects of the air interface of 5G NR as it is getting developed.

Chapter 2 on deployments scenarios contains an introduction of basic deployment scenarios, such as SA and NSA modes, focusing on the spectrum allocation perspective. Target spectrum types and frequency ranges considered for the 5G NR design, including some basic characteristics of the different frequency ranges impacting the design, are described.

Chapter 3 discusses architecture options for 5G including all the architecture options for 5G NSA and SA, the "disaggregated" RAN architecture options and all the options enabling evolved LTE to connect to 5GC. The motivation for introduction of 5GC is explained and an introduction of the main 5GC functionalities is provided. In the end, explanation on the possible migration options for mobile network operators (MNOs) among the different specified 5G architecture options is provided.

Evolution of 5G architecture is addressed in Chap. 4. It introduces the evolution of 5GC in Release 16 providing an overview of the full set of system functionalities such as non-public networks/private networks, V2X, cellular IoT, support for big data, and URLLC. A short introduction of the 5G system features planned for Release 17 is also given.

Chapter 5 is dedicated to introducing numerology and slot structure for 5G. It starts with an introduction of 4G LTE numerology and slot structure, and the lessons learned from 4G standardization, which helped the corresponding 5G standardization. We will discuss the flexible SCS, frequency range, bandwidths, and bands for NR, and the channel bandwidth from the cell perspective and from the UE perspective (also known as a bandwidth part or BWP). In the time domain, we will detail the design considerations for symbols, slots, subframes, and frames for NR. Finally, considerations of the slot structure and forward compatibility are given.

Chapter 6 provides an overview of the procedures related to initial access and mobility. Initial access is the procedure enabling UEs to acquire the system to the point that they can receive the system information from the network, starting from a so-called "cold start" where no synchronization means have been established between the network and the UEs. The acquisition of the system information will, in turn, enable UEs access the network via another procedure, the so-called random access procedure, which is also described in this chapter. Mobility procedures and associated physical layer measurements are introduced and discussed in this chapter.

Downlink control is an integral part of 5G. Chapter 7 first summarizes the Downlink control operation in 4G LTE, necessary to show the evolution of the design considerations from 4G to 5G. For 5G Downlink control, we will detail the control region management, PDCCH (physical Downlink control channel) structure, search space management, DCI (Downlink control information) formats, and the physical layer block diagram for PDCCH. We will also introduce the related power saving considerations, a new feature introduced in Release 16.

Chapter 8 on Downlink data operation contains a comprehensive overview of procedures for DL data transmission and reception on the physical Downlink shared channel (PDSCH). The chapter includes descriptions of aspects specific to the DL and additionally aspects common to the DL and UL, such as channel coding. Design concepts of the PDSCH and the corresponding reference signals are discussed. The chapter also includes descriptions of other signals and procedures that are used in support of PDSCH reception, such as those used for frequency and time tracking, phase noise estimation, and channel state information feedback. Finally, the chapter includes a description of the basic concepts of DL beam management.

Uplink control for 5G is introduced in Chap. 9. Considering the natural evolution, we start by introducing the Uplink control operation in 4G. This is followed by the details of 5G Uplink control information (UCI) types and payload sizes, physical Uplink control channel (PUCCH) formats, PUCCH resource determination, UCI on physical Uplink shared channel (PUSCH), and channel coding for UCI.

Chapter 10 on Uplink data operation includes an overview of procedures for UL data transmission and reception on the PUSCH. Design concepts of the PUSCH and corresponding reference signal structure are discussed. The chapter also contains descriptions of other signals and procedures that are used in support of UL reception, such as phase tracking, power control, and UL timing adjustment. A description of the basic concepts of UL beam management is given as well.

Chapter 11 on coexistence of 4G and 5G includes a description of coexistence scenarios on the system level, such as same channel and adjacent channel coexistence. It also contains an overview of procedures introduced in support of coexistence of LTE and NR in the same device when NSA mode is used, including aspects of power sharing and switched UL operation.

One important aspect for a wireless communication system is the frequency support required for it to operate. While cellular wireless operations have traditionally relied on licensed spectrum, 3GPP introduced the possibility for LTE to operate in unlicensed spectrum in Release 13. 3GPP defined unlicensed operation in LTE is always linked to a cellular anchor which manages the overall connection[4]. This allowed an additional 20 MHz of spectrum to be available, enabling higher Gbps data rates in some and which enabled the proliferation of Gbps LTE in commercial networks. To that end, in Chap. 12 we introduce 5G in Unlicensed and Shared Spectrum which is being enabled in Release 16. Unlike LTE-based access to unlicensed operation, which is based on Licensed Assisted Access (LAA) in the form of carrier aggregation of Scells on unlicensed spectrum linked to a Pcell on licensed spectrum, 5G NR enables more

[4] The MulteFire Alliance (http://www.multefire.org) developed a version that did not require an anchor.

diverse deployment scenarios either license-assistance based, reusing the CA or DC framework, or stand-alone unlicensed not requiring a licensed carrier as an anchor.

As one of the vertical domain expansions, URLLC can be found in many use cases and deployment scenarios, and is covered in Chap. 13. After a brief presentation of the efforts in 4G regarding latency and reliability handling, we provide the detailed considerations for URLLC design in terms of resource management, link efficiency optimization, dynamic Downlink resource sharing, and dynamic Uplink resource sharing for different service types and from both the cell and the UE perspectives. Finally, additional design aspects are introduced including the support of time-sensitive networks (TSN) for IIoT and multi-TRP (Transmit and/or Receive Points).

Chapter 14 focuses on the support of machine type communications (MTC), another vertical domain expansion. After a brief introduction of the efforts in 3GPP on MTC, we provide detailed design considerations for eMTC (evolved MTC), covering the technology enablers to achieve low cost/complexity, extensive power savings, and extreme coverage extension. Similar presentation is also done for another variation of MTC, namely NB-IoT (narrow-band internet of things), by emphasizing the differences in design compared with eMTC. The integration of eMTC and NB-IoT, originally standardized in 4G LTE, into 5GC is then introduced. Finally, we discuss some future trends for MTC.

Chapter 15 covers another important vertical expansion of cellular operations, namely, vehicle-to-everything (V2X) communications. In this chapter, we provide a brief introduction on how V2X was introduced in LTE during Releases 14 and 15. Then, we look into the ways that the Release 16 project on NR V2X complements it. The support of V2X brings the sidelink capability to NR which is expected to go beyond vehicular applications starting in Release 17.

Chapter 16 includes an overview of various broadcast and multicast modes, focusing on multimedia broadcast multicast service (MBMS) content delivery. It also includes a brief introduction of the evolution of physical layer channels used for the delivery of MBMS data. The chapter closes with a discussion of future directions of evolving broadcast multicast in 5G NR.

After the initial version of NR in Release 15, where the entire radio air interface, internode protocols and architecture were developed in a single project, 3GPP has fanned out its evolution of NR into multitude projects from Release 16. While we cover many of the Release 16 projects in either dedicated chapters or embedded into some other fundamental chapters of this book, there are a number of projects that we believe are important to mention and for which we do not have dedicated chapter. To that end, Chap. 17 provides an overview of some miscellaneous topics for 5G which did not get treated in dedicated chapters.

Chapter 18 on typical 5G commercial deployments includes an initial overview of possible topologies for 5G deployments. The description focuses on options for providing coverage and capacity at the same time, such as combining spectrum in different frequency ranges while also providing solutions for mobility.

The final chapter of this book, Chap. 19, attempts to present how the authors see 5G evolving in future 3GPP releases. While we first cover the Release 17 projects that are still ongoing at the time this book goes to press, we also venture into what we expect future releases of 3GPP will focus on and how those areas will bridge into 6G.

We hope this book can provide a good introduction of how and why 5G is standardized, for both the traditional broadband and the expanded industries. We hope this book can be useful for people in both industrial and academic fields, to learn the stories behind the standardization, to always aim for optimizing every single bit or a small fraction of dB while ensuring reasonable complexity and commercial achievability, to be ready for compromise and cooperation with all engineers working toward the same goal, and to get motivated to contributing, either for the first time or with more efforts, to 5G or future generations of standardization for wireless communications.

References

[1] 3GPP RAN Chairman, "Chairman's Summary Regarding 3GPP TSG RAN Workshop on 5G," RWS-150073, RAN-workshop on 5G, Phoenix (US), September 2015.

[2] ITU-R, "IMT Vision: Framework and Overall Objectives of the Future Development of IMT for 2020 and Beyond," M.2083, September 2015.

[3] 3GPP RAN chair, SA chair, "Tasks from Joint RAN-SA Session on 5G Architecture Options," RP-161269, 3GPP TSG RANP#72, June 2016.

[4] Nokia Networks, "Combined NR / NextGen Time Plan," SP-160465, 3GPP TSG SAP#72, June 2016.

[5] Alcatel-Lucent Shanghai-Bell, Alibaba, Apple, AT&T, British Telecom, Broadcom, CATT, China Telecom, China Unicom, Cisco, CMCC, Convida Wireless, Deutsche Telekom, DOCOMO, Ericsson, Etisalat, Fujitsu, Huawei, Intel, Interdigital, KDDI, KT, LG Electronics, LGU+, MediaTek, NEC, Nokia, Ooredoo, OPPO, Qualcomm, Samsung, Sierra Wireless, SK Telecom, Sony, Sprint, Swisscom, TCL, Telecom Italia, Telefonica, TeliaSonera, Telstra, T-mobile USA, Verizon, vivo, Vodafone, Xiaomi, ZTE, "Way Forward on the Overall 5G-NR eMBB workplan," RP-170741, 3GPP TSG RANP#75, March 2017.

[6] 3GPP, "3rd Generation Partnership Project; Technical Specification Group Radio Access Network; NR; NR and NG-RAN Overall Description; Stage 2 (Release 16)," 38.300, v16.0.0, January 2020.

[7] 3GPP, "3rd Generation Partnership Project; Technical Specification Group Radio Access Network; Evolved Universal Terrestrial Radio Access (E-UTRA) and Evolved Universal Terrestrial Radio Access Network (E-UTRAN); Overall description; Stage 2 (Release 16)," 36.300, v16.0.0, January 2020.

About the Authors

Wanshi Chen, Ph.D., Sr. Director, Technology, at Qualcomm Inc., is 3GPP TSG RAN Plenary Chair, elected in March 2021. Formerly 3GPP RAN1 Chair and Vice Chair, he has successfully managed a wide range of RAN1 4G Long Term Evolution (LTE) and 5G New Radio (NR) sessions. He has over 20 years of experience in telecommunications at leading telecom companies, including operators, infrastructure vendors, and chipset vendors.

Peter Gaal, Ph.D., VP, Technical Standards, at Qualcomm's Corporate Standards group, has been with Qualcomm since 1999. Initially, he was involved in cdma2000 standardization. Since 2007, he has been attending 3GPP meetings, first in the RAN4 group and since 2010 in the RAN1 group.

Juan Montojo, Ph.D., VP, Engineering, at Qualcomm's Corporate Standards group, joined Qualcomm in January 1997 and has worked in the system design and standardization of various communication systems, including Globalstar, 3G, 4G, 5G, and WiFi, as part of the corporate R&D and Standards groups.

Haris Zisimopoulos, M.Sc., Sr. Director, Technical Standards, at Qualcomm's Corporate Standards group, joined Qualcomm in September 2012. Since then, he has been participating in 3GPP System Architecture WG2, i.e., the 3GPP group related to 3GPP system architecture. He has been rapporteur of various 3GPP projects, namely Proximity Services/D2D, Next Generation eCall, Unlicensed Spectrum System optimizations, and Radio Capabilities Signaling optimization.

Acronyms

3GPP	3rd Generation Partner Project
5GC	5G Core
5GS	5G System
AC	Authentication Confirmation
AI	Artificial Intelligence
AL	Aggregation Level
AMBR	Aggregate Maximum BitRate
AMF	Access and Mobility Function
ANDSP	Access Network Discovery and Selection Policy
APN	Access Point Name
BG	Base Graph
BLER	BLock Error Rate
BM	Beam Management
BWP	BandWidth Part
C-V2X	Cellular V2X
CA	Carrier Aggregation
CAG	Closed Access Group
CB	Code Block
CBG	Code Block Group
CBR	Channel Busy Ratio
CBRA	Contention-Based Random Access
CC	Component Carrier
CCA	Clear Channel Assessment
CCS	Cross-Carrier Scheduling
CDD	Cyclic Delay Diversity
CDM	Code Division Multiplexing
CF	Correction Field
CFRA	Contention-Free Random Access
CG	Configured Grant
CGS	Computer Generated Sequence
CLI	Cross-Link Interference
CM	Connection Management
CN	Core Network
CoMP	Coordinated Multiple Point (transmission/reception)
CORESET	COntrol REource SET
COT	Channel Occupancy Time

CP	Cyclic Prefix or Control Plane (depending on the context)
CQI	Channel Quality Indicator
CRI	CSI-RS Resource Indicator
CRS	Common Reference Signal
CSI	Channel State Information
CSI-RS	Channel State Information—Reference Signal
CSS	Common Search Space
CW	CodeWord or Contention Window (depending on the context)
D2D	Device-to-Device
DAI	Downlink Assignment Index
DC	Dual-Connectivity
DCI	Downlink Control Information
DCN	Dedicated Core Network
DG	Dynamic Grant (or Dynamically Granted)
DL	DownLink
DM-RS	DeModulation—Reference Signal
DNN	Data Network Name
DPS	Dynamic Power Sharing
DRX	Discontinuous Reception
DSRC	Dedicated Short-Range Communication
DSS	Dynamic Spectrum Sharing
(e)CCE	(enhanced) Control Channel Element
(e)MBB	(enhanced) Mobile BroadBand
(e)MBMS	(enhanced) Multimedia Broadcast/Multicast Services
(e)MTC	(enhanced) Machine Type Communication
(e)PDG	(evolved) Packet Data Gateway
(e)REG	(enhanced) Resource Element Group
(E)-UTRA	(Evolved) Universal Terrestrial Radio Access
EAP	Extensible Authentication Protocol
ECP	Extended Cyclic Prefix
ED	Energy Detection
eIMTA	enhanced Interference Management and Traffic Adaptation
eNB	evolved Node B (an LTE base station)
EPC	Evolved Packet Core
EPF	Network Exposure Function
EPRE	Energy Per Resource Element
EPS	Evolved Packet System
FBE	Frame Based Equipment
FDD	Frequency-Division Duplex
FR1	Frequency Range 1
FR2	Frequency Range 2
GBR	Guaranteed BitRate
gNB	next generation Node B (a 5G base station)
GP	Guard Period
HRLLC	High Reliability Low Latency Communications
HSS	Home Subscriber System
IAB	Integrated Access and Backhaul
IIoT	Industrial Internet of Things
IM	Interference Measurement

IMS	IP Multimedia Subsystem
IoT	Internet of Things
IS	In Service
ITS	Intelligent Transportation System
LAA	License Assisted Access
LADN	Local Area Data Network
LBE	Load-Based Equipment
LBRM	Limited Buffer Rate Matching
LDPC	Low Density Parity Check code
LI	Layer Indicator
LIPA	Local IP Access
LMF	Location Management Function
LTE	Long-Term Evolution
MBR	Maximum BitRate
MBSFN	Multicast Broadcast Single Frequency Network
MCG	Master Cell Group
MCL	Maximum Coupling Loss
MCS	Modulation and Coding Scheme
MDBV	Maximum Data Burst Volume
MIB	Master Information Block
MICO	Mobile Initiated Connectivity Only
MIMO	Multiple-Input-Multiple-Output
MME	Mobility Management Entity
mMTC	massive Machine Type Communication
mmW	milli-meter Wave
MNC	Mobile Network Code
MNO	Mobile Network Operator
MPR	Maximum Power Reduction
MRTD	Maximum Receive Timing Difference
MTTD	Maximum Transmit Timing Difference
N3IWF	Non-3GPP InterWorking Function
NAICS	Network-Assisted Interference Cancellation and Suppression
NAS	Non-Access Stratum
NB-IoT	Narrow-Band Internet of Things
NCP	Normal Cyclic Prefix
NCT	New Carrier Type
NEST	NEtwork Slice Types
NG	Next Generation
NID	NPN ID
NIDD	Non-IP Data Delivery
NOMA	Non-Orthogonal Multiple Access
NPN	Non-Public Networks
NR	New Radio
NR-U	NR access to Unlicensed spectrum
NRF	Network Repository Function
NSA	Non-StandAlone (deployments)
NSSAI	Network Slice Selection Assistance Information
NSSP	Network Slice Selection Policy
NWDAF	NetWork Data Analytics Function

NZP	Non-Zero Power
OCC	Orthogonal Cover Code
OFDMA	Orthogonal Frequency Division Multiple Access
OOS	Out Of Service
OSI	Other System Information
OTA	Over-The-Air
PAPR	Peak-to-Average Power Ratio
PBCH	Physical Broadcast CHannel
Pcell	Primary cell
PCF	Policy Control Function
PCFICH	Physical Control Format Indicator CHannel
PCI	Physical Cell ID
PD	Preamble Detection
PDB	Packet Delay Bound
PDCCH	Physical Downlink Control CHannel (*Note:* ePDCCH stands for enhanced PDCCH; MPDCCH stands for MTC PDCCH; NPDCCH stands for NB-IoT PDCCH)
PDCP	Packet Data Convergence Protocol
PDSCH	Physical Downlink Shared CHannel (*Note:* NPDSCH stands for NB-IoT PDSCH)
PDU	Protocol Data Unit
PER	Packet Error Rate
PHICH	Physical HARQ Indicator CHannel
PI	Pre-emption (or Priority) Indicator
PMI	Precoding Matrix Indicator
PO	Paging Occasion
PPPP	ProSe Per Packet Priority
PRACH	Physical Random Access CHannel (*Note:* NPRACH stands for NB-IoT PRACH)
PRB	Physical Resource Block
PRI	PUCCH Resource Indicator
PSCCH	Physical Sidelink Control CHannel
PSD	Power Spectral Density
PSM	Power Saving Mode
PSS	Primary Synchronization Signal (*Note:* NPSS stands for NB-IoT PSS)
PSSCH	Physical Sidelink Shared CHannel
PT-RS	Phase Tracking Reference Signal
PUCCH	Physical Uplink Control CHannel
PUR	Pre-configured Uplink Resource
PUSCH	Physical Uplink Shared CHannel (*Note:* NPUSCH stands for NB-IoT PUSCH)
QCI	QoS Class Identifier
QCL	Quasi-Co-Location
QFI	QoS Flow ID
QoE	Quality of Experience
QoS	Quality of Service
RAN	Radio Access Network
RAR	Random Access Response

RAT	Radio Access Technology
RB	Resource Block
RBG	Resource Block Group
RE	Resource Element
RI	Rank Indicator
RIM	Remote Interference Management
RLC	Radio Link Control
RLF	Radio Link Failure
RLM	Radio Link Management
RM	Registration Management or Rate Matching or Reed-Muller (depending on the context)
RMSI	Remaining Minimum System Information
RNTI	Radio Network Temporary Identifier
RRM	Radio Resource Management
RSRP	Reference Signal Received Power
RSRQ	Reference Signal Received Quality
RSSI	Received Signal Strength Indicator
RSU	RoadSide Unit
RTT	Round Trip Time
RV	Redundancy Version
SA	StandAlone (deployments)
SC-FDMA	Single-Carrier Frequency Division Multiple Access
SC-PTM	Single Cell Point-To-Multipoint
SCEF	Service Capability Exposure Function
Scell	Secondary cell
SCG	Secondary Cell Group
SCS	SubCarrier Spacing
SD	Slice Differentiator
SD-CDD	Small Delay Cyclic Delay Diversity
SDL	Supplemental DownLink
SFBC	Space-Frequency Block Codes
SFI	Slot Format Indicator
SFN	Single Frequency Network
SFTD	SFN and Frame Timing Difference
SI	Study Item or System Information (depending on the context)
SIB	System Information Block
SIDF	Subscription Identifier De-concealing Function
SINR	Signal to Interference and Noise Ratio
SLA	Service Level Agreements
SLIV	Start and Length Indicator Value
SMF	Session Management Function
SMTC	SS/PBCH block Measurement Time Configuration
SNPN	Standalone Non-Public Networks
SPS	Semi-Persistent Scheduling
SR	Scheduling Request
SRI	SRS Resource Indicator
SRS	Sounding Reference Signal
SS	Synchronization Signals

SSB	Synchronization Signals Block
SSBRI	SSB Resource Indicator
SSC	Session and Service Continuity
SSS	Secondary Synchronization Signal (*Note:* NSSS stands for NB-IoT SSS)
SST	Slice/Service Type
sTTI	short Transmit Time Interval
SUL	Supplemental UpLink
SUPI	SUbscriber's Permanent Identifier
TA	Timing Advance
TAG	Timing Advance Group
TB	Transport Block
TBCC	Tail-Biting Convolutional Code
TBS	Transport Block Size
TCI	Transmission Configuration Indicator
TDD	Time-Division Duplex
TDRA	Time Domain Resource Allocation
TL	Threshold Level
TPC	Transmit Power Control
TPMI	Transmit Precoding Matrix Indicator
TRP	Transmit and/or Receive Point
TRS	Tracking Reference Signal
TSN	Time Sensitive Networking
TTI	Transmit Time Interval
UAC	Unified Access Control
UCI	Uplink Control Information
UDM	Unified Data Management
UDR	User Data Repository
UDSF	Unstructured Data Storage Function
UE	User Equipment
UL	UpLink
ULSUP	UL Sharing from the UE Perspective
UP	User Plane
URLLC	Ultra Reliability Low Latency Communication
URSP	UE Route Selection Policy
USS	UE-specific Search Space
V2I	Vehicle to Infrastructure
V2N	Vehicle to Network
V2P	Vehicle to Pedestrian
V2V	Vehicle to Vehicle
V2X	Vehicle to Everything
VoIP	Voice over IP
VRB	Virtual Resource Block
WI	Work Item
WLANSP	WLAN Selection Policy
WUS	Wake-Up Signal
XR	eXtended Reality
ZP	Zero-Power

CHAPTER 1

5G versus 4G: What's New?

1.1 Overview

From the radio perspective there are many similarities between 4G (LTE) and 5G (NR) but there are some key differences to note. The objective of this chapter is to cover these differences and provide a rationale behind each of them. For the core network architecture, a detailed analysis of the similarities and differences between 4G (EPC) and 5G (5GC) is described in Chap. 3.

5G, from its inception, has been meant to address three major types of applications: enhanced mobile broadband (eMBB), ultra-reliable low-latency communications (URLLC), and massive machine type communications (mMTC).

The first 3GPP release of NR was Release 15 and the main focus was eMBB. In order to address future applications, NR was designed from day one with forward compatibility in mind. Forward compatibility is addressed in multiple ways. First, NR introduced the concept of reserved resources which are meant to be ignored by the UEs. NR design has also drastically reduced reliance on "always on" channels and signals, which largely reduces the network requirements for having to be "ON" to transmit some determined channels or signals in certain parts of the frame. Lastly, unlike LTE, the underlying frame and slot structure, as well as the timing relationships involving action/reaction in NR, are flexible, enabling the possibility of moving transmissions/receptions around for the potential future introduction of new channels and/or signals.

While URLCC support was in the scope of Release 15, only a limited set of features were introduced motivated by this type of application, namely:

- Low latency: mini-slot scheduling (scheduling or mapping type B in the specifications), Capability 2 for shorter processing times at the UE
- High reliability: set of MCS and CQI tables tailored for 10^{-5} error rate operation

Chapter 13 provides more details on how URLLC is enabled by NR in both Release 15 and Release 16. By way of a high-level introduction, it is important to note that URLCC is not enabled by one particular feature or functionality in the NR specifications. Rather, different levels of low-latency and reliability are enabled by different functionalities. In addition, there are features introduced to enable the coexistence in a network and within a terminal of services of different latency and reliability requirements.

On mMTC, it is important to realize that mMTC has not been explicitly addressed by NR Release 15, Release 16, or even Release 17. 3GPP made the conscious decision to avoid further fragmenting the IoT market by developing yet another Low Power Wide Area (LPWA) solution just a couple of releases after the initial introduction of eMTC and NB-IoT in Release 13. Instead, mMTC communications are supported by the LTE developed features eMTC (or LTE-M) and NB-IoT, which can be fully embedded inside an NR carrier and connected to the 5G Core (5GC) from Release 16. Chapter 14, in turn, gives an overview of these systems and how they have evolved over the various releases from their inception in Release 13. Chapter 4 provides an overview of the mMTC functionalities provided by 5GC.

1.2 LTE: A Success Story

There is no doubt that LTE has been a success story for wireless communications. It is fair to say that smartphones proliferated with LTE networks. The overall user experience for data communications increased considerably with the advent of these networks. LTE enabled high speed mobile broadband performance away from home, which – together with new phone form factors, with larger and touch screens – changed forever the way people interact with their mobile devices. Having access to web browsing, video streaming, video conferencing, gaming, and other endless number of "Apps" at your fingertips inside and outside home was hard to imagine a few years earlier.

At the same time all operators provided services that were using the circuit switched (CS) domain (such as voice, SMS) and regulatory services (such support for E911 emergency calls, public warning system, etc.) have been successfully migrated through the evolutions of LTE and EPC to the packet switched (PS) domain and IP Multimedia Subsystem (IMS). One can say that eventually LTE/EPC in 2020 in most operator networks is a fully packet switched cellular system.

When LTE was designed and standardized, there was a clear intent to "keep it simple" and "try to minimize options." This was after the learnings from 3G UMTS over-configurability and associated complexity. That complexity and tool-box approach became manifested especially at the time of the first system deployments which took a long time to mature. Initial 3G UMTS user equipment implementations also suffered from poor performance and battery life.

As a result, LTE is a fairly lean and simple design avoiding unnecessary options and configurability. Note that these design principles were not present in the development of NR, which is designed for a variety of use cases requiring substantially different configurations.

LTE was introduced in Release 8 of 3GPP and still evolves today with the corresponding Release 17 work. The amount of work and functionality that 3GPP has added over the releases to the LTE platform is phenomenal. It is noticed that there is a large gap with respect to the features offered by the LTE standards and those which got commercially deployed. There are some very important additions after Release 8 which saw broad commercial success, voice over IMS (Release 9), E911 and LCS for E911 (Release 9), Transmission Mode 8 (TM8 in Release 9), carrier aggregation (Release 10), eMTC and NB-IoT (Release 13), LAA (Release 13), cellular-V2X (Release 14) to name a few.

There were some revolutionary changes in the migration from 3G networks to 4G networks. The most relevant change from the radio perspective was the change of waveform from CDMA to (a) OFDMA in the Downlink and (b) SC-FDMA in the Uplink. Note that the LTE waveforms never changed after the original waveform choice in Release 8 despite the failed attempt to introduce OFDMA in the Uplink during Release 10 discussions in the context of LTE-advanced. Figure 1.1 shows a high-level comparison of CDMA, OFDMA, and SC-FDMA transmitters.

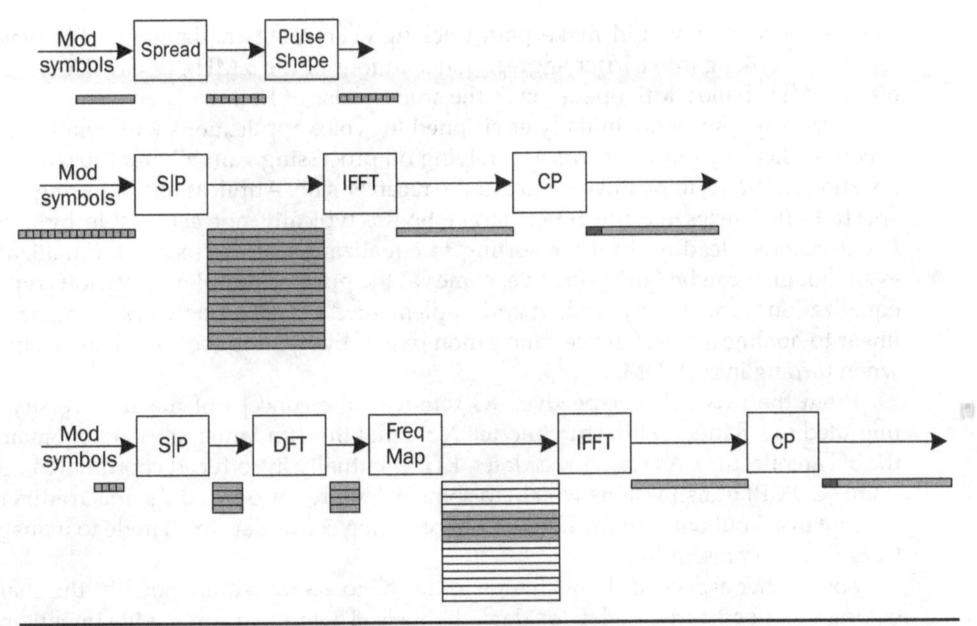

Figure 1.1 CDMA/OFDMA/SC-FDMA transmitter comparison.

CDMA transmitters expand the transmission bandwidth via a spreading function which typically leads to increased inter symbol interference (ISI). The spreading function enables a processing gain at the receiver which makes CDMA communications particularly adequate for low or very low SNR regime. At high SNRs, CDMA systems need to resort to equalization techniques, which are typically of high complexity. In contrast, OFDMA and SC-FDMA parallelize the transmission into multiple frequency subchannels, effectively elongating the symbol duration and, hence, becoming more resilient to channel dispersion and reducing the need for complex equalization.

The main reasons for the change of waveforms in the 3G to 4G transition were the scalability to larger bandwidths, the reduced receiver complexity, and the increased suitability for multiple-input, multiple output (MIMO) operation.

Receiver complexity for CDMA receivers stems from two major reasons:

- Finger management of RAKE receiver
- Equalization complexity for higher spectral efficiencies

RAKE receivers resort to multiple "fingers" latching to individual paths. The path resolution for a CDMA system is inversely proportional to the channel bandwidth. Hence, the path resolution for W-CDMA 3.84 Megachips per second (Mcps) is roughly three times that of IS-95 or cdma2000 1.2288 Mcps system. For CDMA systems with increased bandwidth, the multi-path resolution increases and, hence, more fingers are required for tracking. Also, since there are more resolvable paths, the energy per path drops making it harder to track them. One time-sample (or chip) duration in W-CDMA is 260.41 ns which makes path tracking already challenging. Scaling up the channel bandwidth from the 5 MHz of W-CDMA to say 20 MHz via direct up-sampling of the

CDMA transmitter would make path tracking even more challenging. That was the reason for looking into carrier aggregation solutions of the 5 MHz W-CDMA system to offer 20 MHz bandwidth operation in the study phase of LTE.

CDMA systems were initially envisioned for voice applications with relatively low spectral efficiency requirements and relying on processing gain inherent to CDMA for boosting the SNRs to positive values at the receiver side. Applications requiring higher spectral efficiencies require much larger SNRs, typically not achievable by a mere RAKE receiver leading to the resorting to equalization at the receiver. Equalization, even though it can be fairly effective, comes at the price of complexity. Various types of equalization were heavily studied and implemented in CDMA networks ranging from linear to nonlinear interference cancelation based. Everything looked so much simpler when turning into OFDM.

From the network perspective, 4G removed the concept of macro-diversity and migrated to a flatter architecture model. Note that this fundamental change remains in the 5G architecture. As we will see later, LTE eventually introduced coordinated multi-point (COMP) transmissions which, to some extent, re-introduced the macro-diversity concept in 4G but without the need of a base station controller (BSC) node to manage it. COMP is also present in 5G.

Some other aspects did not change in the 3G to 4G transition; notably, the channel coding remained turbo codes for data channels. There were some adjustments made to the coding of control channels where the regular Convolutional Codes of 3G were replaced by tail-biting convolutional codes in LTE.

From core network and system perspective overall as mentioned previously, EPS is the first fully packet switched system designed by 3GPP. The original goal was to provide all services in PS domain. This goal was watered down due to the introduction of CS Fallback (CSFB) in order to provide support for voice and SMS through transitioning to 2/3G CS. Progressively the majority of operators migrated to support Voice over IMS. In order to support this goal, every 4G LTE UE when it attaches to the network gets an IP address, unlike UMTS where usually the IP address was obtained after application/user's interaction. This "always on" IP connectivity is one of the major enablers of the internet connected "apps" that we are all enjoying in our phones every day.

Another major evolutionary step in EPS was a great simplification in the NAS and RRC protocols Finite State Machine (FSM) through, e.g., removal of URA-PCH and "suspended" state that existed in 2/3G. The QoS framework was also greatly simplified since the numerous QoS parameters that existed in UMTS QoS causing interoperability issues in real-life networks were replaced with only a handful. The main one is the QoS class identifier (QCI) that is a scalar value, which maps to the packet delay budget (PDB), packet error rate (PER), and priority of the specific EPS bearer.

1.3 Physical Layer Changes in 5G

At high level, Table 1.1 summarizes the differences at the physical layer between NR and LTE.

1.3.1 Frame Structure

LTE originally defined two frame structures: frame structure type 1 for FDD operation and frame structure type 2 for TDD operation. In Release 13, with the standardization

Concept	4G LTE	5G NR
Frame structure (FS)	FS type 1: Frequency division duplexing (FDD) FS type 2: Time division duplexing (TDD) FS type 3: Unlicensed operation	Single and highly configurable frame structure for all use cases. Notion of DL, UL, and Flexible resources.
Scheduling flexibility	1ms subframe based Release 15 introduced sTTI for shorter scheduling units	Slot based (type A scheduling) Mini-slot based (type B scheduling)
Waveform	DL: OFDM UL: SC-FDMA 15 kHz SCS	DL: OFDM UL: SC-FDMA* & OFDM FR1: 15 kHz, 30 kHz, 60 kHz (data) 15 kHz, 30 kHz (SSB) FR2: 60 kHz, 120 kHz (data) 120 kHz, 240 kHz (SSB) (*) SC-FDMA for single-layer transmissions.
Forward compatibility	Limited via MBSFN subframes and almost blank subframes (MBSFN subframes with very limited control information and including CRS)	Reserved resources
"Always on" signals	CRS (every DL slot) SSB (5 ms period)	No CRS SSB (20 ms period default)
Transmission modes	TM 1-7 (Release 8) TM8 (Release 9) TM9 (Release 10) TM10 (Release 11) SFBC TxDiv for control channels	Single-transmission mode for data channels Transparent TxDiv scheme
Multi-beam operation	N/A	FR1: Up to 8 SSB beams FR2: Up to 64 SSB beams
Channel coding	Data: Turbo codes Control: TBCC, as well as repetition, simplex, and Reed-Mueller coding for small payload sizes.	Data: LDPC codes Control: Polar codes Coding for control channels with less than 13 bits of information remain the same as in LTE, i.e., repetition, simplex, and Reed-Mueller.
Bandwidths & bandwidth part concept	BWs: 1.4, 3, 5, 10, 15, 20 MHz BWP: N/A	BWs: FR1: 5, 10, 15, 20, 25, 30, 40, 50, 60, 80, 90, 100 MHz FR2: 50, 100, 200, 400 MHz BWP: Up to 4 configured BWPs. Single-active BWP. Additional BWs could be added in future releases of NR.

Table 1.1 Summary of Physical Layer Differences Between NR and LTE

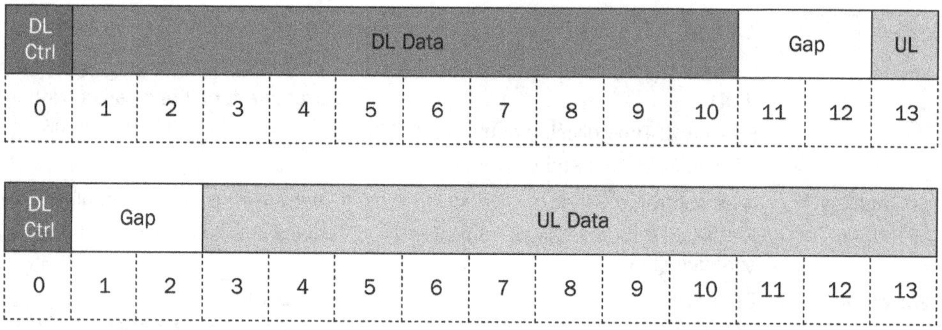

DL Ctrl	DL Data										Gap		UL
0	1	2	3	4	5	6	7	8	9	10	11	12	13

DL Ctrl	Gap		UL Data										
0	1	2	3	4	5	6	7	8	9	10	11	12	13

Figure 1.2 Exemplary NR slot configurations enabling DL and UL transmission in a same slot.

of unlicensed operation for LTE (LAA), frame structure type 3 was introduced in the LTE specifications.

Frame structure type 1 consisted of all DL subframes for the DL frequency and all UL subframes for the UL frequency. Frame structure type 2 introduced the concept of special subframes on top of the DL subframes and UL subframes. The special subframes had a duration of 1 ms and consisted of a portion of DL symbols at the beginning of the subframe, a portion of UL symbols at the end of the subframe, and a guard period in between. Release 8 introduced seven different frame structure type 2 configurations with 5 ms and 10 ms DL/UL switching periodicities. In addition, Release 8 introduced nine possible configurations of the special subframe with various durations of the DL portion (3, 9, 10, 11, 12 symbols), the UL portion (1 or 2 symbols), and the guard period (minimum 1 symbol).

In contrast, NR defines a single-frame structure without distinction of FDD, TDD, or unlicensed operation. Indeed, the definition of the slot format is so flexible in NR that a single definition enables a large variety of frame and slot formats. Figure 1.2 shows an illustration of two possible NR slot configurations.

NR slot format configuration is achieved by semi-static configuration (common plus possibly dedicated RRC configuration) in conjunction with the possibility of dynamic configuration for a given period of time. The designation of resources can be Downlink (D), Uplink (U), or Flexible (F) resources. One or two patterns can be created from a sequence of DL only slots at the beginning of the pattern (possibility for 0 DL only slots is also possible), a sequence of UL only slots (possibility for 0 UL slots is also possible) at the end of the pattern, and a slot with individual symbol configuration for D, U, and F resembling the special subframes in LTE. One important difference is that LTE originally had nine special subframes and NR has defined a table with 256 entries in Release 15 with "only" 56 being defined for Release 15 [1] (more details can be found in Chap. 5).

In case of two patterns configured, their concatenation and subsequent repetition creates the overall frame and slot configuration.

Dedicated semi-static configuration can only override Flexible resources. Dynamic slot format indication (SFI) is conveyed via Downlink control information (DCI) and consists of a pointer to a semi-statically configured slot configuration. This dynamic SFI is not expected to conflict with the semi-static configuration. Note that a link direction

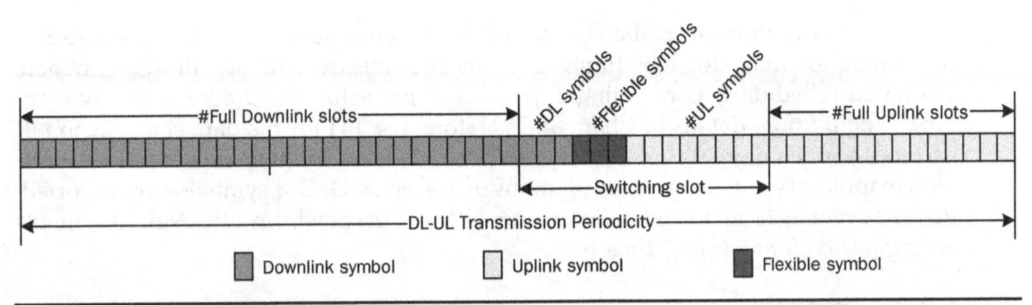

FIGURE 1.3 Exemplary frame and slot configuration.

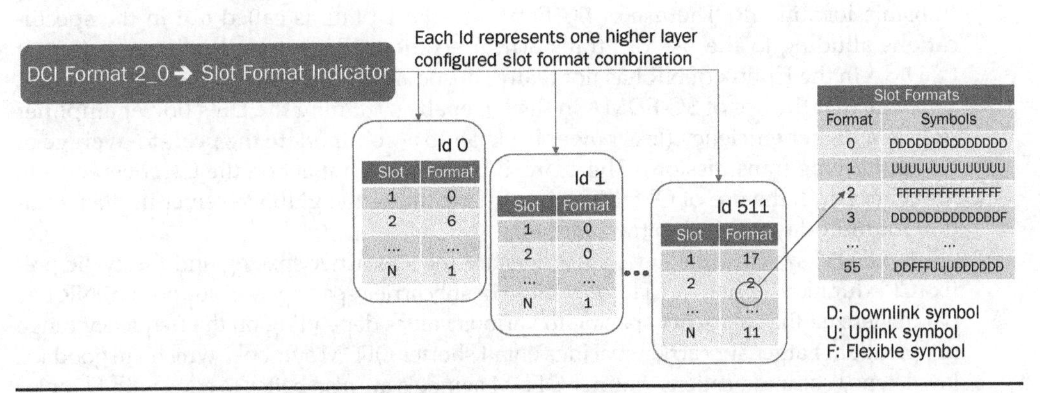

FIGURE 1.4 Exemplary slot format indication.

denoted as "F" in dynamic SFI is not deemed as in conflict with "D" or "U" in the semi-static configuration.

Figure 1.3 illustrates an exemplary frame and slot configuration resorting to full DL slots, full UL slots, and a slot with one switching point between DL and UL with a guard in-between.

Figure 1.4, in turn, illustrates an exemplary slot format indication by the combination of a RRC configuration and an indication by means of DCI format 2_0 of the corresponding slot format.

1.3.2 Scheduling Flexibility

Another aspect that NR has introduced relates to the scheduling flexibility. The scheduling unit or transmission time interval (TTI) was a very rigid quantity in WCDMA (3G) and LTE (4G). In 3G days, the Release 5 High Speed Downlink Packet Access (HSDPA) definition introduced 2 ms TTI for the "HS" channels, while Release 6 High Speed Uplink Packet Access (HSUPA) resorted to 2 ms and 10 ms TTIs for the related UL channels. Note that 10 ms TTI for the UL was to match the link budget of the circuit switch channels. In LTE Release 8, a single TTI of length 1 ms was defined for all the Physical layer channels. LTE Release 15 introduced the notion of short TTI (sTTI), enabling for the first time in LTE the scheduling units of less than 1 ms.

NR, in turn, introduces the concept of slot-based scheduling and non-slot–based scheduling, or mini-slot scheduling from its first release. The specifications denote slot-based scheduling as mapping Type A, and mini-slot–based scheduling as mapping Type B (more details in Chap. 8). Therefore, the TTI for the data channels in NR becomes a much more flexible concept. While in Release 15 there are some limitations with mapping Type B, namely only lengths of 2, 4, and 7 OFDM symbols are supported, this restriction gets alleviated in Release 16 with all duration from 2 through 13 symbols are supported for mapping Type B.

1.3.3 Waveform

OFDMA is introduced for the Uplink of NR on top of the LTE-defined SC-FDMA. Note that, SC-FDMA, also known as DFT-spread-OFDMA, is only supported for single-layer transmissions in NR. The use of SC-FDMA in the Uplink is called out in the specifications alluding to the use of "transform precoding." Therefore, the introduction of OFDMA in the Uplink of NR has not really caused much difficulty in the specifications.

Note that the use of SC-FDMA in the UL enables running the UE's power amplifier (PA) at a higher efficiency (less power back-off to accommodate the peak-to-average of the underlying transmission). Therefore, it has a direct impact on the UL coverage. On the other hand, the use of OFDM waveform facilitates the gNB's receiver implementation for multi-layer (MIMO) transmissions.

In OFDM systems, the key parameters are the subcarrier spacing and the cyclic prefix (CP) duration. While in LTE, only 15 kHz subcarrier spacing was supported, NR has parameterized the subcarrier spacing to various values depending on the frequency range of operation. Larger subcarrier spacings entail shorter OFDM symbols, which are good for low-latency communications. Shorter OFDM symbols are also better to cope with Doppler and the increased phase noise (PN) of transmissions at higher frequencies especially mmWave. Figure 1.5 illustrates the various subcarrier spacings introduced in NR Release 15.

NR has introduced two frequency ranges for operation:

- **Frequency Range 1** (FR1): for NR operations between 425 MHz and 7.125 GHz. This is often denoted as "sub 6" because the original upper bound was set to be 6 GHz which got increased to 7.1 GHz to include the 6 GHz unlicensed band. The data channels in FR1 can use 15, 30, or 60 kHz subcarrier spacing, whereas the sync channels are defined for 15 and 30 kHz.
- **Frequency Range 2** (FR2): for NR operations between 24.25 GHz and 52.6 GHz. This frequency range is often denoted as "mmWave" or, simply, "mmW." The data channels in FR2 can use 60 kHz or 120 kHz subcarrier spacing, whereas the sync channels are defined for 120 kHz and 240 kHz.

LTE defined normal CP and extended CP. The normal CP in LTE has an average duration of 4.76 μs and incurs a 6.66% CP overhead, whereas the extended CP has a duration of 16.66 μs and incurs a 20% CP overhead. In NR, the normal CP duration applies to all the subcarrier spacings with a duration scaled with the symbol duration so that the same CP overhead is incurred for all the different numerologies, i.e., 6.66%. In NR, the extended CP duration applies only to the 60 kHz subcarrier spacing duration.

There is no known LTE deployment employing the extended CP, and the extended CP duration for NR has been designated as an optional feature for the UE.

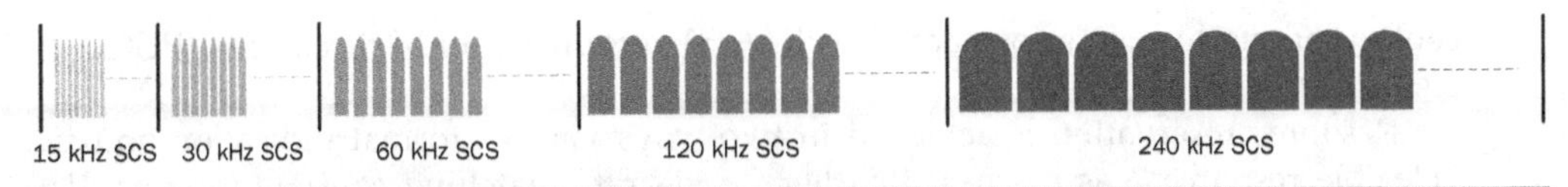

FIGURE 1.5 Illustration of various NR subcarrier spacings.

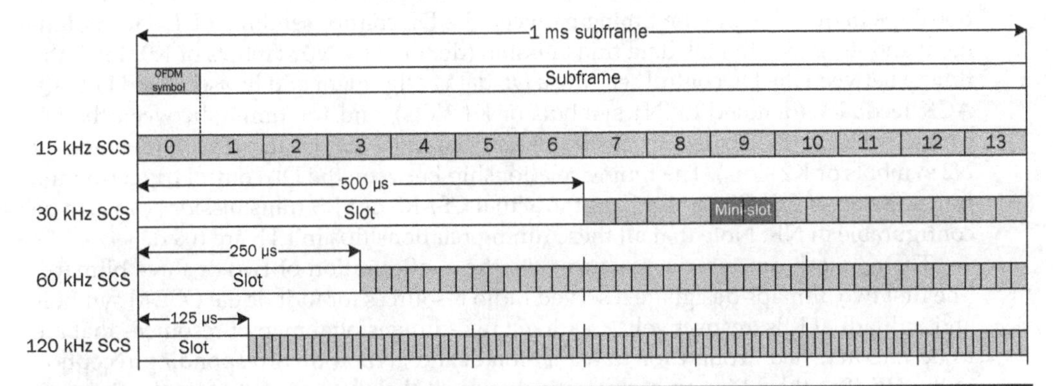

FIGURE 1.6 Illustration of various NR slot durations corresponding to different numerologies.

The scheduling of LTE, from its inception, was centered around the 1 ms *subframe* consisting of 14 OFDM symbols. NR reuses the concept of *subframe* to designate a duration of 1 ms. However, it introduces the concept of 14 OFDM symbol *slot* as the default scheduling unit across numerologies. As a result, the slot duration for 15, 30, 60, and 120 kHz corresponds to 1 ms, 500 µs, 250 µs, and 125 µs slot durations, respectively. Chapter 5 provides a more in-depth treatment of the NR numerology and slot structure.

NR also reuses the concept of resource block (RB) introduced in LTE. One RB consists of 12 subcarriers in a given numerology or subcarrier spacing. As a result, the bandwidth spun by an RB goes from 180 kHz for 15 kHz subcarrier spacing and scales in powers of 2 for increased subcarrier spacings. Figure 1.6 introduces the concept of radio frame, subframe, slot, and mini-slot which will be elaborated in more detail in Chap. 5.

1.3.4 Forward Compatibility

As discussed earlier, forward compatibility was a design goal for 5G NR from the start. Forward compatibility is achieved by making sure that the inclusion of features in the future will not adversely degrade operation or performance of legacy devices. This is achieved by defining reserved resources in the time-frequency grid of resources. The basic **time unit** for reservation is one OFDM symbol while the basic **frequency unit** for reservation is typically one resource block. Note that a finer frequency granularity of resource reservation corresponding to individual subcarriers or resource elements (REs) can be achieved for some limited patterns and only for 15 kHz subcarrier spacing, as discussed later.

Reserved resources should be discarded and not processed (received) at the UE receiver. It is assumed that UL transmissions by the UE will be naturally governed by the corresponding dynamic or configured grants, as well as by semi-static RRC configurations [e.g., random access (PRACH) resources] and UL control (PUCCH) resources.

Resource reservation is achieved in two ways via slot format indication and its "Flexible resources" as discussed earlier, or via rate matching configuration as discussed hereafter. Additionally, timing relationships in NR can be flexible/configurable, making it easier to schedule transmissions around certain resources. The timing relationships in question are the timing between the DL control sending a DL data assignment and its associated DL data transmission (denoted by N0 symbols or K0 slots), the timing between the DL control sending a DL data assignment and its associated HARQ-ACK feedback (denoted by N1 symbols or K1 slots), and the timing between the DL control sending the UL data grant and its associated UL data transmission (denoted by N2 symbols or K2 slots). The timing relationship between the DL control triggering the transmission of CSI-RS or SRS [and the actual CSI-RS or SRS transmission] can also be configurable in NR. Note that all these timing relationships in LTE are fixed.

The network can reserve resources via the configuration of two or three bitmaps. The first two bitmaps designate reserved radio resources for individual OFDM symbols and individual RBs, respectively, creating a two-dimensional map of resources that are to be rate-matched around for transmission at the gNB and corresponding reception at the UE. The third bitmap can selectively repeat the pattern created by the first two bitmaps over a longer time duration up to 40 ms. A bit in the DCI scheduling the corresponding PDSCH can selectively enable/disable a given reservation scheme for the associated scheduled DL transmission.

In addition, for 15 kHz subcarrier spacing, an additional RE level pattern can be configured with the objective to rate match NR DL data (PDSCH) transmissions around LTE CRS resource elements. The configuration enables the possibility to configure the number of CRS ports (up to 4), the frequency shift for the transmission of CRS, and the center carrier frequency and bandwidth.

Note that in addition to the forward compatibility, the resource reservation schemes introduced by NR enable backwards compatibility with LTE and what is being broadly denoted as dynamic spectrum sharing (DSS). Chapter 11 presents this feature of NR in detail.

Looking into the opportunities for forward compatibility that LTE has, the story is quite different because of the rigidity of LTE and the inability to completely blank out portions of the frame for legacy operation to ignore, while enabling new implementations to use those portions of the frame for something else. Indeed, LTE only has the possibility to configure the so-called MBSFN subframes without actually transmitting eMBMS (broadcast), which was the original intent for this type of subframes in LTE. The subframes configured as MBSFN subframes (and without the corresponding eMBMS transmission) will only have the control region corresponding to the first one or two OFDM symbols of the corresponding MBSFN subframe occupied, while the corresponding data portion of the subframe will be left empty. Note that in the control region of the MBSFN subframes, at the very least the REs corresponding to CRS will have to be transmitted. In addition, that control region could carry UL grants for corresponding UL subframes.

1.3.5 Minimizing "Always On" Channels

LTE Downlink has the common reference signals (CRS) transmitted in every Downlink slot. This physical signal is used by the UE receiver for multiple purposes: time and frequency tracking, channel state information measurement, and mobility (RRM) measurements, in addition to providing the phase reference for the demodulation of the DL control and data channels (Transmission Modes 1 through 6). There is no CRS equivalent in NR. Instead, NR resorts to channel specific demodulation reference signals (DMRS) which were also introduced in LTE for Transmission Modes 7 through 10. DMRS transmission is embedded within the resource allocation of the corresponding NR physical layer channel [physical broadcast channel (PBCH), physical Downlink control channel (PDCCH), physical Downlink shared channel (PDSCH), physical Uplink control channel (PUCCH), physical Uplink shared channel (PUSCH)]. Note that incurring upfront the overhead associated with the CRS transmission diminishes the benefits of having a yet additional reference signal for demodulation. There are several benefits associated with DMRS operation. DMRS-based operation enables beamforming of reference signal in the same way as the associated control or data transmission, making the application of the channel estimates obtained from DMRS to demodulate the corresponding control or data straight forward. Also, the reference signal overhead for higher rank transmission in multi-layer MIMO transmissions is only incurred when needed (unlike CRS which would require multiple antenna-port transmission all the time incurring always the higher overhead). Also, not having to transmit CRS brings multiple benefits, especially to the network, as it enables lower network energy consumption and reduced inter-cell interference.

In LTE, the synchronization signals are transmitted every 5 ms. In NR, in contrast, they are transmitted every 20 ms by default (longer and shorter periodicities are allowed by the specifications). More details on the synchronization signals will be covered in Chap. 6.

As a result, transmission of "always on" signals in NR has been reduced considerably. This is positive from the gNB energy efficiency perspective, as well as from the generation of inter-cell interference perspective (oftentimes dominated by CRS interference in lightly loaded LTE networks). However, the absence of a signal for the UE to latch on makes it more difficult for it to track and measure. As a result, NR operation is expected to possibly degrade UE power consumption and increase initial cell search times versus regular LTE operation.

1.3.6 Transmission Modes

The first release of LTE defined seven transmission modes (TMs), TM1 through TM7, for the transmission of the Downlink data channel (PDSCH) semi-statically configured by higher layers. Release 9 introduced TM8 enabling dual-layer beamforming for LTE TDD systems and which has been extensively used in LTE TDD deployments in China.

Two additional transmission modes were introduced in Release 10 and Release 11, respectively. Table 1.2 provides a high-level view of the 10 transmission modes of LTE.

The real necessity of all the10 transmission modes for LTE was tested by what was eventually adopted in actual network deployments. There are some transmission modes that have not seen the light of a broad deployment or have never been deployed. Only TM3, TM4, and TM8 have been broadly deployed in LTE networks.

This is probably one of the reasons where NR went for a radically simpler approach and a single-transmission mode is defined for the Downlink data channel (PDSCH).

In addition, LTE introduced transmit diversity for DL control channels based on space frequency block codes (SFBC). NR, in turn, has resorted to transparent transmit

LTE transmission mode #	Description
1	Release 8 CRS based. Single-antenna port, port 0.
2	Release 8 CRS based. Transmit diversity (TxDiv).
3	Release 8 CRS based. Large delay CDD (TM1 fallback)
4	Release 8 CRS based. Closed-loop spatial multiplexing (TM1 fallback)
5	Release 8 CRS based. Multi-user MIMO (TM1 fallback)
6	Release 8 CRS based. Closed-loop spatial multiplexing with single layer (TM1 fallback)
7	Release 8 DMRS based. Single-antenna port, port 5 (TM1 or TM2 fallback).
8	Release 9 DMRS based. Dual-layer beamforming
9	Release 10 DMRS based. Multi-layer transmission (up to 8 layers)
10	Release 11 for support of coordinated multi-point (CoMP)

TABLE 1.2 LTE Transmission Modes

diversity schemes for control channels and sync signals. The transparency is from the specification and receiver design perspective. The diversity, in turn, is achieved by the use of small cyclic delay diversity schemes not requiring the specification for interoperability between a transmitter and a receiver.

1.3.7 Multi-Beam Operation

Highly directional transmissions have been always used in mmWave systems to cope with the harsh propagation conditions of this frequency regime. The high directionality is typically achieved by an antenna array with a large number of elements, e.g., 128, 256, or even more. These transmissions achieve a respectable antenna and array gain which will help "closing the loop" for the establishment of a communication link. In a system beyond a single point-to-point link, a "codebook of beams" is necessary in order to cover a given geographical coverage area.

In order to maximize the antenna and array gain of a given transmission, all antenna elements contribute to the creation of the "codebook of beams" as illustrated in Fig. 1.7. RFIC implementations of these large antenna arrays typically provide per-element amplitude and phase control to yield the codebook of the so-called "analog beams." These beams can be typically very narrow (directive) but, unlike digitally created beams, which can apply to a given limited frequency resource allocation, these narrow beams, which are applied in the analog domain, apply to the entire transmission bandwidth. This is the reason why analog beamforming requires the concept of "beam sweeping" in order to cover the aggregated area of the beams belonging to the "codebook of beams" (see Fig. 1.7).

NR specifications have been developed for the two frequency ranges of operation concurrently. The concept of multi-beam operation, therefore, appears not only in the context of FR2 but also, albeit at a smaller extent, in the context of FR1 as well.

Synchronization signals themselves can be transmitted in up to 64 beams for FR2 and up to 8 beams in FR1. The acquisition of the synchronization signals in a given

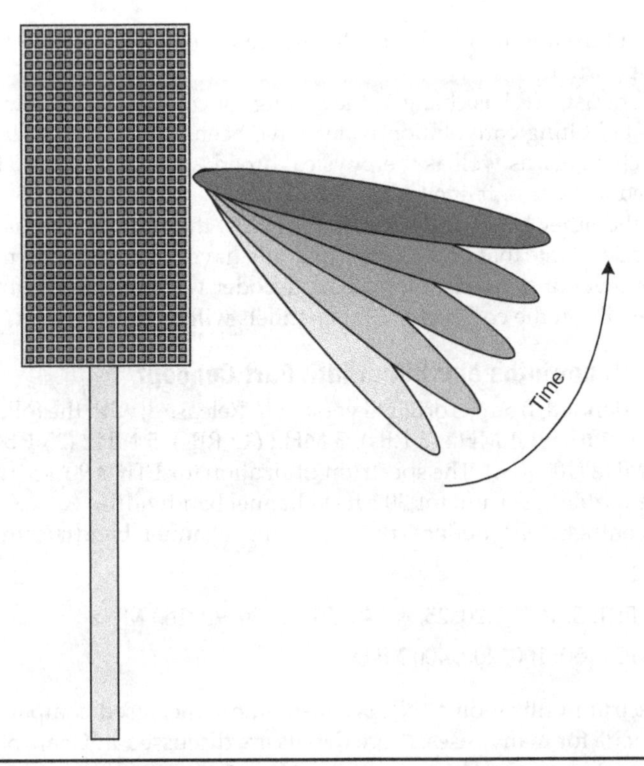

Figure 1.7 Illustration of analog beamforming.

beam direction starts a bootstrap procedure for the UE to choose a given set of resources so that the corresponding gNB's reception is tuned at that corresponding beam direction. Once a connection is established, an entire beam management procedure is introduced in NR, very much similar to the regular mobility procedure with measurements, reports, and failures, but limited to the serving cell and staying at medium access control (MAC) level (unlike the regular mobility procedure which takes place at RRC).

1.3.8 Channel Coding

LTE (4G) did not change the channel coding for the data channels from those used for 3G (WCDMA and cdma2000) reusing the same turbo codes which had been introduced in 3G UMTS. One change that LTE introduced was the simplification of the rate matching procedure with the introduction of the "circular buffer" concept to determine the coded bits to transmit for the different redundancy versions of the transmission. Note that this circular buffer concept still applies to NR.

For control information, LTE introduced tail-biting convolution codes (TBCC) replacing the regular convolutional codes that were in use in 3G to save the overhead incurred by the tail bits, which were used to take the trellis for Viterbi decoding to the all-zero state. TBCC is used in LTE for control payload sizes of more than 22 bits. For smaller control payloads, repetition coding (for 1-bit of information), simplex code (for

2-bit of information), and Reed-Mueller codes (for 3 through 22 bits of information) were supported.

In contrast, NR has changed the coding for control and data transmissions.

The tail-biting convolution codes have been replaced by polar codes. DL and UL control channels, as well as the physical broadcast channel (PBCH) use polar codes as described in more detail in Chap. 7.

On the other hand, for the data channels, the turbo codes have been replaced by LDPC codes. Note that LDPC codes not only have superior performance but also enable a higher level of parallelization at the decoder which is critical for Gbps applications. More details on the coding for data channels will be presented in Chap. 8.

1.3.9 Bandwidths and Bandwidth Part Concept

LTE was defined, from its original version in Release 8, with the following possible channel bandwidths: 1.4 MHz (6 RBs), 3 MHz (12 RBs), 5 MHz (25 RBs), 10 MHz (50 RBs), and 20 MHz (100 RBs). The spectrum utilization for LTE is 90% across bandwidths, e.g., 18 MHz usable spectrum for 20 MHz channel bandwidth.

In contrast, NR defines the following channel bandwidths for FR1 and FR2, respectively:

- FR1: 5, 10, 15, 20, 25, 30, 40, 50, 60, 80, 90, 100 MHz
- FR2: 50, 100, 200, 400 MHz

The spectrum utilization of NR is considerably increased compared to LTE, i.e., levels around 95% for many cases. More details are discussed in Chap. 5.

DL control transmission (in PDCCH) in LTE spans the entire system bandwidth. This was the main obstacle to enable reduced bandwidth operation in LTE when it was being devised for MTC applications. Indeed, the introduction of ePDCCH (for relay operation) and MPDCCH (for eMTC operation) was mainly motivated to have the opportunity to confine the transmission of the DL control channel within certain (narrow) transmission bandwidth. In contrast, in NR, from the start, the transmission of PDCCH is confined within a so-called CORESET or COntrol REsource SET. The bandwidth of the CORESET is fully configurable as discussed in Chap. 7. In addition, NR introduces the notion of bandwidth part (BWP) for DL and UL. The BWP concept enables operation confined to the bandwidth of the BWP. In NR, there are up to four configured BWPs but only one active BWP. Chapter 5 provides more details of the BWP operation and how it enables reduced bandwidth operation at the terminal side, as well as enabling load balancing at the network side.

1.4 Protocol Changes in 5G

Figures 1.8 and 1.9 illustrate the protocol stack of LTE for user plane and control plane, respectively [2].

Figures 1.10 and 1.11, in turn, illustrate the protocol stack of NR for user plane and control plane, respectively [3].

Clearly, the control plane protocol stacks of NR and LTE are, fundamentally, the same. Note that the network nodes are called differently, namely, a base-station in LTE is called an eNB, while in NR it is called a gNB. Also, NAS protocol is terminated at the mobility management entity (MME) in LTE, while it is terminated at the access and

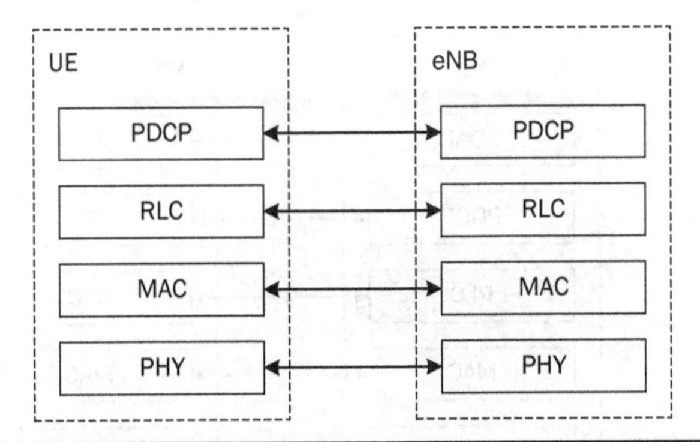

FIGURE 1.8 LTE user plane protocol stack. (PDCP: packet data convergence protocol; RLC: radio link control; MAC: medium access control; PHY: physical)

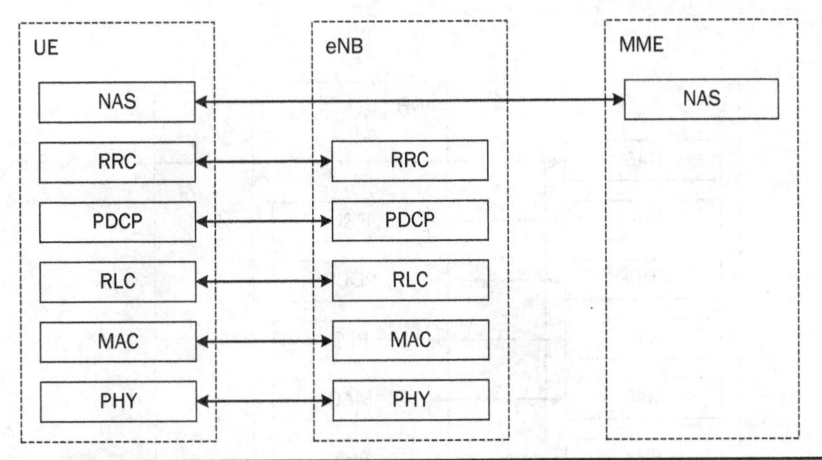

FIGURE 1.9 LTE control plane protocol stack. (NAS: non-access stratum; RRC: radio resource control)

mobility management function (AMF) in NR. More details on the internal split in NAS in mobility management (MM) and session management (SM) and the CN termination are provided in Chap. 3.

For the user plane, the NR protocol stack introduces a new layer above packet data convergence protocol (PDCP), namely, service data adaptation protocol (SDAP). The main services and functions of SDAP include:

- Mapping between a QoS flow and a data radio bearer
- Marking QoS flow ID (QFI) in both DL and UL packets

A single-protocol entity of SDAP is configured for each individual packet data unit (PDU) session.

An example of the layer 2 data flow is depicted in Fig. 1.12 [3], where a transport block is generated by MAC by concatenating two radio link control (RLC) PDUs from

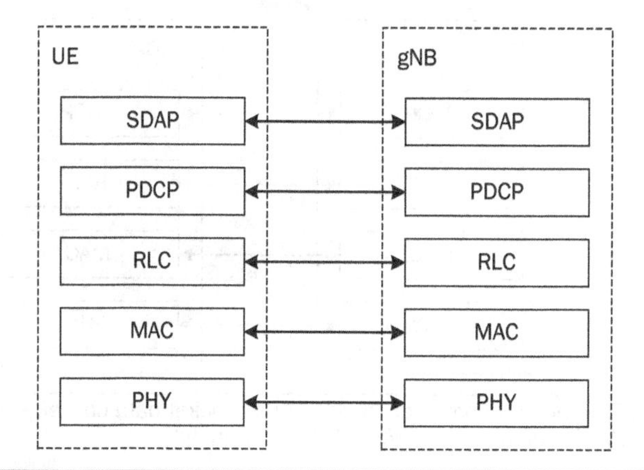

FIGURE 1.10 NR user plane protocol stack. (SDAP: service data protocol)

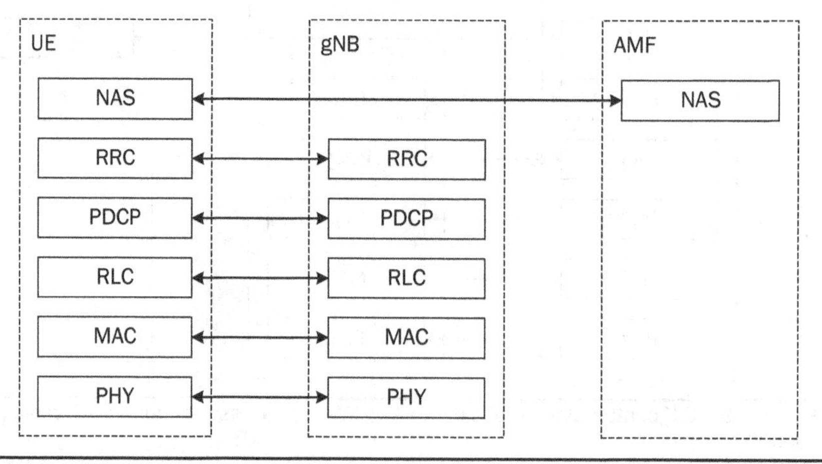

FIGURE 1.11 NR control plane protocol stack.

radio bearer "RB_x" and one RLC PDU from Radio Bearer "RB_y". The two RLC PDUs from RB_x each corresponds to one IP packet (n and $n+1$) while the RLC PDU from RB_y is a segment of an IP packet (m). Note, H depicts the headers and subheaders.

1.5 Main Physical Layer Features of LTE over Releases

Despite the accelerated schedule of the first release of NR, one can say that many of the functionalities provided by seven releases of LTE (Release 8 through Release 14) went into the first release of NR, namely, Release 15.

LTE grew in complexity over time with the addition of features in the course of the seven releases which preceded the first release of NR, i.e., Release 8 through Release 14.

Table 1.3 summarizes the main features of LTE introduced after the first version in Release 8.

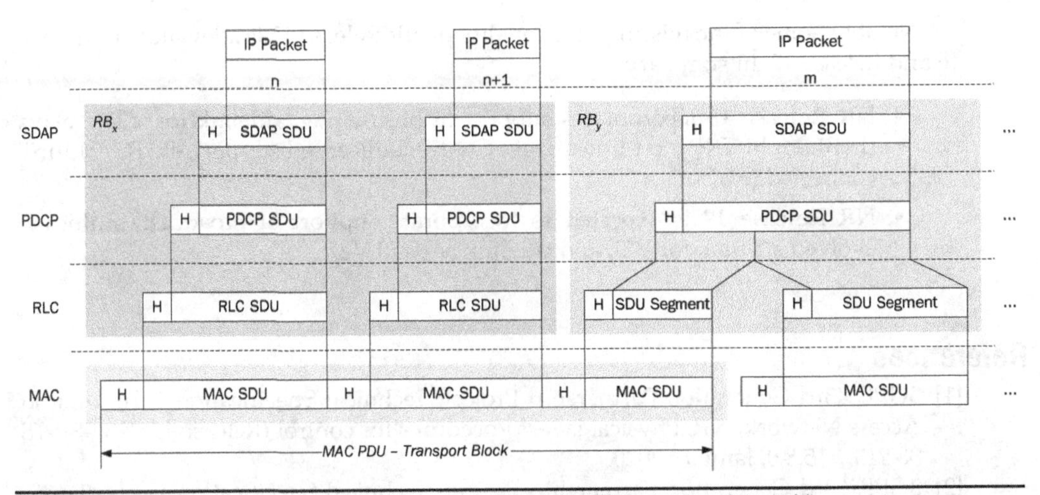

FIGURE 1.12 A data flow example. (PDU: packet data unit; SDU: service data unit)

Release	Modulation		# Layers		#CCs	Salient features of the release
	DL	UL	DL	UL		
Rel-8	64-QAM	64-QAM	1	1	1	First release
Rel-9	64-QAM	64-QAM	1	1	1	Positioning, TM8, eMBMS
Rel-10	64-QAM	64-QAM	5	4	5	TM9, CA, UL-MIMO, eICIC, Relays, HeNBs
Rel-11	64-QAM	64-QAM	5	4	5	TM10 (CoMP), FeICIC, EPDCCH, IDC
Rel-12	256-QAM	64-QAM	5	4	5	D2D, MTC, FDD-TDD CA, DC, eIMTA, NAICS, WLAN traffic offloading
Rel-13	256-QAM	64-QAM	32	4	32	D2D+, LAA, LWA, eMTC, NB-IoT, FD-MIMO
Rel-14	256-QAM	256-QAM	32	4	32	eLAA, eMTC+, NB-IoT+, V2x, eFD-MIMO, MUST, EnTV
Rel-15	1024-QAM	256-QAM	32	4	32	sTTI, drones, evolutions of other items

Note: CC: component carrier; CA: carrier aggregation; eMBMS: evolved multimedia broadcast/multicast service; ICIC: inter-cell interference coordination; eICIC: enhanced ICIC; FeICIC: further enhanced ICIC; HeNB: home eNB; EPDCCH: enhanced PDCCH; IDC: in-device coexistence; D2D: device-to-device; DC: dual connectivity; eIMTA: enhanced interference mitigation and traffic adaptation; NAICS: network-assisted interference cancellation and suppression; WLAN: wireless local area network; LAA: license-assisted access; eLAA: enhanced LAA; LWA: LTE WiFi aggregation; eMTC: enhanced MTC; NB-IoT: narrow-band internet of things; FD-MIMO: full dimension MIMO; V2x: vehicle to everything; eFD-MIMO: enhanced FD-MIMO; MUST: multi-user superposition transmission; EnTV: enhanced television; sTTI: short TTI.

TABLE 1.3 Summary of Physical Layer LTE Features over 3GPP Releases

Notably, some of the missing functionality of NR Release 15 is alleviated in Release 16 and Release 17. In summary:

- **NR Release 16** incorporates support of positioning, sidelink for V2X, relays (IAB), Multi-TRP (non-transparent CoMP), unlicensed support, eMTC/NB-IoT connected to 5GC.
- **NR Release 17** is expected to incorporate support of broadcast/multicast, evolved sidelink for V2X and sidelink beyond V2X.

References

[1] 3GPP, "3rd Generation Partnership Project; Technical Specification Group Radio Access Network; NR; Physical layer procedures for control (Release 15)," 3GPP TS 38.213, v15.8.0, January 2020.
[2] 3GPP, "3rd Generation Partnership Project; Technical Specification Group Radio Access Network; Evolved Universal Terrestrial Radio Access (E-UTRA) and Evolved Universal Terrestrial Radio Access Network (E-UTRAN); Overall description; Stage 2 (Release 15)," 3GPP TS 36.300, v15.8.0, January 2020.
[3] 3GPP, "3rd Generation Partnership Project; Technical Specification Group Radio Access Network; NR; NR and NG-RAN Overall Description; Stage 2 (Release 15)," 3GPP TS 38.300, v15.8.0, January 2020.

CHAPTER 2

Deployment Scenarios

With progressing generations of telecommunication standards, the required bandwidth for higher data rates dictates the use of wider and wider swath of spectrum, while at the same time finding new spectrum for use has become more challenging. To answer this challenge, the 5G standard from the start targeted utilizing an array of different spectrum types. In terms of frequency range, the target spectrum extends from 410 MHz to 52.6 GHz. This wide range of spectrum consists of two noncontiguous frequency ranges, as shown in Table 2.1. Frequency Range 1 (FR1) [1] was originally called sub-6GHz and it ranged from 410 MHz to 6 GHz. Later, the upper end of FR1 was extended to 7.125 GHz. Frequency Range 2 (FR2) [2] is also called mmWave, and it ranges from 24.25 to 52.6 GHz.

The intermediate range between FR1 and FR2 was reserved for future standardization.

Another dimension of spectrum planning is the type of spectrum, which could be, for example, licensed, lightly licensed, unlicensed, or shared. While the importance of using all these types of spectrum was recognized early on, the first release of NR targeted using only licensed spectrum. Of course, the specified operation can be applicable to other spectrum types as well but more customized operation for other spectrum types, i.e., other than licensed, was planned for later releases starting from Release 16.

2.1 LTE-NR Spectrum Sharing

As early as at the start of the spectrum scenario discussions, various concerns were raised. One of the concerns was the time required to implement and deploy the new NR core network. Therefore, an intermediate step—so-called non-standalone (NSA) operation—was introduced, wherein the RRC functionality, initial access, idle mode would be all performed in LTE, while in connected mode, the UE is configured in an LTE-NR dual connectivity mode (EN-DC) [3]. The comparison of standalone (SA) and non-standalone (NSA) modes is depicted in Fig. 2.1.

Frequency range designation	Corresponding frequency range (MHz)
FR1	410–7125
FR2	24250–52600

TABLE 2.1 Frequency Ranges

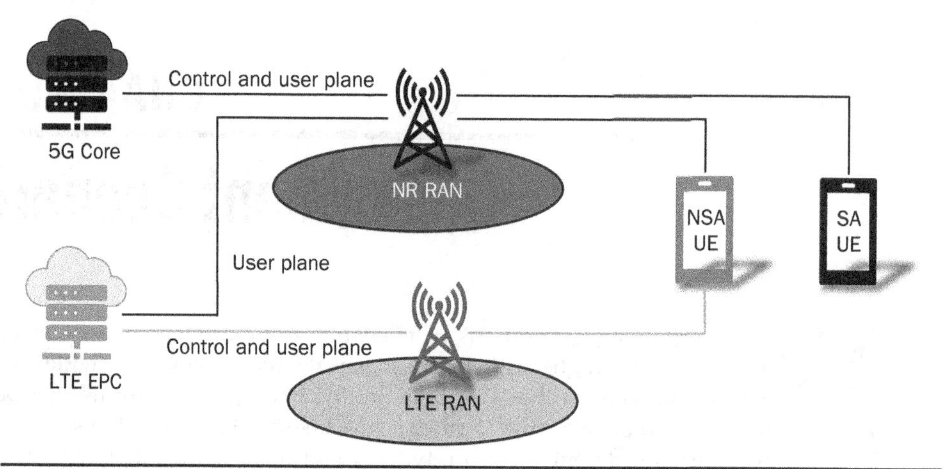

Figure 2.1 5G standalone (SA) and non-standalone (NSA) comparison.

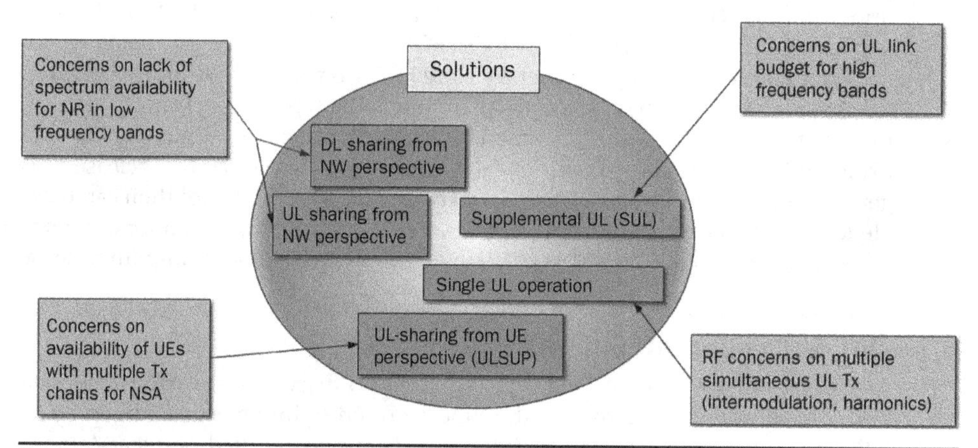

Figure 2.2 Problem statement and corresponding solutions for spectrum sharing.

The NSA and, to some extent, the SA modes of operation in turn raised various new concerns and corresponding solutions, depicted in Fig. 2.2.

In the following, we discuss the details of the various solutions mentioned in Fig. 2.2.

An important dimension of spectrum planning is the various flavors of spectrum aggregation that is also closely related to the concerns on UL link budget mentioned in Fig. 2.2. Using mmWave spectrum can satisfy the need for new wide spectrum; however, it is difficult to achieve this goal while providing universally wide coverage. In order to provide broad spectrum with mmWave while providing wide coverage with another spectrum type, combination of FR2 with FR1 frequencies can be utilized. The combination can be achieved with carrier aggregation (CA) or with dual connectivity (DC). Note that NR was designed with all flavors of CA in mind, but DC was originally not

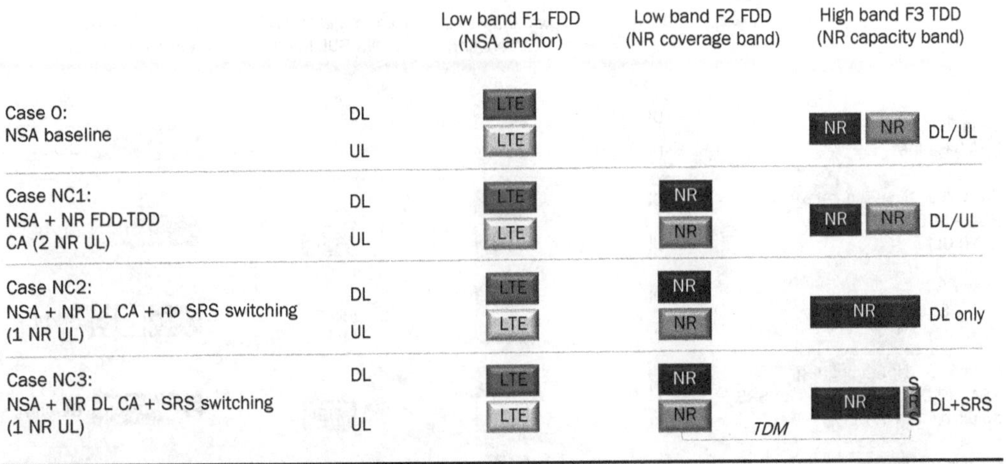

		Low band F1 FDD (NSA anchor)	Low band F2 FDD (NR coverage band)	High band F3 TDD (NR capacity band)
Case 0: NSA baseline	DL	LTE		NR NR DL/UL
	UL	LTE		
Case NC1: NSA + NR FDD-TDD CA (2 NR UL)	DL	LTE	NR	NR NR DL/UL
	UL	LTE	NR	
Case NC2: NSA + NR DL CA + no SRS switching (1 NR UL)	DL	LTE	NR	NR DL only
	UL	LTE	NR	
Case NC3: NSA + NR DL CA + SRS switching (1 NR UL)	DL	LTE	NR	NR SRS DL+SRS
	UL	LTE	NR	TDM

Figure 2.3 Example scenarios for NSA with NR CA.

intended to be part of the first NR release. This was changed when DC was also added as a Release 15 late drop feature, together with other architecture options. However, the Release 15 NR-DC version was limited to the following only:

- Only SA (i.e., there are no three cell groups supported)
- Only FR1+FR2 DC (i.e., the MCG is fully confined in FR1 and SCG is fully confined in FR2)
- Only synchronous DC (in order to make the physical layer operation in the UE the same as for CA)

Other DC operating modes will be introduced in Release 16.

Single-carrier operation is supported in both frequency division duplex (FDD) and time division duplex (TDD) duplex modes. In addition, for multi-carrier operation, additional carried types, Supplemental Downlink (SDL), and Supplemental UL (SUL) were introduced. These are not intended to be used as standalone carriers supporting unicast operation but rather they are attached to another carrier that has both DL and UL. The other carrier can be either FDD or TDD.

Some of the aggregation types for NSA operation are summarized in Fig. 2.3. Note that in Fig. 2.3, vertical separation in each case signifies frequency separation (FDM) within a band, while horizontal separation signifies time separation (TDM) within a band.

Figure 2.4 summarizes various scenarios where SUL is used.

Note that certain cases are marked as not supported in Fig. 2.4. This is because these cases would assume aggregating an SDL and an SUL carrier, with the SDL carrier in a TDD band, which is not supported in the specification. As it was mentioned earlier, either SDL or SUL were meant to be associated with another carrier that has both DL and UL. In addition, case NS1 in Fig. 2.4 is marked as not supported because simultaneous UL transmission in the SUL and in the other NR UL carrier (except for SRS transmission) is precluded.

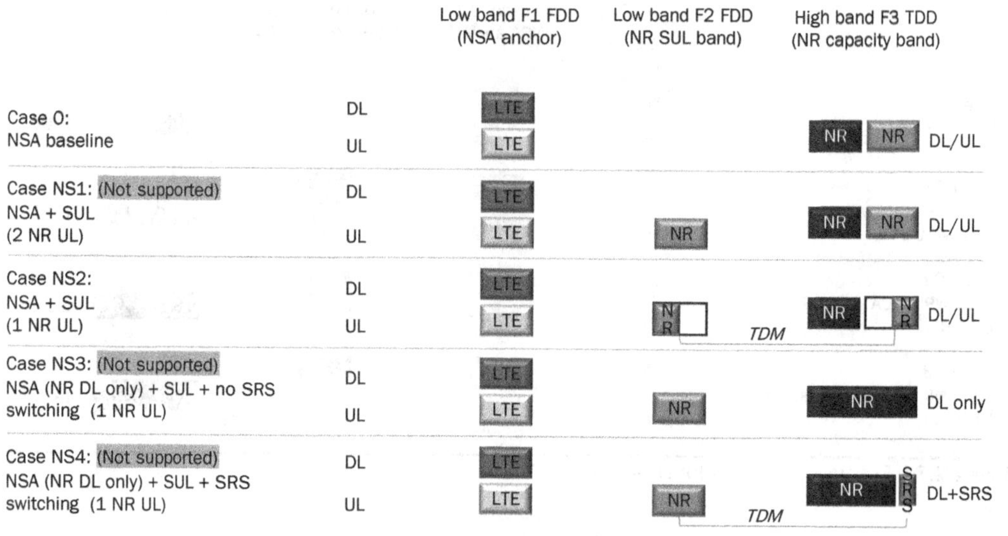

		Low band F1 FDD (NSA anchor)	Low band F2 FDD (NR SUL band)	High band F3 TDD (NR capacity band)
Case 0: NSA baseline	DL	LTE		
	UL	LTE		NR NR DL/UL
Case NS1: (Not supported) NSA + SUL (2 NR UL)	DL	LTE		
	UL	LTE	NR	NR NR DL/UL
Case NS2: NSA + SUL (1 NR UL)	DL	LTE		
	UL	LTE	NR (TDM)	NR NR DL/UL
Case NS3: (Not supported) NSA (NR DL only) + SUL + no SRS switching (1 NR UL)	DL	LTE		
	UL	LTE	NR	NR DL only
Case NS4: (Not supported) NSA (NR DL only) + SUL + SRS switching (1 NR UL)	DL	LTE		
	UL	LTE	NR (TDM)	NR SRS DL+SRS

FIGURE 2.4 Example scenarios for NSA with NR SUL.

Another aspect mentioned in Fig. 2.2 is that in certain regions it was seen problematic to rely on "green-field" FR1 spectrum available for an operator to deploy NR, so various sharing schemes were introduced, where the spectrum used for LTE is reused for NR while not removing LTE users. The techniques used to achieve coexistence of LTE and NR users are comprehensively called dynamic spectrum sharing (DSS). It was envisioned that the refarming from LTE to NR would be carried out in a multi-step manner, with DSS as intermediate step, as depicted in Fig. 2.5.

The detailed mechanisms of supporting LTE-NR coexistence will be discussed in Chap. 11.

As it was mentioned, DSS can be seen as a transitionary phase, since once all UEs are NR UEs (i.e., all UEs support NR via either single mode or part of multi-mode capabilities) there should be no motivation to maintain the LTE network. Even when the ratio of NR UEs to the total is not yet close to 100 percent but achieves, say, 50 percent, it is already reasonable to deploy the NR UEs on fully refarmed carriers without LTE UEs. This results in so-called hard spectrum partitioning, where carriers are designated as either LTE or NR according to the need of the supported UE population. This represents certain suboptimality, since the partitioning cannot respond to dynamic variation, such as when the active UE population in a certain cell at a certain time instance happens to shift toward more NR or more LTE at a given time. But in exchange for that suboptimality, hard partitioning is much simpler and it allows utilizing NR features to their full extent. To contrast this with so-called soft partitioning, soft partitioning allows for better adaptation to temporal traffic needs but it has major downsides in having to support the overhead of both systems, e.g., both LTE PSS/SS/PBCH and NR SSB have to be transmitted, and the NR and LTE transmissions have to be scheduled around each other that results in increased latency in general.

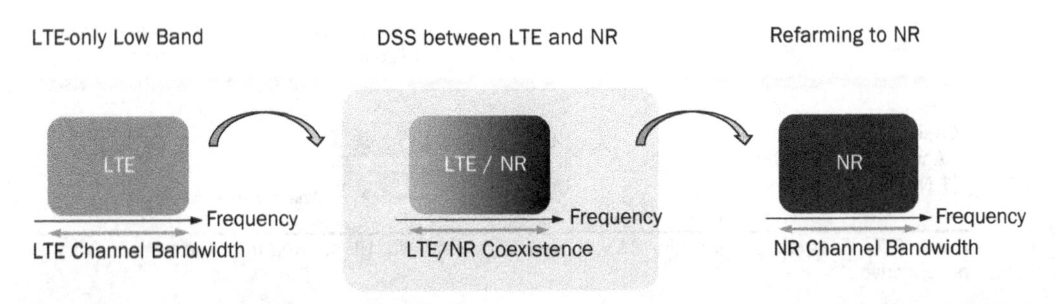

FIGURE 2.5 Transitioning through dynamic spectrum sharing.

		Low band F1/F2 FDD (NR coverage band)	High band F3 TDD (NR capacity band)
Case SC1: SA NR FDD-TDD CA (2 NR UL)	DL	LTE / NR	NR NR DL/UL
	UL	LTE / NR	
Case SC2: SA NR DL CA + no SRS switching (1 NR UL)	DL	LTE / NR	NR DL only
	UL	LTE / NR	
Case SC3: SA NR DL CA + SRS switching (1 NR UL)	DL	LTE / NR	NR SRS DL+SRS
	UL	LTE / NR	TDM

FIGURE 2.6 Example scenarios for SA with DL and UL sharing from the network perspective.

In Fig. 2.6, example scenarios with DL and UL sharing are shown.

Note that in the scenarios in Fig. 2.6, a carrier is shown as shared between LTE and NR; however, this is from the viewpoint of the network only; hence, this scheme is called sharing from the network perspective. The base station serves both LTE and NE UEs on the same carrier at the same time. However, there is no sharing in the above scenarios from the UE's perspective. Each UE is configured to operate with LTE only or NR only on a given carrier.

A further variation of sharing is shown in Fig. 2.7. Case SC4 is UL sharing from the network perspective, without DL sharing. The SA NR UL is paired with a DL in an SDL band. There has been considerable debate about whether to define the NR operation in this scenario as an aggregation of an SDL and an SUL carrier, or more simply just as a new FDD band. It had been argued that aggregation of an SDL and an SUL carrier can be more straightforward from the regulatory perspective. However, after some information exchange, this was found not to be the case, and it was also found that defining a new FDD band for this type of operation was more appropriate, hence the decision was to define this scenario as a new FDD band.

Although, in theory, one could take any pair of FDD bands—taking the DL from one and pairing it with the UL of the other—in order to create a new FDD band, it was not the intent to allow this in the general case. In order to form a new FDD band, the DL

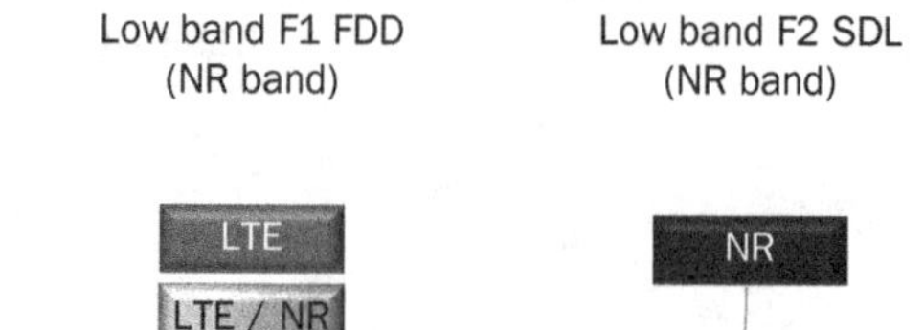

FIGURE 2.7 Example scenario for SA with new FDD band with UL sharing from the network's perspective.

NR band	Channel bandwidths for UL (MHz)	Channel bandwidths for DL (MHz)
n66	5, 10	20, 40
	20	40
n70	5	10, 15
	5, 10, 15	20, 25
n71	5	10
	10	15
	15	20

TABLE 2.2 FDD Asymmetric UL and DL Channel Bandwidth Combinations

NR band	Channel bandwidths for UL (MHz)	Channel bandwidths for DL (MHz)
n50	60	80

TABLE 2.3 TDD Asymmetric UL and DL Channel Bandwidth Combinations

part needs to come from an existing SDL band, and the UL part needs to come from an LTE band that was targeted for UL sharing from the network's perspective.

The new FDD bands to be defined based on Case SC4 have a couple of additional attributes. The first is that the duplex separation, i.e., the frequency offset between the center of the DL channel and UL channel allocated to a UE must be flexible. This is because the DL and UL portions of such bands have been obtained independently by the operator; therefore, maintaining a fixed defined separation between DL and UL would be impractical. The second attribute is that the channel bandwidth of the DL and UL is typically asymmetric, for similar reasons. Note that neither flexible duplex separation nor asymmetric DL/UL channel bandwidth is unique to the newly defined FDD bands, they do exist for some pre-existing FDD bands. However, they exist more on an exceptional basis for pre-existing bands, while they are expected to be the typical case to these new FDD bands. As examples, the following cases of asymmetric DL and UL bandwidth designations are defined [1] for FDD and TDD shown in Tables 2.2 and 2.3, respectively.

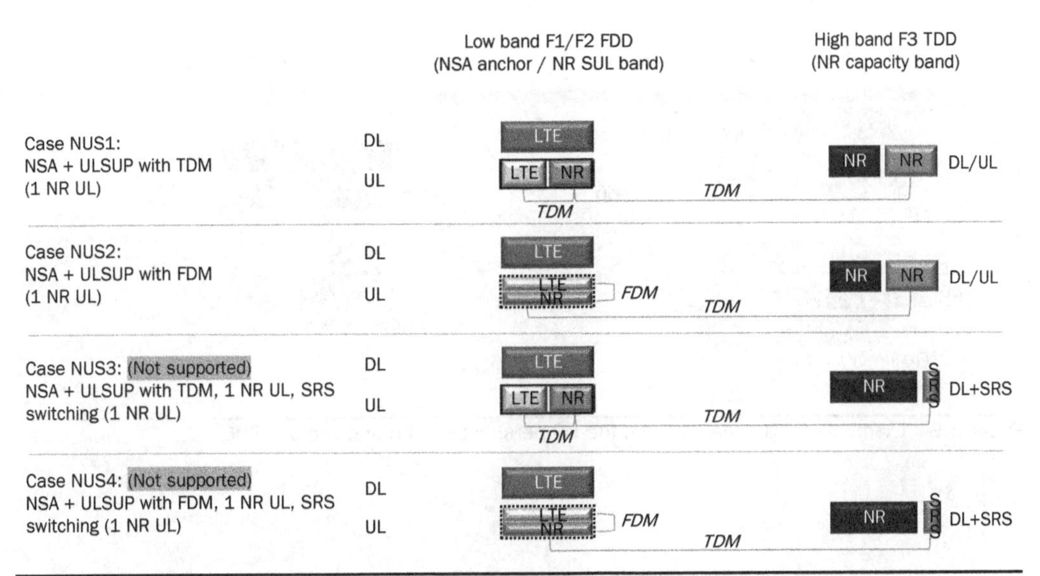

Figure 2.8 Example scenarios for NSA with UL sharing from the UE perspective.

Another newly introduced scenario is UL sharing from the UE perspective (ULSUP) [3]. What distinguishes this from the sharing scenarios from the network perspective is that, as the name implies, the UE is configured to operate both LTE and NR on the same carrier at the same time. There are two ULSUP multiplexing modes, FDM and TDM, each making sure that actual transmission allocations do not overlap. The mechanisms to enable these modes are similar to sharing from the network's perspective. However, UL sharing from the UE perspective is only allowed when LTE and NR use the same subcarrier spacing, which must be 15 kHz.

We also note that sharing from the UE perspective always necessarily involves sharing from the network's perspective; while the converse is not true.

Example scenarios for ULSUP are shown in Fig. 2.8.

Note that certain cases are marked as not supported in Fig. 2.8. This is because these cases would assume aggregating an SDL and an SUL carrier, with SDL in a TDD band, which is not supported in the specification.

We note that one limitation of the existing NR specification is that while UL sharing from the UE perspective is defined, DL sharing from the UE perspective is not. The reason for this is that supporting all forms of UL sharing was seen higher priority than DL sharing, given the perception that the FDD DL is more congested with LTE traffic already than the FDD UL. The lack of DL sharing from the UE perspective also leads to an apparent logical conclusion that ULSUP must operate in SUL. The explanation for this would be the following. Assuming the opposite, i.e., that ULSUP operates in non-SUL, which would mean that the UE operates with LTE and NR in one UL carrier and, since there is no SUL involved, there must be similarly a paired LTE and paired NR DL somewhere. Given that the LTE and NR bands are defined with the same duplex separation, the paired LTE and NR DL would be necessarily overlapping, presuming DL sharing from the UE perspective. But DL sharing from the UE perspective is not supported, leading to an apparent contradiction. However, flexible duplex separation

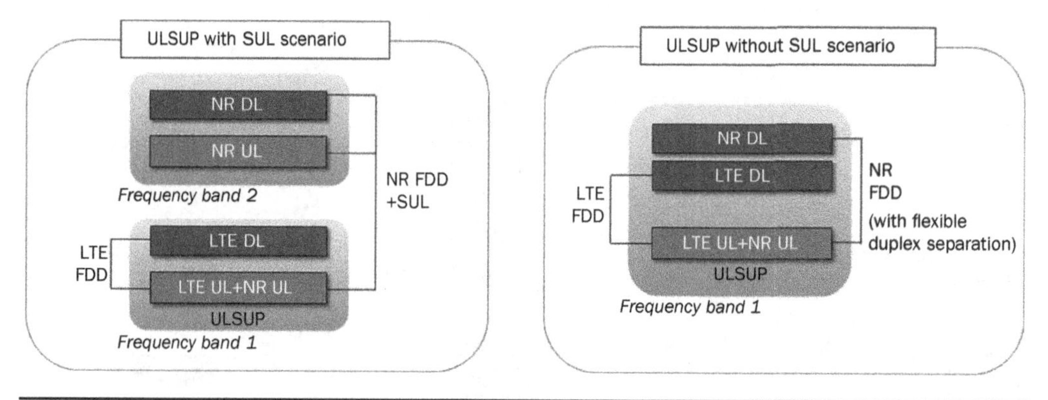

FIGURE 2.9 Examples for UL sharing from the UE perspective with and without SUL.

already exists in LTE and NR and can resolve the apparent contradiction. Although adding more flexible duplex separation cases in LTE is less likely, they can be easily added as needed in NR. Accordingly, an example for UL sharing from the UE perspective without SUL can be constructed from the example in Fig. 2.9 showing the scenario for SA with new FDD band. For this, we can convert the scenario to NSA mode by allocating an LTE carrier also for the same UE. Further examples of ULSUP without SUL are shown in Fig. 2.9.

Note that the example of ULSUP with SUL in Fig. 2.9 is the same as Cases NUS1 and NUS2 in Fig. 2.8.

2.2 Switched NR UL Carrier Aggregation Enhancements

The switched UL operating modes mentioned so far (as mentioned in Sec. 2.1) realized discrete switching, meaning that the UE would be transmitting only on one carrier at a time among two carriers. It is possible, however, that the UE is capable of simultaneous transmission but only with a limited number of transmit chains. This can occur when considering the existence of MIMO operation. Assume for example that the UE is configured with NR FDD-TDD UL CA with two carriers, where the FDD carrier supports non-MIMO operation, while the TDD carrier supports 2×2 MIMO operation. An UL CA capable UE can support simultaneous transmission, requiring up to three transmit chains. A UE that has only two transmit chains requires special accommodation, which is being defined in Release 16.

An example of a UE with three Tx chains not relying on switched UL CA enhancements, and an example of a UE with two Tx chains relying on switched UL CA enhancements are shown in Figs. 2.10 and 2.11, respectively.

The UE with Release 16 switched UL CA support will switch between two states, as given in Table 2.4.

Transitioning between Case 1 and Case 2 states requires a transient period, during which the UE is not expected to transmit. For the duration of those transient periods, the gNB is expected to schedule transmission gaps. Due to the flexible nature of NR

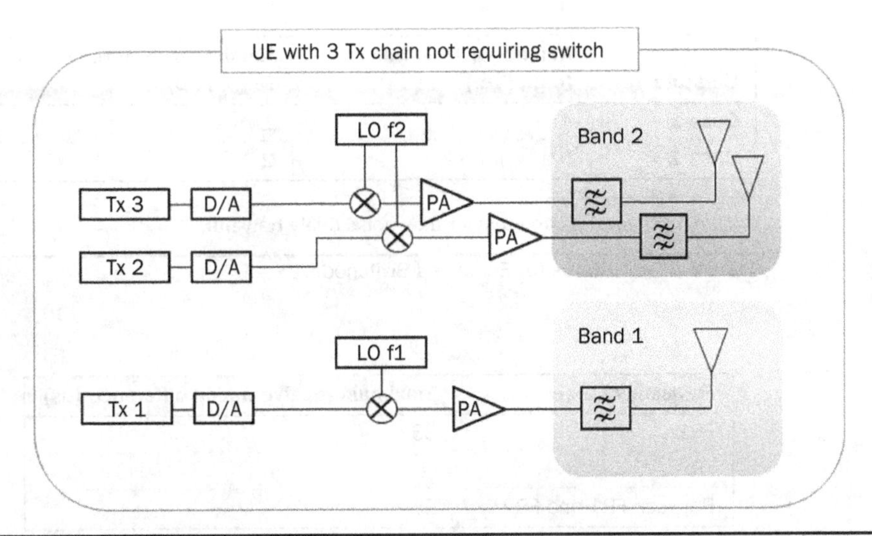

Figure 2.10 UE with Release 15 UL CA with MIMO.

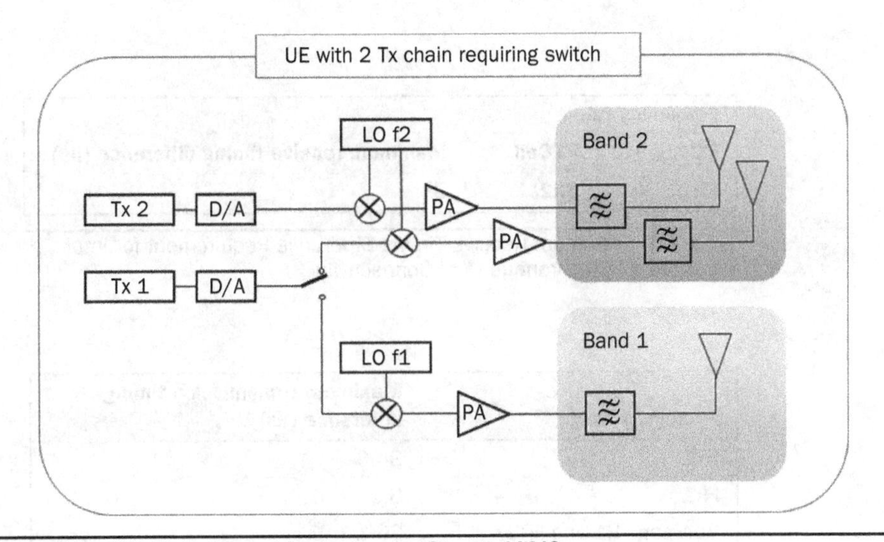

Figure 2.11 UE with Release 16 switched UL CA with MIMO.

scheduling (PUSCH mapping types A and B, various PUCCH formats), such gaps can be created in a transparent manner.

Certain transmission states, e.g., when the UE transmits neither in Band 1 nor in Band 2 could be categorized either as Case 1 or Case 2. It is beneficial to define which case these states belong to, because that way the gNB can know when state transitions requiring transmission gap will happen in the UE.

At the time of writing, the specification work for switched UL CA enhancements was still on going.

UE state	Number of Tx in Band 1 (e.g., FDD)	Number of Tx in Band 2 (e.g., TDD)
Case 1	1	1
Case 2	0	2

Note: A Tx chain in a band may or may not actively transmit.

Table 2.4 UE States for Enhanced Switched UL

Frequency range	Maximum receive timing difference (μs)
FR1	33
FR2	8
Between FR1 and FR2	25

Table 2.5 Maximum Receive Timing Difference Requirement for Inter-Band NR Carrier Aggregation

Frequency range		Maximum receive timing difference (μs)
PCell	PSCell	
FR1	**FR2**	33

Table 2.6 Maximum Receive Timing Difference Requirement for Inter-Band NR-NR Synchronous Dual Connectivity

Frequency range	Maximum transmission timing difference (μs)
FR1	34.6
FR2	8.5
Between FR1 and FR2	26.1

Table 2.7 Maximum Transmission Timing Difference Requirement for Inter-Band NR Carrier Aggregation

2.3 Nonaligned Carrier Aggregation Operation

In Release 15 NR, same as earlier in LTE, the system time for both the DL and UL is aligned across Component Carriers (CCs). Note that alignment is not assumed to be perfect and time misalignment is allowed up to a certain limit. The limits typically ensured that the maximum misalignment is not greater than about an OFDM symbol in most cases for typical subcarrier spacing (SCS) values. The actual maximum receive

Frequency range		Maximum transmission timing difference (μs)
PCell	PSCell	
FR1	FR2	34.1

Table 2.8 Maximum Transmission Timing Difference Requirement for Inter-Band NR-NR Synchronous Dual Connectivity

timing difference (MRTD) and maximum transmit timing difference (MTTD) values are given in Tables 2.5 through 2.8, based on subclauses 7.5 and 7.6 in [4].

When the allowed misalignment is subtracted; however, the alignment is complete in the sense that the CCs are both slot and frame aligned and even the system frame number (SFN) is aligned.

It turned out that in some specific deployments, the frame boundary alignment is too limiting. This may occur, for example, in some inter-band TDD-TDD CA deployments. Due to coexistence requirements with incumbent systems, different DL/UL configurations may have to be used in different bands. NR gives much more flexibility for DL and UL slot arrangement than LTE; nevertheless, there are still certain restrictions. For example, SSB can occur only with certain slot index, which necessarily must be designated as DL. Therefore, even NR does not offer full flexibility. In order to circumvent this, it was decided that Release 16 would introduce relaxed nonaligned timing across aggregated carriers in CA. The introduced nonaligned CA definition specified timing between two CCs, say CC1 and CC2 in a particular way:

- Similar MRTD and MTTD values are allowed between CC1 and CC2 as in aligned CA.
- If MRTD and MTTD were set both to zero, CC1 and CC2 are slot aligned.
- If MRTD and MTTD were set both to zero, the maximum frame boundary misalignment is ±2.5 ms.
- If the ±2.5 ms misalignment is ignored, CC1 and CC2 are SFN aligned.

Given that the time offset between CC1 and CC2 is an integer number of slots, it is natural to define their relative time in terms of a slot number. However, the fact that CC1 and CC2 may use different SCS and the fact that the slot duration for SCS 60 kHz and above is not uniform, led to several complications. Most of the problems could have been resolved, if the slot offset granularity were limited to multiples of 0.5 ms only. However, for 120 kHz SCS, this would have meant a unit of four slots, which was perceived as too coarse granularity. Therefore, it was decided that fine granularity, i.e., any slot offset is supported up to ±2.5 ms.

Firstly, the notion of slot alignment is needed to be clarified. Two CCs are slot aligned if the beginning of slot 0 of the CC with lower SCS coincides with a slot boundary of the CC with higher SCS.

Second, the slot offset is defined for the lowest SCS combinations of other than 60 kHz / 60 kHz and 120 kHz / 120 kHz as follows: For slot offset N, for the CA and DC cases, the beginning of slot #0 of the CC with lower SCS (or PCell/PScell for equal SCS) coincides with the beginning of slot #(qN mod M) of the CC with higher SCS (or SCell for equal SCS), where

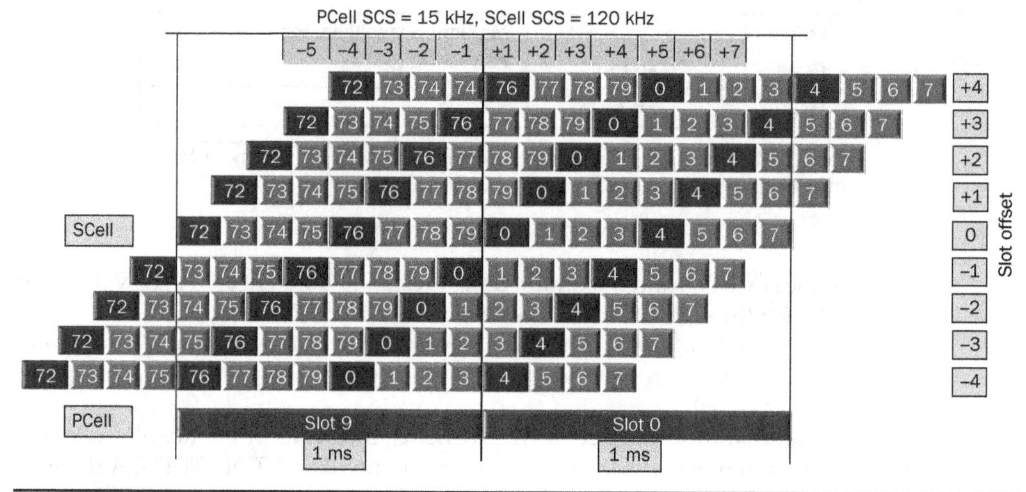

FIGURE 2.12 Nonaligned CA with SCS 15 kHz / 120 kHz.

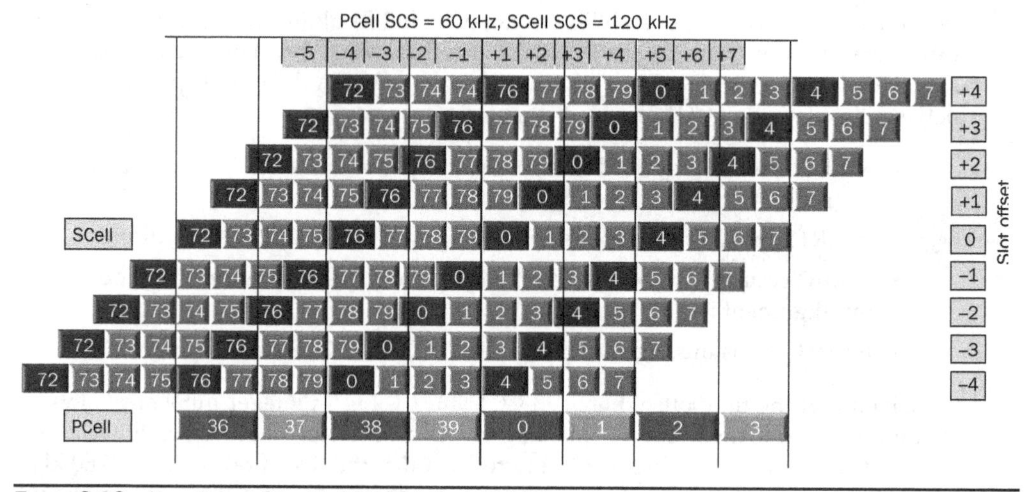

FIGURE 2.13 Nonaligned CA with SCS 60 kHz / 120 kHz.

- $q = -1$, if lowest SCS of PCell/PScell is smaller than or equal to lowest SCS of SCell
- $q = 1$, otherwise
- M is the number of slots per frame in the CC with higher SCS

In Figs. 2.12 through 2.16, we give illustrations of the various nonaligned CA slot offsets. Note that slots with different color represent different slot length. In the case of SCS 120 kHz, every slot with index equal zero modulo 4 is 52.0833 ns longer than the

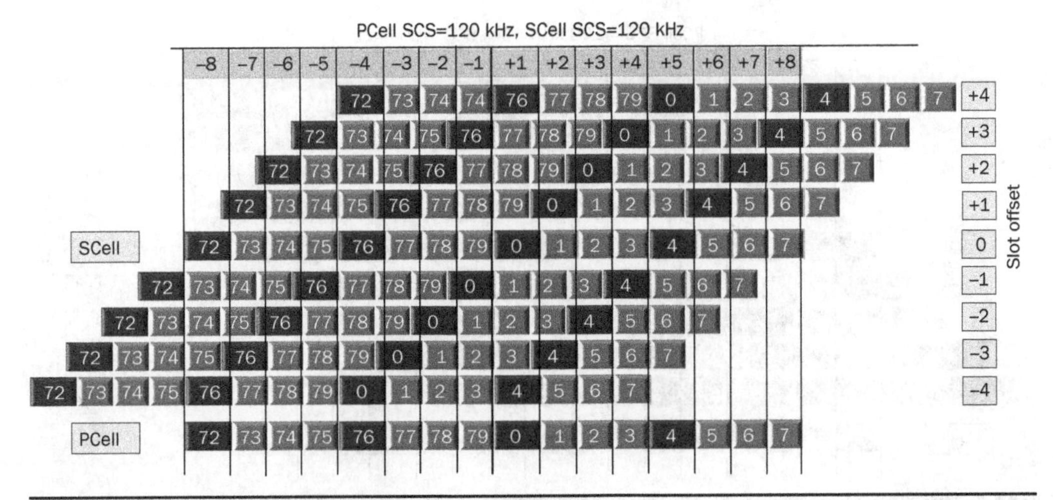

Figure 2.14 Nonaligned CA with SCS 120 kHz / 120 kHz.

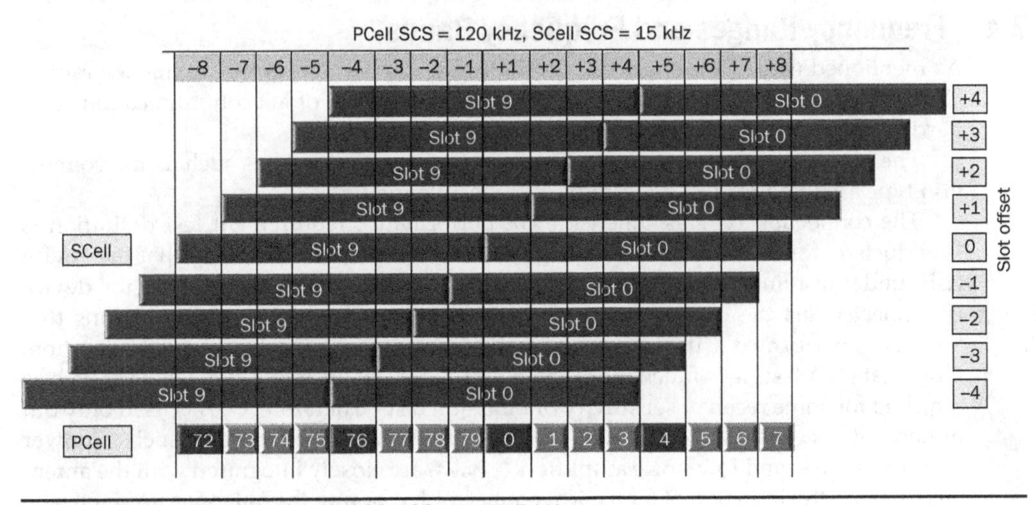

Figure 2.15 Nonaligned CA with SCS 120 kHz / 15 kHz.

other slots. In the case of SCS 60 kHz, every even indexed slot is 52.0833 ns longer than the odd indexed slots. In the case of SCS 15 kHz and 30 kHz, all slots are of equal length.

Finally, we note that in Release 15, NR-NR DC is defined as either SFN aligned synchronous, or slot aligned synchronous. Referring to Fig. 2.15 with SCS 15 kHz / 120 kHz, the SFN aligned synchronous NR-NR DC case is represented by the middle row with slot offset = 0, while slot aligned synchronous NR-NR DC case is represented by any other row. In the slot aligned NR-NR DC case, the ±2.5 ms limitation does not apply.

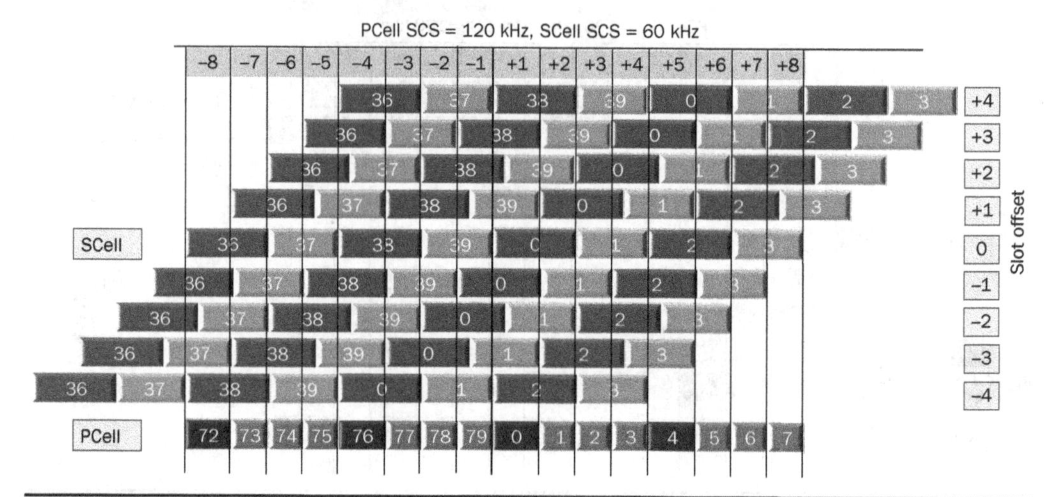

Figure 2.16 Nonaligned CA with SCS 120 kHz / 60 kHz.

2.4 Frequency Ranges and Frequency Bands

As mentioned earlier, the spectrum used for NR was divided in two frequency ranges [1, 2]. A notable difference compared to earlier generation of telecommunication standards is the inclusion of mmWave frequencies in NR.

The two different frequency ranges have additional attributes such as the connection type assumed for test definitions and the antenna types.

The connection type assumed for UE performance requirement test definition is "conducted" for FR1 and over-the-air (OTA) for FR2. Conducted means that the device (UE) under test must have a physical connector bypassing the antenna, and the device is connected via the connector with a cable to the test equipment. OTA means that the test is performed with using the physical antennas as used in normal operation. Note that OTA testing is much more challenging, especially at low signal levels, such as required for some receiver sensitivity or emission tests. Therefore, OTA is used only out of necessity because at the high frequencies of FR2, the RF components such as power amplifiers (PAs) and low-noise amplifiers (LNAs) are closely integrated with the antennas, making the insertion of a physical connector bypassing the antennas impractical.

The assumed antenna type is also different between FR1 and FR2. In FR1, each antenna element used in a multi-antenna (MIMO) set up is connected to a separate transmitter/receiver; therefore, beamforming is performed with adjusting digital phase coefficients in the transmitter/receiver. In FR2, so-called analog beamforming is used. This means that a single receiver/transmitter is connected to multiple antenna elements and the beamforming is performed by amplitude adjustment and phase delays within the antenna elements. The reason for this differentiation lies with the physical characteristics of the radio waves. In FR1, the antenna elements are placed at a distance from each other to provide diversity or spatial multiplexing gain. As the frequency increases, the size of an antenna element is inversely proportionally getting smaller. This means that the area of the antenna gets smaller, which in turn means that the antenna will interact with a smaller cross-section of the impeding electromagnetic flux, which would diminish antenna gain. To counteract this, multiple closely spaced antenna elements are integrated to form an antenna panel. The antenna panel is integrated with low-noise amplifiers on the receiver side and power amplifiers on the transmitter side in order to eliminate cable losses. The closely spaced antenna elements do not provide diversity or spatial multiplexing gain, but they do provide beamforming gain. It should be also noted that an antenna panel may include two sets of elements arranged into a cross-polarized pair, in which case they do provide polarization diversity of multiplexing capability.

A crucial difference between analog and digital beamforming is that, in the case of the receiver, with analog beamforming, the beamforming has to be set before the actual receive operation happens, while with digital beamforming, the beamforming can be adjusted based on the observed signal itself and it can be refined even after the physical signal has been received and stored. Another difference is that with digital beamforming, different subbands can be beamformed differently, while with analog beamforming, all subbands must be beamformed the same way.

The frequency bands for NR operation in FR1 are defined in [1], as shown in Table 2.9. Note that the entry in the last column signifies the duplex type of the band:

- FDD = Frequency Division Duplex
- TDD = Time Division Duplex
- SDL = Supplemental Downlink
- SUL = Supplemental Uplink

The frequency bands for NR operation in FR2 are defined in [2], as shown in Table 2.10.

Note that in FR2 all bands are currently defined as TDD, although in the future, other types such as SDL could be possible.

2.4.1 UE Channel Bandwidth

The band definitions mentioned thus far reflect the availability of frequency ranges to use globally; however, it does not describe the partition of the bands for use by different operators. Of course, the actual allocation of spectrum blocks to operators within the band is highly variable, even before taking into account the possibility of spectrum aggregation and disaggregation. Therefore, the channel bandwidths defined are not exhaustive but are a result of a tradeoff between flexibility and limiting the number of different cases to be implemented and tested. The defined channel bandwidth cases are all multiple of 5 MHz and at present include the following cases [1, 2]:

- FR1: 5, 10, 15, 20, 25, 30, 40, 50, 60, 80, 90, 100 MHz
- FR2: 50, 100, 200, 400 MHz

In Release 15 of NR, the UE is mandated to support 100 MHz in FR1 bands in which 100 MHz is defined, and the UE is mandated to support 200 MHz in FR2 bands in which 200 MHz is defined.

Note the following consideration, which was at least in part the reason for selecting 100 MHz and 400 MHz as the maximum channel bandwidth for FR1 and FR2, respectively. At the time of standards development, it was discussed whether a maximum

Operating band	Uplink (UL) operating band BS receive, UE transmit $F_{UL\_low} - F_{UL\_high}$ (MHz)	Downlink (DL) operating band BS transmit, UE receive $F_{DL\_low} - F_{DL\_high}$ (MHz)	Duplex mode
n1	1920–1980	2110–2170	FDD
n2	1850–1910	1930–1990	FDD
n3	1710–1785	1805–1880	FDD
n5	824–849	869–894	FDD
n7	2500–2570	2620–2690	FDD
n8	880–915	925–960	FDD
n12	699–716	729–746	FDD
n14	788–798	758z–768	FDD
n18	815– 830	860–875	FDD
n20	832–862	791–821	FDD
n25	1850–1915	1930–1995	FDD
n28	703–748	758–803	FDD
n29	N/A	717–728	SDL
n30	2305–2315	2350–2360	FDD
n34	2010–2025	2010–2025	TDD
n38	2570–2620	2570–2620	TDD
n39	1880–1920	1880–1920	TDD
n40	2300–2400	2300–2400	TDD
n41	2496–2690	2496–2690	TDD
n48	3550–3700	3550–3700	TDD
n50	1432–1517	1432–1517	TDD
n51	1427–1432	1427–1432	TDD
n65	1920–2010	2110–2200	FDD
n66	1710–1780	2110–2200	FDD
n70	1695–1710	1995–2020	FDD
n71	663–698	617–652	FDD
n74	1427–1470	1475–1518	FDD
n75	N/A	1432–1517	SDL
n76	N/A	1427–1432	SDL
n77	3300–4200	3300–4200	TDD
n78	3300–3800	3300–3800	TDD
n79	4400–5000	4400–5000	TDD
n80	1710–1785	N/A	SUL
n81	880–915	N/A	SUL
n82	832–862	N/A	SUL
n83	703–748	N/A	SUL
n84	1920–1980	N/A	SUL
n86	1710–1780	N/A	SUL
n89	824–849	N/A	SUL
n90	2496–2690	2496–2690	TDD

Table 2.9 NR Operating Bands in FR1

Operating band	Uplink (UL) operating band BS receive, UE transmit	Downlink (DL) operating band BS transmit, UE receive	Duplex mode
	$F_{UL\_low} - F_{UL\_high}$ (MHz)	$F_{DL\_low} - F_{DL\_high}$ (MHz)	
n257	26500–29500	26500–29500	TDD
n258	24250–27500	24250–27500	TDD
n260	37000–40000	37000–40000	TDD
n261	27500–28350	27500–28350	TDD

Table 2.10 NR Operating Bands in FR2

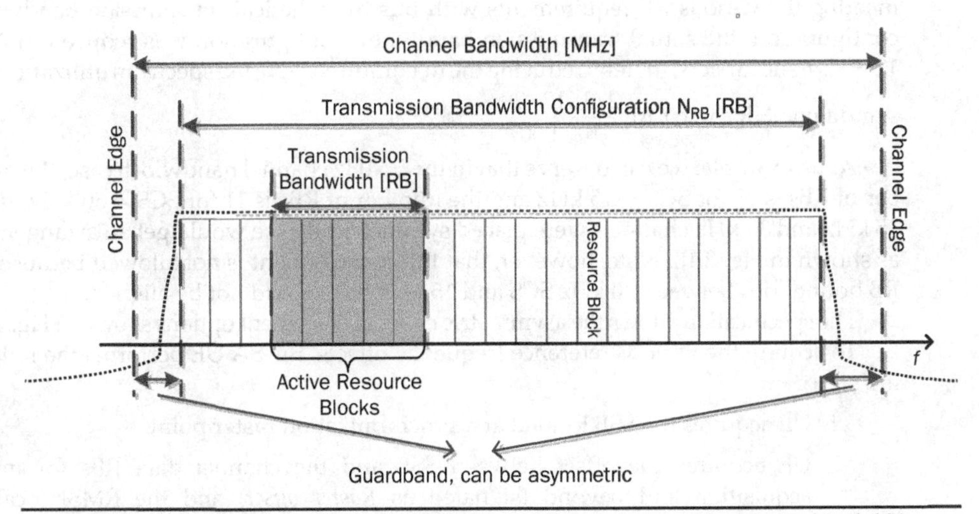

Figure 2.17 Definition of the channel bandwidth and the maximum transmission bandwidth configuration for one NR channel [1].

FFT size of 2048 should be assumed as in LTE, or a larger FFT size. The compromise was to assume 4096 as the maximum FFT size. In FR1, where 30 kHz SCS was assumed as typical target numerology in the bands where large channel bandwidth is available, the maximum bandwidth with 4096 subcarriers would be 4096 times 30 kHz, which is 122.88 MHz. This, however, would not have allowed for anti-alias filtering with any reasonable filter roll-off. To mitigate this, it was decided that using 100 MHz out of the 122.88 MHz total would leave sufficient guard for filtering. In FR2, where 120 kHz SCS was assumed as typical target numerology, the same calculation can be repeated with scaling up by four, resulting in 400 MHz maximum bandwidth.

Another important decision point was defining the guardband allowance between the so-called maximum transmission bandwidth configuration and the channel bandwidth. The maximum transmission bandwidth configuration defines the range of RBs on which the UE is required to transmit or receive, while the channel bandwidth defines the frequency boundaries beyond which the UE is required to meet emission requirements and blocking requirements for the transmitter and receiver, respectively. These concepts, together with the guardband definition, are depicted in Fig. 2.17 from [1].

In LTE, the guardband was assumed to be always 10%, irrespective of the channel bandwidth (with the exception of 1.4 MHz bandwidth). This was seen as wasteful, since the guardband required to achieve the required close-in emission levels is a certain multiple of the subcarrier spacing, rather than a fixed fraction of the channel bandwidth. In response to these considerations, the maximum transmission bandwidth configurations for FR1 and FR2 were defined in Tables 2.11 and 2.12, based on [1] and [2], respectively.

Note in the case of FR2, the spectrum utilization is mostly linear, except for the 50 MHz case with 120 kHz SCS.

We note that the original intent was to enable 3300 out of 4096 subcarriers as useful carriers with 30 kHz SCS in FR1, which would have given a theoretical 99 percent spectrum utilization (considering that 100 percent spectrum utilization in 100 MHz with 30 kHz SCS would be 3333.33 subcarriers). However, due to the difficulties of meeting the various RF requirements with this hypothetical transmission bandwidth configuration, the actual transmission bandwidth configuration was reduced to $273 \cdot 12 = 3276$ subcarriers, thereby reducing the maximum supported spectrum utilization in a standalone NR carrier to $\frac{273 \cdot 12 \cdot 30}{100{,}000} = 98.28\%$.

As an example, we can observe that in the 5 MHz channel bandwidth case, the number of RBs is 25 for SCS = 15 kHz and the number of RBs is 11 for SCS = 30 kHz. If the 15 kHz and 30 kHz channels were placed symmetrically, we would get an arrangement as shown in Fig. 2.18. Note, however, that this arrangement is not allowed because the RB boundaries between 30 kHz SCS and 15 kHz SCS would not be aligned.

The specification allows the asymmetric channel placement options shown in Fig. 2.19.

To acquire the various reference frequency offsets, the SA UE performs the following steps:

1. UE acquires the SSB located at a synchronization raster point.
2. UE acquires the offset between SSB and the channel data RBs for initial acquisition and beyond (signaled as *Raster offset*) and the RMSI location (signaled as *RMSI config*) from the master information block (MIB) in the SSB [5]. Note that the term RMSI refers to the remaining essential system information outside of MIB.
3. The lowest subcarrier of the first RB in the RMSI control resource set (CORESET #0) is used as RE #0 to generate sequences for reference signals and scrambling for RMSI.
4. UE acquires RMSI and extracts the following parameters that are signaled in RMSI for each supported SCS (each SCS is a different CC).
    a. Absolute frequency of Reference point A (*absoluteFrequencyPointA in FrequencyInfoDL,* signaled as *ARFCN-ValueNR*) [5]
    b. Subcarrier spacing (*subcarrierSpacing* in *SCS-SpecificCarrier*)
    c. Offset in PRB unit from Reference point A to the first usable PRB (*offsetToCarrier* in *SCS-SpecificCarrier*)
    d. Carrier bandwidth in PRB unit (*carrierBandwidth* in *SCS-SpecificCarrier*)
5. Reference point A is RE #0 used to generate sequences for reference signals and scrambling for channels other than those used for initial acquisition.

SCS (kHz)	5 MHz	10 MHz	15 MHz	20 MHz	25 MHz	30 MHz	40 MHz	50 MHz	60 MHz	80 MHz	90 MHz	100 MHz
	N_{RB}	N_{RB}	N_{RB}	N_{RB}	N_{RB}	N_{RB}	N_{RB}	N_{RB}	N_{RB}	N_{RB}	N_{RB}	N_{RB}
15	25	52	79	106	133	160	216	270	N/A	N/A	N/A	N/A
	90.00%	93.60%	94.80%	95.40%	95.76%	96.00%	97.20%	97.20%	N/A	N/A	N/A	N/A
30	11	24	38	51	65	78	106	133	162	217	245	273
	79.20%	86.40%	91.20%	91.80%	93.60%	93.60%	95.40%	95.76%	97.20%	97.65%	98.00%	98.28%
60	N/A	11	18	24	31	38	51	65	79	107	121	135
	N/A	79.20%	86.40%	86.40%	89.28%	91.20%	91.80%	93.60%	94.80%	96.30%	96.80%	97.20%

Table 2.11 FR1 Maximum Transmission Bandwidth Configuration N_{RB}

SCS (kHz)	50 MHz	100 MHz	200 MHz	400 MHz
	N_{RB}	N_{RB}	N_{RB}	N_{RB}
60	66	132	264	N/A
	95.04%	95.04%	95.04%	N/A
120	32	66	132	264
	92.16%	95.04%	95.04%	95.04%

Table 2.12 FR2 Maximum Transmission Bandwidth Configuration N_{RB}

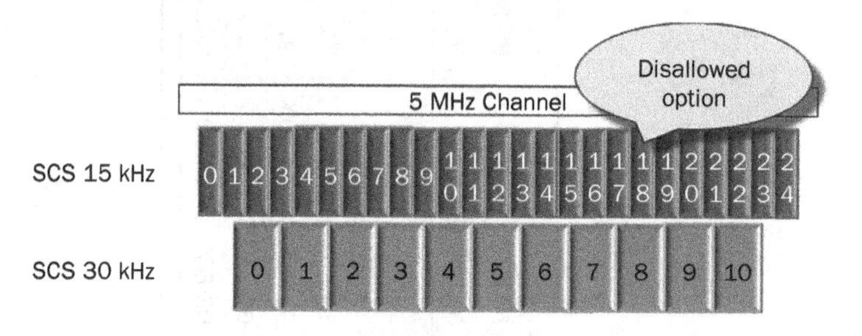

Figure 2.18 Disallowed symmetric channel placement.

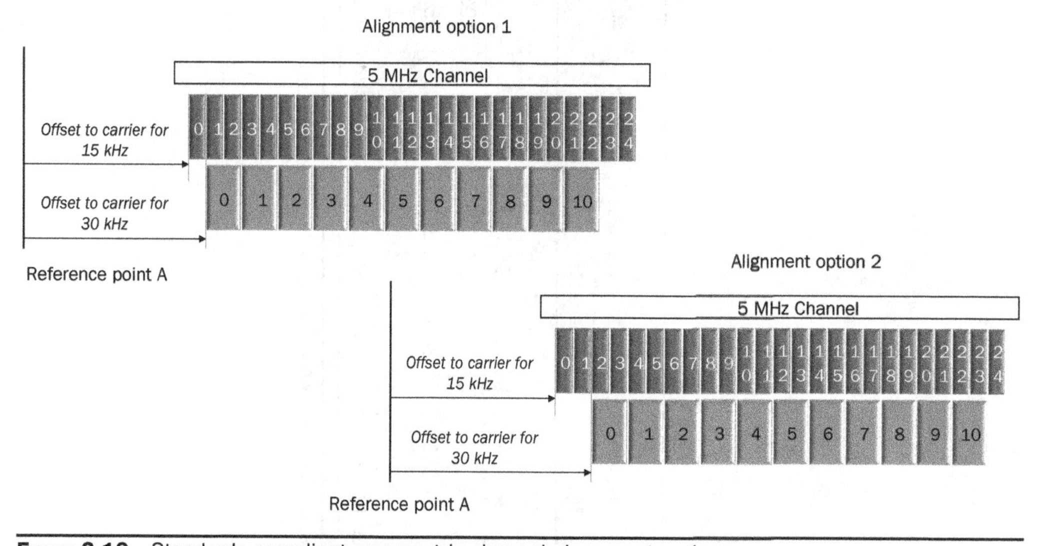

Figure 2.19 Standards compliant asymmetric channel placement options.

Note that NSA UEs can directly start at Step #5 in the sequence above because all parameters can be provided to the UE before cell acquisition.

The various configured frequency offsets are shown in Fig. 2.20.

Note that "Reference point A" is a concept introduced not only for the above scenarios but in general to enable multiplexing UEs having different notions of channel

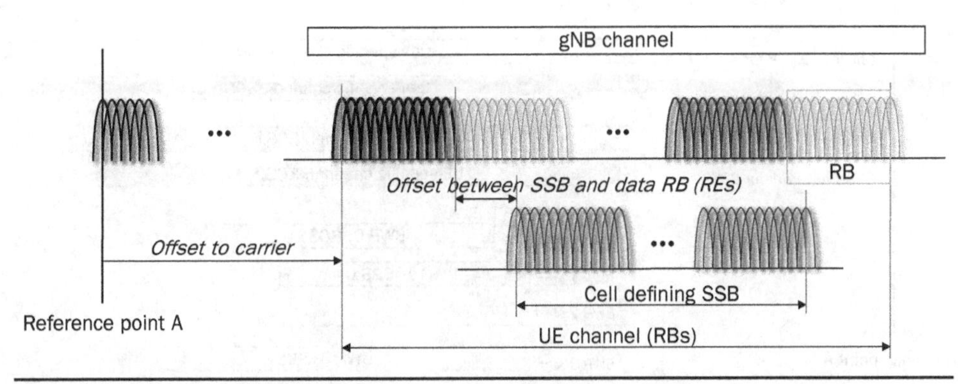

FIGURE 2.20 Configuration of reference frequency offsets.

placement but with a common notion of reference signal definitions. This will be explained further in Sec. 2.4.2.

2.4.2 Bandwidth Migration Scenarios

One of the lessons learned from LTE was that designing for a fixed channel bandwidth is quite limiting. This came to light when in a later LTE release eMTC was added, which necessitated to operate a 1.08 MHz channel bandwidth (6 RB) system within a wider bandwidth, e.g., 10 MHz (50 RB) system. Even though the numerology, waveform, and subframe structure designs of LTE and eMTC were identical, the reference signal design and control channel design made coexistence cumbersome. One of the design goals of NR was to avoid these future problems from the start. In order to achieve this goal, the following mechanisms were introduced:

- Bandwidth Part (BWP) designated within the UE-specific channel bandwidth.
- UE-specific channel bandwidth embedded in a cell-specific channel bandwidth.
- Control resource set (CORESET) bandwidth configured separately from channel bandwidth.
- Reference signal bandwidth configured separately from channel bandwidth.
- Common reference point (so-called Reference Point A) from which reference signal scrambling, resource block group (RBG) numbering starts.

In Fig. 2.21, an illustration is given of different bandwidths being configured for a UE.

As it was mentioned, the goal of the NR design was to allow multiplexing different UEs with different notions of channel bandwidth to coexist and even use the same reference signals. An example illustration for this is shown in Fig. 2.22.

In Fig. 2.22, sharing the same transmitted channel state information reference signal (CSI-RS) by two UEs with different notions of channel is shown. This is made possible by the CSI-RS scrambling sequence definition, where the scrambling starts from Reference point A, which is common among UEs. The same principle applies to other reference signals. In the case of demodulation reference signal (DM-RS), the actual transmitted DM-RS for different UEs is typically different due to the UE-specific

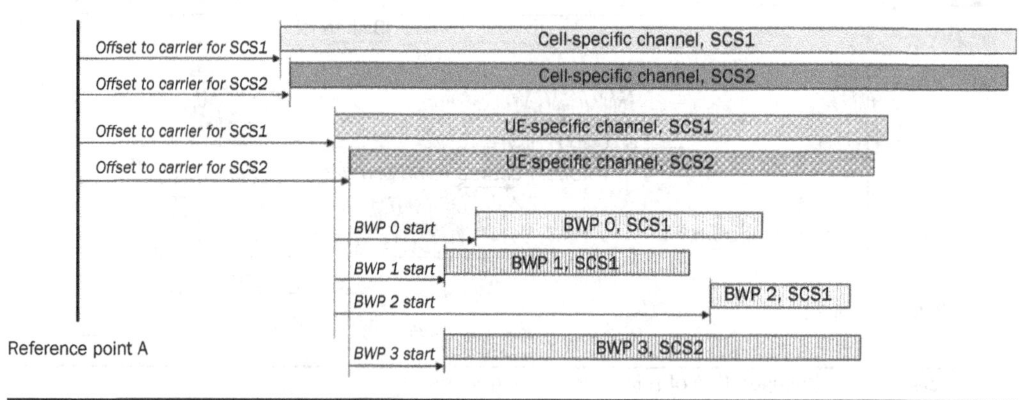

FIGURE 2.21 Bandwidth configuration options.

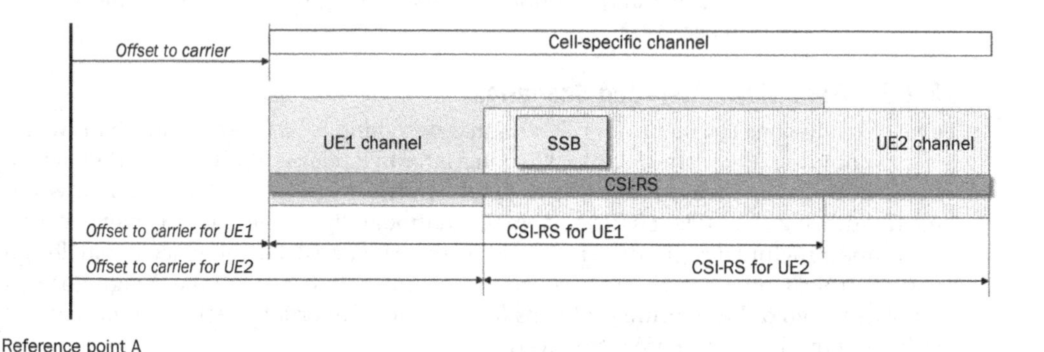

FIGURE 2.22 Common reference signals for UEs with different channels.

beamforming; nevertheless, the use of common Reference point A helps with making orthogonal multiplexing of the DM-RS easier.

As it was mentioned before, the maximum bandwidth supported in Release 15 is 100 MHz in FR1 and 400 MHz in FR2. The channel structure design of NR makes it possible to introduce and deploy future wider bandwidths while not removing the earlier UEs. An example for this is shown in Fig. 2.23.

In Fig. 2.23, two 400 MHz channels are shown. Different UEs can be allocated in each, or UEs can use 2×400 MHz carrier aggregation. Therefore, 800 MHz, 2×400 MHz, and 400 MHz UEs can operate at the same time in the same channel, even though when the 400 MHz or 2×400 MHz UEs were designed, 800 MHz channel bandwidth may not even have existed.

A similar mechanism can be used to introduce nonstandard channel bandwidth in a transparent manner. Transparent here means that the gNB must implement a new channel bandwidth but UEs do not need to be aware of the new channel bandwidth. An example for this is shown in Fig. 2.24.

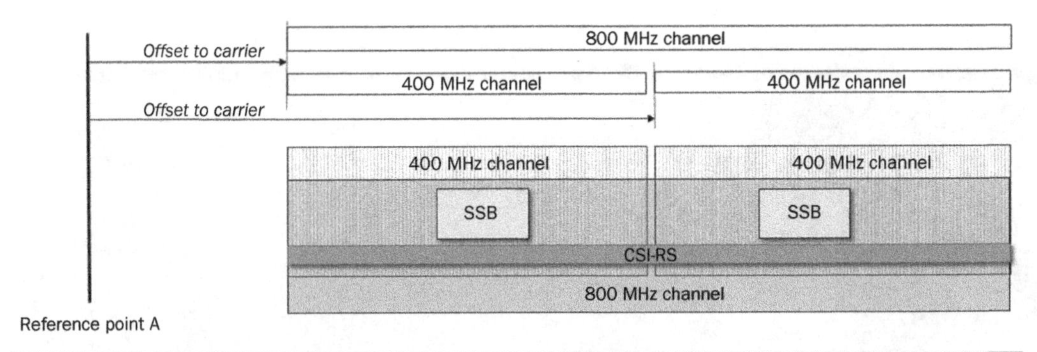

FIGURE 2.23 Example for introduction of future wider bandwidth.

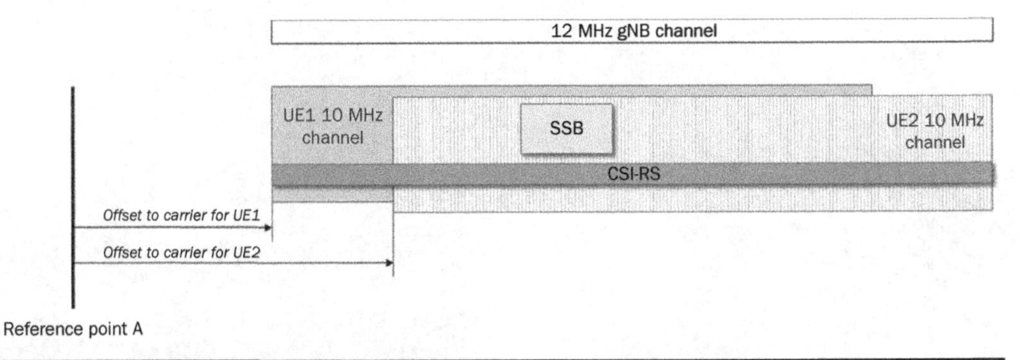

FIGURE 2.24 Example for introduction of nonstandard channel bandwidths.

In Fig. 2.24 an example of 12 MHz gNB channel bandwidth is shown but other gNB channel bandwidth cases can be similarly supported. Note also that while in Fig. 2.24, DL only is shown, the same solution can apply to the UL as well.

References

[1] 3GPP, "3rd Generation Partnership Project; Technical Specification Group Radio Access Network; User Equipment (UE) Radio Transmission and Reception; Part 1: Range 1 Standalone (Release 16)," 38.101-1, v16.2.0, December 2020.

[2] 3GPP, "3rd Generation Partnership Project; Technical Specification Group Radio Access Network; User Equipment (UE) Radio Transmission and Reception; Part 2: Range 2 Standalone (Release 16)," 38.101-2, v16.2.0, December 2020.

[3] 3GPP, "3rd Generation Partnership Project; Technical Specification Group Radio Access Network; User Equipment (UE) Radio Transmission and Reception; Part 3: Range 1 and Range 2 Interworking Operation with Other Radios (Release 16)," 38.101-3, v16.2.1, December 2020.

[4] 3GPP, "3rd Generation Partnership Project; Technical Specification Group Radio Access Network; Requirements for Support of Radio Resource Management (Release 15)," 38.133, v15.8.0, December 2020.

[5] 3GPP, "3rd Generation Partnership Project; Technical Specification Group Radio Access Network; NR; Radio Resource Control (RRC) Protocol Specification (Release 15)," 38.331, v15.8.0, December 2020.

CHAPTER 3

Architecture Options for 5G

3.1 Introduction

Unlike the mono-dimensional migration option defined in previous "G transitions" (e.g., from 3G to 4G) where the transition was happening in tandem in Radio Access Network (RAN) and core network (CN), for the transition from 4G to 5G, for a variety of commercial and technical reasons, 3GPP defined a variety of options that can be categorized in two distinct paradigms:

- Paradigm 1: 5G NR added as secondary radio access type (RAT) to the evolved packet core (EPC).
- Paradigm 2: Legacy radio (LTE) and new radio (5G NR) connect to the new 5G core (5GC).

Paradigm 1 was defined in the so-called "Option 3 variants" as depicted in generic form in Fig. 3.1 that requires minimal EPC changes or enhancements. 5G NR is added as secondary RAT in dual connectivity configuration with three variants (3: split bearer, 3a: SCG, 3x: SCG with split bearer).

From a UE perspective, three bearer types exist: Master cell group bearer, secondary cell group bearer, and split bearer. These three bearer types are depicted in Fig. 3.2 for Option 3 with EPC (aka EN-DC).

Some small EPC enhancements have been introduced in 3GPP in order to support dual connectivity with NR using EPC (aka Option 3 variants):

- Subscription-based access restriction for NR as secondary RAT
- NAS indicator "support DC with NR" used for S/P-GW selection
- NAS and SIB indicator (NR is available to use) (aka used for 5G indicator in UEs user interface)
- Data volume counting and reporting in RAN for NR as secondary RAT
- New QCIs
- Security algorithm negotiation
- Extension to the bitrate values [e.g., for guaranteed bitrate (GBR), maximum bitrate (MBR), AMBR]

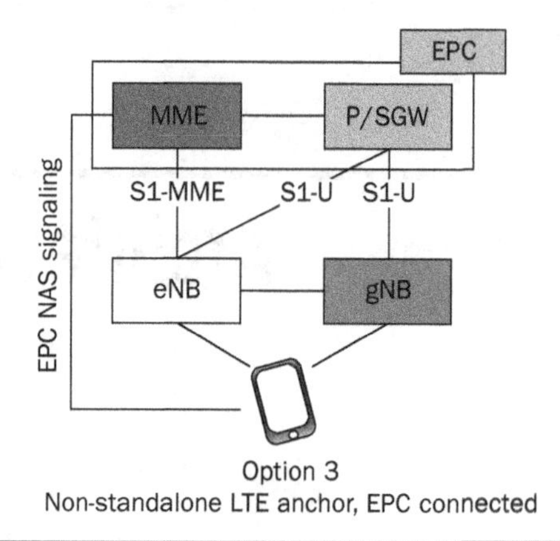

FIGURE 3.1 General Option 3 architecture.

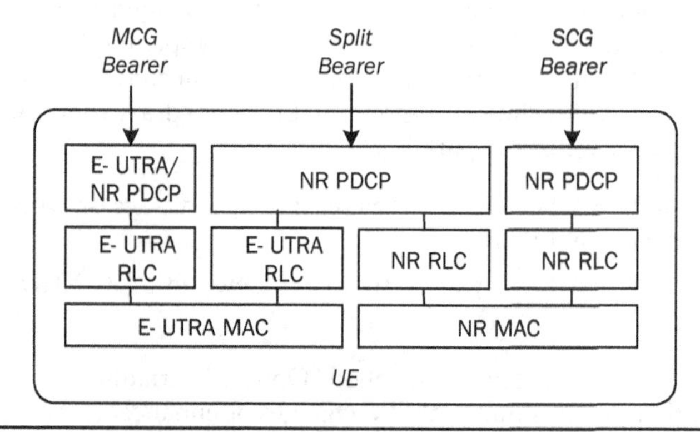

FIGURE 3.2 Radio protocol architecture for master cell group (MCG), secondary cell group (SCG), and split bearers from a UE perspective with EPC (Option 3 family).

These enhancements are considered "optional," i.e., one can launch basic NR NSA "Option 3" without these enhancements but limited functionality.

Paradigm 2 has a variety of options that involve LTE (called evolved LTE/E-UTRA) and NR connecting in non-standalone and standalone configuration connecting to 5GC that are all supported in 5G system architecture and related system procedure (see TS 23.501 [1] /TS 23.502 [2]). The combination of evolved 3GPP RATs (evolved E-UTRA and NR) connecting to 5GC is called next generation radio access network (NG-RAN).

The choice of deployment of these options that are depicted collectively in Fig. 3.3, i.e., if one or more will be deployed by an MNO in the same network depends on the availability of greenfield spectrum in different regions, also more are provided in Sec. 3.15. From CN point of view the common denominator of these options is that both LTE and NR connect to 5GC.

When 5GC is used in NSA variants, NR PDCP is always used for all bearer types. In Option 7 as depicted in Fig. 3.4, E-UTRA RLC/MAC is used in the master node while NR RLC/MAC is used in the secondary node. In Option 4, on the other hand, NR RLC/MAC is used in the master node while E-UTRA RLC/MAC is used in the secondary node.

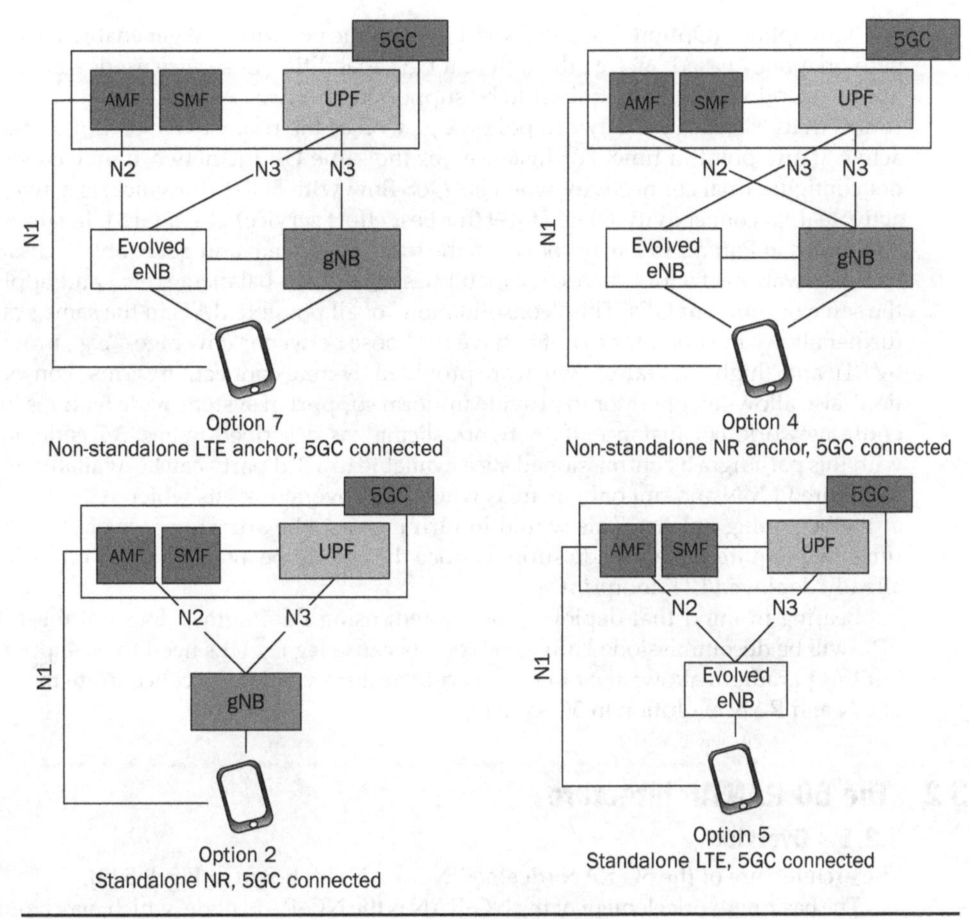

FIGURE 3.3 5GC-based architecture options (Option 7/4/2/5).

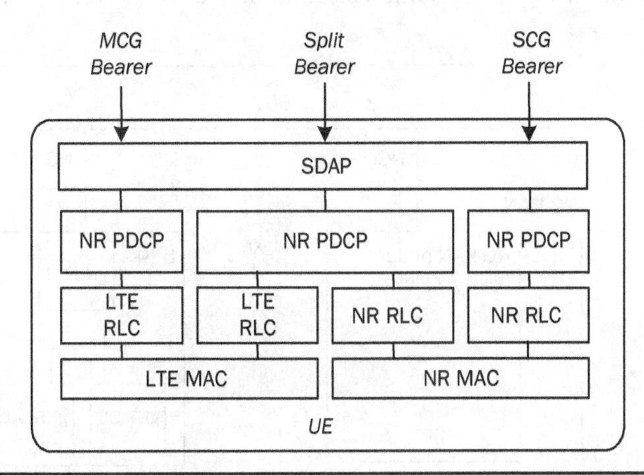

FIGURE 3.4 Radio protocol architecture for master cell group (MCG), secondary cell group (SCG), and split bearers from a UE perspective with 5GC using option 7 E-UTRA RLC/MAC.

NSA options (Option 7/4) can coexist in the same network and get enabled from the network policy based on, e.g., the different UE capabilities, or service used, whereas SA options (Option 2/5) are required to be supported in any case since activation of dual connectivity is based on network policy, e.g., QoS of the respective QoS flows that are active at any point in time. For instance, for the same UE the network may choose to not configure dual connectivity when 5G QoS flow with 5QI-1 (for voice) is active, but activate dual connectivity when 5QI-9 (for best effort service) is activated. In summary all options in Paradigm 2 may coexist in the same 5G system and get enabled/disabled based on various factors such UE capabilities, QoS, load balancing, etc., and apply to the same or different UEs. This "consolidation" of all possible RATs in the same system further allows the operator to not to have to choose between "coverage," e.g., provided by LTE and "high data rates" which are provided by dual connectivity. This "consolidation" also allows an operator to provide uniform support of system-wide features in the entire network. For instance, if "network slicing" as described in Sec. 3.7 is deployed with this paradigm a commissioned slice available to a 3rd party can be available across the entire PLMN and not only in areas where NR coverage exists which will be limited at the beginning at least. This would in turn increase the attractiveness of the 5G features that require system-wide support since they could be available from day 1 in the already deployed LTE footprint.

Bearing in mind that deploying 5G system using Paradigm 2 does not mean that EPC will be decommissioned any time soon because legacy UEs need to be supported, but this paradigm allows the operator and their infra vendors to concentrate the effort of CN and RAN evolution in 5G system.

3.2 The 5G RAN Architecture

3.2.1 Overview

The architecture of the 5G RAN (deemed NG-RAN) is shown in Fig. 3.5 [3].

The basic network element of the NG-RAN is the NG-RAN node, which may be either a gNB (hosting NR access), or a ng-eNB (hosting LTE access). NG-RAN nodes connect to the 5G core using the NG interface, and may be inter-connected by the Xn interface,

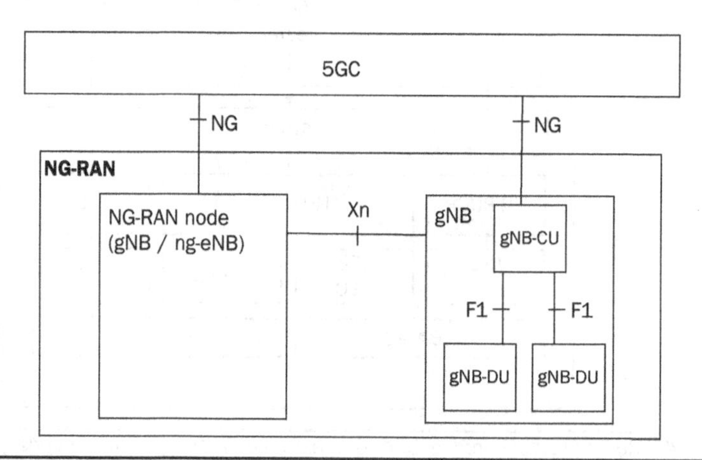

Figure 3.5 Overall NG-RAN architecture.

which supports inter-node functionality and notably mobility and dual connectivity operation. Both the interfaces comprise separate control and user plane components.

Figure 3.5 also shows a split gNB which is further discussed in the following section. For now, it should be noted that the internal nodes and interfaces of a gNB are not visible to other NG-RAN nodes, UEs, or the 5G Core.

3.2.2 Split RAN Architecture in 3GPP

Growing interest in the possibility of introducing RAN virtualization led to work in 3GPP at the outset of 5G. The initial goal was to define a functional split between a "centralized unit" (or CU) and a "distributed unit" (or DU), such the CU could be centralized and potentially virtualized. Eight potential splits were defined as shown in Fig. 3.6 [4], and these were subsequently analyzed in respect of associated requirements and system advantages. For example, key aspects include required backhaul latency and throughput.

It quickly became apparent that the full set of splits covered a very wide range of potential deployment scenarios and product options, and, therefore, it was appropriate to consider separately "higher layer" and "lower layer" splits. Consequently, 3GPP decided to support split Option 2 as the higher layer split in Release 15, and specify a new interface (F1 interface [5]) between newly defined nodes, e.g., gNB-CU and gNB-DU for this purpose. There are strong similarities with dual connectivity since in both cases PDCP PDUs are transported in the user plane, and latency/throughput requirements are also similar, and not over-demanding.

In parallel, 3GPP also considered the possibility of a horizontal split between user plane and control plane in the RAN. After considerable discussion on whether such a split would apply across the full layer model, it was decided to separate the gNB-CU itself into a control plane entity (gNB-CU-CP) and a user plane entity (gNB-CU-UP) [6]. The two entities need not be co-located and may also scale independently. Furthermore, the gNB-CU-CP may be seen as the core of the gNB since it hosts RRC and it terminates the control plane part of NG-RAN interfaces that connect it to other nodes (e.g., NG-C toward the core network, and Xn-C toward neighbor gNBs).

Figure 3.7 shows the resulting overall architecture of the gNB [3], illustrating also the new interfaces and scalability aspects. There is a single gNB-CU-CP per gNB, and in addition

- There may be multiple gNB-CU-UPs, and each may be connected to one or more gNB-DUs through the user plane F1-U interface.

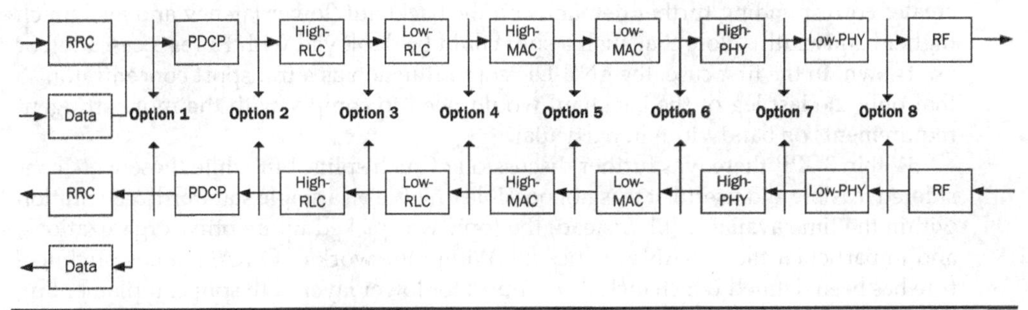

FIGURE 3.6 Potential functional splits between centralized and distributed units.

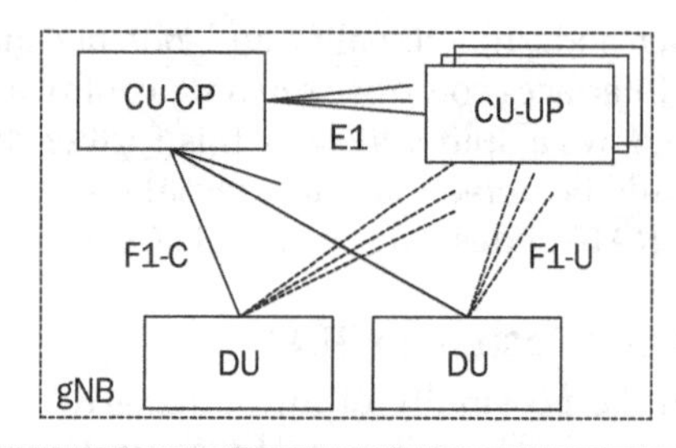

FIGURE 3.7 gNB architecture showing potential splits and interfaces.

- Since a gNB-DU may be connected to multiple gNB-CU-UPs, it is possible to have different PDU sessions / DRBs processed in different gNB-CU-UPs (for the same UE).
- The E1 interface [7] is a pure control plane interface, used to configure the functionality and resources of the gNB-CU-UP (i.e., SDAP, PDCP, and transport resources for F1-U and NG-U).
- The F1-C interface is used to configure the functionality and resources of the gNB-DU; create, modify, and delete UE contexts in the gNB-DU; and also transport RRC messages.

This architecture enables a wide range of deployments; it allows the virtualization of all CU components or, alternatively, some of them. For example, it is possible to centralize the gNB-CU-CP only, leaving the F1-U as an internal interface of a reduced base station. It is also possible to locate specific gNB-CU-UPs according to latency requirements of PDU sessions, and related UPF topology.

The above architecture is defined in 3GPP [3], and, as new 3GPP features are added, the design work also covers the new RAN nodes, so that new functionality is being added to both E1 and F1 interfaces within the cycle of 3GPP releases.

3.2.3 Lower Layer Splits

The higher layer split is suitable for wide area deployments with F1 backhaul latency in the order of several milliseconds – possibly even tens of milliseconds. In return, there is still considerable processing to be performed at the base site. With a lower layer split, it is possible to reduce further such processing close to the transmitting points, thus potentially reducing the size and power drain of the corresponding units; of course this creates corresponding further demands on the backhaul (lower latency and also much higher bandwidth). Note that such a split could be deployed with F1 (as a cascade), or on its own. In the first case, the gNB-DU could function as a transport concentrator, so that only the last leg of the backhaul would need to comply with the more stringent requirements on bandwidth in particular.

Within 3GPP, there was further discussion of such splits, but while these were considered feasible and useful, it was not possible to settle on a single standardized solution within the time available [8]. Instead, the topic was picked up by other organizations, and in particular the O-RAN Alliance [9]. Within the work of O-RAN, a new architecture has been defined which includes support for lower layer gNB split complementing the 3GPP architecture of Fig. 3.8. The architecture also includes a number of aspects

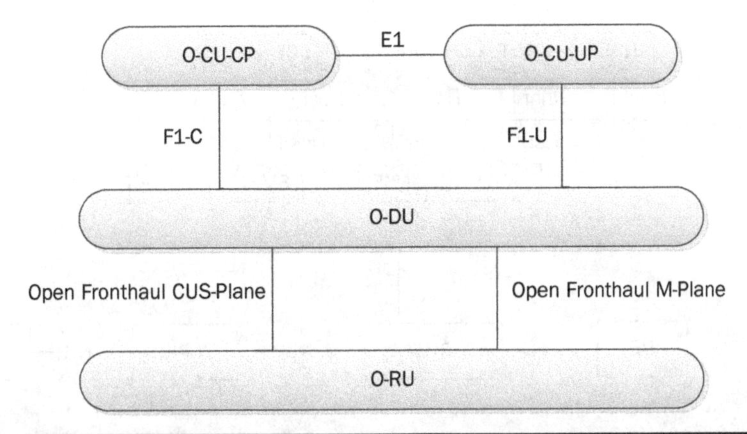

FIGURE 3.8 Partial view of the logical architecture of O-RAN (split aspects only).

related to management and orchestration of the split RAN network elements. A reduced version of the O-RAN architecture (focusing on the split aspects) is shown in Fig. 3.8.

As shown in Fig. 3.8, a new RAN node has been defined by O-RAN (O-RU, or Open – Radio Unit), which terminates the Open Fronthaul or LLS Interface. This interface supports the Option 7 split where the PHY functionality is divided into high PHY (hosted by the gNB-DU, or O-DU in O-RAN's architecture), and low PHY (hosted by the O-RU). There are several option 7 variants, depending on the exact functional mapping to low/high PHY. The O-RAN split architecture was based on a split developed by x-RAN known as 7-2x but enables some limited flexibility in functional mapping (e.g., precoding may be hosted in either O-DU or O-RU). It has been recognized that other functional splits could be of interest to operators, which might include options 6, 8 or even different variants of option 7, and such additional splits or split variants may be supported in the future.

3.3 The 5G Core

The first version of the 5G system architecture has been specified by 3GPP in Release 15 and the stage-2 architecture is frozen in December 2017 in TS 23.501 [1] for the general architecture description and TS 23.502 [2] for the stage-2 system procedures.

The 5G system architecture, as depicted in Fig. 3.9, is defined to support data connectivity and services enabling deployments to use techniques such as network function virtualization and software defined networking. The 5G system architecture leverages service-based interactions between control plane (CP) network functions where identified. Some key concepts and goals are the following:

- Separate the user plane (UP) functions from the control plane (CP) functions, allowing independent scalability, evolution, and flexible deployments, e.g., centralized location or distributed (remote) location.
- Modularize the function design, e.g., to enable flexible and efficient network slicing.
- Define the procedures (i.e., the set of interactions between network functions) as services, so that their re-use is possible.

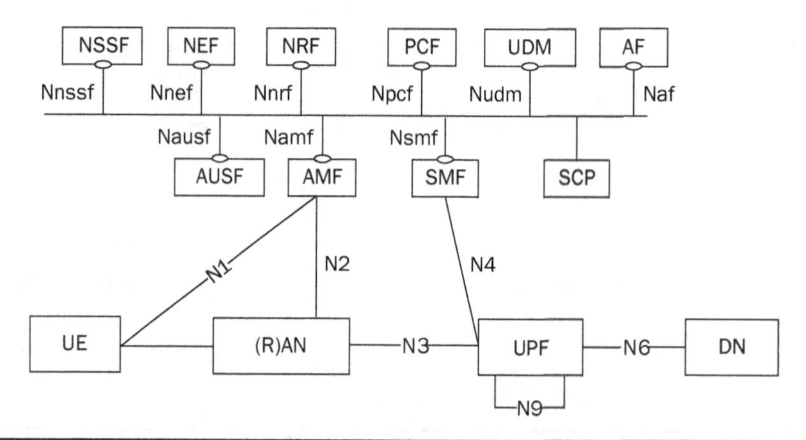

FIGURE 3.9 5G System architecture (non-roaming) (source TS 23.501).

- Enable each network function to interact with other network functions directly if required. The architecture does not preclude the use of an intermediate function to help route control plane messages.
- Minimize dependencies between the access network (AN) and the core network (CN). The architecture is defined with a converged core network with a common AN–CN interface which integrates different access types, e.g., 3GPP access and non-3GPP access such as Wi-Fi or fixed.
- Support a unified authentication framework using in the future non-UICC credentials.
- Support "stateless" network functions, where the "compute" resource is decoupled from the "storage" resource.
- Support network capability exposure via a set of APIs usable by external parties.
- Support concurrent access to local and centralized services. To support low latency services and access to local data networks, UP functions can be deployed close to the Access Network.
- Support roaming with both home routed traffic as well as local breakout traffic in the visited PLMN.

Key new features of 5G system that are not supported by EPS in addition to the above mentioned general principles are:

- Support for network slicing providing the option of segmenting the network to multiple logical networks that provide specific network capabilities and network characteristics.
- A more flexible QoS model allowing flow differentiation within the same PDU session and support of reflective QoS with no need to provide UL packet filters
- Support for RRC-INACTIVE procedures keeping the UE always "connected" in CN while allowing RAN to keep the UE in power saving mode
- Combination of functionalities relevant for reducing end-to-end latency while maintaining service continuity

- Support for new session and service continuity modes allowing support for multiple user plane anchors either using IPv6 multi-homing or multiple user plane anchors (so-called "SSC mode 3")
- Support of Uplink classifier that provides user plane functionality that aims at "offloading" Uplink traffic, based on filter rules provided by SMF, toward data network within the same session.

- Native support for ethernet traffic with QoS differentiation

More details about these functionalities are described in subsequent sections.

3.4 EPC versus 5GC (What Is 5GC For?)

Despite the fact that most MNOs had to roll out 5G NR using EPC/Option 3 at the beginning of their 5G deployment, they are pretty much universally interested to migrate to 5GC in the next few years.

The 5GC architecture network functions and the new services that come with it are summarized in the following sections. It is though important to highlight that comparing the 5GS features with EPS features in Release 15 timeframe, is not the right approach because operators not only want to benefit from new features available in 5GC, but also more importantly the prospects for future evolution since this will become the platform of core network evolution for the years to come. Also in which "system" to invest the needed CAPEX for costs for proprietary features such as charging or O&M.

It is reasonable to expect that the R&D resources of network infrastructure vendors will shift toward the new generation core and will not evolve EPC which will soon be considered legacy. For instance, from the 10 ongoing 3GPP Release 16 "architecture evolution" studies, 8 are for 5GC and 2 for EPC. In this sense the right question is not to ask how many features 5GC can support that EPC cannot, but rather how the two will compare in 3 to 5 years from now and which will be the platform for further R&D in the system area.

Although plans are by no means finalized, there are some common themes that have emerged in terms of why 5GC roll out intentions are being accelerated.

Operators seem to want to move to cloud RAN architecture as soon as possible for the economic and flexibility benefits that this entails. For example, while Cloud RAN is available today for LTE, MNOs prefer to do this in combination with 5GC. This seems to be because of a combination of infra vendors pushing the idea, as they are naturally placing all their new development on 5GC, and MNOs not wanting to keep investing on EPC which is increasingly being considered as a dead-end.

Introduction of 5GC then is required, and it could come as soon as 2020, which means that Options 2, 4, 5, 7 are under consideration by all major MNOs. In the next section we look at which of these options are preferable from MNO perspective and why.

3.5 Main Functional Entities of the 5G Core

Due to the goal of disaggregation of the Core Network functionality the number of (logical) network functions in 5G Core (5GC) has increased. All these functions offer a set of service-based interfaces documented in TS 23.501 [1], the letter N and the name of the corresponding network function, e.g., Namf for AMF, Nsmf for SMF, etc. Using these service-based interfaces any other network function is possible to communicate and retrieve data from the said network function.

The "entry" function to the 5GC is **access and mobility function (AMF)** and connects to the 5G Access Network for 3GPP access or Non-3GPP interworking function (N3IWF) for non-3GPP. **AMF** terminates the UEs NAS session, performs access authentication and mobility management. It is also responsible to route session management messages to the right session management function (SMF). If non-3GPP access is supported, a UE simultaneously connected to the same 5GC Network of a PLMN over a 3GPP access and a non-3GPP access shall be served by a single AMF in this 5GC Network.

Session management function (SMF) is responsible to terminate UEs session management sessions. In addition, the main other functions include, UE IP address allocation and management (including optional authorization), selection and control of UP function, including controlling the UPF. Termination of interfaces toward policy control functions. SMF also determines the session and service continuity (SSC) mode of a session (see more below).

User Plane Function (UPF) implements the packet forwarding and routing for user plane data in the role of the inter-RAT and intra-RAT anchor in the 5GC. In addition, certain "control plane" functions such as allocation of IP address when requested from SMF, policy enforcement such as gating and downlink data buffering, and generation of notification requests.

Unified data management (UDM) is the equivalent of home subscriber system (HSS) in EPS. There was though more emphasis put in separating storage of subscriber data from actual business logic. UDM implements the application logic and does not require an internal user data storage and then several different UDMs may serve the same user in different transactions. **User data repository (UDR),** on the other hand, implements the actual storage.

Policy control function (PCF) is the enabler of the unified policy framework and provider of policy rules to the different control functions to enforce them. A very fundamental difference with EPS is the ability to distribute and enforce rules also at the UE, related to route and slice selection but also to the AMF for access subscription information.

Network exposure function (NEF) exposes APIs that allow 3rd party application functions (AF) to either gather data from the 5G System (e.g., location data) or instantiate certain actions in the 5G system. It expands the role of the service capability exposure function (SCEF) that was defined in EPS for internet of things (IoT) in TS 23.682 [10]. Many more exposure APIs have been defined and documented in TS 23.501 [1] and TS 23.502 [2].

Network repository function (NRF) concentrates all the selection and discovery procedures inside the core network. For instance, CN function selection such as SMF or network function instance selection based on, e.g., the slicing information is performed by the NRF.

Non-3GPP interworking function (N3IWF) allows NAS signaling to be used over non-3GPP access such as Wi-Fi or wireline. In such a way, it enables the use of 5GC for UEs that access via Wi-Fi and wireline access and provides a converged core network architecture. More details can be found in Sec. 3.9.

Unstructured data storage function (UDSF) was one of the most controversial functions in the study phase of the 5G system. Given the stated goals of separation of storage and compute the network functions such as AMF and SMF were not supposed to maintain state information. In the system though mobility management and session management protocols need to maintain state such as timers, connection associations, etc. These had to be stored somewhere, but 3GPP SA WG2 was not possible to get consensus on the format of the data stored. It was therefore agreed to have UDSF as the storage place for "unstructured" data that can be shared between the function. In an implementation it is possible to collapse the UDSF and UDR as common storage function.

As mentioned at the beginning one of the main design goals was the disaggregation of the different functionalities and ability to separate with more granularity the network logic to allow more modular implementations. The list of functions is therefore quite extensive (more than 20 in total). An avid reader can look at TS 23.501 [1] Sec. 6.2 where the 5GC functions are defined in detail.

3.6 High-Level Features of 5G Core

3.6.1 Mobility Management (aka Registration and Connection Management)

3.6.1.1 General

The principles of mobility management have not changed much between EPS and 5GS. Namely the procedures are governed by Registration and Connection Management.
For registration management (RM) there are two states in the UE and for the UE in AMF:

- RM-DEREGISTERED
- RM-REGISTERED

as depicted in Figs. 3.10 and 3.11.

When the UE in RM-DEREGISTERED state is not registered with the network. This effectively means that UE has no state in AMF, and AMF will reject any request, e.g., mobile terminated request for this UE. In order for the UE to go into RM-REGISTERED state it needs to successfully complete a registration procedure. Afterwards it needs to keep this registration active, through periodic registration update procedure triggered by expiration of the periodic update timer. When either the UE or the network triggers deregistration it moves to RM-DEREGISTERED state.

Connection management (CM) controls the rules and procedures for controlling a NAS signaling connection between a UE and the AMF over the NAS interface (N1). The construct of the NAS signaling connection enables the exchange of NAS signaling between the UE and the core network. It is therefore the union of the UEs RRC

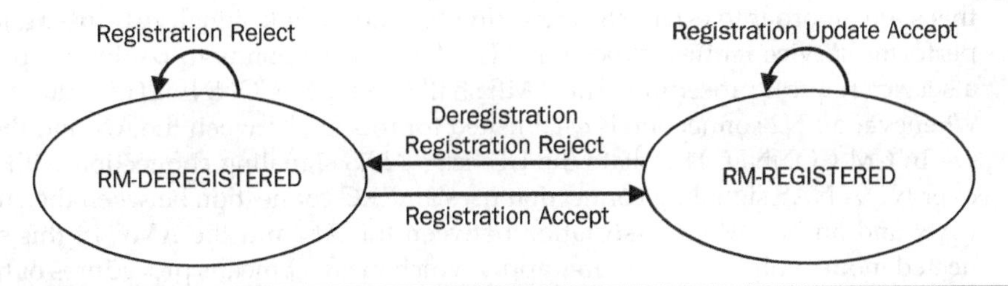

FIGURE 3.10 RM state model in UE (source TS 23.501).

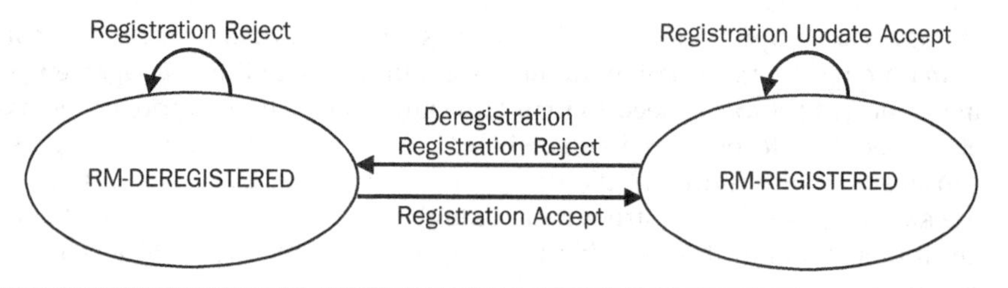

FIGURE 3.11 RM state model in AMF (source TS 23.501).

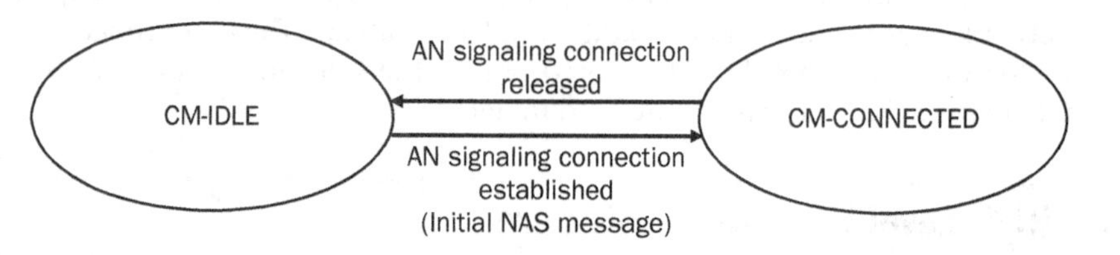

FIGURE 3.12 CM state transition in UE (source TS 23.501).

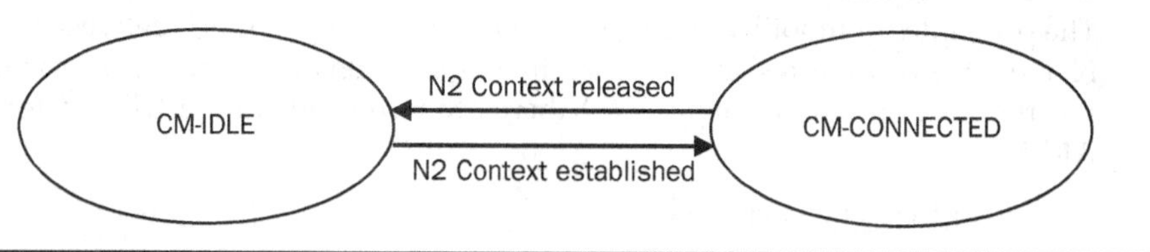

FIGURE 3.13 CM state transition in AMF (source TS 23.501).

connection between the UE and access network, and between the access network and AMF.

Two CM states are used to reflect the NAS Signaling connection state in the UE and of the UE with the AMF:

- CM-IDLE
- CM-CONNECTED

as depicted in Figs. 3.12 and 3.13.

When the UE is CM-IDLE state it has no NAS signaling connection with AMF. When the UE is in this state it performs cell selection and reselection according to TS 38.304 [11] and performs PLMN selection procedures as defined in TS 23.122 [12]. In this state, in order to establish connection to send uplink signaling or user data the UE performs service request procedure. The UE also responds to paging by performing a service request procedure. The AMF shall enter CM-CONNECTED state for the UE whenever an N2 connection is established for this UE between the AN and the AMF.

In CM-CONNECTED state the UE has a NAS signaling connection with the AMF over N1. A NAS signaling connection uses an RRC connection between the UE and the RAN and an NGAP UE association between the AN and the AMF. In this state connected mode mobility procedures apply which in effect means procedures of handover, RRC redirection are used where the UE has state in the RAN.

A UE in CM-CONNECTED state can be in RRC inactive state, which is a new functionality defined in 5G system and described further below.

It is important to highlight that 5G system can support common tracking areas between NR and E-UTRA connected to 5GC. When a UE registers with the network over any 3GPP access, the AMF allocates a set of tracking areas in TAI List to the UE. If the UE and the network support interworking with non-3GPP access, a single TAI dedicated to non-3GPP access is allocated, and this TAI is applicable for the whole PLMN.

3.6.1.2 RRC Inactive

In a nutshell RRC inactive is a new RRC state that is applicable to all the 3GPP RATs that connect to 5GC except NB-IoT. The intention is to keep the UE in connected state in the core network (CM-CONNECTED) but in a state where procedures similar to idle will apply in the RAN. In RRC inactive state, 5G core forwards all downlink for the UE to the UE and the RAN performs RAN paging within the RAN notification area. When the UE performs mobility across areas where the mobility of the UE is tracked by RAN (called RAN notification areas), the UE's context in RAN is transferred across using Xn procedures defined in TS 38.300 [13].

Before RRC Inactive was introduced in 5G Core, SA WG2 had to do quite extensive study for the similar functionality under LTE/EPS (code named LTE Light Connection). SA2 captured the system impacts in TR 23.723 [14] but eventually the impacts were

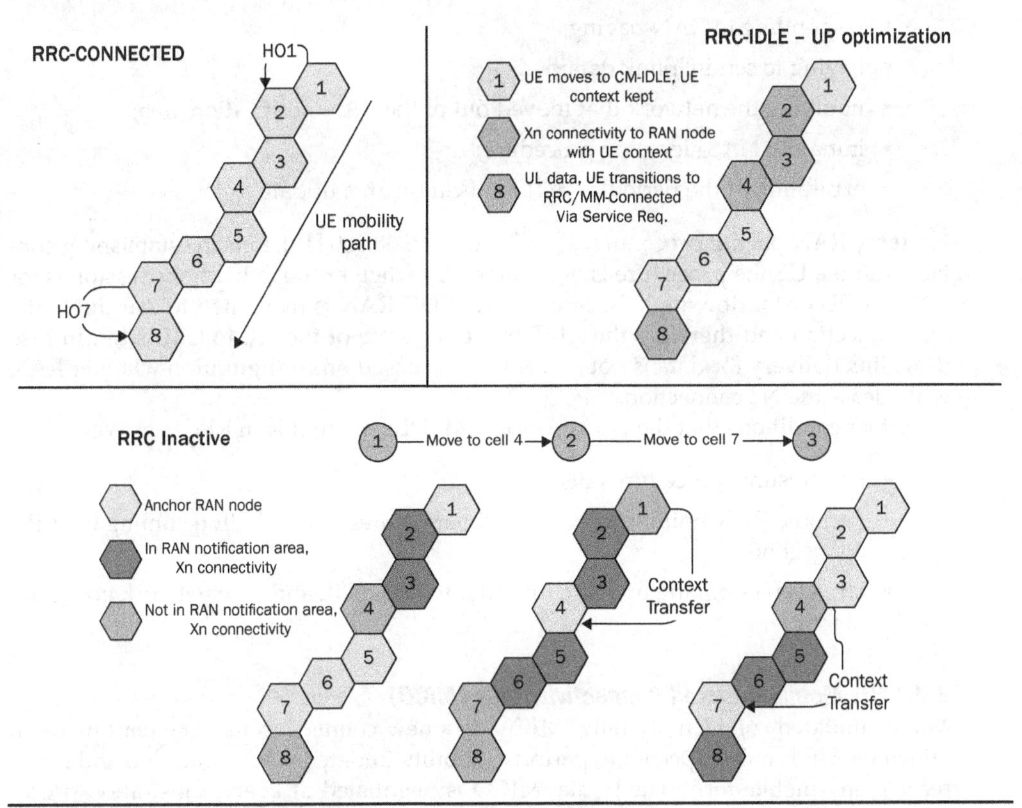

FIGURE 3.14 Interactions of the different states of RRC with mobility.

considered extensive for a legacy system and the feature was not introduced in LTE connected to EPS. In LTE/EPS, though since Release 13 there is the ability to keep UE state in RAN and resume the connection through RRC resume and not requiring service request. The functionality is called user plane optimization and is also introduced in 5G system as part of cellular IoT enhancements (see Chap. 4). The main difference though is that in this case the UE is in CM-IDLE state in the core network. The interactions between these different states of RRC mobility are depicted in Fig. 3.14.

For RRC inactive, the core network assists the RAN by providing parameters that can be used in order to decide the activation of RRC Inactive. These parameters are:

- the UE specific DRX values;
- the NAS periodic registration update timer;
- the registration area provided to the UE;
- Information from the UEs temporary identifier for E-UTRA connected to 5GC that allows the RAN to calculate the UE's RAN paging occasions.

At transition into RRC inactive state, the RAN node configures the UE with a periodic RAN notification area update timer and uses a guard timer with a value longer than the RAN notification area update timer value provided to the UE.

When the UE is in RRC inactive state, the UE may resume the RRC connection due to:

- responding to RAN paging;
- having to send Uplink data;
- notifying the network that moved out of the RAN notification area;
- initiating NAS signaling procedure;
- expiration of the periodic RAN notification area update timer

If the RAN paging procedure, as defined in TS 38.300 [13], fails in establishing contact with the UE the procedure is determined by whether the failed transmission is for a NAS PDU or for downlink. In case of NAS PDU RAN is mandated to tear down the N2 connection and therefore the AMF move the state of the UE to CM IDLE. In case downlink delivery for data is not successful it is based on configuration whether RAN will release the N2 connection.

Other conditions that the UE moves to CM IDLE when it is in RRC inactive:

- RRC resume procedure fails;
- Periodic RAN notification update timer expires without UE resuming the RRC connection;
- UE receives paging by AMF (in order to avoid UE and core network going out of sync);

3.6.1.3 Mobile Initiated Connectivity Only (MICO)

Mobile initiated connectivity only (MICO) is a new connection management mode in which the UE is only expected to perform mobility initiated communication and never receive any mobile terminated data. MICO is negotiated at every registration procedure, and if the UE indicates preference for MICO, the AMF based on local policy and subscription assigns a registration area that is not restricted in size, i.e., the UE will have to perform less often mobility management registration procedures. Alternatively, the AMF can assign a "whole PLMN" registration area and therefore the mobility management procedures do not apply. In either case the UE continues to perform periodic registration management procedure. MICO is meant to be useful for IoT UEs and for this reason in Release 16 it was combined also with the ability to assign an extended connected time, if the AMF knows that the UE has some pending mobile terminated data, the UE is kept in RRC-connected state for whole duration of the extended connected time and active time, in which case the UE is kept in CM-IDLE state and therefore is pageable for the duration of active, but not thereafter. In order to avoid paging in the entire PLMN, it is recommended against assigning "whole PLMN" registration areas for UEs assigned MICO with active time.

3.6.2 Session Management

Packet data unit (PDU) session is providing a virtual pipe between the UE and a data network that is identified by a data network name (DNN). The equivalent in EPS is packet data network (PDN) connection and the DNN is the equivalent of access point name (APN). There are though some additional properties in PDU session compared to the ones considered in EPS: S-NSSAI, i.e., the slicing indication (see more in Sec. 3.7) is used as part of the PDU session parameters. This means in effect that a UE can establish two PDU sessions with the same DNN but different S-NSSAIs (e.g., DNN="internet", S-NSSAI=V2X+SD and DNN="internet", S-NSSAI=eMBB+SD) which is not possible in EPS. This allows for another degree of freedom in terms of PDU session management in the UE and network. The PDU session types in 5GS are: (1) IPv4, (2) IPv6, (3) IPv4v6, (4) Ethernet, (5) Unstructured. Ethernet was not supported in EPS in Release 15 but was added in Release 16. Unstructured can be considered functionally equivalent to "non-IP" in EPS, i.e., a session that can use any type of data. Unstructured (similar to non-IP of EPS) cannot support QoS because obviously there is no way to describe the packet filters in the UE and the core network.

The other novelty of session management in 5GS is the definition of three session and service continuity (SSC) modes:

- SSC mode 1, the connectivity anchor for the UE is preserved. For instance, for all IP type PDU session types, the IP address of the UE is preserved.
- SSC mode 2, could release the connectivity service to the UE upon mobility to a new anchor. For instance, for all IP type PDU sessions the IP address of the UE is released when the UE moves to a new anchor.
- SSC mode 3, the network ensures that the UE suffers no loss of connectivity in a "make before break" fashion but the user plane changes are indicated to the UE. For instance, for all the IP type PDU sessions, the IP address is not preserved in this mode when the PDU session anchor changes, but a new one is assigned to the UE before the "old" one is released.

SSC mode selection is done by the SMF based on the allowed SSC modes including the default SSC mode in the user subscription as well as the PDU session type and, if present, the SSC mode requested by the UE.

The UE may be provisioned with an SSC mode selection policy (SSCMSP) as part of the UE route selection policy (URSP) (see Sec. 3.10). The SMF receives from the UDM the list of allowed SSC modes and the default SSC mode per DNN per S-NSSAI as part of the subscription information. When a UE signals an SSC mode at the establishment of a new PDU session, the SMF can agree with the SSC mode requested by the UE or reject the establishment of the PDU session.

5GC supports multi-homed PDU sessions, effectively allowing a single PDU session to be associated with multiple IPv6 prefixes. Multi-homing of a PDU session applies only for PDU sessions of IPv6 type. When the UE requests a PDU session of type "IPv4v6" or "IPv6" the UE also provides an indication to the network whether it supports a multi-homed IPv6 PDU session. When IPv6 multi-homing is used, then the different user plane paths merge into a "common" UPF referred to as a "branching point" functionality. The branching point provides forwarding of UL traffic toward the session anchors and merge of DL traffic to the UE. The UPF that supports the branching point functionality may also be controlled by the SMF to support traffic measurement for charging, traffic replication for LI, and bit rate. IPv6 multi-homing can also be used to facilitate local breakout as explained in the next section.

3.6.3 5GS Session Management and Enablers for Edge Computing Services

Edge computing is one of the promises of 5G. It will allow hosting of services much closer to the radio sites and therefore reducing the overall end-to-end delay. According to latest GSMA survey "Edge computing in the 5G era: Technology and market developments in China", almost 90% of mobile ecosystem players in China identified edge computing as a major revenue opportunity in the 5G era [15]. In order to achieve these goals 5G system session management contains some "native" functionalities that can be used as enablers in order to facilitate deployments for edge computing services:

1. Uplink Classifier

   Uplink classifier (UL CL) as seen in Fig. 3.15, solves a problem of EPS with local breakout. In EPS local break out required the configuration of a specific APN in order to create a PDN connection that is "locally" broken out, e.g., in a PDN GW close to the RAN and therefore achieving lower end-to-end delay. This required binding of this APN in specific application in the UE that made management of the additional APN rather cumbersome. In 5G System this

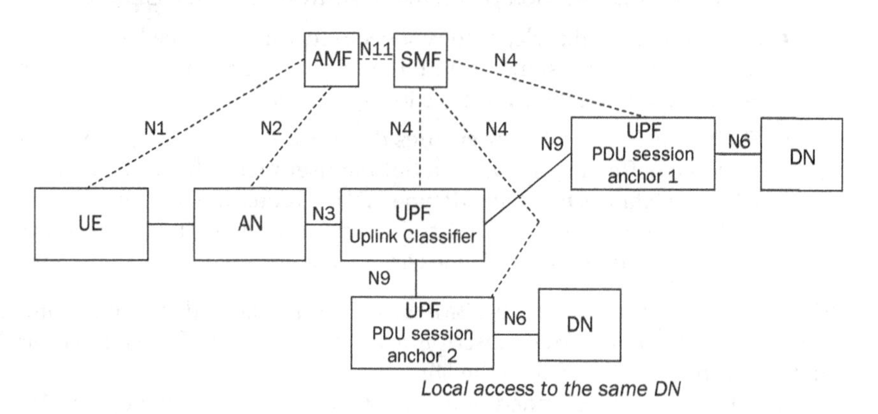

Figure 3.15 Uplink Classifier in 5GS architecture (source TS 23.501).

problem is alleviated with the introduction of an Uplink classifier that is an additional UPF that can be configured by the SMF with specific rules that can offload certain traffic locally, but the rest of the traffic of a given PDU session is still anchored in a different UPF potentially in the back end of the system. The addition of an Uplink classifier and the configuration of rules that govern the offloading of traffic are transparent from the UE that still uses the same PDU session irrespective of whether certain portion of this traffic is locally offloaded.

2. Local Area Data Network

   Local area data network (LADN) is functionally equivalent to local IP access (LIPA) in EPS. Basically, it is a data network that is only accessible from a restricted geographical area. In effect the PDU session to this specific DNN can only be successfully established when the UE is located in given area. The UE is configured to know whether a specific DNN is an LADN DNN and performs binding to the application layer accordingly. Some functionality that is not supported in EPS is that the AMF may provide to the UE based on its subscription and local configuration information about the LADNs that are available in the area when the UE performs registration procedure. Based on this information the UE may establish PDU session to the specific LADN DNN.

3. Local Access to a DN Using Multi-Homed PDU Sessions

   As described in Sec. 3.6.2 IPv6 multi-homed PDU sessions can be supported in 5GS. An assigned UPF as depicted in Fig. 3.16, can act as "branching point" and in similar manner to UL classifier can also provide access to local data network. In effect multi-homed PDU sessions can provide the same functionality by assigning to the UE an IPv6 prefix for local data network.

4. Influencing UPF Selection from AF

   With this functionality the AF can send requests to the SMF in order to influence routeing decisions for traffic of PDU sessions. This functionality only applies to local breakout and not home-routed traffic. The requests from the AF use N5

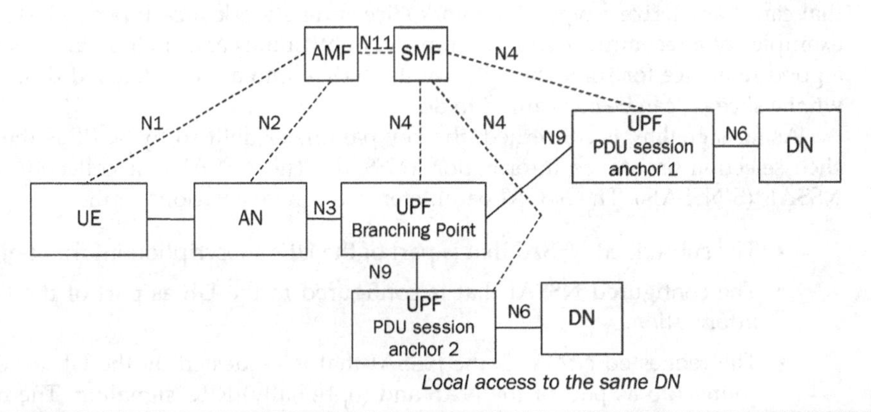

FIGURE 3.16 Access to local DN for multi-homed PDU sessions (source TS 23.501).

(for on-going PDU sessions of individual UE(s) or via the NEF. When the future PDU sessions of multiple UE(s) are supposed to be influenced the requests from the AF are sent via the NEF and may target multiple PCF(s).

5. Session and Service Continuity

   As described in Sec 3.6.2 5GS session management supports three session and service continuity modes. SSC mode 2 and 3 allow the session anchor to be relocated and in case of SSC mode 3 also while the UE can offer "make before break" service continuity to the application.

3.7 Network Slicing

Network slicing is certainly one of the most discussed about features of 5GC. Any publication, presentation, or marketing material about 5GC will undoubtably mention network slicing as one of the main new areas of innovation for the 5GC. In 3GPP defined telecoms system there are most of the times predecessor features with the same or similar goals. In this case the antecedent feature was dedicated core network (DCN) that was defined in Releases 13 and 14. The main difference though is that while DCN only exists in the core network (EPC and GPRS), network slicing is an end-to-end system feature. Another major difference is that DCN allowed each UE to connect to only one dedicated core network.

Network slicing overall provides the ability to segment the network control plane and user plane functions, RAN and CN, in several logical ones that potentially share the same physical infrastructure but can have autonomous operation and control. For example, it gives the ability for the MNO to dedicate a slice of the network to a 3rd party vertical, e.g., automotive company or specific service provided by the operator such as V2X or URLLC. This gives the ability to control, generate statistics, and manage these parts of the network separately establishing easier service level agreements (SLAs) either between the verticals and the MNOs but also across the MNOs in roaming scenarios. While 3GPP defines the signaling framework in order to allow a UE and several of the network functions (in RAN and CN) to assign specific policies to a specific slice, the actual policies and how a slice is used is out of scope of 3GPP and up to the specific MNOs. GSMA has defined in NG.116 the generic slice template, i.e., attributes that can characterize a type of network slice and network slice Types (NEST) that give examples of a recommended minimum set of attributes and their suitable values. It is a good reference for someone who wants to dive into a more detailed description on what a slice is "can be configured to do."

As far signaling is concerned, the key parameter defined by 3GPP is the network slice selection assistance information (NSSAI). The NSSAI is a collection of single NSSAIs (S-NSSAIs). The NSSAI parameter that can take various forms:

- The subscribed NSSAI that is part of the UEs subscription information.
- The configured NSSAI that is configured in the UE as part of the UE policy information.
- The requested NSSAI is the NSSAI that is requested by the UE to register or connect to as part of the NAS and (optionally) RRC signaling. The requested NSSAI signaled by the UE to the network allows the network to select the serving AMF, network slice(s) and network slice instance(s) for this UE.

- The Allowed NSSAI that is provided by the network (AMF) to the UE as part of the mobility management signaling when the UE performs registration to the serving PLMN and indicates the NSSAIs that the UE is allowed to connect to by this specific serving PLMN.

There can be at most eight S-NSSAIs in allowed and requested NSSAIs sent in signaling messages between the UE and the network.

As defined in TS 23.501 [1], the NSSAI is split in two parts:

- A slice/service type (SST), which refers to the expected network slice behavior in terms of features and services;
- A slice differentiator (SD), which is optional information that complements the slice/service type(s) to differentiate among multiple network slices of the same slice/service type.

The slice/service type (SST) has standardized values (at the time of writing this book the following SST values have been defined in TS 23.501, enhanced mobile broadband SST=1, URLLC SST=2, massive IoT SST=3, V2X SST=4). As defined in TS 23.003 [16] the SST information element is 8 bits and the SD information element is 24 bits.

In the UE as depicted in Fig. 3.17, what binds together the applications to specific NSSAIs and PDU session is the network slice selection policy (NSSP) that is part of the UE route selection policy (URSP) (see Sec. 3.10 for general description of UE policies in 5GS).

The network slice configuration information in the UE contains one or more configured NSSAI(s). A configured NSSAI may either be configured by a serving PLMN and apply to the serving PLMN only, or alternatively may be a default configured NSSAI configured by the home PLMN and that applies to any PLMNs for which no specific configured NSSAI has been provided to the UE. There is at most one configured NSSAI per PLMN. The default configured NSSAI, if it is configured in the UE, is used by the UE in a serving PLMN only if the UE has no configured NSSAI for the serving PLMN.

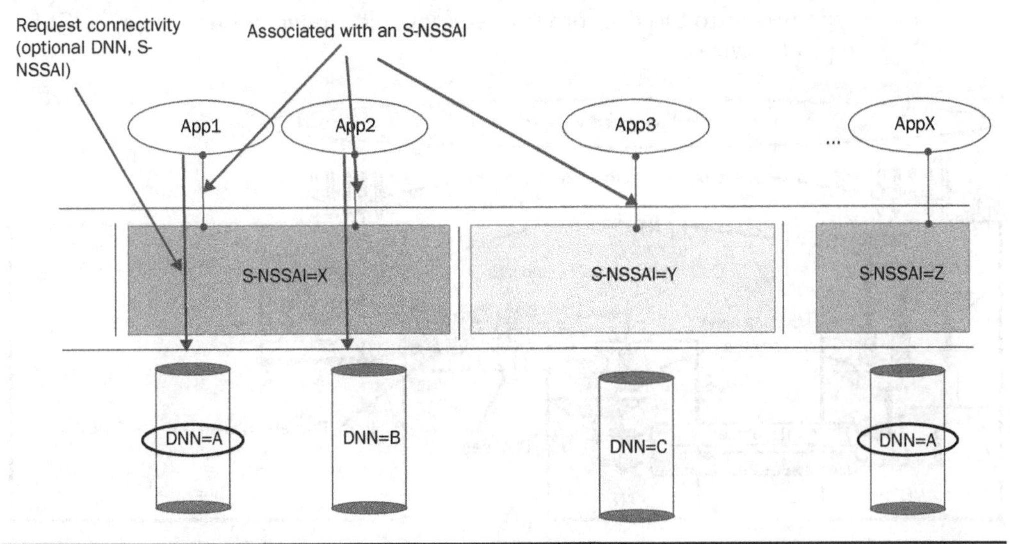

Figure 3.17 Mapping of applications to S-NSSAI and DNNs using the network slice selection policy part of URSP.

On the network side, the network slice selection function (NSSF) is responsible to select the set of network slice instances serving the UE, determine the allowed NSSAI and, if needed, the mapping to the subscribed S-NSSAIs (in case of roaming), determine the configured NSSAI and, if needed, the mapping to the subscribed S-NSSAIs. At last the NSSD is responsible to select the AMF set that is used to serve the UE when it is requesting a specific NSSAI.

3.8 QoS

The QoS fundamentals of 5GS resemble a lot those of EPS. The general QoS architecture is defined in TS 23.501 [1] where the lowest granularity of QoS differentiation within the PDU session is the QoS flow which has similar properties to the EPS bearer in EPS. The UPF in Downlink and the UE in Uplink map the service data flows (SDFs) to the QoS flows based on Uplink (UL) and Downlink (DL) packet filters provided by the policy control function (PCF) to the session management function (SMF) and signaled either at PDU session establishment or dynamically after application interaction, e.g., at IMS call setup. The concept of QoS flow is depicted in Fig. 3.18.

Each QoS flow is identified by a QoS flow ID (QFI) and has a QoS rule that contains the QoS profile which is basically the set of QoS parameters and the QoS flow template that is the set of UL and DL packet filters that represent the service data flows that map to the given QoS flow. There are three "resource types" of QoS flows:

1. Guaranteed bitrate QoS flows, where there is admission control and the signaled bitrate of the QoS flow in GFBR parameter is guaranteed,
2. Non-guaranteed bitrate QoS flows where as the name suggests the bitrate is not guaranteed and therefore there is no guaranteed flow bitrate (GFBR) parameter,
3. Delay critical GBR, where the bitrate signaled by GFBR is guaranteed but in addition any packets delayed above the limit indicated by the packet delay budget (PDB) are counted as "lost." The delay critical GBR QoS flows resource type was introduced in order to support ultra reliable low latency (URLLC) type of services.

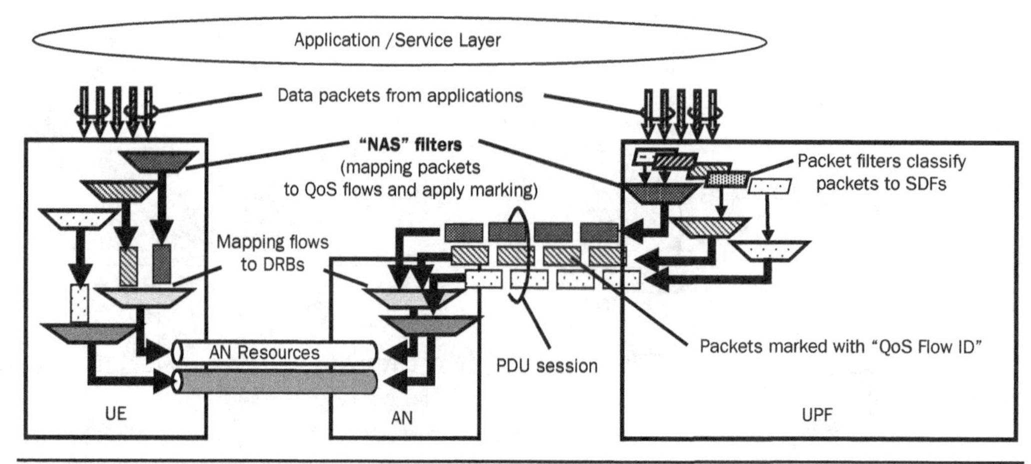

Figure 3.18 The QoS flow.

The QoS parameters signaled are the 5G QoS indicator (5QI), and allocation and retention priority (ARP), for every QoS flow and guaranteed flow bitrate (GFBR), and maximum flow bitrate (MFBR) for GBR and delay critical GBR QoS flows. From these parameters the 5QI of 5GS and QoS class identifier (QCI) of EPS are functionally equivalent scalar values that are point to combination of values of QoS parameters: packet delay budget (PDB), packet error rate (PER), averaging window, and priority for all types of 5QIs. For delay critical GBR 5QIs in addition the maximum data burst volume (MBDV) is also defined.

The packet delay budget (PDB) represents the upper bound for the time that a packet may be delayed between the UE and the UPF that terminates the N6 interface. The packet error rate (PER) represents the upper bound for the rate of packet data units (PDUs) that have been processed by the sender of a link layer protocol (e.g., RLC in RAN) but that are not successfully delivered by the corresponding receiver to the upper layer. The maximum data burst volume (MDBV) represents the largest amount of data that the 5G RAN node is required to serve within a period of the "access stratum" component of the PDB (i.e., excluding the core network delay). MDBV is only used for delay critical GBR QoS flows. The averaging window (which has default value 2 s) is the time defined in order to calculate the rest of the QoS characteristics of 5QI.

The table containing the latest defined standardized 5QI is in TS 23.501 [1] Sec. 5.7.4. Given new 5QIs are defined in every release of the specification in order to support new example services it is better to refer there for the latest set of standardized values. The hierarchy of QoS rule and corresponding parameters is depicted in Fig. 3.19.

There are at some new developments in 5GS QoS architecture compared to EPS:

1. The ability to explicitly signal QoS parameters that are part of the 5QI, e.g., packet delay budget and packet error rate in addition or instead of the 5QI. For example, it is possible to signal a standardized 5QI, but separate value of the PDB if only one of the QoS characteristics cannot fulfill the service's requirements. This allows more flexibility compared to EPS.
2. Reflective QoS, which is an optional feature, allows the UE to derive the mapping of the uplink traffic to QoS flows even when there are no QoS rules signaled. Reflective QoS works with the UE creating QoS rules corresponding to the received downlink traffic.

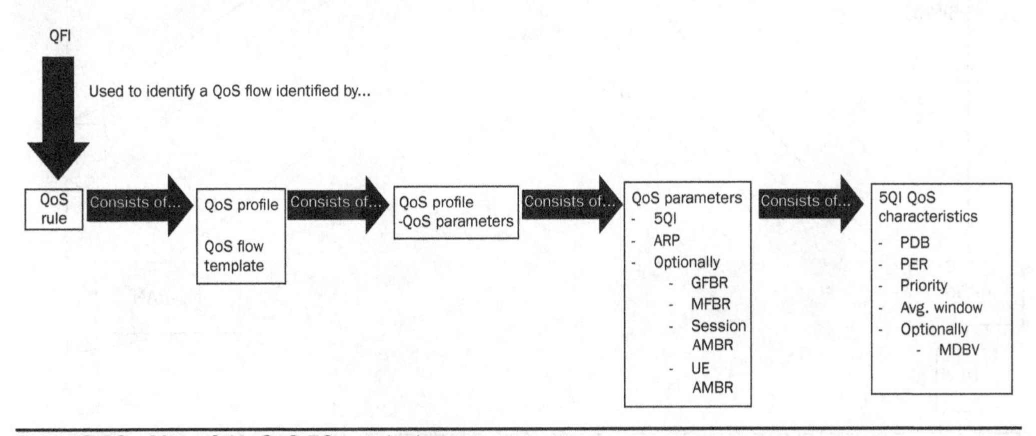

FIGURE 3.19 Map of the QoS 5G terminology.

3.9 Interworking with Non-3GPP Access Technologies

While 3GPP is well known for developing radio access technologies (starting from UMTS all the way to NR) and in addition providing the corresponding system integration of those (starting from GPRS all the way to 5GS), another less known expertise is the support of non-3GPP defined access technologies in the 3GPP systems. The term "non-3GPP access" when used in 3GPP specification is a codename for any access technology (wireless or fixed) that is not defined by 3GPP (radio) standards. Examples include Wi-Fi, CDMA 2000, cable access, or DSL access. In general, the procedures that 3GPP defines for this type of interworking are access agnostic; hence, usually the term "non-3GPP access" is used in the specifications. In previous release a lot of different flavors of non-3GPP interworking were defined starting from TS 23.234 [17] in Release 6 to TS 23.402 [18] for EPS.

This "tradition" continued with the 5G system and in TS 23.501 [1] there are two basic mechanisms in 5G system procedures of Release 15 to connect non-3GPP defined accesses in the 5G system:

- Use evolved packet data gateway (ePDG) that was defined for EPS in TS 23.402 [18] but can continue to be used for 5GS. The UE in that case supports the same protocol stack and procedures to connect to ePDG as defined in TS 23.402 [18] but the S2b termination happens to a "combo" SMF+PGW-C and UPF. This architecture option is depicted in Fig. 3.20.

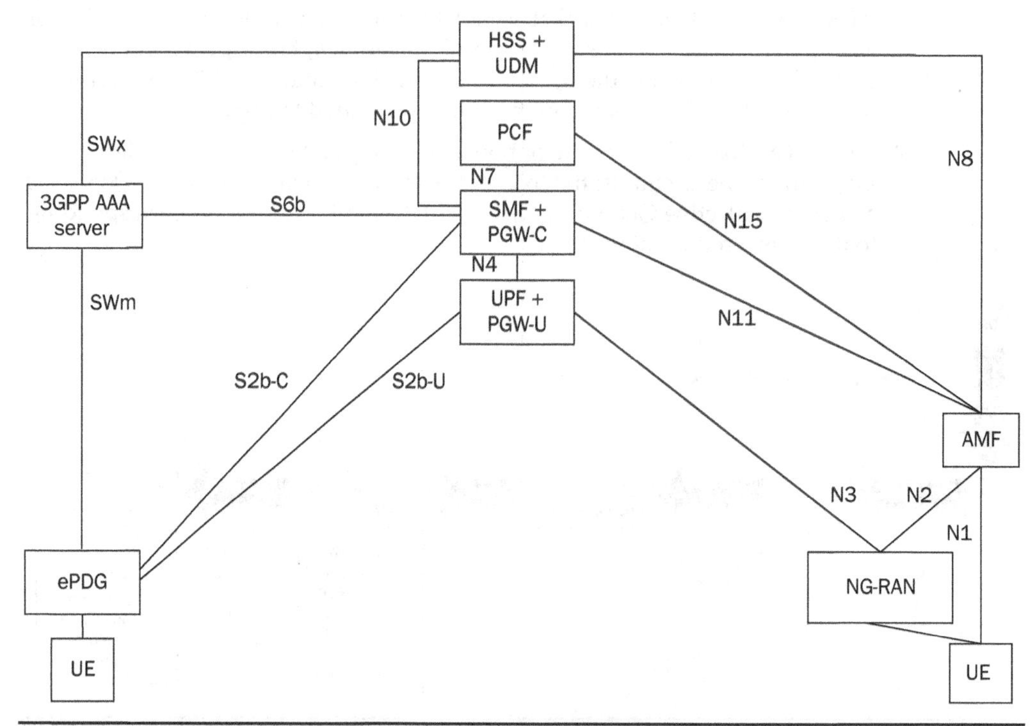

FIGURE 3.20 Architecture for interworking between ePDG/EPC and 5GS (source TS 23.501).

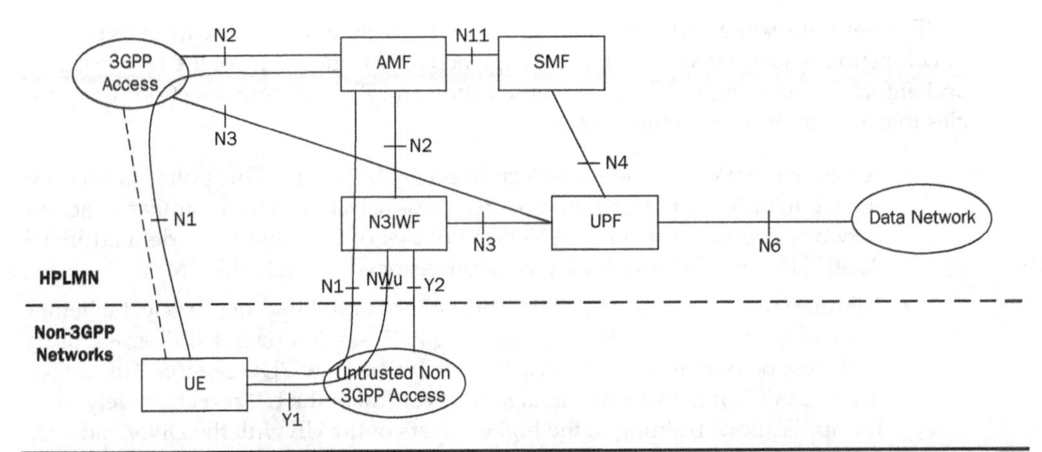

FIGURE 3.21 Architecture for interworking between N3IWF for non-3GPP accesses and 5GS (source TS 23.501).

- Use the newly defined Non-3GPP interworking function (N3IWF). This option introduces a novel way for interconnecting non-3GPP accesses into a 3GPP system, since the UE continues to use NAS and connect to AMF and the rest of control plane of the 3GPP protocol stack on top of IKEv2/IPsec based secure tunnel(s) established between the UE and the N3IWF. This architecture option is depicted in Fig. 3.21.

The UE is selecting non-3GPP accesses network using the access network discovery and selection policy (ANDSP) that is signaled to the UE using the UE policy content with mechanisms described in Sec. 3.10. The contents of the ANDSP are defined in TS 23.503 [19] for Wi-Fi selection policy, the WLAN selection policy (WLANSP) parameters are identical to those defined in TS 23.402 [18].

Release 15 supports only untrusted non-3GPP access while addition methods and procedures are defined in Release 16 to support wireline access and access transfer selection and splitting (ATSSS). A brief description of those functionalities is provided in Chap. 4.

3.10 Policy Control

Policy control was introduced for first time in 3GPP systems in Release 5 with IMS policy control initially only for control of QoS. In later releases of EPS, policy control and charging were combined and a new logical function (policy and charging rules function) – PCRF, was introduced. The system functionality for policy control and charging in EPS has seen many new features across the years and has evolved into a standalone architecture documented in TS 23.203 [20] for EPS. The policy control architecture of EPS was mostly dealing with session management–related procedures.

In 5GS, the new control function was named policy control function (PCF) and what was defined already for session management policy control was almost entirely transferred to the new architecture in order to allow QoS and charging control for IMS and non-IMS sessions. The role of the PCF in 5GS is identical to that of PCRF in EPS and signals traffic rules that are used in order to enforce policies in the user plane. The role of the policy control enforcement function (PCEF) is split between the SMF and UPF, due to the control and user plane split defined in the new 5GS architecture.

Furthermore, some additional new functionalities were added. The most important new function is not related to session management and is the ability of the PCF to signal and enforce policies in the UE for access selection and PDU session selection. The policies that are provisioned to the UE are:

- Access network discovery & selection policy (ANDSP): This policy is used by the UE in order to perform non-3GPP access selection. This is similar to access network selection function (ANDSF) policies of EPS that were defined in TS 23.402 [18]. In EPS though they were provisioned using OMA DM.
- UE route selection policy (URSP): This policy is used in order to select whether some outgoing traffic will use an existing PDU session over 3GPP access, use a PDU session over non-3GPP access or establish a new PDU session. This policy is the "brain" of the session management control of the UE and effectively links the applications residing in the higher layers of the UE with the DNN and slice information (NSSAI). Effectively, it determines DNN and slices each application in the UE uses (see Fig. 3.22).

The 5G policy control architecture as depicted in Fig. 3.23 is defined in TS 23.503 [19] and as is common for 5GS it uses service-based interfaces.

In order for the ANDSP and URSP to be signaled to the UE, the Namf interface is used and the policies are signaled using NAS (TS 24.501 [21], clause D.2.6) and the UE configuration update procedure. The encoding for the ANDSP and URSP policies in stage-3 is defined in TS 24.526 [22].

Another innovation of the 5GS policy control architecture that started in Release 15 is the introduction of the network data analytics function (NWDAF). This function is responsible for data collection and generation of network analytics that can be used to automate the network operation. While NWDAF was introduced first in TS 23.503 [19]

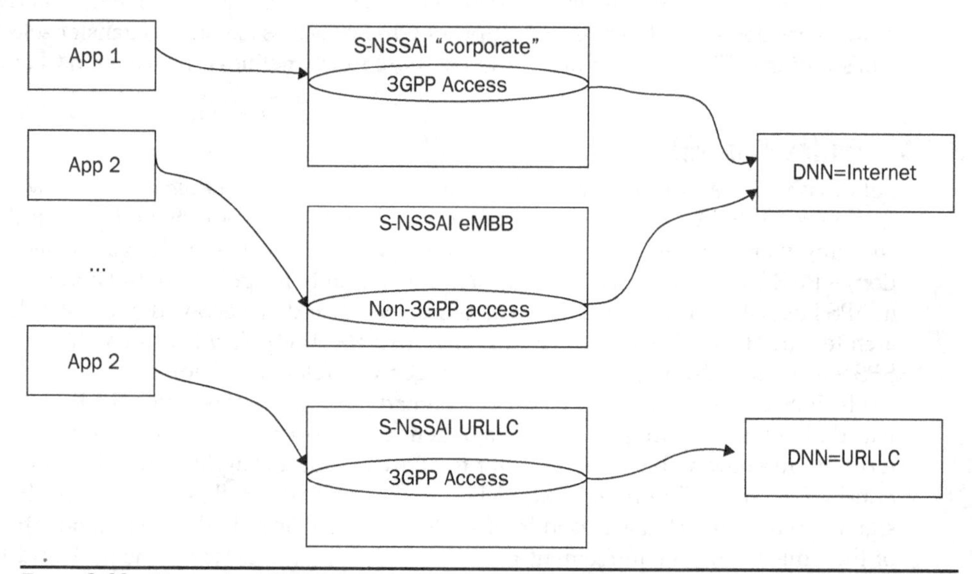

FIGURE 3.22 Illustration of URSP.

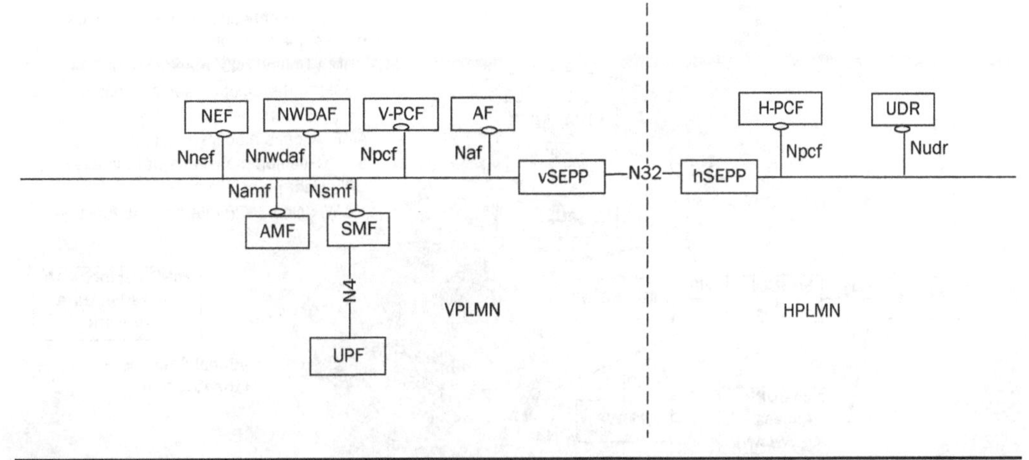

FIGURE 3.23 5GS policy control and charging architecture (service-based representation) (source TS 23.503).

in Release 15, in Release 16 it has evolved in a new architecture in TS 23.288 [23]. More details on the network data analytics services are defined in Chap. 4.

3.11 5G Security

Enhancement to the security architecture is one of the major enablers for operators to deploy 5G core network. The opportunity of introducing a new architecture and therefore new set of signaling procedures allows operators to introduce new security functionalities into the system without the concern of maintaining backwards compatibility.

The overall security architecture is depicted in Fig. 3.24 and the main new security features of the 5G system include:

- **Privacy of the UE's permanent identifier**: The goal is to prevent identification of a subscriber based on capturing the subscriber's permanent identifier sent over the air, e.g., by using IMSI catcher. In order to achieve that, the Subscription Permanent Identifier (SUPI) is never sent over the air, but an encrypted SUPI – SUbcription Concealed Identifier (SUCI) is sent over the air when necessary (e.g., during the initial registration). The SUCI is constructed such that only the MSIN/username part of the SUPI is encrypted while the Home Network identifier is encoded in clear for the routing purpose. The Subscription Identifier De-concealing Function (SIDF) that is part of the UDM is used to decrypt the MSIN/username part of the SUCI.
- **Initial NAS Protection**: For initial access to LTE/EPC, sensitive information elements that could potentially leak information about the user are at best only integrity protected. In 5G System as depicted in Fig. 3.25, care has been taken for the UE to send only essential information for negotiating security capabilities and identifying the security context in clear (i.e., unciphered) in the initial NAS message. The rest of the information elements are always ciphered. In order to achieve this, partial ciphering was introduced to the initial NAS signaling procedures of 5G System.

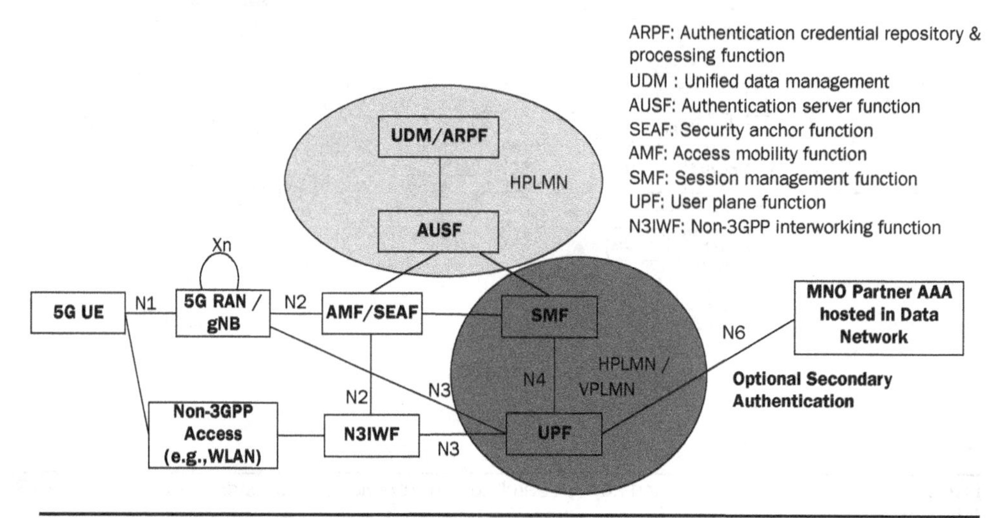

Figure 3.24 5G system security functions. (Note: Only 5G System security relevant functions and interfaces are shown.)

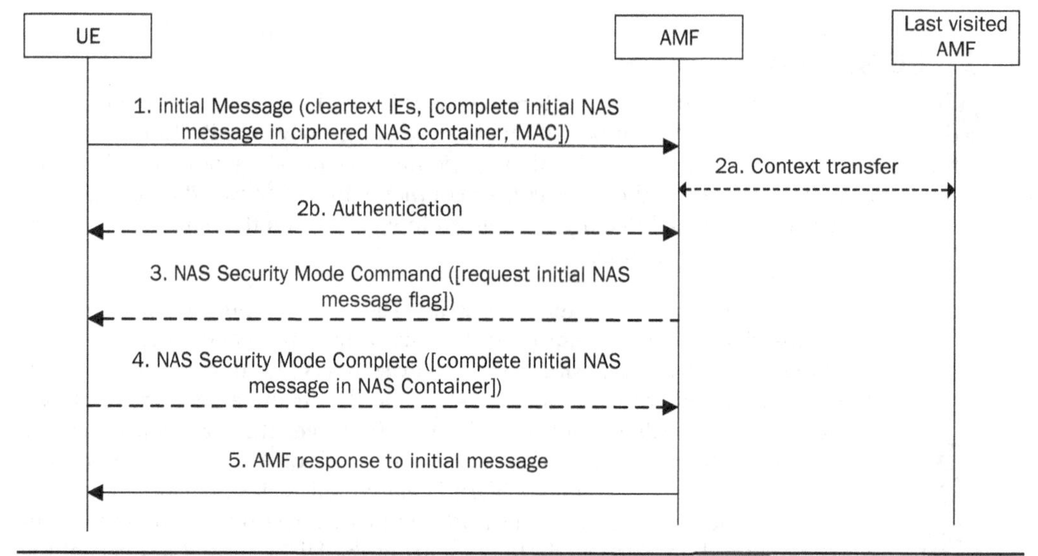

Figure 3.25 Initial NAS message protection/partial ciphering (source TS 33.501).

- **Authentication and Key Agreement:** It is mandatory for the UE and the 5G serving networks to support both 5G AKA and EAP-AKA′. The 5G AKA enhances the EPS AKA with authentication confirmation (AC) from SEAF to AUSF for enhanced home control (i.e., confirmation that the UE is actually present in the serving network) in order to address certain issues during roaming scenarios, e.g., billing fraud. EAP-AKA′ (and other EAP methods) natively provides this control as the EAP authentication is performed between the UE and the AUSF in the home network. It is the home network that decides whether to use 5G AKA or EAP-AKA′. Both 5G AKA and EAP-AKA′ can be used over 3GPP and untrusted non-3GPP accesses. In the 5G EAP framework, the UE takes the role of EAP peer and SEAF takes the role of EAP pass through authenticator, while AUSF takes the role of back-end authentication server.

- **User Plane Integrity Protection:** Integrity protection in the user plane is mandatory to be supported by all UEs that support NR connected to 5GC. Integrity protection is used in the user plane between the UE and gNB. There are two supported data rates (i.e., 64 Kbps or full rate) and the control is negotiated between the UE and SMF in 5GC per PDU session granularity.
- **5G Key Hierarchy:** Key hierarchy is depicted in Fig. 3.26 and two new keys are defined in the 5G key hierarchy: K_{AUSF} and K_{SEAF}. For 5G AKA, K_{AUSF} is derived by the ARPF and sent to the AUSF. For EAP-AKA′ (and other EAP method), AUSF sets the K_{AUSF} to the first 256 bits of the EMSK. This allows for introduction of any EAP authentication method in the future (e.g., EAP-TLS) without impacting the serving network. K_{SEAF} is the anchor key derived by the AUSF and sent to the SEAF/AMF in the serving network. K_{AMF} is the key derived from K_{SEAF} (based on a fresh authentication) or from another K_{AMF} (in case of mobility scenarios). Once K_{AMF} is derived, K_{SEAF} is no longer needed. ngKSI is

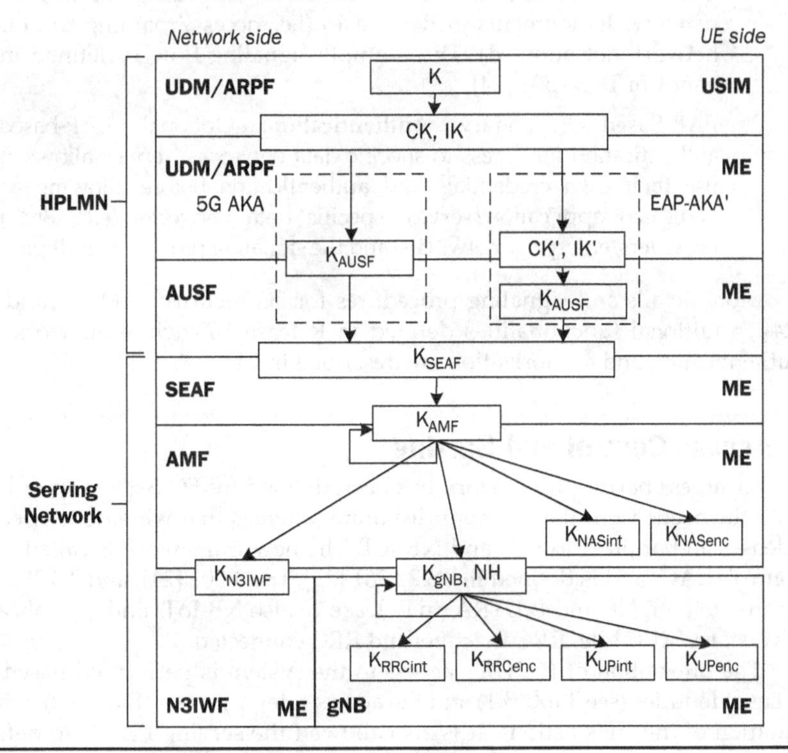

FIGURE 3.26 5G key hierarchy (source TS 33.501).

used to identify the 5G security context in 5GS. If the USIM supports storage of security context (e.g., Release 15 USIM), it is stored on the USIM (e.g., UE power off/detach); otherwise, it is stored on the mobile entity (ME) non-volatile memory (NVM). Note that pre-release 15 USIM can still be used to access the 5G system.

- **5G UE Credential Storage Requirements:** Allows the use of secure element integrated in system on a chip (SoC) ("iUICC" / "ieUICC"). As defined in TS 33.501 [24], the long-term key(s) of the subscription credential(s) shall never be available in the clear outside of the tamper-resistant secure hardware component and the authentication algorithm(s) that make use of the subscription credentials shall always be executed within the tamper-resistant secure hardware component.
- **Primary Authentication over Untrusted Non-3GPP Access:** For Non-3GPP access (i.e., Wi-Fi or wireline using Non-3GPP interworking function (N3IWF) as defined in Sec. 3.9), an IPsec tunnel is established between UE and N3IWF. N3IWF plays similar role as ePDG in EPC (defined in TS 23.402 [18]). AMF takes N3IWF equivalent to "5G RAN or gNB" in 3GPP access. EAP-5G (a new 3GPP specified EAP method) is used as a transport to carry all NAS messages between UE and the AMF.
- **EAP-TLS Use for Non-Public Network Deployments** (see details in Sec. 4.2): UE and AUSF can be pre-configured with certificates that would be used for EAP-TLS authentication. This method is only used for isolated non-public network deployments in Release 15 (i.e., access/roaming from public mobile network not allowed). The example signaling flow is defined in informative annex of TS 33.501 [24].
- **EAP-Based Secondary Authentication:** Optional EAP-based secondary authentication for access to specific data networks/service allows applications to use their own credentials and authentication before allowing access to their dedicated application/service specific data networks (e.g., 3rd party service providers, enterprise networks) and the signaling procedure is depicted in Fig. 3.27.

More details and signaling procedures for 5G security can be found in TS 33.501 [24]. Additional functionalities defined in Release 16 such as network slice specific authentication and authorization are described in Chap. 4.

3.12 Access Control and Barring

A new access barring framework has been defined for 5G System in order to consolidate the needs from the different disparate schemes that were developed in previous releases and applied to LTE and NB-IoT. The new framework is called unified access control (UAC) and is defined in TS 22.261 [25], TS 24.501 [21], and TS 38.331 [26]. UAC applies to both NR and E-UTRA (in Release 16 also NB-IoT) and is applicable to all UE RRC states: RRC Idle, RRC Inactive, and RRC connected.

The prevention of the UE's access to the system is performed based on the UE's access identities (see Table 3.1) and the access categories (see Table 3.2), which is a combination of the UE's actions. It is also allowed the serving PLMN to define operator-defined access categories (not specified by 3GPP) in order for example to control access

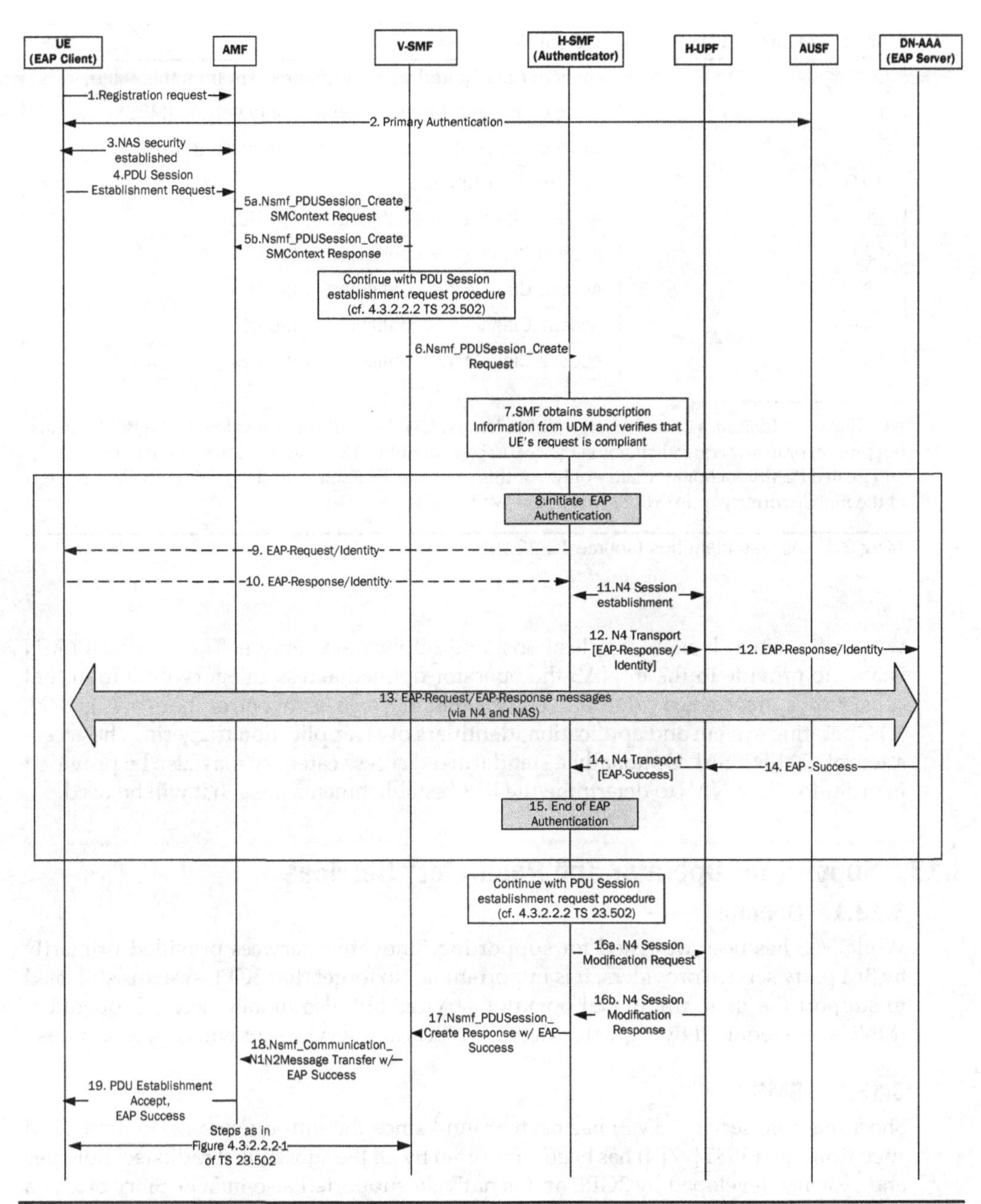

FIGURE 3.27 EAP secondary authentication (source TS 33.501).

Access identity number	UE configuration
0	UE is not configured with any parameters from this table.
1	UE is configured for multimedia priority service (MPS).
2	UE is configured for mission critical service (MCS).
3-10	Reserved for future use
11	Access Class 11 is configured in the UE.
12	Access Class 12 is configured in the UE.
13	Access Class 13 is configured in the UE.
14	Access Class 14 is configured in the UE.
15	Access Class 15 is configured in the UE.

NOTE: Access Identities 11 and 15 are valid in home PLMN only if the equivalent home PLMN list is not present or in any equivalent home PLMN. Access Identities 12, 13, and 14 are valid in home PLMN and visited PLMNs of home country only. For this purpose, the home country is defined as the country of the mobile country code (MCC) part of the IMSI.

TABLE 3.1 Access Identities (source TS 22.261)

to specific network slices, applications, and application servers. The serving PLMN is able to provide to the in NAS the operator-defined access category definition that consists of a precedence value, and criteria that may consist of one or more of (1) DNN, (2) Operating system and application identifiers of an application triggering the access attempt, (3) S-NSSAI. Optionally a standardized access category may also be provided in order to allow NAS to determine the RRC establishment cause that will be used.

3.13 Support for Operator and Regulatory Services

3.13.1 General

While 5GS has been designed for supporting innovative services provided primarily by 3rd party service providers, it is important not to forget that 3GPP systems still need to support the more traditional operator services but also mobile network operators (MNOs) are required by regulators to support certain services for emergency services.

3.13.2 SMS

Short message service (SMS) has been around since the introduction of the first GSM specification in 1985 [27]. It has been supported by all the subsequent radio technologies and systems developed by 3GPP and is natively supported also in 5GS. Since 5GS is a packet switched (PS) only system and has no interconnection with circuit switched (CS) domain, SMS is provided only in packet switched. There are two main mechanisms to transport SMS, one is using IMS where SMS messages are encapsulated in SIP/IP messages and transported over IMS network functions (this architecture option can be used also in 5GS and is documented in TS 23.204 [28]), the other architecture option is to transport the SMS messages over NAS (as defined specifically for 5GS and documented in TS 23.501 [1]). For this latter architecture option in order to route the SMS messages

Example access category number	Conditions related to UE	Type of access attempt
0	All	Mobile originated signaling resulting from paging
1	UE is configured for delay tolerant service and subject to access control for Access Category 1, which is judged based on relation of UE's HPLMN and the selected PLMN.	All except for emergency
2	All	Emergency call
3	All except for the conditions in Access Category 1.	Mobile originated signaling on NAS level resulting from other than paging
4	All except for the conditions in Access Category 1.	Multimedia telephony voice
5	All except for the conditions in Access Category 1.	Multimedia telephony video
6	All except for the conditions in Access Category 1.	SMS
7	All except for the conditions in Access Category 1.	Mobile originated data that do not belong to any other access categories
8	All except for the conditions in Access Category 1	Mobile originated signaling on RRC level resulting from other than paging
9-31		Reserved for other standardized access categories
32-63	All	Operator-defined access categories

Table 3.2 Example Access Categories (source TS 22.261)

to and from the UE via NAS and to the legacy SMS infrastructure (SMS-GMSC/SMS-router) a new function (SMS Function) was defined specifically for this role.

A side effect of the ability of 5G NAS to be transported also over non-3GPP access (see Sec. 3.9) and N3IWF is that SMS can also be supported over non-3GPP access (Wi-Fi, fixed) using also NAS transport as well as SIP/IP transport and IMS.

3.13.3 Voice Services

Support of voice services was one of the biggest challenges the EPS architecture had to accomplish. Initially the view in the 3GPP community was that LTE/EPS will be a packet switched only system and voice services will be offered using IMS. At some point though later in Release 8 several operators and the technical community, at large, realized that the challenge to upgrade both the core network and also deploy IMS in order to offer LTE/EPS connectivity to their customers was a very big challenge. As a result, introduced an interim mechanism to support voice services in CS domain with "CS Fallback" (CSFB). CSFB effectively triggers a handover or redirection of the UE to

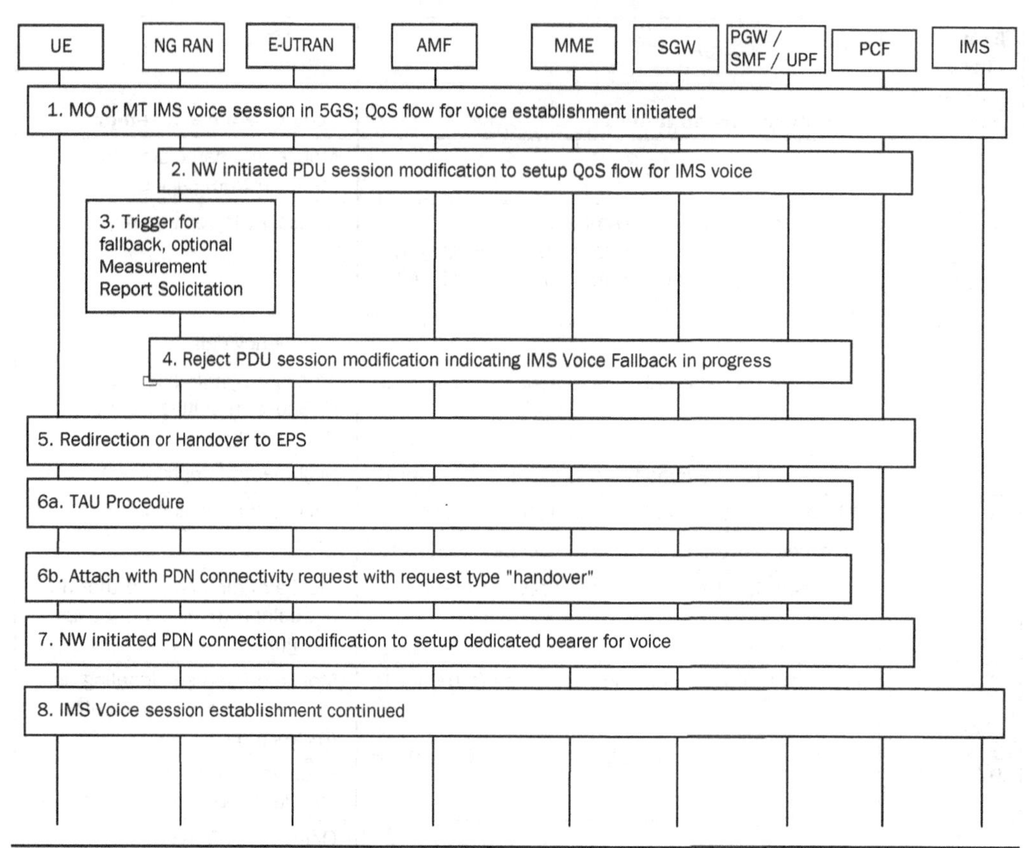

FIGURE 3.28 EPS fallback for voice (source TS 23.502).

2/3G CS (GSM/UMTS or 1x RTT) in order to perform a mobile originating or terminating voice call. Providing voice using CSFB is quite common even today (10 years after the initial launch of LTE). In the meantime, though the majority of MNOs have either launched or are in the process of launching VoLTE which is the native support of voice service over LTE and using IMS for the call setup and control signaling. Almost all of the new LTE handsets that come to the market in the last 2 to 3 years support both CSFB and VoLTE.

5GS is the first 3GPP system that originally did not have support even for interworking with CS domain, even though support for single radio voice call continuity to UMTS CS was eventually added in Release 16.. It is now commonly expected that all MNOs will migrate to supporting IMS by the time of deploying 5GS. What was though also considered is whether NR radio access technology will be mature enough to support the same degree of spectral efficiency and QoS as LTE for voice services. In simple terms the development and testing in order to support "Voice over NR" (VoNR) was not first priority for MNOs in the initial release of 5GS. For this reason, a new signaling procedure was defined in TS 23.501 [1] and TS 23.502 [2] to trigger handover or redirection of the UE to LTE (either connected in EPC or 5GC) when there is mobile originating or terminating call. This procedure is called "EPS/RAT fallback." Unlike though CSFB, mentioned above, voice is still using the packet switched domain and signaling control is still using IMS. The main difference is that radio interface uses LTE that is already optimized and well-tested for voice services. This signaling procedure is transparent from the UE (has no UE impacts) and the decision whether to use VoNR or EPS/RAT fallback is taken entirely on the network.

3.13.4 Emergency Services

In EPC, support for emergency services was defined one release after the initial introduction of EPC (Release 9). In 5G System though the support for emergency services was added from the first release of the specifications. The reason is that it did not require too many changes and the functionalities are mostly a carbon copy of what is supported in EPC. As stated in Sec. 3.13.3, in 5G system there is only support of voice services using IMS and the same applies also to emergency services which are also supported in IMS. The main signaling procedures are defined in TS 23.167 [29] and TS 23.501 [1]. Support for emergency services allows also unauthenticated access when the UE is in limited service state (as defined in TS 23.122 [12]) and is not able to register normally in 5G System.

The only new functionality that was defined specifically for 5GS is the support of emergency fallback which serves similar purpose to EPS/RAT fallback (see Fig. 3.28), i.e., to allow a smooth selection of RAT connected to EPC (LTE) if there is no support for emergency services provided in 5GC. There is a difference between emergency fallback as depicted in Fig. 3.29 and EPS/RAT fallback defined for normal voice service, since for emergency fallback the procedure is UE initiated.

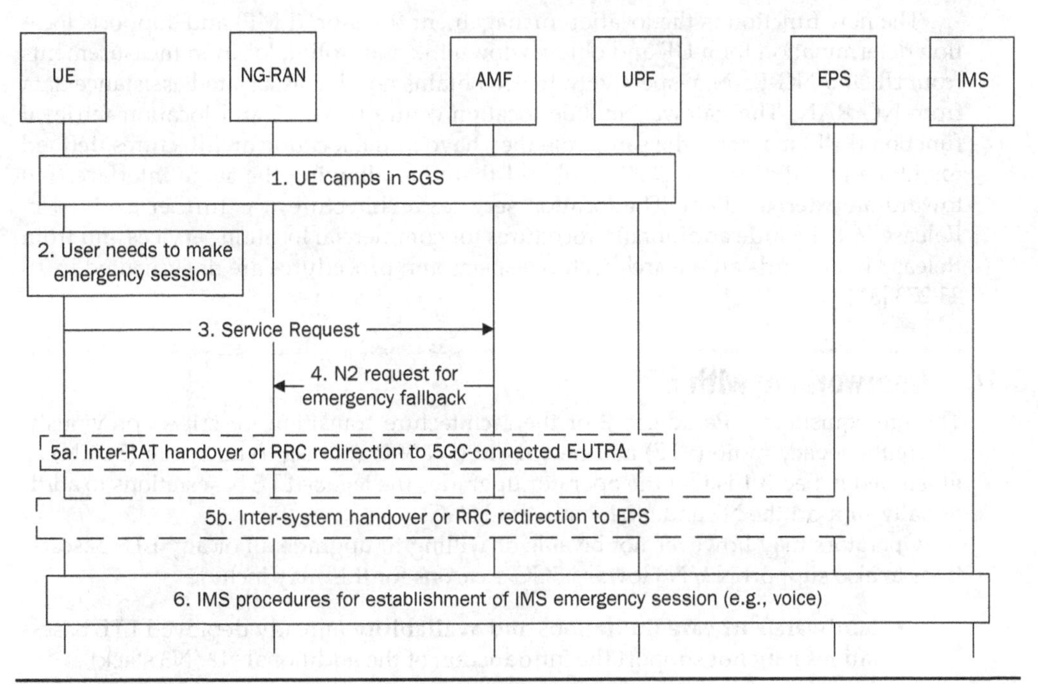

FIGURE 3.29 Emergency services fallback (source TS 23.502).

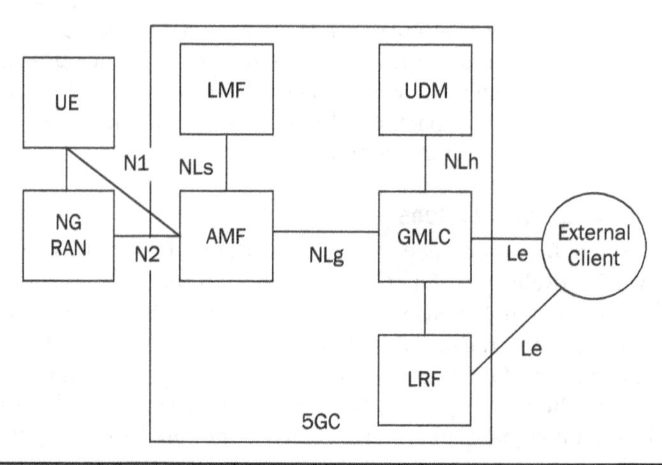

Figure 3.30 Location services architecture (source TS 23.501).

3.13.5 Location Services

While there is no support for NR-based positioning in Release 15 in order to support emergency services and fulfill the related regulatory the 5GS supports basic procedures for location services using RAT-independent position methods (e.g., A-GNSS, WLAN, Sensors), cell ID–based methods and methods defined for E-UTRA. The architecture for Release 15 is depicted in Fig. 3.30 and is defined in TS 23.501 [1] and the procedures supported are limited to the ones strictly necessary for regulatory services are defined in TS 23.502 [2].

The new function is the location management function (LMF) and supports location determination for a UE and obtains downlink and uplink location measurements from UE and NG-RAN, respectively. It also obtains non-UE associated assistance data from NG RAN. The gateway mobile location centre (GMLC) and location retrieval function (LRF) maintain the same role they have in the location architectures defined for EPS and GPRS in TS 23.271 [30] and therefore allowing the same interface (Le) toward an external client. The location services architecture was further evolved in Release 16 to include additional procedures for commercial location services and from Release 16 onwards all the architecture aspects and procedures are documented in TS 23.273 [31].

3.14 Interworking with EPC

The prerequisite for Paradigm 2 of the architecture transition described previously, where the legacy radio (LTE) and new radio (5G NR) connect to new core (5GCN) as illustrated in Sec. 3.1 is that the operator upgrades the legacy LTE basestations to additionally support the N2 and N3 interfaces to 5GC.

Operators may however not be able or willing to upgrade all or any LTE basestations to also support N2/N3 toward 5GC. Reasons for this may include

- **hardware/software limitations and availability** (already deployed LTE basestations may not support the introduction of the additional N2/N3 stack),

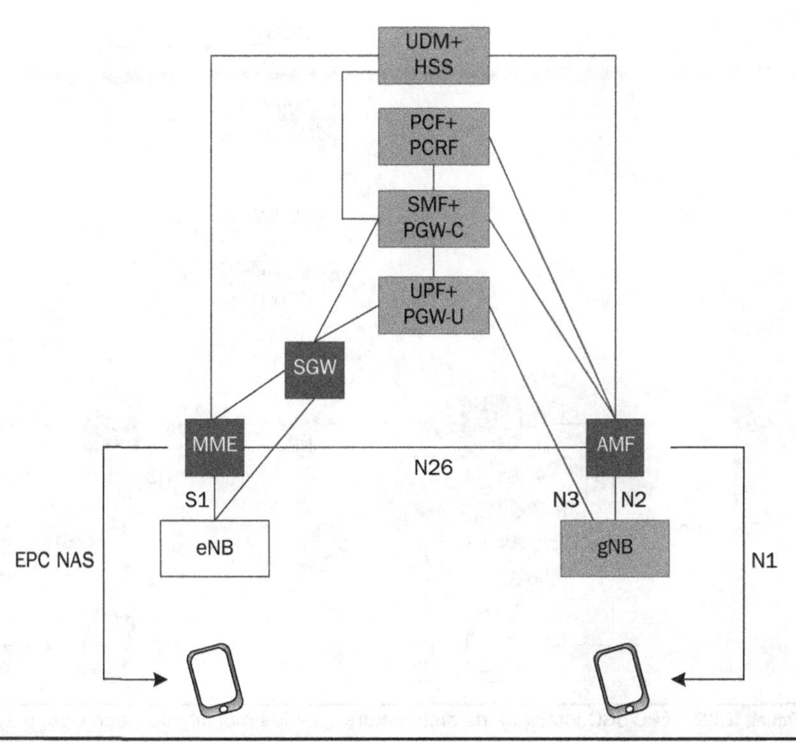

Figure 3.31 EPC-5GC interworking architecture between Option 1 (E-UTRA connected to EPC) and Option 2 (NR connected to 5GC).

- **cost** (licensing cost for N2/N3 support on LTE basestations and migration cost),
- **timing** (an operator may not be able to complete the upgrade of its existing RAN before the NR launch date due to competitive pressure), and
- **migration strategy** (an operator may want to minimize risk when introducing NR connected to 5GC by not unnecessarily destabilizing the existing LTE RAN).

The prerequisite for Paradigm 2 (LTE basestations will be able to connect to 5GC) will therefore not be fulfilled in some networks. To support seamless mobility between NR and LTE in those networks an interworking architecture has been defined in 3GPP (see Fig. 3.31).

The key idea of the EPC-5GC interworking architecture is that the UE is served by 5GC while in NR coverage and by EPC while in LTE coverage. The architecture supports session continuity when the UE changes between NR and LTE (and consequently between 5GC and EPC).

It is worth emphasizing that the EPC-5GC interworking architecture does not only support mobility between E-UTRA connected to EPC and NR connected to 5GC. As depicted in Fig. 3.32 the architecture also enables mobility between the Option 3 variants (non-standalone 5G NR added as a secondary RAT in dual connectivity

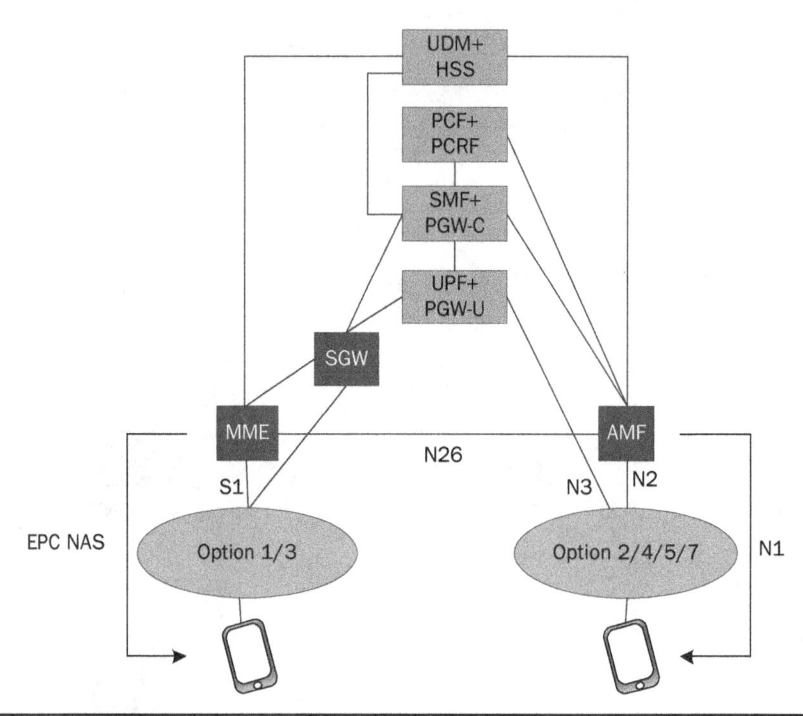

Figure 3.32 EPC-5GC interworking architecture enables mobility between Option 1/3 connected to EPC and Options 2/4/5/7 connected to 5GC.

configuration connected to EPC) and Options 2, 4, 5, and 7 connected to 5GC. This is relevant for operators for various migration scenarios, which will be discussed in detail in Sec. 3.15.

A related aspect worth highlighting is that the different interworking flavors that are enabled also require UE support, e.g., Option 5 and 7 require UE support for E-UTRAN connected to 5GC.

As far interworking is concerned N26 interface connects the MME and AMF and is possible to be supported with no mandatory impacts in legacy EPC. The procedures look like inter-MME handover to the MME. There are though some possible enhancements that have been introduced e.g., a UE capability to allow the MME to select the "combo" SMF+PGW-C node, ability in the MME to select target AMF based on 3-byte tracking area code that is supported in 5GC but not in EPC. In a nutshell though it is possible for an operator to deploy EPC to 5GC interworking with no impacts to legacy E-UTRAN, MME, and SGW and only upgrading the PGW to support SMF functionalities.

3.15 EPC to 5GC Migration

In the initial 5G deployments most MNOs start from NSA EPC-based architecture, i.e., some variant of Option 3. In this section we examine the logic behind this decision and then anticipate what are the next steps MNOs are likely to take in terms of evolving their 5G rollout.

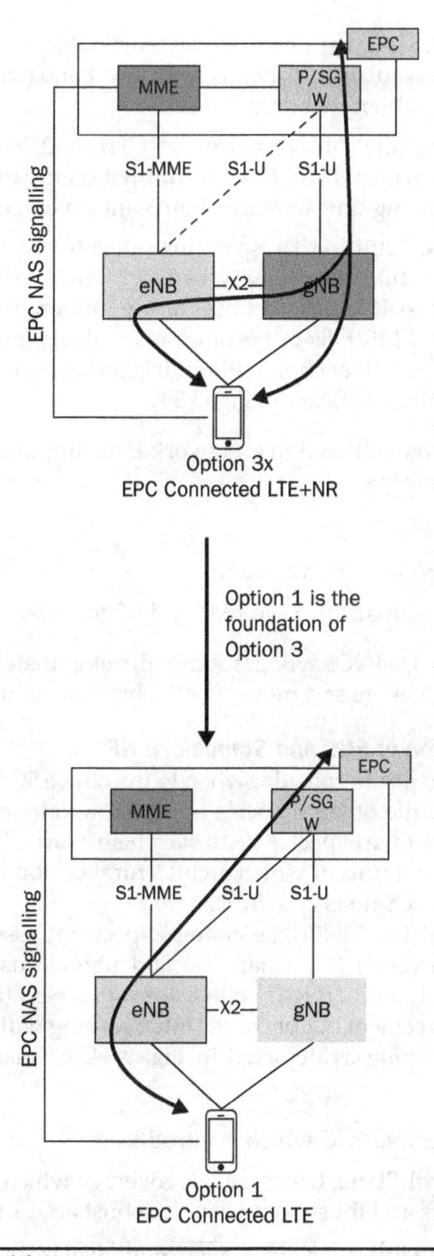

Figure 3.33 Options 3 and 1.

Step 1: Introduction of NR in the RAN

The selection of Option 3 as the initial 5G architecture is based on the following:

- **EPC Core:** Leverage of EPC core for fast time to market and risk reduction, given the expected immaturity of 5GC implementations in early years.

- **LTE Control Plane:** Leverage of nearly-ubiquitous LTE coverage as LTE has been deployed in mid bands and low bands covering from urban to rural, providing robust backdrop to NR.
- **NR RAN:** Utilize NR, primarily 3.5 GHz, and 26 GHz when available. Initially these deployments will be concentrated on urban areas only in 2019/2020 and will take a long time to reach significant coverage.
- **LTE RAN:** Utilize LTE spectrum to enhance throughput and capacity by leveraging dual connectivity between NR and LTE (targeting 2x peak Tput compared to LTE-only). Importantly Option 3 offers robust behavior in the sense that if NR fails or becomes unavailable, the UE can drop NR link while maintaining LTE anchor, without triggering a mobility event, i.e., by switching to Option 1 as depicted in Fig. 3.33.

Note that to turn on Option 3 in a network that supports Option 1 the MNO needs to take the following steps:

- No change to EPC
- Upgrade eNBs with X2 to gNBs
- Introduce gNBs with S1 to EPC and X2 to eNBs

As part of Option 3, MNOs will gradually develop their NR coverage layer by expansion of 3.5 GHz coverage and more effectively by introducing NR on mid/low bands.

Step 2: Introduction of 5GC and Standalone NR

The next logical step is to consider when to introduce 5GC. Note that there is no reason to consider any of the other available architecture options unless the MNO decides to turn on 5GC. In fact we expect 5GC to start being turned on in MNOs as soon as 2020. This is important in terms of which architectural option is chosen from those available in Paradigm 2 (i.e., Options 2/4/5/7).

At some point the MNO has enough spectrum assets on NR or using dynamic spectrum sharing (DSS) (see Chap. 11) and ubiquitous enough coverage that it no longer needs to rely on LTE RAN either for robustness (control layer/mobility) or for throughput enhancement but of course interworking with Option 1 is possible through 5GC-EPC interworking as depicted in Fig. 3.34. At that point the MNO can enable Option 2:

- **5GC Core:** Use 5GC which is introduced.
- **NR Control Plane:** Leverage NR coverage which at this point is assumed to be ubiquitous and thus it can provide robust access to terminals.
- **NR RAN:** Utilize NR, in 3.5GHz and 26GHz, but importantly also on in mid and low bands providing high throughput via CA combinations and high capacity utilizing available bandwidth using dynamic spectrum sharing. By 2020/2021 NR is likely to start getting deployed in a low band, e.g., 700 MHz in Europe when available or 900 MHz if refarmed from 2/3G.
- **LTE RAN:** No need for LTE RAN.

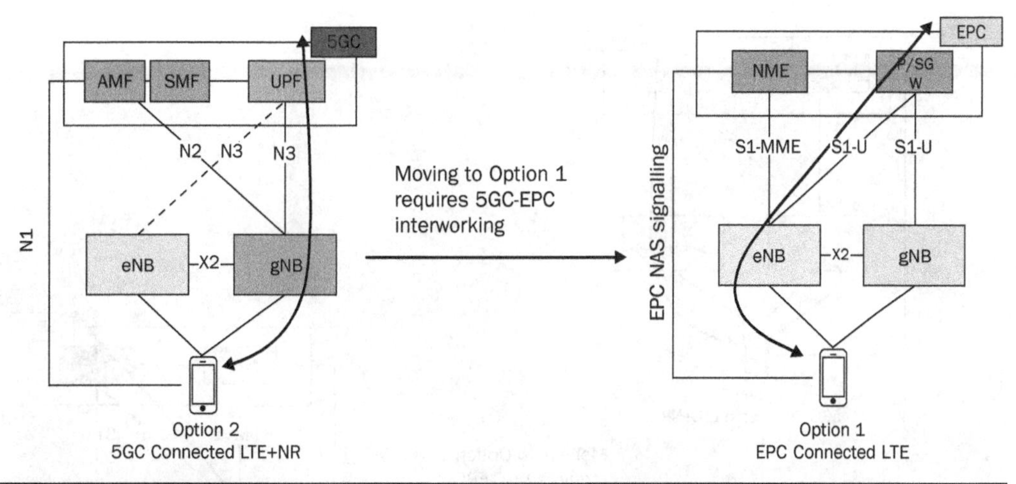

Figure 3.34 Options 2 and 1.

What About Options 5 and 7?

Given use of Option 3 in Step 1 above, another option of choice for introduction of 5GC is Option 7. This is based on the following:

- **5GC Core:** Introduce 5GC when is considered stable (as soon as 2020) to take advantage of new features (e.g., slicing, clean Cloud RAN, etc.) and all additional forward looking developments from infra vendors.
- **LTE Control Plane:** Leverage of near-ubiquitous LTE coverage and the fact that it is unlikely NR is nearly as ubiquitous at that point in time. Introduce "evolved" LTE versions of RRC, MAC.
- **NR RAN:** Utilize NR, primarily when available.
- **LTE RAN:** Utilize LTE spectrum to enhance throughput and capacity by leveraging dual connectivity between NR and LTE. Importantly Option 7 offers robust behavior in the sense that if NR fails or becomes unavailable, the UE can drop NR link while maintaining LTE anchor, without triggering a mobility event, i.e., by switching to Option 5. Introduce NR user plane protocols over LTE, i.e., NR PDCP and SDAP.

Note that to turn on Option 7 in a network that supports Option 1/3 the MNO needs to take the following steps:

- Introduce 5GC
- Upgrade eNBs with N2/N3 to 5GC
- Upgrade gNBs with N3 to 5GC
- Upgrade Xn between eNB and gNB for N3 procedures
- Introduce 5GC-EPC interworking

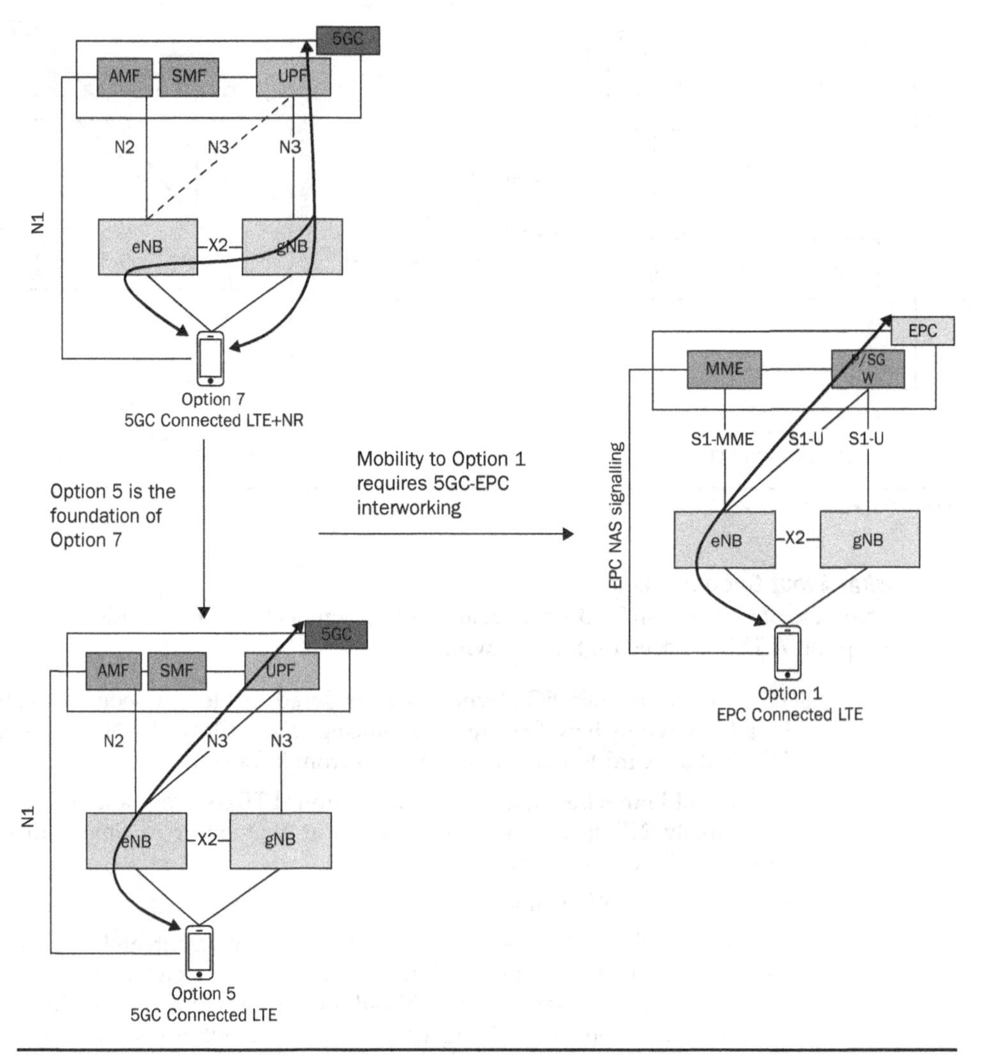

FIGURE 3.35 Options 7 and 5.

As shown in Fig. 3.35, Option 7, by definition implies support for Option 5 as the latter is a foundational part of the former similar to how Option 1 is foundational to Option 3. Option 5 is what happens when an Option 7 device attaches to the network to begin with, before it brings NR carrier up and move to Option 7. It is also what happens if the NR link breaks for whatever reason and the UE has to rely only on its LTE anchor. Finally it is a necessary step during 5GC to EPC interworking if the UE moves between areas where eNBs have been upgraded to Option 7 and eNBs that have not been upgraded to Option 7.

What About Option 4?

Option 4 is essentially the mirror image of Option 3. The reason to do Option 4 would be similar to those of utilizing Option 3. Another way to look at it is that Option 3 introduces NR RAN in the LTE worlds. Option 4 instead, maintain LTE RAN in a future NR dominated world.

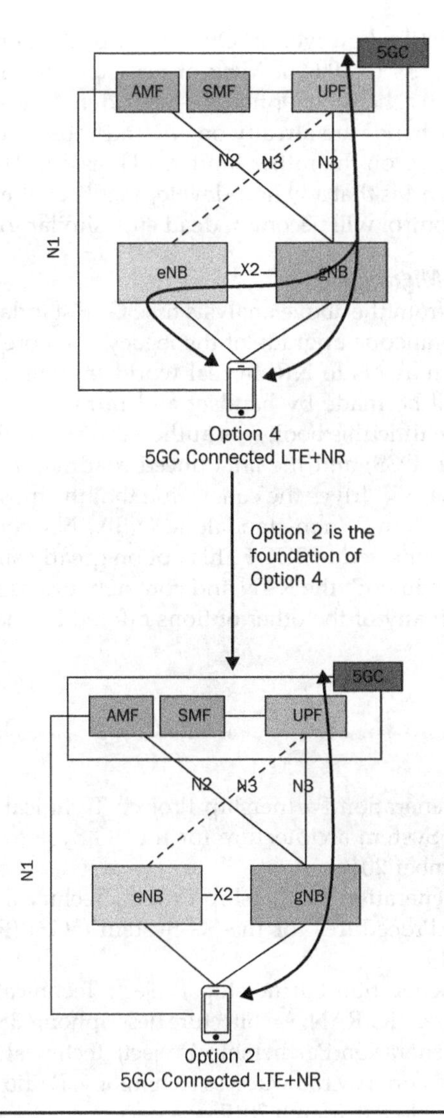

FIGURE 3.36 Options 4 and 2.

To justify Option 4 support the following logic will have to be used:

- **5GC Core:** Utilize 5GC features, etc.
- **NR Control Plane:** Leverage of near-ubiquitous NR coverage, i.e., NR is deployed in low bands at ubiquitous coverage even more than LTE coverage.
- **NR RAN:** Utilize NR, available across multiple bands.
- **LTE RAN:** Utilize LTE when available to boost throughput and capacity.

Note that Option 4 is the foundation to Option 2 as depicted in Fig. 3.36, e.g., when LTE link is not available, again emphasizing how Option 4 is only relevant when

NR dominates. While, however, in Option 3 introduction of NR RAN makes a huge difference (i.e., access to 100s of MHz of new spectrum), this does not apply in this case. Presumably by the time Option 4 is considered most of the spectrum assets and certainly any new bands are already on NR and thus there is no throughput/capacity advantage of adding on the rather limited LTE assets. The only remaining reason for considering Option 4 is that any new developments on the control plane are going to be in NR, and LTE control will become a dead end, similar to EPC.

Prediction on 5G Migration

As it can be seen from the above analysis the 3GPP standards offer a variety of options for the contemporaneous upgrade of the legacy and core or only migrate the existing operator spectrum assets to NR. In real world though a lot depends on whether all these options will be made by handset and infra vendors. From publicly available information at the time this book was authored it seems that the promises of dynamic spectrum sharing (DSS) and the announced roadmap of infrastructure and handset vendors [32, 33, 34, 35] drives the conclusion that the most probable migration is from the architecture option 3 (non-standalone (NSA) NR connected to EPC) to option 2 (standalone NR connected to 5GC). This option greatly simplifies the future evolution of the 5G platform in both the RAN and core network as shown previously. Time will only tell though if any of the other options offered by the standards will be commercially deployed.

References

[1] 3GPP, "3rd Generation Partnership Project; Technical Specification Group System Architecture; System architecture for the 5G System (5GS) (Release 15)," 23.501, v15.8.0, December 2019.

[2] 3GPP, "3rd Generation Partnership Project; Technical Specification Group System Architecture; Procedures for the 5G System (5GS) (Release 15)," 23.502, v15.8.0, December 2019.

[3] 3GPP, "3rd Generation Partnership Project; Technical Specification Group Radio Access Network; NG-RAN; Architecture description," 38.401, v15.7.0, September 2019.

[4] 3GPP, "3rd Generation Partnership Project; Technical Report Group Radio Access Network; Study on new radio access technology: Radio access architecture and interfaces," 38.801, v14.0.0, March 2017.

[5] 3GPP, "3rd Generation Partnership Project; Technical Specification Group Radio Access Network; NG-RAN; F1 general aspects and principles," 38.470, v15.7.0, December 2019.

[6] 3GPP, "3rd Generation Partnership Project; Technical Specification Group Radio Access Network; Study of separation of NR Control Plane (CP) and User Plane (UP) for split option 2," 38.806, v15.0.0, January 2018.

[7] 3GPP, "3rd Generation Partnership Project; Technical Specification Group Radio Access Network; NG-RAN; E1 general aspects and principles," 38.460, v15.4.0, June 2019.

[8] 3GPP, "3rd Generation Partnership Project; Technical Report Group Radio Access Network; Study on CU-DU lower layer split for NR," 38.816, v15.0.0, December 2017.

[9] O-RAN Alliance, https://www.o-ran.org

[10] 3GPP, "3rd Generation Partnership Project; Technical Specification Group System Architecture; Architecture enhancements to facilitate communications with packet data networks and applications (Release 15)," 23.682, v15.10.0, December 2019.

[11] 3GPP, "3rd Generation Partnership Project; Technical Specification Group Radio Access Network; NR; User Equipment (UE) procedures in idle mode and in RRC Inactive state," 38.304, v15.6.0, December 2019.

[12] 3GPP, "3rd Generation Partnership Project; Technical Specification Group Core Network and Terminals; Non-Access-Stratum (NAS) functions related to Mobile Station (MS) in idle mode," 23.122, v15.7.0, December 2019.

[13] 3GPP, "3rd Generation Partnership Project; Technical Specification Group Radio Access Network; NR; Overall description; Stage-2," 38.300, v15.8.0, December 2019.

[14] 3GPP, "3rd Generation Partnership Project; Technical Report Group System Architecture; Study of system level impacts due to introduction of light connection for LTE in EPS," 23.723, v0.2.0, January 2017.

[15] https://www.rcrwireless.com/20200304/5g/5g-driving-edge-computing-momentum-china-gsma

[16] 3GPP, "3rd Generation Partnership Project; Technical Specification Technical Specification Group Core Network and Terminals; Numbering, addressing and identification," 23.003, v15.8.0, September 2019.

[17] 3GPP, "3rd Generation Partnership Project; Technical Specification Group System Architecture; 3GPP system to Wireless Local Area Network (WLAN) interworking; System description; Stage 2," 23.234, v13.1.0, March 2017.

[18] 3GPP, "3rd Generation Partnership Project; Technical Specification Group System Architecture; Architecture enhancements for non-3GPP accesses," 23.402, v15.3.0, March 2018.

[19] 3GPP, "3rd Generation Partnership Project; Technical Specification Group System Architecture; Policy and charging control framework for the 5G System (5GS); Stage 2," 23.503, v15.8.0, December 2019.

[20] 3GPP, "3rd Generation Partnership Project; Technical Specification Group System Architecture; Policy and charging control architecture," 23.203, 15.5.0, October 2019.

[21] 3GPP, "3rd Generation Partnership Project; Technical Specification Group Core Network and Terminals; Non-Access-Stratum (NAS) protocol for 5G System (5GS); Stage 3," 24.501, v15.6.0, December 2019.

[22] 3GPP, "3rd Generation Partnership Project; Technical Specification Group Core Network and Terminals; User Equipment (UE) policies for 5G System (5GS); Stage 3," 24.526, v15.3.0, June 2019.

[23] 3GPP, "3rd Generation Partnership Project; Technical Specification Group System Architecture; Architecture enhancements for 5G System (5GS) to support network data analytics services," 23.288, v16.2.0, December 2019.

[24] 3GPP, "3rd Generation Partnership Project; Technical Specification Group System Architecture; Security architecture and procedures for 5G System," 33.501, v15.7.0, December 2019.

[25] 3GPP, "3rd Generation Partnership Project; Technical Specification Group System Architecture; Service requirements for the 5G system," 22.261, v15.8.0, December 2019.

[26] 3GPP, "3rd Generation Partnership Project; Technical Specification Group Radio Access Network; NR; Radio Resource Control (RRC); Protocol specification," 38.331, v15.8.0, December 2019.

[27] https://en.wikipedia.org/wiki/SMS
[28] 3GPP, "3rd Generation Partnership Project; Technical Specification Group System Architecture; Support of Short Message Service (SMS) over generic 3GPP Internet Protocol (IP) access; Stage 2," 23.204, v15.0.0, March 2018.
[29] 3GPP, "3rd Generation Partnership Project; Technical Specification Group System Architecture; IP Multimedia Subsystem (IMS) emergency sessions," 23.167, v15.6.0, December 2019.
[30] 3GPP, "3rd Generation Partnership Project; Technical Specification Group System Architecture; Functional stage 2 description of Location Services (LCS)," 23.271, v15.2.0, December 2019.
[31] 3GPP, "3rd Generation Partnership Project; Technical Specification Group System Architecture; 5G System (5GS) Location Services (LCS); Stage 2," 23.273, v16.2.0, December 2019.
[32] https://www.ericsson.com/en/ericsson-technology-review/archive/2018/simplifying-the-5g-ecosystem-by-reducing-architecture-options
[33] https://www.ericsson.com/en/news/2019/9/ericsson-spectrum-sharing
[34] https://www.mobileeurope.co.uk/press-wire/swisscom-ericsson-and-qualcomm-make-dynamic-spectrum-sharing-5g-call
[35] https://www.qualcomm.com/news/onq/2019/08/19/key-breakthroughs-drive-fast-and-smooth-transition-5g-standalone

CHAPTER 4

Evolution of 5G Architecture

4.1 Introduction

One key point worth keeping in mind is that any new generation of cellular technology systems becomes a platform for continuous evolution with new features added at every 3GPP release. It is therefore not appropriate to see the first release of 5G and the functionalities described in previous sections as static. Instead they should be considered as a snapshot of a continuously evolving landscape. Evidence of that is the extent of additional features that are added in Release 16. While Release 16 was ongoing at the time of writing this book, with the goal of completing all the specifications by 2Q 2020 (as shown in Fig. 4.1), the set of features that are going to be included are already clear.

In subsequent sections we provide a short overview of the additional functionalities that are under specification for the 5G system in Release 16 of 3GPP specifications.

4.2 Non-Public Networks

Support for non-public networks (NPNs) using the 5G system architecture is a major evolutionary step for the 3GPP-based cellular ecosystem. The feature set developed under the umbrella of non-public networks provides the opportunity to deploy 5G without the need for a PLMN and therefore allows entities other the traditional mobile network operators to deploy networks for various purposes and use cases. NPNs or private networks are effectively 5G system–based networks that do not need to offer the traditional MNO services like voice, SMS, etc., to general subscribers. NPNs can operate in any spectrum, licensed or unlicensed (more on that in Chap. 12) that can be owned by any entity that has access to this spectrum, e.g., enterprise, factory, etc. Several regulators in various countries are already acting to assign spectrum for private networks for specific uses (see [1], [2]). From 3GPP point of view, the main use case that drove the design for NPNs in Release 16 is industrial automation or also known as industrial internet of things (IIoT), which was based on use cases developed in TS 22.804 [3] for cyber-physical control usage and requirements for ultra-low latency, high reliability, and high communication service availability. The final design though is not limited solely to this use case. Effectively NPNs can be deployed for a multitude of other use cases, e.g., NPNs for audio-video production, NPNs used for data access as replacement of Wi-Fi in public venues like stadia, etc.

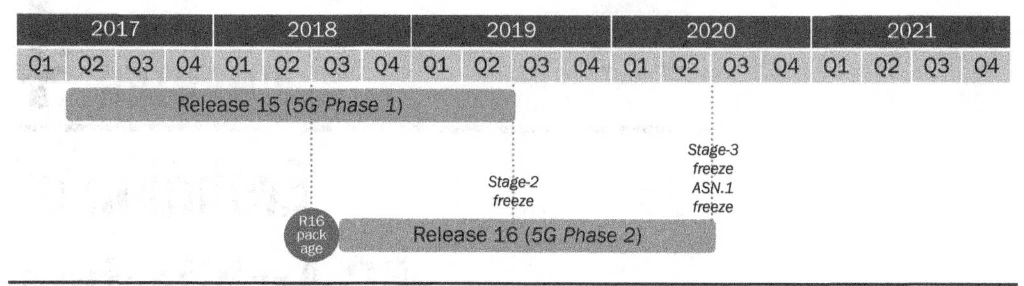

Figure 4.1 3GPP timescales for Release 16.

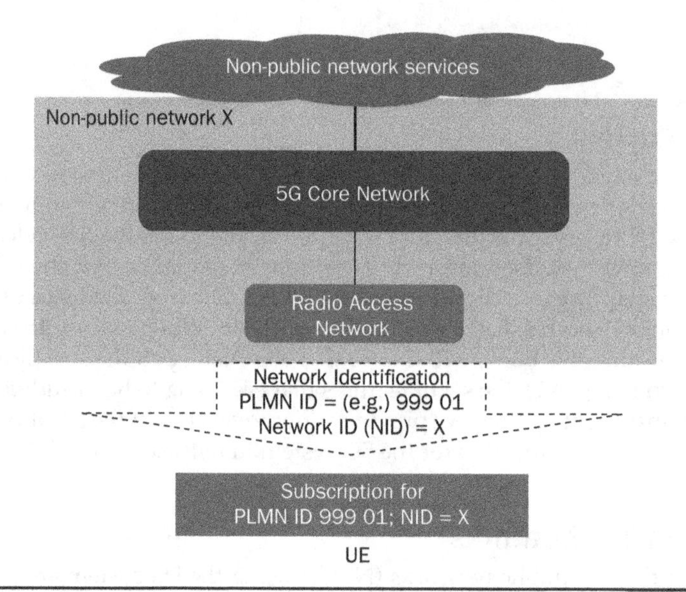

Figure 4.2 Standalone non-public network.

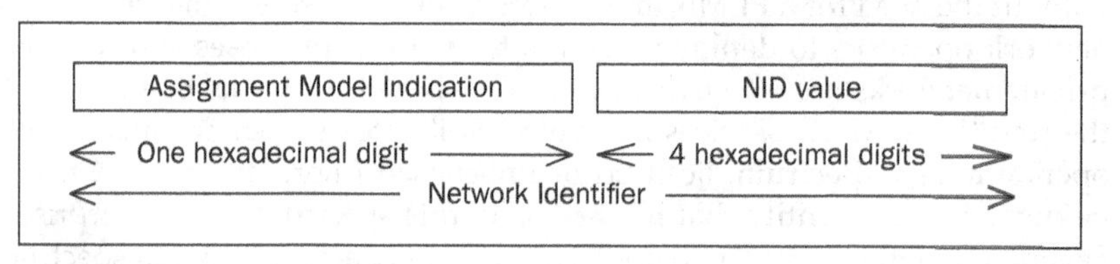

Figure 4.3 SNPN network ID (source TS 23.003).

Non-public networks are split in two main categories: **Standalone non-public networks (SNPNs)** do not require any interaction with PLMNs (even though they can support RAN sharing with PLMNs). SNPNs can use a non-unique PLMN Id (e.g., MCC = 999 assigned by ITU for private networks [4] and any MNC to identify the cell as part of a non-public network) and a network Id (NID) as defined in TS 23.003 [5].

The NID shall support two assignment models:

1. **Self-assignment**: NIDs are chosen individually by SNPNs at deployment time (and may therefore not be unique) but use a different numbering space than the coordinated assignment NIDs.
2. **Coordinated assignment**: NIDs are assigned using one of the following two options:
   a. The NID is assigned such that it is globally unique independent of the PLMN ID used. This is achieved by allowing the use of a private enterprise number (PEN) as NID (see TS 23.003 [5]). PENs are issued by the Internet Assigned Numbers Authority (IANA) [6].
   b. The NID is assigned such that the combination of the NID and the PLMN ID is globally unique. This implies that the NIDs for a given PLMN ID are centrally assigned by or on behalf of the PLMN ID owner. One example is the citizens broadband radio service (CBRS) alliance in the United States [7].

The other category of NPNs is **PLMN integrated NPNs (PNI-NPNs)** as shown in Fig. 4.4. In this case the NPN is logically part of a PLMN. A new identifier Closed Access Group (CAG) is used in order to restrict access to cells from UEs that are not member of the Closed Access Group. The selection of CN nodes is based on the usual techniques, i.e., use of S-NSSAI and DNN. Given that slicing isolation is supported by default it is possible to also achieve dedicated CN nodes for the PNI-NPN using a dedicated S-NSSAI. The UEs that are required to get access to the PNI-NPN are provisioned with an allowed CAG list and optionally an indication that the UE is only allowed to access CAG cells. The UE only selects cells that broadcast in SIB a CAG ID that is part of the UE's allowed CAG list. If the UE is only allowed to access CAG cells, it will not attempt access on cells that do not broadcast a CAG ID. The network checks the UEs access against the subscription from the UDM.

In this case the UE is operating PLMN selection procedures as normal and there is support for emergency calls from CAG cells.

As mentioned earlier, the main motivation for NPNs is the use for industrial automation and the requirements for ultra-low latency, ultra-reliable time-sensitive communication. In order to do that 3GPP had to adapt the 5G system architecture to support time sensitive networking (TSN). The work that was done in Release 16 allows the 5G system to be integrated transparently as a bridge in an IEEE TSN network as shown in Fig. 4.5.

To enable TSN synchronization, two key enhancements were specified for 5GS: (1) support of time synchronization using generalized precision time protocol (gPTP) as defined in IEEE Std 802.1AS [8] and (2) support of time-deterministic ethernet frame forwarding as specified in IEEE 802.1Qbv [9].

To enable time synchronization, the variable delays that result from forwarding gPTP messages as defined in [8] via 5GS had to be mitigated. This was achieved by time-stamping gPTP frames when entering 5GS and by correcting the gPTP time information based on the time that a gPTP frame spent within 5GS before forwarding the gPTP messages to downstream nodes connected to the UE.

The 5GS projects itself as one or more TSN bridges of the TSN network. As depicted in Fig. 4.5, the 5GS bridge is composed of ethernet ports on a network-side TSN

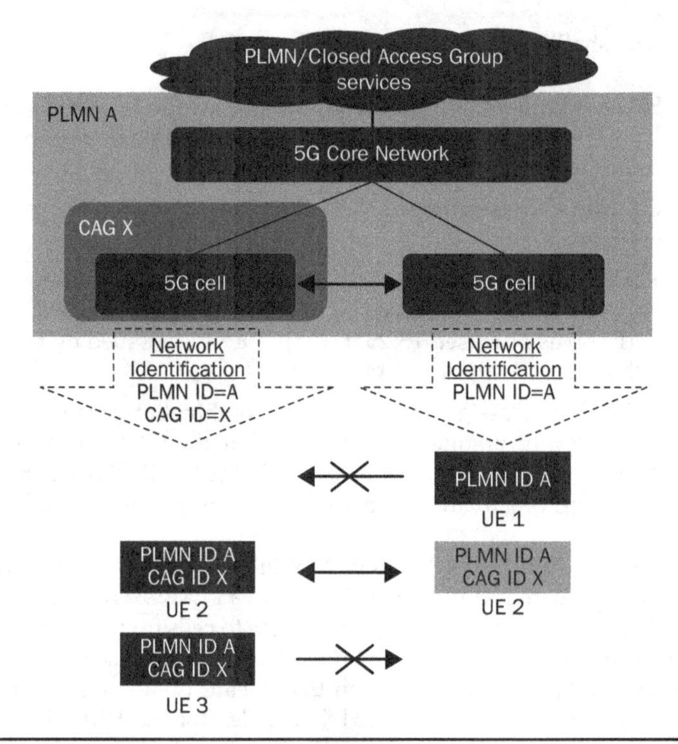

FIGURE 4.4 PLMN integrated non-public networks (PNI-NPNs): While UE 1 is not allowed to access CAG X, UE 2 is allowed to access both CAG X and non-CAG cells. UE 3 is restricted to CAG cells.

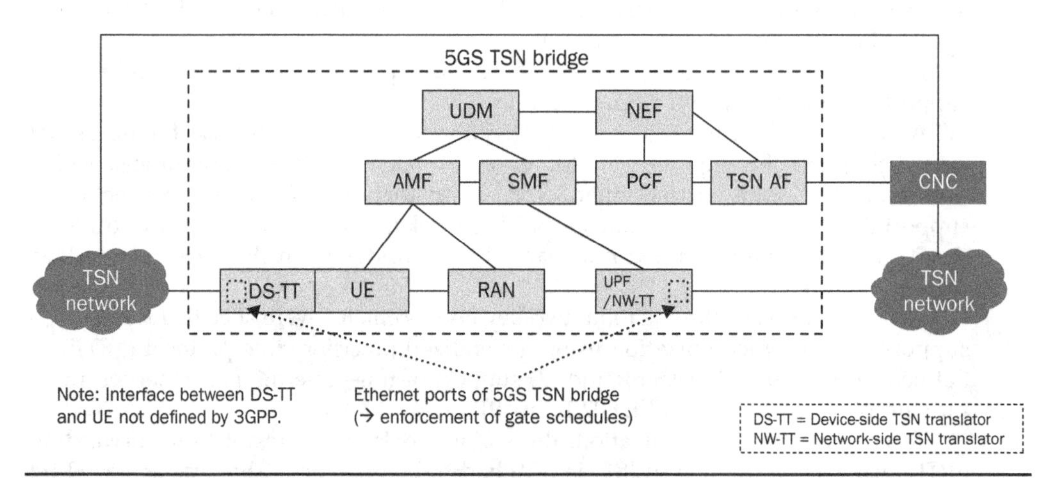

FIGURE 4.5 5GS adaptation with TSN (5GS as a logical TSN bridge).

translator, which is co-located with a single UPF, the user plane tunnel between the UE and UPF, and ethernet ports on the device-side TSN translator (DS-TT). For each 5GS bridge of a TSN network, the ports on UPF support the connectivity to the TSN network, the ports on the DS-TT are associated with the PDU session providing connectivity to the TSN network.

To enable support of time-deterministic ethernet frame forwarding as specified in IEEE 802.1Qbv [9], the 5GS bridge is required to provide 5GS bridge delays for each DS-TT and NW-TT port pair and traffic class to the IEEE TSN system, specifically the centralized network configuration (CNC) entity. This information is used by the CNC to calculate forwarding path and scheduling information. CNC then sends frame scheduling information as defined in IEEE 802.1Qbv [9] per traffic class and egress port to all bridges in a TSN network. 5GS enforces this scheduling information by hold and forward functionality in DS-TT and NW-TT.

The overall system functionality for adaptation of 5GS in a TSN system and the related signaling procedures are defined in Release 16 versions of TS 23.501 [10] and TS 23.502 [11], respectively.

4.3 Cellular V2X

Cellular vehicle to anything (C-V2X) communication is not new as far as cellular systems are concerned. In fact, 3GPP in the previous release before Release 15 (i.e., the initial release of 5G) has completed the second version of the LTE-based V2X specification

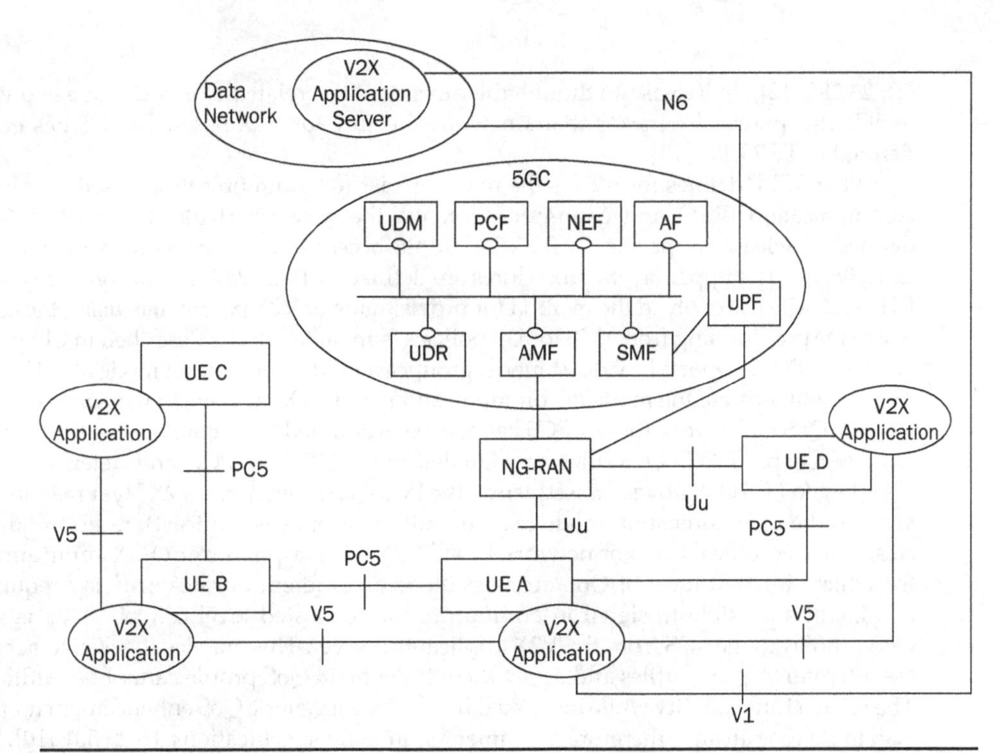

FIGURE 4.6 5GS-based architecture for V2X (source TS 23.287).

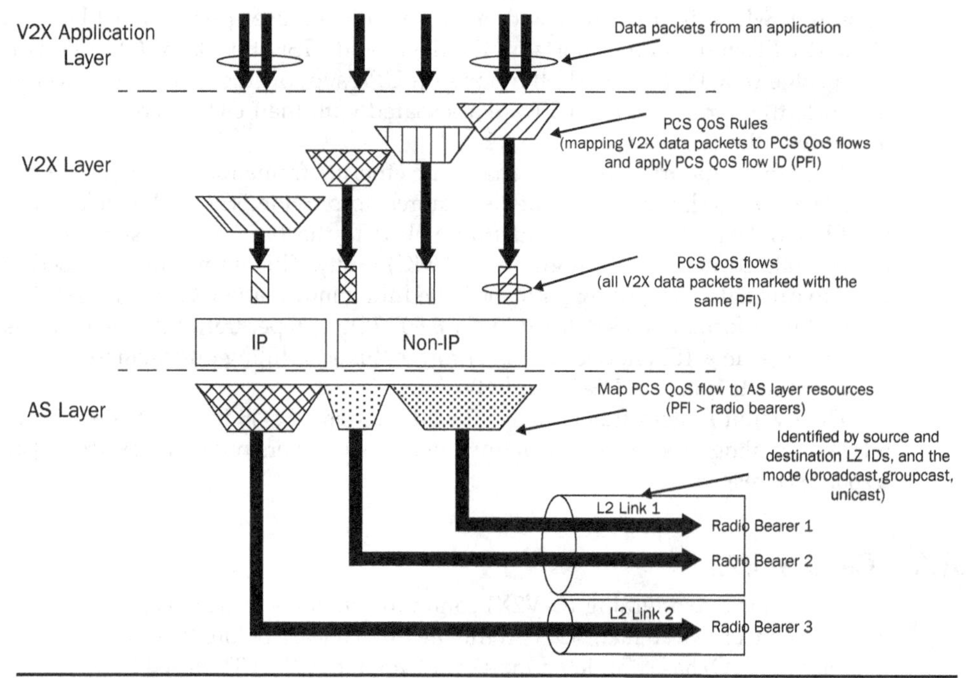

Figure 4.7 Handling of PC5 QoS flows based on PC5 QoS rules (source TS 23.287).

(TS 23.285 [12]). In Release 16 though the functionalities related to C-V2X have evolved to 5G. This practically means that a new architecture for V2X based on 5GS has been defined in TS 23.287 [13].

What 3GPP defines for V2X is the main enabler for communication based on direct communication (PC5) and communication via the network (Uu). NR-based PC5 is defined in Release 16 specifically for V2X and the access stratum procedures are defined in TS 38.300. The upper layers procedures are defined in TS 23.287 [13]. In comparison to LTE V2X 5GS also evolved the method for provisioning of V2X parameters using the control plane provisioning from PCF for UE policies (similar to what is described in Chap. 3). NR-based PC5 supports broadcast mode, groupcast mode, and unicast mode at AS layer. The UE will indicate the mode of communication for a V2X message to the AS layer.

The QoS framework for NR PC5 has also been adapted to resemble more that of NR Uu (see Chap. 3). PC5 QoS rules are signaled to the UE from PCF, and determine the mapping to PC5 QoS flows. Subsequently the PC5 QoS flows at the V2X layer map in AS layer to the communication mode (e.g., broadcast, groupcast, unicast), radio frequencies, and Tx profile used. For network-based V2X (Uu) inspired from V2X applications for which different level of QoS are possible (e.g., different bitrates or delay requirements), it is possible to signal in addition to the requested level of QoS. Alternative QoS profiles to the 5GS from the V2X application server. This enables the 5GS to act on the alternative QoS profiles and apply them if the main QoS profile cannot be fulfilled. The related functionality while instigated from V2X is a generic QoS enhancement for the 5GS in Release 16 and, therefore, documented in main specifications TS 23.501 [10], TS 23.502 [11], and TS 23.503 [14].

4.4 Cellular IoT

Various industry reports predict a huge increase in the number of connections for IoT devices. For example, Qualcomm report [15] suggests that the number of low power wide area (LPWA) IoT connections will be approximately 6 billion. In this context a major development in Release 16 is the integration of support for massive MTC (mMTC) radio technologies (namely NB-IoT and eMTC) in the 5GC. In essence this includes all the system wide features that have been worked out under the banner of "cellular IoT" (CIoT) for EPC that are now provided by 5GC. This offers two major advantages to operators: (1) it allows all the radio technologies to be provided under one core network, especially as explained in Chap. 14 when NB-IoT is supported inband in an NR carrier and, as explained in Sec. 4.2, it allows 5GC to be used as a single evolution platform in future releases, and (2) allows features that are inherently available in 5GC, e.g., the use of EAP authentication that allows possible use of non-SIM credentials, slicing, and support for APIs in the core network to be used in order to fulfill new use cases and development models.

The following main new features have been added in 5GC:

- Support power saving optimizations: Power saving mode that allows the UE to go to "sleep" for extended periods of time except from a short "active time" that is negotiated and known to the network. Support for extended discontinuous reception (eDRX) that extends beyond values of 2.56 s that are allowed in Release 15. In order to do that procedures for negotiation of extended DRX values and handling of mobile terminated data (storage, paging, and UE reachability) have been defined.

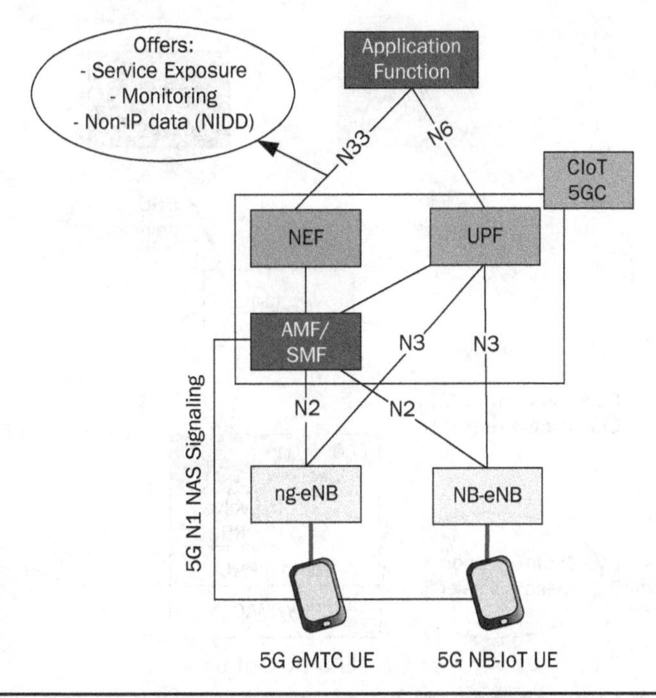

FIGURE 4.8 Cellular IoT in 5GC.

- Support for procedures for UE sending and receiving small data in the form of control plane optimization which allows the use of data over NAS without the need for data radio bearers in RAN and therefore without access stratum security context. This method of sending and receiving data is mandatory for NB-IoT UEs to support. Another procedure which resembles the use of RRC-inactive is user plane optimization which allows RAN to store the UE's access stratum context while the UE is in RRC-Idle/CM-Idle state. This method is optional to be supported in both eMTC and NB-IoT UEs.
- Related to above non-IP data delivery procedures and reliable data service (RDS) have been defined in 5GC.

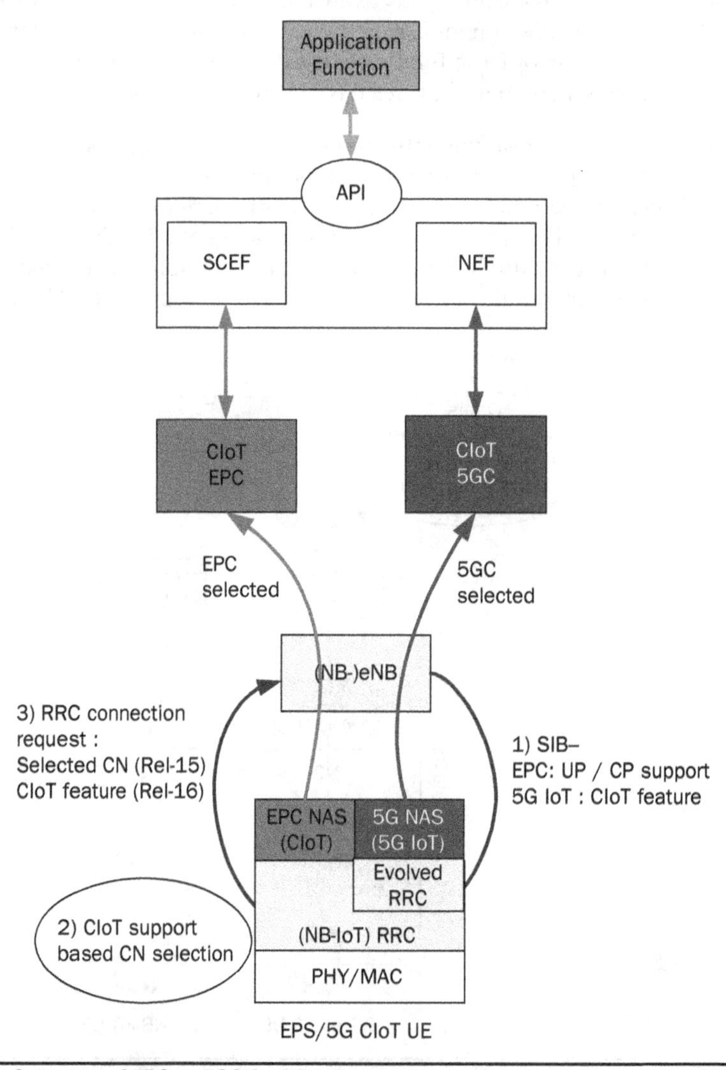

Figure 4.9 Selection of EPC or 5GC for NB-IoT.

- A variety of exposure services have been defined on the new N33 reference point via NEF that allows, for example, external application to perform monitoring for the state and location of the UE, and support for non-IP data delivery (NIDD).
- Additional support for congestion and overload control in the core network and RAN, like integration of NB-IoT access barring in unified access control (UAC).
- Support for paging with enhanced coverage indications for eMTC.

One important aspect is that the same NB-IoT cell can connect to both EPC and 5GC, and as depicted in the Fig. 4.9 the UE can select whether to connect to EPC or 5GC based on system information broadcast (SIB) indication that also indicates the support for features such as user plane CIoT optimization, control plane CIoT optimization in EPC and 5GC. The UE further indicates in RRC CIoT feature in order for NB-eNB to select the appropriate CN.

All the above features plus some others, for instance, selection of EPC or 5GC from a UE that supports both "stacks" are documented in Release 16 versions of TS 23.501 [10] and TS 23.502 [11].

4.5 "Big Data" Collection (Enhanced Network Automation)

By 2025, the addressable market for mobile network operators in IoT derived big data beyond connectivity is estimated at $386 billion [16]. Part of this is expected to be due to the promises of artificial intelligence (AI) use of networks' own analytics in order to simplify tasks and network decisions. This is the goal of the enhanced network automation (eNA). The "heart" of the data analytics collection and exposure is the network data analytics function (NWDAF) that is responsible to collect data by any network function in the 5GC (e.g., AMF, SMF, PCF, etc.), application functions, and the OAM system. NWDAF can also perform on demand provision of analytics to network functions of the 5GC, AF, and OAM and data repositories as is shown in Fig. 4.10.

An exemplary use of NWDAF for adjusting the UEs QoS is shown in Fig. 4.11.

- An application function (AF) provides the quality of experience (QoE) parameter (e.g., mean opinion score) of the particular application to NWDAF.
- 5GC provides the QoS information for the specific PDU session used for this particular application to the NWDAF.
- NWDAF subscribes the network data from 5GC and service data from application function (AF).
- NWDAF offline trains a MOS model for the application which will be used to determine/estimate the service experience for the application.
- NWDAF provides the data analytics to the PCF. PCF can use this information to determine the used QoS parameters for the service.

Further details on the functionality of enhanced network automation (eNA) are defined in TS 23.288 [17].

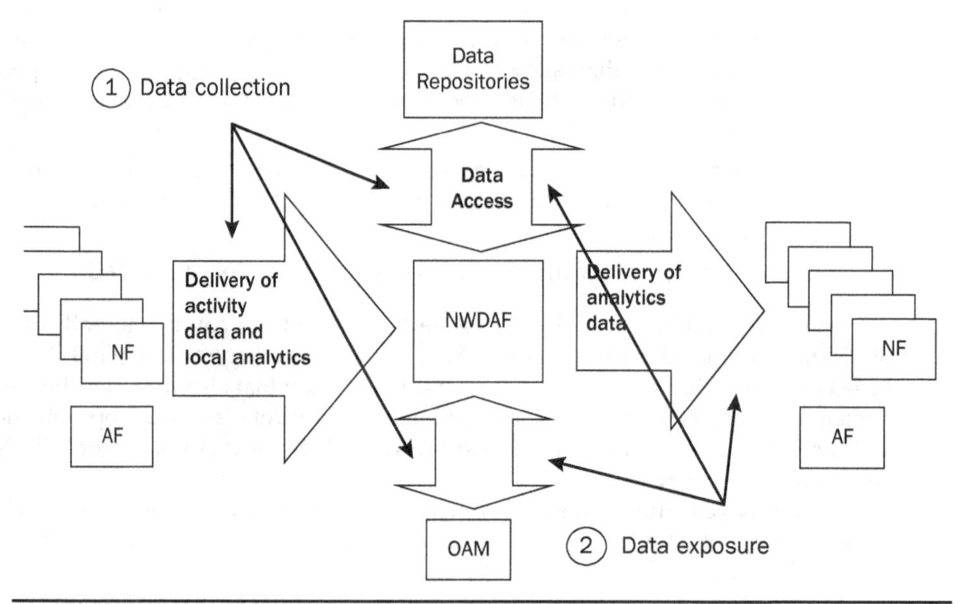

Figure 4.10 Network data analytics function for data collection and exposure.

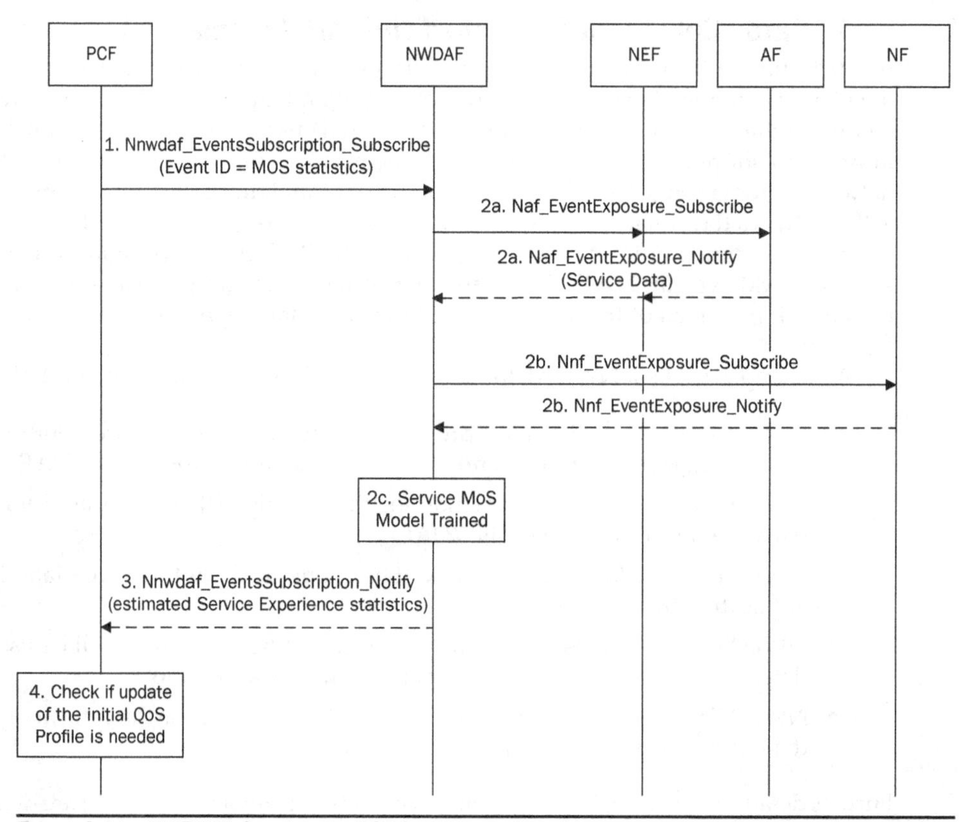

Figure 4.11 NWDAF informing the PCF for UEs quality of experience (QoE) [18].

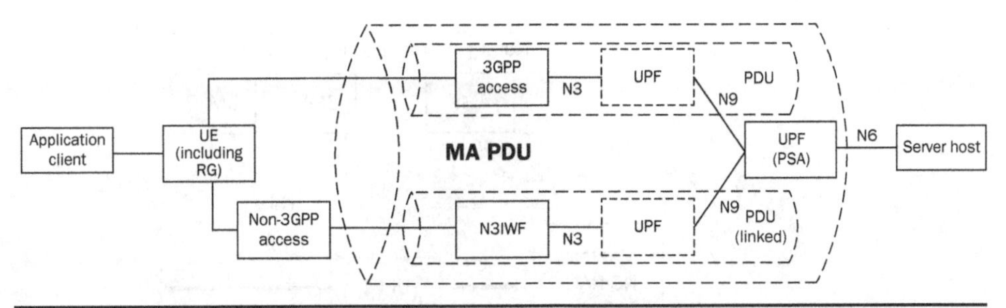

Figure 4.12 Multi-access PDU session.

4.6 Enhancements to Interworking with Non-3GPP Accesses

As it was already explained in Chap. 3, 5G system already from Release 15 supports interworking with non-3GPP access technologies via N3IWF. In Release 16 this basic functionality is further enhanced to support:

- **Access Traffic Steering, Switching, and Splitting:** This Release 16 feature called "ATSSS" allows to select Wi-Fi or any 3GPP access (NR or LTE) in order to send a particular type of traffic, split traffic across 3GPP access, and Wi-Fi achieving aggregation using Multipath-TCP (MP-TCP) or lower layer aggregation. The main construct to allow the ATSSS functionality is the use of multi-access PDU sessions (MA-PDU session) that allows the UE to exchange packets with the data network by simultaneously using one 3GPP access network and one non-3GPP access network and may have user-plane resources on two access networks as is shown in Fig. 4.12.

  Architecture and signaling procedures for ATSSS are defined in TS 23.501 [10] and TS 23.502 [11].

- **Wireless-Wireline Convergence:** A number of mobile network operators (MNOs) that deploy 5G also wireline access networks, and therefore access of wireline access network and fixed wireless access to 5G system is an important feature. Access from fixed network residential gateways (FN-RG) and 5G residential gateways (5G-RG) is defined in new Release 16 specification TS 23.316 [19]. This allows existing fixed network standards defined by broadband forum (BBF) or CableLabs to interwork with 5G system.

4.7 URLLC

Ultra-reliable low-latency communication (URLLC) is one of the main unique selling points for 5G. In addition to the radio enhancements defined in Chap.14, there are some further system functionalities that have been defined in order to enable better reliability. There are various forms of redundant transmissions that are defined in TS 23.501 [10], for example:

- Dual connectivity based end-to-end redundant user plane paths as is shown in Fig. 4.13.

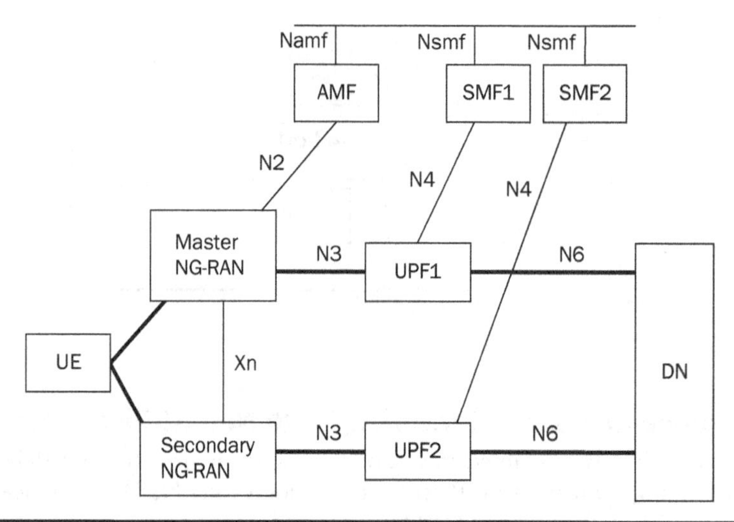

FIGURE 4.13 Example scenario for end-to-end redundant user plane paths using dual connectivity (source TS 23.501).

In order to improve reliability, a UE may set up two redundant PDU sessions and as a result two distinct user plane paths of the two redundant PDU sessions are established, the PDU sessions in return are expected to be supported by the master and secondary RAN nodes, respectively. The user's subscription determines whether it is allowed to have redundant PDU sessions and this indication is provided to SMF from UDM. It is out of scope of 3GPP specifications how applications make use of these redundant PDU sessions.

- Support of redundant transmission on N3/N9 interfaces

Using this functionality, at the time or after a URLLC QoS flow establishment, if it is decided that redundant transmission shall be performed by the SMF based on, e.g., authorized 5QI, the SMF informs the UPF and RAN to perform redundant transmission via N4 interface and N2 information accordingly. In this case, also on the transport path these tunnels should be mapped to disjoint paths according to configuration.

If duplication transmission is performed on the path between UPF and RAN and between two UPFs, for each Downlink packet of the QoS flow the UPF received from data network, the UPF duplicates the packet and assigns the same GTP-U sequence number to them for the redundant transmission. The RAN node that receives the duplicated packets eliminates them based on the GTP-U sequence number and then forwards the packet to the UE.

For each Uplink packet of the QoS flow the RAN node received from UE, it duplicates the packet and assigns the same GTP-U sequence number to them for redundant transmission. These packets are transmitted to the UPF via two RAN-to-UPF tunnels separately. The UPF eliminates the duplicated packets based on the GTP-U sequence number accordingly.

More URLLC 5G system functionalities like duplication at the transport paths only and QoS monitoring are defined in TS 23.501 [10].

4.8 Slice Authentication

As described in Chap. 3, network slicing is one of the pillars of 5G system. It enables use cases where "virtual" networks can be defined that are used to provide dedicated services (e.g., V2X) or dedicated to 3rd party enterprises. In order to further enable this deployment model where a 3rd party is assigned a slice, slice authentication has been defined in Release 16. The network slice authentication is defined in TS 23.501 [10], TS 23.502 [11], and TS 33.501 [20] and is when the S-NSSAI in the UE subscription is indicated to require slice authentication. In this case the AMF decides to initiate slice authentication with an authentication, authorization, and accounting (AAA) server that belongs to a 3rd party. The method for this authentication using EAP and the credentials are out of scope of 3GPP. If the UE is successful it can get access to the specific network slice, otherwise access to this network slice is denied.

4.9 Other Release 16 Features

Other system features defined in Release 16 that enhance the basic functionality of 5G system are:

- Radio capabilities signaling optimization optimize the signaling of radio capabilities since primarily due to the increase of band combinations have significantly increased in size and exceed 10s Kbytes. The basic idea is to replace explicit signaling of the UEs radio capabilities with an assigned identifier called UE radio capability ID and is signaled in NAS. Related functionalities are defined TS 23.501 [10] and TS 23.502 [11].
- 5G SRVCC from NR to UMTS is defined in TS 23.216 [21] in order to facilitate voice handover from NR to UMTS for operators that do not have excessive LTE coverage. The procedures only allow handover for the voice QoS flow to UMTS CS and not PS handover. The architecture and signaling procedures are defined in TS 23.216 [21].

References

[1] https://www.ofcom.org.uk/__data/assets/pdf_file/0033/157884/enabling-wireless-innovation-through-local-licensing.pdf

[2] https://venturebeat.com/2019/11/21/germany-opens-door-to-private-5g-networks-with-3-7-3-8ghz-licenses/

[3] 3GPP, "3rd Generation Partnership Project; Technical Report Group System Architecture; Study on Communication for Automation in Vertical domains (CAV)," 22.804, v16.2.0, December 2018.

[4] International Telecommunication Union (ITU), Standardization Bureau (TSB): "Operational Bulletin No. 1156." http://handle.itu.int/11.1002/pub/810cad63-en (retrieved October 5, 2018).

[5] 3GPP, "3rd Generation Partnership Project; Technical Specification Group Core Network and Terminals; Numbering, addressing and identification (Release 16)," 23.003, v16.1.0, December 2019.

[6] https://www.iana.org/assignments/enterprise-numbers/enterprise-numbers

[7] https://www.cbrsalliance.org/specifications/

[8] IEEE Std 802.1AS-Rev/D7.3, August 2018: "IEEE Standard for Local and metropolitan area networks—Timing and Synchronization for Time-Sensitive Applications."
[9] IEEE 802.1Q: "Standard for Local and metropolitan area networks—Bridges and Bridged Networks."
[10] 3GPP, "3rd Generation Partnership Project; Technical Specification Group System Architecture; System architecture for the 5G System (5GS) (Release 16)," 23.501, v16.3.0, December 2019.
[11] 3GPP, "3rd Generation Partnership Project; Technical Specification Group System Architecture; Procedures for the 5G System (5GS) (Release 16)," 23.502, v16.3.0, December 2019.
[12] 3GPP, "3rd Generation Partnership Project; Technical Specification Group System Architecture; Architecture enhancements for V2X services (Release 15)," 23.285, v15.4.0, December 2019.
[13] 3GPP, "3rd Generation Partnership Project; Technical Specification Group System Architecture; Architecture enhancements for 5G System (5GS) to support Vehicle-to-Everything (V2X) services (Release 16)," 23.287, v16.1.0, December 2019.
[14] 3GPP, "3rd Generation Partnership Project; Technical Specification Group System Architecture; Policy and charging control framework for the 5G System (5GS); Stage 2 (Release 16)," TS 23.503, v16.3.0, December 2019.
[15] https://www.qualcomm.com/media/documents/files/leading-the-lte-iot-evolution-to-connect-the-massive-internet-of-things.pdf
[16] https://www.gsma.com/iot/wp-content/uploads/2019/10/The-IoT-Big-Data-revenue-opportunity-for-operators_GSMA_IoT.pdf
[17] 3GPP, "3rd Generation Partnership Project; Technical Specification Group System Architecture; Architecture enhancements for 5G System (5GS) to support network data analytics services (Release 16)," 23.288, v16.2.0, December 2019.
[18] S2-1901080, TS 23.288 NWADF-assisted QoS profile provision, China Mobile, Huawei, Ericsson.
[19] 3GPP, "3rd Generation Partnership Project; Technical Specification Group System Architecture; Wireless and wireline convergence access support for the 5G System (5GS) (Release 16)," 23.316, v16.2.0, December 2019.
[20] 3GPP, "3rd Generation Partnership Project; Technical Specification Group System Architecture; Security architecture and procedures for 5G System (Release 16)," 33.501, v16.1.0, December 2019.
[21] 3GPP, "3rd Generation Partnership Project; Technical Specification Group System Architecture; Single Radio Voice Call Continuity (SRVCC); Stage 2 (Release 16)," 23.216, v16.3.0, December 2019.

CHAPTER 5

Numerology and Slot Structure

From the very beginning of 5G standardization, it was envisioned that 5G would support diverse services, a wide range of spectrum, and flexible deployments. In addition, it is necessary to be future proof, such that it can readily accommodate an introduction of any new services in the future, with expansion to vertical domains (see Chaps. 13, 14, 15, and 16), while causing zero or minimal impact to legacy operations. To achieve this, among all the technology enablers, the most fundamental one is the numerology and slot structure, with all other technology enablers built on top of it.

As discussed in Chap. 1, both 4G LTE and 5G NR are based on CP-OFDM and DFT-s-OFDM waveforms. To be more specific, both 4G LTE Downlink and 5G NR Downlink are based on CP-OFDM; while 4G LTE Uplink is based on DFT-s-OFDM only, 5G NR Uplink supports both DFT-s-OFDM- and CP-OFDM-based operations, with the former to ensure a good Uplink coverage and the latter for more scheduling flexibility and to exploit the subband scheduling gain.

5.1 Numerology and Slot Structure in 4G LTE

In 4G LTE, both frequency-division duplexing (FDD) and time-division duplexing (TDD)-based frame structure are supported [1]. At the beginning of 4G LTE standardization, one contentious issue was the possible subframe duration [2], e.g.,

- A 0.5-ms subframe only, or
- A 1-ms subframe only, or
- Both a 0.5-ms subframe and a 2-ms subframe.

There was also a possibility of supporting an additional 0.675-ms subframe duration specific to TDD. Considering the latency and coverage tradeoff and in order to keep reasonable standardization efforts, a 1-ms subframe was eventually adopted [3] for both FDD and TDD. Some key numerology parameters for LTE include [1]:

- Sub-carrier spacing (SCS): 15 kHz only
    - Note that there are also other SCS values for different services such as broadcast (7.5 kHz), PRACH (1.25 kHz, 1.875 kHz, 7.5 kHz), etc.

- System bandwidths: 1.4 MHz, 3 MHz, 5 MHz, 10 MHz, 15 MHz, and 20 MHz
  - A maximum fast fourier transform (FFT) size of 2048 is assumed.
- CP lengths: normal (NCP) and extended (ECP), intended for different channel conditions and deployment scenarios
- Frame, Subframe, Slot, and Symbol
  - Each frame is fixed at 10 ms or 10 subframes.
  - Each subframe consists of two slots, each of 0.5 ms.
  - For NCP, there are 14 symbols per subframe or 7 symbols per slot.
    - The CP length is 160 T_s or 5.21 μs for symbols 0 & 7, and 144 T_s or 4.69 μs for other symbols. The parameter T_s is the base time unit for LTE, defined as 1 ms/(2048 × 15000) = 0.032552 μs.
  - For ECP, there are 12 symbols per subframe or 6 symbols per slot.
    - The CP length is 512 T_s or 16.67 μs for all symbols.
- Resource Element (RE): a resource grid formed by one tone in frequency and one symbol in time
- Resource block (RB): 12 consecutive tones in frequency and 7 consecutive symbols in time, a minimum resource allocation granularity in frequency (except for MTC, see Chap. 14)
  - For system bandwidths 1.4 MHz, 3 MHz, 5 MHz, 10 MHz, 15 MHz, and 20 MHz, the numbers of RBs that can be used for resource allocation are 6, 15, 25, 50, 75, and 100 RBs, respectively. Correspondingly, given a 180-kHz bandwidth of one RB, we have 1.08-MHz, 2.7-MHz, 4.5-MHz, 9-MHz, 13.5-MHz, and 18-MHz usable bandwidths, respectively. As a result, the usable system bandwidth is roughly 90%, where a 10% overhead is assumed for bandwidth gaps necessary for interference protection, performance, and regulation satisfaction.
- Minimum resource allocation granularity in time, also known as a TTI (transmit time interval): 1 ms
  - The 1-ms TTI consists of a PRB pair, where a first RB is in the first slot of a subframe and a second RB is in the second slot of the subframe. The first RB and the second RB may have the same physical RB index (which implies no frequency hopping) or different physical RB indices (which implies frequency hopping).
  - Note that starting from LTE Release 15, shortened TTI (sTTI) is supported, where the time domain resource allocation granularity can be as small as 2 symbols. This helps reduce the over-the-air latency for a transmission. More details can be found in Chap. 13.

Figure 5.1 provides an illustration of the key numerology for 4G LTE, using the FDD Downlink as an example. The synchronization signals and PBCH, crucial for the initial access of 4G LTE systems (see Chap. 6), are also shown.

Before the introduction of supporting unlicensed spectrum in Release 13 [4], 4G LTE only supported FDD and TDD, also known as frame structure 1 and frame structure 2, respectively. In FDD, the paired spectrum consists of a Downlink carrier dedicated to

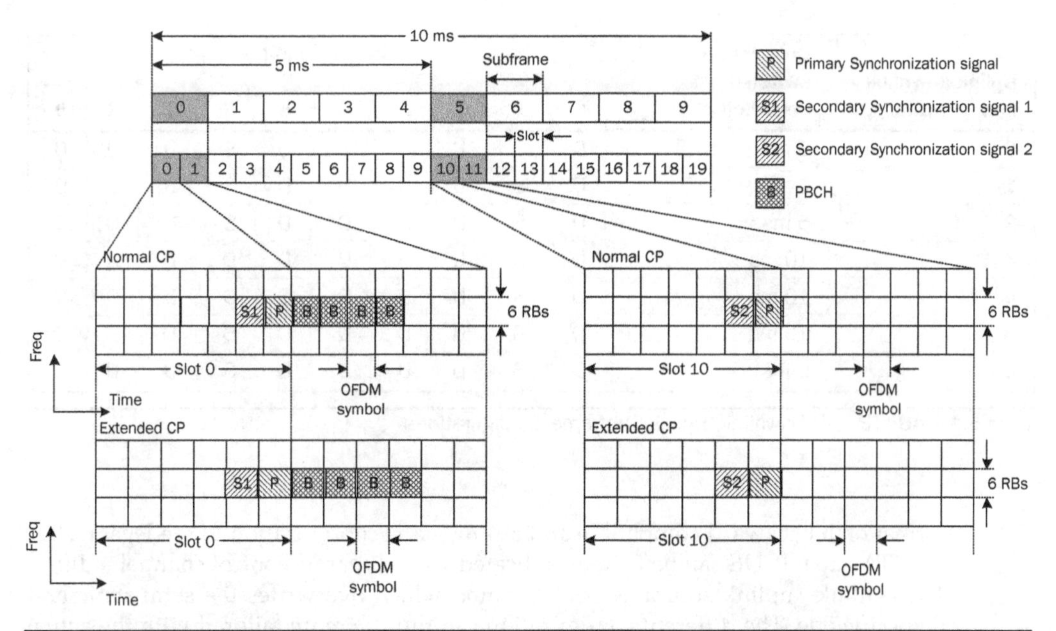

Figure 5.1 Illustration of 4G LTE numerology in Downlink, including synchronization signals and PBCH (using FDD as an example).

Downlink traffic only and an Uplink carrier dedicated to Uplink traffic only. In TDD, there are seven predefined Downlink-/Uplink-subframe configurations [1], as shown in Table 5.1, where "D" stands for a Downlink subframe, "U" stands for an Uplink subframe, and "S" stands for a special subframe which consists of a set of Downlink, guard, and Uplink symbols. The seven configurations provide the possibility for operators to choose a configuration to best match with the actual deployment need particularly in terms of the expected *long-term* Downlink traffic versus the *long-term* Uplink traffic. For instance, if it is expected that for several hours or longer, the traffic is Uplink heavy, the Downlink-/Uplink-subframe configuration can be configuration #0, which is the most Uplink-heavy configuration. It roughly offers a 2 (DL subframes) to 8 (UL subframes) ratio. If after a while, the traffic turns out to be Downlink heavy, configuration #2 (roughly an 8:2 ratio) may become a better choice. If the Downlink traffic is extremely heavy, configuration #5 (roughly a 9:1 ratio) may be the best. Although such a semi-static switching of a Downlink-/Uplink-subframe configuration is feasible, the configuration in commercial deployments, once chosen, is generally fixed for an indefinite amount of time, primarily due to concerns over the cross-link (Downlink to Uplink, or Uplink to Downlink) interference.

The introduction of evolved interference management and traffic adaptation (eIMTA) in Release 12 in 4G LTE [5] was intended to address the restriction in updating the Downlink-/Uplink-subframe configuration in response to bursty traffic needs. With eIMTA, the change from a Downlink-heavy configuration to an Uplink-heavy configuration (or vice versa) can be dynamic in a *UE-specific* manner. To be fully backward compatible, the possible Downlink-/Uplink-subframe configurations are still chosen from the existing seven configurations. While an eIMTA-enabled cell broadcasts in its system

Uplink-downlink configuration	Downlink-to-uplink switch-point periodicity	Subframe index									
		0	1	2	3	4	5	6	7	8	9
0	5 ms	D	S	U	U	U	D	S	U	U	U
1	5 ms	D	S	U	U	D	D	S	U	U	D
2	5 ms	D	S	U	D	D	D	S	U	D	D
3	10 ms	D	S	U	U	U	D	D	D	D	D
4	10 ms	D	S	U	U	D	D	D	D	D	D
5	10 ms	D	S	U	D	D	D	D	D	D	D
6	5 ms	D	S	U	U	U	D	S	U	U	D

Table 5.1 4G LTE TDD Downlink/Uplink-Subframe Configurations

information a backward compatible and a semi-static configuration to the legacy UEs, an eIMTA-capable UE can be further indicated via a dynamic control channel a different Downlink-/uplink-subframe configuration which overwrites the semi-static and cell-specific one. The different configuration can thus be more tailored with the actual needs for the UE.

Note that although eIMTA can dynamically update the Downlink-/Uplink-subframe configuration, the update can only be as frequently as up to one frame (10 ms). This is due to the need for backward compatibility and to ensure that inter-cell interference and/or inter-carrier interference (e.g., carriers of the same or adjacent bands) can be properly managed especially for certain subframes. In particular, it is important that a set of subframes may not experience cross-link directions within the same or across adjacent carriers so that critical communications such as for signaling and control can be ensured. For any given 10 ms, one of the seven configurations as in Table 5.1 has to be used. This implies that even under such a dynamic subframe configuration, it is still impossible to achieve an over-the-air latency of 1 ms or less, since some subframes have to be designated as either a Downlink or an Uplink subframe. Within a subframe (including the special subframes), there is *no dynamic adaption* of the *subframe structure*. This restriction has been addressed in 5G NR.

5.2 Lessons Learned from 4G LTE and 5G Considerations

When 4G LTE was standardized during the first release, the primary focus was on the traditional mobile broadband, with the need to optimize for VoIP as well [6, 7]. When 4G LTE was further evolved, additional services and scenarios were gradually introduced. In particular, in Release 12, small cell densification was introduced [8]. Machine type communication (MTC) communication was first studied in Release 12 [9], with more dedicated enhancements starting from Release 13 focusing on low cost/complexity, increased coverage and improved power consumption. Two versions of MTC specifications were added, one is called enhanced MTC (eMTC) and the other is called narrow-band internet-of-things (NB-IoT). While NB-IoT emphasizes more regarding extremely low data rates and extreme coverage extension, eMTC emphasizes more regarding relatively higher data rates and sufficient coverage extension.

Both eMTC and NB-IoT continued evolving in Releases 14, 15, 16, and 17. Vehicle to everything (V2X)–type communications in LTE started in Release 14 [10], while high-reliability and low latency communication (HRLLC) services were studied in Release 14 [11] and specified in Release 15. Chapter 13 has a high-level introduction of 4G LTE HRLLC.

As can be seen, the additional services and new spectrum (e.g., unlicensed spectrum as in [4]) for 4G LTE NR were not considered in the first 4G LTE release, but only got introduced *gradually* later. However, addition of new features in 4G LTE in general was not straightforward, primarily due to the need to maintain backward compatibility. In particular, it is necessary to ensure *zero* impact on the *always-on* common reference signals (CRS), since legacy 4G LTE UEs rely on the CRS for initial access, mobility measurement, demodulation, etc. It is worth noting that there was careful discussion of possibly reducing or completely removing the CRS in 4G LTE, creating a so-called new carrier type (NCT) [8] so that new services/requirements (e.g., green base stations [8], MTC, etc.) can be more readily accommodated. However, NCT was not adopted in 4G LTE primarily because it cannot be accessed by legacy LTE UEs.

The lessons learned in 4G LTE are valuable for 5G NR standardization. 5G NR was duly developed in response to the continuing evolution of mobile broadband, the continuing expansion of wireless communications into other fields, the ever-increasing desire to multiplex various kinds of services with different requirements into the same spectrum and/or the same cell, the need to be scalable and flexible so that a new spectrum can be integrated into the same framework, the necessity of being future proof, and the readiness in introducing new features that the future may offer. Reference [12] provides a nice summary of some system design principles for 5G NR.

In particular, in terms of numerology and slot structure for 5G NR, it is crucial to take the following into account:

- 4G LTE versus 5G NR deployments:
  - It is critical to facilitate migration from 4G LTE networks to 5G NR networks, as discussed in Chap. 2 regarding possible deployment options.
  - NR and LTE may have to coexist, either on the same channel (but may be of different bandwidths), or adjacent channels, for some time for an operator or across operators in a given region, as discussed in Chap. 11 regarding the coexistence between 4G LTE and 5G NR networks.
- Spectrum for 5G NR
  - The carrier frequency can be as low as hundreds of MHz, and can be as high as tens of GHz or even higher.
  - It may be paired or unpaired.
  - It may be licensed, unlicensed, or shared.
- It is possible that
  - A cell may need to provide two or more types of services with different requirements (e.g., one for eMBB and the other for URLLC), e.g., in terms of latency, reliability, and coverage.
  - A UE may need to support two or more types of services with different requirements as well.

- The deployments can be
  - Macro cells or small cells.
  - Non-standalone or standalone.
  - Outdoor or indoors (office-like indoors, industrial indoors [13]).
  - Stationary or mobile.
  - With or without relays [14], where the relays can be a gNB-like or a UE-based node, one-hop or multiple hops, etc.
- The link type can be Downlink, Uplink, sidelink, backhaul link, satellite link, etc.
- The necessity of being future proof, which can be realized by flexibly managing resources as reserved or blanked without causing issues to backward compatibility, by minimizing or removing transmissions of always-on signals, and by confining physical layer signals and channels within a configurable or an allocable time and frequency resource at a gNB's disposal.

5.3 SCSs for 5G NR

The foremost issue for 5G NR numerology is to choose a set of SCSs for 5G NR. There is no doubt that the number of SCSs for 5G NR has to be more than one due to a wide range of spectrum and services. That is, the SCS should be scalable in order to support different carrier frequencies and different latency requirements. Note that 4G LTE has a fixed SCS of 15 kHz, and a 14-symbol based 1-ms subframe duration (for NCP). The minimum scheduling unit in the time domain in 4G LTE was originally fixed at a 1-ms TTI, until when the shortened TTI (sTTI) was introduced (more details can be found in Chap. 13). The sTTI duration can be either one-slot (or 7 symbols for NCP), or 2-symbol based. However, since there are 14 symbols per subframe and 7 symbols per slot, it is rather cumbersome to have a *flexible* and a *scalable* TTI length considering the need to efficiently multiplex different TTI lengths together in one cell. Obviously, it would be much easier if the number of symbols for possible TTI lengths can be in the form of 2^{μ} ($\mu = 0, 1, 2, 3, \ldots$).

Indeed, at the beginning of 5G NR study, the following options for the "base" 5G NR SCS f_0 were considered:

- Alternative 1: $f_0 = 15$ kHz (i.e., 4G LTE–based numerology).
- Alternative 2: $f_0 = 17.5$ kHz with a uniform symbol duration including the CP length [15].
- Alternative 3: $f_0 = 17.07$ kHz with a uniform symbol duration including the CP length [16].
- Alternative 4: $f_0 = 21.33$ kHz with a uniform symbol duration including the CP length [16].

Table 5.2 summarizes some key parameters associated with the above design alternatives using a 1-ms TTI as the reference.

While Alternative 1 (LTE) and Alternative 2 maintain the same property of a power of 2 of the FFT size, Alternative 3 and Alternative 4 keep the same sampling rate as

Parameters	LTE NCP	LTE ECP	Alternative 2 – Scalable CP		Alternative 3	Alternative 4
f_0 [kHz]	15	15	17.5	35	17.07	21.33
OFDM Symbol Length	66.67	66.67	62.5	31.25	58.6	46.87
CP Length [μs]	5.2/4.7	16.67	5.36	2.68	3.9	15.63
# of Symbols/1 ms	14	12	16	32	16	16
Sampling Rate [MHz]	30.72	30.72	35.84	71.68	30.72	30.72
CP Overhead	7%	20%	8.6%	8.6%	6.0%	25.0%
FFT Size	2048	2048	2048	2048	1800	1440

TABLE 5.2 Comparison of Some Key Parameters for Different SCS Design Alternatives, Assuming a 1-ms TTI

that of LTE, but the FFT size is no longer a power of 2 (1800 and 1440, respectively). To achieve the different CP lengths in order to combat different channel and operating conditions, Alternative 2 relies on scalable SCS values resulting in scalable CP lengths, while Alternative 3 and Alternative 4 are based on "hand-picked" SCS values. Note that Alternative 1 (LTE) chooses the fixed SCS but varies the number of FFT samples to realize NCP and ECP.

Alternatives 2, 3, and 4 have one common philosophy – within a pre-defined TTI (e.g., 1 ms), the number of symbols is 2^{μ}, one key component to enable flexibility and scalability. Scaling the TTI length is crucial in meeting tight latency requirements, e.g., for URLLC (more details can be found in Chap. 13). These alternatives also make it easier to manage and multiplex different TTI lengths scheduled within a same time instance. This is quite different from Alternative 1, as will be shown in Chap. 13, where special handling must be done considering the interactions among *incompatible* TTI lengths such as 2, 7, and 14 symbols.

Despite the clear benefits of 2^{μ}-based SCS proposals, there is one major drawback. These new SCS values and hence the new symbol durations imply that when 5G NR networks and 4G LTE networks coexist in the same cell or adjacent cells using the same carrier frequency or adjacent frequencies, there would be mutual interference. Therefore, managing the coexistence is not straightforward. Rather, extra care has to be taken particularly for the case when 4G LTE and 5G NR need to be supported in the same cell or adjacent cells. For instance, dynamic spectrum sharing (DSS) may be highly desirable for some operators (see Chap. 11 for more details). In many cases, the deployment of 5G NR base stations and UEs has to be done in a gradual manner due to time and budget constraints and the natural penetration rate of a new radio access technology. Such coexistence or multiplexing may be there for a long time, depending on the deployment need and the adoption rate of 5G NR. In addition, it is also important for 4G LTE and 5G NR to possibly aggregate together, forming the so-called EN-DC or NE-DC (see Chap. 11 for more details). The term EN-DC refers to the case when 4G LTE and 5G NR are configured as dual-connectivity for a UE, with one LTE carrier as the anchor or the primary carrier. The term NE-DC is similar to EN-DC but with one NR carrier as the anchor or the primary carrier. In both EN-DC and NE-DC, the Uplink transmission power by a UE has to be shared by LTE and NR carriers, either in a semi-static or dynamic manner. In both cases, it is easier for the power sharing if the numerologies are such that the two radio access technologies (RATs) can be synchronized at every certain number of symbols, although the symbol durations for the two RATs may be different.

With the above considerations, 5G NR eventually adopted Alternative 1. That is, **15-kHz SCS is used as the foundation for 5G NR numerology**, which can be further scaled to achieve flexible and scalable SCSs.

In terms of SCS scaling, there were different alternatives as well, although less controversial. To be more specific, on a high level, two sets of scalable SCSs were considered:

- Alternative 1: $f_0 * 2^{\mu}$ based, where μ is an integer, and
- Alternative 2: $f_0 * U$ based, when U is an integer chosen from a set of possible positive values.

Alternative 2 can be designed to be a superset or a subset of Alternative 1, depending on the possible set of values of U. As stated earlier, one important metric in determining the set of possible SCSs is to allow *convenient multiplexing* of different services (or different SCSs) in the same cell or for the same UE. A power of 2 based set of SCSs obviously make it easier to ensure the frequency-alignment and time-alignment of different SCSs, and to construct a nested structure in both the frequency domain and the time domain. For instance, under Alternative 1, when $\mu = 0, 1, 2,$ and 3, we can have SCSs of 15 kHz, 30 kHz, 60 kHz, and 120 kHz. The nested structure in the frequency domain is shown in Fig. 5.2. Although not shown, it should be straightforward to see that a similar nested structure holds for the time domain as well. As a result, Alternative 1 was adopted for 5G NR.

Table 5.3 summarizes the supported SCSs ($\Delta f = 15 * 2^{\mu}$ kHz) in 5G NR [17]. It is noted that ECP is only supported for the 60-kHz SCS. In this case, the CP length is 4.17 μs, comparable to that of the 15-kHz SCS with NCP (for both LTE and NR). The support of ECP is primarily driven by URLLC-related considerations – a combination of a long CP duration with the 60-kHz SCS makes it possible to utilize a large SCS (for low latency) in deployment scenarios with a large range of delay spreads. This of course comes at the cost of additional CP overhead.

Similar to 4G LTE, the minimum resource allocation granularity is also called an RB. In order to ensure the nested structure, each RB should contain a fixed number of consecutive tones, *irrespective of the SCS*. At the early stage of 5G NR standardization, two possible values were studied: 12 tones/RB or 16 tones/RB. The former is the same

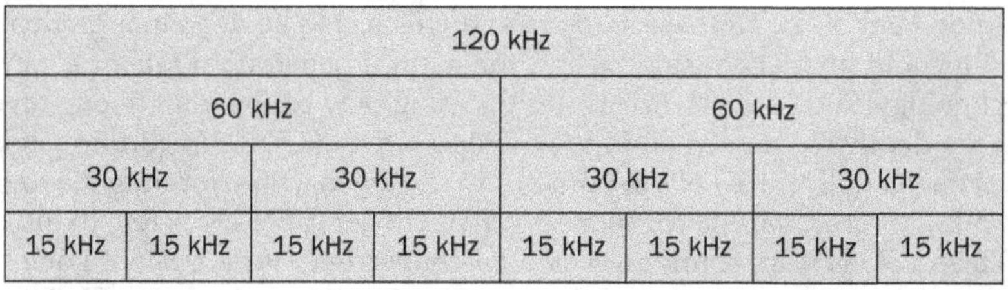

<table>
<tr><td colspan="8">120 kHz</td></tr>
<tr><td colspan="4">60 kHz</td><td colspan="4">60 kHz</td></tr>
<tr><td colspan="2">30 kHz</td><td colspan="2">30 kHz</td><td colspan="2">30 kHz</td><td colspan="2">30 kHz</td></tr>
<tr><td>15 kHz</td><td>15 kHz</td><td>15 kHz</td><td>15 kHz</td><td>15 kHz</td><td>15 kHz</td><td>15 kHz</td><td>15 kHz</td></tr>
</table>

FIGURE 5.2 The nested structure of f_0*2^{μ} [kHz]–based SCSs in the frequency domain.

μ	$\Delta f = 15 * 2^{\mu}$ [kHz]	Cyclic prefix
0	15	Normal
1	30	Normal
2	60	Normal, extended
3	120	Normal
4	240	Normal

TABLE 5.3 Supported NR SCSs in 5G NR [17]

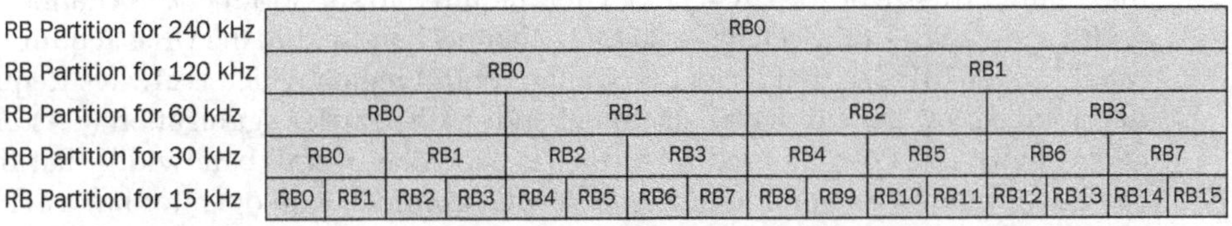

RB Partition for 240 kHz: RB0
RB Partition for 120 kHz: RB0 | RB1
RB Partition for 60 kHz: RB0 | RB1 | RB2 | RB3
RB Partition for 30 kHz: RB0 | RB1 | RB2 | RB3 | RB4 | RB5 | RB6 | RB7
RB Partition for 15 kHz: RB0 | RB1 | RB2 | RB3 | RB4 | RB5 | RB6 | RB7 | RB8 | RB9 | RB10 | RB11 | RB12 | RB13 | RB14 | RB15

FIGURE 5.3 The nested RB structure via a relatively fixed grid for all SCSs.

as LTE, while the latter is intended to address some limitations that the 12 tones/RB has, especially in terms of the design and the processing of the reference signals (e.g., for demodulation, CSI, etc.). This is because 16 tones/RB makes it easier to scale a reference signal pattern with the number of ports (1, 2, 4, 8, etc.) for the reference signal and to be compatible with FFT sizes of power of 2 [18]. The 12 tones/RB was the eventual winner for 5G NR, for a better alignment with the LTE numerology.

In addition to a fixed number of tones/RB, additional conditions are necessary in order to satisfy the nested structure. That is, the RBs for different numerologies have to be located on a fixed grid relative to each other. In other words, the definition of RBs for different numerologies is not supposed to be staggered in the frequency domain. To that end, for the set of SCSs of $\Delta f = 15 * 2^{\mu}$ kHz, $\mu = 1, 2, 3$, and 4, the RBs are defined as a subset of the RB grids for the 15-kHz SCS in a nested manner in the frequency domain. This is illustrated in Fig. 5.3 where the RB is indexed for each SCS separately.

The nested RB structure makes it easier to schedule services with different numerologies simultaneously in a cell with minimal scheduling blocking. To better understand the benefit, Fig. 5.4 provides an example. In this example, a first UE is scheduled with a 15-kHz SCS using RB0, RB1, and RB2, a second UE is scheduled with a 30kHz SCS using RB2 and RB3, and a third UE is scheduled with a 60-kHz SCS using RB2. With the nested structure, all three UEs can be scheduled without overlapped resources in frequency (or in other words, blocking/interfering each other). This also makes it easier for a gNB to choose to use or not use a guardband among UEs with different SCSs. In this example, there is a guardband (RB3 in the 15-kHz SCS) between the first UE and the second UE. However, there is no guardband between the second UE and the third UE, which may be possible if the performance degradation due to the mutual interference is tolerable or manageable, e.g., if the two UEs are both scheduled with low MCSs.

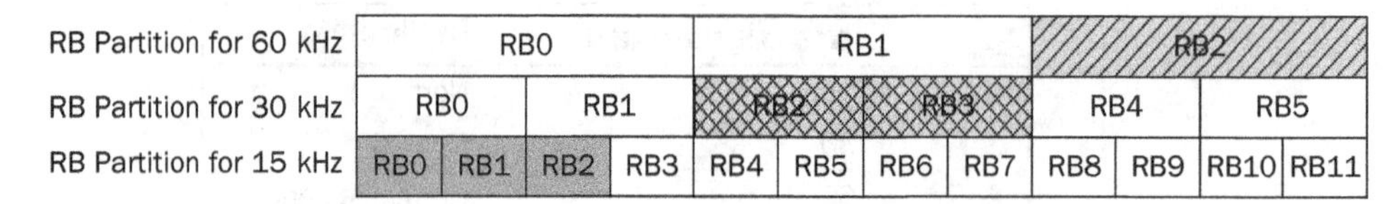

FIGURE 5.4 Facilitation of scheduling of UEs with different SCSs via the nested RB-structure.

5.4 Frequency Ranges, Bandwidths, and Bands for 5G NR

As discussed earlier, each 4G LTE carrier has a maximum bandwidth of 20 MHz, and a maximum FFT size of 2048. A 4G LTE UE is required to support the *same* channel bandwidth of an eNB – if the cell has a 20 MHz channel bandwidth, the UE is required to be capable of 20 MHz so that it can decode the control region which is always frequency-distributed over the entire eNB channel bandwidth. Carrier aggregation (CA) or dual connectivity (DC) can be used to further expand the overall bandwidth that the UE can be configured for communications. As one example, if an operator has a 100-MHz bandwidth, it can be divided into five component carriers (CCs), each of 20 MHz. A UE can be configured with the five CCs as CA so that it can be scheduled over the entire 100-MHz bandwidth. It is important to note that each CC needs to be scheduled by its own control channel. In addition, as discussed earlier, the guardband defined for a 4G LTE CC is about 10 percent. In other words, for a 20-MHz channel bandwidth, the schedulable bandwidth for 4G LTE communications is about 18 MHz (100 RBs). These factors (separate control channels and guardbands) contribute to a sizeable overhead for 4G LTE.

In light of the shortcomings in 4G LTE, there are a couple of questions related to channel bandwidths for 5G NR:

- How much should the maximum channel bandwidth be for a gNB?
- How much should a guardband be for a carrier?

Note that 5G NR can be deployed in a wide range of carrier frequencies. In different carrier frequencies, the available bandwidths for an operator, license or unlicensed, certainly vary. Generally speaking, the available bandwidth increases along with the carrier frequency – for carrier frequencies of several GHz, the available bandwidths may be in order of tens or a few hundred MHz. For carrier frequencies of tens of GHz, the available bandwidths can be as high as 1 GHz or more.

To accommodate carrier-frequency-dependent channel bandwidths, it is necessary to classify or group different carrier frequencies. To that end, 5G NR adopts a terminology called frequency range or FR [19], as summarized in Table 2.1 in Chap. 2. Reference [19] also provides the corresponding frequency bands defined for FR1 and FR2, as also shown in Tables 2.9 and 2.10, respectively, where the prefix "n" is used for any NR band. Note that for some frequency bands, instead of FDD or TDD, either supplemental DL or supplemental UL is defined. It should be noted that additional bands for FR1 and FR2 may be introduced in the future, primarily driven by operators' needs. In addition, additional frequency ranges may also be introduced in the future. For example, in Release 17, scaling up 5G NR to frequency ranges of 52.6 to 71 GHz is to be studied [20] and specified [21].

To determine the maximum channel bandwidth for a carrier frequency, the following factors need to be considered:

- The available spectrum (licensed/unlicensed/shared),
- The target maximum FFT size,
- Tradeoff between single-carrier versus CA operations,
- Suitable SCSs for a carrier frequency.

It is envisioned that 5G NR needs to provide means for a UE to be able to operate with an aggregated bandwidth larger than 1 GHz. Note that the more than 1 GHz aggregated bandwidth may not be contiguous, which implies that a single FFT operation may not be desirable. Consequently, there is always a need for carrier aggregation, which is a very useful feature supported in 4G LTE from Release 10. In 4G LTE, up to 32 carriers can be configured for a UE for carrier aggregation, resulting in a maximum of 32 (CCs) $\times$20 (MHz/CC) = 640 MHz aggregation bandwidth.

A wider bandwidth for a CC is overhead-friendly, as per earlier discussion due to factors such as control overhead and guardbands. However, a wider bandwidth makes it more difficult to ensure sufficient synchronization within the wideband. Often, multiple copies of synchronization signals may become necessary. A wider bandwidth may also imply a larger FFT size. As an example, an FFT size of 2048 is necessary for a 15-kHz SCS and a 20-MHz channel bandwidth as in 4G LTE. Assuming a 160-MHz channel bandwidth with a 15-kHz SCS, an FFT size of $8 \times 2048 = 16{,}384$ would become necessary. Such an FFT size may be too large for implementation, especially considering low latency–related applications. A larger bandwidth under a certain maximum transmit power constraint leads to a smaller power spectral density (PSD), which may result in less coverage. It is also important to note that regardless of the maximum gNB channel bandwidth, it is also a question whether a UE's channel bandwidth can be allowed to be smaller, which is to be further discussed in the next section.

In practical implementation, there is always a phase noise from implementation, e.g., due to the impact of reference clock, loop filter noise and voltage-controlled oscillator sub-components, etc. The impact of phase noise to 5G NR performance requires specific consideration especially in high-carrier frequencies. Reference [22] contains several models for evaluating the impact of phase noise. Generally speaking, a large SCS size helps mitigate the impact of phase noise. A large SCS size also reduces the FFT size for a given channel bandwidth, crucial for wide contiguous available bandwidths at higher carrier frequencies.

In light of the above considerations, 5G NR finally adopts the following:

- Maximum channel bandwidth
  - For FR1, up to **100 MHz**,
  - For FR2, up to **400 MHz**.
- Each UE can be aggregated, using CA or DC, with up to 16 component carriers (CCs).
- SCS:
  - FR1: 15 kHz, 30 kHz, 60 kHz,

- FR2: 60 kHz, 120 kHz.
    - Note that 240-kHz SCS is not supported for data transmissions – it is only used for synchronization signals (details can be found in Chap. 6).

Therefore, the maximum aggregated bandwidth for a UE in 5G NR is

- FR1: 100 (MHz/CC) × 16 (CCs) = **1.6GHz**,
- FR2: 400 (MHz/CC) × 16 (CCs) = **6.4GHz**.

It should be noted that the maximum aggregation bandwidth for a UE in 5G NR can be further increased in future releases, e.g., by aggregating more CCs (e.g., up to 32 CCs), by introducing new SCSs and/or a larger maximum channel bandwidth. Such an extension would be straightforward by using the existing numerology framework, thanks to its future compatibility.

For a given band, the Downlink channel bandwidth and the Uplink channel bandwidth can be different, as shown in Table 2.9 in Chap. 2.

A minimum channel bandwidth is also defined, namely, 5 MHz for FR1 and 50 MHz for FR2. The introduction of the minimum channel bandwidth is necessary to achieve a good tradeoff between the standardization efforts (as the minimum channel bandwidth has impact on some physical layer channels/signals, e.g., synchronization signals, to be detailed in Chap. 6) and the necessary facilitation of re-farming the existing 4G LTE to NR 5G for FR1 in commercial deployments.

Similar to 4G LTE, the bandwidth available in a cell for Downlink or Uplink transmissions is always less than the channel bandwidth, due to the necessary guardband reserved to handle emissions and regulatory requirements. Compared with a roughly constant 10 percent guardband defined for 4G LTE, the guardband for 5G NR is more carefully chosen, allowing more efficient usage of frequency resources. The maximum transmission bandwidth (denoted as *TxBw*, in RBs) and the ratio of transmission bandwidth to channel bandwidth (denoted as *Ratio*, in %) are shown in Table 2.11 for FR1 [19] and in Table 2.12 for FR2 [23] in Chap. 2. Obviously, the higher the ratio, the more efficient the utilization of the channel bandwidth. From Tables 2.11 and 2.12, it can be seen that the ratio of the maximum transmission bandwidth to channel bandwidth is generally more than 90%, particularly for the case of large channel bandwidths. The amount of minimum guardband is not linearly increasing with the channel bandwidth. Rather, it is tailored with the combinations of the SCS and the channel bandwidth. For 15-kHz SCS, the minimum guardband (the two bandwidth edges combined) is roughly within a range of 0.5 to 1.4 MHz for the channel bandwidths ranging from 5 to 50 MHz. For 30-kHz SCS, the minimum guardband is roughly within a range of 1 to 2 MHz for the channel bandwidths ranging from 5 to 100 MHz. For 60-kHz SCS, the minimum guardband is roughly within a range of 2 to 3.2 MHz for the channel bandwidths ranging from 10 to 100 MHz.

Therefore, it is clear that the utilization of channel bandwidths for both Downlink and Uplink transmissions in 5G NR increases substantially compared with 4G LTE.

5.5 gNB Channel Bandwidth versus UE Channel Bandwidth

As mentioned earlier, in 4G LTE, the UE channel bandwidth is *required* to be the same as the eNB channel bandwidth. For a maximum channel bandwidth of 20 MHz, such a requirement is reasonable considering the need to fully utilize the frequency diversity when possible and feasible and the simplification of standardization efforts.

However, for 5G NR, it seems to be too much of a requirement if all UEs are still required to be capable of supporting the same maximum channel bandwidth as the gNB, i.e., up to 100 MHz for FR1 and up to 400 MHz for FR2. After all, there are different kinds of devices that 5G NR aims to support: some are highly capable, while others may be not. Devices such as MTC, wearables, sensors in an industrial setting, etc., may require limited data rates, limited bandwidths, limited transmit power, and limited number of transmit antennas. Complexity and cost may be crucial for this kind of devices. Some devices may be sensitive to power consumption. Operating with a narrower bandwidth helps reduce power consumption. Therefore, from the beginning of 5G NR standardization, it was agreed that as a design principle:

- A UE's channel bandwidth can be the same as or different than a gNB's channel bandwidth,
- Regardless of a gNB's channel bandwidth, UEs of different bandwidth capabilities can be efficiently multiplexed in the gNB,
- For a UE, the set of bandwidths supported for Downlink are not necessarily the same as those supported for Uplink,
  - In particular, in a given cell, the Bandwidth for Downlink operation is not necessarily to be the same as that for Uplink operation for the UE.

Based on the reported capability by a UE, a gNB can configure the UE to operate within a part of the gNB's channel bandwidth, which is commonly known as a **bandwidth part (BWP)** for the UE. Several use cases for UE BWPs are presented in Fig. 5.5. Case (a) and Case (b) were discussed earlier and are rather straightforward. Case (c) shows the case that with the BWP framework, a UE can operate in a first BWP with a first numerology (e.g., targeting eMBB traffic), and in a second BWP with a second numerology (e.g., targeting URLLC traffic). In case (d), although it is a single carrier from the gNB's perspective, a UE can be configured with multiple carriers as part of CA.

Therefore, the introduction of UE BWP not only provides the necessary accommodation of different UE capabilities, but also facilitates support of different services, different deployments, improved power savings, etc. A BWP can be activated or deactivated via a Downlink control channel. A UE can be configured with up to four BWPs in a cell. For simplicity, only one BWP may be active at any time. It should be noted that there are no specific requirements on UE reception/transmission outside of an active BWP.

Although the signaling in 5G NR can configure a UE with any channel bandwidth for a BWP, the possible practical UE BWP sizes are specified in [19] and [23]. Generally, the possible sizes are confined to possible gNB channel bandwidths, i.e., {5, 10, 15, 20, 25, 30, 40, 50, 60, 80, 90, 100} MHz for FR1, and {50, 100, 200, 400} MHz for FR2.

5.6 Symbol, Slot, Subframe, and Frame for 5G NR

As discussed earlier, the set of SCSs for 5G NR are defined as $\Delta f = f_0 * 2^{\mu}$ (kHz), where $f_0 = 15$ kHz, $\mu = 0, 1, 2, 3$, and 4. When $\mu = 0$, $\Delta f = f_0$, the resulting OFDM symbol

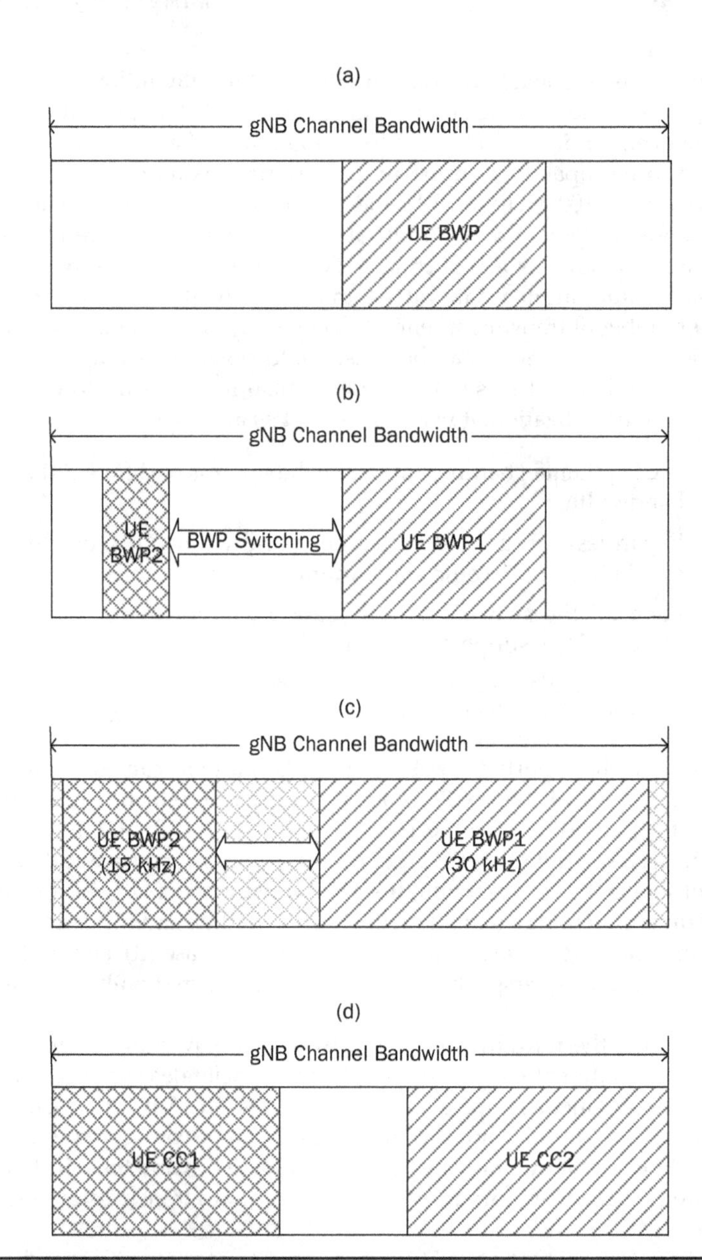

FIGURE 5.5 Illustration of some use cases for UE BWPs: (a) reduced UE bandwidth capability; (b) BWP switching for UE power savings; (c) support of different services with different SCS (15 kHz and 30 kHz); (d) CA for a UE within a gNB (where the middle part is unknown to the UE).

boundaries are completely aligned with that of 4G LTE (in case of NCP). For other values of μ, the resulting OFDM symbol boundaries are generally a scaled version of $\mu = 0$. However, similar to the frequency domain, it is also important to ensure the nested structure in the time domain to facilitate multiplexing of different SCSs.

Note that in 4G LTE, for NCP, in each slot (7 symbols) we have the following CP lengths:

- 1st symbol: 160 T_s
- 2nd to 7th symbols: 144 T_s

The special handling for the 1st symbol every slot is due to the fact that 1024 (T_s) is not divisible by 7 (symbols). Implementation-friendly numbers (160 and 144) were thus chosen. For 5G NR, similar special handling is done as well in order to align with 4G LTE and to ensure the nested structure in the time domain. In particular, a special CP duration is specified for **the first symbol every 0.5 ms** for all possible values of μ for NCP in 5G NR:

- For any $\mu = 0, 1, 2, 3$, and 4, every 0.5 ms (in total there are $7 \times 2^{\mu}$ symbols), the CP lengths are:
  - 1st symbol: $(144 \times \kappa \times 2^{-\mu} + 16 \times \kappa) \times T_c = 144 \times \kappa \times 2^{-\mu} \times T_c + 16 \times T_s =$ $\mathbf{144 \times 2^{-\mu} \times T_s + 16 \times T_s}$
  - All other symbols ($7 \times 2^{\mu} - 1$ symbols): $144 \times \kappa \times 2^{-\mu} \times T_c = 144 \times 2^{-\mu} \times T_s$

where T_c is the basic time unit for NR defined as $1/(\Delta f_{max} \times N_f)$, where $\Delta f_{max} = 480 \times 10^3$ (Hz) and $N_f = 4096$, and $\kappa = T_s/T_c = 64$. In other words, all symbols in each 0.5 ms have a duration linearly scaled (by $2^{-\mu}$) from those of the 15-kHz SCS, except for the first symbol which always has an extra duration of $16 \times T_s$ or 0.52 μs.

For ECP (60-kHz SCS only), the CP duration is a straightforward linearly scaled version of the CP duration from 4G LTE ECP, i.e., $512 \times \kappa \times 2^{-\mu} \times T_c = 512 \times 2^{-\mu} \times T_s$, where $\mu = 2$, for all symbols.

In 5G NR, DL and UL can be configured separately with different CP types (similar to 4G LTE) and different SCSs (e.g., a 30-kHz NCP Downlink associated with a 60-kHz ECP Uplink). The configured CP type is applied to all the applicable physical layer channels and signals in the given link direction.

Similar to 4G LTE, Downlink and Uplink transmissions are organized in terms of subframes and frames. In particular,

- Subframe: fixed at 1 ms or $(\Delta f_{max} \times N_f/1000) \times T_c$
- Frame: fixed at 10 ms or $(\Delta f_{max} \times N_f/100) \times T_c$

However, it should be noted that while the definition of subframes and frames in 5G NR are necessary in establishing the physical time, useful for synchronization across cells or between a gNB and a UE, it is NOT used as a TTI for scheduling Downlink or Uplink transmissions. This is different from LTE, where the 1-ms based TTI is the most dominant resource allocation duration.

Instead, 5G NR utilizes the **slot** concept [17] for resource scheduling for both Downlink and Uplink transmissions. The slot duration depends on the SCS, as summarized in Tables 5.4 and 5.5.

From Tables 5.4 and 5.5, it is clear the slot duration depends on the SCS. For instance, for the 15-kHz SCS, a slot has a duration of 1 ms, while for the 30-kHz SCS, a slot is 0.5 ms. The slot definition is thus different from that of 4G LTE.

To better illustrate the concepts of symbol, slot, subframe, frame, and the nested structure in the time domain, Fig. 5.6 provides an example for SCSs 15 kHz, 30 kHz, and 60 kHz, where the special handling of symbol 0 every 0.5 ms or $7 \times 2^{\mu}$ symbols is highlighted.

μ	N_{symb}^{slot}	$N_{slot}^{frame,\mu}$	$N_{slot}^{subframe,\mu}$
1	14	20	2
2	14	40	4
3	14	80	8
4	14	160	16

TABLE 5.4 Number of OFDM Symbols per Slot, Slots per Frame, and Slots per Subframe for NCP

μ	N_{symb}^{slot}	$N_{slot}^{frame,\mu}$	$N_{slot}^{subframe,\mu}$
0	14	10	1
2	12	40	4

TABLE 5.5 Number of OFDM Symbols per Slot, Slots per Frame, and Slots per Subframe for ECP

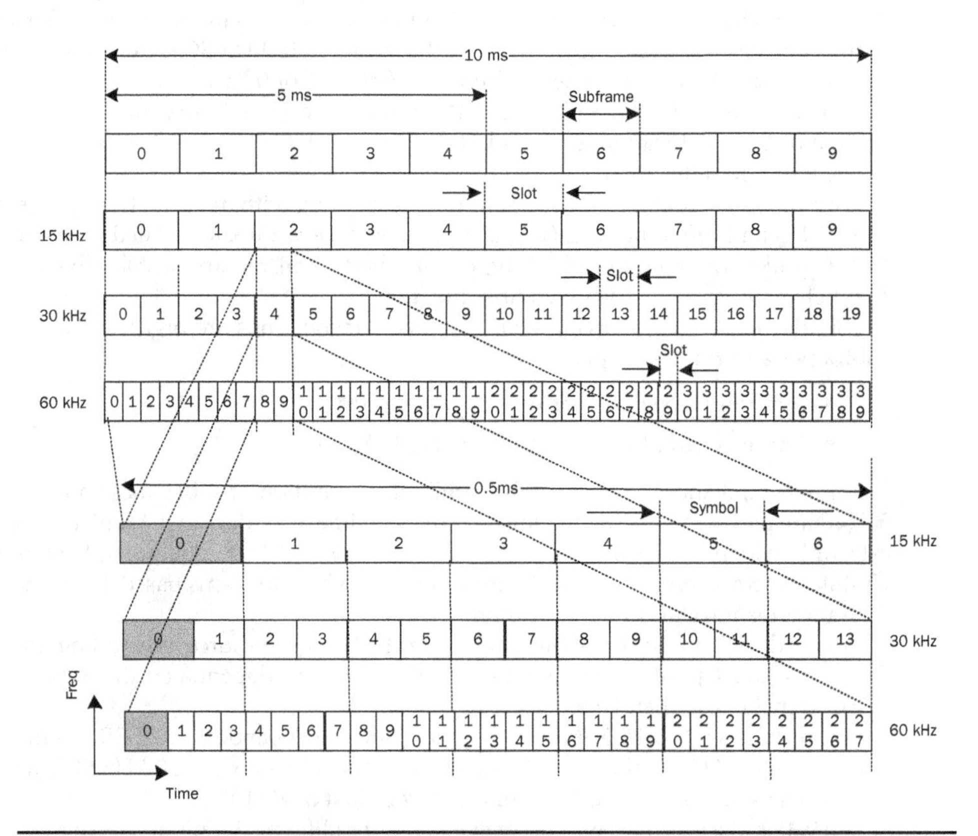

FIGURE 5.6 Illustration of the nested structure in the time domain, including the special handling of symbol 0 every 7 × 2^{μ} symbols for symbol-level alignment, where μ = 0, 1, and 2 (15/30/60-kHz SCS).

Besides the slot-based scheduling, the *mini-slot* based scheduling is possible with 5G NR. The concept of mini-slot is similar to the sTTI concept in 4G LTE, where the TTI can be in units of symbols. The mini-slot based scheduling offers the necessary low latency operation for traffic types such as URLLC. More details can be found in Chap. 13. In addition, slot (or mini-slot) aggregation–based scheduling is also supported, where the data transmission can be repeated in multiple slots (or mini-slots), in order to further improve reliability or coverage. Note that mini-slot–based aggregation is a feature introduced in Release 16.

5.7 Slot Structure for 5G NR and Forward Compatibility Considerations

One lesson learned from 4G LTE is its inflexibility in future compatibility. There is an always-on cell-specific reference signal, which is present in every Downlink subframe except when in the multicast broadcast single-frequency network (MBSFN) region in an MBSFN subframe [1]. However, in a backward compatible 4G LTE carrier, only up to six subframes in a frame can be possibly configured as MBSFN subframes. Indeed, in the early standardization of 4G LTE, there was a proposal to possibly configure all subframes as *blank* subframes [24], but was not adopted. Blank subframes enable forward compatibility, since a legacy UE is not expected to make any assumptions about the blank subframes while new services can be introduced to new UEs in a backward-compatible manner.

5G NR is designed to have a flexible and forward-compatible slot structure. A slot can support:

- Zero DL/UL switching points
  - 14 "**DL**" symbols (D), or 14 "**flexible**" symbols (X), or 14 "**UL**" symbols (U)
- One DL/UL switching point of all combinations
  - Starting with zero or more DL symbols, ending with zero or more UL symbols, and with "flexible" symbols in between, where there is at least one "flexible" symbol and one DL or UL symbol
- Two DL/UL switching points within a slot
  - The first 7 symbols start with zero or more DL symbols, and end with at least one UL symbol at symbol #6 with zero or more "flexible" symbols in between.
  - The second 7 symbols start with one or more DL symbols and end with zero or more UL symbols with zero or more "flexible" symbols in between.

Note that in an FDD system, the Downlink can have either DL symbols or flexible symbols while the Uplink can have either UL symbols or flexible symbols, but not symbols of a conflicting link direction. However, such restriction may be revisited in future 5G NR releases.

For flexible symbols, unless explicitly instructed, a UE is not expected to make any assumption for these symbols, which is essential for forward compatibility. It is worth noting that a flexible symbol for a UE means it is *reserved* for some other uses and its usage is transparent to the UE. Another possibility is to further introduce an *empty* symbol, where a UE may assume that the symbol is not used for any transmission. Therefore, the UE can use the empty symbol for purposes such as interference measurement. However, the concept of empty symbol was not adopted for 5G NR. Instead, resources for interference measurement were introduced separately for a UE, as detailed in Chap. 8.

More specifically, the indication of the slot structure/format, also known as slot format indicator (SFI), is done via three types of signaling:

- Cell-specific semi-static SFI,
- UE-specific semi-static SFI, and
- Group-specific dynamic SFI.

Cell-specific semi-static SFI is indicated by SIB1 [25] on a per cell basis. The indication has a configured periodicity, with potential values dependent on the SCS value, which is also called the reference numerology for SFI, denoted as μ_{ref}. To be more specific, the following periodicity values are supported [25]:

$$\{0.5, 0.625, 1, 1.25, 2, 2.5, 3, 4, 5, 10\} \text{ ms}$$

where the value of 0.625 ms is valid only for $\mu_{ref} = 3$ (120 kHz), the value of 1.25 ms is valid only for $\mu_{ref} = 2$ (60 kHz) or $\mu_{ref} = 3$ (120 kHz), and the value of 2.5 ms is valid only for $\mu_{ref} = 1$ (30 kHz), $\mu_{ref} = 2$ (60 kHz), or $\mu_{ref} = 3$ (120 kHz). As a result, the number of slots within a period can be as small as a single slot, and as large as 80 slots (a periodicity of 10 ms with a 120-kHz SCS).

For each period, the configuration provides fully flexible combinations of all symbol types, via the following parameters:

- From the beginning of the period, a number of $d_{slot} \geq 0$ consecutive slots with only Downlink symbols,
- Subsequent to d_{slot} Downlink slots, a number of $d_{sym} \geq 0$ consecutive Downlink symbols,
- From the end of the period and counting backward, a number of $u_{slot} \geq 0$ slots with only Uplink symbols, and
- Preceding to the d_{slot} Uplink slots, the number of $u_{sym} \geq 0$ consecutive uplink symbols.

Any symbols in between the DL portion (governed by d_{slot} and d_{sym}) and the UL portion (governed by u_{slot} and u_{sym}) are flexible symbols. This is illustrated in Fig. 5.7. In the extreme case, if d_{slot}, d_{sym}, u_{slot}, and u_{sym} all configured to be zeros, all symbols are flexible symbols.

Furthermore, the flexibility in configuring d_{slot}, d_{sym}, u_{slot}, and u_{sym} also makes it easier for 5G NR to coexist with 4G LTE TDD, either for co-channel or adjacent-channel deployments. As one example, an operator may have a 4G LTE TDD deployment with a Downlink/Uplink subframe configuration of #1 (DSUDDDSUDD) (see Table 5.1), and decides to gradually deploy 5G NR on an adjacent carrier with a 30-kHz SCS. To minimize the mutual interference between 4G LTE TDD and 5G NR, the operator can choose a cell-specific SFI for 5G NR as

- A 5-ms SFI periodicity with the 10 slots indexed by [0, 1, 2, 3, 4, 5, 6, 7, 8, 9] along with the configuration [D, D, D, D/X, U, U, D, D, D, D], respectively.

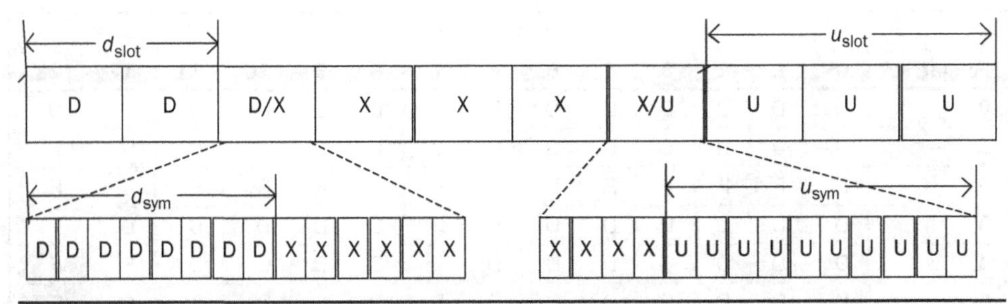

FIGURE 5.7 Illustration of cell-specific semi-static SFI indication, a periodicity of 10 slots with $\boldsymbol{d}_{slot} = 2$, $\boldsymbol{d}_{sym} = 8$, $\boldsymbol{u}_{slot} = 3$, and $\boldsymbol{u}_{sym} = 10$.

The above 5-ms periodicity–based SFI can match well with the 4G LTE configuration in each 5 ms, and thus minimizes the mutual interference.

For additional flexibility, a UE can be configured with a second cell-specific pattern with a separate periodicity and a set of DL/UL/flexible symbols, which is concatenated with the first pattern to form a dual-periodicity configuration, as long as the combined periodicity can divide 20 ms [2].

For each UE, it can be further configured via dedicated signaling a configuration regarding how to utilize the X symbols in the cell-specific configuration. In other words, the X symbols in the cell-specific configuration can be over-written by a UE-specific semi-static configuration. It should be noted that the Downlink symbols and the Uplink symbols in the cell-specific configuration can NOT be over-written by the UE-specific configuration.

The UE-specific configuration can be done separately for each of the slot containing the X symbols for the one or more slots in the cell-specific configuration. For each slot, the UE-specific configuration can indicate whether it is all Downlink symbols, all Uplink symbols, or an explicit indication of Downlink, Uplink, and/or flexible symbols with zero or one DL/UL switching point in the slot, where all combinations of the three types of symbols are possible. For instance, in the example shown in Fig. 5.7, the *UE-specific configuration* can only apply to the X symbols in slots 2, 3, 4, 5, and 6, where the configuration can be done separately for each of the five slots. For instance, a gNB can choose to configure:

- For a first UE: all X symbols in slot 4 are Downlink symbols while other X symbols in the cell-specific configuration still remain as X symbols.
- For a second UE: X symbols in slot 5 are configured to have a pattern of DDDDDDDXXXXXXU, and all X symbols in slot 6 are configured to Uplink symbols. Other X symbols in the cell-specific configuration remain as X symbols.

Both the above cell-specific and UE-specific configurations are *semi-static*. If a new configuration is deemed necessary, a re-configuration can be performed but it takes some time, e.g., tens of milliseconds. To address the relatively long latency, a dynamic SFI indication is also supported in 5G NR.

The dynamic SFI indication is done via PDCCH using a Downlink control information (DCI) format called 2-0. The details regarding how the PDCCH is transmitted can

Format	Symbol index in a slot													
	0	1	2	3	4	5	6	7	8	9	10	11	12	13
0	D	D	D	D	D	D	D	D	D	D	D	D	D	D
1	U	U	U	U	U	U	U	U	U	U	U	U	U	U
2	F	F	F	F	F	F	F	F	F	F	F	F	F	F
3	D	D	D	D	D	D	D	D	D	D	D	D	D	F
4	D	D	D	D	D	D	D	D	D	D	D	D	F	F
5	D	D	D	D	D	D	D	D	D	D	D	F	F	F
6	D	D	D	D	D	D	D	D	D	D	F	F	F	F
7	D	D	D	D	D	D	D	D	D	F	F	F	F	F
8	F	F	F	F	F	F	F	F	F	F	F	F	F	U
9	F	F	F	F	F	F	F	F	F	F	F	F	U	U
10	F	U	U	U	U	U	U	U	U	U	U	U	U	U
11	F	F	U	U	U	U	U	U	U	U	U	U	U	U
12	F	F	F	U	U	U	U	U	U	U	U	U	U	U
13	F	F	F	F	U	U	U	U	U	U	U	U	U	U
14	F	F	F	F	F	U	U	U	U	U	U	U	U	U
15	F	F	F	F	F	F	U	U	U	U	U	U	U	U
16	D	F	F	F	F	F	F	F	F	F	F	F	F	F
17	D	D	F	F	F	F	F	F	F	F	F	F	F	F
18	D	D	D	F	F	F	F	F	F	F	F	F	F	F
19	D	F	F	F	F	F	F	F	F	F	F	F	F	U
20	D	D	F	F	F	F	F	F	F	F	F	F	F	U
21	D	D	D	F	F	F	F	F	F	F	F	F	F	U
22	D	F	F	F	F	F	F	F	F	F	F	F	U	U
23	D	D	F	F	F	F	F	F	F	F	F	F	U	U
24	D	D	D	F	F	F	F	F	F	F	F	F	U	U
25	D	F	F	F	F	F	F	F	F	F	F	U	U	U
26	D	D	F	F	F	F	F	F	F	F	F	U	U	U
27	D	D	D	F	F	F	F	F	F	F	F	U	U	U
28	D	D	D	D	D	D	D	D	D	D	D	D	F	U
29	D	D	D	D	D	D	D	D	D	D	D	F	F	U
30	D	D	D	D	D	D	D	D	D	D	F	F	F	U
31	D	D	D	D	D	D	D	D	D	D	D	F	U	U
32	D	D	D	D	D	D	D	D	D	D	F	F	U	U
33	D	D	D	D	D	D	D	D	D	F	F	F	U	U
34	D	F	U	U	U	U	U	U	U	U	U	U	U	U
35	D	D	F	U	U	U	U	U	U	U	U	U	U	U
36	D	D	D	F	U	U	U	U	U	U	U	U	U	U

TABLE 5.6 Slot Formats for Dynamic SFI for Normal CP

Format	Symbol index in a slot													
	0	1	2	3	4	5	6	7	8	9	10	11	12	13
37	D	F	F	U	U	U	U	U	U	U	U	U	U	U
38	D	D	F	F	U	U	U	U	U	U	U	U	U	U
39	D	D	D	F	F	U	U	U	U	U	U	U	U	U
40	D	F	F	F	U	U	U	U	U	U	U	U	U	U
41	D	D	F	F	F	U	U	U	U	U	U	U	U	U
42	D	D	D	F	F	F	U	U	U	U	U	U	U	U
43	D	D	D	D	D	D	D	D	D	F	F	F	F	U
44	D	D	D	D	D	D	F	F	F	F	F	F	U	U
45	D	D	D	D	D	D	F	F	U	U	U	U	U	U
46	D	D	D	D	D	F	U	D	D	D	D	D	F	U
47	D	D	F	U	U	U	U	D	D	F	U	U	U	U
48	D	F	U	U	U	U	U	D	F	U	U	U	U	U
49	D	D	D	D	F	F	U	D	D	D	D	F	F	U
50	D	D	F	F	U	U	U	D	D	F	F	U	U	U
51	D	F	F	U	U	U	U	D	F	F	U	U	U	U
52	D	F	F	F	F	F	U	D	F	F	F	F	F	U
53	D	D	F	F	F	F	U	D	D	F	F	F	F	U
54	F	F	F	F	F	F	F	D	D	D	D	D	D	D
55	D	D	F	F	F	U	U	U	D	D	D	D	D	D
56 – 254	Reserved													
255	Special entry													

Table 5.6 Slot Formats for Dynamic SFI for Normal CP (*Continued*)

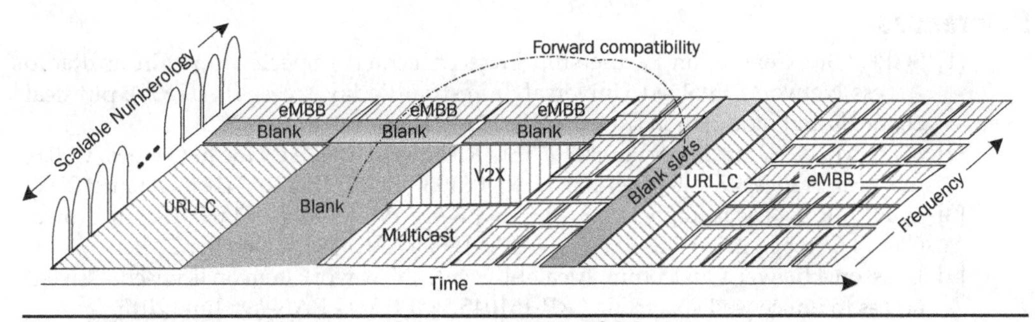

Figure 5.8 Illustration of flexible and forward-compatible 5G NR slot structure.

be found in Chap. 7. The PDCCH for SFI can only be present at the beginning of a slot, and is valid starting from the same slot.

The dynamic SFI is used to indicate to a UE (or a group of UEs) how to use the *leftover X symbols* after applying the cell-specific and the UE-specific semi-static configurations. However, it can NOT update any Downlink or Uplink symbols in the above semi-static configuration(s). The dynamic SFI can be sent on a periodic basis with a period chosen from 1, 2, 5, 10, and 20 slots, where the slot duration is determined based on the corresponding PDCCH SCS. The SFI indication can be transmitted on a cell for the same cell (i.e., same-cell SFI indication). The PDCCH can also be sent on one cell indicating SFI for another cell (i.e., cross-carrier SFI indication).

The possible slot structure indicated by the dynamic SFI for a slot is determined by a pre-determined table, as shown in Table 5.6 [26], where a UE can be indicated using a subset of the entries from the table. The table entries are indexed by an 8-bit indicator, where only 56 entries provide the explicit slot structure. Entry 255 is a special entry, which indicates to the UE that the dynamic SFI is not used on top of the cell-specific and/or UE-specific semi-static configurations (i.e., no over-writing of the left-over X symbols from the semi-static configurations). All other entries are reserved, which can be further defined for new UEs in future releases. The table provides entries with all DL symbols (format 0), all UL symbols (format 1), all flexible symbols (format 2), short consecutive DL symbols (e.g., formats 16–18), short UL consecutive symbols (formats 8–9), DL dominant entries (e.g., formats 3, 4, etc.), UL dominant entries (e.g., formats 10, 11, etc.), a mixture of DL, UL, and flexible symbols (e.g., formats 20, 21, etc.), entries with two DL/UL switching points in a symmetric manner (e.g., formats 46–53), etc.

As can be seen, the numerology and slot structure for 5G NR are designed to be flexible and future-proof, which establishes a solid ground for efficient and effective Downlink/Uplink transmissions accommodating different services in a same cell using a wide range of spectrum under various deployment scenarios. To get a better understanding, Fig. 5.8 provides an illustration where in a same cell, different services may be multiplexed, such as eMBB, URLLC, V2X, Multicast, etc. The slot duration and the SCS can be scaled flexibly. Blank resources can be reserved in the time domain and/or in the frequency domain for forward compatibility. The subsequent chapters will provide more details describing various aspects of 5G NR.

References

[1] 3GPP, "3rd Generation Partnership Project; Technical Specification Group Radio Access Network; Evolved Universal Terrestrial Radio Access (E-UTRA); physical channels and modulation (Release 16)," 36.211, v16.1.0, March 2020.

[2] 3GPP MCC, "Draft Report of 3GPP TSG RAN WG1 LTE Ad Hoc in Cannes, v1.0.0, (Cannes, France, 27 – 30 June, 2006)," R1-061937, 3GPP TSG RAN1#46, August 2006.

[3] 3GPP MCC, "Draft Report of 3GPP TSG RAN WG1 #46bis v.1.0.0, (Seoul, Korea, 09 – 13 October, 2006)," R1-063021, 3GPP TSG RAN1#46bis, October 2006.

[4] Ericsson, Huawei, Qualcomm, Alcatel-Lucent, "New work item on licensed-assisted access to unlicensed spectrum," RP-151045, 3GPP TSG RAN#68, June 2015.

[5] CATT, "New work item proposal for further enhancements to LTE TDD for DL-UL interference management and traffic adaptation," RP-121772, 3GPP TSG RAN#58, December 2012.

[6] 3GPP, "3rd Generation Partnership Project; Technical Specification Group Radio Access Network; requirements for evolved UTRA (E-UTRA) and evolved UTRAN (E-UTRAN) (Release 7)," 25.913, v7.3.0, March 2006.

[7] 3GPP, "3rd Generation Partnership Project; Technical Specification Group Radio Access Network; requirements for further advancements for evolved universal terrestrial radio access (E-UTRA) (LTE-Advanced) (Release 8)." 36.913, v8.0.1, March 2009.

[8] 3GPP, "3rd Generation Partnership Project; Technical Specification Group Radio Access Network; small cell enhancements for E-UTRA and E-UTRAN – Physical layer aspects (Release 12),"36.872, v12.1.0, December 2013.

[9] 3GPP, "3rd Generation Partnership Project; Technical Specification Group Radio Access Network; study on provision of low-cost machine-type communications (MTC) user equipments (UEs) based on LTE (Release 12),"36.888, v12.0.0, June 2013.

[10] 3GPP, "3rd Generation Partnership Project; Technical Specification Group Radio Access Network; study on LTE-based V2X services; (Release 14),"36.886, v14.0.0, July 2016.

[11] Ericsson, Huawei, "New SI proposal: Study on latency reduction techniques for LTE," RP-150465, 3GPP TSG RAN#67, March 2015.

[12] N. Bhushan, T. Ji, O. Koymen, J. Smee, J. Soriaga, S. Subramanian, and Y. Wei, "Industry Perspective: 5G Air Interface System Design Principles," *IEEE Wireless Communications*, vol. 24, no. 5, pp. 6-8, October 2017.

[13] 3GPP, "3rd Generation Partnership Project; Technical Specification Group Radio Access Network; study on channel model for frequencies from 0.5 to 100 GHz (Release 15)," 38.901, v15.1.0, September 2019.

[14] 3GPP, "3rd Generation Partnership Project; Technical Specification Group Radio Access Network; NR; study on integrated access and backhaul; (Release 16)," 38.874, v16.0.0, January 2019.

[15] Qualcomm Inc., "NR Numerology Design Principles,." R1-165112, 3GPP TSG RAN1#85, May 2016.

[16] ZTE, ZTE Microelectronics, "Overview of numerology candidates," R1-164271, 3GPP TSG RAN1#85, May 2016.

[17] 3GPP, "3rd Generation Partnership Project; Technical Specification Group Radio Access Network; NR; physical channels and modulation (Release 15)," 38.211, v15.8.0, January 2020.

[18] Qualcomm Inc., "NR RB Size Design: 12 vs 16," R1-1610125, 3GPP TSG RAN1#86bis, October 2016.

[19] 3GPP, "3rd Generation Partnership Project; Technical Specification Group Radio Access Network; NR; User Equipment (UE) radio transmission and reception; Part 1: Range 1 Standalone (Release 15)," 38.101-1, v15.6.0, July 2018.

[20] Intel Corporation, "Study on supporting NR from 52.6 GHz to 71 GHz," RP-193259, 3GPP TSG RAN#86, December 2019.

[21] Qualcomm, "Extending current NR operation to 71GHz," RP-193229, 3GPP TSG RAN#86, December 2019.

[22] Nokia, Alcatel-Lucent Shanghai Bell, NTT DoCoMo, Samsung, Qualcomm, Ericsson, Intel, InterDigital, "WF on phase noise modeling," R1-165685, 3GPP TSG RAN1#85, May 2016.

[23] 3GPP, "3rd Generation Partnership Project; Technical Specification Group Radio Access Network; NR; user equipment (UE) radio transmission and reception; Part 2: Range 2 Standalone (Release 15)," 38.101-2, v15.6.0, July 2018.

[24] Qualcomm Europe, "Specifying blank subframes for efficient support of relays," R1-083817, 3GPP TSG RAN1#54bis, September 2008.
[25] 3GPP, "3rd Generation Partnership Project; Technical Specification Group Radio Access Network; NR; Radio Resource Control (RRC) protocol specification (Release 15)," 38.331, v15.8.0, January 2020.
[26] 3GPP, "3rd Generation Partnership Project; Technical Specification Group Radio Access Network; NR; physical layer procedures for control (Release 15)," 38.213, v15.8.0, January 2020.

CHAPTER 6

Initial Access and Mobility

6.1 Overview

This chapter covers the physical layer procedures associated with initial acquisition and random access. It also covers the aspects related to the transmission of the system information and paging messages. Finally, we present measurements related to mobility and radio link monitoring.

6.2 Initial Access

Initial access is the mechanism for the UE to acquire synchronization, cell identification, and system information of the cell providing NR coverage. The following signals and channels are relevant for initial access:

- **Synchronization signals** (SS) provide OFDM symbol synchronization. They also provide the physical cell ID which is relevant for the detection of control and data channels of the corresponding cell. There are two types of synchronization signals: primary synchronization signal (PSS) and secondary synchronization signal (SSS).
- **Physical broadcast channel** (PBCH) carries the master information block (MIB). PBCH provides the slot and (10 ms) radio-frame synchronization, as well as the identification of the transmitted beam direction. Note that the PBCH has its own reference signals, DMRS, for demodulation at the receiver.
- **Physical downlink control channel** (PDCCH) and **physical downlink shared channel** (PDSCH) carry the system information (SI) in various System Information Blocks (SIBs). Reception of **SIB1** at the UE enables it to perform random access. Note that SIB1 is sometimes called remaining minimum system information (RMSI).

The combination of synchronization signals and PBCH constitute the so-called SS/PBCH block. The SS/PBCH block is transmitted at fixed locations in a half-frame. As discussed in Chap. 1, the sync signals in LTE are transmitted every 5 ms while in NR the default SS/PBCH block periodicity for initial cell selection is 20 ms.

The SS/PBCH block can be transmitted in different beam directions to improve its associated coverage. The maximum number of beam directions [or number of SS/PBCH block (SSB) indices] in which the transmission of the SS/PBCH block can take place depends on the frequency regime (FR1 vs. FR2) and within FR1, on the carrier frequency of the corresponding transmission. In all cases, the transmission of the SS/PBCH in all defined directions is always confined within a half-frame, i.e., within 5 ms.

6.2.1 Synchronization Signal and PBCH Block

The SS/PBCH block is constituted by

- Primary sync signal (PSS)
- Secondary sync signal (SSS)
- Physical broadcast channel (PBCH)

The SS/PBCH block spans four OFDM symbols in time domain. The SS/PBCH block can be transmitted at multiple beam directions in what we designate as multiple sync signal block index (SSB index). The SS/PBCH block is transmitted at fixed locations within a 5-ms half-frame. While it can be transmitted at various periodicities, the UE at initial cell selection shall assume, by default, that the SS/PBCH block is transmitted every 20 ms.

The SS/PBCH block spans 20 RBs in the frequency domain and is transmitted on the sync raster of the corresponding frequency band [1, 2]. The sync raster is a new feature of NR which is discussed in more detail in Chap. 2. The main purpose for the sync raster is to reduce the search complexity of a finer granularity raster, e.g., a coarser granularity raster of 100 kHz is always assumed in LTE.

In LTE systems, the UE implementations rely on a quick frequency scanning to look for transmissions which are always present, e.g., common RS (CRS) energy, in order to narrow down the initial cell search frequency hypotheses. The considerable reduction of always-on signals in NR (Chap. 1) renders frequency scanning inadequate for initial acquisition in NR. Hence, it becomes important in NR to sparse out the raster for initial acquisition and, hence, introduce the sync raster.

Figure 6.1 from Ref. [3] illustrates the SS/PBCH block structure in time and frequency domains. As it can be observed in the figure, there are a number of empty REs around PSS and SSS. Note that no other transmission can take place in the empty REs within the SS/PBCH block.

Table 6.1 provides a high-level overview of some of the properties of the SS/PBCH block.

The location of the SS/PBCH within a frame depends on the transmit SCS. Figure 6.2 illustrates the SS/PBCH block locations for 15 kHz SS/PBCH subcarrier spacing. For FR1 bands below 3 GHz, a maximum number of four SSB indices can be transmitted. For FR1 bands above 3 GHz, a maximum number of eight SSB indices can be transmitted. Note that Fig. 6.2 illustrates the location of the eight SSB indices.

Figure 6.3 illustrates the SS/PBCH block locations for 30 kHz SS/PBCH subcarrier spacing. Note that there are two possible band-dependent mappings.

For the first mapping (top), for FR1 bands below 3 GHz, a maximum number of 4 SSB indices can be transmitted. For FR1 bands above 3 GHz, a maximum number of eight SSB indices can be transmitted.

For the second mapping (bottom), for FR1 FDD bands below 3 GHz, a maximum number of four SSB indices can be transmitted, while for FR1 FDD bands above 3 GHz, a maximum number of eight SSB indices can be transmitted. For FR1 TDD bands below 1.88 GHz, a maximum number of four SSB indices can be transmitted, while for FR1 TDD bands above 1.88 GHz, a maximum number of eight SSB indices can be transmitted.

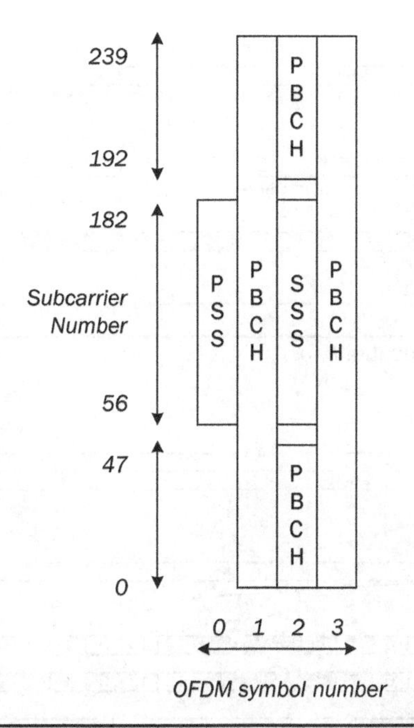

Figure 6.1 SS/PBCH block structure.

Concept	Notes
Block structure	Time: four consecutive symbols [PSS][PBCH][SSS|PBCH][PBCH] Frequency: 20 consecutive RBs
Numerology	FR1: 15 kHz, 30 kHz FR2: 120 kHz, 240 kHz
Frequency raster	Sync raster. Indication of offset from RB grid in PBCH.
Hypotheses on SSB numerology	Single except n5, n41, and n66 [1] which have two
Cyclic Prefix	Normal (NCP)
Antenna ports	Single antenna port (same for entire SS/PBCH block transmission)
Periodicity	{5, 10, 20, 40, 80, 160} ms Default: 20 ms (UE assumed for initial cell selection) Indicated in SIB1
Max number of SSB indices "L_{max}"	Number of beam directions for SSB transmission FR1: L_{max} = 4 or L_{max} = 8 depending on carrier frequency FR2: L_{max} = 64

Table 6.1 High-Level Properties of SS/PBCH Block

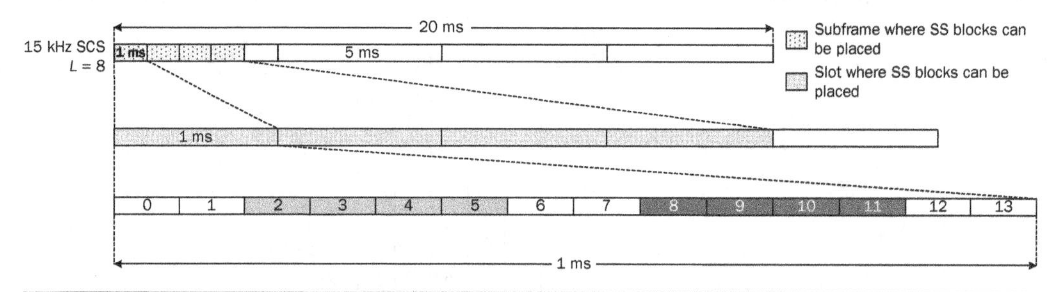

FIGURE 6.2 SS/PBCH block locations for 15 kHz (L_{max} = 8).

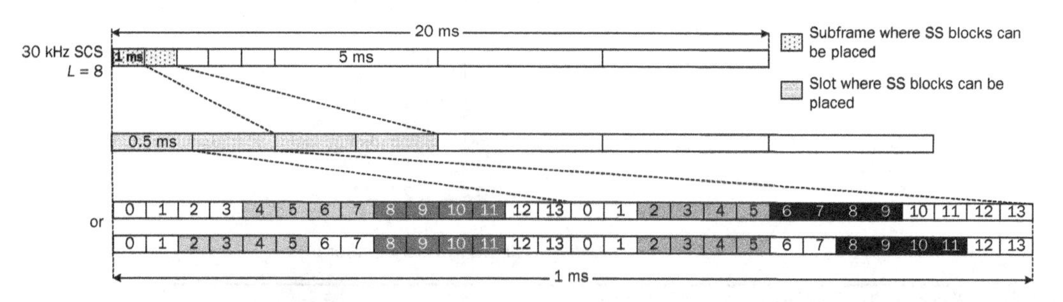

FIGURE 6.3 SS/PBCH block locations for 30 kHz (L_{max} = 8) – two possible mappings.

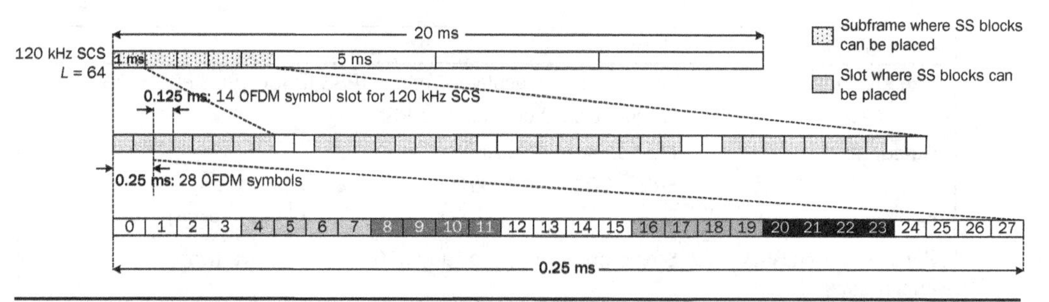

FIGURE 6.4 SS/PBCH block locations for 120 kHz (L_{max} = 64).

Note that Fig. 6.3 illustrates the location of the eight SSB indices.

Figure 6.4 illustrates the SS/PBCH block locations for 120 kHz SS/PBCH subcarrier spacing applicable to FR2 bands and where a maximum of 64 SSB indices can be transmitted.

Figure 6.5 illustrates the SS/PBCH block locations for 240 kHz SS/PBCH subcarrier spacing applicable to FR2 bands and where a maximum of 64 SSB indices can be transmitted.

There can be multiple SS/PBCH blocks transmitted within the bandwidth of a wideband carrier. However, only the *cell defining SS/PBCH block* has an associated SIB1. The frequency location of additional SSBs may or may not be on the sync raster. These additional SSBs can be used for the purpose of measurements.

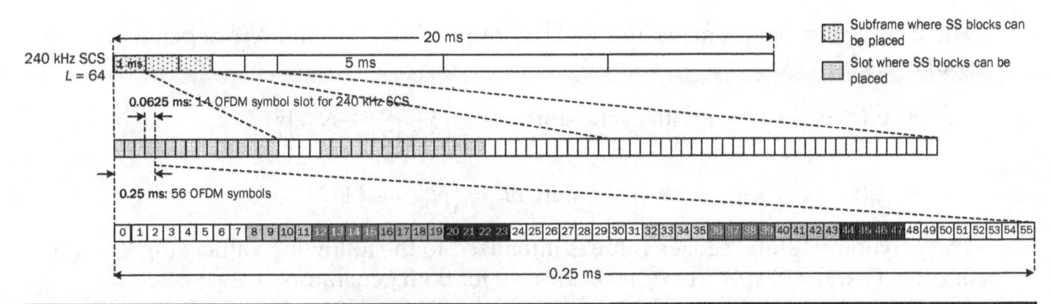

FIGURE 6.5 SS/PBCH block locations for 240 kHz (L_{max} = 64).

In addition, to assist the UEs in gaining OFDM symbol synchronization, the sync signals (SS) provide the physical identity of the cell, cell ID, which is denoted as N_{ID}^{cell} in the 3GPP specifications. The cell ID is relevant because it governs the scrambling of PBCH DMRS and PBCH itself.

Indeed, there are 1008 unique physical cell IDs in NR, unlike LTE where there are 504. The cell ID is defined as $N_{ID}^{cell} = 3N_{ID}^{(1)} + N_{ID}^{(2)}$ with $N_{ID}^{(2)} = \{0,1,2\}$ is carried by PSS and $N_{ID}^{(1)} = \{0,1,...,335\}$ is carried by SSS.

A UE may assume that antenna ports used for transmissions of SS/PBCH blocks with same SSB index are quasi co-located (QCL'd) with respect to spatial, average gain, delay, and Doppler parameters.

By default, a UE may not assume that antenna ports used for transmissions of SS/PBCH blocks with different SSB index are QCL'd with respect to spatial, average gain, delay, and Doppler parameters.

In the next subsections we will provide a few more details about PSS, SSS, and PBCH.

6.2.1.1 Primary Synchronization Signal (PSS)

This signal is an M-sequence with three possible values. As discussed above, PSS provides the parameter $N_{ID}^{(2)} = 0, 1, 2$ which is part of the cell ID ($N_{ID}^{cell} = 3N_{ID}^{(1)} + N_{ID}^{(2)}$).

- PSS sequence length: 127 mapped to consecutive subcarriers
- PSS sequence details: BPSK M-sequence with generator polynomial $g(x) = x^7 + x^4 + 1$
- Three cyclic shifts in frequency domain (0, 43, 86) to get the three PSS signals

The polynomial shift register value is initialized to the following value: x(0) = 0, x(1) = 1, x(2) = 1, x(3) = 0, x(4) = 1, x(5) = 1, x(6) = 1.

6.2.1.2 Secondary Synchronization Signal (SSS)

This signal consists of an M-sequence scrambled with another M-sequence. There is a total number of 1008 SSS signals after scrambling.

- SSS sequence length: 127 mapped to consecutive subcarriers
- SSS sequence details: BPSK M-sequence with scrambling, i.e., two M-sequences XOR'd in binary (0/1) representation or multiplied in decimal (+1/−1) representation.

The 2 generator polynomials and their corresponding, $N_{ID}^{(1)}$ and $N_{ID}^{(2)}$ dependent, cyclic shifts are given by:

- $g_0(x) = x^7 + x^4 + 1$ with cyclic shift $m_0 = \left(3\left\lfloor \frac{N_{ID}^{(1)}}{112} \right\rfloor + N_{ID}^{(2)}\right)5$
- $g_1(x) = x^7 + x + 1$ with cyclic shift $m_1 = \left(N_{ID}^{(1)} \; mod \, 112\right)$

The polynomial shift register value is initialized to the following value: x(0) = 1, x(1) = 0, x(2) = 0, x(3) = 0, x(4) = 0, x(5) = 0, x(6) = 0 for both generators.

SSS carries the entire cell ID information, i.e., $N_{ID}^{cell} = 3N_{ID}^{(1)} + N_{ID}^{(2)}$.

6.2.1.3 Physical Broadcast Channel (PBCH)

The PBCH carries the master information block (MIB) and has the following physical layer properties:

- BW: 20 RBs, i.e., 240 subcarriers
- Time-span: Two full OFDM symbols plus eight RBs on the SSS symbol
- Phase reference: dedicated DMRS (PBCH-DMRS)

Table 6.2 presents the fields of MIB and the contents of PBCH. As can be seen from Table 6.2, a number of bits are appended at the physical layer outside the MIB but nonetheless very relevant for the PBCH transmission and for the entire SS/PBCH block structure and the information it provides.

MIB TTI is 80 ms although it can be transmitted more often than that, e.g., every 20 ms, and the PHY layer fields mapped to PBCH change every radio-frame.

PBCH-DMRS A single, long sequence is mapped to all PBCH-DMRS REs within a SS/PBCH block with QPSK modulation. The PBCH DMRS sequence is generated in the same fashion as the scrambling for other physical layer channels of NR. It uses as Gold Code with underlying polynomials: $x^{31} + x^3 + 1$, $x^{31} + x^3 + x^2 + x + 1$.

The sequence initialization depends on the cell ID and three LSBs of the SSB index. For the case of $L_{max} = 4$, the half-frame indicator, in addition to the 2-bits to indicate the SSB index, is used for the initialization. The remaining bits for SSB index applicable for FR2 are explicitly carried by a field in PBCH as shown in Table 6.2.

The UE may assume the same energy per resource element (EPRE) is used for SSS, PBCH-DMRS, and PBCH-data. On the other hand, PSS EPRE can either have the same EPRE value or be boosted by 3dB.

The DMRS is mapped on every PBCH symbol with 1/4 frequency density and same positions on both full symbols and on the RBs where PBCH is transmitted in SSS symbol. The sequence mapping is frequency-first, time-second in increasing frequency order.

A cell ID–based frequency shift for PBCH-DMRS RE locations is applied, i.e., $v_{shift} = N_{ID}^{cell} \bmod 4$.

PBCH Data PBCH is scrambled twice: before CRC and encoding, and after encoding. The same Gold sequence generation used for other channels is used for both scrambling operations.

Field	MIB (number of bits)	PBCH (number of bits)	Notes
System frame number (SFN)	6	6	6 MSBs of SFN.
Default subcarrier spacing	1	1	SCS used for SIB1, Msg. 2/4 for initial access, paging, and broadcast SI-messages. 15 or 30 kHz for FR1, 60 or 120 kHz for FR2.
SSB subcarrier offset	4	4	MSBs of k_{SSB} which is the frequency domain offset between SSB and the overall RB grid in number of subcarriers (Chap. 2). The value range of this field may be extended by an additional MSB encoded within PBCH (later field). This field may indicate that this cell does not provide SIB1 and that there is hence, no CORESET#0 configured in MIB. In this case, the field PDCCH configuration SIB1 may indicate the frequency positions where the UE may (not) find a SS/PBCH with a CORESET and search space for SIB1.
DMRS Type A position	1	1	Position of 1st DMRS symbol DL & UL (Chaps. 8 and 10).
PDCCH config SIB1	8	8	Time/freq location of CORESET#0: 4-bits. PDCCH monitoring occasions: 4-bits. If the field SSB subcarrier offset indicates that SIB1 is absent, this field indicates the frequency positions where the UE may find SS/PBCH block with SIB1 or the frequency range where the network does not provide SS/PBCH block with SIB1.
Cell barred	1	1	Barred, not barred.
Intra frequency reselection	1	1	Controls cell selection/reselection to intra-frequency cells when the highest ranked cell is barred.
Spare	1	1	Reserved bit for future use.
Future extension	1	1	Bit for possible future MIB extension.
Physical layer added fields:			
SFN LSBs		4	To obtain 10-bit SFN.
Half-frame index		1	First or second half-frame in radio-frame.
Additional PHY layer info		3	If L_{max} = 64, three MSBs of SSB index. Note that PBCH DMRS scrambling carries three LSBs of SSB index. Else, 1st bit is MSB of k_{SSB}, 2nd/3rd bits are reserved.
CRC		24	
Duration			
Total #Bits	24	56	

TABLE 6.2 MIB and PBCH Contents [4, 5, 6]

The following info bit mapping before 1st PBCH scrambling and CRC encoding is applied to PBCH:

- Let $a_0, a_1, a_2, \ldots, a_{31}$ denote the input bits to 1st PBCH scrambling.
- Timing related bits, ($s_9, s_8, s_7, s_6, s_5, s_4, s_3, s_2, s_1, s_0, c_0, b_5, b_4, b_3$), are mapped to ($a_{16}, a_{23}, a_{18}, a_{17}, a_8, a_{30}, a_{10}, a_6, a_{24}, a_7, a_0, a_5, a_3, a_2$), respectively, where s_i indicates SFN bits, c_0 indicates the half-frame indicator, and b_i indicates the 3-bits denoted as *Additional PHY layer info* in Table 6.2.
- The remaining info bits are mapped to ($a_1, a_4, a_9, a_{11}, a_{12}, a_{13}, a_{14}, a_{15}, a_{19}, a_{20}, a_{21}, a_{22}, a_{25}, a_{26}, a_{27}, a_{28}, a_{29}, a_{31}$).

D-CRC interleaving effect is taken into account in the above info bit mapping design.

The 1st scrambling, initialized by $c_{init} = N_{ID}^{cell}$, is applied to PBCH payload prior to CRC and the encoding process excluding the following fields: SS/PBCH block index, half radio-frame, and 2nd and 3rd LSBs of SFN [5].

PBCH is appended a 24-bit CRC increasing the overall payload size from 32- to 56-bits before encoding.

Polar coding is applied to PBCH using the same coding scheme as for PDCCH (Chap. 7).

The 2nd scrambling, also initialized by $c_{init} = N_{ID}^{cell}$, is applied to the PBCH after CRC attachment and encoding [7].

The PBCH coded bits are mapped across REs in the three symbols with PBCH.

Note that PBCH content, except the SSB index, is the same for all SS/PBCH blocks within a half-frame for the same center frequency. Combining PBCH at the receiver is assumed to be a UE implementation.

SS/PBCH block enables UEs to acquire time synchronization at OFDM symbol, slot, and radio-frame level. Table 6.3 provides a high-level view about how each level of synchronization is obtained at the UE receiver.

6.2.2 SIB1

The SIB1 in NR is equivalent to SIB1+SIB2 of LTE. SIB1 is also known as the remaining minimum system information (RMSI) as it will provide UEs with sufficient information for them to be able to access the system. The PDSCH carrying the SIB1 is scheduled by PDCCH.

As described in Sec. 6.2.1, PBCH provides the configuration information for that PDCCH providing the necessary information for the corresponding CORESET (Type0-PDCCH CSS [6]) and the search space therein. PDCCH scheduling SIB1 and PDSCH carrying SIB1 are on the RB grid. The parameters k_{SSB}, *PDCCH config SIB1*, and *Default*

Level	Notes
OFDM symbol	PSS
Slot	PBCH DMRS (3 LSBs of SSB index known)
Half-frame	PBCH DMRS if $L_{max} \leq 4$ PBCH payload else
Radio-frame	PBCH payload (SFN in PBCH)

Table 6.3 Time Synchronization Levels

Subcarrier Spacing provided in PBCH enable UEs to detect the PDCCH in the so-called CORESET#0 with the corresponding SCS.

6.2.2.1 SIB1 CORESET Configuration

SS/PBCH block and SIB1 multiplexing can take one of the following forms:

"**Pattern 1**" refers to the multiplexing pattern where SS/PBCH block and SIB1 CORESET occur at **different** time instances (TDM), and SS/PBCH block transmit bandwidth and initial active DL BWP containing SIB1 CORESET overlap.

"**Pattern 2**" refers to the multiplexing pattern where SS/PBCH block and RMSI CORESET have different numerologies, SS/PBCH block and RMSI PDSCH occur at the **same** time instances, and SS/PBCH block transmit bandwidth and initial active DL BWP containing SIB1 CORESET do **not overlap** (FDM).

- Applicable for {SSB SCS, SIB1 PDSCH SCS} = {120, 60}, {240, 120} kHz.
- The starting symbol index R for the RMSI CORESET monitoring window occurs earlier than the SSB symbols in the same slot or one slot before.
- The duration of the monitoring window is 1 slot.

"**Pattern 3**" refers to the multiplexing pattern where SS/PBCH block and RMSI CORESET have the same numerology, occur at **same** time instance, and SS/PBCH block transmit bandwidth and initial active DL BWP containing SIB1 CORESET do **not overlap** (FDM)

- Applicable for {SSB SCS, RMSI PDSCH SCS} = {120, 120} kHz. Two symbols for PDCCH, two symbols for PDSCH.
- The starting symbol index R for the RMSI CORESET monitoring window is the same as the starting symbol of the SSB.
- The duration of the monitoring window is 1 slot.

The three SS/PBCH block and SIB1 multiplexing patterns are illustrated in Fig. 6.6.

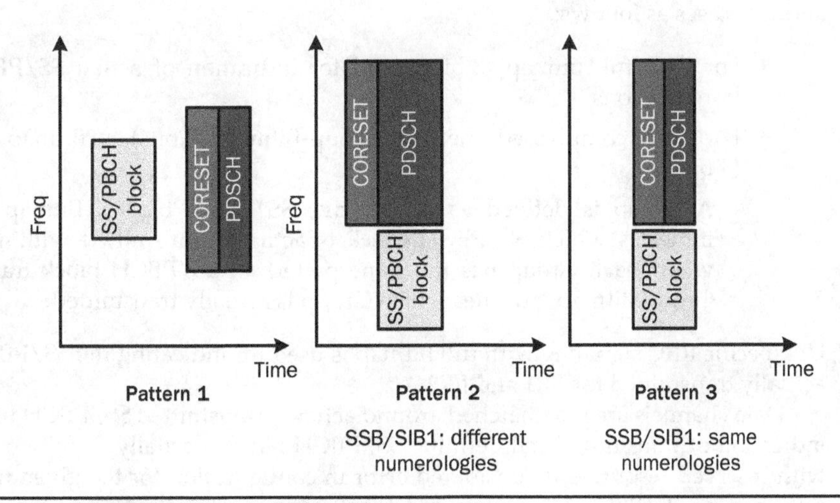

FIGURE 6.6 SS/PBCH block and SIB1 multiplexing patterns.

The SIB1 CORESET configuration is provided as part of the MIB field "*PDCCH config SIB1*" with 4-bits. Multiple 16-entry tables are defined in Ref. [6] for various combinations of SCS for SS/PBCH block and SIB1, as well as for some tables, the applicable frequency band's corresponding minimum channel bandwidth (e.g., 5/10 MHz or 40 MHz). These tables indicate the SIB1 CORESET configuration in terms of:

- Pattern number (1, 2, or 3).
- CORESET BW in number of PRBs (24, 48, 96).
- CORESET time-duration in number of OFDM symbols (1, 2, or 3).
- Frequency offset in number of PRBs defined as the frequency difference from the lowest PRB of RMSI and the lowest PRB of SS/PBCH block.
  - Note: The offset in subcarriers between the edge of the SS/PBCH block and SIB1 CORESET PRB grid is indicated by k_{SSB} in PBCH (5-bits for FR1 and 4-bits for FR2).

FR1 only uses Pattern 1 multiplexing (TDM). FR2 has SS/PBCH block and SIB1 multiplexing configurations for Pattern 1, 2, and 3.

There is an RMSI PDCCH monitoring window associated with an SSB recurring periodically. Each window has duration of x consecutive slots. The period, y, of the monitoring window can be the same as or different from the period of the SSB burst set. When the SS/PBCH blocks and corresponding RMSI CORESETs occur in different time instances, the UE assumes that the SIB1 CORESET monitoring window corresponding to an SS/PBCH block in the radio-frame satisfying the condition mod(SFN,2) = 0. Note: RMSI scheduling periodicity is up to the gNB implementation. The monitoring window is associated with one SS/PBCH block in a burst set.

6.2.2.2 SIB1 SSB Transmission Indication

An indication of the SS/PBCH blocks actually transmitted is sent in SIB1 for both FR1 and FR2 cases as follows:

- For **FR1**: full bitmap (8-bits) used for indication of actual SS/PBCH block transmissions
- For **FR2**: compressed method Group-Bitmap (8-bits) + Bitmap in Group (8-bits)
  - A Group is defined as consecutive SS/PBCH blocks. Bitmap in Group indicates which SS/PBCH block is actually transmitted within a Group, where each Group has the same pattern of SS/PBCH block transmission, Group-Bitmap indicates which Group is actually transmitted.

UE-specific RRC signaling with full bitmap is used for indicating the SS/PBCH blocks actually transmitted for FR1 and FR2.

Data channels are rate matched around actually transmitted SS/PBCH blocks. gNB indication of transmitted cell-defining SS/PBCH blocks partially or fully overlapping with reserved resources is considered error in configuration for the given transmitted cell-defining SS/PBCH blocks and no UE behavior is specified.

6.2.3 Other System Information

The broadcast delivery of other system information (OSI) is supported by PDSCH. The scheduling information of broadcast PDSCH is carried by PDCCH. The numerology for on-demand OSI via broadcast delivery is the same as used for SIB1.

For broadcast OSI CORESET, the SI monitoring window configuration, e.g., time offset, duration, periodicity, is explicitly signaled in corresponding SIB1. For broadcast OSI CORESET configuration, we reuse the same configuration as for SIB1 CORESET indicated in PBCH. PDCCH configuration which gives search space configuration includes monitoring occasions within the SI monitoring window (same PDCCH configuration assumed for all SIs in Release 15).

For non-broadcast, on-demand (i.e., dedicated) OSI transmission for connected mode UEs, scheduling is up to gNB implementation, i.e., no specific handling for non-broadcast on-demand (i.e., dedicated) OSI CORESET.

6.3 Random Access

The random access procedure enables the UE to access the system. Release 15 defined random access procedure with the classical 4-step procedure depicted in Fig. 6.7. Note that Release 16 had a project on 2-step RACH which is discussed in some depth in Chap. 17.

In this section we will introduce the 4-step random access procedure and will dwell into each of the steps. Specifically, we will cover the details of the physical random access channel (PRACH) which takes the form of a random access preamble.

We will also briefly discuss the concept of scheduling request (SR) as a form for the UEs to access the UL with dedicated resources when up-to-date timing advance is available. The SR is transmitted via the PUCCH, which is discussed in detail in Chap. 9.

We introduce the following acronyms, which will be heavily used in this section:

- CBRA: Contention-based random access
- CFRA: Contention-free random access

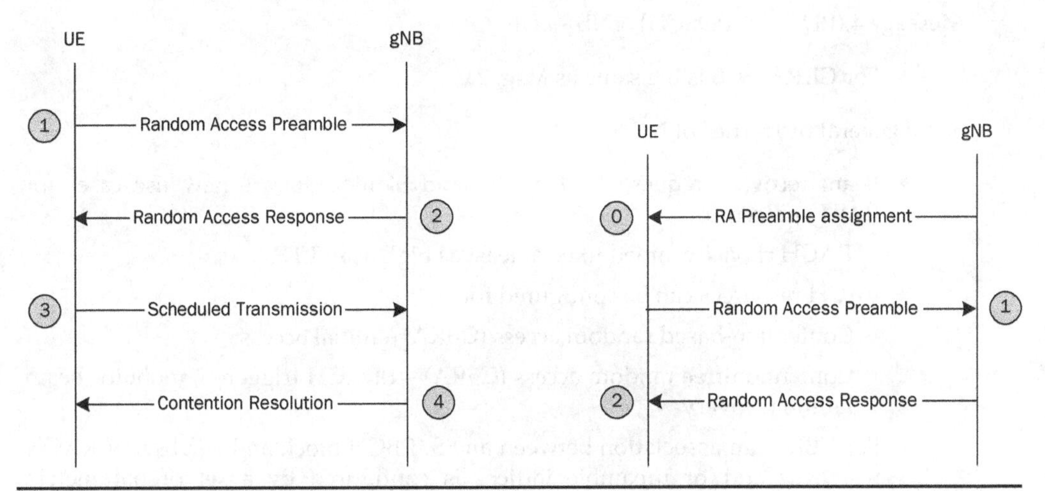

FIGURE 6.7 4-step CBRA (left) and CFRA (right) random access procedure.

6.3.1 Random Access Procedure

Release 15 defined 4-message–based random access procedure for NR as in LTE. Note that Release 16 introduces 2-step random access procedure as introduced in Chap. 17.

Message 1 (RACH preamble): UE → gNB

- Transmitted without applying any Timing Advance (TA)
- Zadoff-Chu sequence based with two sequence lengths (L):
  - Long sequences: $L = 839$: SCS = {1.25, 5} kHz.
  - Short sequences: $L = 139$: SCS = {15, 30, 60, 120} kHz. Note that longer sequences are introduced in Release 16 motivated by NR-U as discussed in Chap. 12.
  - For long/short sequence, root Zadoff-Chu sequence order (logical to physical root mapping) as in LTE given by [8 (Tables 5.7.2-4 and 5.7.2-5)].
- For contention-based RA (CBRA): SCS configured in RACH configuration (SIB1)
- For contention-free RA (CFRA): For PDCCH-triggered CFRA, SCS is same as for CBRA (SIB1 configured). For mobility CFRA, SCS provided in Handover command.

Message 2 (Random access response: RAR on PDCCH/PDSCH): gNB → UE

- For **CBRA**: SCS is the same as used in SIB1.
- For **CFRA**: SCS provided in Handover command.

Message 3 (first PUSCH transmission): UE → gNB

- For **CBRA**: SCS is configured in the RACH configuration (SIB1) separately from SCS for Msg. 1 (1-bit). In FR1, SCS of Msg. 3 can be either 15 or 30 kHz while in FR2, SCS of Msg. 3 can be either 60 or 120 kHz.

Message 4 (PDCCH/PDSCH): gNB → UE

- For **CBRA**: SCS is the same as Msg. 2.

Some general properties of NR RACH:

- Beam recovery requests and on-demand SI requests are new use cases for RACH procedure.
  - RACH capacity aimed to be at least as high as in LTE.
- RACH resources can be configured for
  - Contention-based random access (CBRA) – initial access
  - Contention-free random access (CFRA) – PDCCH triggered, mobility, beam failure recovery
- For CBRA, an association between an SS/PBCH block and a subset of RACH resources and/or preamble indices is configured by a set of parameters in SIB1.

- Same number of PRACH preambles for all actually transmitted SS/PBCH blocks:
  - 4 * N (where $1 \leq N \leq 16$). Preamble indices not used for CBRA in a RACH occasion can be reserved for CFRA as in LTE.
- After the UE selects one PRACH transmission occasion for Msg. 1 transmission, the UE is not allowed to select another one before the expiration of RAR window for the same Msg. 1 transmission in Release 15.

6.3.2 Preamble Formats

6.3.2.1 Long-Sequence-Based (L = 839)

Long-sequence–based preambles are formats enabling long transmissions (1 ms or longer) with a relatively narrow bandwidth. Table 6.4 presents the relevant parameters of the preamble formats, namely, Format 0, 1, 2, and 3, based on long sequences of length 839.

Figure 6.8 illustrates the various long-sequence–based PRACH preambles showing the span in time and frequency.

6.3.2.2 Short-Sequence-Based (L = 139)

Short-sequence–based preambles are formats enabling short transmissions of 2, 4, 6, or 12 OFDM symbols with CP aggregated at the beginning of the burst and with or without guard time (GT) at the end. Table 6.5 presents the relevant parameters of the preamble formats, namely, Format A1, A2, A3, B1, B2, B3, B4, C0, and C2 based on short sequences of length 139 for SCS = 15 kHz. For 30, 60, and 120 kHz subcarrier spacing, the preamble format can be scaled according to the subcarrier spacing. $T_s = 1/(2*30720)$ ms for 30 kHz subcarrier spacing, $T_s = 1/(4*30720)$ ms for 60 kHz subcarrier spacing, $T_s = 1/(8*30720)$ ms for 120 kHz subcarrier spacing. Note that some of the formats may not be applicable to all subcarrier spacings.

Concept	Format 0	Format 1	Format 2	Format 3
L	839	839	839	839
N	864	864	864	864
SCS (kHz)	1.25	1.25	1.25	5
Bandwidth (MHz)	1.08	1.08	1.08	4.32
1 seq length (ms)	0.8	0.8	0.8	0.2
N_{SEQ}	1	2	4	4
Tot seq length (ms)	0.8	1.6	3.2	0.8
T_{CP} (ms)	0.103	0.684	0.153	0.103
T_{GT} (ms)	0.097	0.713	0.147	0.097
Total length (ms)	1	3	3.5	1
Use case	LTE refarming	Large cell (100 km)	Coverage enhancement	High-speed case

TABLE 6.4 Long-Sequence–Based Preamble Formats (*L* = 839)

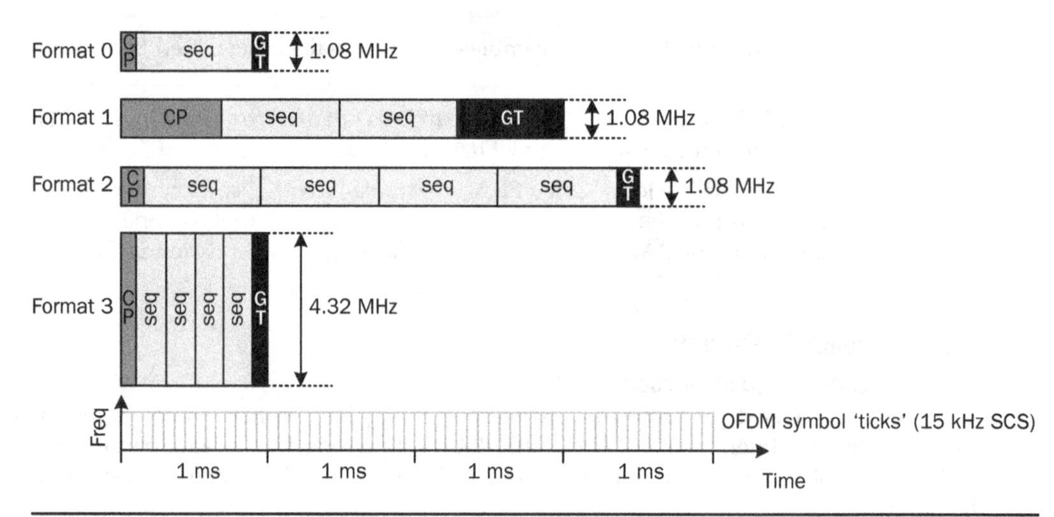

FIGURE 6.8 Illustration of long-sequence–based PRACH preambles.

Concept	A1	A2	A3	B1	B2*	B3*	B4	C0	C2
L	139	139	139	139	139	139	139	139	139
N	144	144	144	144	144	144	144	144	144
SCS (kHz)	15	15	15	15	15	15	15	15	15
Bandwidth (MHz)	2.16	2.16	2.16	2.16	2.16	2.16	2.16	2.16	2.16
1 seq length (μs)	66.67	66.67	66.67	66.67	66.67	66.67	66.67	66.67	66.67
N_{SEQ}	2	4	6	2	4	6	12	1	4
Tot seq length (μs)	133.33	266.67	400	133.33	266.67	400	800	66.67	266.67
T_{CP} (μs)	9.375	18.75	28.125	7.031	11.719	16.406	30.469	40.36	66.67
T_{GT} (μs)	0	0	0	2.344	7.031	11.719	25.781	35.677	94.922
Total length (μs)	142.71	285.42	428.13	142.71	285.42	428.125	826.25	142.71	428.26
# OFDM symbols	2	4	6	2	4	6	12	2	6

Note: (*) Formats B2 and B3 are only supported in combination with formats A as follows: A1/B1, A2/B2, A3/B3.

TABLE 6.5 Short-Sequence–Based Preamble Formats (L = 139)

Further, note that additional preambles of this kind are introduced in Release 16 motivated by NR-U (wider bandwidth transmission) as discussed in Chap. 12.

Figure 6.9 illustrates the various short-sequence–based PRACH preambles for 15 kHz SCS showing the span in time and frequency.

PRACH preambles are aligned with OFDM symbol boundaries for data with same numerology. Additional 16Ts for every 0.5 ms should be included in T_{CP} when RACH preamble is transmitted across 0.5-ms boundary or from 0.5-ms boundary. For format A,

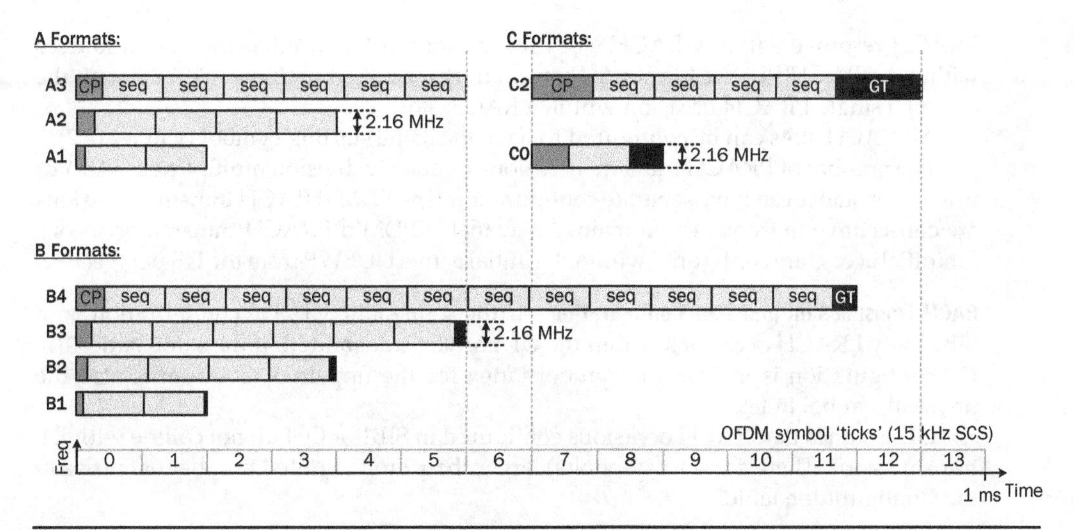

Figure 6.9 Illustration of short-sequence–based PRACH preambles.

GP can be defined within the last RACH preamble among consecutively transmitted RACH preambles.

6.3.2.3 RACH Configuration

Random access (RA) configuration is included in SIB1.

- All RA config information is broadcasted in all beams used for SIB1 within a cell, i.e., SIB1 information is common for all beams.
- For initial access, either long-sequence–based preamble or short-sequence–based preamble is configured.
- RACH configurations are specified using a table similar to LTE, this table is indexed by the *PRACH Config Index*. Frequency multiplexed PRACH transmission occasions use the same *PRACH Config Index*. The pattern given by the *PRACH Config Index* repeats every RACH Configuration Period. RACH Configuration Period values supported are 10, 20, 40, 80, and 160 ms.

The slot duration for PRACH resource mapping for

- short preamble formats (i.e., $L = 139$) is based on the RACH Msg. 1 numerology, i.e., SCS.
- long preamble formats (i.e., $L = 839$) is based on 15 kHz SCS, i.e., 1 ms.

RMSI can indicate PDCCH configuration giving search space configuration for RACH procedure (before RRC connection setup is complete).

A UE is not expected to monitor more than one Msg. 2/Msg. 3/Msg. 4 search space in one slot. Starting symbol of Msg. 2/Msg. 3/Msg. 4 search space is the same in every slot.

NR defines the pattern of slots that contain PRACH resources into a larger time interval. Consecutive RACH resources can be configured within a RACH slot.

One PRACH format is configured for a cell. For PRACH formats based on short sequence length: formats A&B are considered a single format, i.e., PRACH configuration, configures either format A/B or format C. If format A/B is configured, the last

PRACH resource within a RACH slot uses the format B and other PRACH resources within the RACH slot use format A. Support only format B4 within a RACH slot, in the case of a single PRACH occasion within a RACH slot.

All RACH slots can be configured to have the same starting symbol (values: 0, 2).

The number of PRACH transmit occasions frequency division multiplexed (FDM'd) in a time instance can have separate configuration. The FDM'd RACH transmit occasions are consecutive in frequency domain. All available FDM'd PRACH transmit occasions for initial access are configured within the initial active UL BWP from the UE perspective.

RACH Transmission and TDD Configuration If the semi-static UL/DL configuration is in SIB1, only PRACH occasions within the UL part are transmitted. If the semi-static UL/DL configuration is in OSI, the symbol index for the unpaired spectrum is also the physical symbol index.

UE assumes that RACH occasions configured in SIB1 or OSI do not collide with DL transmissions. There are start symbol(s) larger than 2 for a limited number of entries in the Configuration table.

RACH Procedure (Multi-Beam Handling) The association between one or multiple occasions for SSB and a subset of RACH resources and/or subset of preamble indices is communicated to the UE via broadcast SI.

Preamble indices for CBRA and CFRA are mapped consecutively for one SSB in one RACH transmit occasion. The association of CFRA preambles with SSBs can be reconfigured through UE-specific RRC signaling. Note: this does not preclude the gNB to possibly configure that the number of CFRA preambles per RACH occasion is smaller than the number of actually transmitted SSBs configured in SIB1.

NR supports gNB configuring the number of preambles for actually transmitted SS/PBCH blocks. The gNB configures in SIB1 the number of CBRA preambles per SSB per RACH transmit occasion and the number of SSBs per RACH occasion.

The number of PRACH transmission occasions in the RACH config period which are associated (by the SSB to RACH resource association rule) can be different than the number of actual transmitted SSBs in the SS burst set period.

The mapping from actually transmitted SS/PBCH blocks to RACH occasion/preamble index is in increasing order of preamble indices in a single RACH occasion, then in increasing order of frequency multiplexed RACH occasions, then in increasing order of the number of time multiplexed RACH occasions within a RACH slot, and then in increasing order of the number of RACH slots.

For initial access, the threshold for SS/PBCH block selection for RACH resource association is configurable by network, where the threshold is based on RSRP (*RSRPThreshold-SSB*). The value rage of RSRPThreshold-CSI-RS is the same as *RSRPThershold-SSB*.

RACH configuration maps RACH resources onto slots irrespective of the time locations of actually transmitted SSBs for FDD and TDD.

6.3.3 Additional Information about RACH Message 1

For every time period, the first actually transmitted SSB index is mapped to the first PRACH occasion.

Assuming N preamble indices are available in one RACH transmission occasion, if only one SSB is mapped to only one RACH transmission occasion, each RACH transmission occasion has preamble index 0 to N-1. For Zadoff-Chu type RACH preamble sequence, RACH preamble indices are in order of increasing cyclic shifts of a root sequence with logical root index and then increasing logical root index.

If multiple SSBs are mapped to one RACH transmission occasion, the preamble indices for CBRA for each SS/PBCH block are mapped consecutively.

The relative frequency offset of Msg. 1 defines the offset of the lowest PRACH transmission occasion in frequency domain with respective to PRB 0 of the initial active UL BWP(s).

6.3.4 Additional Information about RACH Message 2 (RAR)

At least for initial access, the PDSCHs for Msg. 2 and Msg. 4 are confined within UE minimum DL BW for a given frequency band.

As in LTE, the PRACH preamble is identified via random access preamble identifier (RAPID). The bit field length of RAPID is fixed in the specifications to 6-bits. The preambles contained within each RACH transmission occasion are indexed from 0 to 63, which is also used for RAPID.

Regarding timing advance (TA) in Msg. 2, the maximum size of TA command for RAR is 12-bits. The granularity of timing advance in RAR depends on the subcarrier spacing of the first uplink transmission after RAR, i.e., Msg. 3.

The maximum size of TA command for MAC-CE is 6-bits. For the timing advance in MAC-CE, the granularity depends on the SCS of the UL BWP in the TA Group (TAG) that TA in MAC-CE applies to.

6.3.5 RACH Message 3

Msg. 3 is scheduled by the uplink grant in RAR. The UE adjusts its power setting for Msg. 3 using the TPC command in Msg. 2 and the transmit power of the latest PRACH preamble. NR supports asynchronous adaptive HARQ retransmissions for Msg. 3. The HARQ process ID for Msg. 3 PUSCH re-transmission is fixed in the specification to a value of 0.

Msg. 3 is transmitted after a minimum time gap from the end of Msg. 2 over-the-air reception. gNB has the flexibility to schedule the transmission time of Msg. 3 while ensuring the minimum time gap.

The minimum time gap between Msg. 2 and Msg. 3 assuming Msg. 2 and Msg. 3 have the same numerology is N1 + N2 + L2 + TA, where N1 refers to ACK turnaround time with front loaded plus additional DMRS and UE capability #1, N2 refers to UL-grant turnaround time for UE capability #1, TA = max timing advance value that the 12-bit TA command can provide with respect to the SCS of Msg. 3. L2 refers to the MAC processing latency (L2 = 500 μs) and does not depend on SCS.

The network signals the waveform for Msg. 3 in SIB1 as one bit. Note that the waveform for Msg. 3 can be DFT-S-OFDM or CP-OFDM.

SIB1 or handover command provides the SCS of Msg. 3 depending on the reason for the transmission.

6.3.6 Other RACH Details

RACH QCL Relationships The UE may assume the following:

- DMRS of PDCCH and DMRS of PDSCH conveying Msg. 2 are QCL'd with the SS/PBCH block that the UE selected for RACH association and transmission.

- DMRS of PDCCH conveying Msg. 3 re-transmission grant is QCL'd with the SS/PBCH block that the UE selected for RACH association and transmission.
- DMRS of PDCCH and DMRS of PDSCH conveying Msg. 4 are QCL'd with the SS/PBCH block that the UE selected for RACH association and transmission.

Slot/Mini-Slot-Based Transmission of Msg. 2/3/4 NR supports both slot-based PDCCH, PDSCH, and PUSCH, and mini-slot–based PDSCH/PUSCH transmissions for Msg. 2, Msg. 3, and Msg. 4. For the mini-slot–based transmission, 2, 4, and 7 OFDM-symbol durations for PDSCH/PUSCH are supported in Release 15.

6.3.7 Scheduling Request (SR)

Used by synchronized UEs in RRC connected state to access the UL on dedicated resources. SR in NR has the following properties:

- Single-bit transmitted on PUCCH format 0 (short-duration) or PUCCH format 1 (long-duration) as in LTE.
- Note: Up to 12 SRs per PRB can be semi-statically configured.
- In case of SR only, the physical layer can only transmit one SR at any given time.
- Multiplexing of SR and HARQ feedback is supported on short and long PUCCH formats.
- For each SR configuration, the following is indicated via RRC:
  - Periodicity and offset which identify the slots/symbols to be used for SR
  - One configured SR can be associated with either short or long PUCCH
- Periodicities for SR resources are summarized in Table 6.6.

6.4 Paging

For paging, the same subcarrier spacing is used for data and control channels. The subcarrier spacing of data and control channel for paging transmission is the same as SIB1 (default SCS indicated in MIB).

Subcarrier spacing (kHz)	Supported periodicities [#OFDM symbols in given numerology] or [ms]
15	2 symbols, 7 symbols, 1, 2, 5, 10, 20, 40, 80
30	2 symbols, 7 symbols, 0.5, 1, 2, 5, 10, 20, 40, 80
60	2 symbols, 7 symbols (6 symbols for ECP), 0.25, 0.5, 1, 2, 5, 10, 20, 40, 80
120	2 symbols, 7 symbols, 0.125, 0.25, 0.5, 1, 2, 5, 10, 20, 40, 80

TABLE 6.6 PUCCH-Supported Periodicities

The following parameters for paging are explicitly signaled in the corresponding OSI/SIB1:

- Paging occasion configuration, e.g., time offset, duration, periodicity
- PDCCH configuration which gives search space configuration including monitoring occasions within the paging occasion

For paging CORESET configuration, we reuse the same configuration as for SIB1 CORESET indicated in PBCH.

6.4.1 Paging Mechanism

Paging scheduling DCI and paging Message may be sent in the same slot. NR supports paging mechanism sending short messages, e.g., *systemInfoModification, etwsAndCmasIndication*. Short messages can be transmitted on PDCCH using P-RNTI with or without associated paging message using short message field (8-bits) in DCI format 1_0.

In Release 15, 2-bits are designated for indicating *systemInfoModification*, i.e., indication of a BCCH modification other than SIB6, SIB7, and SIB8, and *etwsAndCmasIndication*, i.e., indication of an earthquake and tsunami warning system (ETWS) primary notification and/or an ETWS secondary notification and/or a commercial mobile alert system (CMAS) notification.

6.4.2 Other Paging Characteristics

QCL Assumptions The UE may assume QCL between SSBs, Paging DCIs and Paging Messages. UEs are not required to soft combine multiple Paging DCIs within one Paging Occasion (PO).

Mini-Slot-Based SIB1, Broadcast OSI, and Paging NR supports both slot-based PDCCH/PDSCH, and mini-slot–based PDSCH transmission for SIB1, broadcast OSI, and paging. For the mini-slot–based transmission, 2, 4, and 7 OFDM-symbol durations for the transmission of SIB1, broadcast OSI, and paging PDSCH are supported in Release 15.

6.5 Mobility

NR mobility is very much the same as for LTE. Release 15 NR mobility is based on the traditional single serving cell framework. The UE carries out radio resource management (RRM) measurements and sends reports based on various mobility triggers. The source cell manages the handover to a target cell. Upon indication from the source cell, the UE attempts to access the target cell via random access (CFRA) and the reception of the confirmation from the target cell successfully completes the handover procedure.

Note that a project to enhance NR mobility has been carried out in Release 16, as presented and discussed in Chap. 17.

At least for handover case, a source cell can indicate in the handover command.

- Association between RACH resources and CSI-RS configurations.
- Association between RACH resources and SS/PBCH blocks.
- A set of dedicated RACH resources.
- Note that above CSI-RS configuration is UE-specifically configured.

6.5.1 Radio Resource Management (RRM)

RRM is based on DL measurements as 3G and 4G. RRM can be based on **SS/PBCH block** measurements or **CSI-RS** measurements. The following RRM-related measurements are defined in Release 15:

- SS/PBCH block RSRP, CSI-RS RSRP
- SS/PBCH block RSRQ, CSI-RS RSRQ
- SS/PBCH block SINR, CSI-RS SINR

It is up to UE implementation how to select a set of receive beams to perform RRM measurements on a carrier with the following guidelines:

- Different sets of receiving beams can be used in measurements for different measurement objects.
- Same set of receiving beams shall be used in measurement of each transmit beam for a measurement object.
- If receiver diversity is in use by the UE, the reported *measurement quantity* (i.e., RSRP, RSRQ, SINR) value shall not be lower than the corresponding *measurement quantity* of any of the individual receiver branches.

6.5.1.1 SS/PBCH Block Based RRM

SS/PBCH block based RRM is configured via the SS/PBCH block measurement timing configuration (SMTC):

- SMTC window duration = {1, 2, 3, 4, 5} ms.
- SMTC window timing offset = {0, 1, …, SMTC periodicity −1} ms.
  - SMTC window timing reference for the timing offset is SFN#0 of the corresponding serving cell (or cell UE is camped on in RRC Idle).
- SMTC periodicity = {5, 10, 20, 40, 80, 160} ms.
- For intra-frequency measurements, SMTC configuration is signaled in SIB2 for IDLE mode, and RRC for connected mode.
- For inter-frequency measurements, SMTC configuration is signaled per frequency, in SIB4 in cell UE is camped on for IDLE mode, and RRC for connected mode.

Multiple SSB-based RRM measurement for a wide-band carrier is not supported in Release 15.

Measurements in Idle Mode No network indication of set of SSBs to be measured, i.e., all SSBs within SMTC measurement duration are used. Single SMTC is configured per frequency carrier.

Measurements in Connected Mode The network indicates a set of SS/PBCH blocks to be measured within the SMTC measurement duration. The indication is per frequency layer and the UE is not required to measure SS/PBCH blocks not indicated as transmitted.

RRC signaling via full bitmap applicable to both intra- and inter-frequency measurements. If there is no indication, the default value is that all SS/PBCH blocks within the SMTC measurement duration are to be measured.

For intra-frequency measurement, up to two SMTC periodicities can be configured. The UE can be informed of which cells are associated with which measurement window periodicity. For not listed cells, longer measurement window periodicity is used. Single measurement window offset and duration are configured per frequency carrier.

For inter-frequency measurements, a single SMTC periodicity is configured per frequency carrier. At least for inter-frequency measurement, SS/PBCH block–based RSSI and interference measurement resources are confined within at most the measurement gap duration, e.g., 6 ms.

6.5.1.2 CSI-RS Based RRM

The CSI-RS for L3 mobility is separately configured from that for beam management. The CSI-RS framework is based on the CSI-RS framework for beam management (Chap. 8). Only single port and periodic CSI-RS resources are configured. The periodicity is configurable.

For each CSI-RS resource, at most one associated SSB can be configured. If associated SSBs are configured for CSI-RS, maximum N1 = 96 CSI-RS resources can be configured per frequency layer. $M \geq 1$ number of CSI-RS resources per associated SSB can be configured. If associated SSBs are not configured for CSI-RS, a maximum $N2 \geq 1$ CSI-RS resources can be configured per frequency layer. In this case, UE may assume that the carrier is synchronized with the serving cell.

A UE is not required to measure CSI-RS configured for L3 mobility outside the **active time where** the **active time** relates to the time when UE is monitoring PDCCH in *onDuration* or due to any timer triggered by gNB activity, i.e., when any of *onDurationTimer*, *drx-InactitivityTimer*, or *drx-RetransmissionTimer* is running.

6.5.1.3 PHY Measurements [9]

Reference Signal Received Power (RSRP)

SS/PBCH block RSRP (SS-RSRP):

- It is up to UE to use SSS only or SSS + PBCH DMRS for RSRP measurement
- Applicable for RRC_IDLE intra-frequency, RRC_IDLE inter-frequency, RRC_INACTIVE intra-frequency, RRC_INACTIVE inter-frequency, RRC_CONNECTED intra-frequency, RRC_CONNECTED inter-frequency

CSI-RS RSRP (CSI-RSRP):

- Measured RSRP from CSI-RS configured for RRM in CONNECTED mode
- CSI-RS RSRP measurement quantity should be defined per configured CSI-RS port
- Applicable for RRC_CONNECTED intra-frequency and RRC_CONNECTED inter-frequency

Received Signal Strength Indicator (RSSI)

NR carrier RSSI:

NR carrier RSSI comprises the linear average of the total received power [in (W)] observed only in certain OFDM symbols of measurement time resources, in the measurement bandwidth over N RBs from all sources, including co-channel serving and non-serving cells, adjacent channel interference, thermal noise, etc.

- Measurement time resources are confined within SMTC window durations.
- In Release 15, the same measurement bandwidth for SS-RSSI and SS-RSRP is assumed.
- A set of slots for RSSI time-domain measurement resource can be explicitly configured per frequency carrier by OSI for IDLE, by RRC for CONNECTED.
- In CONNECTED mode, the set of slots (based on SSB numerology) is configured by a bitmap with each bit corresponding to each slot of the slots within the SMTC window.
- A set of OFDM symbols in the configured slot are used from symbol 0 to a configured ending symbol.

CSI-RS RSSI:
CSI-RSSI comprises the linear average of the total received power [in (W)] observed only in OFDM symbols of measurement time resources, in the measurement bandwidth, over N RBs from all sources, including co-channel serving and non-serving cells, adjacent channel interference, thermal noise, etc.

- Measurement time resources for CSI-RSSI correspond to OFDM symbols containing L3 mobility CSI-RS.
- In Release 15, the same measurement bandwidth for CSI-RSSI and CSI-RSRP is assumed.

Reference Signal Received Quality (RSRQ) RSRQ based on SS/PBCH block and CSI-RS [SS-RSRQ and CSI-RSRQ] supported in both IDLE and CONNECTED mode.

- SS-RSRQ defined as (N×SS-RSRP)/(NR carrier RSSI)
  - where N is the number of RBs of NR carrier RSSI measurement bandwidth. The measurement in the numerator and the denominator is made over the same set of RBs.
- CSI-RSRQ defined as (N×CSI-RSRP)/CSI-RSSI
  - where N is the number of RBs in CSI-RSSI measurement bandwidth. Measurement in the numerator and the denominator is made over same set of RBs.

For RSRQ measurement, same receive beam shall be applied between RSSI measurement and RSRP measurement.

Signal-to-Noise and Interference Ratio (SINR) SS-SINR–based CSI-SINR is supported in CONNECTED mode.

- It is up to UE implementation to use SSS only or SSS + PBCH DMRS as interference measurement resources.
- Interference measurement resources for CSI-SINR are the CSI-RS REs used for the RSRP measurement.

SFN and Frame Timing Difference (SFTD) The observed SFN and frame timing difference (SFTD) between an E-UTRA PCell and an NR PSCell is defined as comprising the following two components:

- SFN offset = $(SFN_{PCell} - SFN_{PSCell})$ mod 1024, where SFN_{PCell} is the SFN of a E-UTRA PCell radio-frame and SFN_{PSCell} is the SFN of the NR PSCell radio-frame of which the UE receives the start closest in time to the time when it receives the start of the PCell radio-frame.
- Frame boundary offset = $\lfloor(T_{FrameBoundaryPCell} - T_{FrameBoundaryPSCell})/5\rfloor$, where $T_{FrameBoundaryPCell}$ is the time when the UE receives the start of a radio-frame from the PCell, $T_{FrameBoundaryPSCell}$ is the time when the UE receives the start of the radio-frame, from the PSCell, that is closest in time to the radio-frame received from the PCell. The unit of $(T_{FrameBoundaryPCell} - T_{FrameBoundaryPSCell})$ is Ts.
- Applicable for: RRC_CONNECTED intra-frequency.

6.5.2 Radio Link Monitoring (RLM)

NR Release 15 supports RLM on PCell and PSCell only. Similar to RRM, NR supports **SS/PBCH block**–based RLM and **CSI-RS**–based RLM. Therefore, NR has two types of reference signals (RS) for RLM (RLM-RS), i.e., SS/PBCH block and CSI-RS.

A hypothetical PDCCH BLER (based on RLM-RS SINR) is the metric for determining in-sync (IS) and out-of-sync (OOS) as done for LTE.

The UE assumes the same antenna port is used between the hypothetical PDCCH and the RS used for RLM. Signal and interference measurement for a given CORESET may be performed using the same receive beam. It is up to the UE implementation how interference and noise measurement can be performed, i.e., no explicit resources are defined and indicated to the UE for interference measurement resources for RLM. It is understood that the UE may perform interference measurements on any resource (excluding SS/PBCH resources) with a known signal, i.e., a known reference signal, a transmission the UE can decode, or a resource element the UE knows is empty. The UE is not required to perform RLM measurements outside the active DL BWP.

For a cell group, a single IS or OOS indication is reported by the UE (regardless of the number of available beams in a cell). A single IS BLER threshold (**Q_in**) is configured for a UE at a time. Also, a single OOS BLER threshold (**Q_out**) is configured for a UE at a time. Those values are configurable from two pairs of values for IS/OOS BLERs explicitly indicated by RRC.

An IS/OOS threshold pair index corresponds to a specific IS/OOS threshold pair. The default value for IS/OOS threshold pair indication corresponds to the LTE-like IS/OOS threshold of 2% and 10% BLER.

NR provides only periodic IS/OOS indications. There was no consensus in 3GPP to provide aperiodic indication(s) based on beam failure recovery procedure to assist radio link failure (RLF) procedure despite the possibility that the same RS is used for beam failure recovery and RLM procedures.

Periodic IS is indicated if the estimated link quality (i.e., hypothetical PDCCH BLER) of at least one RLM-RS resource among all configured RLM-RS resource(s) is above Q_in threshold.

Periodic OOS is indicated if the estimated link quality (i.e., hypothetical PDCCH BLER) of all configured RLM-RS resource(s) is below Q_out threshold.

L_{max}	$N_{LR\text{-}RLM}$	N_{RLM}
4	2	2
8	6	4
64	8	8

Table 6.7 $N_{LR\text{-}RLM}$ and N_{RLM} as a Function of L_{max}

6.5.2.1 RLM-RS Resource Configuration

The UE can be configured with up to $N_{LR\text{-}RLM}$ *RadioLinkMonitoringRS* for link recovery procedures and for radio link monitoring. From the $N_{LR\text{-}RLM}$ *RadioLinkMonitoringRS*, up to N_{RLM} *RadioLinkMonitoringRS* can be used for radio link monitoring depending on a maximum number L_{max} of candidate SS/PBCH blocks per half-frame as described in Sec. 6.2, and up to two *RadioLinkMonitoringRS* can be used for link recovery procedures. Table 6.7 provides the values of $N_{LR\text{-}RLM}$ and N_{RLM} as a function of L_{max}.

One RLM-RS resource can be either one SS/PBCH block or one CSI-RS resource. NR supports configurability of different RLM-RS types for each RLM-RS. The RLM-RS resources are UE-specifically configured by RRC. RLM-RS is undefined until explicitly configured. Note, this implies that network needs to configure RLM-RS for UE to perform RLM.

For SS/PBCH block–based RLM, the SS/PBCH blocks to be used for RLM are individually indicated. The RLM-RS resources for CSI-RS–based RLM can be separately configured from CSI-RS for beam management albeit reusing the same CSI-RS design/configuration framework:

- Only single-port CSI-RS resources supported
- CSI-RS frequency density D = {1, 3}
- *RLM-CSI-RS-timeConfig*:
  - Periodicity (P) = {5 ms, 10 ms, 20 ms, 40 ms}.
  - Slot offset: {0, …, Ps-1} slots, where Ps is number of slots within period P in the CSI-RS numerology
- *RLM-CSI-RS-FreqBand*: same as for beam management but with a minimum number of PRB is 24

References

[1] 3GPP, "3rd Generation Partnership Project; Technical Specification Group Radio Access Network; NR; User Equipment (UE) radio transmission and reception; Part 1: Range 1 Standalone (Release 15)," 3GPP TS 38.101-1, v15.8.2, January 2020.

[2] 3GPP, "3rd Generation Partnership Project; Technical Specification Group Radio Access Network; NR; User Equipment (UE) radio transmission and reception; Part 1: Range 2 Standalone (Release 15)," 3GPP TS 38.101-2, v15.8.0, January 2020.

[3] 3GPP, "3rd Generation Partnership Project; Technical Specification Group Radio Access Network; NR; NR and NG-RAN overall description; Stage 2 (Release 15)," 3GPP TS 38.300, v15.8.0, January 2020.

[4] 3GPP, "3rd Generation Partnership Project; Technical Specification Group Radio Access Network; NR; Radio Resource Control (RRC) protocol specification (Release 15)," 3GPP TS 38.331, v15.8.0, January 2020.
[5] 3GPP, "3rd Generation Partnership Project; Technical Specification Group Radio Access Network; NR; multiplexing and channel coding (Release 15)," 3GPP TS 38.212, v15.8.0, January 2020.
[6] 3GPP, "3rd Generation Partnership Project; Technical Specification Group Radio Access Network; NR; physical layer procedures for control (Release 15)," 3GPP TS 38.213, v15.8.0, January 2020.
[7] 3GPP, "3rd Generation Partnership Project; Technical Specification Group Radio Access Network; NR; physical channels and modulation (Release 15)," 3GPP TS 38.211, v15.8.0, January 2020.
[8] 3GPP, "3rd Generation Partnership Project; Technical Specification Group Radio Access Network; Evolved Universal Terrestrial Radio Access (E-UTRA); physical channels and modulation (Release 15)," 3GPP TS 36.211, v15.8.1, January 2020.
[9] 3GPP, "3rd Generation Partnership Project; Technical Specification Group Radio Access Network; NR; physical layer measurements (Release 15)," 3GPP TS 38.215, v15.6.0, January 2020.

CHAPTER 7

Downlink Control Operation

Downlink control is an integral component for wireless communications. Due to the complexity involved in Downlink data decoding and the necessary variations for PDSCH in terms of resource allocation, transport block sizes, etc., blind decoding of Downlink data channels without any control channel assistance is prohibitively expensive. Non-preconfigured or non-managed Uplink data transmission are not desirable as well due to the resulting decoding complexity and the non-controllable Uplink interference.

Efficient Downlink control design is crucial for wireless communications. Downlink control should provide sufficient information for Downlink data channel decoding or Uplink data channel transmission for a UE. At the same time, the Downlink control overhead has to be carefully managed. Blind decodes of Downlink control channels are often necessary, since a gNB often has to schedule multiple UEs simultaneously which necessitates flexibility in control channel transmissions. The performance of Downlink control, including both the detection performance and the false alarm probability, should be satisfactory, especially for ultra-reliable communications. Frequent control channel monitoring may also be necessary for low latency communications, while at the same time, excessive control channel monitoring by a UE is not friendly to UE power savings.

In this chapter, we will first introduce the Downlink control design in 4G LTE, followed by the detailed Downlink control design for 5G NR.

7.1 Downlink Control in 4G LTE

In 4G LTE, Downlink control has the following channels [1]:

- Physical control format indicator channel (PCFICH), which provides dynamic indication of a time span of the control region in every subframe.
- Physical HARQ indicator channel (PHICH), which provides HARQ feedback for PUSCH.
- Physical Downlink control channel (PDCCH), which provides DL and UL assignments, TPC commands for a group of users, DL data arrival indication, paging indication, SPS activation/release, etc.

For these Downlink control channels, one fundamental building block for resource mapping and transmission is called a resource element group (REG), which consists of either four or six REs. A four REs/REG is for the case when there is no CRS in the OFDM symbol (hence three REGs/RB, see symbol 2 in Fig. 7.1), while the six REs/REG is for the case when there is CRS in the OFDM symbol (hence two REGs/RB, see symbols 0 and 1 in Fig. 7.1). In both cases, there are always four effective REs/REG *available* for the control channel transmission, which is friendly for space-frequency block codes (SFBC)–based transmit diversity scheme for LTE Downlink control channel transmissions. For PDCCH, the resource mapping granularity is a group of REGs forming a control channel element (CCE), which consists of nine REGs or 36 effective REs/CCE available for PDCCH transmissions. A PDCCH may aggregate 1, 2, 4, or 8 CCEs for a transmission, also known as aggregation levels, effectively resulting in different coding rates for PDCCH.

Figure 7.1 illustrates the Downlink control region in 4G LTE. A four-port CRS is assumed (represented by R_0, R_1, R_2, and R_3), and there are three control symbols assumed in the example. It is noted that the management of the Downlink control region is in units of a full OFDM symbol. In other words, Downlink data cannot be scheduled to use any resources in the control region.

The size of the control region (in units of symbols) of a subframe is *dynamically* signaled by PCFICH in the subframe, where 1, 2, or 3 OFDM symbols can be signaled for system bandwidths larger than 10 RBs, and 2, 3, or 4 OFDM symbols can be signaled for system bandwidths of 6 to 10 RBs. PCFICH is transmitted in the first OFDM symbol in every Downlink subframe, using four REGs (or 16 effective REs) fully distributed over the entire eNB channel bandwidth.

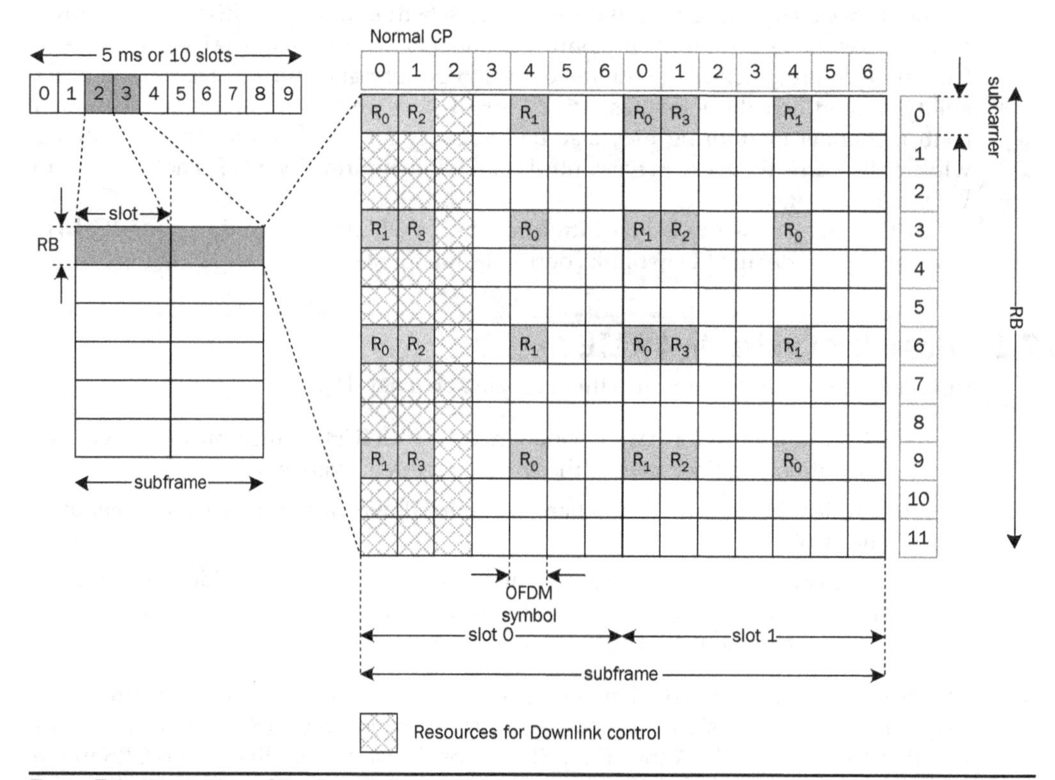

Figure 7.1 Illustration of the Downlink control region in 4G LTE by assuming a four-port CRS and three OFDM symbols for control.

In LTE, UL data transmission is based on synchronous HARQ, where after a PUSCH transmission in a subframe, a UE looks for a HARQ re-transmission indication or a scheduling grant in a subsequent Downlink subframe based on a pre-defined timing (e.g., a 4-ms gap). The HARQ for PUSCH is supported via PHICH, which may be transmitted in the first symbol only or over several symbols (in order for additional power boosting and time diversity), depending on the configuration. The PUSCH re-transmissions can be done by either PHICH or PDCCH. PDCCH triggers the so-called adaptive PUSCH re-transmissions (since many PUSCH parameters can be updated via PDCCH, e.g., modulation order, resource allocation, etc.), while PHICH triggers the so-called non-adaptive PUSCH re-transmissions (although hopping is possible if pre-defined). PHICH can be more efficient in terms of the associated Downlink control overhead, especially when there are many re-transmissions (by a same UE or by a large number of UEs).

PDCCH provides scheduling for a wide range of use cases, such as:

- DL scheduling, including scheduling unicast data, paging messages, system information blocks (SIBs), random access responses (RAR), etc.
- UL grants, including scheduling unicast data, requesting for aperiodic CSI reports, triggering of SRS transmissions, etc.
- Multi-user TPC commands, for PUCCH or PUSCH for a group of UEs
- DL data arrival, intended for UEs not synchronized with an eNB
- SPS activation/release

The PDCCH has a 16-bit CRC protection, where the CRC is scrambled by an RNTI (radio network temporary identifier). The RNTI type depends on the purpose of the PDCCH transmission, e.g., SI-RNTI for system information scheduling, P-RNTI for paging scheduling, RA-RNTI for RAR scheduling, C-RNTI for unicast data scheduling, SPS C-RNTI for SPS activation/release or re-transmissions, TPC PUCCH RNTI and TPC PUSCH RNTI for the TPC commands for PUCCH and PUSCH transmissions, respectively, etc. Some key parameters for PDCCH include [1]:

- Coding: Tail-biting convolutional coding (TBCC) is used for PDCCH (same as PBCH)
- Set of aggregation levels: {1, 2, 4, 8} CCEs
- Reference signal: the wide-band CRS
- Transmission scheme: SFBC based transmit diversity
- Search space: a set of logically consecutive CCEs containing PDCCH candidates for a UE to perform blind decoding
    - Common search space (CSS): for broadcast/paging/RAR and limited unicast data scheduling, common for all UEs served by the same cell
    - UE-specific search space (USS): for unicast data
- Number of blind decodes: up to 44 or 60
    - $44 = 2 \times 6$ (CSS) $+ 2 \times 16$ (USS), where the factor 2 is the number of distinct DCI sizes, 6 is the number of decoding candidates in CCS ({4, 2} decoding

candidates for aggregation levels {4, 8}, respectively) and 16 is the number of decoding candidates in USS ({6, 6, 2, 2} decoding candidates for aggregation levels {1, 2, 4, 8}, respectively).

- ◦ 60 = 2 × 6 (CSS) + 3 × 16 (USS), where in this case, there are 3 different DCI sizes in USS (due to UL MIMO).

PDCCH search space follows a tree-structure, where each aggregation level L (={1, 2, 4, 8}) always starts with CCE indices of integer multiples of L. That is, for PDCCH:

- Aggregation Level 1: may start with any CCE indices.
- Aggregation Level 2: may start with CCEs 0, 2, 4, …
- Aggregation Level 4: may start with CCEs 0, 4, 8, …
- Aggregation Level 8: may start with CCEs 0, 8, 16, …

This is illustrated in Fig. 7.2. The tree-structure helps alleviate PDCCH blocking probability, hence improving the overall PDCCH resource management. For example, under the tree-structure, it is possible that two PDCCH transmissions of aggregation Level 2 only block one PDCCH of aggregation Level 4.

It is worth noting that the nine REGs for each CCE for a PDCCH are fully distributed in the frequency domain in order to maximize the frequency diversity gain. In other words, PDCCH resource mapping is designed to be in a *frequency-distributed* manner. The mapping is also dependent on the physical cell ID (PCI) of the cell, in order to improve the interference randomization across cells.

Enhanced PDCCH (ePDCCH) was later introduced in 4G LTE [2 (Sec. 8)], primarily motivated by the following:

- PDCCH capacity may be limited,
- PDCCH is not friendly to inter-cell coordination (since it is fully frequency-distributed), and
- PDCCH cannot benefit from any beamforming gain (since it is based on CRS, which is not precoded).

ePDCCH is FDM based. That is, each ePDCCH occupies one or more PRB pairs, which has a time span over the entire subframe. This facilitates inter-cell coordination on a per

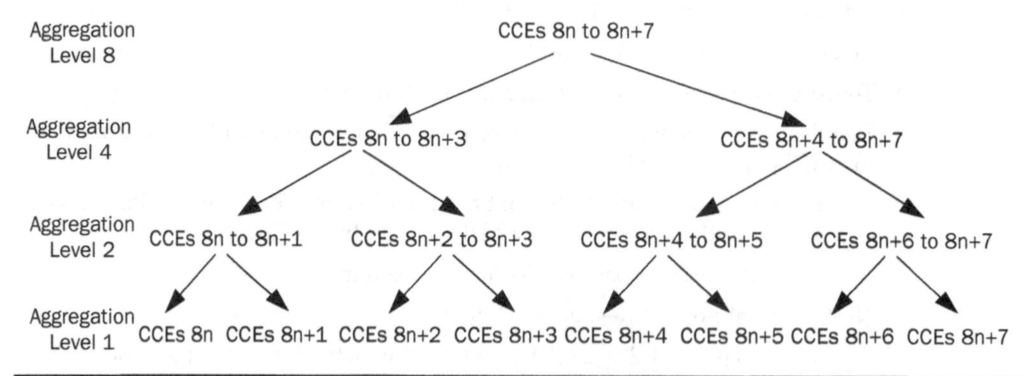

Figure 7.2 Illustration of the tree-structure of PDCCH search space, where n is an integer ≥ 0.

PRB-pair basis. ePDCCH demodulation is based on DM-RS, which makes it possible to exploit the beamforming gain. The construction of ePDCCH is based on the enhanced REG (eREG) and the enhanced CCE (eCCE). The eREG is defined within a PRB pair by excluding 24 DM-RS REs (for NCP) or 16 DM-RS REs (for ECP) in a PRB pair, although an ePDCCH may be associated with a smaller number of DM-RS REs (e.g., 12 DM-RSs for NCP) for demodulation. The remaining REs in a PRB pair, including the REs used by the legacy control, are *always* evenly and sequentially (frequency-first-time-second) divided into 16 eREGs. In particular,

- For NCP, each eREG has $[(14\times12)-24]/16 = 9$ REs, where 14 is the number of symbols in a PRB pair, 12 is the number of tones per RB, and 24 is the assumed number of REs excluded for DM-RS.
- For ECP, each eREG has $[(12\times12)-16]/16 = 8$ REs, where the first 12 is the number of symbols in a PRB pair in the extended CP case and the first 16 is the assumed number of REs excluded for DM-RS.

This is shown in Fig. 7.3 for NCP, where the numbering in each RE represents the associated eREG index, and the shaded REs without numbering represent the assumed REs excluded for DM-RS. For instance, eREG 0 has the following nine REs ({tone index, symbol index}): {0, 0}, {4, 1}, {8, 2}, {0, 4}, {8, 5}, {8, 7}, {0, 9}, {4, 10}, and {8, 11}.

Note that some REs in an eREG may be used by other channels/signals (e.g., legacy control, CRS, etc.). An ePDCCH transmission has to rate match around these REs occupied by other channels/signals. It implies that the number of *available* REs for an eREG (and hence for an ePDCCH) is a variable, which is different from that of PDCCH.

The number of eREGs (denoted by N) for an eCCE depends on many factors [1], but in most cases, $N = 4$ (thus, 36 nominal REs/eCCE). The construction of an eCCE is a function of the configured ePDCCH mode(s), where two modes are defined:

- Localized mode, where a single precoder is applied for each PRB pair and eREGs from a same PRB pair form a same eCCE (so that subband-based precoding can be exploited), and
- Distributed mode, where two precoders cycle through the allocated resources within each PRB pair, and eREGs from different PRB pairs (typically frequency-distributed) form a same eCCE (so that frequency diversity can be exploited).

	Symbol Index													
Tone Index	0	1	2	3	4	5	6	7	8	9	10	11	12	13
0	0	12	8	4	0			8	4	0	12	8		
1	1	13	9	5	1			9	5	1	13	9		
2	2	14	10	6	2	12	2	10	6	2	14	10	4	10
3	3	15	11	7	3	13	3	11	7	3	15	11	5	11
4	4	0	12	8	4	14	4	12	8	4	0	12	6	12
5	5	1	13	9	5			13	9	5	1	13		
6	6	2	14	10	6			14	10	6	2	14		
7	7	3	15	11	7	15	5	15	11	7	3	15	7	13
8	8	4	0	12	8	0	6	0	12	8	4	0	8	14
9	9	5	1	13	9	1	7	1	13	9	5	1	9	15
10	10	6	2	14	10			2	14	10	6	2		
11	11	7	3	15	11			3	15	11	7	3		

FIGURE 7.3 Illustration of eREG definition for ePDCCH in 4G LTE (NCP).

(a)

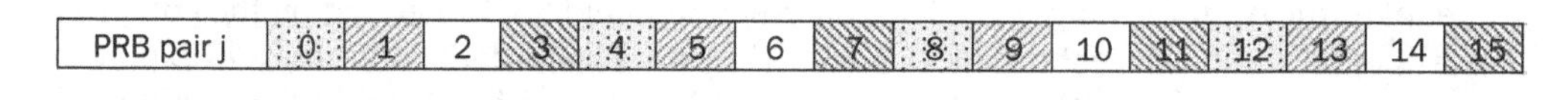

(b)

	eCCE indexing															
	0	1	2	3	4	5	6	7	8	9	10	11	12	13	14	15
PRB pair j_1	0	12	8	4	1	13	9	5	2	14	10	6	3	15	11	7
PRB pair j_2	4	0	12	8	5	1	13	9	6	2	14	10	7	3	15	11
PRB pair j_3	8	4	0	12	9	5	1	13	10	6	2	14	11	7	3	15
PRB pair j_4	12	8	4	0	13	9	5	1	14	10	6	2	15	11	7	3

FIGURE 7.4 Illustration of eCCE to eREG mapping, (a) localized mode; (b) distributed mode, NCP, 4 eREGs/eCCE.

To illustrate, Fig. 7.4 (a) provides one example of eCCE to eREG mapping for the localized ePDCCH, where the numbering in each box represents the *logical* eREG index (the mapping of eREG to REs is shown in Fig. 7.3), while eREGs of the same pattern are associated with the same eCCE. For example, eREGs 0, 4, 8, and 12 belong to the same eCCE. Figure 7.4 (b) provides one example for the distributed ePDCCH, where in this case, eREGs of one eCCE are distributed across different PRB pairs (indexed by j_1, j_2, j_3, and j_4). For example, eCCE0 consists of eREG 0 of PRB pair j_1, eREG 4 of PRB pair j_2, eREG 8 of PRB pair j_3, and eREG 12 of PRB pair j_4. Note that the four PRB pairs are not necessarily to be consecutive in frequency.

Compared with the wide-band CRS-based PDCCH, the narrow-band DM-RS–based ePDCCH may not achieve better performance even with the help of beamforming gain. This is due to the channel estimation loss when comparing the narrow-band DM-RS with the wide-band CRS. Figure 7.5 provides some evaluation results of the performance gap between ePDCCH and PDCCH. A DCI format 1A [1] of 43 bits is assumed. A positive gap means that the required SNR to achieve a 1% PDCCH miss-detection probability for ePDCCH is higher than that of PDCCH. As can be seen, for a 2 (transmit antennas) by 2 (receive antennas) configuration (denoted as 2 × 2), both localized and distributed ePDCCH (over four frequency-distributed PRB pairs) are generally worse than the legacy PDCCH, primarily due to the channel estimation loss and the resource-dimension loss (<36 available REs/eCCE). However, under a 4×2 system, when there is increased beamforming gain as shown in Fig. 7.5 (c), the localized ePDCCH may outperform the legacy PDCCH especially when the aggregation levels are low where the beamforming gain can be readily exploited. Note that the larger loss at 1 eCCE (compared to the cases of two or more eCCEs) is primarily due to the resource-dimension loss and the resulting coding loss (the coding rate is much higher than the mother coding rate of 1/3). Therefore, in order to compensate the potential performance loss compared with the legacy PDCCH, in addition to aggregation levels {1, 2, 4, 8}, ePDCCH can additionally take two larger aggregation levels: 16 and 32.

Cross-carrier scheduling (CCS) can be configured for both PDCCH and ePDCCH. CCS refers to the case where a first carrier (also known as a scheduling carrier or cell) transmits a (e)PDCCH scheduling a PDSCH or a PUSCH transmission on a second

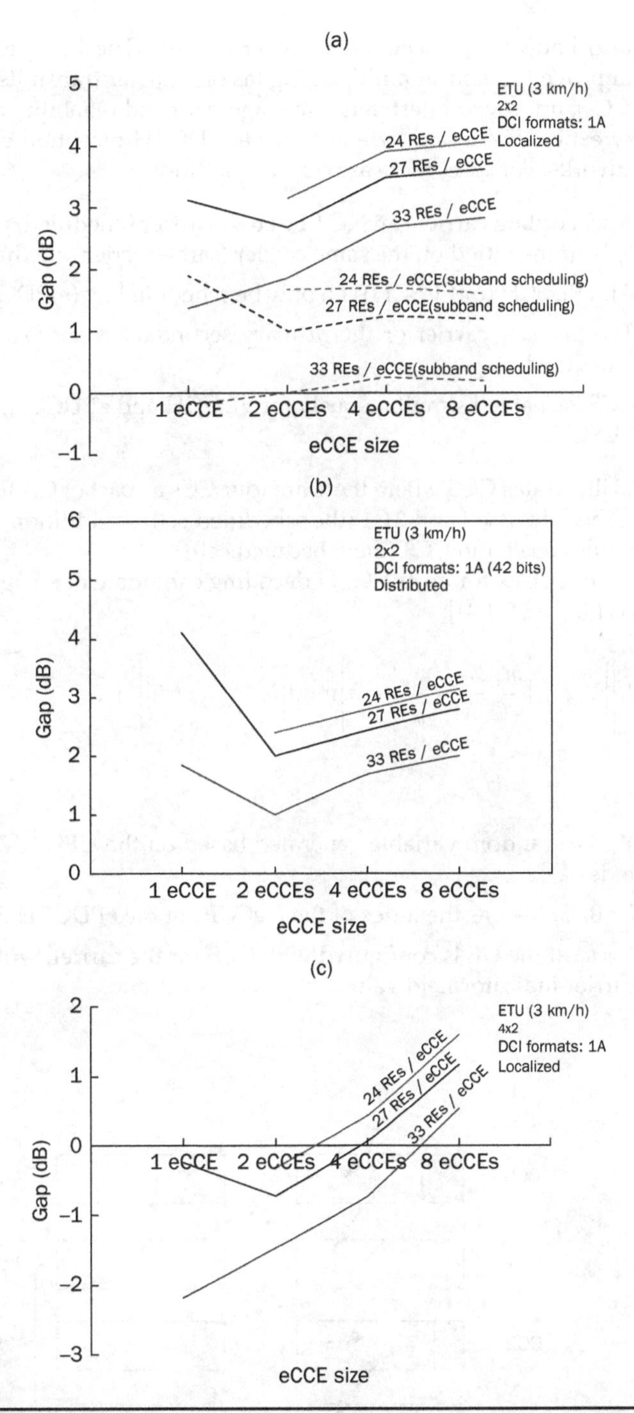

FIGURE 7.5 Performance gap between ePDCCH and PDCCH, ETU, DCI format 1A [1] of 43 bits, (a) Localized ePDCCH, 2×2; (b) distributed ePDCCH, 2×2; (c) localized ePDCCH, 4×2.

carrier (also known as a scheduled carrier or cell). The benefits of supporting CCS include improved statistical multiplexing (as one carrier transmits (e)PDCCHs for two or more CCs), improved interference management and reliability for (e)PDCCH where a UE may experience less interference for (e)PDCCH reception especially in heterogenous networks. For CCS, there are certain conditions:

- A scheduling carrier can NOT be cross-carrier scheduled, i.e., its (e)PDCCH has to be transmitted on the same carrier (same-carrier scheduling).
- Any PDSCH (or PUSCH) can only be scheduled by (e)PDCCH from one carrier.
- The primary carrier or the primary secondary carrier cannot be cross-carrier scheduled.
- CCS is enabled simultaneously for PDCCH and ePDCCH, and for PDSCH and PUSCH.

Figure 7.6 illustrates CCS where there are four CCs as part of CA for a UE. The primary cell CC0 schedules itself and CC1 (the scheduled cell). In addition, CC2 (the scheduling cell) schedules itself and CC3 (the scheduled cell).

The set of eCCEs for an ePDCCH decoding candidate of an aggregation level L are defined as [3, (Sec. 9.1.4)]:

$$L\left\{\left(Y_{p,k}+\left\lfloor\frac{m\cdot N_{\mathrm{eCCE},p,k}}{L\cdot M_{p,full}^{(L)}}\right\rfloor+b\right)\operatorname{mod}\left\lfloor N_{\mathrm{eCCE},p,k}/L\right\rfloor\right\}+i \tag{7.1}$$

where

- $Y_{p,k}$ is a random variable generated based on the UE's RNTI and the subframe index k,
- $i = 0, \ldots, L-1$ as the index of the L eCCEs of the ePDCCH decoding candidate,
- $b = n_{CI}$ if the UE is configured with CCS for the carrier, with n_{CI} being the cross-carrier indicator field value, otherwise $b = 0$, and

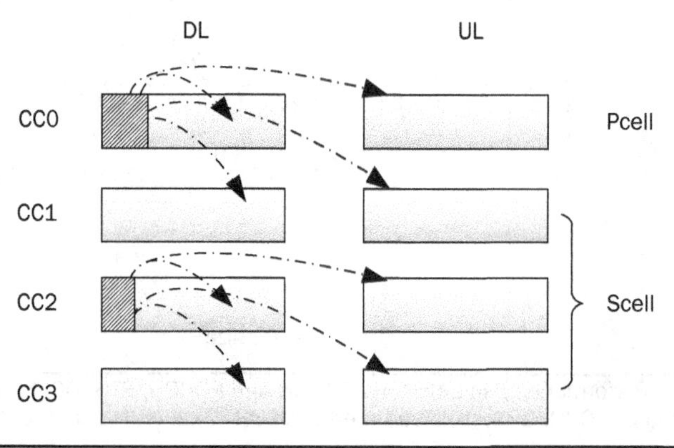

FIGURE 7.6 Illustration of CCS for a UE in CA.

eCCE Index	0	1	2	3	4	5	6	7	8	9	10	11	12	13	14	15	16	17	18	19	20	21	22	23	24	25	26	27	28	29	30	31
L=1																																
L=2																																
L=4																																
L=8																																

FIGURE 7.7 Illustration of ePDCCH decoding candidates for a UE, distributed within the eCCE domain. The same distribution is also adopted for 5G NR PDCCH.

- $M^{(L)}_{p,full}$ is the number of ePDCCH candidates derived based on a configuration with p as an index of the configured ePDCCH resource set, $m = 0, 1, \ldots, M_p^{(L)} - 1$, where the resource set has $N_{eCCE,p,k}$ eCCEs.

To better understand how the eCCEs of an ePDCCH for a UE are mapped within a configured ePDCCH resource set, let us take a look at one specific example. Assume $Y_{p,k} = 0$, $b = 0$, $N_{eCCE,p,k} = 32$, $M^{(L)}_{p,full} = \{6, 6, 2, 2\}$ for $L = \{1, 2, 4, 8\}$, respectively. Figure 7.7 illustrates the mapping of the ePDCCH decoding candidates for each aggregation level to the eCCE domain for a UE. As can be seen, the multiple decoding candidates for each aggregation level are generally evenly distributed in the eCCE domain. For instance, for aggregation Level 1, the six decoding candidates have eCCE indices of 0, 5, 10, 16, 21, and 26. The distribution of eCCEs for the multiple decoding candidates of a given aggregation level helps reduce the ePDCCH blocking probability among different UEs. As an example, if one UE's ePDCCH of aggregation Level 8 is transmitted using eCCEs 16 to 23, these eCCEs are no longer available for other UEs (unless MU-MIMO is used for ePDCCH, which is supported but may not be desirable). If aggregation Level 1 ePDCCH decoding candidates are designed to map to six consecutive eCCEs, e.g., eCCEs 16–21 for a UE, the UE can no longer be scheduled with aggregation Level 1. Distributed eCCE mapping makes it possible for an eNB to transmit an ePDCCH of aggregation Level 1 for the UE using other decoding candidates, e.g., eCCEs 0, 5, 10, or 26.

As will be detailed later, the same hash function is also adopted for 5G NR PDCCH.

7.2 Control Region Management in 5G NR

In 4G LTE, the size of the legacy control region is dynamically indicated. It is noted that the legacy control region is managed in a *cell-specific* manner and cannot be used for data transmissions (i.e., no FDM between control and data). On the other hand, ePDCCH is managed in a *UE-specific* manner, with the set of RBs for ePDCCH configured for a UE semi-statically.

As discussed in Chap. 5, in 5G NR, UE BWP is introduced so that a UE can operate in a gNB with a UE channel bandwidth less than the gNB's channel bandwidth. In designing the control region for 5G NR, the following needs to be considered:

- Should the control region be managed in a cell-specific or a UE-specific manner?
- Should the control region be dynamically or semi-statically indicated to a UE?
- Should the control region be allowed to be FDM with data channels or even possibly be re-used for data channels?

To answer these questions, we have to first differentiate the control region for broadcast from that for unicast. Obviously, it is more sensible to have the control region managed in a cell-specific manner at least for initial access, since the control operation during initial access is generally a broadcast transmission in nature. However, for unicast, UE-specific management of control region is preferable due to the following:

- The channel bandwidths for UEs can be different.
- The required frequency of time instances of control channel monitoring can be different for different UEs, e.g., due to different latency requirements and different power saving needs.
- The required reliability of control channel transmissions can be different for different UEs, e.g., due to different QoS requirements.

Therefore, the control region for a UE is preferably defined with both time-domain and frequency-domain restrictions, ideally in response to the UE's channel bandwidth restriction, the need for power saving, the need for possibly multiplexing with different UEs, and the performance requirements. In addition, the control region should not be limited to only the beginning of a slot (as for LTE PDCCH). Otherwise, certain latency requirement may not be met. In other words, in addition to *slot-level* PDCCH opportunities, *mini-slot-level* PDCCH opportunities should be provided for a UE when necessary. Lastly, given different service requirements, more than one control region should be available in a gNB for a same UE or different UEs.

The usage of UE-specific management of the control region for unicast traffic not only provides the necessary tool to address UE-specific scheduling needs, but also brings flexibility in control channel capacity. As a result, it becomes questionable whether it is necessary to dynamically indicate the size of the control region via a physical layer signal or channel as in the 4G LTE case. Indeed, 5G NR does not have a 4G LTE PCFICH like physical layer channel. Rather, it relies on a semi-static indication in the physical broadcast channel (PBCH) in an implicit manner. The parameter is called *dmrs-TypeA-Position* [4] (Sec. 6.2.2). It specifies that, for both type A PDSCH assignments and type A PUSCH assignments (i.e., slot-based PDSCH/PUSCH scheduling, see Chap. 8), the first DM-RS symbol starts from either symbol 2 (the 3rd symbol) or symbol 3 (the 4th symbol). This also provides an upper limit to the control region located at the beginning of the slot, i.e., no more than two symbols or three symbols, respectively.

A **CO**ntrol **Re**source **SET** configured for a UE is denoted as a **CORESET** in 5G NR. Considering the need for broadcast control and UE-specific control, the following CORESETs are defined:

- CORESET #0, indicated via *pdcch-ConfigSIB1* in MIB [5] (Sec. 6.2.2), with detailed parameters specified in [6 (Sec. 13)],
- An additional CORESET configured in SIB1, with an CORESET ID > 0 whose resources are contained in the bandwidth of CORESET#0, and
- Additional CORESETs configured UE-specifically, up to 3 CORESETs per BWP.

In particular, to better match with practical deployments, the number of RBs and the number of symbols for CORESET #0 depend on the combinations of the SCS of

SSB and the SCS of CORSET#0, and the minimum channel bandwidths. Specifically, the possible numbers of RBs for CORESET #0 are:

- For FR1
  - For 15 kHz PDCCH SCS: {24, 48, 96} RBs or {4.32, 8.64, 17.28} MHz
  - For 30 kHz PDCCH SCS: {24, 48} RBs or {8.64, 17.28} MHz
- For FR2
  - For 60 kHz PDCCH SCS: {48, 96} RBs or {34.56, 69.12} MHz
  - For 120 kHz PDCCH SCS: {24, 48} RBs or {34.56, 69.12} MHz

The number of symbols for CORESET #0 may take a value of {1, 2, 3} symbols for {15, 30, 60} kHz, and {1, 2} for 120 kHz PDCCH SCS, but the selection of the value depends on the number of RBs for CORESET #0 as well [6 (Tables 13-1 to 13-10)]. Note that CORESE #0 is intended for the so-called initial DL BWP (BWP #0), which is the BWP associated with the initial access.

Additional CORESET configuration in SIB1 is optional. If configured, it provides the flexibility of using a different CORESET (other than CORESET #0) for the reception of other system information (OSI), paging, and messages related to the RACH procedure. The additional CORESET configuration in SIB1 can also be used for UE-specific control transmissions.

For UE-specifically configured CORESETs, the configuration in the frequency domain is in units of a group of six consecutive RBs. A bitmap of 45 bits can be used to indicate up to 45 groups, or up to 45 × 6 = 270 RBs. The number of symbols can be {1, 2, 3}. Note that since a CORESET is configured within a BWP for a UE, it has to be fully contained within the BWP.

Besides the absence of a 4G LTE PCIFCH like physical channel, 5G NR does not specify a 4G LTE PHICH like physical channel. The reason is very simple – in 5G NR, HARQ operation for PUSCH HARQ is asynchronous and, consequently, there is no fixed timing between a current PUSCH transmission and the next re-transmission opportunity.

7.3 PDCCH Structure in 5G NR

As discussed earlier, 4G LTE PDCCH and ePDCCH are constructed based on REG/CCE and eREG/eCCE, respectively. Each REG always has four available REs for PDCCH, which is carefully chosen to facilitate SFBC-based transmission scheme using CRS [1]. However, the number of available REs per eREG (hence per eCCE) for ePDCCH is a *variable*, which is not necessarily a multiple of 4. This is feasible since an ePDCCH transmission is based on DM-RS with beamforming, where it can be easily specified for any number of REs/eREGs.

For both PDCCH and ePDCCH, the introduction of REG/CCE and eREG/eCCE is critical, providing the necessary construction units which can be easily aggregated to achieve a good tradeoff between performance (for each control channel and across different control channels, e.g., the blocking probability as shown in Fig. 7.2) and the associated overhead (for each control channel and the resulting impact on the overall system resource utilization, e.g., resource fragmentation due to control channel transmissions).

Therefore, it is straightforward for 5G NR to also adopt the concepts of REG and CCE as the construction units for NR PDCCH. However, for NR PDCCH structure, we need to answer the following questions:

- What is the transmission scheme for NR PDCCH?
- How should REG and CCE be defined for NR PDCCH?

Different transmission schemes for PDCCH were proposed and studied during the process of 5G NR standardization. In particular, two transmission schemes were standing out, namely:

- A two-port–based SFBC transmit diversity scheme and
- A one-port–based transmit diversity scheme.

Note that in LTE, SFBC for LTE PDCCH can be based on a two-port CRS or a four-port CRS, while a localized LTE ePDCCH is based on a single-port DM-RS and a distributed LTE ePDCCH is based on a two-port DM-RS for precoding cycling on a per RE basis. The performance comparison was discussed earlier (e.g., see Fig. 7.5). For 5G NR, extensive simulations were carried out as well. Some key evaluation assumptions include [7]:

- Aggregation levels: 1, 2, 4, 8 (other values such as 16, 32 can also be simulated),
- DCI size: 32 bits and 76 bits including CRC,
- CCE size: 4 to 8 REGs per CCE, where each REG is defined as one PRB (including DM-RS tones),
- Practical channel estimation,
- Carrier frequency: 4 GHz and 30 GHz,
- DMRS density: 33%,
- Channel model: TDL-A, TDL-C [8 (Sec. 7.7)]:
  - Delay spread of 30 ns, UE speed of 3 km/h (may also evaluate 70 and 500 km/h),
  - Delay spread of 300 ns, UE speed of 3 km/h,
  - Delay spread of 1000 ns, UE speed of 3 km/h.

Some evaluation results can be found in Refs. [9–14]. Although the evaluation results from different sources naturally differ to some extent, it is generally observed that the two schemes offer similar performance in most cases. However, the 2-port based SFBC transmit diversity scheme requires a set of REs to be grouped into pairs, where the two REs in the same pair have to be close to each other to ensure good performance. Due to the presence of other signals (e.g., DM-RS), it is likely that some REs may not be feasible or straightforward to be paired with other REs in proximity, leading to the so-called *orphan* REs. Handling orphan REs for an SFBC-based transmission would cause more standardization efforts. Therefore, the one-port–based transmit diversity scheme was eventually adopted.

How to choose an appropriate REG size for 5G NR control? On a high level, we need to consider:

- How to achieve sufficient transmit diversity to ensure good PDCCH performance?
- How to minimize resource fragmentation resulted from PDCCH? Or, in other words, how to multiplex control and data together for efficient system resource utilization?

In 4G LTE, one transmit antenna may still be deployed by some eNBs, while a UE may be equipped with at least two receive antennas. To achieve sufficient diversity for LTE PDCCH, a CCE is designed to have nine fully-frequency-distributed REGs. In 5G NR, it is envisioned that a gNB is typically equipped with multiple transmit antennas, especially at higher carrier frequencies. Indeed, as detailed in Chap. 8, massive antenna technique is one key technology enabler for 5G NR. In addition, a UE may typically be equipped with four receive antennas.

On the other hand, different from LTE PDCCH, where control and data are TDM, 5G NR is designed to have more efficient multiplexing of control and data in a same symbol. Since a PDSCH transmission is scheduled with a resource granularity of one RB, any finer granularity of PDCCH other than one RB would inevitably create some difficulty in multiplexing PDCCH and PDSCH together without resource fragmentation.

Moreover, in an RB containing PDCCH, it may have to carry both the DM-RS and the corresponding control data. Since an RB is defined to have 12 tones, the split of resources between the control DM-RS and the control data implies that the available REs for the control data are less than 12. For example, assuming a DM-RS density of 1/4, the number of available REs/RB for the control data is nine. Considering the performance comparison between LTE PDCCH and LTE ePDCCH as discussed earlier and the need to ensure 5G NR to have at least the same coverage as 4G LTE, it is sensible to define the following in 5G NR:

- **REG**: one RB in one symbol (12 × 1 REs)
- **CCE:** 6 REGs

In a CORESET, the REG is numbered in an increasing order following a time-first manner, starting with 0 for the first OFDM symbol and the lowest numbered resource block in the CORSET.

Another important concept is an **REG bundle**, which is defined as a number of REGs – either consecutive in frequency only or in frequency and time – for which a same precoder may be assumed by a UE. This is a necessary definition to balance the need of coherent channel estimation and a diversity level for PDCCH transmissions. The possible REG bundle sizes for NR PDCCH are {2, 3, 6} REGs.

Similar to LTE ePDCCH (see Fig. 7.4), two operation modes for NR PDCCH are also necessary, namely:

- Non-interleaved CCE-to-REG mapping
- Interleaved CCE-to-REG mapping

The interleaver is based on a row/column rectangular interleaver, defined as [15 (Sec. 7.3.2.2)]:

$$
\begin{aligned}
&f(x) = (rC + c + n_{\text{shift}}) \bmod (N_{\text{REG}}^{\text{CORESET}}/L) \\
&x = cR + r \\
&r = 0, 1, ..., R - 1 \\
&c = 0, 1, ..., C - 1 \\
&C = N_{\text{REG}}^{\text{CORESET}}/(LR)
\end{aligned}
\tag{7.2}
$$

where $R \in \{2,3,6\}$ is a configured interleave size, $N_{\text{REG}}^{\text{CORESET}} = N_{\text{RB}}^{\text{CORESET}} N_{\text{symb}}^{\text{CORESET}}$ with $N_{\text{RB}}^{\text{CORESET}}$ being the number of resource blocks and $N_{\text{symb}}^{\text{CORESET}} \in \{1,2,3\}$ being the number of symbols in the CORESET, and L is the REG bundle size. The parameter n_{shift} is either the same as the PCI of the cell or configurable with a value range of {0, 1, …, 274}. Essentially, the interleaving operation consists of writing into a rectangular interleaver by rows and reading by columns, along with a configurable cyclic shift to the interleaved REG bundles.

For the non-interleaved CCE-to-REG mapping, the REG bundle size is fixed at six REGs, in order to better exploit the beamforming gain and the subband scheduling gain. For the interleaved CCE-to-REG mapping, the REG bundle size can be {2, 6} REGs for a 1-symbol CORESET, {2, 6} REGs for a 2-symbol CORESET, and {3, 6} REGs for a 3-symbol CORESET, such that a trade-off between coherent channel estimation and a diversity level can be achieved. Note that for CORESET #0, interleaved mapping is always supported, where the REG bundle size L is fixed at 6 while the interleaver size R is fixed at 2.

Figures 7.8 to 7.10 provide examples of both non-interleaved and interleaved CCE-to-REG mapping for a 1-symbol CORESET, a 2-symbol CORESET, and a 3-symbol CORESET, respectively.

As mentioned earlier, within an REG bundle, the UE may assume the same precoding is used so that coherent channel estimation can be performed within a narrowband REG bundle (up to six REGs). For LTE PDCCH, the wideband and non-precoded CRS is used for channel estimation, implying that the effective REG bundle size is all the REGs across the *entire cell bandwidth*. Similar to the comparison between LTE PDCCH and LTE ePDCCH, it is expected that, relatively speaking, there is some channel estimation loss for NR PDCCH due to a limited number of DM-RS REs in a narrow-band REG bundle, although the loss may be compensated by the precoding gain by a UE-specific precoding operation. However, for scenarios when precoding is not effective (e.g., for broadcast PDCCH, or a UE under high mobility, etc.), it is necessary to still possibly use a *wide-band RS* for NR PDCCH channel estimation. To that end, a parameter called *precoderGranularity* can be used to indicate to a UE that instead of being restricted within an REG bundle, all contiguous RBs within a CORESET can be assumed to have the same precoding. For example, if a CORESET is configured to map to two sets of contiguous RBs, namely, from $[N_1, N_1 + 1, \ldots, N_1 + M - 1]$ and $[N_2, N_2 + 1, \ldots, N_2 + M - 1]$, each of the M contiguous RBs can be assumed to have the same precoding by the UE.

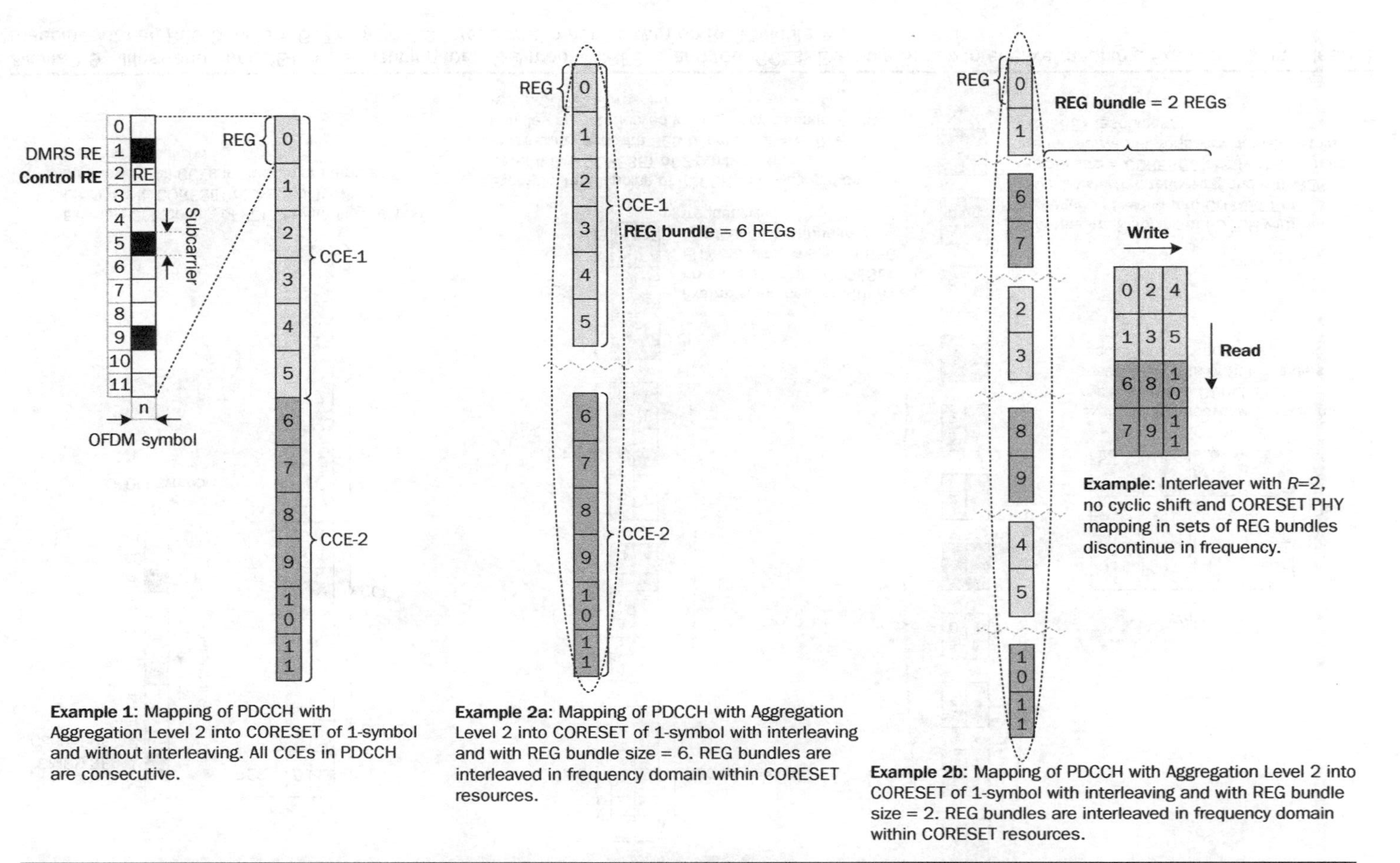

Figure 7.8 Illustration of CCE-to-REG mapping for a 1-symbol CORESET and two CCEs. Example 1: non-interleaved mapping; Example 2a: interleaved mapping with an REG bundle = 6; Example 2b: interleaved mapping with an REG bundle = 2.

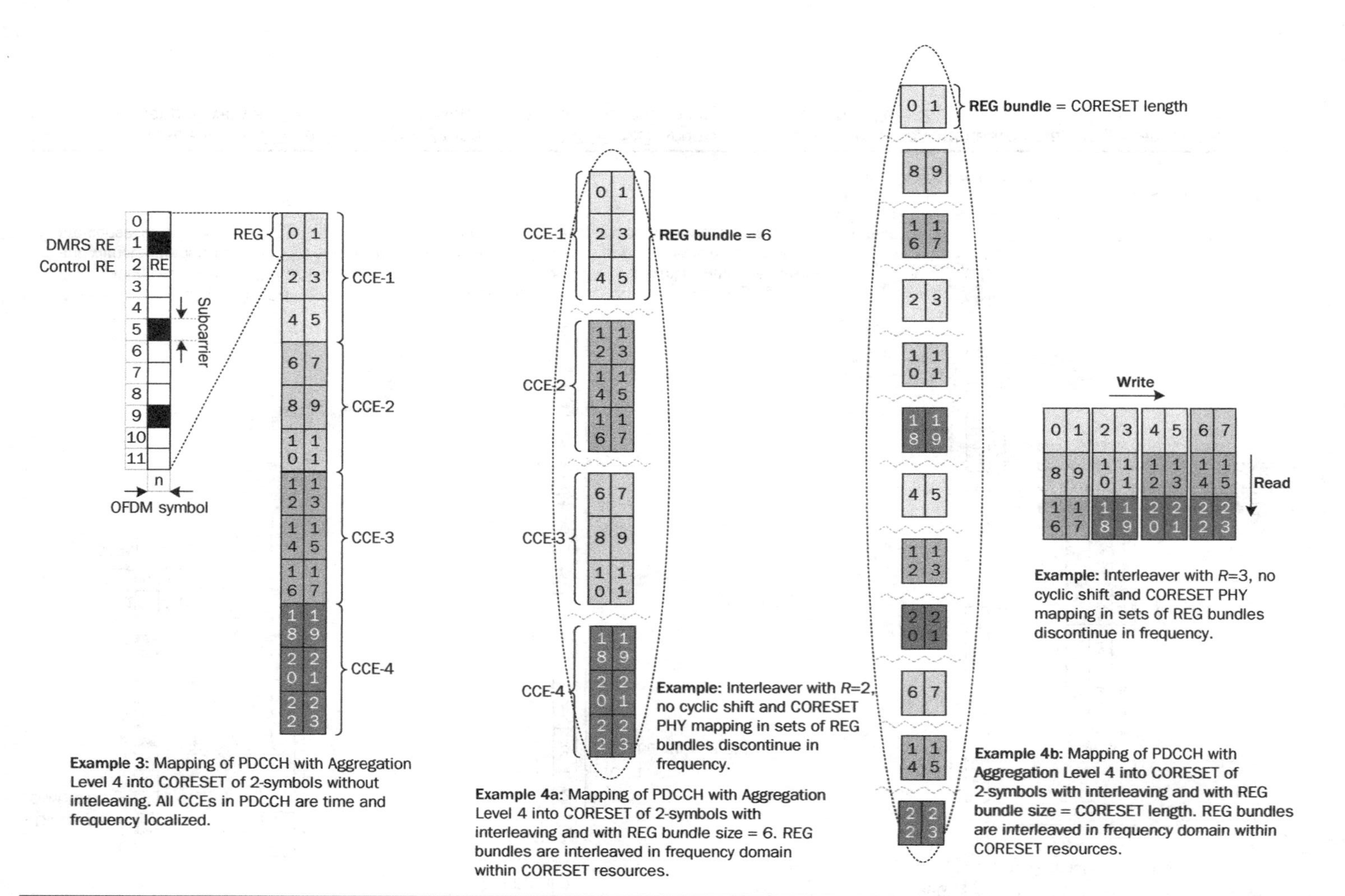

FIGURE 7.9 Illustration of CCE-to-REG mapping for a 2-symbol CORESET and four CCEs. Example 3: non-interleave mapping; Example 4a: interleaved mapping with an REG bundle = 6; Example 4b: interleaved mapping with an REG bundle = 2.

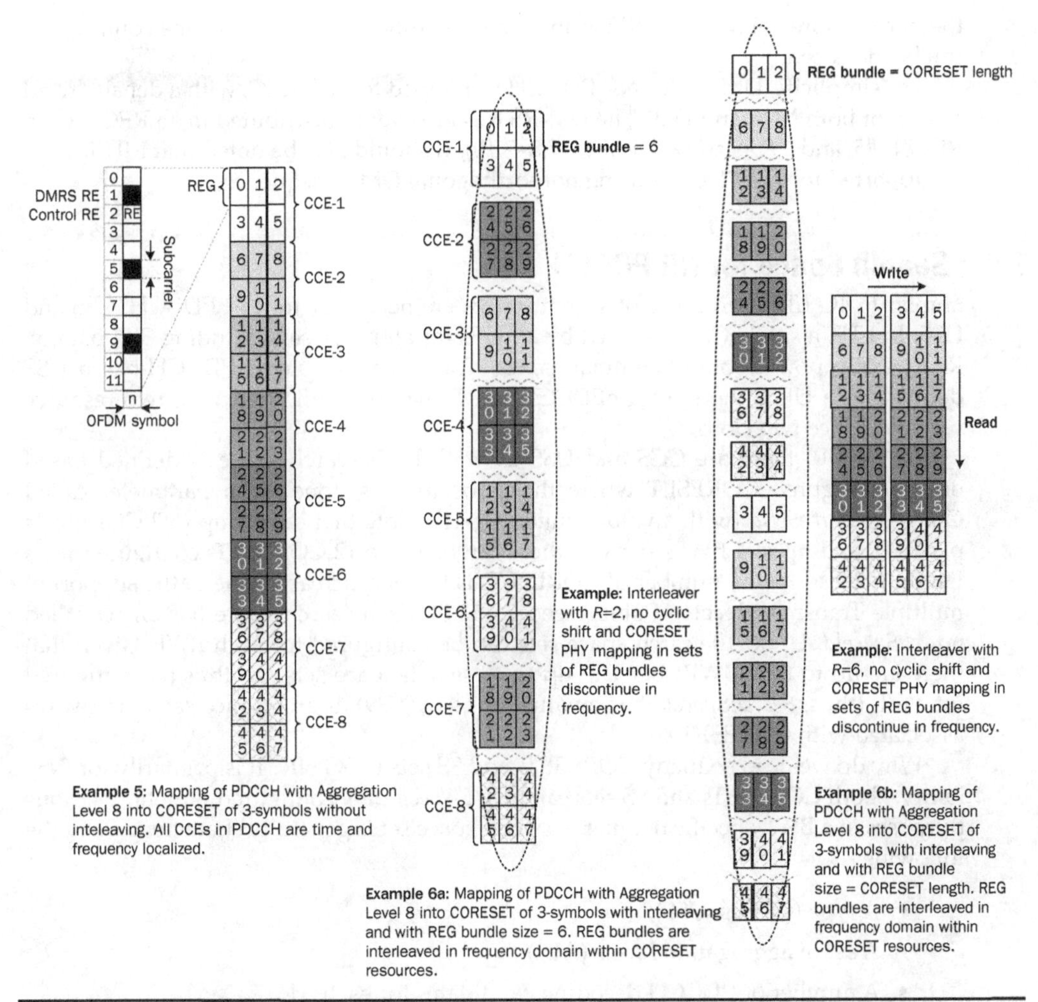

FIGURE 7.10 Illustration of CCE-to-REG mapping for a 3-symbol CORESET and eight CCEs. Example 5: non-interleave mapping; Example 6a: interleaved mapping with an REG bundle = 6; Example 6b: interleaved mapping with an REG bundle = 3.

The finalization of DM-RS pattern for PDCCH was also a result of evaluating different design alternatives. In particular,

- How much should the DM-RS density for NR PDCCH be within an REG?
- Should the DM-RS be present in every REG?

The density of DM-RS depends on many factors. Naturally, a higher density helps improve channel estimation at the expense of a resource dimension loss for the control data. Having DM-RS in every single REG makes it easier for both standardization and implementation efforts. However, it does not take advantage of the potential usage of DM-RS in one REG for one or more neighboring REGs so that DM-RS may be omitted in the one or more neighboring REGs and consequently, the resource for the control data can be increased.

For simplicity, DM-RS for NR PDCCH is mapped to each REG, with a density fixed at 1/4 for both NCP and ECP. The DM-RS REs are evenly distributed in an REG, i.e., in REs #1, #5, and #9 (as shown in Figs. 7.8–7.10). It should also be noted that MU-MIMO is supported for NR PDCCH using non-orthogonal DMRS sequences.

7.4 Search Space for NR PDCCH

Similar to 4G LTE, two types of search spaces are necessary for NR PDCCH: CSS and USS. In LTE, a single CSS is shared by all UEs for all purposes, including SIB, paging, RAR, group power control commands, etc. There is also a single PDCCH-based USS defined for a UE, although for ePDCCH, a UE may be configured with two resource sets and hence two USSs.

In 5G NR, there are CCS and USS as well. Each search space is defined based on a configured CORESET where the association is done via a parameter called *ControlResourceSetId* (with a value range of 0–11). Note that due to up to 3 CORESETs per BWP and up to 4 BWPs per UE, there can be up to 12 CORESETs configured for a UE (in Release 16, the number of CORESETs is further increased due to the support of multiple Transmit/Receive Points, see Sec. 13.8). Each search space has an ID called *SearchSpaceId*. Up to 10 search space sets can be configured for each BWP. Given that there are up to four BWPs per UE, up to 40 search space sets can thus be configured for a UE (thus, *SearchSpaceId* has a value range of 0–39). Search space set #0 is always associated with CORESET #0.

Why do we need so many CORESETs and SS sets in 5G NR? It is primarily for flexibility – both CORESETs and SS sets can be *UE-specifically* configured targeting various purposes in a BWP specific manner. A search space set is generally characterized by the following:

- A type (CSS vs. USS),
- A set of aggregation levels (ALs),
- A number of PDCCH decoding candidates for each AL,
- PDCCH monitoring occasions for the set of PDCCH decoding candidates,
- A duration in terms of a number of consecutive slots to monitor, and
- The associated DCI format(s).

For CSS, the definition is closely tied with the purposes, namely, for scheduling SIB1, OSI, paging and RAR, and group-specific control (e.g., SFI, pre-emption indicator or PI, TPC, SRS switching, etc.). Accordingly, there are five types of CCS defined:

- Type0-PDCCH CSS,
- Type0A-PDCCH CSS,
- Type1-PDCCH CSS,
- Type2-PDCCH CSS, and
- Type3-PDCCH CSS.

Type0-PDCCH CSS is for SIB1 reception and is configured only on the primary cell of the master cell group or MCG (recall that a UE may be configured with two groups, one MCG and the one is a secondary cell group or SCG. Each group may have one or more CCs.). This implies that the UE is only required to monitor SIB1 using the type0-PDCCH CSS on the primary cell of the MCG. For other secondary cells, the SIB1 can be delivered to the UE by dedicated messaging. The Type0-PDCCH CSS set can be configured:

- By *pdcch-ConfigSIB1* in MIB, or,
    - Note that the search space configured by *pdcch-ConfigSIB1* is always search space #0.
- By *searchSpaceSIB1* in *PDCCH-ConfigCommon*, or
- By *searchSpaceZero* in *PDCCH-ConfigCommon*,

for a DCI format with CRC scrambled by SI-RNTI. The information element *PDCCH-ConfigCommon* contains a set of parameters related to PDCCH and is used to configure cell-specific PDCCH parameters by SIB1 or dedicated signaling. The information element (IE) of *pdcch-ConfigSIB1* is specified in Ref. [5], as shown below:

```
PDCCH-ConfigSIB1 ::=                    SEQUENCE {
    controlResourceSetZero                  ControlResourceSetZero,
    searchSpaceZero                         SearchSpaceZero
}
```

In the initial DL BWP, the search space is always search space #0. In other DL BWPs, either search space #0 or searchSpaceSIB1 can be used. For simplicity and flexibility, similar to CORESET #0, search space #0 can be used in an active DL BWP other than #0 but only if the active DL BWP and the initial DL BWP have the same SCS and the same CP length, and the active DL BWP includes all RBs of CORESET #0. Note that a gNB may choose to NOT configure *searchSpaceSIB1* or *searchSpaceZero* in a DL BWP. In this case, the UE is not required to monitor Type0-PDCCH CSS for SIB1 reception.

Type0A-PDCCH CSS and Type2-PDCCH CSS are for OSI and paging reception, respectively. Both are configured only on the primary cell of the MCG. Accordingly, the UE is required to monitor OSI using Type0A-PDCCH CSS (or Paging using Type2-PDCCH CSS) on the primary cell of the MCG only. The configuration of Type0A-PDCCH-CSS and Type2-PDCCH-CSS is done via *searchSpaceOtherSystemInformation* and *pagingSearchSpace* in *PDCCH-ConfigCommon* for a DCI format with the CRC scrambled by SI-RNTI and P-RNTI, respectively. Note that such a configuration can be done for any DL BWP configured for the UE, up to a gNB's implementation.

The set of aggregation levels and the maximum number of decoding candidates for Type0-PDCCH CSS, Type0A-PDCCH CSS, and Type2-PDCCH CSS are defined in Table 7.1. Note that the support of aggregation Level 16, not in LTE PDCCH, is to improve NR PDCCH coverage for broadcast traffic.

For monitoring of paging messages, a UE needs to further determine a paging occasion (PO), which is defined as a set of PDCCH monitoring occasions and can consist of multiple time slots where a paging DCI can be sent. Within one paging frame (PF), which is defined as one radio frame (10 ms), there may be one or multiple PO(s). The

CCE aggregation level	Number of candidates
4	4
8	2
16	1

Table 7.1 CCE Aggregation Levels and the Maximum Number of PDCCH Candidates for CSS Sets Configured by *searchSpaceSIB1*

SFN for the PF (denoted by SFN_{paging}) and the index of the PO (denoted by i_s) for paging for a UE are determined by the following formulae [16 (Sec. 7.1)]:

$$\mathrm{mod}((SFN_{paging} + PF_{offset}), T) = (T/N)^{*}\mathrm{mod}(UE_{ID}, N) \tag{7.3}$$

$$i_s = \mathrm{mod}(\mathrm{floor}\ (UE_{ID}/N), N_s)$$

where:

- T: the dis-continuous reception (DRX) cycle of the UE,
- N and PF_{offset}: a number of total paging frames in T and an offset used for PF determination, respectively, which are derived based on the configuration parameter *nAndPagingFrameOffset*, which may take values of $\{T, T/2, T/4, T/8, T/16\}$,
- N_s: a number of paging occasions for a PF, which is configurable with values of 1, 2, and 4,
- UE_{ID} = mod(5G-S-TMSI, 1024), where 5G-S-TMSI is a 48-bit string defined in Ref. [17].

For paging, PDCCH without a corresponding PDSCH can also be used in order to convey *Short Messages*. Two types of short messages are defined [16]:

- A 1-bit indication of a system information update,
- A 1-bit indication of an earthquake and tsunami warning system (ETWS) primary notification and/or an ETWS secondary notification and/or a commercial mobile alert system (CMAS) notification.

Type1-PDCCH CSS is for the random access response related operation, and is present only on the primary cell of the MCG or the SCG. The configuration of Type1-PDCCH CSS is done via *ra-SearchSpace* in *PDCCH-ConfigCommon* for a DCI format with CRC scrambled by RA-RNTI or TC-RNTI on the primary cell.

Type3-PDCCH CSS is intended for operations for a *group* of UEs. Configuration of Type3-PDCCH CSS can be on any cell (primary or secondary) configured for a UE. The group-specific operations may include:

- Slot format indication (see Sec. 5.7), using a DCI format called DCI format 2-0 with the CRC scrambled by an SFI-RNTI,
- DL pre-emption indication (see Sec. 13.5), using a DCI format called DCI format 2-1 with the CRC scrambled by an INT-RNTI,

- Group power control for PUSCH [6], using a DCI format called DCI format 2-2 with the CRC scrambled by a TPC-PUSCH-RNTI,
- Group power control for PUCCH [6], using a DCI format called DCI format 2-2 with CRC scrambled by TPC-PUCCH-RNTI,
- Group power control for SRS [6], using a DCI format called DCI format 2-3 with the CRC scrambled by a TPC-SRS-RNTI, for SRS operation under carrier switching.

Note that additional group-specific DCI formats were introduced in Release 16, to be discussed in Sec. 7.5.

For additional scheduling flexibility, a CSS may also carry UE-specific DCIs with the CRC scrambled by UE specific RNTIs (e.g., C-RNTI, MCS-C-RNTI, SP-CSI-RNTI, or CS-RNTI).

For USS, it is configured by *SearchSpace* in PDCCH-Config with *searchSpaceType = ue-Specific* for DCI formats with the CRC scrambled by a UE-specific RNTI. The possible aggregation levels include 1, 2, 4, 8, and 16, while the number of decoding candidates for an aggregation level is configurable (e.g., 0, 1, 2, 3, 4, 5, 6, and 8).

Different from 4G LTE where PDCCH is required to be monitored in every subframe when the UE is active, an NR UE can be configured with a PDCCH monitoring periodicity larger than one slot so that the UE is only required to monitor PDCCH every $N > 1$ slots, where N can be up to 2560. This provides the necessary flexibility for agNB to manage PDCCH transmissions and for potential power savings for a UE. It is worth noting that the set of possible periodicity values (including the maximum configurable periodicity value) depend on a DCI format, tailored with the actual needs. As an example, for DCI format 2-1, only values of {1, 2, 4} (slots) can possibly be configured for a UE; for DCI format 2-0, only values {1, 2, 4, 5, 8, 10, 16, 20} (slots) can possibly be configured for a UE.

To accommodate different service needs (e.g., eMBB vs. URLLC), different cases of PDCCH monitoring are introduced, namely:

- Case 1: A fixed location in a slot with each monitoring occasion up to three consecutive symbols.
  - Case 1-1: Up to first three symbols in a slot. This is most applicable for scheduling of eMBB type of traffic.
  - Case 1-2: Up to three symbols in other locations of a slot. This provides more flexibility in PDCCH transmissions in a slot.
- Case 2: other than Case 1.
  - This is particularly useful for ultra-low latency services such as URLLC. As an example [18], a UE may report whether it can support (X, Y) equal to:
    - (7, 3) only, or
    - (4, 3) and (7, 3) only, or
    - (2, 2), (4, 3) and (7, 3),

    where X represents the minimum time separation between two consecutive transmissions of PDCCH, and Y represents the span of the PDCCH monitoring (2 or 3 symbols). A (7, 3) combination implies that the UE is capable of monitoring PDCCH in every seven symbols with a span of up to three symbols in each monitoring occasion. This makes it possible to schedule URLLC for the UE as frequently as every seven symbols. A (2, 2) combination implies that a UE can be scheduled with URLLC as frequently as every two symbols, necessary for some time-stringent URLLC services [see Chap. 13].

For both CSS and USS, for a search space set s associated with CORESET p, the CCE indexes for aggregation level L corresponding to PDCCH candidate $m_{s,n_{CI}}$ of the search space set in slot $n^{\mu}_{s,f}$ for an active DL BWP of a serving cell corresponding to a carrier indicator field value n_{CI} are given by [6] (Sec. 10.1).

$$L \cdot \left\{ \left(Y_{p,n^{\mu}_{s,f}} + \left\lfloor \frac{m_{s,n_{CI}} \cdot N_{\text{CCE},p}}{L \cdot M^{(L)}_{s,\max}} \right\rfloor + n_{CI} \right) \bmod \lfloor N_{\text{CCE},p}/L \rfloor \right\} + i \tag{7.4}$$

Compared with Eq. (7.1) for LTE ePDCCH, it can be observed that the hashing function of Eq. (7.4) for NR PDCCH is the same as that of LTE ePDCCH. Therefore, Fig. 7.7 also provides an illustration of the hashing function for NR PDCCH. Note that although the parameter b as in Eq. (7.1) is replaced by n_{CI} in Eq. (7.4), it still follows the same definition. Indeed, the management of cross-carrier scheduling in NR is the same as that in LTE, except that in NR, it is possible that the scheduling CC may have a different SCS from that of the scheduled CC.

As can be seen, monitoring of NR PDCCH is much more flexible than that of LTE PDCCH/ePDCCH, as all NR PDCCH parameters are configurable (CORESETs, SS, periodicity, aggregation levels, number of decoding candidates, periodicity of monitoring, monitoring symbols within a slot, etc.). This flexibility is necessary to facilitate scheduling of different service types, adaptation to a given slot structure, adaptation of a UE's channel conditions and traffic needs, etc.

On the other hand, it is important to ensure a reasonable *PDCCH false alarm probability* and manageable *implementation complexity* at the UE side. Generally, the PDCCH false alarm probability can be derived from three factors: an effective CRC length for PDCCH (N_1), the number of blind decodes (N_2), and the number of RNTIs associated with each blind decode (N_3). Note that a PDCCH CRC is always scrambled by an RNTI in order to identify the intended UE or the purpose. The number of blind decodes is determined by the number of PDCCH decoding candidates in combination with the number of DCI sizes associated with each PDCCH decoding candidate. In PDCCH decoding, a UE may perform de-scrambling for RNTI-matching by using multiple possible RNTIs based on the configuration, e.g., by using three different RNTIs including C-RNTI, MCS-C-RNTI, and CS-RNTI. The detailed RNTI values and usage can be found in Table 7.8. The PDCCH false alarm probability can be approximated by:

$$N_2 \times N_3 \times 2^{-N_1} \tag{7.5}$$

Note that the above formula is simplified, e.g. since the number of RNTIs for descrambling may be different for different decoding candidates. Assuming $N_1 = 21$ (bits), $N_2 = 44$, and $N_3 = 1$, the false alarm probability is roughly 2×10^{-5}.

A falsely detected PDCCH may result in unintended transmissions, thus creating unintended interference. If the PDCCH is for DL scheduling, a UE may transmit an intended PUCCH carrying HARQ for a corresponding non-existing PDSCH

SCS (kHz)	Maximum number of monitored PDCCH candidates
15	44
30	36
60	22
120	20

TABLE 7.2 Maximum Number of Monitored PDCCH Candidates per Slot for a Serving Cell

transmission. If the PDCCH is for UL scheduling, a UE may transmit an intended PUSCH. For both cases, if an SRS is triggered in the falsely detected PDCCH, a UE may transmit the unintended SRS as well. In addition to the interference aspect, unintended transmissions would also cause more UE power consumption, which is a critical factor for battery-powered devices.

By implementation, a UE may be able to perform some pruning to alleviate the issue of PDCCH false alarm. Pruning can be done by checking the integrity of the DCI contents. This is because some information fields in the DCI may not fully utilize all the entries provided by the respective bit-width. In addition, some combinations of information fields may be invalid.

Implementation complexity of PDCCH mainly comes from two aspects: channel estimation and blind decoding. Since PDCCH parameters are configurable, the resulting PDCCH implementation complexity can be overwhelmingly large, especially when considering the case when the UE is configured with multiple carriers (e.g., up to 16 carriers). To address this, there are two parameters to manage NR PDCCH implementation complexity:

- A maximum number of monitored PDCCH candidates, as shown in Table 7.2, which is related to the maximum number of blind decodes, and
- A maximum number of non-overlapped CCEs, as shown in Table 7.3, which is related to the maximum channel estimation complexity.

When the UE is configured with more than four carriers, the UE may additionally report a limit for a total maximum number of blind decodes over all the configured carriers. Such limit has to be taken into account by the gNB in determining the maximum number of blind decodes for the UE on a per carrier basis. Detailed procedure in handling the above two parameters aiming for reasonable NR PDCCH implementation complexity can be found in Sec. 10 of Ref. [6].

7.5 DCI Formats for NR PDCCH

Table 7.4 summarizes the DCI formats defined for NR PDCCH [19 (Sec. 7.3.1)]. DCI formats 0_0 and 1_0 are commonly called the *fallback* DCIs or the *compact* DCIs, only scheduling rank 1 transmissions. The non-fallback DCI formats include 0_1/0_2/1_1/1_2, where 0_2 and 1_2 were introduced in Release 16 intended for URLLC-related scheduling. The group DCI format 2_4 was newly introduced in Release 16 for URLLC as well (see Sec. 13.6), intended for UL URLLC-related operations (similar to DCI format 2_1,

SCS (kHz)	Maximum number of non-overlapped CCEs per slot
15	56
30	56
60	48
120	32

Table 7.3 Maximum Number of Non-Overlapped CCEs per Slot for a Serving Cell

DCI format	Usage
0_0	Scheduling of PUSCH in one cell
0_1	Scheduling of one or multiple PUSCH in one cell, or indicating Downlink feedback information for configured grant PUSCH (CG-DFI)
0_2	Scheduling of PUSCH in one cell
1_0	Scheduling of PDSCH in one cell
1_1	Scheduling of PDSCH in one cell, and/or triggering one shot HARQ codebook feedback
1_2	Scheduling of PDSCH in one cell
2_0	Notifying a group of UEs of the slot format, available RB sets, COT duration and search space set group switching
2_1	Notifying a group of UEs of the PRB(s) and OFDM symbol(s) where UE may assume no transmission is intended for the UE
2_2	Transmission of TPC commands for PUCCH and PUSCH
2_3	Transmission of a group of TPC commands for SRS transmissions by one or more UEs
2_4	Notifying a group of UEs of the PRB(s) and OFDM symbol(s) where UE cancels the corresponding UL transmission from the UE
2_5	Notifying the availability of soft resources
2_6	Notifying the power saving information outside DRX Active Time for one or more UEs
3_0	Scheduling of NR sidelink in one cell
3_1	Scheduling of LTE sidelink in one cell

Table 7.4 List of DCI Formats for NR PDCCH

which is intended for DL URLLC-related operations). DCI format 2_5 is for integrated access and backhaul (IAB) operations (see Chap. 17), while DCI format 2_6 is used as a wake-up signal (WUS) for UE power savings (see Sec. 7.7). DCI formats 3_0 and 3_1 were introduced in Release 16 for V2X operations (see Chap. 15).

Information element	Bitwidth (bits)	Description
Identifier	1	Always set to 1 to indicate a DL DCI format
FDRA	$\lceil \log_2(N_{RB}^{DL,BWP}(N_{RB}^{DL,BWP}+1)/2) \rceil$	Frequency domain resource assignment; where $N_{RB}^{DL,BWP}$ is the BWP bandwidth in RBs
TDRA	4	Time domain resource assignment; Bitwidth depends on configuration
VRB-to-PRB mapping	1	Whether or not to apply virtual RB to physical RB mapping (interleaved vs. non-interleaved PDSCH mapping)
MCS	5	Modulation and coding scheme
NDI	1	New data indicator
RV	2	Redundancy version
HARQ process number	4	An index of the HARQ process for the PDSCH
DAI	2	To indicate an accumulative index of the current PDSCH assignment (i.e., a counter)
TPC for PUCCH	2	2 bits power control commands for PUCCH
PUCCH resource indicator	3	An indicator of which PUCCH resource to use
PDSCH-to-HARQ-feedback timing indicator	3	An indicator of which slot HARQ feedback is due

Table 7.5 DCI Contents for DCI Format 1_0 When the CRC Is Scrambled by C-RNTI

The two fallback DCI formats have a same DCI size, where the size is typically invariant to the UE-specific configurations, as shown in Table 7.5 for DCI format 1_0 when the CRC is scrambled by the C-RNTI. Such invariance is important since if there is any ambiguity, e.g., due to a potential misalignment between a gNB and a UE during a new RRC (re)-configuration, the gNB can still use the fallback DCI formats to communicate with the UE (hence the name "*fallback*"). This also implies that the fallback DCI formats are not intended to be used for scheduling the UE with advanced features (which are typically UE-specifically configured). In contrast, the non-fallback DCI formats 0_1/0_2 and 1_1/1_2 have their size(s) depend on the UE-specific configurations. To illustrate, Table 7.6 shows the contents of DCI format 1_1. As can be seen, many information elements in the DCI have variable bit-widths, which are determined by the UE-specific configurations. It is worth noting that while the UE-specific configurations provide gNB scheduling flexibility, it imposes some burden on the UE side as the UE has to support all possible configurations resulting in a large number of possible DCI sizes. Some DCI sizes (including the 24-bit CRC), e.g., for DCI format 1_1, can exceed 100 bits.

Information element	Bitwidth (bits)	Description
Identifier	1	Always set to 1 to indicate a DL DCI format
Carrier indicator	0 or 3	3 bits when the UE is configured with cross-carrier scheduling; 0 otherwise
BWP indicator	0, 1, or 2	To indicate which BWP; Bitwidth depends on the number of configured BWPs for the UE
FDRA	Variable	Frequency domain resource assignment; bitwidth depends on resource allocation type (for details, refer to Chap. 8), BWP bandwidth, etc.
TDRA	0, 1, 2, 3, or 4	Time domain resource assignment; bitwidth depends on configuration
VRB-to-PRB mapping	0 or 1	Whether or not to apply virtual RB to physical RB mapping (interleaved vs. non-interleaved PDSCH mapping)
PRB bundle size Indicator	0 or 1	1 bit if configured to indicate one of two PRB bundle size (with the same precoding)
Rate matching indicator	0, 1, or 2	Indicate which rate matching pattern group(s) (up to 2 groups) to use
ZP CSI-RS trigger	0, 1, or 2	Indicate zero-power CSI-RS for rate matching
MCS of TB1	5	Modulation and coding scheme for transport block 1
NDI for TB1	1	New data indicator for TB1
RV for TB1	2	Redundancy version for TB1
MCS of TB2	5	Modulation and coding scheme for TB2, only present if configured with a maximum of 2 codewords
NDI for TB2	1	New data indicator for TB2, only present if configured with a maximum of 2 codewords
RV for TB2	2	Redundancy version for TB2, only present if configured with a maximum of 2 codewords
HARQ process number	4	An index of the HARQ process for the PDSCH
DAI	0, 2, or 4	2 bits for dynamic HARQ codebook without CA/DC; 4 bits for dynamic HARQ codebook with CA/DC; 0 otherwise. To indicate an accumulative index of the current PDSCH assignment (i.e., a counter) and a total number of PDSCH assignments so far (i.e., a total)
TPC for PUCCH	2	2 bits power control commands for PUCCH
PUCCH resource indicator	3	An indicator of which PUCCH resource to use
PDSCH-to-HARQ-feedback timing indicator	0, 1, 2, or 3	An indicator of which slot HARQ feedback is due
Antenna port(s)	4, 5, or 6	Antenna port(s) for PDSCH
TCI	0 or 3	Transmission configuration indication
SRS request	2 or 3	To request for aperiodic SRS transmissions
CBGTI	0, 2, 4, 6, or 8	Code-block-group transmission information
CBGFI	0 or 1	CBG flushing out information
DM-RS sequence initialization	1	A value of 0 or 1 for PDSCH DM-RS sequence initialization

Table 7.6 DCI Contents for DCI Format 1_1

For both DL and UL, the unicast DCI formats 0_0/0_1/0_2/1_0/1_1/1_2 can be used for configured grant (CG) activation and release. Note for UL, there are two types of CGs, where type 1 UL CG is completely managed by an RRC configuration, while type 2 UL CG is subject to DCI-based activation and release. In this case, the DCI formats have some information fields which are set to pre-defined values, with the CRC scrambled by CG (configured grant)-RNTI. In particular, we have:

- NDI = 0,
- HARQ process number = 0000, and
- RV = 00.

In addition, to validate the release of a configured grant, the following fields are additionally pre-defined:

- MCS = 11111 and
- FDRA = all 1's.

For improved Downlink control overhead efficiency, it is also possible to use a single DCI to release two or more UL configured grants, a new feature introduced in Release 16 [6].

The predefined setting of the values discussed above not only helps identify activation or release of a configured grant, but also helps reduce the false alarm probability of the corresponding PDCCH. This is because a falsely detected PDCCH has to have the same predefined values for these information fields in order for it to be possibly considered as a valid PDCCH. Such improved false alarm probability for activation or release of configured grants is very important. This is because if an activation or a release is falsely detected, the resulting unintended transmissions are no longer a one-shot transmission, but may prolong for a while.

7.6 Physical Layer Block Diagram for NR PDCCH

Figure 7.11 illustrates the physical layer diagram for NR PDCCH. The information bits for NR PDCCH are used to first generate a 24-bit CRC for protection, followed by interleaving, coding, and rate matching. The coding for NR PDCCH is also known as a 24-bit CRC-assisted (CA) Polar coding with interleaving, with additional details shown in Fig. 7.12.

A 16-bit RNTI is used to provide a mask (by a bitwise XOR operation) onto the last 16 CRC bits, where the possible RNTI values and usage are summarized in Tables 7.7 and 7.8 [20], respectively. The 24-bit CRC is calculated using the following generator polynomial [19]:

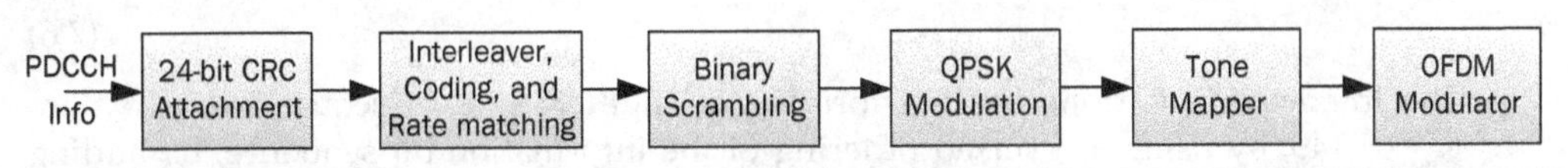

FIGURE 7.11 Physical layer block diagram for NR PDCCH.

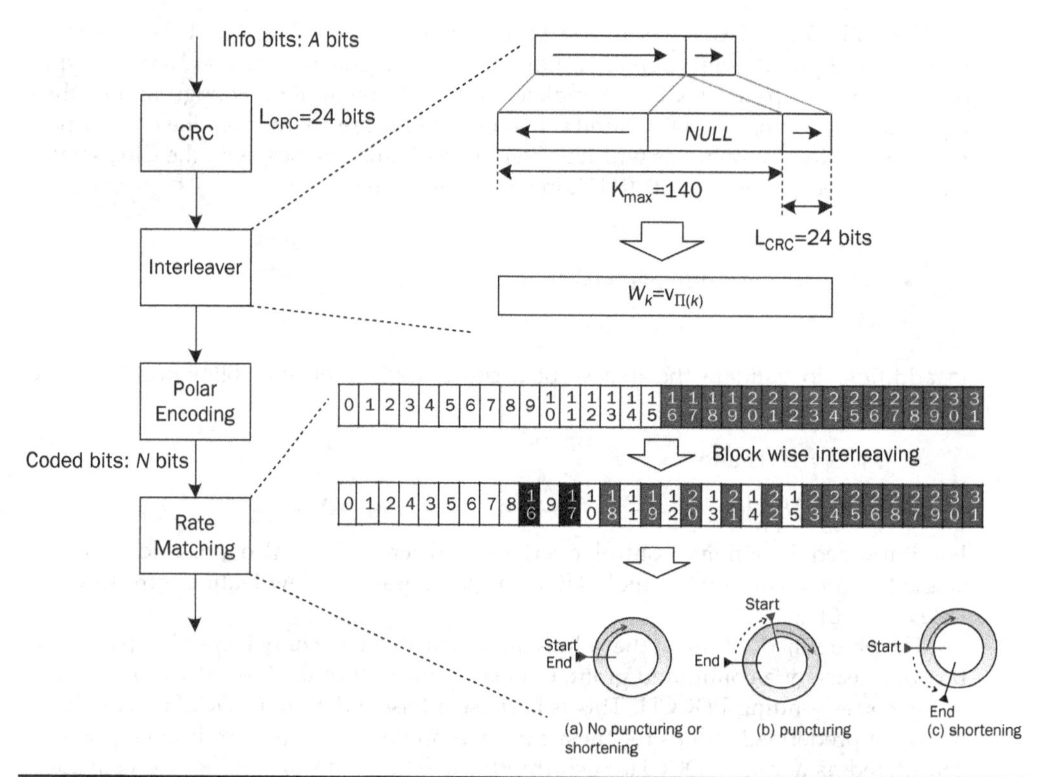

FIGURE 7.12 Polar codes channel coding chain for NR PDCCH.

Value (hexa-decimal)	RNTI
0000	N/A
0001–FFEF	RA-RNTI, Temporary C-RNTI, C-RNTI, MCS-C-RNTI, CS-RNTI, TPC-PUCCH-RNTI, TPC-PUSCH-RNTI, TPC-SRS-RNTI, INT-RNTI, SFI-RNTI, and SP-CSI-RNTI
FFF0–FFFD	Reserved
FFFE	P-RNTI
FFFF	SI-RNTI

TABLE 7.7 RNTI values in 5G NR

$$g_{CRC24C}(D) = [D^{24} + D^{23} + D^{21} + D^{20} + D^{17} + D^{15} + D^{13} + D^{12} + D^{8} + D^{4} + D^{2} + D + 1] \tag{7.6}$$

The interleaving is done by first forming a length K_{max} + 24 sequence bits, where K_{max} = 140, by using a reversed ordering of the information bit sequence, by adding NULL bits when necessary, and by appending the 24 CRC bits. The formed K_{max} + 24

RNTI	Usage
P-RNTI	Paging and system information change notification
SI-RNTI	Broadcast of system information
RA-RNTI	Random access response
Temporary C-RNTI	Contention resolution (when no valid C-RNTI is available)
Temporary C-RNTI	Msg3 transmission
C-RNTI, MCS-C-RNTI	Dynamically scheduled unicast transmission
C-RNTI	Dynamically scheduled unicast transmission
MCS-C-RNTI	Dynamically scheduled unicast transmission
C-RNTI	Triggering of PDCCH ordered random access
CS-RNTI	Configured scheduled unicast transmission (activation, reactivation, and retransmission)
CS-RNTI	Configured scheduled unicast transmission (deactivation)
TPC-PUCCH-RNTI	PUCCH power control
TPC-PUSCH-RNTI	PUSCH power control
TPC-SRS-RNTI	SRS trigger and power control
INT-RNTI	Indication pre-emption in DL
SFI-RNTI	Slot format indication on the given cell
SP-CSI-RNTI	Activation of semi-persistent CSI reporting on PUSCH

Table 7.8 Usage of RNTIs

sequence bits are then interleaved by using an interleaving pattern $\Pi_{IL}^{max}(m)$, where the pattern is defined by Table 5.3.1.1-1 in Ref. [19]. This interleaving would place some CRC bits in earlier positions in the sequence of bits, so that in some cases, the earlier CRC bits can help the decoding process via CRC-checking for early decoding termination before reaching the end of the payload.

Polar encoding is performed using a Polar code of length $N = 128$, 256, or 512 bits as determined by the DCI size and the aggregation level for the DCI. The code's generator matrix is recursively constructed from a 2×2 kernel using the Kronecker product until the target length is reached: $G_N = \begin{bmatrix} 1 & 0 \\ 1 & 1 \end{bmatrix}^{\otimes \log_2 N}$. The DCI payload is arranged with 0-valued (frozen) bits according to a reliability order and the rate-matching scheme to obtain the encoder input vector u, which is multiplied by the generator matrix, providing the Polar code block. The reliability order is provided by Table 5.3.1.2-1 in Ref. [19] using an ordered sequence of length 1024 (the maximum value of 1024 is due to the Polar coding for Uplink control). A Polar code of length N is a subset of the length-1024 Polar sequence, whose elements of values from 0 to $N - 1$, ordered in ascending order of reliability, are used to form the length N Polar code.

After Polar encoding, rate matching interleaving is performed to further enhance performance, where block-wise interleaving is used. The sequence of N encoded bits is divided into 32 sub-blocks, each of $N/32$ bits, with the sub-block interleaving pattern as shown in Fig. 7.12. After the rate-matching interleaving, the output is written into a length-N circular buffer, where one of the following modes is taken depending on the number of coded bits for transmission (M), the number of information bits, etc.:

- Puncturing: realized by selecting bits from position ($N-M$) to position ($N-1$) from the circular buffer, or
- Shortening: realized by selecting bits from position 0 to position $M-1$ from the circular buffer, or
- Repetition: realized by selecting all bits from the circular buffer, and additionally repeating ($M-N$) consecutive bits with the smallest index bit from the circular buffer.

Note that punctured and shortened bit locations are marked as frozen bits and are not used to carry the encoder input mapped from the DCI payload.

Scrambling for NR PDCCH is done in a bit-wise manner, i.e.:

$$\tilde{b}(i) = \left(b(i) + c(i)\right) \bmod 2 \tag{7.7}$$

where $b(0), \ldots, b(M_{bit}-1)$ is the block of M_{bit} bits after rate matching, and $c(i)$ is the binary scrambling sequence based on a length-31 Gold sequence [1], which is initialized as follows:

- For CSS: based on the physical cell ID of the serving cell, so that the PDCCH can be accessed by all UEs in the serving cell, while providing the inter-cell interference randomization, and
- For USS: either based on the physical cell ID of the serving cell, or a configurable ID and C-RNTI of the UE so that the UE-specific randomization can be realized.

7.7 Power Saving Considerations

Compared with 4G LTE, 5G NR may pose a more challenging task for battery-powered UEs, considering the following factors:

- Higher throughput (at least one magnitude higher): resulting in higher baseband processing requirement;
- Lower latency: also leading to higher baseband power due to shorter slot durations, more frequent PDCCH monitoring, etc.;
- Higher band (mmW) operations: requiring higher RF power due to the necessity of separating RF and IF (intermediate frequency) chips, and the beamforming operation in Downlink and Uplink using a massive number of antennas.

To reduce power consumption, a wake-up signal (WUS) was introduced in Release 16, where the WUS is a special PDCCH sent by a gNB before a DRX ON duration to indicate whether the UE should stay active to receive new data or skip the current DRX ON duration till the next DRX ON duration. This is illustrated in Fig. 7.13. The WUS is sent outside of the DRX active time, using a group-common DCI with CRC scrambled by a new RNTI called PS-RNTI. The DCI format is called DCI format 2_6 (see Table 7.4), which carries one or more information blocks, each intended for a specific UE:

- block number 1, block number 2,..., block number N,

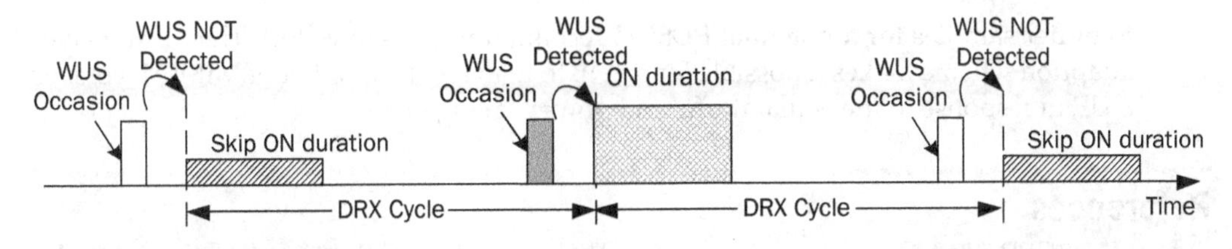

FIGURE 7.13 The WUS for UE power savings in 5G NR.

where the starting position of a block for a UE is determined by a higher-layer parameter. The block information contains the following:

- A 1-bit wake-up indication, and
- A Scell dormancy indication, 0 to 5 bits.

Each bit in the Scell dormancy indication corresponds to one of the Scell group(s) configured by higher layers, signaling the corresponding Scells in the Scell group to transition to/from dormancy. Dormancy is a new "sub-state" for a Scell for the purpose of power saving, where a UE is not required to monitor PDCCH for the Scell when the Scell is in this sub-state. This is illustrated in Fig. 7.14. Note also that while the group-common WUS DCI is used for the state transition outside the DRX active time, a UE-specific DCI (based on the existing DCI formats, e.g., 0_1 and 1_1) can be used for the state transition within the DRX active time.

Note that there are also other UE power saving schemes, e.g., cross-slot scheduling, MIMO layer adaptation, etc. The cross-slot scheduling scheme provides a relaxation on the UE side so that the UE knows that any PDSCH scheduled by a PDCCH has a minimum delay of one slot. Such knowledge enables the UE to have the so-called "micro-sleep," since it can go to sleep if there is no PDCCH detected in a slot (e.g., no need

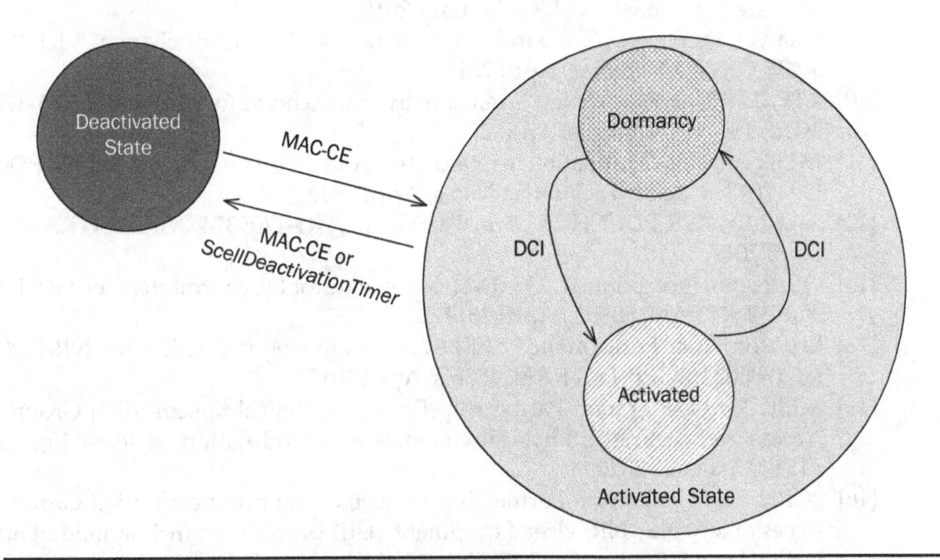

FIGURE 7.14 Scell state transitions for UE power savings in 5G NR.

to buffer samples for a potential PDSCH reception in the same slot). The MIMO layer adaption scheme makes it possible for a gNB to manage the number of MIMO layers for a UE in response to the actual traffic and power saving needs.

References

[1] 3GPP, "3rd Generation Partnership Project; Technical Specification Group Radio Access Network; Evolved Universal Terrestrial Radio Access (E-UTRA); physical channels and modulation (Release 16)," 36.211, v16.1.0, March 2020.

[2] 3GPP, "3rd Generation Partnership Project; Technical Specification Group Radio Access Network; Evolved Universal Terrestrial Radio Access (E-UTRA); Downlink multiple input multiple output (MIMO) enhancement for LTE-advanced (Release 11)," 36.871, v11.0.0, December 2012.

[3] 3GPP, "3rd Generation Partnership Project; Technical Specification Group Radio Access Network; Evolved Universal Terrestrial Radio Access (E-UTRA); physical layer procedures (Release 11)," 36.213, v11.13.0, September 2017.

[4] 3GPP, "3rd Generation Partnership Project; Technical Specification Group Radio Access Network; NR; physical layer procedures for data (Release 15)," 38.214, v15.8.0, January 2020.

[5] 3GPP, "3rd Generation Partnership Project; Technical Specification Group Radio Access Network; NR; Radio Resource Control (RRC) protocol specification (Release 15)," 38.331, v15.8.0, January 2020.

[6] 3GPP, "3rd Generation Partnership Project; Technical Specification Group Radio Access Network; NR; physical layer procedures for control (Release 16)," 38.213, v16.0.0, January 2020.

[7] 3GPP MCC, "Final Report of 3GPP TSG RAN WG1 #88 v1.0.0. (Athens, Greece, 13th – 17th February 2017)," R1-1704172, 3GPP TSG RAN1#88bis, April 2017.

[8] 3GPP, "3rd Generation Partnership Project; Technical Specification Group Radio Access Network; Study on channel model for frequencies from 0.5 to 100 GHz (Release 14)," 38.901, v14.3.0, January 2018.

[9] Huawei, HiSilicon, "Transmission schemes of DL control channel," R1-1704200, 3GPP TSG RAN1#88bis, April 2017.

[10] ZTE, ZTE Microelectronics, "Transmit diversity scheme for NR control," R1-1704368, 3GPP TSG RAN1#88bis, April 2017.

[11] LG Electronics, "Evaluation results on transmit diversity schemes for NR-PDCCH," R1-1704898, 3GPP TSG RAN1#88bis, April 2017.

[12] Samsung, "NR-PDCCH Tx Diversity Scheme," R1-1705374, 3GPP TSG RAN1#88bis, April 2017.

[13] Qualcomm Incorporated, "Tx diversity scheme for DL control channel," R1-1705601, 3GPP TSG RAN1#88bis, April 2017.

[14] Ericsson, "On Performance of SFBC and Pre-Coder Cycling for NR-PDCCH," R1-1706024, 3GPP TSG RAN1#88bis, April 2017.

[15] 3GPP, "3rd Generation Partnership Project; Technical Specification Group Radio Access Network; NR; Physical channels and modulation (Release 15)," 38.211, v15.8.0, January 2020.

[16] 3GPP, "3rd Generation Partnership Project; Technical Specification Group Radio Access Network; NR; User Equipment (UE) procedures in Idle mode and RRC Inactive state (Release 15)," 38.304, v15.6.0, January 2020.

[17] 3GPP, "3rd Generation Partnership Project; Technical Specification Group Services and System Aspects; System architecture for the 5G System (5GS); Stage 2 (Release 15)," 23.501, v15.8.0, December 2019.
[18] 3GPP, "3rd Generation Partnership Project; Technical Specification Group Radio Access Network; NR; User Equipment (UE) radio access capabilities (Release 15)," 38.306, v15.8.0, January 2020.
[19] 3GPP, "3rd Generation Partnership Project; Technical Specification Group Radio Access Network; NR; Multiplexing and channel coding (Release 16)," 38.212, v16.0.0, January 2020.
[20] 3GPP, "3rd Generation Partnership Project; Technical Specification Group Radio Access Network; NR; Medium Access Control (MAC) protocol specification (Release 15)," 38.321, v15.8.0, January 2020.

CHAPTER 8

Downlink Data Operation

Only one transmission scheme is defined in NR for the PDSCH that is used for all PDSCH transmissions [1]. In contrast, LTE has defined 10 transmission modes. One lesson learned from LTE was that introducing many transmission modes was unnecessary, practical deployments will only focus on the most essential ones.

The NR DL transmission scheme is closest to Transmission Mode 10 in LTE. It relies on UE-specific DM-RS. Another lesson from LTE was that "always on" signals should be avoided. Therefore, NR does not use any common reference signals. The only guaranteed periodic signal in NR is the sparse SSB discussed in Chap. 6. For channel estimation for channel state information (CSI) feedback, CSI reference signals (CSI-RS) are transmitted. Although the number of different virtualized ports that can be represented via CSI-RS to different UEs is virtually unlimited, the number of data ports used for one UE in one PDSCH transmission in one carrier is limited up to eight. When co-scheduling more than one UE, up to 12 port DM-RS can be orthogonally multiplexed in a non-transparent fashion using the DM-RS structure described in Sec. 8.12.

First, we discuss the NR data channel coding in Sec. 8.1, which is common between DL and UL.

8.1 Channel Coding for Data

This section applies to both the DL and UL.

For the channel coding, a significant debate occurred about which coding scheme to select. The proposed candidates were

- Enhanced Turbo code
- Low density parity check code (LDPC)
- Polar code

The main benefit of an enhanced Turbo code was its familiarity, as it had been used for LTE. The main drawback is that the decoding complexity (in terms of the number of arithmetic operations in the decoder) did not scale with the number of parity bits. That is, even if the data was transmitted at very high code rate, i.e., very few parity bits were included, the decoding complexity was almost the same as if all possible parity bits had been included. This is particularly costly for cellular systems, where the highest achievable peak rates, which are the envelope defining cases for decoding complexity, are typically with high code rates.

LDPC did not suffer from this drawback. At the highest code rates, the computational overhead of LDPC decoding operation is at its lowest. Another benefit of LDPC codes, in particular quasi-cyclic LDPC codes, which are lifted with permutations matrices from a common base graph, is that the larger code block lengths could be decoded with large degrees of parallelism [2]. This was more natural and efficient to implement compared to parallelized turbo decoders. One perceived drawback of LDPC was that for dual mode devices employing LTE and NR, the decoding hardware could not be re-used. However, given the much larger throughput and lower latency for NR relative to LTE, the relative amount of hardware re-use may not be significant in any case.

Finally, polar code in theory offered good performance, but to achieve performance comparable with Turbo and LDPC codes, one would have to implement polar codes with a list decoder [3] and an outer parity check (or CRC) code to prune the candidates of the list decoder. Such structures do not scale well at very high throughput and low latency [4] and can only be partially parallelized. As a compromise, even a combination was considered, where polar code would be used only for small code block sizes and LDPC for larger code block sizes. But having very different channel coding schemes on the hardware data path would have incurred a higher complexity compared with a single code solution. Moreover, at the time of technology selection, there was no proven polar code HARQ scheme presented that would have robust decoding performance for data operation, which was seen as a drawback for polar codes for data in general.

In the end, LDPC was selected for any code block size for NR data [5].

There was some debate on whether PBCH should be categorized as data cannel or control channel for the purpose of coding scheme selection. In the end, it was decided that PBCH uses polar code since the payload size for MIB was small.

In the remainder of this section, we discuss channel coding with LDPC codes.

A few of the general characteristics of the data channel coding are the following:

- LDPC coding is used for all block lengths.
- The same coding chain is used for DL and UL.
- Four redundancy versions are available.
- Intra-code-block bit-interleaving is applied after rate matching.

Much of the design philosophy for the specific LDPC structure used in NR is described in Ref. [2].

As it was mentioned, quasi-cyclic LDPC codes support different code block sizes by using a common base graph with variable lifts to the final binary parity matric. However, at the medium and small code block lengths, smaller lifts would be employed, and so any hardware designed for lift-parallel processing would be used less often. Additionally, smaller code blocks could be further optimized for better performance with a separate base graph, including support for lower code rate. After further study in 3GPP, it was found that close to optimum performance can be achieved with just two basic parity check protograph matrices. These are called base graph 1 (BG1) and base graph 2 (BG2). In general, base graph 1 is for larger CB and larger code rate, while base graph 2 is for smaller CB and lower code rates. In more detail, the BG is selected according to the following:

- If $A \leq 292$, or if $R \leq 0.25$, or if $A \leq 3824$ and $R \leq 0.67$, base graph 2 is used
  - where A is the number of information bits (without CRC) in the transport block (TB)
  - and R is the code rate of the initial transmission;
- otherwise base graph 1 is used.

The same base graph is used for initial transmission and retransmissions of the same TB. Since the code rate is reduced for retransmissions, this means that the above determination should be expressly based on the initial transmission only. This has a potential to create problems when the grant for the initial transmission was lost at the receiver. A solution considered for this problem was to add a dedicated DCI bit in grants to explicitly indicate the base graph. This was viewed wasteful and not adopted. Instead, the assumed solution is that the transmitter applies restrictions to the MCS of all retransmissions to ensure that the TBS calculation results in the same BG selection as for the initial transmission. The TBS determination procedure provides sufficient flexibility to find the same TBS for a retransmission.

The region of applicability for each base graph is shown in Fig. 8.1.

Figure 8.2 shows the general steps of encoding from the transmitter's perspective.

The decoding operation in the receiver is generally a sequence of the inverse operations performed in the reverse order relative to the steps in Fig. 8.2.

8.1.1 CRC Attachment

The transport block (TB) CRC is 16 or 24 bits based on TB size. For TB size (without CRC) $A \leq 3824$, CRC length 16 is used, and for $A > 3824$, CR length 24 is used. We note for comparison that for NR control, the CRC length is 24 bits. The smaller CRC size in some cases for data does not necessarily reflect any more tolerance to false pass of mis-decoded data compared to control. Rather, it reflects the fact that LDPC codes provide

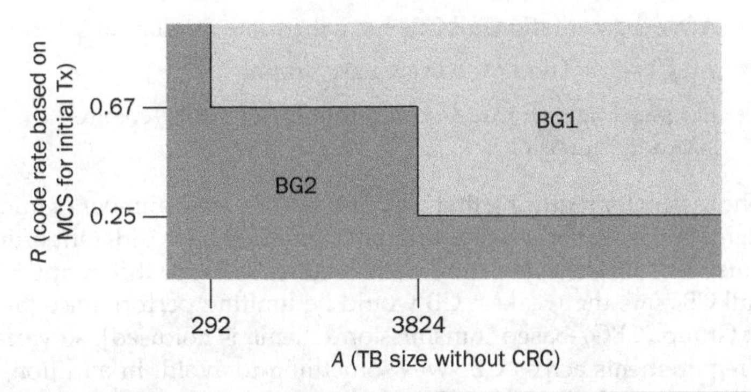

FIGURE 8.1 Region of applicability for LDPC base graph 1 and base graph 2.

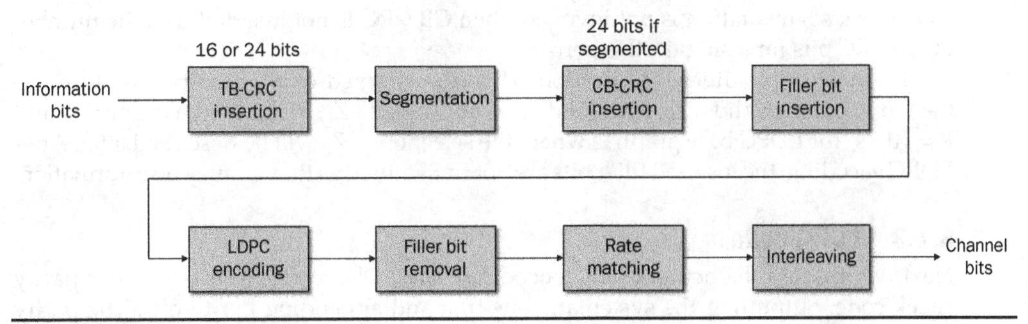

FIGURE 8.2 Channel coding for data.

some baseline protection against mis-decoding without any CRC already, because the LDPC decoding itself fails to converge when the decoding output would be incorrect. With added 16-bit CRC, the error detection performance can be raised to adequate levels already with relatively low overhead. For large TB sizes, where the CRC overhead is negligible, the added 24-bit CRC improves error detection performance even further.

The CRC scheme for data is the common type defined by a cyclic generator polynomial, one for each CRC length. TB CRC is appended to the end of the information bits.

8.1.2 Code Block Segmentation

The next step is code block (CB) segmentation. For the two different base graphs, the code block segmentation is invoked at different TB size thresholds.

- For LDPC base graph 1, the maximum input size without segmentation is: $K_{cb} = 8448$ (8424 information bits + 24-bit TB CRC)
- For LDPC base graph 2, the maximum input size without segmentation is: $K_{cb} = 3840$ (3824 information bits + 16-bit TB CRC)

The CB segmentation is invoked whenever the number of input bits, including TB CRC, is greater than K_{cb}. The CB segmentation algorithm was designed to ensure the following:

- CB size is byte-aligned.
- After segmentation, all CBs have the same number of information bits.
- All CBs in a TB use the same base graph.
- No padding is needed with either base graph (ensured by the method of TB size determination).

The reason for requiring that all CBs have the same number of information bits after segmentation is that otherwise different CBs could have different lifting sizes, resulting in somewhat different decoding SNR requirements for different CBs. Since if one CB fails all CBs fail, the weakest CB would be limiting performance [at least when Code Block Group (CBG)–based transmission scheme is not used], so variation in decoding SNR requirements across CBs was something to avoid. In addition, having the same lifting size through the entire TB makes the implementation more efficient.

If code block segmentation was invoked, the next step is 24-bit CB CRC insertion for each CB segment. CB CRC is appended to the end of the information bits. When the code block segmentation is not invoked, then CB CRC is not inserted, i.e., the number of CB CRC bits for a single CB is zero.

The next step is filler bit insertion. This is performed only when needed to make the CB to be of a valid size. The valid sizes are $K = 22 \cdot Z_C$ for LDPC base graph 1 and $K = 10 \cdot Z_C$ for LDPC base graph 2; where the selection of Z_C will be described later. After LDPC encoding, the inserted filler bits are not transmitted as they convey no information.

8.1.3 LDPC Code

Next, we discuss the actual LDPC encoding. The LDPC code itself is a linear parity check code, outputting the systematic bits first and appending parity bits. The parity bits are generated as the modulo-2 sum of a subset of information bits, where the subset is relatively small (hence the name low density) for most parity bits, and a different subset is used for each parity bit. The LDPC code is formally described in the specification [5] by its parity check matrix **H**. When the parity check matrix is multiplied with any valid encoded codeword, the result is an all-zero column vector, which can be written as

$$\mathbf{H} \times \begin{bmatrix} \mathbf{c} \\ \mathbf{w} \end{bmatrix} = \mathbf{0} \tag{8.1}$$

where **c** is the set of information bits and **w** is the set of parity bits, both organized as column vectors.

The key of the actual LDPC realization is how the parity check matrix is defined. As mentioned, the NR LDPC defines two base graphs. The first base graph, BG1 has a protograph size of 22 information nodes. The second base graph, BG2 has a protograph size of 10 nodes. Then the full size of the actual parity check matrix is achieved by so called lifting, which is replicating the protograph Z_C times, where Z_C is of the form $Z = a \cdot 2^{i_{LS}}$, and cyclically shifting connections between those replicas. The possible values of a, the set index i_{LS} and the lifting size Z are shown in Table 8.1 based on Ref. [5]. One important aspect of the design is that only a coarse set of liftings with exponential scaling was sufficient to cover the entire byte-level granularity for code blocks [2].

The transmitter determines Z_C as the smallest possible lifting size Z in Table 8.1, with which the lifting results in a size that is at least as great as the number of input bits, i.e., $K_b \cdot Z_C \geq K'$ where K' is the number of input bits without filler bits, and K_b is determined as $K_b = 22$ for base graph 1 and K_b is between 6 and 10, depending on the CB size, for base graph 2, with larger K_b values assigned to larger CB sizes. Note that irrespective of the actual value of K_b, the LDPC coding for base graph 2 assumes $K = 10 \cdot Z_C$, so the smaller K_b values increase the number of filler bits.

We note that the maximum lifting size Z is 384, which refers back to $K_{cb} = 8448 = 384 \cdot 22$ for base graph 1 and $K_{cb} = 3840 = 384 \cdot 10$ for base graph 2. A fully parallel decoding architecture at the largest lift could then be realized at the smaller block length of 3840, and overall decoding throughput was much improved from multiple base graphs.

The code rate (before rate matching) for base graph 1 is close to 1/3 and the code rate (before rate matching) for base graph 2 is close to 1/5. Therefore, the codeword length N is $N = 3 \cdot 22 \cdot Z_C = 66 \cdot Z_C$ for base graph 1 and is $N = 5 \cdot 10 \cdot Z_C = 50 \cdot Z_C$ for base graph 2.

Z		A							
		2	3	5	7	9	11	13	15
i_{LS}	0	2	3	5	7	9	11	13	15
	1	4	6	10	14	18	22	26	30
	2	8	12	20	28	36	44	52	60
	3	16	24	40	56	72	88	104	120
	4	32	48	80	112	144	176	208	240
	5	64	96	160	224	288	352		
	6	128	192	320					
	7	256	384						

TABLE 8.1 Possible Lift Sizes Z

With this, the size of the parity check matrix **H** should be $(66-22)\cdot Z_C = 44\cdot Z_C$ rows by $66\cdot Z_C$ columns for base graph 1 and it should be $(50-10)\cdot Z_C = 40\cdot Z_C$ rows by $50\cdot Z_C$ columns for base graph 2. However, there is a peculiar design detail of the NR LDPC codes that makes the size of **H** slightly larger. Namely, the first $2\cdot Z_C$ systematic bits are punctured and not transmitted. The reason for this is that these bits are part of the information bit subset of many parity bits, i.e., these are high degree systematic bits, therefore can be reliably recovered from the other bits. As a replacement for these bits, $2\cdot Z_C$ extra parity bits are generated and transmitted. With this, the actual size of the NR LDPC parity check matric **H** becomes the following:

- $46\cdot Z_C$ rows by $68\cdot Z_C$ columns for base graph 1
- $42\cdot Z_C$ rows by $52\cdot Z_C$ columns for base graph 2

The NR LDPC code is defined by a 46-by-68 protograph base graph matrix ($\mathbf{H}_{BG}$) for base graph 1 and a 42-by-52 protograph base graph matrix ($\mathbf{H}_{BG}$) for base graph 2. The matrix is lifted to the required size of $46\cdot Z_C$ - by - $68\cdot Z_C$ and $42\cdot Z_C$ - by - $52\cdot Z_C$, respectively, by replacing each entry of the protograph base graph matrix by a Z_C-by-Z_C size submatrix. Whenever the entry in the base graph matrix is zero, the replacement is a Z_C-by-Z_C all-zero submatrix. Whenever the entry in the base graph matrix is not zero, the replacement is a cyclically shifted version of a Z_C-by-Z_C identity matrix (which is a permutation matrix). The shift value is specified for each entry and each value of set index i_{LS}, individually. Protograph base graph matrix ($\mathbf{H}_{BG}$) for base graph 1 and base graph 2 are given in Tables 5.3.2-2 and 5.3.2-3 in Ref. [5], respectively. We note that a value 0 in those tables does not mean zero entry in the base graph. Rather, it means a non-zero entry in the base graph is to be replaced with a zero-shifted Z_C-by-Z_C identity matrix. Zero entry is specified for those entries whose joint row and column index does not appear in Tables 5.3.2-2 and 5.3.2-3 for base graph 1 and base graph 2, respectively. In Fig. 8.3, we give another illustration of the general structure of the 46-by-68 protograph base graph matrix ($\mathbf{H}_{BG}$) for base graph 1. The differently colored rectangular parts show the building blocks that the LDPC code design was focusing on.

The next step in the channel coding process is filler bit removal. As it was mentioned earlier, filler bits were added when needed to make the CB to be of a valid size. After encoding, these bits are removed and not transmitted.

8.2 Channel Code Rate Matching

The next step in the transmit data processing is rate matching. Rate matching is of a circular buffer type, similar to that used in LTE. The circular buffer is filled with the sequence of systematic bits and parity bits, N bits in total. For HARQ operation, four redundancy versions (RVs) can be read from the circular buffer, denoted with RV_i, $i = 0, \ldots, 3$. RV_i is assigned a starting bit location S_i in the circular buffer. To get the contents of RV_i, the coded bits are read out sequentially from the circular buffer, starting with the bit location at S_i. Note that, ideally, if the start of a HARQ retransmission could be at any point, it would be best to start at immediately the next consecutive bit in the soft buffer after where the previous HARQ transmission ended. This would require, however, that the receiver knows where the previous transmission ended, which in turn assumes that the receiver decoded the grant of all previous transmissions of the same TB. The latter is not a robust assumption. It was decided that the receiver must be

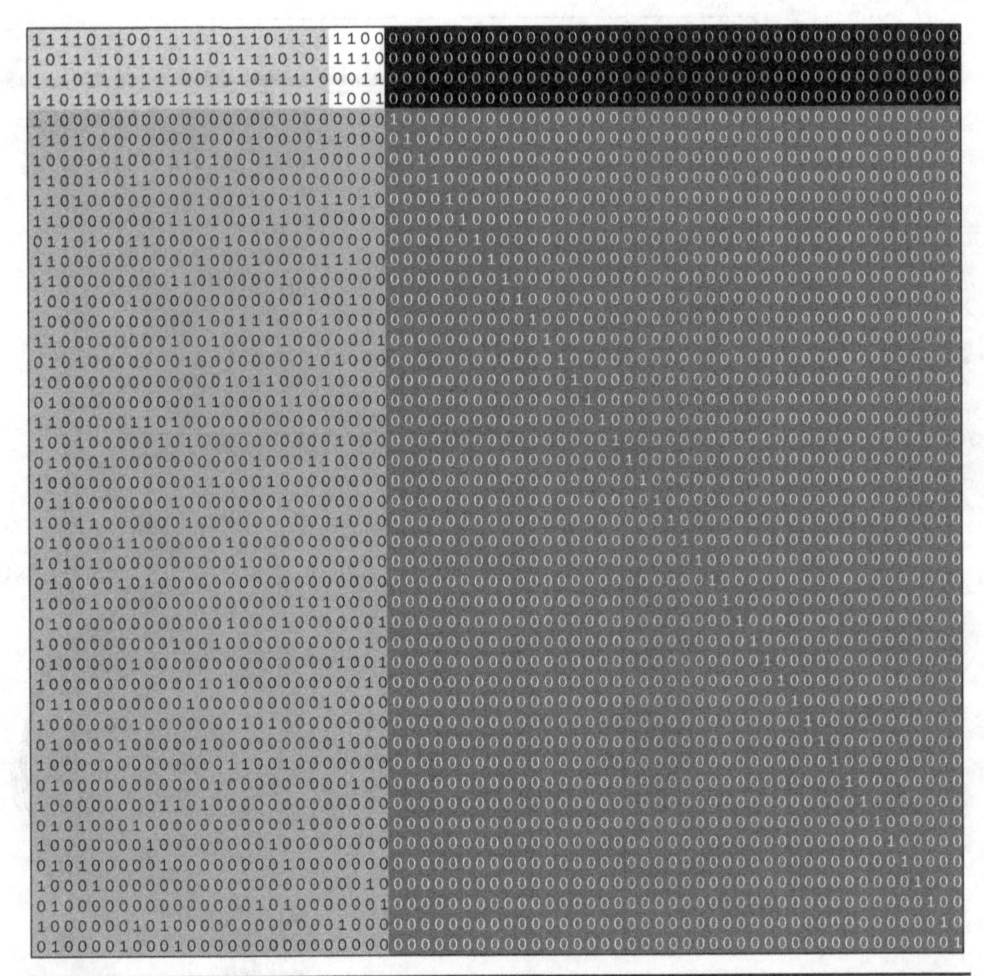

FIGURE 8.3 General structure of the 46-by-68 protograph base graph matrix (H_{BG}) for base graph 1.

able to determine the starting position from the grant of the current transmission alone. Therefore, the starting position should depend only on rv_{id} and the TB size. The starting positions, S_i for RV_i are as shown in Table 8.2 based on Ref. [5].

When limited buffer rate matching (LBRM) is not applied, $N_{cb} = 66 \cdot Z_C$ for base graph 1 and $N_{cb} = 50 \cdot Z_C$ for base graph 2; therefore, the starting positions simplify to $\{0, 17, 33, 56\} \cdot Z_C$ and to $\{0, 13, 25, 43\} \cdot Z_C$ for base graph 1 and base graph 2, respectively. LBRM will be discussed in Sec. 8.3. As it was mentioned in Sec. 8.1.3, the first $2 \cdot Z_C$ bits are punctured and not entered in the circular buffer with or without LBRM.

Note that the first three RV_i starting points are approximately uniformly distributed around the circular buffer but the fourth is closer to the end. The reason for this is that RV_3 can include more systematic bits this way; therefore, the "self-decodability" of RV_3 is increased. The same holds also for the case when LBRM is applied. Figure 8.4 shows the approximate starting positions in the circular buffer. The length of each RV in Fig. 8.4 is variable, depending on the amount of resources allocated for transmission.

$i = rv_{id}$	S_i	
	LDPC base graph 1	LDPC base graph 2
0	0	0
1	$\left\lfloor \frac{17 \cdot N_{cb}}{66 \cdot Z_c} \right\rfloor \cdot Z_c$	$\left\lfloor \frac{13 \cdot N_{cb}}{50 \cdot Z_c} \right\rfloor \cdot Z_c$
2	$\left\lfloor \frac{33 \cdot N_{cb}}{66 \cdot Z_c} \right\rfloor \cdot Z_c$	$\left\lfloor \frac{25 \cdot N_{cb}}{50 \cdot Z_c} \right\rfloor \cdot Z_c$
3	$\left\lfloor \frac{56 \cdot N_{cb}}{66 \cdot Z_c} \right\rfloor \cdot Z_c$	$\left\lfloor \frac{43 \cdot N_{cb}}{50 \cdot Z_c} \right\rfloor \cdot Z_c$

Table 8.2 Starting Position of Different Redundancy Versions, S_i

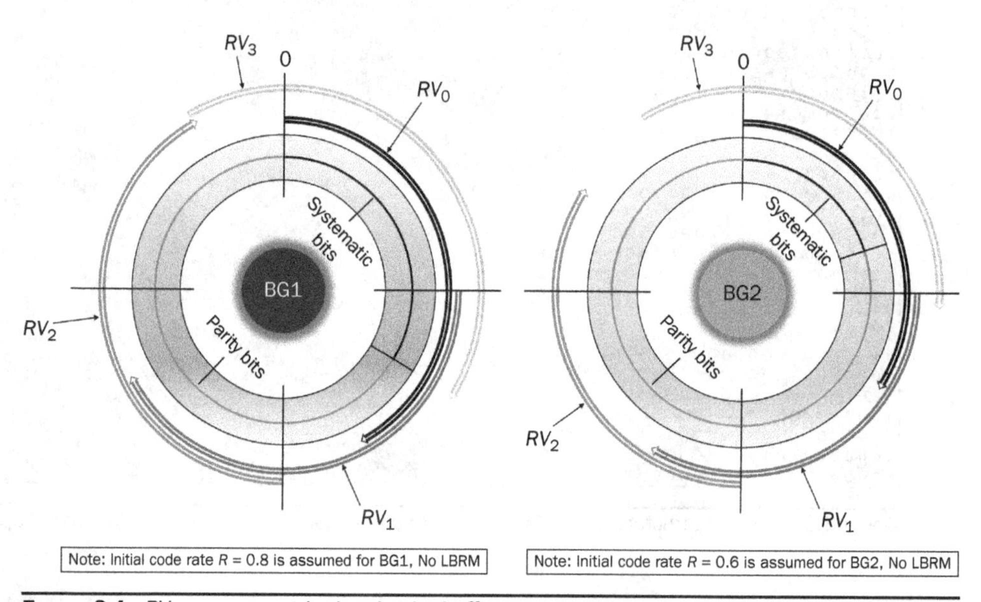

Figure 8.4 RV arrangement in the circular buffer.

The default *RV* order is {0, 2, 3, 1} for both base graph 1 and base graph 2 in cases where RV index is not explicitly signaled and there is no ambiguity about which instance of a transmission occurred. This solution is the same as used for the LTE UL. The rationale for using the sequence {0, 2, 3, 1} instead of the more natural {0, 1, 2, 3} can be explained as follows. In nominal conditions, the probability of successive retransmissions forms an approximately geometric decreasing series. For example, if we assume that initial transmission and all retransmissions are with the same SNR, and they contain the same number of bits, then the probability of decoding after one transmission can be 90%, after two transmissions greater than 99%, after three transmissions greater than 99.9% and so on. Therefore, it is important to optimize HARQ performance for one, two, etc., transmissions in that order. To optimize decoding performance for the first transmission, obviously, the best is to start at position $S_0 = 0$. This will maximize the number of systematic bits included for any initial code rate. Note that the small starting bit position offset used in LTE for Turbo codes was not carried over to NR because it did not provide any discernible benefits. For the second transmission (i.e., first retransmission), it is best to start at the 50% point of the soft buffer, which corresponds to RV_2 instead of RV_1, as this will minimize the channel bit repetitions (overlap) across the first and second transmissions.

It is important to note that this is not universally the best starting point selection. For example, as it was mentioned already, if the start of the second transmission could be at any point, it would be better to start at immediately the next consecutive bit in the soft buffer after where the first transmission ended. To make things even more complicated, the selection of the ideal starting point and of the ideal rv_{id} sequence also depends on the SNR statistical model. The discussion so far assumed uniform SNR across retransmissions, as already mentioned. Another common model is bursty interference, where large interference, and correspondingly low SNR, occurs independently at random retransmission occasions. To maximize HARQ performance in this scenario, the ideal starting point selection for the second transmission would be actually $S_i = 0$, i.e., the same as for the first transmission. In order to take advantage of the transmitter possibly knowing the SNR statistics, the transmitter can dynamically signal rv_{id} for each transmission. Even when rv_{id} for a sequence of transmissions is not explicitly signaled, such as in the case of slot aggregation or configured grant transmission, NR allows to semi-statically configure a different signal rv_{id} sequence, different from the default RV order of {0, 2, 3, 1}.

Another reason for using {0, 2, 3, 1} as the default RV order instead of {0, 1, 2, 3} was that, as it was mentioned earlier, self-decodability of RV_3 was enhanced by delaying its starting position in the circular buffer, therefore increasing the number of systematic bits included. If the first transmission (RV_0) in the sequence was lost then with the sequence {0, 1, 2, 3}, it would take a total of four transmissions to get good self-decodability. With {0, 2, 3, 1}, it only takes three. To maximize early self-decodability, one could choose sequences {0, 3, 2, 1} or even {0, 3, 0, 3}. However, these are also not universally optimum solutions, since {0, 3, ...} has much less incremental coding gain after the very important first two transmissions than {0, 2, ...}, as it was already mentioned. In the case of slot repetition with configured grant in the UL, the RV sequence is configurable, it can be configured to use one of three options {0, 2, 3, 1, ...}, {0, 3, 0, 3, ...}, or {0, 0, 0, 0, ...}.

The length C of the rate-matching output sequence E_r, $r = 0, 1, ..., C - 1$ is derived as follows. The first $C' - \gamma$ transmitted code blocks have $E_r = E^- = N_L \cdot Q_m \cdot \lfloor G' / C' \rfloor$ coded bits. The last γ transmitted code blocks have $E_r = E^+ = N_L \cdot Q_m \cdot \lceil G' / C' \rceil$ coded bits, where

- C' is the number of scheduled code blocks for the TB.
- $\gamma = G' \bmod C'$
- $G' = \dfrac{G}{(N_L \cdot Q_m)}$
- G is the total number of bits available for the scheduled transmission of the TB.

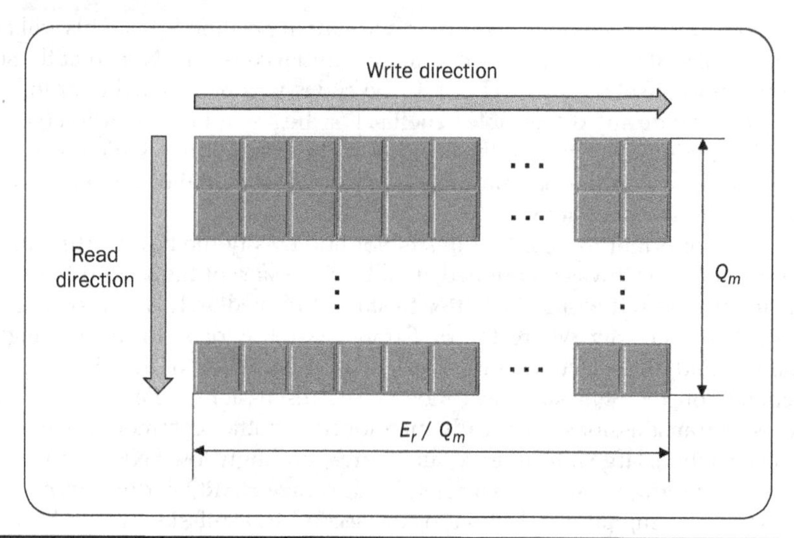

Figure 8.5 Intra-code-block interleaving.

- Q_m is the number of bits in a modulation symbol.
- N_L is the number of layers for the TB.

The final step in the channel coding process is interleaving. A bit-level interleaver is applied within each code block individually, at the output of the rate matching. The interleaver is a row-column interleaver with the number of rows equal to the modulation order. The operation is writing rows first, columns second and reading columns first, rows second. This is important, as it achieves that systematic bits in RV_0, after being written in the lower number rows are mapped to high-reliability locations in QAM symbols. The interleaver structure is illustrated in Fig. 8.5.

The number of coded bits in a code block is an integer multiple of the modulation order, so no filler bits are needed for the interleaving.

Note that the interleaving described above is external to the LDPC channel coding process, i.e., the interleaving is performed on the already encoded (and rate matched) channel bits. The LDPC channel coding does not require the use of an internal interleaver like Turbo codes.

8.3 DL Soft Buffer Management

In LTE, soft buffer management was introduced to control the memory requirements for HARQ buffering at the receiver, including accounting for memory across all HARQ processes. Over time, this became complicated, and in some cases unnecessary.

In NR, soft buffer management was introduced for a different purpose. Specifically, the decoding throughput is a function of the amount of parity, and at peak rates the decoding throughput could be significantly slowed down if a re-transmission of the largest TB size was ever needed at the lowest code rate. Therefore, the support of limited buffer rate matching (LBRM) was added in NR. Aside from not accounting for the number of HARQ processes, the principle of LBRM in NR is the same as in LTE.

For small TB size, the receiver is assumed to be able to store and process all coded bits, which is three times the number of information bits for base graph 1. For the largest TB size, the supported minimum code rate is approximately 2/3. This means the maximum number of parity bits is limited to be no more than 50% of the number of information bits. LBRM operation is applied independently in each HARQ process. Irrespective of the actual TB size, the circular buffer size is fixed and determined with assuming the maximum possible TB size and $R_{LBRM} = 2/3$, meaning maximum TB size times 1.5 bits.

Since it is critical that the transmitter and receiver use the same LBRM assumption, care must be taken that the maximum possible TB size is determined in a well-determined manner, due to the dependency of LBRM operation on this. On the other hand, in order to exploit the opportunity for receiver complexity reduction, the maximum possible TB size should take into account the receiver capabilities, at least when the receiver is the UE, i.e., when the LBRM is in the DL.

Rate matching circular buffer size is determined as follows. If no LBRM is used, $N_{cb} = N$. If LBRM is used, $N_{cb} = min(N, N_{ref})$, where $N_{ref} = \left\lfloor \frac{TBS_{LBRM}}{C \cdot R_{LBRM}} \right\rfloor$ and where C is the number of code blocks for the TB. The TB size TBS_{LBRM} assumed for LBRM operation for a serving cell is determined as the maximum possible according to the general TBS determination rules specified in subclause 5.1.3.2 in Ref. [1] for the DL and in subclause 6.1.4.2 for the UL, with the following important modifications.

- The number of MIMO layers for one TB is determined as *maxMIMO-Layers* for the DL and *maxRank* for the UL when they are configured; otherwise, it is determined as the maximum supported by the UE according to the UE capabilities.
- The maximum modulation order is $Q_m = 8$ when 256QAM is configured and $Q_m = 6$ otherwise.
- Maximum code rate is 948/1024 = 0.9257…
- $N_{RE} = 156 \cdot n_{PRB,LBRM}$, where $n_{PRB,LBRM}$ is given in Table 8.3.

The quantization of the maximum number of PRBs as shown in Table 8.3 ensures that the number of cases for LBRM operation, which the gNB and UE need to be able to handle, is limited. There is no further quantization of TBS_{LBRM} in the specification.

All the parameters in Table 8.3, namely, the maximum number of PRBs, maximum number of layers, maximum modulation order, can be different across different

Maximum number of PRBs	$n_{PRB,LBRM}$
<33	32
33–66	66
67–107	107
108–135	135
136–162	162
163–217	217
>217	273

TABLE 8.3 Value of $n_{PRB,LBRM}$

bandwidth parts (BWPs) configured for a UE in a serving cell. When this is the case, the maximum value across all BWPs is used in the TBS_{LBRM} determination. Note that the maximum is taken for each individual parameter, as opposed to taking the maximum of hypothetical TBS_{LBRM} values calculated for each individual BWP. With this, it is enough to calculate TBS_{LBRM} only once. The reason for taking the maximum in general is to avoid having to resize the soft buffer for any dynamic BWP change. This enables faster BWP transitions. At the same time, RRC reconfiguration of any BWP may result in soft buffer corruption for all BWPs. This, however, is expected to be an infrequent event.

The above definition creates a further complication in that before the initial UE capability exchange, the gNB does not know some of the DL LBRM parameters. The assumed solution is that in these cases, the gNB limits TB sizes to relatively small values where LBRM does not take effect even for the lowest UE capability. This in turn necessitates that some minimum UE capabilities are assumed by the gNB not only for current UEs but also for future UEs.

The general philosophy of making LBRM operation dependent on configured parameters when the configuration is lesser than what the UE can support is for optimizing resource usage and for providing forward compatibility. In the future, the allowed range of UE capabilities can be extended, while dependency on configured parameters allows the gNB to limit the possible parameter combinations.

Running LBRM independently for each HARQ process assumes no capability in the receiver to reallocate soft buffer dynamically across HARQ processes in a carrier, or across carriers, on an as-needed basis. Nevertheless, a receiver that does have this capability can freely use dynamic soft buffer reallocation to optimize its decoding performance, as long as it is done transparently to the transmitter. The specification does allow this operation, but it neither mandates this operation nor does it define specific features to further enhance it.

8.4 DL MCS and TBS Determination

Similar to LTE, the UE has to determine the transport block size (TBS) (note that we use the terms TBS and TB size interchangeably) from the modulation and coding scheme index (I_{MCS}) information received in a DL grant [1]. Some details of this procedure are different from LTE. There are three main reasons for this difference, as follows:

- The maximum number of RBs has increased from 100 in LTE to 275 in NR; therefore, the RB-indexed-table-based approach used in LTE would have been unwieldy.
- The number of symbols for a PDSCH is made much more variable in NR (2...14 symbols) compared to LTE (10...13 symbols).
- There was a desire to make the TBS determination ensure that after code block segmentation, each segment is byte aligned without the use of any additional filler bits.

The above differences motivated a TBS determination method in NR that is more "equation-based," compared to the "table-lookup-based" TBS determination method in LTE. Another benefit of the equation-based TBS determination method is reducing specification impact when new TBS values become possible, e.g., no need to make changes to TBS determination if 1024QAM is introduced in a future release.

One of the design goals for the NR TBS determination was to limit the total number of possible transport block sizes. The result was that there are 471 unique sizes covering all scheduling possibilities. This helps limit the total number of cases that the decoder and encoder in the UE need to be tested for. Another benefit of TBS quantization is that it increases the flexibility of changing MCS between retransmissions while maintaining the same TBS.

The UE receives within the DL grant the following information that is going to be used in the TBS determination:

- N_{symb}^{sh} : number of OFDM symbols allocated for the PDSCH (in the range of 2 to 14)
- n_{PRB}: number of PRBs allocated for the PDSCH (in the range of 1 to 275)
- v: number of MIMO layers for the TB (in the range of 1 to 4)
- N_{DMRS}^{PRB} : number of REs per PRB used for DM-RS (in the range of 4 to 48) including actual DM-RS REs and REs used for rate matching around DM-RS of other UEs (DM-RS CDM groups without data)
- I_{MCS}: modulation and coding scheme (MCS) Index (in the range of 0 to 31)

The UE uses the above information in the following steps to determine TBS:

1. Based on I_{MCS}, determine the modulation order Q_m and code rate R
2. Based on N_{symb}^{sh}, n_{PRB}, and N_{DMRS}^{PRB}, determine N_{RE} the total number of REs allocated for PDSCH (note that N_{RE} is a nominal number of available REs, not the actual number of available REs because it does not yet account for rate matching of the PDSCH around certain signals)
3. Based on N_{RE}, Q_m, and R, determine an intermediate number of information bits N_{info}
4. Based on N_{info}, determine the transport block size TBS

In the following, we give descriptions of each of the above steps.

8.4.1 Determination of Modulation Order and Code Rate

Modulation order Q_m and code rate R are determined with a simple look up from one of the three MCS tables based on I_{MCS}, where I_{MCS} is a 5-bit value. The three MCS tables are:

- MCS table 1: 64QAM MCS table (same as in LTE)
- MCS table 2: 256QAM MCS table (same as in LTE)
- MCS table 3: low spectral efficiency MCS table, used for high reliability, e.g., URLLC (new in NR)

The selection between the 64QAM MCS table and 256QAM MCS table is based on the RRC configured maximum modulation order. Note that 64QAM is actually not explicitly configured but instead it is implied by the omission of the field *mcs-Table* parameter from the PDSCH configuration. The selection between the low spectral efficiency MCS table and the other two MCS tables can be either dynamic based on the scrambling of the CRC of the PDCCH carrying the grant or it can be RRC configured.

The default MCS table to be used before RRC configuration is the 64QAM MCS table. Even after RRC configuration, the 64QAM MCS table can be used when a "fallback DCI" is used to schedule the PDSCH. For the specific purposes of fallback determination, DCI format 1_0 is fallback with respect to the use of 256QAM because the

256QAM MCS table is only used with DCI format 1_1. Similarly, DCI in the common search space is fallback with respect to the use of the low spectral efficiency MCS table, because this table is only used for DCI in the UE-specific search space and not for DCI in the common search space.

In Tables 8.4 to 8.6, MCS table 1, 2, and 3 are shown, respectively, based on Ref. [1].

MCS index I_{MCS}	Modulation order Q_m	Target code rate $R \times 1024$	Spectral efficiency
0	2	120	0.2344
1	2	157	0.3066
2	2	193	0.3770
3	2	251	0.4902
4	2	308	0.6016
5	2	379	0.7402
6	2	449	0.8770
7	2	526	1.0273
8	2	602	1.1758
9	2	679	1.3262
10	4	340	1.3281
11	4	378	1.4766
12	4	434	1.6953
13	4	490	1.9141
14	4	553	2.1602
15	4	616	2.4063
16	4	658	2.5703
17	6	438	2.5664
18	6	466	2.7305
19	6	517	3.0293
20	6	567	3.3223
21	6	616	3.6094
22	6	666	3.9023
23	6	719	4.2129
24	6	772	4.5234
25	6	822	4.8164
26	6	873	5.1152
27	6	910	5.3320
28	6	948	5.5547
29	2	Reserved	
30	4	Reserved	
31	6	Reserved	

Table 8.4 MCS Table 1 for PDSCH (64QAM)

MCS index I_{MCS}	Modulation order Q_m	Target code rate $R \times 1024$	Spectral efficiency
0	2	120	0.2344
1	2	193	0.3770
2	2	308	0.6016
3	2	449	0.8770
4	2	602	1.1758
5	4	378	1.4766
6	4	434	1.6953
7	4	490	1.9141
8	4	553	2.1602
9	4	616	2.4063
10	4	658	2.5703
11	6	466	2.7305
12	6	517	3.0293
13	6	567	3.3223
14	6	616	3.6094
15	6	666	3.9023
16	6	719	4.2129
17	6	772	4.5234
18	6	822	4.8164
19	6	873	5.1152
20	8	682.5	5.3320
21	8	711	5.5547
22	8	754	5.8906
23	8	797	6.2266
24	8	841	6.5703
25	8	885	6.9141
26	8	916.5	7.1602
27	8	948	7.4063
28	2	Reserved	
29	4	Reserved	
30	6	Reserved	
31	8	Reserved	

TABLE 8.5 MCS Table 2 for PDSCH (256QAM)

MCS index I_{MCS}	Modulation order Q_m	Target code rate $R \times 1024$	Spectral efficiency
0	2	30	0.0586
1	2	40	0.0781
2	2	50	0.0977
3	2	64	0.1250
4	2	78	0.1523
5	2	99	0.1934
6	2	120	0.2344
7	2	157	0.3066
8	2	193	0.3770
9	2	251	0.4902
10	2	308	0.6016
11	2	379	0.7402
12	2	449	0.8770
13	2	526	1.0273
14	2	602	1.1758
15	4	340	1.3281
16	4	378	1.4766
17	4	434	1.6953
18	4	490	1.9141
19	4	553	2.1602
20	4	616	2.4063
21	6	438	2.5664
22	6	466	2.7305
23	6	517	3.0293
24	6	567	3.3223
25	6	616	3.6094
26	6	666	3.9023
27	6	719	4.2129
28	6	772	4.5234
29	2	reserved	
30	4	reserved	
31	6	reserved	

Table 8.6 MCS Table 3 for PDSCH (Low Spectral Efficiency)

Note that the last column (Spectral Efficiency) entry in the above mentioned three tables is not an independent parameter but can be obtained as the product of second column (Q_m) and third column ($R \cdot 1024$) divided by 1024.

The last three rows corresponding to I_{MCS} = 29, 30, 31 in MCS tables 1 and 3 are not used directly to determine TBS. These values can be used in retransmissions to implicitly derive the TBS as the TBS of the initial transmission. The three different values enable changing the modulation order to be different from the initial transmission. I_{MCS} = 29, 30, 31 indicates QPSK, 16QAM, and 64QAM, respectively. The same procedure applies to MCS table 2 as well, but here there are four I_{MCS} values reserved, since there are four possible modulation order values: I_{MCS} = 28, 29, 30, 31 indicates QPSK, 16QAM, 64QAM, and 256QAM, respectively. Note that the same implicit determination was used in LTE already. We note again that the 64QAM and 256QAM MCS tables are the same as in LTE.

8.4.2 Determination of the Nominal Number of REs Allocated to PDSCH

Based on N_{symb}^{sh}, n_{PRB}, and N_{DMRS}^{PRB}, the UE determines the total number of REs allocated to PDSCH, N_{RE}, as: $N_{RE} = n_{PRB} \cdot \min\left(156, N_{sc}^{RB} \cdot N_{symb}^{sh} - N_{DMRS}^{PRB} - N_{oh}^{PRB}\right)$, where $N_{sc}^{RB} = 12$ is the number of subcarriers in a physical resource block, N_{symb}^{sh}, n_{PRB}, N_{DMRS}^{PRB} are dynamically signaled parameters that were described earlier, and N_{oh}^{PRB} is an additional parameter configured by RRC. N_{oh}^{PRB} is the estimated overhead expressed as number of REs per PRB and can take on the values 0, 6, 12, 18. When not configured, zero is assumed. The reason for using this parameter is the following. The calculated number of REs allocated for PDSCH, N_{RE}, is not precise because it does not account for rate matching for SSB, CSI-RS, TRS, etc., which further reduces the number of REs available for data. Of course, the gNB knows exactly the number of REs lost and it can lower the signaled I_{MCS} values if needed to reduce the TBS and approximately maintain the desired effective code rate (which will no longer be the same as the R parameter derived based on the I_{MCS}). However, in case the I_{MCS} value indicates an entry just above a modulation order switch point, i.e., the lowest I_{MCS} of a particular Q_m modulation order, then lowering the I_{MCS} will result in switching to a lower modulation order. This is undesirable, since rate matching the PDSCH around some additional signals does not change the SNR of the PDSCH and therefore should not change the modulation order use either. If the gNB sets N_{oh}^{PRB} to match the exact additional overhead, then this undesirable effect can be avoided. Unfortunately, the actual overhead varies from slot to slot and can also vary based on which symbols within the slot are allocated for PDSCH, so the correction factor N_{oh}^{PRB} is somewhat imprecise in practice.

8.4.3 Determination of the Intermediate Number of Information Bits

The intermediate number of information bits, N_{info}, is obtained as $N_{info} = N_{RE} \cdot R \cdot Q_m \cdot v$, where the parameters N_{RE}, R, Q_m, and v have been described earlier.

8.4.4 Determination of the Transport Block Size

The transport block size (TBS) is determined differently when $N_{info} \leq 3824$ and when $N_{info} > 3824$.

When $N_{info} \leq 3824$, code block segmentation cannot occur, and the TBS determination only has to make sure that the TBS is byte aligned. First, a quantized version of intermediate number of information bits is determined as $N'_{info} = max\left(24, 2^n \cdot \left\lfloor \frac{N_{info}}{2^n} \right\rfloor\right)$, where $n = max\left(3, \lfloor log_2(N_{info}) \rfloor - 6\right)$. Second, Table 8.7 below is used to find TBS as the closest number in the table not less than N'_{info}.

When $N_{info} > 3824$, code block segmentation can occur, and the *TBS* determination must make sure that each code block segment is byte aligned without any padding.

Index	TBS	Index	TBS	Index	TBS	Index	TBS
1	24	**31**	336	**61**	1288	**91**	3624
2	32	**32**	352	**62**	1320	**92**	3752
3	40	**33**	368	**63**	1352	**93**	3824
4	48	**34**	384	**64**	1416		
5	56	**35**	408	**65**	1480		
6	64	**36**	432	**66**	1544		
7	72	**37**	456	**67**	1608		
8	80	**38**	480	**68**	1672		
9	88	**39**	504	**69**	1736		
10	96	**40**	528	**70**	1800		
11	104	**41**	552	**71**	1864		
12	112	**42**	576	**72**	1928		
13	120	**43**	608	**73**	2024		
14	128	**44**	640	**74**	2088		
15	136	**45**	672	**75**	2152		
16	144	**46**	704	**76**	2216		
17	152	**47**	736	**77**	2280		
18	160	**48**	768	**78**	2408		
19	168	**49**	808	**79**	2472		
20	176	**50**	848	**80**	2536		
21	184	**51**	888	**81**	2600		
22	192	**52**	928	**82**	2664		
23	208	**53**	984	**83**	2728		
24	224	**54**	1032	**84**	2792		
25	240	**55**	1064	**85**	2856		
26	256	**56**	1128	**86**	2976		
27	272	**57**	1160	**87**	3104		
28	288	**58**	1192	**88**	3240		
29	304	**59**	1224	**89**	3368		
30	320	**60**	1256	**90**	3496		

Table 8.7 *TBS* for $N_{info} \leq 3824$

First, a quantized version of intermediate number of information bits is determined as $N'_{info} = \max\left(3840, 2^n \cdot \text{round}\left(\frac{N_{info} - 24}{2^n}\right)\right)$, where $n = \lfloor \log_2(N_{info} - 24) \rfloor - 5$. The purpose of the quantization is to control the total number of *TBS* cases. This limits the cases an encoder and decoder must be tested for. The term (–24) was added to N_{info} in the above formula because it had been shown that for small payload size this rate back-off term minimized the average overhead. Here overhead means the ratio of padding bits needed to round up the number of input bits to the closest *TBS* larger or equal.

Second, the number of code block segments C is calculated based on the procedures for channel coding. Code block segmentation occurs when $N_{info} > 3824$ and $R \leq 1/4$ or in all cases when $N_{info} > 8424$; therefore, these cases are treated separately;

- When $N_{info} > 3824$ and $R \leq 1/4$, the number of code blocks C is given as $C = \left\lceil \frac{N'_{info} + 24}{K_{CB} - L} \right\rceil$, where $K_{CB} = 3840$ is the maximum code block size for LDPC base graph 2 and $L = 24$ is the code block CRC length. With this, we have $C = \left\lceil \frac{N'_{info} + 24}{3816} \right\rceil$.
- When $N_{info} > 8424$, the number of code blocks C is given as $C = \left\lceil \frac{N'_{info} + 24}{K_{CB} - L} \right\rceil$, where $K_{CB} = 8448$ is the maximum code block size for LDPC base graph 1 and $L = 24$ is the code block CRC length. With this, we have $C = \left\lceil \frac{N'_{info} + 24}{8424} \right\rceil$.
- Otherwise (i.e., when $3824 < N_{info} \leq 8424$ and $R > 1/4$), there is code block segmentation and $C = 1$.

Finally, in order to select a TBS value such that the equal code block segments each become byte aligned, the TBS, after adding 24 bits as TB CRC, must be a multiple of $8 \cdot C$. Therefore, TBS is determined as $TBS = 8 \cdot C \cdot \left\lceil \frac{N'_{info} + 24}{8 \cdot C} \right\rceil - 24$. The subtraction of 24 accounts for the TB CRC.

When PDSCH is used for paging or for random access response (RAR) in message 2, the typical payload size is small, and it is beneficial to allow lowering the minimum supportable spectral efficiency. One option would be to use the lower spectral efficiency MCS table (MCS table 3) for these cases. However, neither for paging nor for RAR can it be assumed that the UE targeted by the message supports MCS table 3. Therefore, a different solution was chosen. The intermediate number of information bits N_{info} is further scaled by a factor *S* indicated in the DCI. The value of *S* can be 0.25, 0.5, or 1. Other than applying the scaling factor, the same steps are followed for paging and RAR as for all other PDSCH cases.

8.5 DL Resource Allocation in the Time Domain

The DL resource allocation in the time domain can refer to two PDSCH mapping types: mapping type A and mapping type B [1]. Mapping type A is also known as slot-based scheduling and mapping type B is also known as mini-slot–based scheduling. Such naming is somewhat deceptive because both mapping type A and B can refer to both

PDSCH mapping type	Normal cyclic prefix			Extended cyclic prefix		
	S	L	S+L	S	L	S+L
Type A	{0, 1, 2, 3}	{3, ..., 14}	{3, ..., 14}	{0, 1, 2, 3}	{3, ..., 12}	{3, ..., 12}
Type B	{0, ..., 12}	{2, 4, 7}	{2, ..., 14}	{0, ..., 10}	{2, 4, 6}	{2, ..., 12}

TABLE 8.8 Start and Length of PDSCH

short and long time-domain duration, as a matter of fact, slot-based mapping type A allows for more flexibility in terms of duration than mini-slot–based mapping type B, at least in Release 15. The main difference between the two mapping types is that PDSCH with mapping type A always starts close to the beginning of the slot, more precisely, it starts in one of the first four symbols, while PDSCH mapping type B can start in any symbol, of the slot, as long as it starts and ends in the same slot. Straddling slot boundaries is disallowed for both mapping types in Release 15. Another difference is that for PDSCH mapping type A, the DM-RS symbols are more or less fixed relative to the slot, except the first symbol that adjusts to the start of the PDSCH, while for PDSCH mapping type B, the DM-RS symbols slide together with the PDSCH.

The allowed start *S* and length *L* in units of symbols for PDSCH mapping type A and B are given in Table 8.8 based on Ref. [1].

For PDSCH mapping type B, the only PDSCH lengths allowed in Release 15 are 2, 4, or 7 symbols for normal cyclic prefix. In later releases, other lengths may be introduced.

The DL grant defines the time-domain allocation as a combination of two parameters: K_0 and *SLIV*. The first parameter, K_0, gives a slot index offset between the slot in which the DL grant is received and the slot in which the PDSCH is received. Note that $K_0 = 0$, i.e., same slot scheduling is most typical, at least in scenarios not targeting power savings. The second parameter, the so-called start and length indicator value (*SLIV*) gives the exact combination of start *S* and length *L* values, for which the allowed values were described in Table 8.8.

The DL grant actually does not include K_0 and *SLIV* explicitly but rather it includes a single time-domain resource assignment field value m that provides a row index $m + 1$ to an allocation table. The allocation table defines a specific K_0, *SLIV*, and PDSCH mapping type (A or B) combination for each row indexed by $m + 1$. The entries are configurable with RRC parameter *pdsch-TimeDomainAllocationList*, which itself has the following format [6].

```
PDSCH-TimeDomainResourceAllocation ::=    SEQUENCE {
    k0                                  INTEGER(0..32)  OPTIONAL,   -- Need S
    mappingType                         ENUMERATED {typeA, typeB},
    startSymbolAndLength                INTEGER (0..127)
}
```

When K_0 is not included, same slot scheduling is applied. The maximum slot offset is 32.

SLIV itself is a 7-bit value that is mapped to a pair of start *S* and length *L* in terms of symbols according to the following equations:

$$\text{if } (L-1) \leq 7 \text{ then } SLIV = 14 \cdot (L-1) + S \tag{8.2}$$

$$\text{else } SLIV = 14 \cdot (14 - L + 1) + (14 - 1 - S), \text{ where } 0 < L < 14 - S. \tag{8.3}$$

The *SLIV* definition is based on the same principles as the resource indication value (*RIV*) definition in both LTE and NR, which will be discussed in Sec. 8.6.

If there were no restriction, the set of all possible start *S* and length *L* combinations would be $(14 \cdot 13)/2 = 91$, which is a subset of what is possible to be expressed in 7-bits. Therefore, some of the *SLIV* values do not give start *S* and length *L* combinations and even those that do are down selected to the set of allowed *S* and length *L* combinations according to Table 8.8.

In lack of RRC configuration, or before RRC configuration during initial access, one of a set of default tables is used. An example default table for normal CP is given in Table 8.9.

Note that in the default tables, such as in the example given in Table 8.9, *S* and *L* are explicitly included, instead of being jointly signaled via *SLIV*.

Row index	*dmrs-TypeA-Position*	PDSCH mapping type	K_0	S	L
1	2	Type A	0	2	12
	3	Type A	0	3	11
2	2	Type A	0	2	10
	3	Type A	0	3	9
3	2	Type A	0	2	9
	3	Type A	0	3	8
4	2	Type A	0	2	7
	3	Type A	0	3	6
5	2	Type A	0	2	5
	3	Type A	0	3	4
6	2	Type B	0	9	4
	3	Type B	0	10	4
7	2	Type B	0	4	4
	3	Type B	0	6	4
8	2,3	Type B	0	5	7
9	2,3	Type B	0	5	2
10	2,3	Type B	0	9	2
11	2,3	Type B	0	12	2
12	2,3	Type A	0	1	13
13	2,3	Type A	0	1	6
14	2,3	Type A	0	2	4
15	2,3	Type B	0	4	7
16	2,3	Type B	0	8	4

TABLE 8.9 Example Default PDSCH Time-Domain Resource Allocation for Normal CP

Release 15 also introduced some restriction on possible "negative" offsets between DL grant and corresponding PDSCH. Negative offset means that the first symbol of the PDSCH starts earlier than the end of the last symbol of the DL grant. The restrictions are the following:

- The UE is not expected to receive a PDSCH with mapping type A in a slot, if the PDCCH scheduling the PDSCH was received in the same slot and was not contained within the first three symbols of the slot.
- The UE is not expected to receive a PDSCH with mapping type B in a slot, if the first symbol of the PDCCH scheduling the PDSCH was received in a later symbol than the first symbol indicated in the PDSCH time-domain resource allocation.

The motivation for avoiding large negative offset is to relieve the UE from having to do excessive buffering. Even without negative offset, the UE has to buffer potential PDSCH symbols during the PDCCH decoding time, even if it turns out that there was no DL grant in the PDCCH. Large negative offsets would increase the buffering requirement even further. In addition, negative offsets would increase the PDSCH processing time.

The same limitation on negative DL grant to PDSCH time offset applies to the case of cross-carrier scheduling where the DL grant and PDSCH are on different carriers. In Release 15, cross-carrier scheduling only applies when the scheduling and scheduled carriers have the same SCS. In Release 16, the feature of cross-carrier scheduling with different SCS was introduced. For these cases, further buffering relaxation was also introduced. For example, when the scheduling carrier SCS is lower than the scheduled SCS, and, therefore, it is expected that the DL grant decoding takes more PDSCH symbols, a positive threshold was introduced as a minimum offset between DL grant and the granted PDSCH.

8.5.1 PDSCH Slot Repetition

Another aspect of the DL time-domain resource allocation is PDSCH slot repetition [1], also called PDSCH slot aggregation or PDSCH slot bundling. The motivation for introducing slot bundling is improving DL coverage and link budget. The same TB is repeated in the bundled slots. PDSCH slot bundling is enabled when the UE is configured with *pdsch-AggregationFactor*, in which case the same symbol allocation is applied across the *pdsch-AggregationFactor* consecutive slots with *pdsch-AggregationFactor* set to 2, 4, or 8, in response to a single received DL grant. All the other resource allocation parameters are identical across the aggregated slots, except rv_{id}, which is taking on consecutive values from the default sequence {0, 2, 3, 1, 0, 2, 3, 1, 0, ...}. The starting point in the sequence is the rv_{id} indicated in the DL grant. To clarify this further, rv_{id} selection is also described in Table 8.10 based on Ref. [1].

Unlike in the UL for PUSCH, there is no inter-slot frequency hopping defined in the DL for PDSCH. The reason for this will be discussed in Sec. 8.6.

Note that while it is possible for the UE to perform joint channel estimation and/or joint decoding across the aggregated slots, the baseline assumption is that the UE does not have the capability to perform this, because it could demand significantly increased buffering to perform this type of operation. The baseline assumption is that the UE decodes the repeated slots as if they were separate HARQ retransmissions.

rv_{id} indicated by the DCI scheduling the PDSCH	rv_{id} to be applied to n^{th} transmission occasion			
	n mod 4 = 0	*n* mod 4 = 1	*n* mod 4 = 2	*n* mod 4 = 3
0	0	2	3	1
2	2	3	1	0
3	3	1	0	2
1	1	0	2	3

TABLE 8.10 Applied Redundancy Version with PDSCH Slot Aggregation

PDSCH slot repetition can be dynamically switched off if DCI format 1_0 is used, since slot repetition is only used with DCI format 1_1. Switching on or off by DCI format change allows for the base station scheduler to be more responsive to channel condition changes. When PDSCH slot repetition is used, the PDSCH is limited to single-layer transmission. This is motivated by the fact that the use of PDSCH slot aggregation is primarily a tool to improve DL link budget, where spatial multiplexing would not be relevant.

PDSCH slot repetition can be applied on top of any PDSCH symbol allocation within the slot. The typical application, however, would be using some of the PDSCH time-domain allocation cases that have more symbols allocated within the slot, e.g., more than seven, as a basis for the repetition. Although repeating short PDSCH, e.g., two-symbol mini-slots (PDSCH mapping type B) across slots is supported, its practical applicability is limited. Given that slot repetition with short PDSCH does not address link budget limitations, individual DL grants per slot could be sent instead of slot repetition, retaining full scheduling flexibility and with little downside. Repeating mini-slots (i.e., PDSCH mapping type B) within a slot is not supported in Release 15. After some debate in Release 16, it has been decided that Release 16 supports this option. Note that having two PDSCH mapping type A allocations with length 3 repeated at the beginning of a slot would in theory be possible, but it is supported in neither Release 15 nor Release 16.

A further complication with bundling is that some of the repeated slots may not be available, for example, due to a measurement gap or the presence of an UL slot in TDD. In these cases, the interrupted slots are omitted by the UE. The remaining slots are neither postponed nor cancelled. Also, the HARQ ACK feedback timing follows the last nominal repeat of the PDSCH, irrespective of whether the last slot has been omitted or not.

8.6 DL Resource Allocation in the Frequency Domain

In the frequency domain, two resource allocation types, type 0 and type 1, are used in the NR DL [1]. Type 1 is the default, e.g., used with DCI format 1_0.

The applicable resource allocation types can be dynamically indicated in the DL grant when dynamic switch is configured by RRC. When it is configured, one bit is included in DCI format 1_1 to indicate the resource allocation type applicable to the granted PDSCH. Otherwise, the resource allocation type itself is RRC configured. This allows the gNB to choose between maximum scheduling flexibility and DCI payload minimization. When dynamic switching is configured, the maximum of the type 0 and

Bandwidth part size	Configuration 1	Configuration 2
1–36	2	4
37–72	4	8
73–144	8	16
145–275	16	16

Table 8.11 RBG Size *P*

type 1 bit-width is used in the DCI format 1_1 in order to make the payload size invariant. In this case, an additional bit is also added as the switching flag.

The frequency-domain allocation is referenced to, and is valid, within the UE's active BWP. When the UE receives an active BWP change indication in the DL grant, then for the purposes of resource allocation, the indicated new BWP is considered active.

Resource allocation type 0 is bitmap based. Each bit explicitly includes or excludes a resource block group (RBG). If the RBG size P was 1, then resource allocation type 0 would provide full scheduling flexibility. However, this would also mean excessive DCI overhead, because the maximum bitmap size would be 275 bits. In order to manage the overhead, $P > 1$ is defined in general, and P is increasing with increasing BWP bandwidth. The applicable values of P are shown in Table 8.11 based on Ref. [1]. As it can be seen from the table, two configurations are defined, selectable by RRC. This gives the gNB further flexibility to fine tune between scheduling flexibility and DCI overhead. Note that in LTE, there was no configurability in this respect, i.e., there was a single P for a given channel bandwidth.

It is not guaranteed that the BWP bandwidth is an integer multiple of P. Therefore, the last RBG can be fractional to fill in the leftover space in frequency. This is the same as in LTE. But unlike in LTE, the first RBG in the BWP can also be fractional in NR. The reason for this is to allow aligning the RBG grid for different UEs with different BWP allocations. We can note that the P values, defined in Table 8.11, follow a power of 2 progression. This also serves the same purpose, i.e., allows UEs with overlapping BWP with different P to be scheduled on a compatible RBG grid. An example of this is shown in Fig. 8.6.

Resource allocation type 1 is based on a contiguous allocation with a start and end points. The resource allocation is signaled in a resource indication value (RIV) field, which is designed similarly as in LTE. The NR *RIV* is defined [1] as

$$\text{if } (L_{RBs} - 1) \leq \lfloor N_{BWP}^{size} / 2 \rfloor \text{ then } RIV = N_{BWP}^{size}(L_{RBs} - 1) + RB_{start} \tag{8.4}$$

$$\text{else } RIV = N_{BWP}^{size}\left(N_{BWP}^{size} - L_{RBs} + 1\right) + \left(N_{BWP}^{size} - 1 - RB_{start}\right), \text{ where } L_{RBs} \geq 1 \text{ and shall not exceed } N_{BWP}^{size} - RB_{start}. \tag{8.5}$$

Resource allocation type 1, similarly to resource allocation type 0, is also referenced to the UE's active BWP. But unlike resource allocation type 0, resource allocation type 1 can be used with DCI format 1_0, which creates a problem because that makes the size of the *RIV* be based on the size of the initial BWP, which may not be of equal size of the UEs current active BWP. More precisely, the *RIV* size is based on CORESET 0 if CORESET 0 is configured for the cell or based on the size of the initial DL bandwidth part if CORESET 0 is not configured for the cell. This size mismatch makes many allocation cases unusable. To solve this problem, the size of the "building blocks" of the

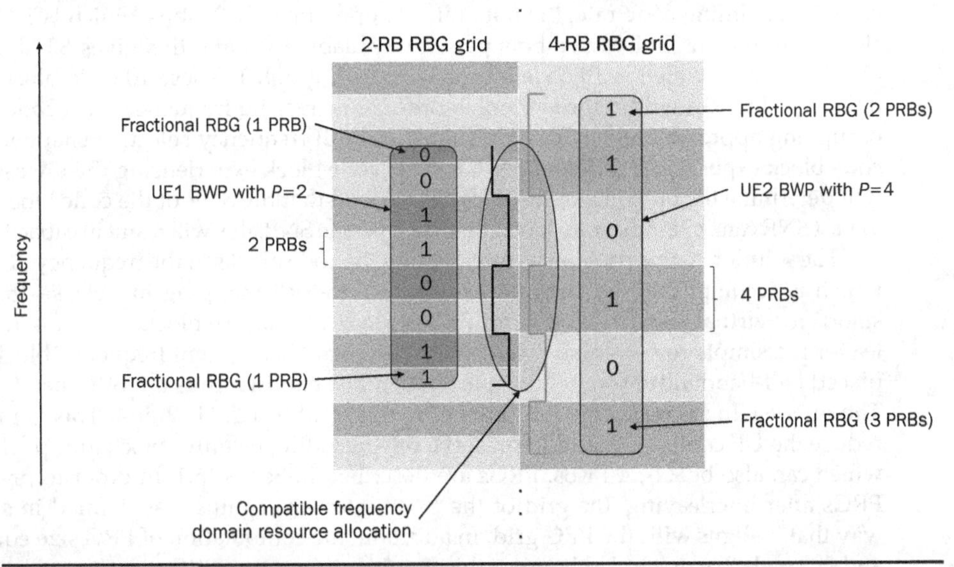

FIGURE 8.6 Example for alignment with RBG grid.

resource allocation is increased from 1 PRB to K PRBs, so that the start of the resource allocation and the length has to be both in K-PRB units. Note that this solution is very similar to what was adopted for the so-called compact DCI formats in LTE. The unit size K is chosen from the set {1, 2, 4, 8}. The largest possible element is chosen for K, for which $K \leq \lfloor N_{BWP}^{active} / N_{BWP}^{initial} \rfloor$. Note that this does not always allow fitting the active BWP size-based allocation in the *RIV*, whose size is determined based on the initial BWP size. For example, if $\lfloor N_{BWP}^{active} / N_{BWP}^{initial} \rfloor = 7$ then $K = 4$ is chosen, instead of $K = 8$. That means that certain allocation cases based on K-PRB building blocks still cannot be signaled. This was not seen as a major limitation though, since even if K were doubled to avoid this, the larger K would introduce similar or worse reduction in scheduling granularity.

As it was mentioned in Sec. 8.5.1, there is no inter-slot frequency hopping defined for the PDSCH in the DL, unlike for the PUSCH in the UL, so in the case of PDSCH slot repetition, the frequency-domain resource allocation remains constant over the aggregated slots. The main reason for this is that in the DL, at least with frequency-domain resource allocation type 0, the scheduler can allocate non-contiguous assignments in frequency in all repeated slots, which can achieve frequency diversity already without hopping. This can be done much more freely in the DL than in the UL due to latter's contiguous or "almost contiguous" resource allocation restriction, which will be discussed in Sec. 10.3.

8.6.1 Interleaved VRB-to-PRB Mapping

In Release 15, a version of frequency-domain interleaving was introduced in the DL for the PDSCH [1]. Similar interleaving was not introduced in the UL. The motivation for this interleaving in the DL was not primarily an intent to improve frequency diversity for the full PDSCH allocation. Rather, the motivation was mostly to improve code block diversity within the PDSCH. This can be explained as follows. In a 100-MHz channel with 30 kHz SCS, four-layer spatial multiplexing, 256QAM modulation order and close to maximum code rate, the data rate is approximately 2 Gbps, which is 1 Mbits/slot. Assuming 12 OFDM symbols per slot available for data, this gives 83 kbits per OFDM symbol, which is 10 code blocks with base graph 1. These 10 code blocks will be mapped to REs within the symbol in order of increasing frequency, each code block occupying approximately a 10-MHz subband. With frequency selective channel, each code block experiences different SNR and the code block experiencing the lowest SNR will determine the outcome of the decoding. Given that the SNR of the code block with lowest SNR can be significantly lower than the average SNR, this will result in capacity loss.

The solution to this problem is interleaving the code blocks in the frequency domain, which is accomplished by the interleaved VRB-to-PRB mapping in Release 15. VRB stands for virtual resource block and PRB is physical resource block. The chosen interleaver is a simple row-column interleaver with depth 2. Adjacent frequency blocks are placed half-bandwidth separation apart, where the bandwidth is the BWP bandwidth. The interleaving is performed in units of L_i RBs, with L_i equals 2 or 4. This is to both reduce the UE complexity and to preserve the precoding resource block groups (PRGs) which can also be 2 or 4 PRBs. PRGs are described in Sec. 8.16.1. In order to preserve PRGs after interleaving, the grid of the L_i RB interleaver units was defined in such a way that it aligns with the PRG grid. In addition, the combination of PRG size equals 4 and L_i equals 2 is precluded because this would not preserve PRGs.

The interleaving can be switched on or off with an indicator bit in the DL grant. The "VRB-to-PRB mapping" indicator is included both in DCI format 1_0 and 1_1.

Figure 8.7 shows an example for the interleaved VRB-to-PRB mapping with $L_i = 2$.

The Release 15 interleaved VRB-to-PRB mapping has a few drawbacks, which can be listed as follows:

- The depth-2 interleaving gives only a moderate improvement in SNR averaging. For example, if there are 10 code blocks in an OFDM symbol in a 100-MHz channel, code block #1 and code block #6 are 25 MHz apart, so their SNR can be still quite different.
- The CQI feedback does not take into account whether interleaved VRB-to-PRB mapping is used; therefore, the rate prediction will have increased error.
- When UEs operate in different BWPs, then with interleaved VRB-to-PRB mapping, resource collision will occur.

A resource collision example is shown in Fig. 8.8. In the example, two UEs operate in overlapping bandwidth parts. UE 1 operates in BWP 1 and UE 2 operates in BWP 2. UE 1 is allocated all RBs in BWP 1 and the gNB intends to allocate all remaining RBs to UE2. This, however, is not possible due to the resource collision shown in Fig. 8.8.

In a future release, an improved interleaver may be introduced that avoids the drawbacks of the Release 15 interleaving scheme.

8.7 DL Rate Matching

Similar to LTE, the PDSCH in NR is also rate matched around certain signals. Rate matching means that during the process of mapping the modulation symbols to resource elements (REs), certain REs that are used for the transmission of other signals

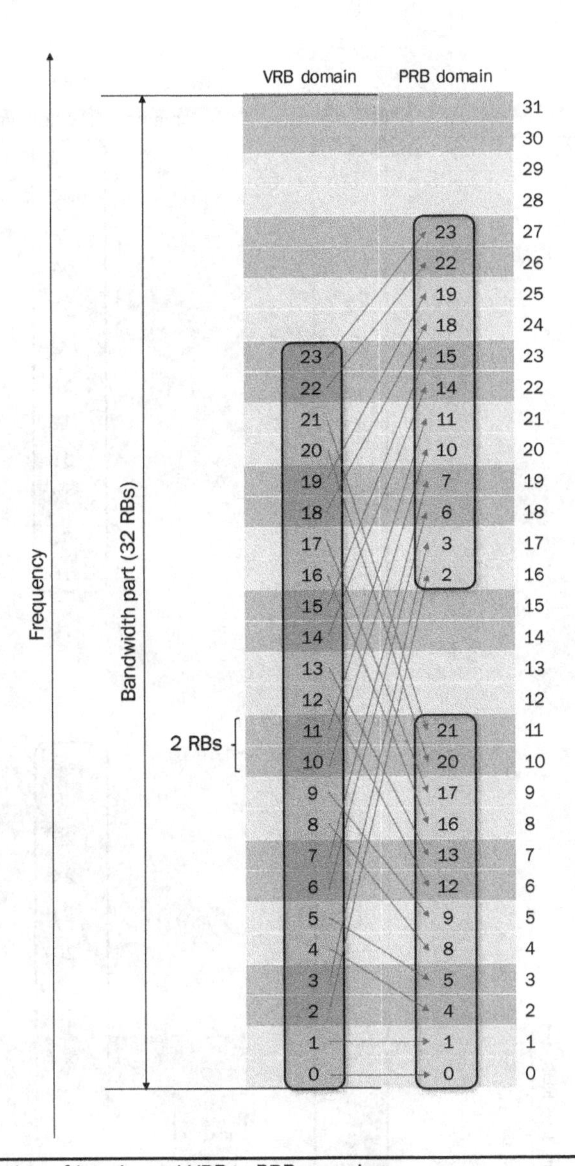

FIGURE 8.7 Illustration of interleaved VRB-to-PRB mapping.

are designated as unavailable and the mapping to these REs is skipped. The skipped REs are subtracted from the number of available REs and this is accounted for in the rate matching mentioned in Sec. 8.1. Another option is so called puncturing. In the case of puncturing, the impacted REs are not skipped, the modulation symbols are still mapped in the sequence, but before transmission, these modulation symbols are replaced with another signal. The impact of puncturing depends greatly on whether the receiver is aware of it and whether it excludes the impacted REs from the data decoding. Even if the receiver is aware of the puncturing, rate matching typically still gives better link level demodulation performance than puncturing because with the proper

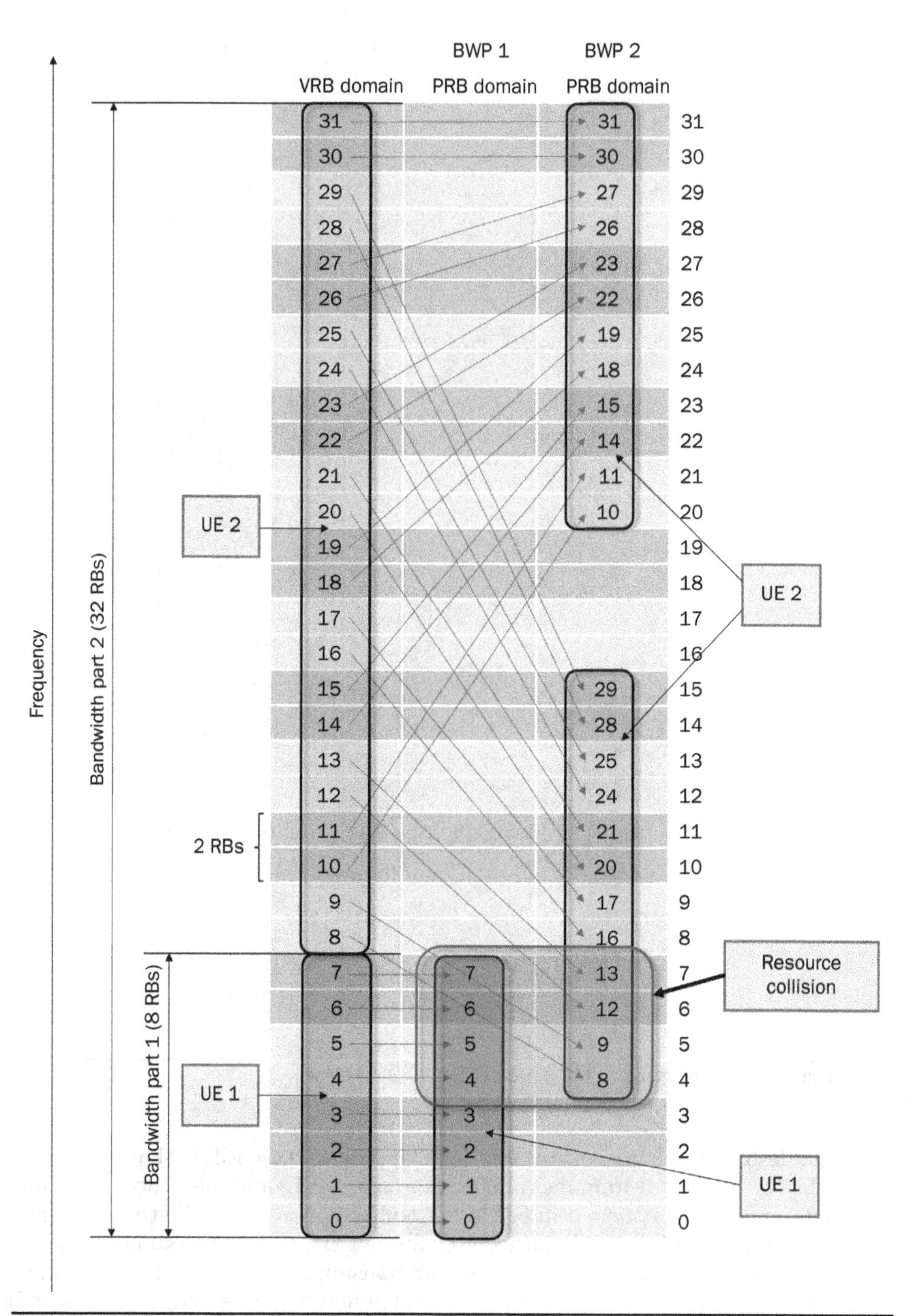

Figure 8.8 Resource collision with interleaved VRB-to-PRB mapping.

rate matching operation, only parity bits will be omitted, the systematic bits will be preserved. For this reason, rate matching operation was prioritized in NR over puncturing for dealing with unavailable REs. One downside of rate matching though is that the transmitter and receiver must be perfectly aligned with respect to exact rate matching operation, otherwise decoding will be impossible in general, since the codewords themselves will be wrong. This is as opposed to puncturing, wherein the receiver not knowing about the puncturing will suffer performance loss but decoding in general is still possible. Due to this, care must be taken so that for the rate matching operation the configuration information exchange is defined in such a way that the transmitter and receiver are always making the same assumption. For this purpose, the rate matching is conveyed in RRC configuration, and until the UE confirms the completion of applying the configuration, all transmission and reception is performed with assuming no configured rate matching. An exception to this is the application of dynamic adaptation of the rate matching pattern based on PDCCH, which will be described later. The impact of mis-applying this PDCCH-based rate matching information is minimal because if the UE cannot decode the grant then the decoding of data will anyhow not be performed. And when the decoded rate matching information is wrong but goes undetected due to false CRC pass (which is already a very low likelihood event with probability less than $\frac{1}{2^{24}} = 6 \cdot 10^{-8}$), then other decoded control information is almost always wrong, so decoding would not have been successful anyway. The application of dynamic rate matching information is memory-less, it is only applied once and does not impact future PDSCH decoding attempts.

Rate matching of PDSCH around the following signals is defined in NR:

- SSB resource blocks according to the configuration provided by *ssb-PositionsInBurst* in *ServingCellConfigCommon* [1]. An RB is unavailable for PDSCH in an OFDM symbol if at least one of its REs overlaps with the SSB.
- DM-RS configured for the PDSCH [7] and DM-RS CDM groups indicated unavailable in the DL grant (CDM groups used for other UEs' DM-RS in MU-MIMO) [1]. This aspect was described in Sec. 8.12.
- PT-RS configured for the PDSCH [7].
- Periodic and semi-persistent non-zero power (NZP) CSI-RS configured for the UE [7] for any uses including TRS, NZP CSI-RS for channel estimation, NZP CSI-RS for interference estimation, NZP-CSI RS for beam management, NZP CSI-RS for pathloss estimation, but excluding NZP CSI-RS for RRM. CSI-RS for RRM can be transmitted by other base stations not serving the UE and the serving base station does not perform rate matching around these in general since time synchronization with other base stations may not be ensured. When synchronization is ensured and the gNB intends to perform rate matching, then corresponding zero power (ZP) CSI-RS can be explicitly configured. Note that rate matching is not performed around aperiodic CSI-RS. This is because the aperiodic CSI-RS is triggered by an UL grant while the PDSCH allocation is provided by a DL grant. Due to PDCCH decoding errors, the UE may receive the DL grant but not the UL grant. In order to perform proper rate matching around the aperiodic CSI-RS, the gNB will trigger an aperiodic ZP CSI-RS in the DL grant. Also note that for V2X in Release 16, an exception was made in that a

PSCCH granting a PSSCH also triggers rate matching in the granted PSSCH. This is because in the V2X sidelink, there are no separate DL and UL grants.

- Periodic, semi-persistent and aperiodic ZP CSI-RS when activated [1]. There can be up to two bits included in the DL grant that can trigger three different ZP CSI-RS patterns. The bit setting "00" indicates no ZP CSI-RS triggering. Since there is a single grant for PDSCH with slot aggregation, the aperiodic ZP CSI-RS triggered in the first slot is going to be rate matched around in all aggregated slots. This typically results in "over rate matching," i.e., rate matching around REs that would be in fact available for PDSCH transmission, causing unnecessary overhead. However, this was seen still better than the alternative of no rate matching in a slot where it would have been needed, in which case the missing rate matching would cause resource collision.
- LTE CRS pattern configured for rate matching [1]. This aspect was described in Sec. 11.2. In Release 15 only one CRS pattern can be configured per NR carrier. In Release 16, up to six CRS patterns can be configured per NR carrier.
- The PDCCH granting the PDSCH [1]. The PDCCH in this context includes the REs used for control information and associated DM-RS. When the PDCCH granting the PDSCH is configured with wideband precoding, then the associated DM-RS includes DM-RS REs in all REGs of the CORESET. Otherwise, the associated DM-RS is the DM-RS in REGs of the PDCCH. When the UE monitors PDCCH with both aggregation level 8 and 16 and the detected PDCCH was aggregation level 8, then the UE is required to assume rate matching around aggregation level 16. Correspondingly, the gNB performs rate matching around aggregation level 16, irrespective of whether the actual PDCCH aggregation level was 8 or 16. The reason for this provision is a particular ambiguity where the aggregation level 16 PDCCH is detected by the UE as aggregation level 8 or vice versa.
- Periodic and dynamic RB-symbol level rate matching [1], which will be described next.

RB-symbol level rate matching is performed around a semi-statically configured resource pattern that can occur either periodically or can be dynamically activated [1]. The granularity of the configured rate matching pattern is a 1RB-by-1symbol block. RB-symbol rate matching was introduced in NR in order to accomplish the following goals:

- Rate matching around CORESETs.
- Rate matching around LTE PSS/SSS/PBCH.
- Forward compatibility rate matching around signals potentially introduced in future releases.

An RB-symbol level rate matching pattern is defined by a triplet of bitmaps:

- First bitmap in the frequency domain indicating, with one bit per RB, the availability of the RB for PDSCH data.
- Second bitmap in the time domain indicating, with one bit per OFDM symbol, the availability of the symbol for PDSCH data. The length of the second bitmap

can be 14 bits in the case of a single-slot pattern or 28 bits in the case of a double-slot pattern.

- Third bitmap also in the time domain indicating, with one bit per slot (if the second bitmap is of length 14) or with one bit per two slots (if the second bitmap is of length 28), the applicability of the second bitmap in the slot. The length of the third bitmap can be 1, 2, 4, 5, 8, 10, 20, or 40 units long, where a unit is one or two slots. For 15 kHz SCS, the 40-unit bitmap is not supported with two-slot unit, in order to limit the maximum duration of the rate matching pattern to no longer than 40 ms.

A set of multiple triplets can be configured for a BWP in a carrier. The final rate matching pattern is determined by combining the patterns given by all triplets with an OR operation.

In Figs. 8.9 to 8.11, an example pattern set is given with two single-slot patterns using the following two example triplets of bitmaps:

- Pattern 1:
  - 1st bitmap: {1, 1, 0, 1, ..., 0, 1}
  - 2nd bitmap = {1, 0, 1, 1, 0, 0, 1, 1, 1, 1, 0, 0, 0, 1}
  - 3rd bitmap = {1, 0, 1, ..., 1}
- Pattern 2:
  - 1st bitmap: {0, 0, 1, 1, ..., 0, 0}
  - 2nd bitmap = {1, 1, 1, 1, 1, 1, 1, 1, 1, 1, 1, 1, 1, 1}
  - 3rd bitmap = {0, 1, 1, ..., 1}

Figure 8.9 shows Pattern 1, Fig. 8.10 shows Pattern 2, and Fig. 8.11 shows the combined rate matching pattern.

The configured rate matching patterns described above can be switched dynamically when pattern groups are also configured. Up to two pattern groups can be configured and each pattern group is dynamically switched with a dedicated bit in DCI format 1_1 granting the PDSCH. When two pattern groups are configured, two bits are included in DCI format 1_1. When a PDSCH is granted, then the pattern groups activated by the granting DCI will indicate resources unavailable for PDSCH. The indication is only valid within the time and frequency resources allocated to the PDSCH. So, for example, if a pattern is activated for a PDSCH in a first slot, where the pattern includes a second bitmap that has a duration of two slots, then only the pattern in the first slot is activated. If there is another PDSCH granted in the second slot, then its PDCCH must also have the dynamic rate matching bit set to indicate that the pattern is activated in the second slot for the rate matching to be effective.

The pattern groups are formed by associating them with configured patterns. As it was mentioned, a set of patterns are semi-statically configured, where each pattern is defined by a triplet of bitmaps. Each individual pattern in the set can have one of the following association types:

- Rate matching pattern not associated with a pattern group
  - This rate matching pattern is always active and indicates unavailable resources.

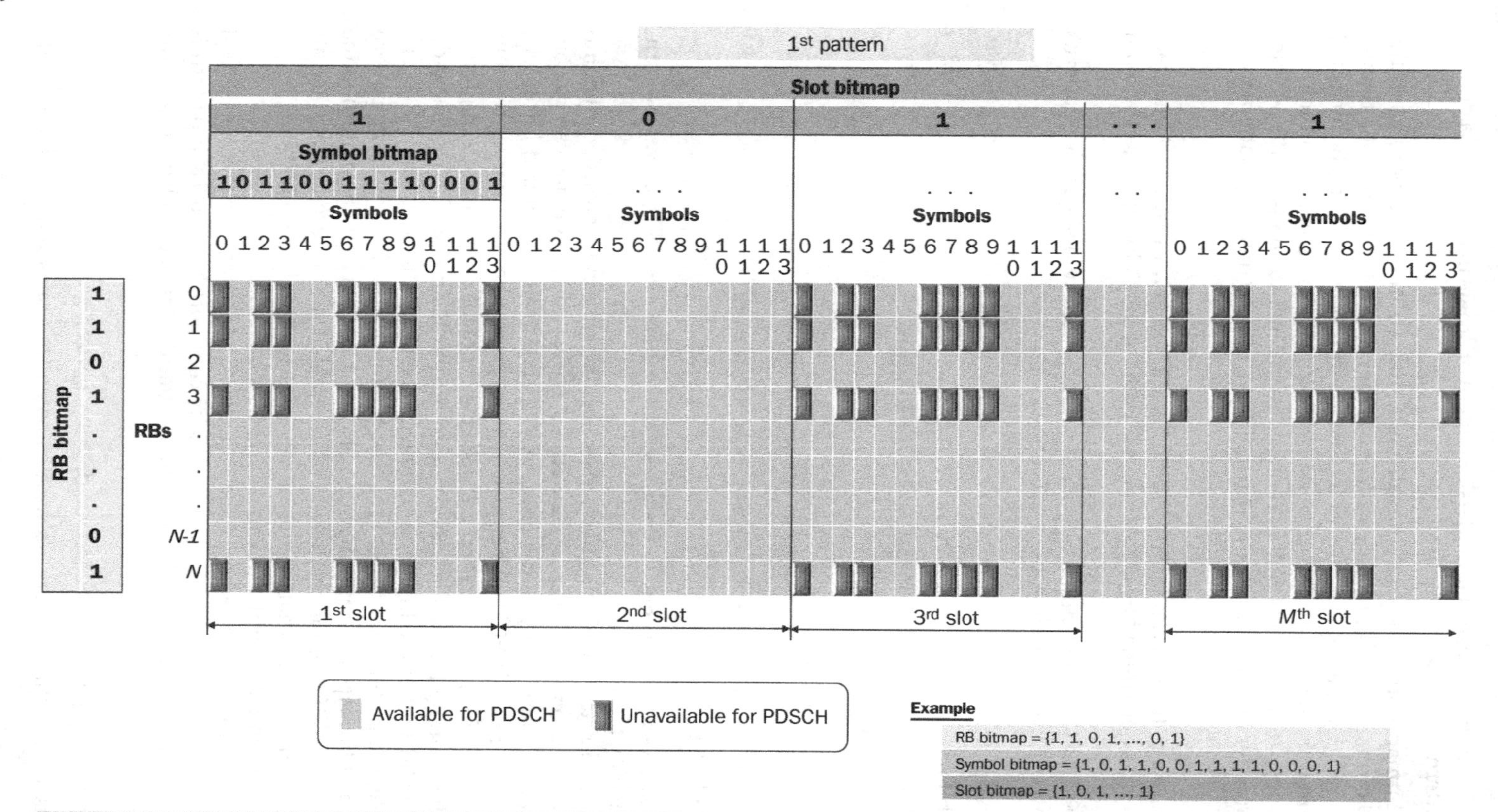

FIGURE 8.9 Example of a first configured rate matching pattern.

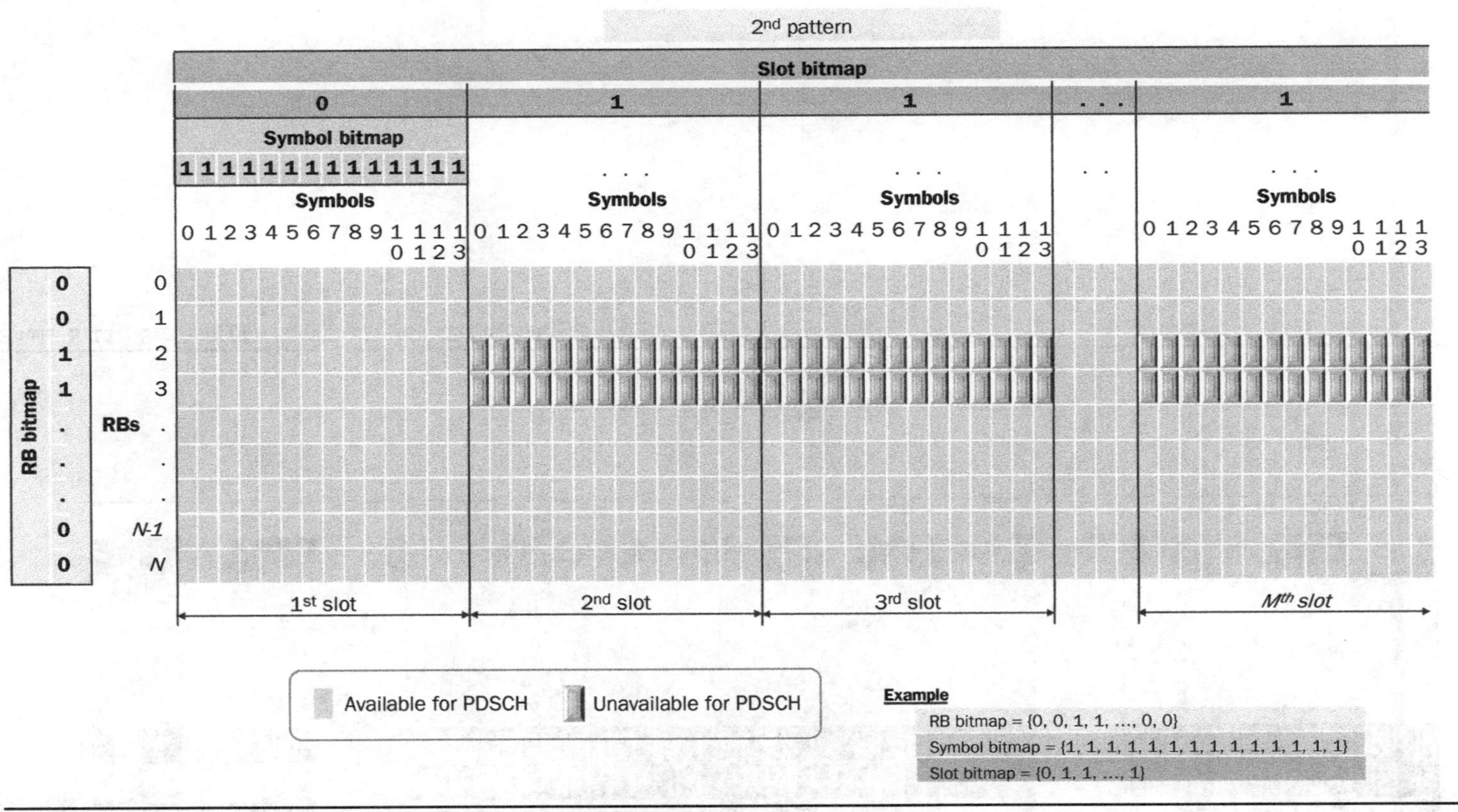

FIGURE 8.10 Example of a second configured rate matching pattern.

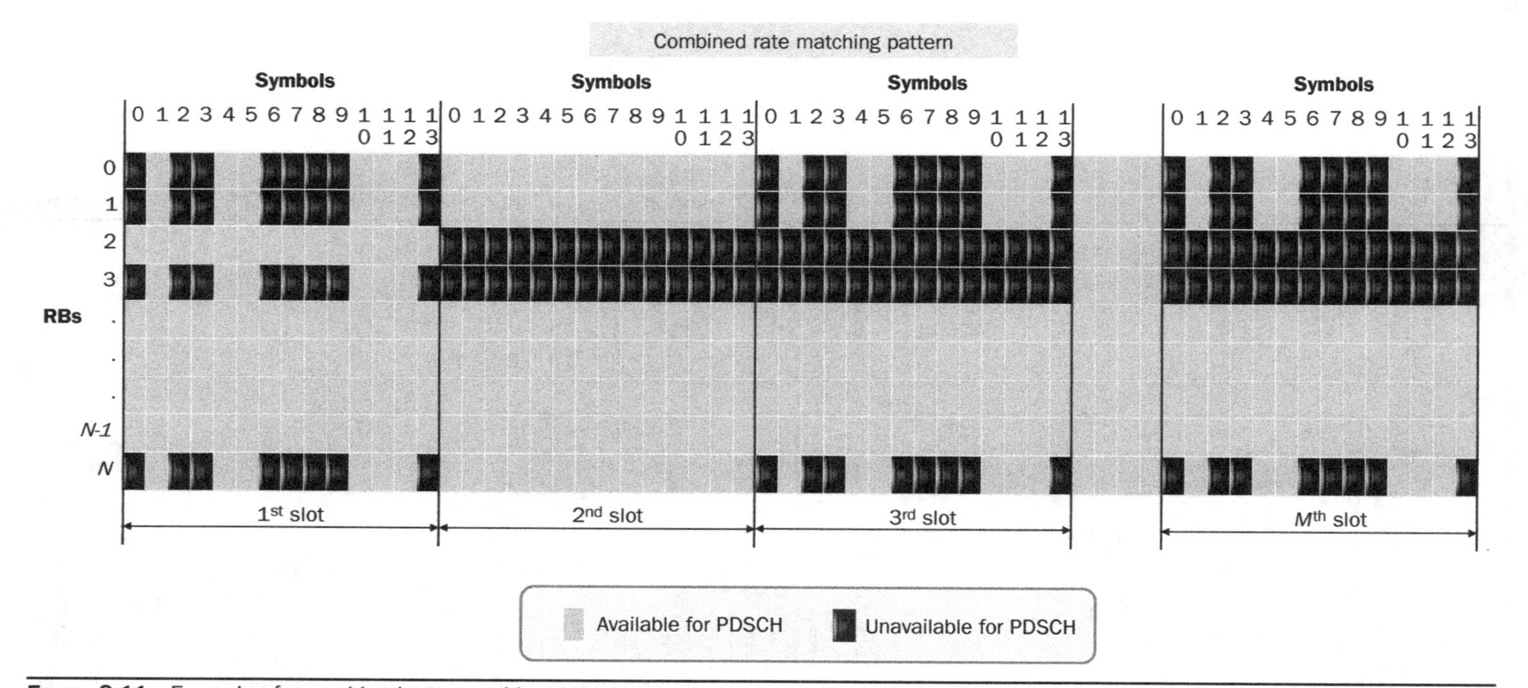

FIGURE 8.11 Example of a combined rate matching pattern.

- Rate matching pattern associated with one of the pattern groups
  - This rate matching pattern is activated by the dedicated dynamic rate matching indication in DCI format 1_1 and indicates unavailable resources only within the allocated PDSCH resources.
- Rate matching pattern associated with both pattern groups
  - This rate matching pattern is activated by either of the dynamic rate matching bits in DCI format 1_1 and indicates unavailable resources only within the allocated PDSCH resources.

The rate matching patterns that are associated with either pattern group are still defined in the same way as other rate matching patterns. This includes the possibility of configuring a 3rd bitmap that is referenced to absolute system time, as opposed to being referenced to the time-domain location of the PDSCH for which the dynamic rate matching indication is valid. Take for example the semi-static pattern shown in Fig. 8.9. Assume that this pattern is associated with the first pattern group and, therefore, it is dynamically switchable by the first indicator bit in DCI format 1_1. If the bit is set to "1" for a PDSCH in the first slot shown in Fig. 8.9, then the pattern gets activated. However, the indicator bit has no effect in the second slot in the given example because the 3rd bitmap, which is referenced to absolute system time, indicated that there are no reserved RB-symbol pairs in this slot. Therefore, setting the indicator bit to "0" or "1" for this slot makes no difference.

The main reason for introducing dynamic rate matching operation was to optimize rate matching for CORESETs. As it was mentioned earlier, the UE always performs PDSCH rate matching around the PDCCH granting the PDSCH. However, the UE is not aware of PDCCH sent to other UEs. A CORESET can be occupied by PDCCH sent to other UEs or it can be available for PDSCH for this UE. The dynamic rate matching operation is able to make this differentiation and enable the CORESET resources used for PDSCH when not occupied by PDCCH sent to other UEs. Note that PDSCH rate matching is not performed around any PDCCH that is received by the UE but not granting the given PDSCH. For example, the UE may receive an UL grant in REs that collide with its PDSCH. Even though the UE knows about the resource collision, the UE is not performing rate matching around these PDCCH REs. The reason for this is that the UE may miss the UL grant, in which case there would be a rate matching error because the gNB performed rate matching, but the UE did not. In order to perform proper rate matching, the gNB can include DL semi-static or dynamic rate matching for the CORSET including the PDCCH carrying the UL grant.

As it was mentioned, when pattern groups are configured, then up to two dedicated dynamic rate matching indication bits are included in DCI format 1_1. But these bits are missing from DCI format 1_0. The rule for handling this case is that the missing bits are assumed to take a default value "1," i.e., both pattern groups get activated. In general, this can result in "over rate matching" causing unnecessary overhead. However, this was seen still better than the alternative of no rate matching for the PDSCH granted by DCI format 1_0 because that could result in resource collisions.

The rate matching operation described in this section is only applicable to PDSCH data. Any overlap between active rate matching patterns and DM-RS for the PDSCH or DM-RS for any other channel or overlap between active rate matching patterns and PDCCH control information remains an error case in Release 15. Also, there is no equivalent rate matching operation in the UL.

8.8 DL HARQ Operation

The HARQ scheme in LTE was defined based on assuming fixed timing relationships. For example, for a DL transmission in FDD, the HARQ feedback is sent always 4 ms after the DL transmission, and in TDD it is sent in the first available UL subframe occurring no earlier than 4 ms after the DL transmission. Similarly, in the UL in FDD, PUSCH is sent 4 ms after the dynamic UL grant, and in TDD, it is sent in the first UL subframe occurring no earlier than 4 ms after the dynamic UL grant. In all cases in LTE, the control transmission in both the DL and UL starts at the beginning of the subframe, so the HARQ delay can be measured in integer number of subframes.

In NR, the HARQ scheme is made more dynamic [1], so that information in the DL grant indicates how much time after the PDSCH the UL feedback should be sent by the UE, and information in the UL grant indicates how much time after the UL grant the UL PUSCH should be sent by the UE. In addition, the DL grant can also indicate a delay between the grant and the corresponding PDSCH. The reasons for adding this flexibility in NR was for the most part lessons learned from LTE. For example, when an UL subframe has to be reserved and, therefore, not used in LTE FDD, the corresponding DL subframe 4 ms earlier also becomes effectively wasted because of the missing HARQ feedback. Conversely, if a DL subframe has to be reserved and therefore not used in LTE FDD, e.g., due to a measurement gap, then the corresponding UL subframe 4 ms later will be wasted because UL grant could not be sent. Both these can be solved if the HARQ delay is made dynamically configurable, and this is what the NR design decided to do. The same examples also hold for TDD, in a sense even more so, because the loss of one UL subframe can mean the loss of multiple DL subframes in a typical DL heavy configuration. Making the HARQ delay dynamically configurable in TDD is also a necessary component of enabling dynamic TDD operation.

For the DL, a maximum of 16 HARQ processes per DL carrier is supported by the UE. The actual maximum number of processes used for the DL is configurable by RRC parameter *nrofHARQ-ProcessesForPDSCH* independently for each carrier. When no configuration is provided, the default number of processes is 8.

The UE is not required to decode dynamically granted PDSCHs overlapped in time in the same carrier, even if they are not overlapped in frequency. The only exception is when one of the PDSCHs is for broadcast information, for example, system information (SI) or paging, and the other PDSCH is for unicast data [8]. In these cases, under certain restrictions, the UE is required to decode both PDSCHs as long as they do not overlap in frequency. The reason for this exception is that the gNB cannot know for sure when the UE decodes broadcast PDSCH, so if the gNB wanted to avoid requiring simultaneous decode for any UE, it would have to refrain scheduling any PDSCH carrying unicast data that overlaps in time with PDSCH carrying broadcast information. This could be a waste of resources because broadcast PDSCH typically occupies less than full bandwidth in the frequency domain. However, opinions were split about this solution, since even if simultaneous decode of broadcast PDSCH and PDSCH carrying unicast data were not required, and the UE would occasionally drop PDSCH carrying unicast data as a consequence, the system impact would have been fairly small, since the decoding of broadcast PDSCH by any given UE is infrequent. As it was mentioned, simultaneous decoding is only required under certain restrictions, for example, the PDSCH carrying unicast data cannot require Capability 2 (faster) UE processing time as defined later in Sec. 8.10.

There were various provisions introduced to exclude so called out-of-order HARQ in the DL. These included the following restrictions:

- For a given HARQ process ID in a carrier, the UE is not required to decode a PDSCH until after the UE completed the HARQ feedback for the previous PDSCH of the same HARQ process ID.
- For two different HARQ process IDs on the same carrier, the PDSCH and the corresponding HARQ feedback cannot be in reverse order. That is, the UE cannot be required to send HARQ feedback later for a first PDSCH than HARQ feedback for a second PDSCH, if the first PDSCH is received earlier than the second PDSCH. Note that sending HARQ feedback at the same time for the two PDSCHs is allowed.
- For two different HARQ process IDs on the same carrier, the PDSCH and the corresponding DL grant cannot be in reverse order. That is, the UE cannot be required to receive a first PDSCH before a second PDSCH, if the DL grant for the first PDSCH was received later than the DL grant for the second PDSCH. For deciding the relative timing of the two DL grants, the end of the last symbol of the PDCCH carrying the grant is counted. Note that receiving the DL grants at the same time for the two PDSCHs is allowed in theory, but it is not mandatory for the UE to receive two DL grants at the same time.

Some of these restrictions may be relaxed in a future release of the specification, for example, in order to enhance eMBB-URLLC multiplexing in a UE in a given carrier.

All the above restrictions apply to the HARQ processes associated with a given carrier, and there is no additional restriction across HARQ processes occurring on different carriers.

8.9 DL Data Rate Capability

In order to provide solutions to a diverse set of use cases, there was a perceived need to define several NR UE capability classes in terms of maximum data rate capability in Release 15. For LTE, this aspect was handled by managing an ever-growing set of UE DL and UL categories. Due to the larger target bandwidths of NR compared to LTE, it seemed that from the start, there may be a need to have many more categories in NR. Also, a lesson learned from LTE was that after introducing advanced features, a particular data rate could be achieved different ways, each of which resulting in slightly different data rates, so the category data rates became widening intervals rather than values. One option considered in NR was that the UE category could be just a grid of quantization points, e.g., the UEs maximum data rate rounded up or down to the closest multiple of a fixed step size (e.g., 10 Mbps or 100 Mbps).

After considerable debate, it has been decided that NR will not introduce the concept of UE categories for the time being. Instead, a standardized formula was provided, by which the UE's DL data rate capability is computed. All input elements to the formula can be extracted from the UE reported capabilities. The performance requirements mandate that the UE can achieve this calculated maximum data rate in a sustained manner, at least under ideal channel conditions.

The formula introduced for calculating the maximum data rate in Ref. [9] is the following:

$$\text{DL data rate (Mbps)} = 10^{-6} \cdot \sum_{j=1}^{J} \left(v_{\text{Layers}}^{BW(j)} \cdot Q_m^{(j)} \cdot f^{(j)} \cdot R_{\max} \cdot \frac{N_{PRB}^{BW(j),\mu} \cdot 12}{T_s^{\mu}} \cdot \left(1 - OH^{(j)}\right) \right) \tag{8.6}$$

where

- J is the number of aggregated DL component carriers in a band or band combination.
- $R_{\max} = 948/1024$ is the nominal maximum code rate.
- For the jth component carrier:
  - $v_{\text{Layers}}^{BW(j)}$ is the maximum number of supported layers given by higher layer parameter *maxNumberMIMO-LayersPDSCH*.
  - $Q_m^{(j)}$ is the maximum supported modulation order given by higher layer parameter *supportedModulationOrderDL*, which can take the following values: $Q_m^{(j)} = \{1,2,4,6,8\}$.
  - $f^{(j)}$ is a UE selected scaling factor given by higher layer parameter *scalingFactor*, which can take the values $f^{(j)} = \{1, 0.8, 0.75, 0.4\}$.
  - μ is the SCS index ($\mu = 0$ for 15 kHz, $\mu = 1$ for 30 kHz, $\mu = 2$ for 60 kHz, $\mu = 3$ for 120 kHz).
  - T_s^{μ} is the average OFDM symbol duration in a subframe for numerology μ, i.e., $T_s^{\mu} = \frac{10^{-3}}{14 \cdot 2^{\mu}}$ for normal cyclic prefix.
  - $N_{PRB}^{BW(j),\mu}$ is the maximum RB allocation in bandwidth $BW^{(j)}$ with numerology μ, where $BW^{(j)}$ is the UE supported maximum bandwidth in the given band or band combination in Ref. [10] or in [11].
  - $OH^{(j)}$ is the nominal overhead and takes the following values: $OH^{(j)} = 0.14$ for DL in FR1 and $OH^{(j)} = 0.18$ for DL in FR2.

In order to determine the UE's peak data rate capability, the data rate is calculated in every DL band combination and feature combination that the UE indicated to support, and the maximum among all those calculated values is selected.

As an example, assume that the UE signals the following capabilities:

- $J = 1$, single carrier only
- $v_{\text{Layers}}^{BW(0)} = 4$, maximum supported rank is four
- $Q_m^{(j)} = 8$, reference modulation order is 256QAM
- $f^{(j)} = 1$, scaling factor of 1
- $\mu = 1$, 30 kHz SCS support
- $N_{PRB}^{BW(0),1} = 273$, 100-MHz channel support in FR1 and 273 RB maximum transmission bandwidth configuration based on Ref. [10] as mentioned in Sec. 2.4.1

With these, the maximum supported data rate will be calculated as:

- DL data rate = 2337 Mbps = 2.337 Gbps

It is important to note that $Q_m^{(j)}$ is not the modulation order the UE supports. It is mandatory for the UE to support 64QAM in all bands in the DL [9]. In addition, it is mandatory for the UE to support 256QAM in the DL in FR1 [9]. Yet, the UE can signal $Q_m^{(j)} = 1$ (Pi/2-BPSK) or $Q_m^{(j)} = 2$ (QPSK), for example. This is because $Q_m^{(j)}$ represents only a scaling factor to be used in the maximum data rate formula, independent of the supported modulation order. For example, in the FR1 DL, a UE can signal $Q_m^{(j)} = 1$ and another UE can support $Q_m^{(j)} = 8$. Both UEs can be configured with 256QAM in the same maximum channel bandwidth. Assuming all other UE capabilities being the same, the first UE supports 1/8 of the maximum supported data rate supported by the second UE. It is interesting to note, that the UE can signal $Q_m^{(j)} = 1$ (Pi/2-BPSK) in the DL, while this modulation order is not even specified in the standard for the DL. There is also another scaling factor, $f^{(j)}$, serving a similar purpose. In the data rate calculation, the product of both applies. Both $Q_m^{(j)}$ and $f^{(j)}$ were introduced with the purpose of allowing UE decoding complexity reduction for the cases of maximum aggregated bandwidth.

It should be also noted that even though $Q_m^{(j)} \cdot f^{(j)}$ is set independent for each band, their effect is aggregated across carriers. Assume for example that the UE supports 256QAM maximum modulation order in two 100-MHz carriers with $J = 2$, and for each of the two carriers, the UE indicates $Q_m^{(j)} \cdot f^{(j)} = 4$, i.e., half the theoretical maximum data rate. Then the UE must support half of the theoretical maximum data rate across the two carriers. But the UE must also support the full theoretical data rate on one of the carriers, as long as the other carrier is not scheduled at the same time. The carrier scheduled can be changed dynamically. To say it differently, the data rate scaling acts on the aggregated date rate, not on the per carrier data rate. However, the following limitations apply to this data rate aggregation: Unused data rate cannot be "transferred" from a cell group (CG) to another CG, i.e., from MCG to SCG or vice versa. Similarly, unused data rate cannot be "transferred" from FR1 to FR2 or vice versa in FR1+FR2 carrier aggregation. For a carrier configured with processing time capability 2, which is discussed in Sec. 8.10, an even stricter restriction applies: The unused data rate in any other carrier cannot be transferred into the carrier configured with processing time capability 2.

In order to further limit UE complexity, a couple of additional restrictions were specified.

The UE may skip decoding a transport block in an initial transmission if the effective channel code rate is higher than 0.95 [1]. Note that the maximum required code rate in NR represents a slight increase compared to LTE, where it was 0.932.

To limit both the soft buffer data transfer rate and the minimum code rate for the largest TB sizes, the following relaxation was defined in Ref. [1]: The UE is not expected to handle any transport blocks (TBs) in any consecutive 14-symbol duration, ending at the last symbol of the latest PDSCH transmission whenever

$$\sum_{i \in S} \left\lceil \frac{C_i'}{L_i} \right\rceil \cdot x_i \cdot F_i > \left\lceil \frac{X}{4} \right\rceil \cdot \frac{TBS_{LBRM}}{R_{LBRM}} \tag{8.7}$$

where

- S is the set of TBs belonging to PDSCH(s) that are partially or fully contained in the consecutive 14-symbol duration.
- for the ith TB:
    - C_i' is the number of scheduled code blocks.

- L_i is the number of OFDM symbols assigned to the PDSCH.
- x_i is the number of OFDM symbols of the PDSCH contained in the consecutive 14-symbol duration.
- F_i is the number of received channel bits per code block that need to be used in the decoding of the ith TB. F_i is calculated as $F_i = \max_{j=0,\ldots,J-1}\left(\min\left(k_{0,i}^{j} + E_i^{j}, N_{cb,i}\right)\right)$, where
  - $J-1$ is the current (re)transmission for the ith TB.
  - $k_{0,i}^{j}$ is the starting location of the RV for the jth transmission of the ith TB.
  - E_i^{j} is the length of the soft buffer segment used for the jth transmission of the ith TB.
  - $N_{cb,i}$ is the circular buffer length.
- $R_{LBRM} = 2/3$.
- TBS_{LBRM} is the quantized maximum TBS assumed for the PDSCH rate matching.
- X is the maximum number of MIMO layers assumed for the PDSCH LBRM determination.

Descriptions of the above parameters can be found in Secs. 8.1.2, 8.2, and 8.3, and in Ref. [5]. With this information, the inequality in the condition for the soft buffer–related data rate relaxation can be explained as shown in Fig. 8.12.

In Fig. 8.13, an example is shown where the soft buffer–related data rate limitation takes effect.

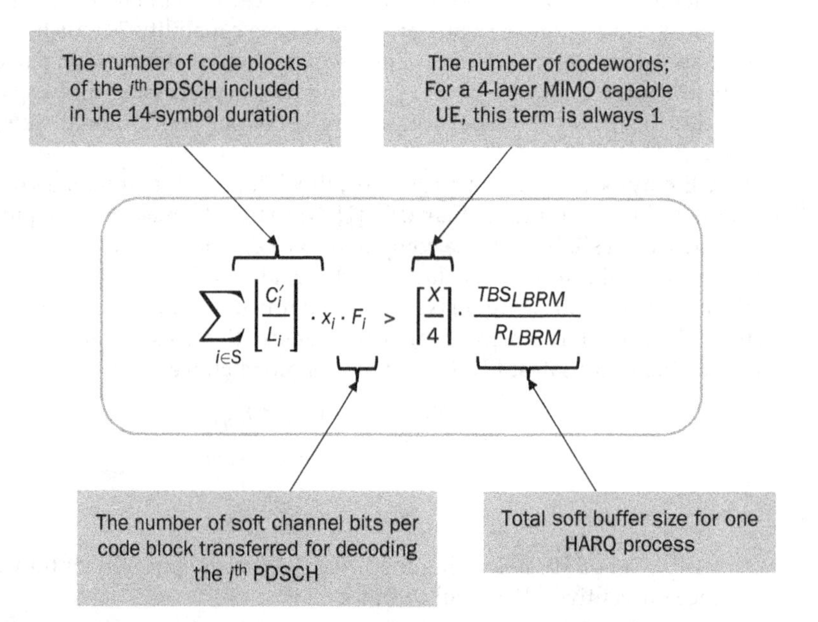

Figure 8.12 Interpretation of the soft buffer–related data rate relaxation inequality.

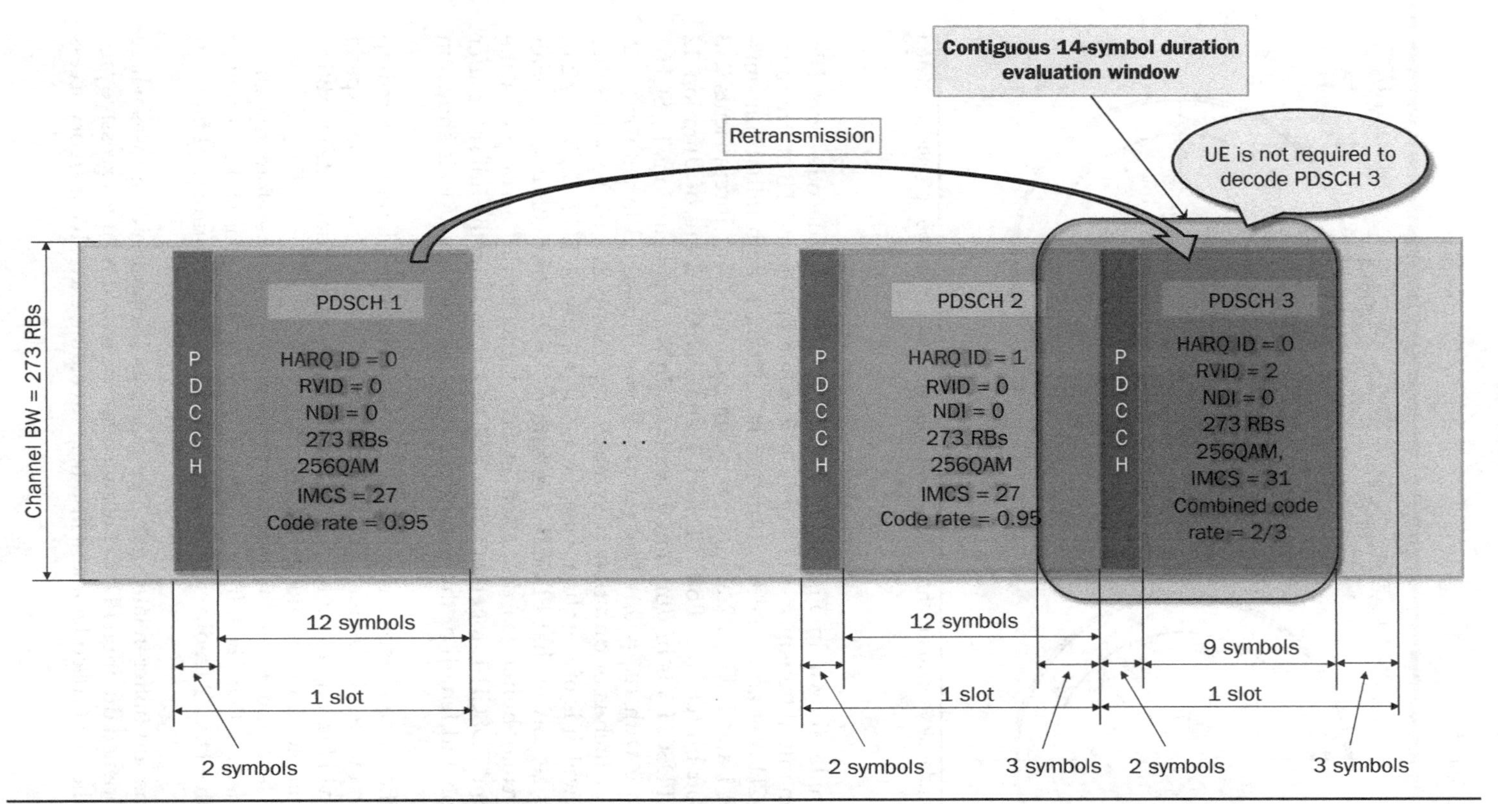

Figure 8.13 Example for the soft buffer–related data rate relaxation.

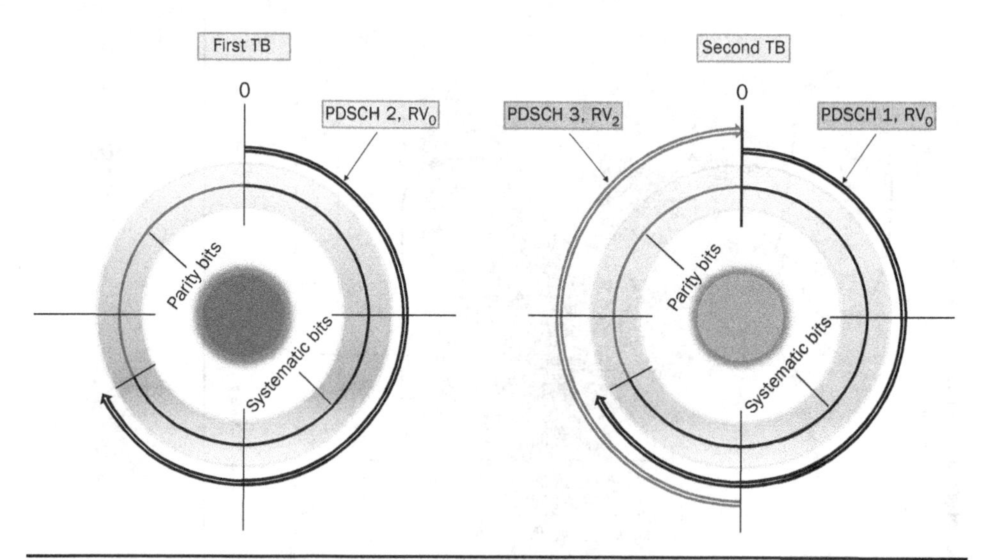

FIGURE 8.14 Redundancy versions in the circular buffer in the example for the soft buffer–related data rate relaxation.

In the example shown in Fig. 8.13, the UE attempts to decode two TBs. The first TB is an initial transmission in PDSCH 2, while the second TB is a retransmission in PDSCH 3 of an initial transmission in PDSCH 1. Both initial transmissions in PDSCH 1 and PDSCH 2 are at maximum data rate, which represents 2/3 of the LBRM soft buffer size. Both PDSCH 1 and PDSCH 2 are of duration 12 symbols, while PDSCH 3 is of duration 9 symbols. Initial transmissions PDSCH 1 and PDSCH 2 are with $rv_{id} = 0$, while retransmission PDSCH 3 is with $rv_{id} = 2$. Since PDSCH 1, which is of duration 12 symbols, represents 2/3 of the soft buffer size, PDSCH 3, which is of duration 9 symbols, will represent $9/12 \cdot 2/3 = 1/2$ of the soft buffer size. Given that PDSCH 3 is with $rv_{id} = 2$, which starts at the half point in the circular buffer (as discussed in Sec. 8.2), PDSCH 3 completes the circular buffer and PDSCH 1, together with PDSCH 3, represents the full soft buffer size. The location and length of the redundancy versions RV_0 and RV_2 is shown in Fig. 8.14.

For the specific example depicted in Fig. 8.13, and for which the placement of redundancy versions in the circular buffer was illustrated in Fig. 8.14, we can further expand the interpretation in Fig. 8.12 to explain how the soft buffer–related data rate relaxation applies in the example as shown in Fig. 8.15.

As it can be seen, the inequality holds in the example case; therefore, the UE is not required to decode PDSCH 3. If the UE skips decoding, the UE is still expected to send back negative HARQ feedback, although there is no clear requirement mandating this.

In general, the soft buffer–related data rate relaxation helps avoiding cases where back-to-back shortened duration PDSCH retransmissions would require excessive processing load on the LDPC decoder and it also helps avoiding excessive memory access data rates.

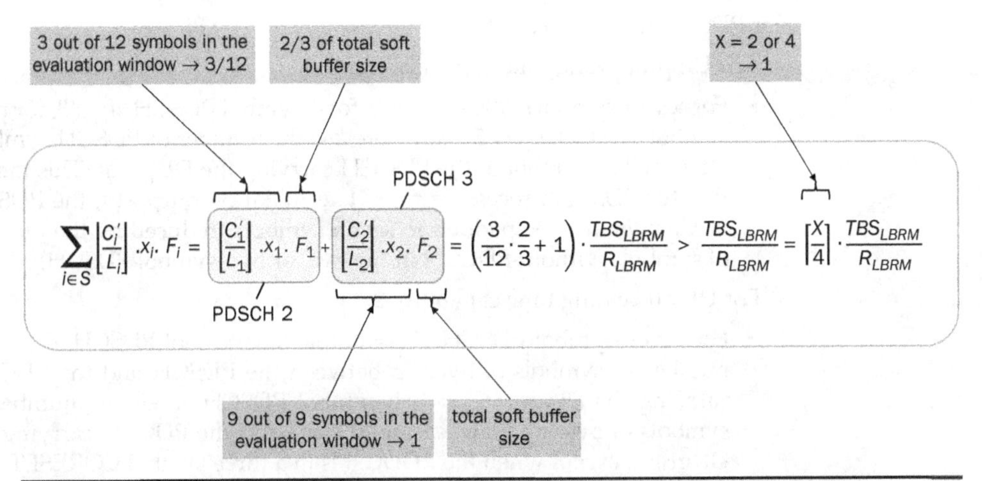

FIGURE 8.15 Evaluation of the soft buffer–related data rate relaxation corresponding to the example.

8.10 Processing Time for DL Data

In LTE, the nominal HARQ feedback delay following a PDSCH reception was 3 ms minus the UL timing advance. Beside the larger delay compared to NR, the LTE processing time definition itself was a bit inefficient because the strictest LTE UE processing time requirement was sized for receiving peak data rate at the cell edge in the maximum supportable cell size, which combination would never occur in practice. In the practical scenarios, the HARQ feedback for the PDSCH at peak date rate is delayed corresponding to the smaller UL timing advance in smaller cells, thereby wasting the UE processing time capability.

One of the main advancements made in NR was reduced latency, which required significantly reducing the DL processing time relative to LTE. In this context, the DL processing time is measured from the end of the PDSCH to the beginning of the PUCCH (or possibly PUSCH) carrying the feedback. During this time, the UE has to decode the PDSCH, encode and multiplex the control information, and generate the UL waveform.

Two UE capability classes were introduced in Ref. [1]: UE DL processing time capability 1 and UE DL processing time capability 2.

The PDSCH HARQ feedback delay requirement denoted with $T_{proc,1}$ was defined as

$$T_{proc,1} = N_1 + d_{1,1} \tag{8.8}$$

where

- $d_{1,1}$ is a term dependent on the length of the PDSCH.
  - For PDSCH mapping type A, if the last symbol of PDSCH is in the ith OFDM symbol of the slot, then $d_{1,1} = 7 - i$ when $i < 7$, otherwise $d_{1,1} = 0$. This means that if the PDSCH finishes before the end of the first half of the slot, then for PDSCH processing time determination purposes the PDSCH is considered to be "padded" to the end of the first half of the slot. PDSCH mapping types were discussed in Sec. 8.5.

- For PDSCH mapping type B,
  - For UE processing time capability 1:
    - For seven-symbol PDSCH, $d_{1,1} = 0$, for 4-symbol PDSCH, $d_{1,1} = 3$, for two-symbol PDSCH, $d_{1,1} = 5 - k$, where k is the number of PDSCH symbols after the last symbol of the PDCCH carrying the DL grant. This means that for PDSCH processing time determination purposes, the PDSCH is considered to be padded to seven symbols (reduced by the number of symbols without PDCCH in the case of two-symbol PDSCH).
  - For UE processing time capability 2:
    - For seven-symbol PDSCH, $d_{1,1} = 0$, for four-symbol PDSCH, $d_{1,1}$ is the number of symbols of overlap between the PDSCH and the PDCCH carrying the DL grant, for two-symbol PDSCH, $d_{1,1}$ is the number of symbols of overlap between the PDSCH and the PDCCH carrying the DL grant, except when the PDDCH is in a three symbol CORESET that ends after the PDSCH, in which case $d_{1,1} = 3$. This means that for PDSCH processing time determination purposes, the number of symbols of overlap between the PDSCH and the PDCCH carrying the DL grant is added to the processing time, to accommodate the additional delay caused for PDCCH decoding, and in the case of three symbol CORESET, an additional symbol may be added.
- N_1 is based on the SCS and is given in Tables 8.12 and 8.13 for UE processing time capability 1 and capability 2, respectively [1]. The SCS corresponds to one of the PDCCH, PDSCH, PUCCH, PUSCH involved in either the DL processing for reception of the data or UL processing for transmission of the corresponding HARQ ACK, with details described later in this section.

Note that Eq. (8.8) is not precise because it omits certain scaling factors, but it suffices for the current description.

The choice between 13 and 14 symbols in the case of processing time capability 1 with 15 kHz SCS and with additional PDSCH DM-RS position shown in Table 8.12 depends on whether the second DM-RS position was delayed due to possible collision with CRS in the shared LTE-NR DL carrier case (as described in Chap. 11).

The advanced UE processing capability 2 was not defined for FR2 in Release 15 or Release 16. It was seen that with 120 kHz SCS, the processing times are relatively

	PDSCH HARQ feedback delay			
	With no additional PDSCH DM-RS position		With additional PDSCH DM-RS position	
SCS (kHz)	N_1 (symbols)	μs	N_1 (symbols)	μs
15	8	570.8	13 or 14	927.6 or 1000
30	10	356.8	13	463.8
60	17	303.3	20	356.8
120	20	178.4	24	214.1

Table 8.12 PDSCH HARQ Feedback Delay for DL Processing Time Capability 1

	PDSCH HARQ feedback delay			
	With no additional PDSCH DM-RS position		With additional PDSCH DM-RS position	
SCS (kHz)	N_1 (symbols)	μs	N_1 (symbols)	μs
15	3	214.1	N/A	N/A
30	4.5	160.6	N/A	N/A
60	9 (for FR1)	160.6 (for FR1)	N/A	N/A

TABLE 8.13 PDSCH HARQ Feedback Delay for DL Processing Time Capability 2

short with UE processing capability 1 already, and in addition, in some cases, the beam switching delays required in FR2 could reduce the benefits of reduced FR2 processing times.

It is important to note that when determining whether there is additional DM-RS position or not in order to determine the applicable columns in Table 8.12 or Table 8.13, the configuration of both PDSCH mapping type A and PDSCH mapping type B together needs to be considered. That is, the column "with additional PDSCH DM-RS position" needs to be selected if additional DM-RS position is configured for either PDSCH mapping type A or B, irrespective of what the mapping type of the currently scheduled PDSCH is. The reason for this is that because the PDSCH mapping type can be dynamically changed from slot to slot, a delay incurred in either mapping type due to configured additional DM-RS position(s) will impact the other mapping type as well when PDSCH with different mapping types are scheduled back to back. With additional DM-RS position(s), the channel estimation induces additional delay. Due to process pipelining, the UE is still decoding the previous PDSCH data while processing the symbols of the current PDSCH. If the previous PDSCH decoding is delayed due to the additional DM-RS position, that will delay the current PDSCH decoding as well.

The values for N_1 were agreed after significant debate. Even though N_1 increases in terms of number of symbols for increasing SCS, it actually decreases in terms of time. On the one hand, an argument can be made that when going to higher SCS, the PDSCH symbol processing, the PUCCH waveform generation, and other aspects have to be sped up anyway with a factor corresponding to the SCS ratio, which would result in a constant processing delay in terms of number of symbols. On the other hand, it can also be argued that many of the decoding aspects are not dependent on SCS; therefore, due to these, the number of symbols should be scaled up with a factor corresponding to the SCS ratio. The actual N_1 selection represented a middle way between these two viewpoints.

The actual time between PDSCH and PUCCH is affected by the UL timing advance. The limits given by $T_{proc,1}$ must be applied after including the effects of timing advance and any relevant DL-to-DL or UL-to-UL timing differences.

As it was noted before, some part of the PDSCH HARQ response time, like PUCCH waveform generation actually depends on the UL SCS, not the DL SCS. Also, it is possible to have processing time capability 2 in the DL carrying the PDSCH and processing time capability 1 in the UL carrying the PUCCH, or vice versa. To cover these cases, the general rule was adopted that in the determination of $T_{proc,1}$, a first value is determined based on the relevant PDCCH parameters (DL SCS and processing time capability of the PDCCH carrier), a second value is determined based on the relevant PDSCH parameters (DL SCS and processing time capability of the PDSCH carrier), a third value is determined based on the relevant PUCCH parameters (UL SCS and processing time capability of the PUCCH carrier), and $T_{proc,1}$ is finally determined by selecting among the first, second, and third value the one that gives the largest delay in absolute terms.

Overall, it can be said that the PDSCH decoding time, which is at most 1 ms, is indeed a significant improvement upon LTE, which had 2.3 ms as the minimum PDSCH decoding time. It can be also noted that with processing time capability 2, NR enables a so-called self-contained slot operation, wherein the DL data reception, decoding of the data and transmitting the corresponding HARQ feedback by the UE all occur within the same slot.

8.11 Demodulation Reference Signals for Data

This section applies to both the DL and UL.

Demodulation reference signals (DM-RS) are used for obtaining channel estimation, and in many cases interference estimation for PDSCH and PUSCH data demodulation. The NR DM-RS design is an evolution of the LTE DM-RS with some significant changes. In LTE, DL DM-RS was not incorporated in the first release, it was added in Release 9 as part of LTE Transmission Mode 8 and a subsequent enhanced version was added in Release 10. In part because of the later addition, the use of DM-RS based DL transmission modes never became widespread, LTE still mostly uses common reference signal (CRS) for DL demodulation. This was all changed in NR. NR does not support CRS or CRS-based demodulation in the DL, it uses DM-RS only. Another important difference between LTE and NR is that because NR introduced CP-OFDM waveform in the UL, it became possible to design a DM-RS pattern that is common between DL and UL, at least for this waveform. Further changes relative to LTE were motivated by an objective to provide increased flexibility to have different DM-RS patterns in both frequency and time in order to address various demands in multiplexing capability, channel estimation capability, time-domain allocation length, time dispersion of the channel, etc. As a matter of fact, the design of various DM-RS patterns to meet the diverse requirements was the main focus of the NR DM-RS design.

8.11.1 DM-RS Design Principles

This section applies to both the DL and UL.

In the following, we give some general requirements/concepts that were adopted and influenced the NR DM-RS design.

1. Within the allocated data RBs, the DM-RS should form a regular pattern design. The frequency-domain pattern must be uninterrupted by rate matching. The purpose of this is to avoid additional complicated filtering needed to extrapolate the channel to missing DM-RS parts. The same also holds for the time-domain placement of DM-RS.
2. The DM-RS patterns should be invariant to SCS. Once the scaling in time and frequency is applied, the DM-RS patterns for different SCSs are identical. The purpose of this is to reduce complexity for both the specification and implementation.
3. The DM-RS placement in time should gravitate toward the beginning of the slot. The purpose of this is to improve data processing timeline. Since the

receiver must wait for the channel estimation results to be available before it can start data demodulation, placing DM-RS later would increase decoding delay.

4. There should be at least one DM-RS pattern, in which there is a single DM-RS symbol in the first available data symbol, i.e., in the first symbol uninterrupted by other signals. This is called "front-loaded" DM-RS. The purpose of this is to provide a format fully optimized for low decoding latency.
5. There should be at least one DM-RS pattern, in which the need for channel extrapolation before the first DM-RS symbol and extrapolation after the last DM-RS symbol are minimized. Note that since channel estimation is needed in every symbol containing data, this requirement is equivalent to saying that the first DM-RS should be close to the beginning of the PDSCH, and the last DM-RS symbol should be close to the end of the PDSCH. The purpose of this is to provide a format with optimized performance in time dispersive, high Doppler channels.
6. Orthogonal multiplexing of at least four DM-RS ports is to be supported by any DM-RS pattern for data. This is to support the nominal minimum four-layer SU-MIMO operation.
7. At least one DM-RS pattern should support multiplexing up to 12 orthogonal DM-RS ports for MU-MIMO purposes.
8. The orthogonalization of DM-RS ports, at least for CP-OFDM waveforms, should use CDM patterns that have narrow frequency span. The purpose of this is to maximize orthogonality by minimizing cross-port interference in frequency selective channels.
9. There should be one family of DM-RS patterns that maximizes consistency of the time-domain placement of the DM-RS symbols among data channels that may have different but at least partially overlapping resource allocations in the time domain. The purpose of this is to maximize MU-MIMO multiplexing opportunities. The DM-RS for PDSCH/PUSCH mapping type A was designed for this purpose. PDSCH/PUSCH mapping type A is discussed in Secs. 8.5 and 10.2.
10. There should be one family of DM-RS patterns that maximizes time-domain allocation flexibility. The purpose of this is to reduce scheduling latency, e.g., for URLLC, by allowing the data transmission to start virtually anytime. The DM-RS for PDSCH/PUSCH mapping type B was designed for this purpose. PDSCH/PUSCH mapping type B is discussed in Secs. 8.5 and 10.2.
11. Dynamic switching between different types of patterns for the same PDSCH/PUSCH mapping type should be minimized. The purpose of this is to alleviate the need for the receiver to dynamically swap filter parameters.
12. The DM-RS for CP-OFDM should have as much commonality between DL and UL as possible. The purpose of this is to enhance demodulation performance and possibly enhance multiplexing capability for cases of interference from a disparate duplex direction in dynamic TDD. An example for this operation is when the UE receives PDSCH DM-RS that is orthogonal to another UE's PUSCH DM-RS transmission in a neighboring cell.

13. There should be at least one DM-RS pattern family whose PAPR matches the data channel PAPR. This necessitated a separate design for DFT-S-OFDM UL. The purpose is to enable transmitting DM-RS at the same power level as data. Because of the filtered Pi/2-BPSK waveform introduced in NR has even lower PAPR than other DFT-S-OFDM UL transmissions, a separate new DM-RS design was needed for this case. However, the enhanced DM-RS for Pi/2-BPSK was only added in Release 16, due to the limited time in Release 15.
14. Orthogonal MU-MIMO multiplexing of DM-RS corresponding to transmissions to/from devices that have different assumption on the configured DL or UL channel. For example, DM-RS can be multiplexed in simultaneous overlapping PDSCHs targeting two different UEs, where one UE assumes that it is communicating with a 20 MHz gNB, the other UE assumes it is communicating with a 100 MHz gNB. This can be achieved by counting scrambling sequences and precoding boundaries from a common frequency, called "Reference Point A," which was described in Chap. 2. The purpose of this is to provide forward compatibility and to provide means of improving coexistence between different device types.
15. Both wideband and narrow band precoding should be supported. This means that the DM-RS should be segmentable according to resource block group (RBG) designation. The purpose of this is to allow the scheduler to choose between maximum MU-MIMO multiplexing flexibility and maximum channel estimation processing gain in low delay spread channels. RBG is described in Secs. 8.16 and 10.12.
16. The DM-RS overhead should be variable but unnecessary overhead should be avoided in all cases. The purpose of variable overhead is to provide flexibility for the gNB to choose between highest peak data rate performance at high SNR in less varying channels and optimum capacity at low SNR in more varying channels.
17. The Tx power in OFDM symbols containing DM-RS should be the same as in other OFDM symbols of the same channel (PDSCH or PUSCH) that contain only data without DM-RS. This should hold irrespective of whether or not the symbol containing DM-RS also contains data. This is an essential requirement in the UL in order to avoid introducing phase discontinuity between the DM-RS and data symbols that would degrade demodulation performance. The negative impact of power variations is described in more detail in Sec. 10.12.

Note that there was no specific design effort to enhance orthogonal multiplexing capability across different SCSs. At the same time some minor modification was applied to accommodate DSS, which was described in Chap. 11.

As it was mentioned, a lot of the design effort went into finding appropriate patterns in the time, frequency, and code domains that meet as much of the requirements listed in (1) to (17) above as possible. As it often happens, there are conflicts between some of the requirements. For example, (3) and (5) cannot be simultaneously satisfied. Or (4), (7), and (8) cannot be simultaneously satisfied. For these cases, either a compromise design choice was taken, or multiple configurable options were introduced. Of course, the latter itself conflicts with a general requirement of not inflating the number of supportable options, so this required additional compromises.

Configurability	Options
PDSCH/PUSCH mapping type	A or B
DM-RS type	1 or 2
Number of consecutive DM-RS symbols	1 or 2
Total number of DM-RS symbol positions	1, 2, 3 or 4
Waveform	Release 15: CP-OFDM, DFT-S-OFDM, or CGS
	Release 16: CP-OFDM, DFT-S-OFDM, Pi/2-BPSK, or CGS
Number of ports	1, 2, ..., 8 for a single UE 1, 2, ..., 12 for multiple UEs
Number of CDM groups without data	1 or 2 for DM-RS type 1 1, 2, or 3 for DM-RS type 2

Table 8.14 Configurability of DM-RS Attributes

Also, a lot of evaluation effort went into determining the optimum DM-RS densities. DM-RS REs need to be spaced close enough both in time and frequency to make the variation of the channel tractable both in time and frequency. At the same time, placing DM-RS REs too close to each other either in frequency or in time results in excessive overhead, degrading achievable capacity.

In broad terms, DM-RS pattern configurability was introduced along different "dimensions." These are described in Table 8.14.

In the following, we list additional comments for the configurability options with the intent of clarifying them further.

- The PDSCH/PUSCH mapping type determines the time-domain placement of DM-RS symbols within the PDSCH/PUSCH and within the slot. This is described in Secs. 8.5 and 10.2.
- The DM-RS type determines the frequency-domain RE pattern, which will be described later.
- The number of consecutive DM-RS symbols designates two options: single symbol or two consecutive symbols. Note that the number of consecutive symbols is only indirectly related to the total number of DM-RS symbol positions. In general, DM-RS symbols within the PDSCH/PUSCH are spread out in time, but two consecutive symbols are used when there is insufficient port multiplexing capability within a single symbol.
- The total number of DM-RS symbols provides tradeoff between overhead and resistance to high Doppler or low SNR. The total number of DM-RS symbols is expressed as first plus additional positions. For example, a four-symbol pattern is called DM-RS with three additional positions. When using the term "number of additional positions," two consecutive DM-RS symbols are counted together as one position.
- The DM-RS waveform options are as listed in Table 8.14. Computer generated sequence (CGS) is used for DFT-S-OFDM UL with one-RB or two-RB allocation

in the frequency domain. In Release 16, an additional waveform type, Pi/2-BPSK was introduced for the DM-RS.

- The number of ports listed in Table 8.14 is not actually an explicit configuration defining a DM-RS pattern, rather the maximum number of ports is an attribute of each pattern. The actual ports scheduled to a UE are signaled for each transmission/reception, which is a subset of the maximum number of ports.
- The number of CDM groups without data is only used for rate matching of data, so it is not a direct attribute of the DM-RS pattern. One of the options listed in Table 8.14 in the row "Number of CDM groups without data" is selected for the UE for every transmission/reception. CDM groups with data apply to CP-OFDM DM-RS only.

For the DFT-S-OFDM waveform, low PAPR DM-RS is needed, requiring special sequence design. This will be discussed in Sec. 10.9.4.

8.12 PDSCH DM-RS

The main components of the DM-RS design are the mapping in frequency, time, and OCC code domains. These determine the location of the DM-RS REs in the two-dimensional frequency-time resource grid, as well as the allocation of the code domain OCC patterns. In the following, we give a description of the DM-RS used for the PDSCH. We will use the terminology of DL and PDSCH but most of the description equally applies to UL PUSCH with CP-OFDM waveforms. The differences between DL and UL will be discussed in Sec. 10.9.

8.12.1 PDSCH DM-RS Frequency Domain and Code Domain Patterns

In the frequency domain, two basic mapping types, DM-RS type 1 and DM-RS type 2 were defined [7]. DM-RS type 1 uses an equidistant comb-2 pattern in the frequency domain. DM-RS type 2 uses 2-RE CDM groups, somewhat resembling pattern used for the CSI-RS. The frequency-domain arrangement in a single OFDM symbol can be replicated in an adjacent symbol to create a double-symbol DM-RS. Both single-symbol and dual-symbol DM-RS is supported together with both type 1 and type 2 frequency-domain patterns. The RRC parameter *maxLength* determines the number of consecutive OFDM symbols for the DM-RS and can be set to one or two symbols. When it is set to one symbol, only the single-symbol DM-RS can be used. However, when it is set to two symbols, both single-symbol or double-symbol DM-RS can be used, based on dynamic scheduling information. When double-symbol DM-RS is used, then all occurrence of the DM-RS will be in two-symbol units.

In Table 8.15, the multiplexing capability of the different options is summarized.

Multiplexing capability, number of ports		Number of consecutive symbols	
		Single	Double
DM-RS type	type 1	4	8
	type 2	6	12

Table 8.15 Multiplexing Capability of the Different DM-RS Options

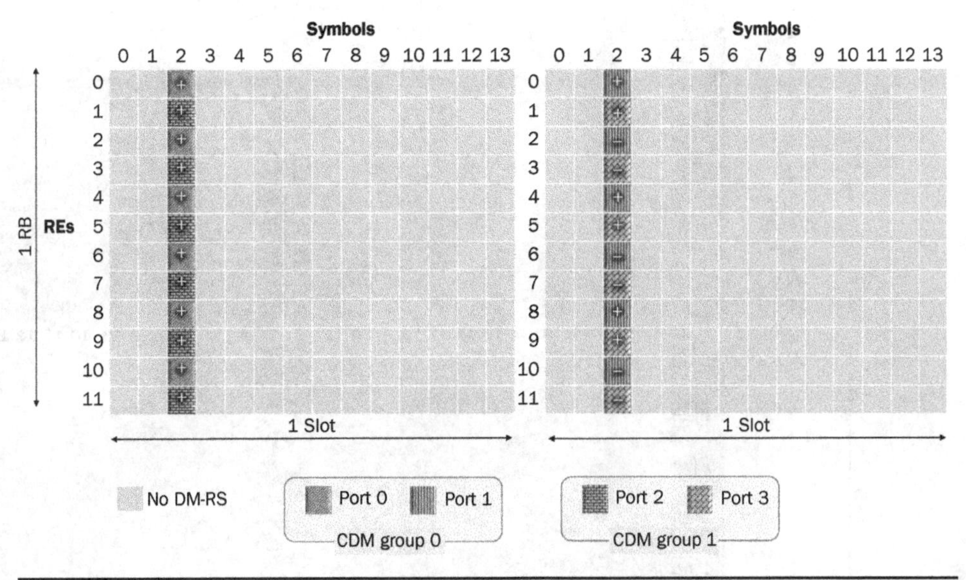

FIGURE 8.16 Frequency-domain mapping for DM-RS type 1, single-symbol.

As it can be seen in Table 8.15, the four combinations of {DM-RS type 1, DM-RS type 2} and {single-symbol, double-symbol} achieve different multiplexing capability, at the cost of different level of immunity to frequency or time-domain variations. In particular, DM-RS type 1 provides a channel observation at every other subcarrier, while DM-RS type 2 has a gap of four subcarriers between observations, as it can be seen in Figs. 8.16 through 8.19.

The frequency and OCC code domain mapping of DM-RS for the four combinations of {DM-RS type 1, DM-RS type 2} and {single-symbol, double-symbol} are shown in Figs. 8.16 through 8.19. These cases are all called front-loaded DM-RS, with no additional DM-RS symbol positions.

The CDM groups are subsets of DM-RS REs that are frequency separated from each other. Each CDM group includes two DM-RS ports for the single-symbol pattern and four DM-RS ports for the double-symbol pattern. The two or four ports within a CDM groups are orthogonalized with a binary Walsh code, which is in a single dimension across frequency for the single-symbol pattern and in two dimensions across both time and frequency for the double-symbol pattern. The OCC patterns apply on top of the DM-RS sequences assigned to the antenna ports.

8.12.2 PDSCH DM-RS Time-Domain Patterns

The discussion of the DM-RS frequency-domain pattern in Sec. 8.12.1 focused on the description of a front-loaded DM-RS pattern. In order to have more observations in the time domain for improving both the channel estimation SNR and the ability to track channel variations, additional DM-RS symbol positions can be configured in the time domain [7]. In these cases, the same frequency-domain arrangement and port allocation gets repeated over time. Note that NR does not support staggering, i.e., alternating the port-to-subcarrier mapping across symbols. The main reason for this was reducing channel estimation complexity in the receiver. With non-staggered patterns, it is much

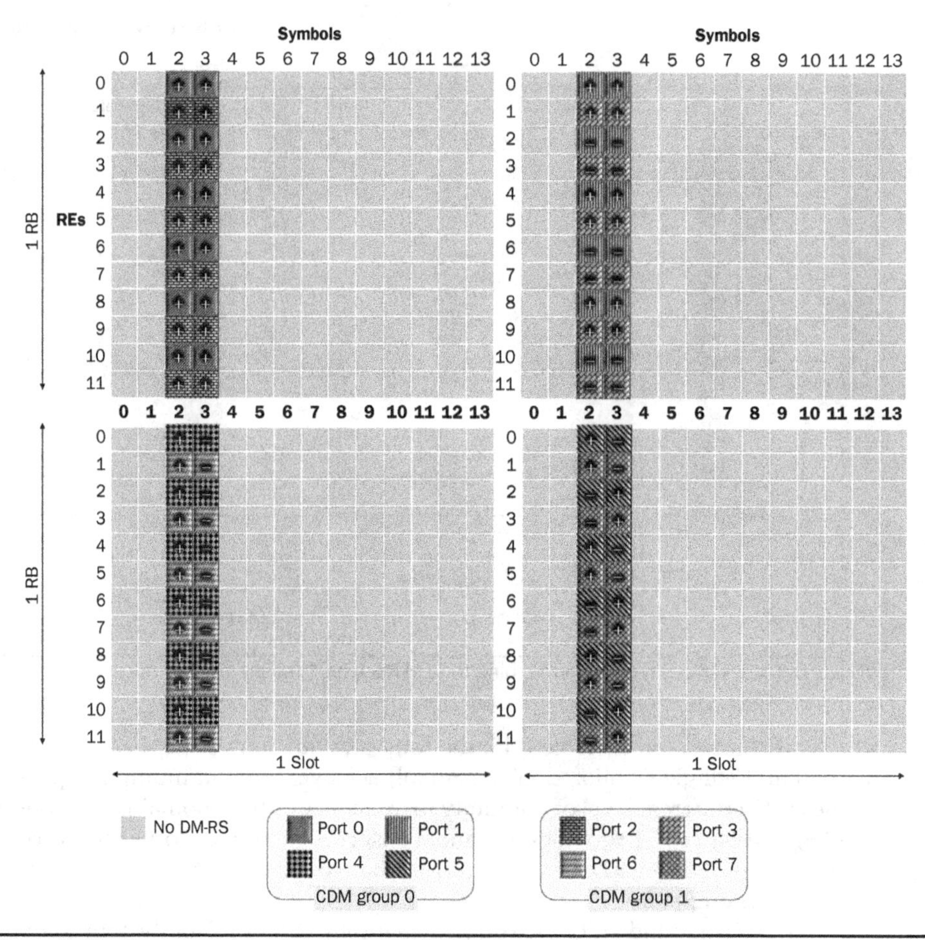

Figure 8.17 Frequency-domain mapping for DM-RS type 1, double-symbol.

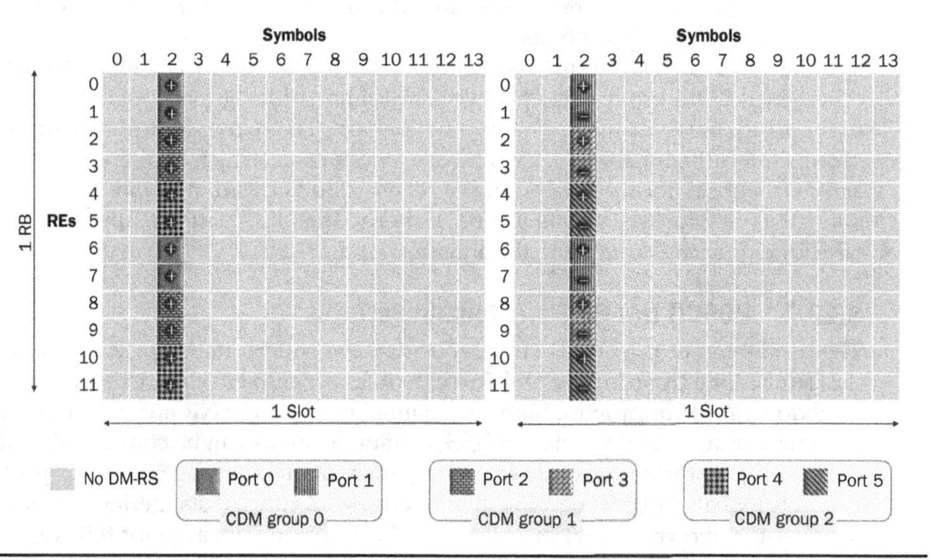

Figure 8.18 Frequency-domain mapping for DM-RS type 2, single-symbol.

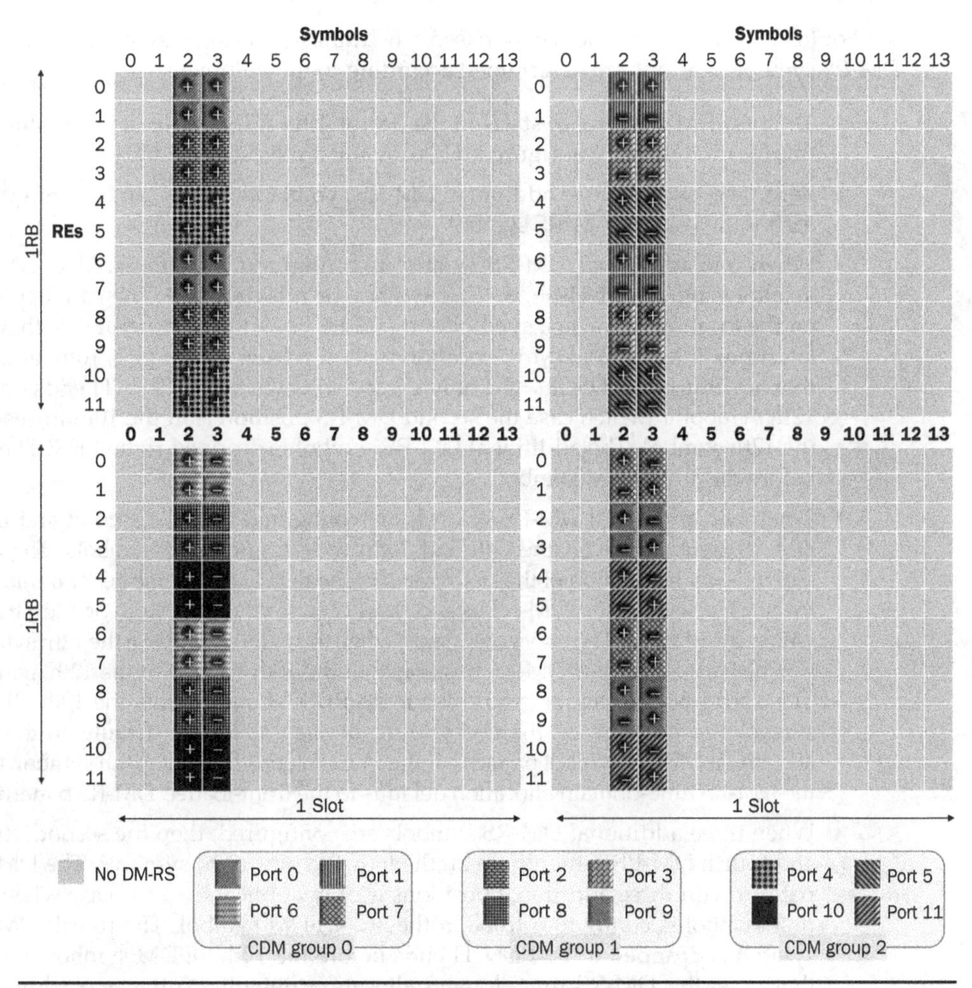

FIGURE 8.19 Frequency-domain mapping for DM-RS type 2, double-symbol.

easier to decompose the channel estimation processing into independent frequency and time-domain filtering with an interchangeable order.

The configuration of additional DM-RS symbol positions is accomplished with a mix of semi-static and dynamic signaling. In particular, an indirect method is provided for performing dynamic switching of the configuration of additional DM-RS symbol positions. This works the following way. The semi-static configuration of the number of additional DM-RS symbol positions can be set differently for mapping types A and B (for either PDSCH or PUSCH). The PDSCH or PUSCH mapping type itself can be dynamically switched by setting the time-domain resource allocation (TDRA) field in the granting DCI, which will then dynamically switch the number of additional DM-RS symbol positions as well. TDRA will be described in more detail in Secs. 8.5 and 10.2. Note that this dynamic switching feature caused some complications in the PDSCH processing time definition, which is described in more detail in Sec. 8.8.

For PDSCH mapping type A (slot-based scheduling) and single-symbol DM-RS, the DM-RS symbol locations are determined according to the following:

- The first DM-RS symbol starts in the 3rd or 4th OFDM symbol in the slot and this location is RRC configured, without any dynamic variability.
- Zero, one, two, or three additional DM-RS symbol locations can be configured, each with a single DM-RS symbol.
- When one additional DM-RS symbol is configured then the second DM-RS symbol is on the 8th, 10th, or 12th symbol, or it is dropped. The case depends on the location of the last symbol of the PDSCH, with the general rule that the latest possible second DM-RS location is chosen from the set {8th, 10th, or 12th} that is within the PDSCH. The only exception is when the PDSCH ends on the 12th symbol in which case the second DM-RS position is in the 10th instead of the 12th symbol. The additional DM-RS symbol is dropped if the PDSCH ends before the 8th OFDM symbol.
- When two additional DM-RS symbols are configured then the second and third DM-RS symbols are on the {7th and 10th} or {8th and 12th} symbols. The case depends on the location of the last symbol of the PDSCH, with the general rule that the latest possible third DM-RS location is chosen from the set (10th or 12th) that is within the PDSCH. The only exception is when the PDSCH ends on the 12th symbol in which case the third DM-RS position is in the 10th instead of the 12th symbol. The third DM-RS symbol is dropped if the PDSCH ends before the 10th OFDM symbol and in that case, the DM-RS time-domain allocation defaults to the one additional DM-RS symbol patterns. If the PDSCH ends before the 8th symbol, then the DM-RS time-domain allocation defaults to the front-loaded DM-RS pattern.
- When three additional DM-RS symbols are configured, then the second, third, and fourth DM-RS symbols are on the {6th, 9th and 12th} symbols. The DM-RS pattern with three additional positions is only applicable to the case when the first symbol is configured to be in the 3rd OFDM symbol. The fourth DM-RS symbol is dropped if the PDSCH ends before the 12th OFDM symbol, and, in that case, the DM-RS time-domain allocation defaults to the two additional DM-RS symbol patterns. If the PDSCH ends before the 10th symbol, then the DM-RS time-domain allocation defaults to the DM-RS pattern with one additional position. If the PDSCH ends before the 8th symbol, then the DM-RS time-domain allocation defaults to the front-loaded DM-RS pattern.

For PDSCH mapping type A (slot-based scheduling) and double-symbol DM-RS, the DM-RS symbol locations are determined according to the following:

- The first DM-RS symbol starts in the 3rd or 4th OFDM symbol in the slot and this location is RRC configured, without any dynamic variability.
- Only zero or one additional DM-RS symbol position can be configured, giving a total of two or four DM-RS symbols. Both the first and the additional position contain two consecutive symbols.
- When one additional DM-RS symbol location is configured then the additional DM-RS symbol starts in the 9th, or 11th symbol, or it is dropped. The case

depends on the location of the last symbol of the PDSCH, with the general rule that the latest possible second DM-RS location is chosen from the set {9th or 11th} so that both of the consecutive DM-RS symbols are still within the PDSCH. The only exception is when the PDSCH ends on the 12th symbol in which case the second DM-RS position is in the 9th instead of the 11th symbol, even though the DM-RS in the 11th and 12th symbol could still be within the PDSCH. The additional DM-RS symbol location is dropped if the PDSCH ends before the 10th OFDM symbol and, in that case, the DM-RS time-domain allocation defaults to the front-loaded DM-RS pattern with two consecutive symbols.

For PDSCH mapping type B (mini-slot–based scheduling) and single-symbol DM-RS, the DM-RS symbol locations are determined according to the following:

- The first DM-RS symbol starts in the 1st OFDM symbol in the PDSCH (mini-slot) or if it collides with PDCCH then it can be moved to the first later symbol after the PDCCH.
- Zero or one additional DM-RS symbol locations can be configured.
- When one additional DM-RS symbol is configured then the second DM-RS symbol is on the 5th symbol, or it is dropped. The additional DM-RS symbol is dropped if the Type B PDSCH length is 2 or 4 OFDM symbols.

For PDSCH mapping type B (mini-slot–based scheduling) and double-symbol DM-RS, the DM-RS symbol locations are determined according to the following:

- The first DM-RS symbol starts in the 1st OFDM symbol in the PDSCH (mini-slot) or if it collides with PDCCH then it can be moved to the first later symbol after the PDCCH.
- No additional DM-RS symbol locations can be configured.

The above rules were captured in table formats in the specification [7], which are repeated in Tables 8.16 and 8.17 for single-symbol and double-symbol DM-RS, respectively.

The value of l_1 in Table 8.16 (in rows corresponding to $l_d = 13$ and $l_d = 14$) depends on whether the additional DM-RS position was shifted due to possible collision with CRS in the shared LTE-NR DL carrier case (as described in Chap. 11). If there is a possible collision, then $l_1 = 12$, otherwise $l_1 = 11$.

The values of $\bar{l}$ in Tables 8.16 and 8.17 are symbol indices, with the first symbol being symbol 0. Therefore, $l = 6$, for example, means the 7th symbol. The symbol index l is defined differently for PDSCH mapping type A and B. For PDSCH mapping type A, $l = 0$ is the first symbol in the slot; therefore, l is symbol index within the slot. For PDSCH mapping type B, $l = 0$ is the first symbol in the PDSCH; therefore, l is symbol index within the PDSCH. Similarly, the duration l_d is counted from the beginning of the slot for PDSCH mapping type A, while it is counted from the beginning of the PDSCH for PDSCH mapping type B.

As it was mentioned earlier, for PDSCH mapping type A, l_0 is either 2 or 3, depending on configuration.

l_d in symbols	DM-RS positions $\bar{l}$							
	PDSCH mapping type A				PDSCH mapping type B			
	dmrs-AdditionalPosition				*dmrs-AdditionalPosition*			
	pos0	*pos1*	*pos2*	*pos3*	*pos0*	*pos1*	*pos2*	*pos3*
2	-	-	-	-	l_0	l_0		
3	l_0	l_0	l_0	l_0	-	-		
4	l_0	l_0	l_0	l_0	l_0	l_0		
5	l_0	l_0	l_0	l_0	-	-		
6	l_0	l_0	l_0	l_0	l_0	l_0, 4		
7	l_0	l_0	l_0	l_0	l_0	l_0, 4		
8	l_0	l_0, 7	l_0, 7	l_0, 7	-	-		
9	l_0	l_0, 7	l_0, 7	l_0, 7	-	-		
10	l_0	l_0, 9	l_0, 6, 9	l_0, 6, 9	-	-		
11	l_0	l_0, 9	l_0, 6, 9	l_0, 6, 9	-	-		
12	l_0	l_0, 9	l_0, 6, 9	l_0, 5, 8, 11	-	-		
13	l_0	l_0, l_1	l_0, 7, 11	l_0, 5, 8, 11	-	-		
14	l_0	l_0, l_1	l_0, 7, 11	l_0, 5, 8, 11	-	-		

Table 8.16 PDSCH DM-RS Positions $\bar{l}$ for Single-Symbol DM-RS

l_d in symbols	DM-RS positions $\bar{l}$					
	PDSCH mapping type A			PDSCH mapping type B		
	dmrs-AdditionalPosition			*dmrs-AdditionalPosition*		
	pos0	*pos1*	*pos2*	*pos0*	*pos1*	*pos2*
< 4				-	-	
4	l_0	l_0		-	-	
5	l_0	l_0		-	-	
6	l_0	l_0		l_0	l_0	
7	l_0	l_0		l_0	l_0	
8	l_0	l_0		-	-	
9	l_0	l_0		-	-	
10	l_0	l_0, 8		-	-	
11	l_0	l_0, 8		-	-	
12	l_0	l_0, 8		-	-	
13	l_0	l_0, 10		-	-	
14	l_0	l_0, 10		-	-	

Table 8.17 PDSCH DM-RS Positions $\bar{l}$ for Double-Symbol DM-RS

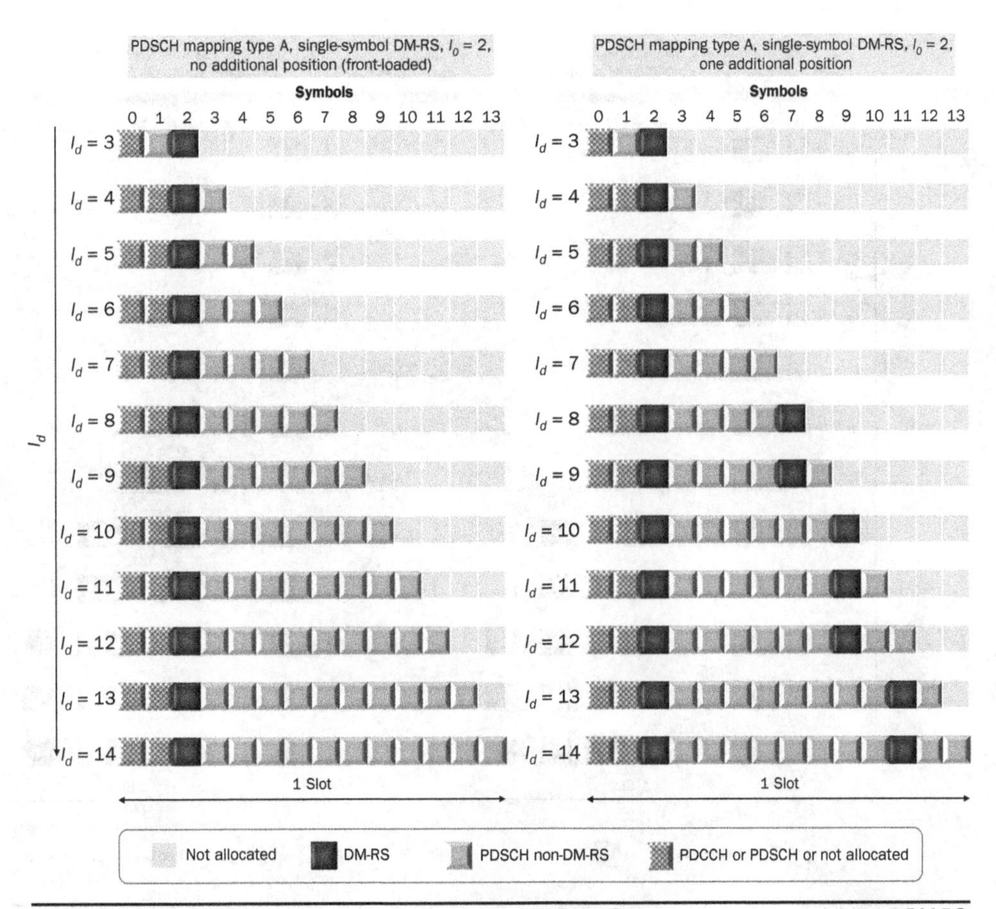

FIGURE 8.20 Time-domain mapping for single-symbol DM-RS with zero and one additional DM-RS position for PDSCH mapping type A.

In Figs. 8.20 to 8.22, we give example illustration for PDSCH DM-RS symbol positions in the time domain for PDSCH mapping type A. Figures 8.20 and 8.21 show DM-RS time-domain patterns with single-symbol DM-RS. Figure 8.22 shows DM-RS time-domain patterns with double-symbol DM-RS.

In Figs. 8.23 and 8.24, we give example illustration for PDSCH DM-RS symbol positions in the time domain for PDSCH mapping type B. Figure 8.23 shows DM-RS time-domain patterns for single-symbol DM-RS. Figure 8.24 shows DM-RS time-domain patterns for double-symbol DM-RS.

As it has been mentioned earlier, the first DM-RS position is determined differently for PDSCH mapping types A and B. For PDSCH mapping type A, the position is configurable but not dynamically variable relative to the slot boundaries; however, it is dynamically variable relative to the PDSCH boundaries because the latter itself is dynamically variable relative to the slot boundaries. For PDSCH mapping type B, it is the other way around, the position of the first DM-RS position is not configurable but dynamically variable relative to the slot boundaries; at the same time, it is fixed relative to the PDSCH boundaries. There are a limited number of exceptions to the last rule.

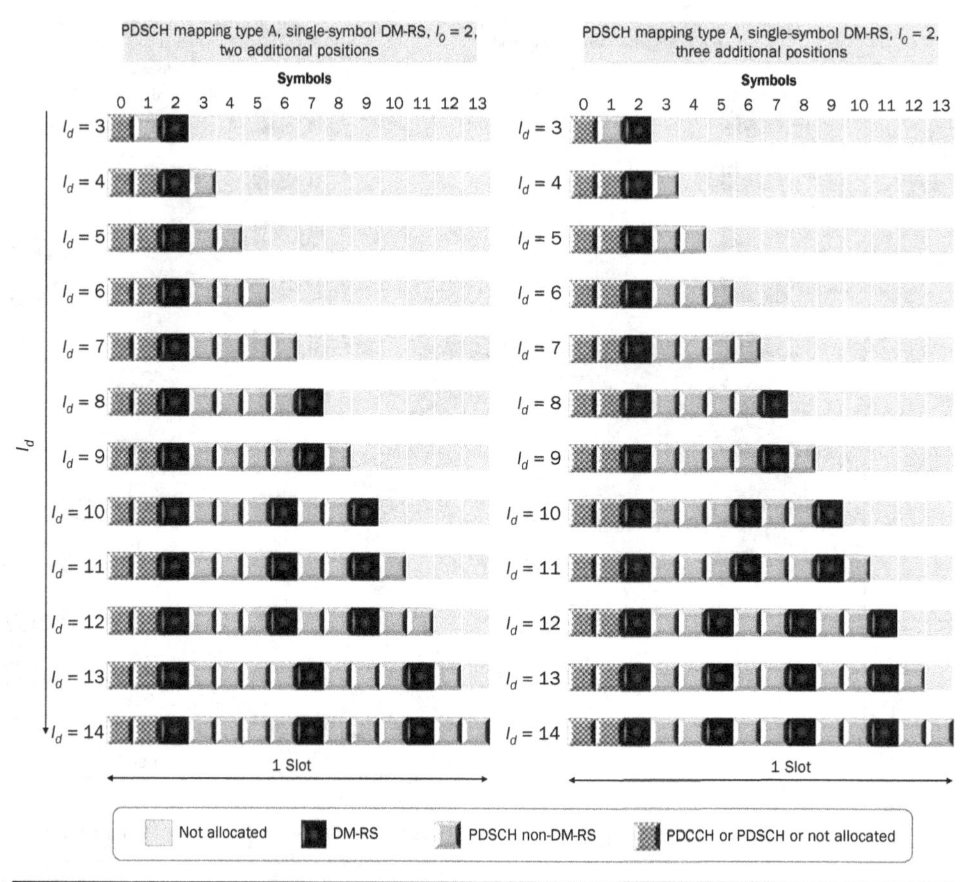

Figure 8.21 Time-domain mapping for single-symbol DM-RS with two and three additional DM-RS positions for PDSCH mapping type A.

When the first DM-RS position would overlap with resources reserved, for example, for a PDCCH CORESET, then the DM-RS position is moved to the next available position. If the next position is also not available, then the DM-RS symbol is moved further. This requirement is necessary to enable the use of PDSCH mapping type B operation in resource collision cases. On other hand, this operation is clearly in conflict with the requirement of placing DM-RS symbols as close to the beginning of the slot as possible (Requirement #3 listed in Sec. 8.11.1), which is needed for controlling PDSCH decoding latency, an important requirement for PDSCH mapping type B. To resolve the conflict, there were limits placed on how many symbols the DM-RS can be shifted by. The allowed cases are depicted in Figs. 8.25 and 8.26. Figure 8.25 shows the non-shifted versus allowed shifted DM-RS positions for PDSCH mapping type B with single-symbol DM-RS. Figure 8.26 shows the non-shifted versus allowed shifted DM-RS positions for PDSCH mapping type B with double-symbol DM-RS.

If the PDSCH duration is $l_d = 2$, the latest allowed DM-RS is the 2nd symbol. This is obvious, since the DM-RS must stay within the PDSCH time-domain resource allocation. If the PDSCH duration is $l_d = 4$, the latest allowed DM-RS symbol is the 3rd symbol

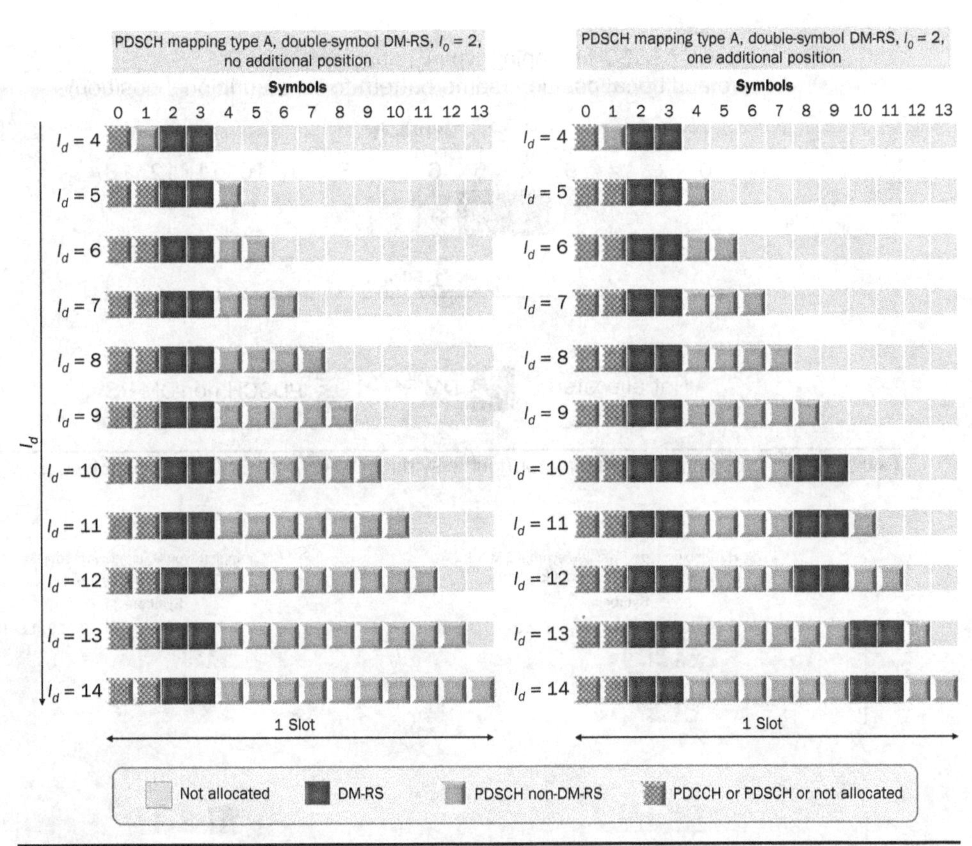

FIGURE 8.22 Time-domain mapping for double-symbol DM-RS with zero and one additional DM-RS position for PDSCH mapping type A.

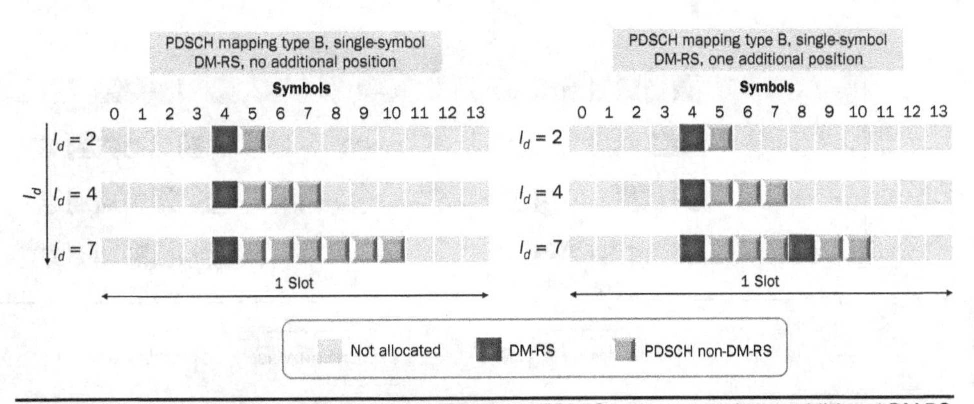

FIGURE 8.23 Time -domain mapping for single-symbol DM-RS with zero and one additional DM-RS position for PDSCH mapping type B.

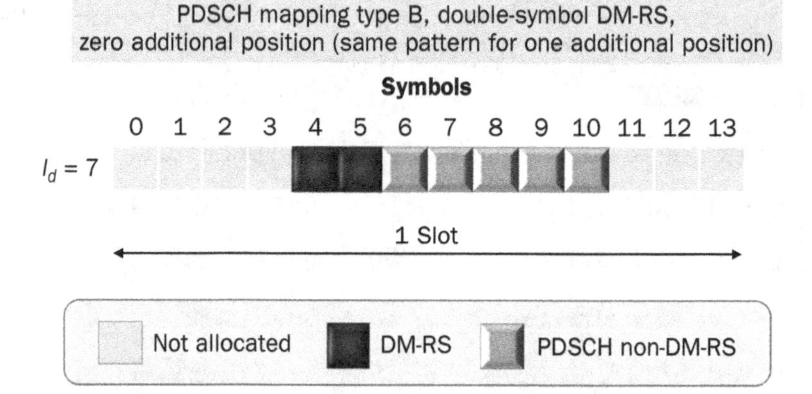

FIGURE 8.24 Time-domain mapping for double-symbol DM-RS for PDSCH mapping type B.

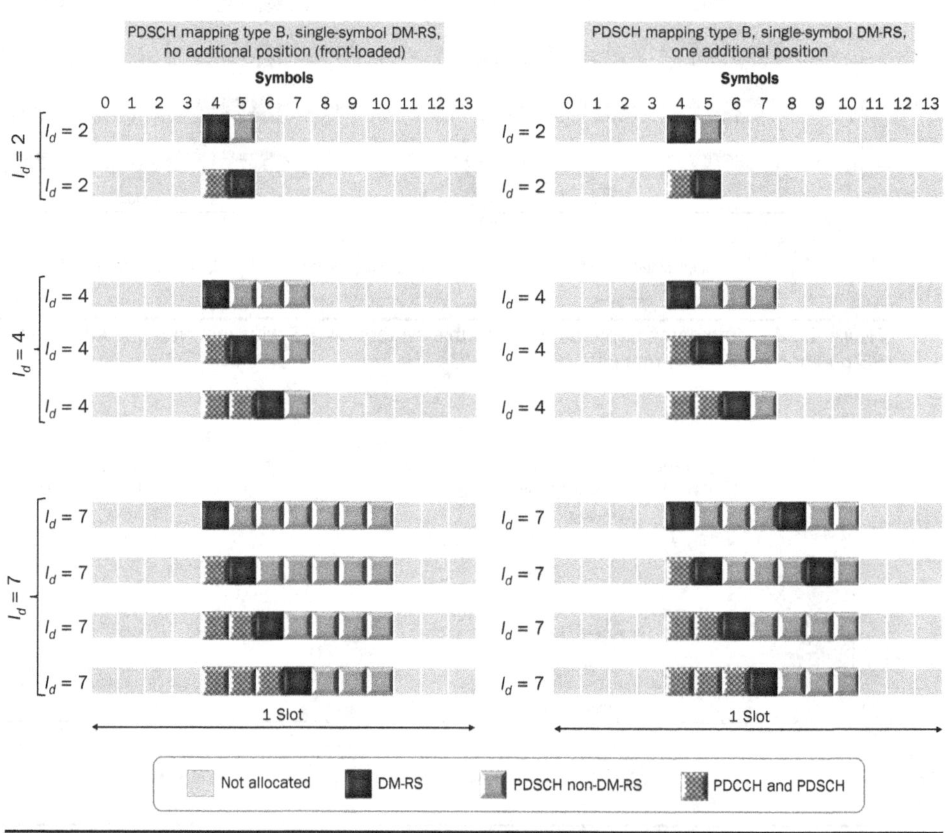

FIGURE 8.25 Time-domain mapping for single-symbol DM-RS with zero and one additional DM-RS position for PDSCH mapping type B with non-shifted and shifted first DM-RS symbol positions.

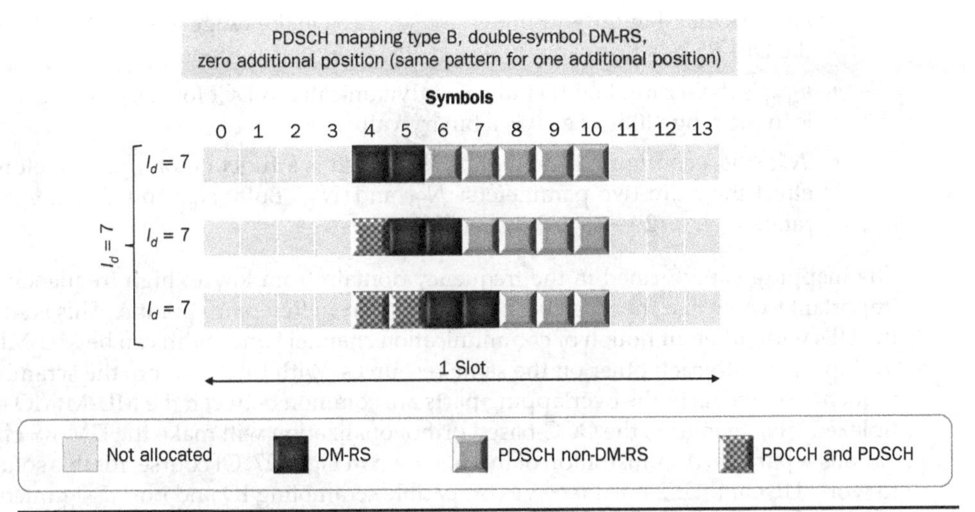

FIGURE 8.26 Time-domain mapping for double-symbol DM-RS for PDSCH mapping type B with non-shifted and shifted first DM-RS symbol positions.

and if the PDSCH duration is $l_d = 7$, the latest allowed position of the first DM-RS symbol is the 4th symbol.

8.12.3 PDSCH DM-RS Scrambling

The mapping in frequency and time domains discussed so far determines the location of the DM-RS REs in the two-dimensional frequency-time resource grid, as well as the allocation of the code domain OCC patterns. The mapping to resources and the application of OCC patterns act on the DM-RS sequences assigned to the antenna ports, which will be discussed next.

The DM-RS for PDSCH and CP-OFDM PUSCH uses pseudo-random QPSK sequences [7], which were defined very similar to the LTE DL DM-RS. The sequence is given as follows:

$$r(n) = \frac{1}{\sqrt{2}}(1 - 2 \cdot c(2n)) + j\frac{1}{\sqrt{2}}(1 - 2 \cdot c(2n + 1)) \tag{8.9}$$

The sequence $c(n)$ is obtained as the output of a Gold-sequence generator defined the same way as in LTE and initialized with the following value:

$$c_{\text{init}} = \left(2^{17}\left(N_{\text{symb}}^{\text{slot}} n_{\text{s,f}}^{\mu} + l + 1\right)\left(2N_{\text{ID}}^{n_{\text{SCID}}} + 1\right) + 2N_{\text{ID}}^{n_{\text{SCID}}} + n_{\text{SCID}}\right) mod 2^{31} \tag{8.10}$$

where

- $n_{\text{s,f}}^{\mu}$ is the slot index within the 10-ms radio frame, which is in the range {0, 1, ..., 9}, for SCS 15 kHz, in the range {0, 1, ..., 19} for SCS 30 kHz, in the range {0, 1, ..., 39} for SCS 60 kHz, and in the range {0, 1, ..., 79}, for SCS 120 kHz.
- $N_{\text{symb}}^{\text{slot}}$ is the number of symbols in a slot, $N_{\text{symb}}^{\text{slot}} = 14$.

- l is the symbol index within the slot, which is in the range {0, 1, ..., 13}, although the DM-RS position cannot take on all values in this range.
- n_{SCID} is the scrambling ID indicated dynamically in DCI format 0_1 or 1_1. n_{SCID} is in the range {0, 1}, i.e., it is a binary value.
- $N_{ID}^{n_{SCID}}$ is a configured scrambling ID, which is a function of n_{SCID}; therefore, in effect there are two parameters: N_{ID}^{0} and N_{ID}^{1}. Both N_{ID}^{0} and N_{ID}^{1} are in the range $\{0, 1, ..., 2^{16} - 1\} = \{0, 1, ..., 65535\}$.

The mapping is performed in the frequency domain from low to high frequency. It is important to note that the mapping always starts from Reference point A. This is so that the UEs with different notion of communication channel bandwidth can be MU-MIMO multiplexed with each other on the same resources. With this solution, the scrambling sequence segments in the overlapping parts are common between the MU-MIMO multiplexed UEs; therefore, the OCC-based orthogonalization will make the DM-RS observations separable. An illustration of this is shown in Fig. 8.27. Of course, for this solution to work, UE1 and UE2 must receive compatible scrambling ID and port assignments.

For channels (PDSCH or PUSCH) used during the initial acquisition, for which information about Reference point A is not yet available, the starting point of the scrambling sequence is the lowest RE of the common CORESET. Because of this, channels used for initial acquisition are not MU-MIMO multiplexable with the other types of channels; however, channels used for initial acquisition in theory could be still multiplexable with each other. The latter does not have much practical significance though and for this reason n_{SCID} is not switchable ($n_{SCID} = 0$ always) for PDSCH carrying broadcast information, and more generally $n_{SCID} = 0$ for any PDSCH granted by DCI format 1_0 and PUSCH granted by format 0_0.

As it has been mentioned, it was a desirable property for the DM-RS design (Requirement #13 listed in Sec. 8.11.1) that the DM-RS PAPR should not be higher than that of the symbols carrying data. For the case of CP-OFDM, this means that the CP-OFDM DM-RS scrambling sequence should ensure Gaussian sample power statistics in the time domain and constant power across OFDM symbols. There were two systematic problems identified with the DM-RS structure that create higher DM-RS PAPR than that of a Gaussian signal.

The first problem is due to the overlapped OCC codes. Assume that one physical antenna port participates in four spatial layers with coefficients $1/2 \cdot \{1, 1, 1, 1\}$ and

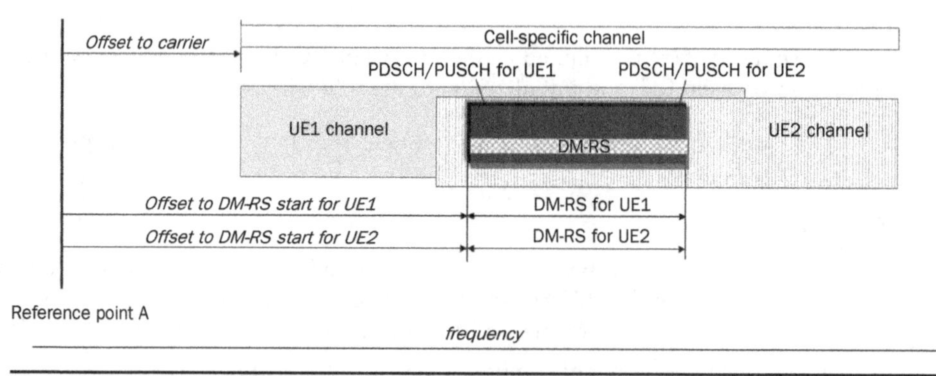

FIGURE 8.27 Orthogonal DM-RS for UEs with different channels.

that the double-symbol DM-RS type 1 configuration is used. The four OCC codes in this case are $\begin{bmatrix} + & + \\ + & + \end{bmatrix}, \begin{bmatrix} + & + \\ - & - \end{bmatrix}, \begin{bmatrix} + & - \\ + & - \end{bmatrix}, \begin{bmatrix} + & - \\ - & + \end{bmatrix}$. With this, the DM-RS pattern before scrambling will be: $1/2 \cdot \left(\begin{bmatrix} + & + \\ + & + \end{bmatrix} + \begin{bmatrix} + & + \\ - & - \end{bmatrix} + \begin{bmatrix} + & - \\ + & - \end{bmatrix} + \begin{bmatrix} + & - \\ - & + \end{bmatrix} \right) = \begin{bmatrix} 2 & 0 \\ 0 & 0 \end{bmatrix}$. As it can be seen, all power will be concentrated in a single RE out of the four, and the other three REs carry no power. Since this is in the frequency domain, uneven distribution across frequency does not in itself increase PAPR in the time domain. (One can think of the example of concentrating all power in the frequency domain into a single tone, which gives perfect 0 dB baseband PAPR in the time domain.) However, in the example with $= \begin{bmatrix} 2 & 0 \\ 0 & 0 \end{bmatrix}$, it can also be observed that the power in the first DM-RS OFDM symbol of the double-symbol allocation, which corresponds to the first column of the matrix is larger than in the second DM-RS symbol corresponding to the second column of the matrix. As a matter of fact, the second symbol carries zero power for this physical antenna port and all power would be concentrated in the first symbol. There was no standards-based solution introduced for this problem because it was identified that there are two standards-transparent solutions that can be deployed without any standard change already. The first is that all four ports need not to be placed in the same CDM group. If we assume the use of two CDM groups then the combined OCC would be $\begin{bmatrix} + & + \\ + & + \\ 0 & 0 \\ 0 & 0 \end{bmatrix}, \begin{bmatrix} + & + \\ - & - \\ 0 & 0 \\ 0 & 0 \end{bmatrix}, \begin{bmatrix} 0 & 0 \\ 0 & 0 \\ + & + \\ + & + \end{bmatrix}, \begin{bmatrix} 0 & 0 \\ 0 & 0 \\ + & + \\ - & - \end{bmatrix}$ and the combined signal would be $1/2 \cdot \left(\begin{bmatrix} + & + \\ + & + \\ 0 & 0 \\ 0 & 0 \end{bmatrix} + \begin{bmatrix} + & + \\ - & - \\ 0 & 0 \\ 0 & 0 \end{bmatrix} + \begin{bmatrix} 0 & 0 \\ 0 & 0 \\ + & + \\ + & + \end{bmatrix} + \begin{bmatrix} 0 & 0 \\ 0 & 0 \\ + & + \\ - & - \end{bmatrix} \right) = \begin{bmatrix} 2 & 2 \\ 0 & 0 \\ 2 & 2 \\ 0 & 0 \end{bmatrix}$. As it can be seen, the power in the frequency domain, i.e., across elements in a column is still uneven, but as it was mentioned this in itself does not increase time-domain PAPR. On the other hand, the power across OFDM symbols is even, as the total power in the first and second columns are equal. For the same reason, there would be no PAPR increase caused by OCC with single-symbol DM-RS in any case.

Another identified mechanism to mitigate the PAPR problem is transparent rotation of the DM-RS constellation. For example, for the physical antenna port that was assumed to participate in four spatial layers with precoding coefficients $1/2 \cdot \{1, 1, 1, 1\}$, the coefficients can be changed to $1/2 \cdot \{-1, 1, 1, 1\}$. Note that this does not change the MIMO precoding matrix, since the coefficients given here act on different spatial layers, across which there is no coherence of data symbols. The MIMO precoding matrix describes the relative phase of physical ports contributing to one spatial layer, not the relative phase of spatial layers contributing to one physical antenna port. Therefore, we can change the coefficients for the latter without changing the precoding or changing link performance. Of course, the same coefficient change applied to the DM-RS must also be applied to the data. With this change, the DM-RS pattern before scrambling will become $1/2 \cdot \left(\begin{bmatrix} + & + \\ + & + \end{bmatrix} - \begin{bmatrix} + & + \\ - & - \end{bmatrix} + \begin{bmatrix} + & - \\ + & - \end{bmatrix} + \begin{bmatrix} + & - \\ - & + \end{bmatrix} \right) = \begin{bmatrix} +1 & -1 \\ +1 & +1 \end{bmatrix}$. As it can be seen, all power will be evenly distributed and the PAPR problem is mitigated. Of course, this hypothetical coefficient change will also apply to contribution of the same spatial layer to the other physical antenna ports; therefore, the coefficient change should involve a joint optimization. But as it was mentioned, it is not necessary the power to be equally distributed across all REs, it is enough if the total power of the antenna is equally distributed across the OFDM symbols.

The second problem turned out to be more serious and required a standard change in Release 16. This problem is caused by the repetition of the scrambling sequence for different antenna ports in different CDM groups in the same OFDM symbol. As before, the two antenna ports can contribute to the transmission on a single physical antenna port (on one PA). Assuming the same OCC code in two CDM groups, the signal transmitted on one physical antenna port can be described as a linear combination on the DM-RS signals in the two CDM groups, where the signals in the two CDM groups are identical except for a frequency shift. With DM-RS type 1, the frequency shift equals the SCS. With DM-RS type 2, the frequency shift is twice the SCS. This can be seen easily from the frequency-domain tone patterns for these two DM-RS types. If we now consider the linear sum of two frequency shifted versions of the same signal, we get an amplitude modulated version of the same signal. When the frequency shift is the same as the SCS, there will be one "peak" of the amplitude modulated signal, where the signal power is twice the average power, and one "valley," where the power drops to zero. When the frequency shift is twice the SCS, there will be two equally spaced "peaks" of the amplitude modulated signal, where the signal power is twice the average power, and two "valleys," where the power drops to zero. In either case, the PAPR increases by 3 dB, which is a significant problem. If we change the relative phase between the two CDM groups contributing to the transmission on the same physical antenna port, we just change the location of the peak and valley in the amplitude modulated signal, without mitigating the 3 dB PAPR degradation.

As mentioned, a Release 16 standards change was adopted to solve this problem. The solution is simply to change the scrambling ID n_{SCID} between the first and second CDM group. Since the n_{SCID} is binary, this can be described as simply flipping the n_{SCID} value from 0 to 1 or from 1 to 0. With this change, the scrambling initialization is

$$c_{init}(\lambda) = \left(2^{17}\left(N_{symb}^{slot} n_{s,f}^{\mu} + l + 1\right)\left(2N_{ID}^{n'_{SCID}(\lambda)} + 1\right) + 2N_{ID}^{n'_{SCID}(\lambda)} + n'_{SCID}(\lambda) + 2^{17}\left(\frac{\lambda}{2}\right)\right) mod 2^{31} \tag{8.11}$$

where λ is the CDM group index, which can take on value {0, 1, 2}, and $n'_{SCID}(0) = n'_{SCID}$, $n'_{SCID}(1) = 1 - n_{SCID}$, $n_{SCID}(2) = n_{SCID}$.

This solution makes sure that the CDM groups do not correspond to identical, but frequency shifted DM-RS signals, thereby mitigating the PAPR problem. The reason why the chosen solution relies on using the n_{SCID} values is that in this way the opportunity to MU-MIMO multiplex Release 15 and Release 16 UEs with each other in the same resources is preserved.

8.12.4 PDSCH DM-RS Power Allocation

It was a desirable property for the DM-RS design (Requirement #17 listed in Sec. 8.11.1) that the Tx power in OFDM symbols containing DM-RS should be the same as in other PDSCH OFDM symbols containing only data. This would not hold without special handling because the number of REs occupied by DM-RS in the OFDM symbol containing DM-RS varies based on the number of co-scheduled PDSCH spatial layers. When the number of allocated DM-RS ports equals the maximum port multiplexing capability of the DM-RS pattern, then all REs will be occupied by DM-RS. However, when the number of ports is less than this, there may be REs unused by DM-RS and these can be either allocated to data or left empty. When REs are left empty, the power in the DM-RS symbol is reduced, creating a symbol-to-symbol power variation. In order to be able to counteract this, a scaling of the energy per resource element (EPRE) was introduced [12]. The scaling factor for ERPE is defined as

$$\frac{\text{DM-RS EPRE}}{\text{Data EPRE}}(\text{dB}) = 10 \cdot \log_{10}(\text{Number of CDM groups without data}) \quad (8.12)$$

Since in the OFDM symbol with DM-RS there is at least one CDM group containing DM-RS, there is at least one CDM group without data. When there is exactly one CDM group without data then from one antenna port's perspective, all REs are uniformly occupied, and in this case, $\frac{\text{DM-RS EPRE}}{\text{Data EPRE}} = 0$ dB. When one additional CDM group is unoccupied by data, then that CDM group carries no power and its power is recycled for the CDM group carrying the DM-RS. In this case, $\frac{\text{DM-RS EPRE}}{\text{Data EPRE}} = 3$ dB. 0 dB or 3 dB are the only possibilities for DM-RS type 1, since there are only two CDM groups. With DM-RS type 2, there are three CDM groups, so in addition, it is also possible to have three CDM groups unoccupied by data, in which case the power of two CDM groups is recycled for the CDM group carrying the DM-RS giving $\frac{\text{DM-RS EPRE}}{\text{Data EPRE}} =$ 4.77 dB.

These are summarized in Table 8.18.

Applying the power scaling given in Table 8.18 ensures that the Tx power in OFDM symbols containing DM-RS is the same as in other OFDM symbols that contain data only without DM-RS.

Note that when the scheduler uses MU-MIMO multiplexing and CDM groups are used for DM-RS of other UEs, those will be counted as CDM groups without data.

Even when CDM groups could be used for data, the scheduler may choose not to use the additional CDM group for data but rather reuse the power for DM-RS as a

Number of DM-RS CDM groups without data	DM-RS type 1	DM-RS type 2
1	0 dB	0 dB
2	−3 dB ($10 \cdot \log_{10}(2)$)	−3 dB ($10 \cdot \log_{10}(2)$)
3	-	−4.77 dB ($10 \cdot \log_{10}(3)$)

Table 8.18 The Ratio of PDSCH EPRE to DM-RS EPRE

way to increase DM-RS SNR and hence improve channel estimation performance. This improvement comes at the cost of reduced coding gain, since the fewer REs available for data will increase the code rate. The optimum choice depends on the SNR operating point and code rate regime.

8.13 DL Phase Tracking Reference Signal

Phase tracking reference signal (PT-RS) is a new concept introduced in NR. The reason for adding this new signal was to help mitigate phase noise. Because phase noise is more prevalent at higher frequencies, the main target was introduction in FR2; however, the use in FR1 is not precluded. The nature of phase noise is such that in the bandwidths relevant to NR, it can be modeled reasonably well with a time varying phase rotation component that is approximately constant across frequency. This means that estimating phase noise can be accomplished with a signal that is reasonably dense in time but can be sparse in frequency. Since the phase noise varies continuously, it is strictly speaking not constant even within one OFDM symbol. Estimating the variation within the symbol is reasonably simple with DFT-S-OFDM because the modulation symbol observations can be viewed as time-domain samples within the OFDM symbol. But doing the same with CP-OFDM waveform would be more involved. There were proposals on defining the PT-RS waveform to better enable estimating variations within an OFDM symbol but, in the end, it was decided that the degree of intra-symbol variation is small enough, so a dedicated signal designed for enabling intra-symbol phase variation was not adopted for CP-OFDM.

The DL PT-RS is only present within allocated PDSCH RBs. If the UE receives non-contiguous PDSCH frequency resource allocation, there is no PT-RS in the non-allocated RBs.

The density of the PT-RS both in time and frequency is a function of other parameters with a configurable dependency. The configuration is part of the *PTRS-DownlinkConfig* record.

In the time domain, the PT-RS density is once per every $L_{PT\text{-}RS}$ OFDM symbols, with $L_{PT\text{-}RS} = 1, 2,$ or 4. Lower $L_{PT\text{-}RS}$ means higher density. The density selection is MCS dependent, as shown in Table 8.19 based on Ref. [1].

The MCS thresholds are configurable, by configuring the three ptrs-MCS$_i$ threshold parameters with $i = 1, 2, 3$. Note that ptrs-MCS$_4$ is not configurable, but rather it represents the lowest reserved I_{MCS} value (either $I_{MCS} = 28$ or $I_{MCS} = 29$). The reserved values are used for retransmissions. When the DL grant includes one of the reserved values, e.g., $I_{MCS} = 29$, the PT-RS density is the same as in the initial transmission. Higher density enables better phase noise tracking, which improves the noise floor, which in turn

Scheduled MCS	Time density ($L_{PT\text{-}RS}$)
I_{MCS} < ptrs-MCS$_1$	PT-RS is not present
ptrs-MCS$_1 \leq I_{MCS}$ < ptrs-MCS$_2$	4
ptrs-MCS$_2 \leq I_{MCS}$ < ptrs-MCS$_3$	2
ptrs-MCS$_3 \leq I_{MCS}$ < ptrs-MCS$_4$	1

TABLE 8.19 Time Density of PT-RS as a Function of Scheduled MCS

enables demodulation with higher MCS. Therefore, the density is expected to increase with higher MCS. When a reserved I_{MCS} value is used, the method of using the initial transmission does not give a match for the modulation order and code rate of the current retransmission, but anyhow the modulation order and code rate of the initial transmission better represents the required SNR, so using that in the PT-RS density is a better match overall. In addition, relying only on the initial transmission parameters gave a simpler rule.

In the frequency domain, the PT-RS density is once per every $K_{\text{PT-RS}}$ RBs, with $K_{\text{PT-RS}} = 2$ or 4. Lower $K_{\text{PT-RS}}$ means higher density. The density selection is dependent on the number of allocated RBs, N_{RB}, as shown in Table 8.20. In the case of non-contiguous allocation, N_{RB} only counts the allocated RBs.

The thresholds are configurable, by configuring two N_{RBi} parameters with $i = 1, 2$. The purpose of this configurability, unlike the PT-RS time-domain density case, is not to tune the density according to SNR requirements but rather to optimize overhead. As it was mentioned earlier, the frequency variation of the phase noise need not be estimated, so the number of frequency-domain observations does not have to grow linearly with increased allocation bandwidth. Therefore, the frequency-domain density is expected to decrease with larger RB allocation. An alternate option could have been to keep the density constant but to omit inserting PT-RS in additional RBs once N_{RB} reaches a certain limit as N_{RB} increases. This option would have achieved similar control of the overhead, but it would not have provided the same level of frequency diversity; therefore, it was not chosen. Although frequency variations of the phase noise need not be estimated, having the observations spread out in frequency still helps avoid situations where the frequency subband containing PT-RS is lost due to channel fade.

With regards to the configuration options in *PTRS-DownlinkConfig* record described above, it is interesting to note that the UE can indicate to the gNB its own preferred version of the configuration tables. For example, a UE with better receiver phase noise characteristic can indicate higher values for ptrs-MCS$_i$, and consequently request sparser PT-RS, than a UE with worse phase noise characteristics. In the end, the gNB is not required to follow the UE recommendation though, and the actual PT-RS density always follows the configuration given by the gNB. For the case when the gNB omits the threshold configuration, the PT-RS density defaults to the following rule:

- If the initial transmission modulation order is QPSK, or if the number of allocated RBs is one or two, PT-RS is omitted.
- In all other cases, PT-RS is included with $L_{\text{PT-RS}} = 1$ (every OFDM symbol), $K_{\text{PT-RS}} = 2$ (every other RB).

Scheduled bandwidth	Frequency density ($K_{\text{PT-RS}}$)
$N_{RB} < N_{RB0}$	PT-RS is not present
$N_{RB0} \leq N_{RB} < N_{RB1}$	2
$N_{RB1} \leq N_{RB}$	4

Table 8.20 Frequency Density of PT-RS as a Function of Scheduled Bandwidth

Note that when the number of allocated RBs is one or two, then anyhow only a single RE per OFDM symbol could carry PT-RS. This provides very limited processing gain, which is part of the reason why the default is not to transmit PT-RS in these cases, unless thresholds are configured indicating otherwise.

PT-RS is always omitted in PDSCH carrying broadcast information, such as system information or paging. In addition, PT-RS can be configured to be not present in any transmission.

The PT-RS signal applicable to the CP-OFDM waveform is special in that it does not have its own scrambling, it is simply a repetition of one of the DM-RS signals within the data channel, more exactly, a repetition of a subset of the DM-RS REs of one DM-RS port. The modulation values of the first DM-RS symbol before applying the Walsh OCC are repeated.

The general philosophy is that DM-RS itself should always serve as part of PT-RS. So, in an OFDM symbol containing DM-RS, PT-RS is never inserted but the DM-RS observation is used instead for phase noise estimation. For the same reason, the symbol counting for time-domain density is reset at every DM-RS symbol. This is shown in an example in Fig. 8.28. For this example, PDSCH allocation type A is assumed with PT-RS time-domain density $L_{\text{PT-RS}} = 2$. In the left part of Fig. 8.28, single-symbol type 1 DM-RS with two additional DM-RS positions and in the right part of Fig. 8.28, double-symbol type 2 DM-RS with one additional DM-RS position is shown. As it can be seen, the PT-RS occupies every other symbol ($L_{\text{PT-RS}} = 2$), skipping DM-RS symbols, and the counting resets at every DM-RS symbol.

When $L_{\text{PT-RS}} = 4$, PT-RS is not included if the length of the PDSCH is four symbols or less. When $L_{\text{PT-RS}} = 2$, PT-RS is not included if the length of the PDSCH is two symbols. These are obvious because after the DM-RS, the first candidate PT-RS symbol would fall outside of the PDSCH in these cases.

In the case of slot repetition, PT-RS insertion is handled for each repeated slot independently.

The RBs selected for PT-RS inclusion have a limited level of randomization. Based on the scrambling ID allocated to a UE (n_{RNTI}), the first RB that includes PT-RS is with index $k_{\text{ref}}^{\text{RB}}$ among the N_{RB} allocated RBs, where $k_{\text{ref}}^{\text{RB}}$ is determined as

$$k_{\text{ref}}^{\text{RB}} = \begin{cases} n_{\text{RNTI}} \bmod K_{\text{PT-RS}} & \text{if } N_{\text{RB}} \bmod K_{\text{PT-RS}} = 0 \\ n_{\text{RNTI}} \bmod (N_{\text{RB}} \bmod K_{\text{PT-RS}}) & \text{otherwise} \end{cases} \tag{8.13}$$

This method ensures that the number of PT-RS REs is not reduced by the randomization. For example, if $N_{\text{RB}} = 20$ and $K_{\text{PT-RS}} = 4$ then the starting RB, $k_{\text{ref}}^{\text{RB}}$, is selected based on n_{RNTI} from the set {0, 1, 2, 3}, and for each case, there are 5 PT-RS REs. On the other hand, if $N_{\text{RB}} = 18$ and $K_{\text{PT-RS}} = 4$ then the starting RB, the set from which $k_{\text{ref}}^{\text{RB}}$, is selected is reduced to {0, 1} because for these, there are still 5 PT-RS REs but if $k_{\text{ref}}^{\text{RB}}$ was 2 or 3, then there would be only four PT-RS REs.

The RE selected for PT-RS inclusion, $k_{\text{ref}}^{\text{RB}}$, within the RB, is nominally either the first or second lowest indexed RE allocated to the given DM-RS port that is associated with the PT-RS. Whether it is the first or second depends on the port index within the DM-RS CDM group. If the PT-RS is associated with the lower port index within the CDM group, then the PT-RS uses the lowest RE of the DM-RS port within the RB. If the PT-RS is associated with the higher port index within the CDM group, then the PT-RS uses the second lowest RE of the DM-RS port within the RB. For example, if DM-RS

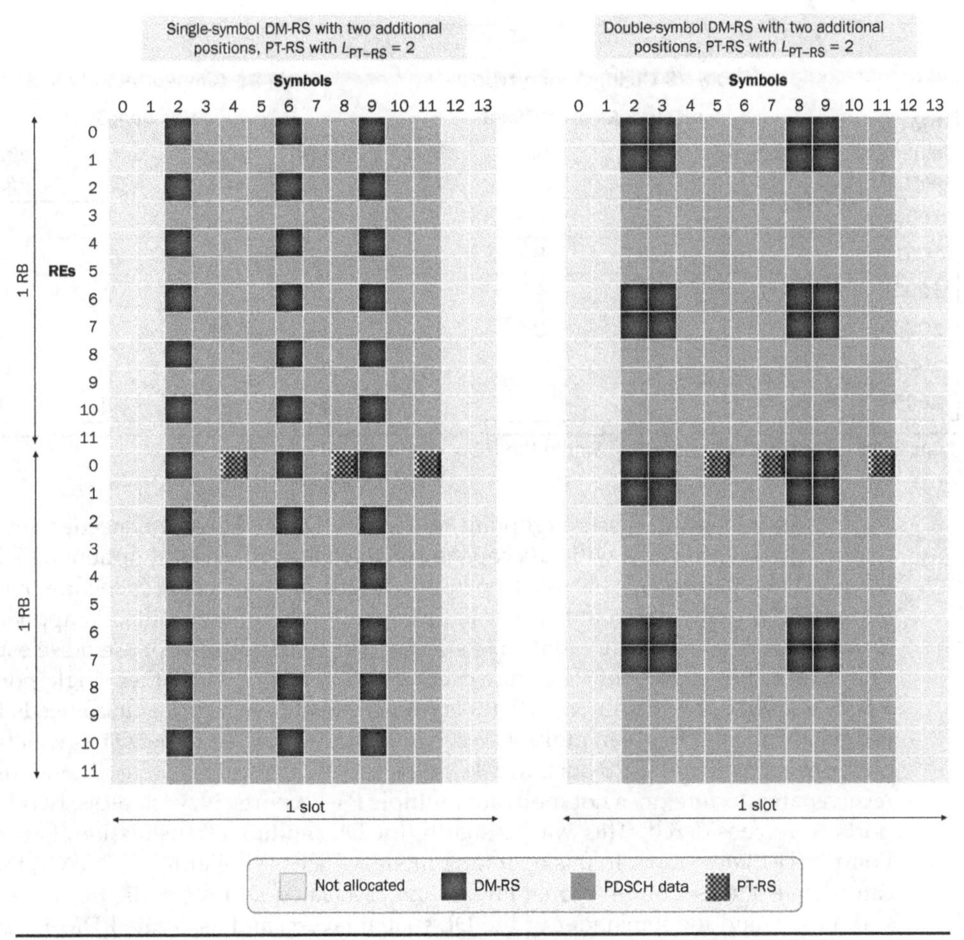

FIGURE 8.28 Example for symbol counting for PT-RS time-domain density determination.

type 1 is used, the first CDM group is shared by DM-RS ports 1000 and 1001, and the first CDM group occupies RE #0 and RE #2 in the RB. Therefore, if the PT-RS port is associated with DM-RS port 1000, then PT-RS is in RE #0 ($k_{\text{ref}}^{\text{RB}} = 0$), while if the PT-RS port is associated with DM-RS port 1001, then PT-RS is in RE #2 ($k_{\text{ref}}^{\text{RB}} = 2$). Similarly, if DM-RS type 2 is used, the first CDM group is shared by DM-RS ports 1000 and 1001, and the first CDM group occupies RE #0 and RE #1 in the RB. Therefore, if the PT-RS port is associated with DM-RS port 1000, then PT-RS is in RE #0 ($k_{\text{ref}}^{\text{RB}} = 0$), while if the PT-RS port is associated with DM-RS port 1001, then PT-RS is in RE #1 ($k_{\text{ref}}^{\text{RB}} = 1$).

The nominal RE location described above applies when there is no *resourceElementOffset* configured in the *PTRS-DownlinkConfig* record. If *resourceElementOffset* is configured, it selects one of the four possible offset values. The first offset value gives the nominal RE allocation without offset. The other three offset values select a different RE, still within the CDM group allocated to the DM-RS port associated with the PT-RS, as given in Table 8.21 based on Ref. [7]. Note that the configurable offset is not used for better randomization but rather to enable avoiding the DC tone location of the UE receiver.

DM-RS antenna port p	k_{ref}^{RB}							
	DM-RS Configuration type 1				DM-RS Configuration type 2			
	resourceElementOffset				*resourceElementOffset*			
	offset 00	offset 01	offset 10	offset 11	offset 00	offset 01	offset 10	offset 11
1000	0	2	6	8	0	1	6	7
1001	2	4	8	10	1	6	7	0
1002	1	3	7	9	2	3	8	9
1003	3	5	9	11	3	8	9	2
1004	-	-	-	-	4	5	10	11
1005	-	-	-	-	5	10	11	4

Table 8.21 The PT-RS RE Index, k_{ref}^{RB}, within the RB

It is an interesting discussion point how many DL PT-RS port is needed for adequate performance. A UE with 2Rx capability, which is a typical assumption in FR2, the UE will have two receive chains. Depending on the UE receiver architecture, namely depending on whether common oscillator signal with common jitter is applied for down-conversion in the different chains or not, single or multiple phase noise estimation need to be carried out. This, however, would not impact whether single-port DL PT-RS is sufficient or multi-port PT-RS is needed. Even when phase noise needs to be estimated independently in multiple receive chains, it could be done just as well based on a single common PT-RS port, which is received by all receive chains. Therefore, UE receiver architecture does not motivate multiple PT-RS ports. Nevertheless, two PT-RS ports were considered. This was primarily for DL multiple Transmission/Reception Point (mTRP) operation. In this operating mode, which is similar to LTE CoMP, the UE can receive a subset of DL spatial layers, and associated DM-RS ports, from one base station site, and the remainder of the DL spatial layers, and associated DM-RS ports, from another base station site. In this case, the original Release 15 design was to have two PT-RS ports, one being transmitted from each of the sites (TRPs). It is important to note though that the motivation for having a separate PT-RS port for each of the DL transmission points is not that a separate phase noise estimation needs to be made for each of the TRPs. It has been always assumed that phase noise is a degradation only within the UE receiver (and transmitter). No inherent phase noise is assumed to be generated in the gNB transmitter (and receiver). This is because the gNB hardware is not limited by the same power and complexity constraints as the UE hardware. Therefore, in the DL, only the UE phase noise needs to be estimated, the phase noise of the signal transmitted by the gNB itself can be assumed to be zero, irrespective of how many TRPs are transmitting. The primary motivation for considering two PT-RS ports for mTRP operation was diversity. Assume, for example, that for high reliability the same data is transmitted to a UE from two TRPs. If one of the transmissions is blocked, the UE can still decode the data from the other transmission received. If there was a single PT-RS port, then it would have to come from one of the TRPs only. If the transmission from the TRP carrying the PT-RS gets blocked then the UE would receive the data from the other TRP but without any phase tracking reference, so the UE would not be able to estimate its own receiver phase noise. Having two PT-RS ports helps avoid this situation. At a later stage of the standards development, non-transparent mTPR support was removed from Release 15, and as part of this, the Release 15 PT-RS definition reverted to supporting single port only. The support of two port PT-RS was added back in Release 16 as part of the non-transparent mTRP operation definition.

When there are multiple DM-RS ports, one of them has to be selected for PT-RS port association. The selected DM-RS port signal will be repeated as PT-RS. Obviously, the best would be to select the DM-RS port with the highest SNR. But since the precoding and power allocation for DM-RS ports is based on the gNBs decision, the UE cannot know beforehand, which DM-RS port will have the highest SNR. Even if the PT-RS processing was delayed in order to enable the UE detecting the port with highest SNR, the choice of DM-RS for PT-RS association would not always be in sync between the gNB and UE, since the SNR estimation has its own uncertainty. In order to avoid the ambiguities caused by this, the selected DM-RS port is always the port with the lowest port index. This is uniquely determined. The gNB is anyhow free to choose the ordering of the precoder matrix rows, it can always apply the precoding vector it expects to result in the highest SNR among the precoder rows to the DM-RS port with the lowest index. A similar rule could have been applied to the case of two codewords (CWs), which is used when the transmission rank is 5 or higher. For example, it could have been decided that the PT-RS is associated with the DM-RS with the lowest port index within the first codeword (CW0). However, in this case, the gNB is not completely free to permute the precoder rows, since permutation should be confined within codewords, and also the two codewords themselves cannot be swapped whenever their numbers of layers are unequal. On the other hand, the two codewords have associated individual I_{MCS} whose values can be used to establish an SNR ranking among the codewords; therefore, in order to deterministically select a DM-RS port with good SNR, the rule was chosen that requires selecting the lowest-indexed DM-RS ports of the codeword with higher individual I_{MCS}.

As with other reference signals, the PDSCH data is rate matched around the REs allocated to PT-RS. Also, on the REs allocated to PT-RS, only the layer of the DM-RS port associated with the PT-RS is transmitted, the other layers are muted. Unlike DM-RS though, PT-RS is allowed to be punctured by other signals like CSI-RS. Since the frequency-domain phase noise variation is not estimated, the loss of observations in certain subcarriers that are subject to such puncturing can be easily recovered as long as the fraction of lost REs is not excessive.

Because PT-RS is transmitted only on one spatial layer, and the other spatial layers are muted, the power of the unused layer can be recycled for power boosting of the PT-RS port. For optimum combining of the DM-RS and PT-RS observations, the UE should know about whether the PT-RS power is boosted or not. To enable this, a configuration parameter *epre-Ratio* is included in the *PTRS-DownlinkConfig* record. There are four possible power offset values but only two are specified, the other two are reserved for future use. The two values indicate whether power boost is applied or not. In the case of no power boost, there may be a slight variation in transmit power across symbols, but this hasn't been seen a problem in the DL.

8.14 Channel State Information Reference Signal

Channel state information reference signal (CSI-RS) is a signal used in the DL for multiple purposes. It is utilized for channel and interference measurement for CSI feedback, measurement for beam management, Rx beam sweeping, frequency and time tracking and mobility measurement for RRM. CSI-RS is only transmitted in the DL.

The CSI-RS can be configured as periodic, semi-persistent or aperiodic. The characteristics of each of these configuration types are the following:

- Periodic CSI-RS:
  - Configured with a periodicity and slot offset
  - Can be used for periodic, aperiodic or semi-persistent reporting
- Semi-persistent CSI-RS:
  - Configured with a periodicity and slot offset
  - Activated and deactivated with a MAC control element (MAC CE) command
  - Can be used for aperiodic or semi-persistent reporting
- Aperiodic CSI-RS:
  - No periodicity is configured.
  - Triggered dynamically with DL or UL DCI.
  - Slot offset is dynamically selected from a configured set.
  - Can be used for aperiodic reporting only.

When CSI-RS is used for CSI feedback, either non-zero power (NZP) CSI-RS or CSI Interference Measurement (CSI-IM) resource is configured. CSI-IM is used for interference measurement only. On CSI-IM resources, other signals designated to the UE are typically muted, so interference can be directly measured. NZP CSI-RS is used primarily for channel measurement, but it can also be used for interference measurement, wherein the channel estimated based on the CSI-RS is included in the assumed interference.

In addition to NZP CS-RS, zero power (ZP) CSI-RS can also be configured. The resource allocation schemes are the same for NZP and ZP CSI-RS. ZP CSI-RS is used for rate matching purposes only, for example, to ensure that there is no PDSCH transmission from the serving cell on CSI-IM resources.

When CSI-RS is used for beam management, it is configured either with *repetition* set to "off" or *repetition* set to "on." The specification [1], somewhat confusingly, identifies a CSI-RS used for beam management as a "CSI-RS configured with higher layer parameter *repetition*". What is confusing is that the CSI-RS configured with repetition in fact may or may not be repeated. The actual repetition is controlled with setting parameter *repetition* to "on" or "off."

When CSI-RS is used for beam management and *repetition* is set to "off," a set of CSI-RS resources can be configured, each with single or two ports. Within the set, all resources must have the same number of ports. Each resource typically corresponds to a different Tx beam direction from the gNB. When receiving these signals, the UE can be configured with *reportQuantity* set to "cri-RSRP". This means that the UE is expected to measure L1-RSRP, which is the CSI-RS power on the CSI-RS port if single port is used or the average power across two ports if two ports are used, and the UE is expected to report the CSI-RS resource index (CRI) of the strongest CSI-RS resources (although selecting the strongest CSI-RS resources to report is not a requirement, the UE is allowed to use other selection criteria as well) within the configured resource set, together with their power (L1-RSRP) sorted in descending order.

When CSI-RS is used for beam management and *repetition* is set to "on," the CSI-RS is intended to be used for Rx beam sweeping. In this case, multiple resources configured have the same Tx beam direction. The UE can vary the analog beamforming direction in its receiver and compare the signal strength of the different directions. The Rx direction with the strongest received signal can be used subsequently for receiving other signals from the same Tx beam direction. This is a transparent operation in that the UE is not expected to report measurement results. Note that when *repetition* is set to "on," there are no actual repetition parameters configured. The repetition simply means that the resources within the resource set have the same repeated Tx beam direction to make the UE's Rx beam sweeping operation meaningful.

When CSI-RS is used for mobility, a set of single-port CSI-RS resources are configured for detection and measurement. When the UE is configured to measure these, the UE reports L3-RSRP, which is the filtered measured power of the CSI-RS resource. Since the CSI-RS signal, unlike the SSB signal, is not designed for efficient blind time hypothesis search, the UE is always given a timing hypothesis where the CSI-RS signal is supposed to be detected and measured. This can be done either via a configuration of *associatedSSB*, in which case the UE first detects the associated SSB to find the timing of the CSI-RS, or when there is no *associatedSSB* configured, the UE is allowed to assume that the timing of the CSI-RS is the same as the timing of the serving cell. Of course, the precise timing of a neighbor cell is usually not exactly the same as the timing of the serving cell due to propagation delay differences but at least in the case of small cell deployments, the timing differences are small enough to make the no *associatedSSB* option viable. When all CSI-RS configured for mobility have associated SSB, the maximum number of configured CSI-RS resources the UE is required to handle on a measurement frequency is 96. When some or all CSI-RS have no associated SSB, the maximum number of configured CSI-RS resources on a measurement frequency is 64. The first number is greater than the second because when there is associated SSB the UE will typically find less than 96 SSBs, and when an SSB is not found the corresponding CSI-RS is not measured either, reducing the actual measurement load. In FR1, the configured set of CSI-RS resources can have a mix of configurations, i.e., some resources may have an associated SSB, while other resources may not. In FR2, such mixing is not allowed, either all or none of the CSI-RS resources have associated SSB.

Whether there is associated SSB or not, the UE is always provided with a list of CSI-RS resources to measure for mobility. This is different from the SSB-based mobility, where the UE can be configured with no neighbor list, in which case the UE must search for all possible SSB IDs at all possible time offset, without any assistance. As mentioned earlier, the reason for this difference is that the SSB signal was designed to make such unassisted search operation feasible with reasonable complexity, while the CSI-RS signal wasn't designed for that.

8.14.1 CSI-RS Frequency and Time-Domain Patterns

Compared to LTE, some new frequency and time-domain resource patterns were introduced for the CSI-RS. When determining the complete set of patterns, the following requirements were adopted:

- Each pattern accommodates X antenna ports, where the X is from the set: $\{1, 2, 4, 8, 12, 16, 24, 32\}$

- A frequency-domain density ρ is configured, which is from the set: {0.5, 1, 3}. Density $\rho = 0.5$ means that CSI-RS is contained only in every other RB. This was introduced to control CSI-RS overhead. The first allocated RB starts at a configured offset (*startPRB* or *startingRB*) relative to Reference point A (described in Sec. 2.4.1). With $\rho = 0.5$, all X ports of a given resource are confined to the same set of RBs (even or odd), but it is possible for different CSI-RS resources to be configured in alternating resources, by setting *startPRB* or *startingRB* to even versus odd value. Density $\rho = 3$ is only used for single port ($X = 1$).
- Each pattern contains a number of equal sized CDM groups, where the group size L is from the set: {1, 2, 4, 8}.
- The number of CDM groups is from the set: {1, 2, 3, 4, 6, 8, 12, 16}.
- Each CDM group comprises a block of contiguous REs, where the block size is one of (1, 1), (2, 1), (2, 2), (2, 4), with the first value representing the number of consecutive REs in frequency, the second value representing the number of consecutive REs in time, with the product of the two being equal to the CDM group size L.
- The L orthogonal cover codes (OCCs) in a CDM group are one- or two-dimensional binary Walsh codes.
- The CDM group size in the frequency domain is never larger than 2. This is in order to limit loss of orthogonality in frequency selective channels. At the same time, for all multi-port ($X > 1$) patterns, there is always frequency-domain CDM2 codes involved, i.e., there is no CDM group size 1 in the frequency domain other than for the single-port case. This is because severe frequency selectivity, where adjacent REs are significantly decorrelated, would be already problematic for the DM-RS frequency-domain patterns, which are all with comb 2 or greater, therefore CSI-RS performance would not be the limiting factor in these cases.
- The CDM group size in the time domain can be 1, 2, or 4. The larger values are more vulnerable to loss of orthogonality in time varying channels, and for this reason, values larger than 4 were precluded.
- For the same number of ports X, multiple pattern options were defined. This is so that different tradeoffs can be realized. As it was mentioned, the CDM group size in the time domain can be up to 4. But the exact same resource utilization and multiplexing capability can be achieved with shorter time-domain CDM group size, and with TDM multiplexing. Some patterns are of this type. While TDM multiplexing is much more resilient to loss of orthogonality, it has a down-side in loss of power when the CSI-RS is non-precoded. Non-precoded CSI-RS typically means that a physical gNB antenna port contributes only to one or few CSI-RS ports. Then with CSI-RS port TDM multiplexing, some physical antenna ports would transmit in a subset of the OFDM symbols and would be muted in the remainder symbols of the CSI-RS pattern. Because the power in the muted OFDM symbol cannot be recycled to boost the power in the unmuted OFDM symbol, power loss occurs. If longer time-domain CDM group is used instead of TDM multiplexing, then such losses can be avoided. Therefore, the pattern selection

for a given number of ports X should in the end depends on a tradeoff between time-domain orthogonality and full power utilization. These in turn depend on the use of CSI-RS precoding, expected UE speed, carrier frequency, etc.

- In order to provide flexibility for multiplexing with other CSI-RS resources in the frequency domain, the frequency-domain location within the PRB is configurable. For multi-port patterns, X > 1, there is a 6-bit bitmap provided. Each bit in the bitmap indicates whether a pair of consecutive REs from the set {(0, 1), (2, 3), (4, 5), …, (10, 11)} is used for the given CSI-RS pattern or not. When there are n frequency-domain CDM groups, there are exactly n bits set to "1" in the 6-bit bitmap. The n bits set to "1" indicate n starting positions: $k_0, k_1, ..., k_{n-1}$. The starting positions $k_0, k_1, ..., k_{n-1}$ together are denoted with $\bar{k}$. Note that the fixed size 6-bit bitmap is not the most optimized way of signaling the frequency-domain allocation, but it is sufficiently simple. As it can be noted, the pair of REs always starts on an even numbered RE index. This enhances the flexibility for multiplexing with other multi-port CSI-RS patterns. There is one exception to the above description, which is captured in row 4 of Table 8.22. In this exception case, the two frequency-domain CDM groups are bundled together without any gap between them and they can be at one of the three positions in the RB, starting at RE index 0, 4, or 8. For single-port patterns, X = 1, with frequency-domain density $\rho = 1$ or $\rho = 0.5$ there are 12 possible frequency-domain positions in the PRB, at any of the 12 REs. For single-port patterns with frequency-domain density $\rho = 3$, there are four possible starting frequency-domain positions in the PRB, starting at any one of the first four REs.
- In order to provide flexibility for multiplexing with other signals in the time domain, the time-domain location of the CDM groups within the slot is configurable. The CDM groups are organized into one or two time-domain bundles. Each bundle comprises consecutive OFDM symbols. The first bundle starts at OFDM symbol l_0 and the second bundle, if used, starts at OFDM symbol l_1. The values of l_0 and l_1 can be any in the range {0, 1, 2, …, 13}, with the limitation that $l_0 < l_1$ and the two bundles can have no overlap in time, which restricts l_1 to certain values, for example, l_1 can never be zero. The starting OFDM symbols l_0, l_1 together are denoted with $\bar{l}$. The same frequency-domain arrangement is repeated across the bundles, and if there are TDM-ed CDM-groups within one bundle, then the same frequency-domain arrangement is repeated in every OFDM symbol within the bundle.

The complete set of CSI-RS patterns in frequency and time are described in Ref. [7] by way of Table 8.22.

The CSI-RS patterns are depicted in Figs. 8.29 to 8.31. The row numbers refer to the row index in Table 8.22. Figure 8.29 includes rows 1 through 6, Fig. 8.30 includes rows 7 through 12, and Fig. 8.31 includes rows 13 through 18.

Note that in Figs. 8.29 to 8.31, the illustrations were made with arbitrary values of positions $k_0, k_1, ..., k_{n-1}$ and l_0, l_1. For example, if two blocks are shown in Figs. 8.29 to 8.31 with a gap in between them in the time domain, then the gap can be of any value, including no gap, as long as blocks remain in the same slot. On the other hand, blocks that are shown in Figs. 8.29, 8.30, or 8.31 as contiguous in the time domain must remain contiguous.

Row	Ports X	Density ρ	CDM-Type	$(\bar{k}, \bar{l})$	CDM group index j	k'	l'
1	1	3	No CDM	(k_0, l_0), $(k_0 + 4, l_0)$, $(k_0 + 8, l_0)$	0, 0, 0	0	0
2	1	1, 0.5	No CDM	(k_0, l_0)	0	0	0
3	2	1, 0.5	FD-CDM2	(k_0, l_0)	0	0, 1	0
4	4	1	FD-CDM2	(k_0, l_0), $(k_0 + 2, l_0)$	0, 1	0, 1	0
5	4	1	FD-CDM2	(k_0, l_0), $(k_0 + l_0, 1)$	0, 1	0, 1	0
6	8	1	FD-CDM2	(k_0, l_0), (k_1, l_0), (k_2, l_0), (k_3, l_0)	0, 1, 2, 3	0, 1	0
7	8	1	FD-CDM2	(k_0, l_0), (k_1, l_0), $(k_0, l_0 + 1)$, $(k_1, l_0 + 1)$	0, 1, 2, 3	0, 1	0
8	8	1	CDM4-(FD2, TD2)	(k_0, l_0), (k_1, l_0)	0, 1	0, 1	0, 1
9	12	1	FD-CDM2	(k_0, l_0), (k_1, l_0), (k_2, l_0), (k_3, l_0), (k_4, l_0), (k_5, l_0)	0, 1, 2, 3, 4, 5	0, 1	0
10	12	1	CDM4-(FD2, TD2)	(k_0, l_0), (k_1, l_0), (k_2, l_0)	0, 1, 2	0, 1	0, 1
11	16	1, 0.5	FD-CDM2	(k_0, l_0), (k_1, l_0), (k_2, l_0), (k_3, l_0), $(k_0. l_0 + 1)$, $(k_1, l_0 + 1)$, $(k_2, l_0 + 1)$, $(k_3, l_0 + 1)$	0, 1, 2, 3, 4, 5, 6, 7	0, 1	0
12	16	1, 0.5	CDM4-(FD2, TD2)	(k_0, l_0), (k_1, l_0), (k_2, l_0), (k_3, l_0)	0, 1, 2, 3	0, 1	0, 1
13	24	1, 0.5	FD-CDM2	(k_0, l_0), (k_1, l_0), (k_2, l_0), $(k_0, l_0 + 1)$, $(k_1, l_0 + 1)$, $(k_2, l_0 + 1)$, (k_0, l_1), (k_1, l_1), (k_2, l_1), $(k_0, l_1 + 1)$, $(k_1, l_1 + 1)$, $(k_2, l_1 + 1)$	0, 1, 2, 3, 4, 5, 6, 7, 8, 9, 10, 11	0, 1	0
14	24	1, 0.5	CDM4-(FD2, TD2)	(k_0, l_0), (k_1, l_0), (k_2, l_0), (k_0, l_1), (k_1, l_1), (k_2, l_1)	0, 1, 2, 3, 4, 5	0, 1	0, 1
15	24	1, 0.5	CDM8-(FD2, TD4)	(k_0, l_0), (k_1, l_0), (k_2, l_0)	0, 1, 2	0, 1	0, 1, 2, 3
16	32	1, 0.5	FD-CDM2	(k_0, l_0), (k_1, l_0), (k_2, l_0), (k_3, l_0), $(k_0, l_0 + 1)$, $(k_1, l_0 + 1)$, $(k_2, l_0 + 1)$, $(k_3, l_0 + 1)$, (k_0, l_1), (k_1, l_1), (k_2, l_1), (k_3, l_1), $(k_0, l_1 + 1)$, $(k_1, l_1 + 1)$, $(k_2, l_1 + 1)$, $(k_3, l_1 + 1)$	0, 1, 2, 3, 4, 5, 6, 7, 8, 9, 10, 11, 12, 13, 14, 15	0, 1	0
17	32	1, 0.5	CDM4-(FD2, TD2)	(k_0, l_0), (k_1, l_0), (k_2, l_0), (k_3, l_0), (k_0, l_1), (k_1, l_1), (k_2, l_1), (k_3, l_1)	0, 1, 2, 3, 4, 5, 6, 7	0, 1	0, 1
18	32	1, 0.5	CDM8-(FD2, TD4)	(k_0, l_0), (k_1, l_0), (k_2, l_0), (k_3, l_0)	0, 1, 2, 3	0,1	0,1, 2, 3

Table 8.22 CSI-RS Locations in Frequency and Time

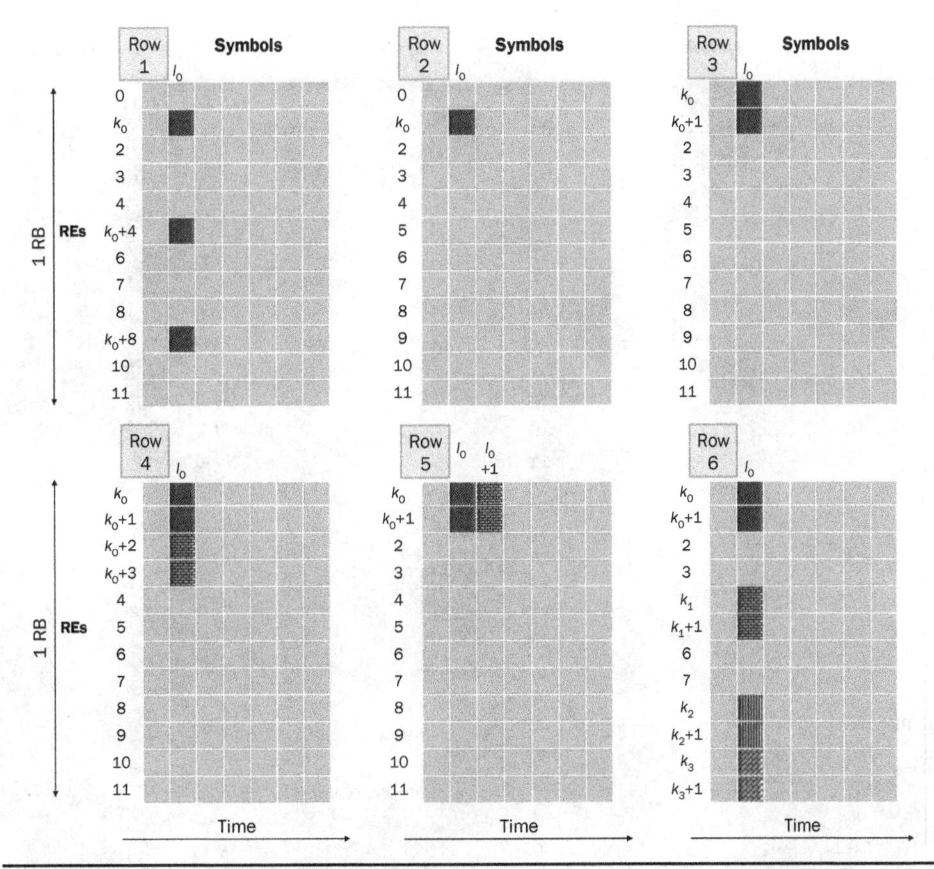

Figure 8.29 Illustration of the CSI-RS patterns in frequency and time, rows 1 through 6.

8.14.2 CSI-RS Scrambling

The CSI-RS signal follows the general OFDM signal generation method. The CSI-RS REs contain signal that reflects the applied OCC, if any, and the pseudo-random scrambling. The scrambling utilizes a pseudo-random QPSK sequence defined [7] as

$$r(m) = \frac{1}{\sqrt{2}}\big(1 - 2 \cdot c(2m)\big) + j\frac{1}{\sqrt{2}}\big(1 - 2 \cdot c(2m+1)\big) \tag{8.14}$$

The sequence $c(n)$ is obtained as the output of a Gold-sequence generator initialized with the following value:

$$c_{\text{init}} = \left(2^{10}\left(N_{\text{symb}}^{\text{slot}} n_{\text{s,f}}^{\mu} + l + 1\right)\left(2n_{\text{ID}} + 1\right) + n_{\text{ID}}\right) \bmod 2^{31} \tag{8.15}$$

where

- $n_{\text{s,f}}^{\mu}$ is the slot index within the 10-ms radio frame, which is in the range {0, 1, ..., 9} for SCS 15 kHz, in the range {0, 1, ..., 19} for SCS 30 kHz, in the range {0, 1, ..., 39} for SCS 60 kHz, and in the range {0, 1, ..., 79} for SCS 120 kHz.
- $N_{\text{symb}}^{\text{slot}}$ is the number of symbols in a slot, $N_{\text{symb}}^{\text{slot}} = 14$ for Normal CP (NCP).

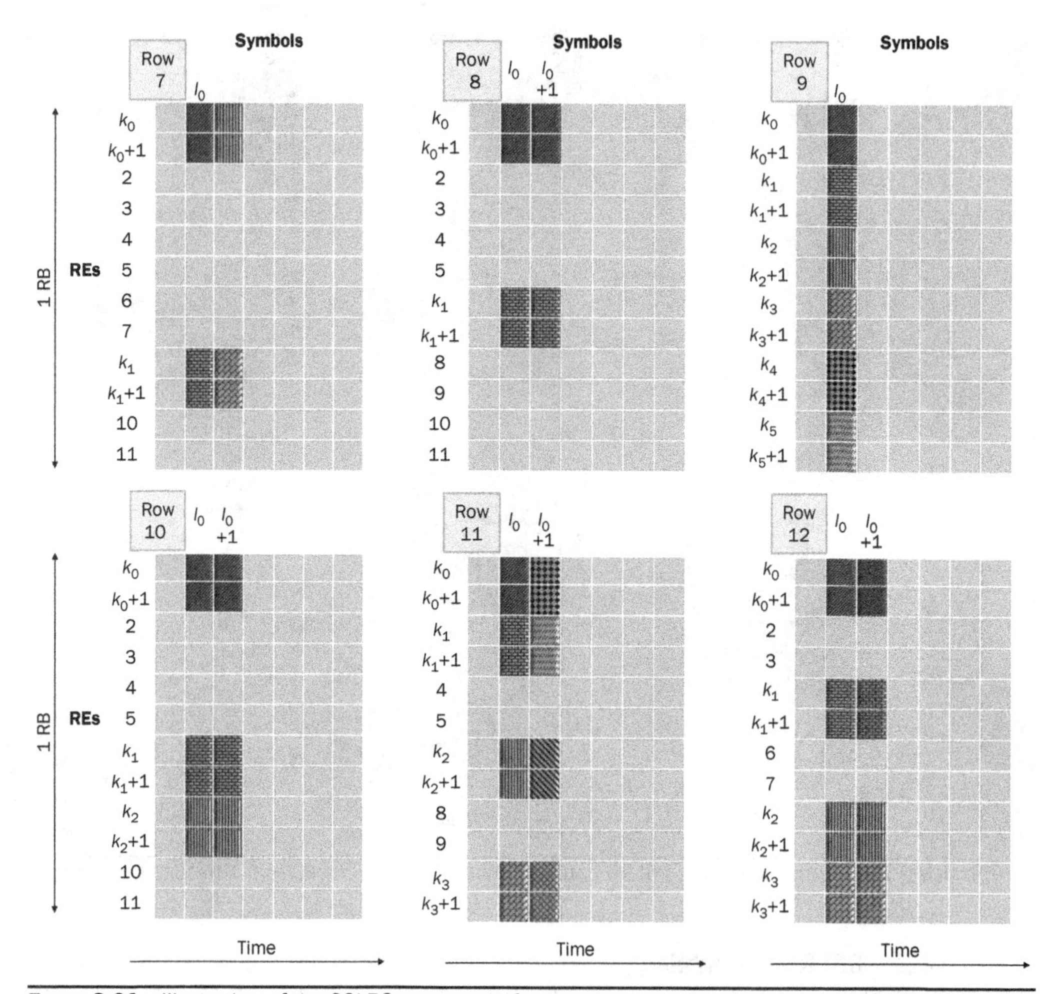

FIGURE 8.30 Illustration of the CSI-RS patterns in frequency and time, rows 7 through 12.

- l is the symbol index within the slot, which is in the range {0, 1, ..., 13}.
- n_{ID} is a configured scrambling ID.

A new scrambling sequence is generated in every symbol. The mapping of the scrambling sequence is based on the following formula from Ref. [7]:

$$
\begin{aligned}
a_{k,l}^{(p,\mu)} &= \beta_{\text{CSIRS}} w_{\text{f}}(k') \cdot w_{\text{t}}(l') \cdot r_{l,n_{s,f}}(m') \\
m' &= \lfloor n\alpha \rfloor + k' + \left\lfloor \frac{\bar{k}\rho}{N_{sc}^{\text{RB}}} \right\rfloor \\
k &= nN_{sc}^{\text{RB}} + \bar{k} + k' \\
l &= \bar{l} + l' \\
\alpha &= \begin{cases} \rho & \text{for } X = 1 \\ 2\rho & \text{for } X > 1 \end{cases} \\
n &= 0,1,\ldots
\end{aligned}
\tag{8.16}
$$

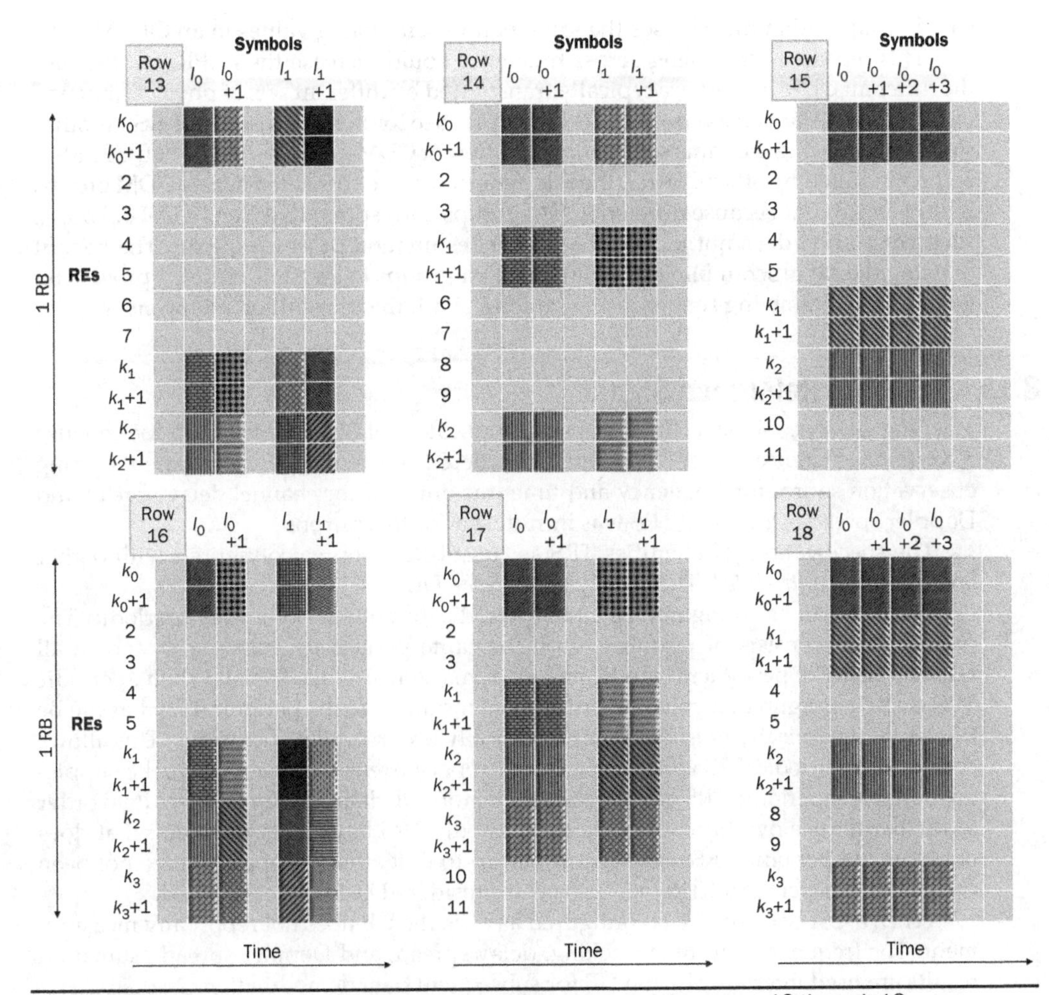

Figure 8.31 Illustration of the CSI-RS patterns in frequency and time, rows 13 through 18.

where $\beta_{\text{CSIRS}} = 0$ for ZP CSI-RS and $\beta_{\text{CSIRS}} > 0$ for NZP CSI-RS. β_{CSIRS} is selected such that when a power offset is specified between the SSB and CSI-RS, the required relative power is provided.

$w_f(k')$, $w_t(l')$ are the frequency and time-domain OCC functions, respectively, and the values k', l' are the frequency and time indices within the CDM group.

The value n is the RB index. It is important to note that the mapping of n always starts from Reference point A, which is necessary to enable UEs with different notion of communication channel BW to be able to utilize the same CSI-RS signal.

The formula is a bit complicated, but it can be simply summarized as follows. The scrambling sequence mapping is performed in the frequency domain from low frequency to high. For a single port $X = 1$ pattern the scrambling sequence is advanced by one element for each RE occupied by the CSI-RS resource. For a multi-port $X > 1$ pattern, the scrambling sequence is advanced by two elements for each RB occupied by the CSI-RS resource. The pair of scrambling values are used for $k' = 0$ and $k' = 1$. Every

CDM group within the RB uses the same pair of scrambling values in an OFDM symbol. This repeated scrambling across frequency could increase the PAPR, but because the different CDM groups are typically transmitted by different sets of physical antenna ports, such PAPR increase does not occur in practice for non-precoded CSI-RS. Another side effect of using the same scrambling in different CDM groups is that when CSI-RS is being interfered by other CSI-RS, the interference will be correlated across CDM groups within an RB. But because different CDM groups correspond to different CSI-RS ports, such correlation does not reduce the channel estimation processing gain. The benefit of using the same scrambling in different CDM groups in an RB is reduced processing load for the scrambling sequence generation in both the transmitter and receiver.

8.15 Tracking Reference Signal

Tracking reference signal (TRS) is a special application of CSI-RS. Since NR does not use a common reference signal (CRS) like LTE, there is a need to provide some recurring observation source for frequency and time tracking and for channel delay spread and Doppler spread estimation. TRS was introduced for this purpose.

The specification [1] identifies TRS as "a CSI-RS resource configured with higher layer parameter *trs-Info*". TRS is only used in the DL.

The TRS can be configured as either periodic, or both periodic and aperiodic. The periodic TRS consists of resources with the same periodicity, bandwidth and in all occurrences. The periodicity can be one of {10ms, 20ms, 40ms, 80ms}. When aperiodic TRS is also configured, it generates additional instances of the periodic TRS that can be triggered dynamically outside of the periodically occurring TRS occasions. This allows configuring periodic TRS with low frequency of occurrence in time that can be supplemented with aperiodic TRS on demand when more DL data activity requires it. In order to further reduce overhead, standalone aperiodic TRS have been proposed that does not have the periodic TRS component, but up to date, these proposals have not been adopted. Semi-persistent TRS has also been considered but not yet adopted.

When a CSI-RS resource is configured as TRS, the UE does not report any measurement. The frequency and time tracking, delay spread, and Doppler spread estimation results are used internally by the UE for subsequent data demodulation.

In order to estimate delay spread, the frequency selectivity of the channel needs to be observed, for which relatively dense CSI-RS pattern is needed in the frequency domain. The definition of density three ($\rho = 3$) single-port pattern was motivated by this goal, and this configuration is used for TRS.

In order to provide Doppler shift and Doppler spread estimate, multiple time observations are needed, that are well separated in time. Although the CSI-RS pattern definitions allow for two blocks of CDM groups to be configured with arbitrary separation between them within the slot, these are only applicable when the duration of the CDM group is two or four. Having two or four closely spaced observations is not beneficial for TRS because observable time variations across adjacent OFDM symbols are often too small. At the same time, having two or four REs is wasteful from the overhead perspective, especially considering that the configured TRS density in the frequency domain may already be high. This motivated introducing specific time-domain patterns for TRS. The target time-domain pattern is four single-symbol CSI-RS locations across two consecutive slots. The OFDM symbol locations in the two slots are identical. There was a concern with TRS overhead in FR2, where periodic TRS may need to be transmitted in 64 or more different beam directions. Due to this concern, a single-slot TRS version was also introduced in FR2, which allows multiplexing twice the TRS in the same resources. The one slot version is simply taking the first half of the two-slot pattern. Later in Release 16, further concerns were raised regarding FR1 TRS, because in certain FR1 TDD deployments, there may not be two consecutive DL slots available. Due to this consideration, the single-slot TRS version was also added as an option in FR1 to address these cases.

The configurable OFDM symbol location l in the slot for TRS was defined [7] as follows:

- For FR1: $l \in \{4, 8\}$, $l \in \{5, 9\}$, or $l \in \{6, 10\}$
- For FR2, all 10 possible pairs with a separation of four symbols between the elements of the pair

Examples of configurable TRS symbol locations are depicted in Fig. 8.32 for FR1.

Figure 8.32 shows the two-slot patterns. The single-slot patterns can be obtained by simply taking the first half.

The FR2 patterns can be similarly constructed with allowing the pattern to start at any OFDM symbol in the range {0, 1, ..., 9}. The additional time allocation options for FR2 were motivated by the fact that TRS corresponding to multiple beams may need to be multiplexed in the same slot in FR2.

The characteristics of the CSI-RS used for TRS are summarized in Table 8.23.

In order to provide adequate delay spread estimation opportunity, the TRS frequency span needs to be reasonably wide. The configurable TRS bandwidth is the full

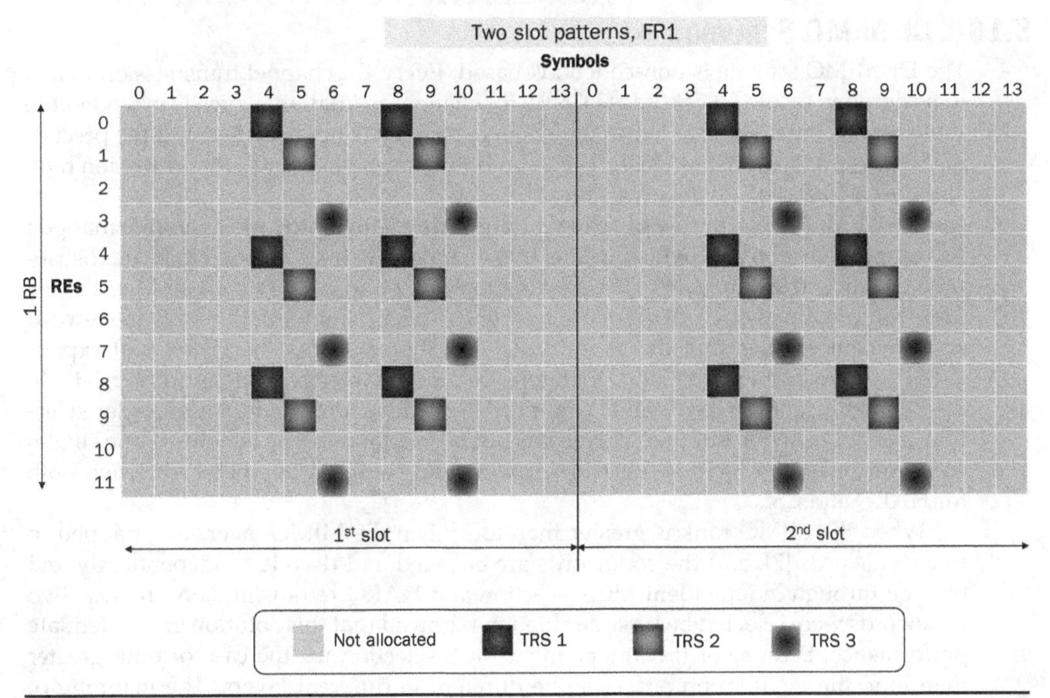

FIGURE 8.32 Illustration of configurable TRS symbol patterns for FR1.

Parameter	Value in FR1	Value in FR2
Number of ports, x	1	1
Frequency density, ρ	3	3
Spacing between adjacent tones	4	4
Spacing between adjacent symbols in slot	4	4
Burst length in Release 15	2 slots	1 slot or 2 slots
Burst length in Release 16	1 slot or 2 slots	1 slot or 2 slots
Periodicity in ms	10, 20, 40, 80	10, 20, 40, 80

Table 8.23 CSI-RS Parameters for TRS

BWP bandwidth, unless the BWP is wider than 52 RBs, in which case the TRS bandwidth can be configured to be either 52 RBs or the full BWP bandwidth. Note that the UE must be able to handle both options and perform appropriate PDSCH rate matching; however, what TRS bandwidth to actually process for tracking is up to UE implementation. The UE can choose to process any bandwidth as long as it enables adequate demodulation performance. The gNB is not allowed to configure the shortest periodicity of 10 ms when the TRS bandwidth is configured to be larger than 52 RBs. This was intended to relax the UE complexity, but as mentioned above, the UE could have anyhow chosen to limit the TRS processing bandwidth without this relaxation.

8.16 DL MIMO Scheme

The DL MIMO scheme is non-codebook based. Every DL channel transmission that is to be decoded by the UE includes DM-RS for channel estimation, which is precoded the same way as data. The manner in which it is precoded is not specified and the precoding itself is transparent to the UE. The UE is only informed about the transmission rank and the DM-RS port-to-layer mapping.

The MIMO layers, in case of spatial multiplexing with up to four layers, are mapped to a single codeword [7], which means that channel bits from a code block are distributed evenly across the layers. The modulation order and code rate on all layers is the same for rank up to 4. The SNR on each layer can be different, but the single-codeword arrangement ensures that the code blocks distributed across the layers will experience the same average SNR and, therefore, using the same code rate is appropriate. In case of significant per layer SNR variation, layer dependent modulation order selection showed benefits, but this scheme has not been adopted. The benefits of the single-codeword mapping include less CSI variation and feedback overhead reduction both for HARQ and CSI.

When the MIMO rank is greater than four, then the MIMO layers are mapped to two codewords [7], and the codewords are encoded and decoded independently and they go through independent MCS selection and HARQ retransmission process. Two codewords were selected because simulations showed that this solution gave adequate performance. Because of the independent MCS selection for the case of rank greater than four, the modulation order can be different in different layers. The mapping of layers to codewords is fixed. In the case of even rank, the layers are distributed equally.

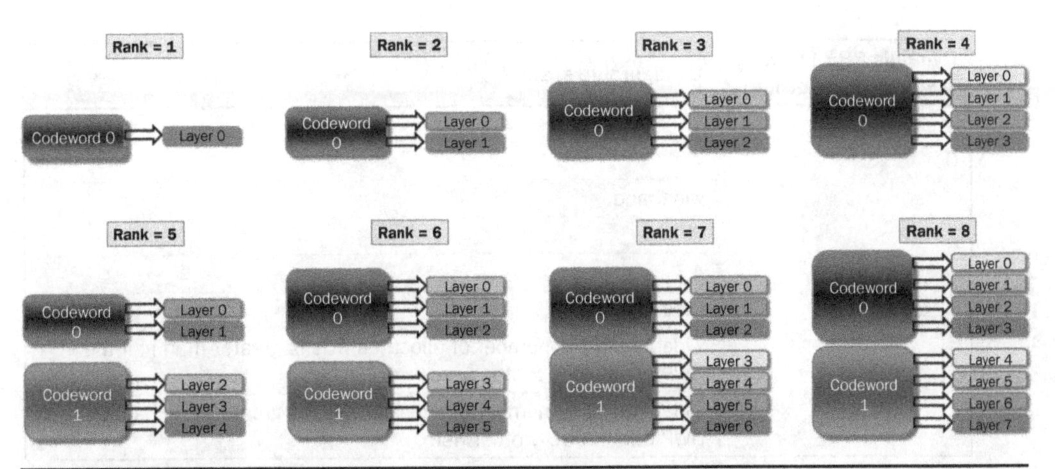

FIGURE 8.33 Illustration of the codeword to layer mapping cases.

In case of odd rank, one more layer is mapped to the second codeword than to the first codeword. The layer mapping cases are shown in Fig. 8.33.

There were a number of DL transmit diversity schemes considered for NR. Both transparent and non-transparent schemes were studied. In the end, transparent transmit diversity scheme was adopted, which can be, for example, PRG-level precoder cycling or small delay cyclic delay diversity (SD-CDD). The concept of PRGs will be discussed in Sec. 8.16.1. SD-CDD means that multiple transmit ports transmit the same signal but with a small time delay difference relative to each other. Although the diversity provided by PRG-level precoder cycling or SD-CDD is not the best, especially when small number of RBs are allocated for data transmission, this is compensated by the fact that both these schemes require only a single DM-RS port (per layer), not two, which reduces overhead. Another advantage of both PRG-level precoder cycling and SD-CDD is simpler interference estimation. When interfered by PDSCH, the interference can always be accurately estimated by observing the DM-RS of the interferer.

8.16.1 PRB Bundling

Similar to LTE, NR also uses the concept of PRB bundling in the DL [1]. PRB bundling means that the UE can assume that the gNB used the same MIMO precoder across a number of contiguous RBs. This increases the available channel estimation processing gain, which in turn improves the channel estimation quality.

The bundling size is $P'_{BWP,i}$ RBs, where $P'_{BWP,i}$ can be 2, 4, or 'wideband'. The wideband bundling option is new in the NR DL and it means that the UE can assume that gNB used the same precoding across all allocated PDSCH RBs. Wideband bundling only applies to contiguous resource allocation in the frequency domain. In the UL, only wideband precoding is supported.

Note that the UE being able to assume that the gNB used the same precoder within the bundle does not mean that the UE can assume the channel is the same across the bundle. Typically, the UE still has to do frequency selective channel estimation but at least the UE does not have to reset the channel estimation at every RB boundary. This improves the channel estimation quality.

Dynamic PRB bundling size indicator setting	Configurable state $P'_{BWP,i}$
0	2
	4
	wideband
1	2
	4
	wideband
	wideband if the number of allocated RBs is greater than half the BWP bandwidth, 2 otherwise
	wideband if the number of allocated RBs is greater than half the BWP bandwidth, 4 otherwise

TABLE 8.24 Configurable States for the Dynamic PRB Bundling Size Indicator

When $P'_{BWP,i}$ is 2 or 4, the bundles are aligned with a fixed grid within the bandwidth part. The purpose of this is that UEs that are MU-MIMO multiplexed with each other have their precoding bundles aligned, so the interference estimation can also benefit from increased processing gain, even if the UEs' frequency-domain resource allocations are not the same. The fixed grid partitions the bandwidth part into precoding resource block group (PRGs) where every PRG is of size $P'_{BWP,i}$, except possibly the first and/or last PRG within the bandwidth part (BWP), which can be of smaller size. It is not guaranteed that the BWP bandwidth is an integer multiple of $P'_{BWP,i}$, therefore, the last PRG can be fractional to fill in the leftover space in frequency. This is the same as in LTE. But unlike in LTE, the first PRG in the BWP can also be fractional in NR. The reason for this is to allow aligning the PRG grid even for UEs with different BWP allocations. For this purpose, there is an overall PRG grid that is common across all possible BWPs and whose numbering starts from Reference point A mentioned in Chap. 2.

In LTE, there was a fixed PRG size for a given channel bandwidth. In NR, the selection of $P'_{BWP,i}$ from the set {2, 4, 'wideband'} is not fixed but configurable to allow for more flexibility [1]. The configuration can be semi-static or dynamic. The dynamic selection allows for more adaptive DL scheduling. Dynamic selection is enabled when higher layer parameter *prb-BundlingType* is set to 'dynamic', and when enabled, a PRB bundling size indicator bit is included in DCI format 1_1. The bit chooses between two states, where the states themselves are configurable according to Table 8.24. For each indicator bit setting (i.e., "0" or "1"), one of the options listed in Table 8.24 is selected by the configuration.

If no PRG configuration is received by the UE, the UE assumes $P'_{BWP,i} = 2$. The same assumption is made also for all PDSCH granted by DCI format 1_0.

8.17 CSI Feedback

The NR CSI feedback framework is significantly more flexible than that in LTE. In NR, reporting settings, resource settings, and list of trigger states are defined. The purpose of this new structure is to enable using the same measurement resources in multiple different measurements. For example, a periodic CSI-RS resource defined in the resource

		CSI Report Configuration		
		Periodic CSI reporting	Semi-persistent CSI reporting	Aperiodic CSI reporting
CSI resource configuration	Periodic CSI-RS	Supported No dynamic triggering/ activation	Supported For reporting on PUCCH, the UE receives an activation command For reporting on PUSCH, the UE receives triggering on DCI	Supported Triggered by DCI
	Semi-persistent CSI-RS	Not supported	Supported For reporting on PUCCH, the UE receives an activation command For reporting on PUSCH, the UE receives triggering on DCI	Supported Triggered by DCI
	Aperiodic CSI-RS	Not supported	Not supported	Supported Triggered by DCI

Table 8.25 Possible CSI Reporting and CSI Resource Configurations

settings can be used in a periodic CSI report in one reporting setting and can be also used in an aperiodic CSI report in another reporting setting, with the latter being also mapped to one or more triggering states.

Table 8.25 summarizes the allowed combinations of CSI resource time-domain behavior and CSI reporting time-domain configuration.

The higher layer configuration records corresponding to each of the setting types are the following:

- Reporting settings is configured in *CSI-ReportConfig*.
- Resource settings is configured in *CSI-ResourceConfig*.
- List of trigger states is configured in *CSI-AperiodicTriggerStateList* or *CSI-SemiPersistentOnPUSCH-TriggerStateList*.

Each trigger state in the list of trigger states contains a list of associated reporting settings.

8.17.1 Reporting Settings

Each reporting setting contains the following parameters:

- Codebook configuration including codebook subset restriction indicated in the record *CodebookConfig*

Bandwidth part (PRBs)	Subband size (PRBs)
<24	N/A
24–72	4, 8
73–144	8, 16
145–275	16, 32

Table 8.26 Configurable Subband Sizes

- Time-domain behavior indicated by *reportConfigType* and can be set to
    - 'aperiodic'
    - 'semiPersistentOnPUCCH',
    - 'semiPersistentOnPUSCH', or
    - 'periodic'
- Frequency granularity for CQI and PMI, which is configurable according to Table 8.26. One of the two entries in the "Subband size" column is selected by RRC configuration.
- CSI band defining the frequency subbands in which the CSI report is valid.
- Measurement restriction configured by *timeRestrictionForChannelMeasurements* and *timeRestrictionForInterferenceMeasurements*, each of which controls whether the UE is allowed to perform time averaging for the relevant quantity.
- CSI-related quantities to be reported by the UE such as the layer indicator (LI), L1-RSRP, CRI, SSBRI (SSB Resource Indicator), RI, PMI, and CQI.

In general, the NR CSI may consist of channel quality indicator (CQI), precoding matrix indicator (PMI), CSI-RS resource indicator (CRI), SS/PBCH block resource indicator (SSBRI), layer indicator (LI), rank indicator (RI), or L1-RSRP. The UE calculates CSI parameters with observing the following dependencies:

- RI is calculated conditioned on the reported CRI.
- PMI is calculated conditioned on the reported RI and CRI.
- CQI is calculated conditioned on the reported PMI, RI, and CRI.
- LI is calculated conditioned on the reported CQI, PMI, RI, and CRI.

Unlike LTE, the NR design for CSI emphasized self-contained reports. Even in the case of periodic CSI reports, the option of reporting different information elements with different periodicities or time offsets is not used.

8.17.2 Resource Settings

Each CSI resource setting is a collection of measurement resources of a given type and contains a list of CSI resource sets given by higher layer parameter *csi-RS-ResourceSetList*. The following are configured for a CSI resource setting:

- NZP CSI-RS resource for channel measurement
- NZP CSI-RS resource for interference measurement
- CSI-IM resource for interference measurement

There can be one, two, or three resource settings configured for each reporting setting.

When a single resource setting is configured, the resource setting (given by higher layer parameter *resourcesForChannelMeasurement*) is for channel measurement for L1-RSRP or L1-SINR computation.

When two resource settings are configured, the first resource setting (given by higher layer parameter *resourcesForChannelMeasurement*) is for channel measurement and the second (given by either higher layer parameter *csi-IM-ResourcesForInterference* or higher layer parameter *nzp-CSI-RS-ResourcesForInterference*) is for interference measurement.

When three resource settings are configured, the first resource setting (given by higher layer parameter *resourcesForChannelMeasurement*) is for channel measurement, the second (given by higher layer parameter *csi-IM-ResourcesForInterference*) is for CSI-IM–based interference measurement and the third (given by higher layer parameter *nzp-CSI-RS-ResourcesForInterference*) is for NZP CSI-RS–based interference measurement.

8.17.3 Codebook Types

The following codebook types are defined in NR [1] in Release 15:

- Type I single-panel codebook (*CodebookType* set to 'typeI-SinglePanel')
- Type I multi-panel codebook (*CodebookType* set to 'typeI-MultiPanel')
- Type II codebook without port selection (*CodebookType* set to 'typeII')
- Type II codebook with port selection (*CodebookType* set to 'typeII-PortSelection')

In the following, we give a brief description of each codebook type.

8.17.3.1 Type I Single-Panel Codebook

The Type I single-panel codebook is very similar to the codebook defined in LTE and supports $P_{CSI-RS} = \{2, 4, 8, 12, 16, 24, 32\}$ Tx ports. The Type I single-panel codebook supports rank up to 8. The codebook assumes certain closely spaced Tx antenna arrangements along linear or rectangular arrays with two cross-polarized antenna elements at each array element. The supported antenna configurations are listed in Table 8.27, where N_1 is the number of antenna elements in the horizontal direction and N_2 is the number of antenna elements in the vertical direction. Note that the notation "horizontal" versus "vertical" represents typical arrangements, the actual orientation used is transparent to the UE.

The principle of the Type I codebook is that a precoder is selected with a W_1 and a W_2 component. W_1 is wideband and selects a single beam direction from a DFT-base codebook with possible oversampling. W_2 defines beam offsets and co-phasing coefficients across polarizations for the different layers. In the Type I codebook, each layer is associated with a single beam direction in each polarization.

8.17.3.2 Type I Multi-Panel Codebook

The Type I multi-panel codebook is a modified version of the Type I single-panel codebook, with added functionality to support multiple Tx panels. The number of supported panels, N_g can be either two or four. The Type I multi-panel codebook supports $P_{CSI-RS} = \{8, 16, 32\}$ Tx ports, i.e., the options are more limited compared to the single-panel codebook. The Type I multi-panel codebook supports rank up to 4. The supported

Number of CSI-RS antenna ports, P_{CSI-RS}	(N_1, N_2)	Number of polarizations
4	(2,1)	2
8	(2,2)	2
	(4,1)	2
12	(3,2)	2
	(6,1)	2
16	(4,2)	2
	(8,1)	2
24	(4,3)	2
	(6,2)	2
	(12,1)	2
32	(4,4)	2
	(8,2)	2
	(16,1)	2

TABLE 8.27 Type I Single-Panel Codebook Supported Configurations of (N_1, N_2)

Number of CSI-RS antenna ports, P_{CSI-RS}	Number of panels, N_g	(N_1, N_2)	Number of polarizations
8	2	(2,1)	2
16	2	(4,1)	2
	4	(2,1)	2
	2	(2,2)	2
32	2	(8,1)	2
	4	(4,1)	2
	2	(4,2)	2
	4	(2,2)	2

TABLE 8.28 Type I Multi-Panel Codebook Supported Configurations of (N_g, N_1, N_2)

antenna configurations are listed in Table 8.28, where N_g is the number of panels, N_1 is the number of antenna elements in the horizontal direction, and N_2 is the number of antenna elements in the vertical direction.

The Type I multi-panel codebook selects a precoder in each panel and in addition selects co-phasing coefficients across the panels for each layer. The co-phasing coefficient is $c_{p,r,l}$, where p is the panel index, r is the polarization index, and l is the layer index. There are two co-phasing modes defined:

- Mode 1, using lower overhead and supporting only wideband co-phasing across panels, applicable to both $N_g = 2$ and $N_g = 4$
- Mode 2, using higher overhead and supporting subband co-phasing across panels, applicable to only $N_g = 2$

The selection between Mode 1 and Mode 2 is based on RRC configuration. The encoding scheme selection for the co-phasing coefficients was based on performance evaluation comparing the gain with more accurate co-phasing versus overhead.

8.17.3.3 Type II Codebook Without Port Selection

The Type II codebook is an evolution of the Type I single-panel codebook, with the major change that each layer can use a linear combination of beam directions.

A comparison between the operations of the Type I single-panel and Type II codebooks is shown in Fig. 8.34.

The Type II codebook supports $P_{\text{CSI-RS}} = \{4, 8, 12, 16, 24, 32\}$ Tx ports. The Type II codebook supports rank up to 2 in Release 15. Support of higher rank was added in Release 16.

The linear combination of beams can be written in the form:

$$\tilde{w}_{r,l} = \sum_{i=0}^{L-1} \boldsymbol{b}_{k_1^{(i)} k_2^{(i)}} \cdot p_{r,l,i}^{(WB)} \cdot p_{r,l,i}^{(SB)} \cdot c_{r,l,i} \tag{8.17}$$

where

- L is the maximum number of beam directions selectable for the linear combination. The value of L is configurable, $L = \{2, 3, 4\}$.
- $\boldsymbol{b}_{k_1,k_2}$ is an oversampled 2-D DFT beam.
- r is the polarization index, $r = \{0, 1\}$.
- l is layer index, $l = \{0, 1\}$.
- $p_{r,l,i}^{(WB)}$ is the wideband (WB) beam amplitude scaling factor for beam i on polarization r and layer l.

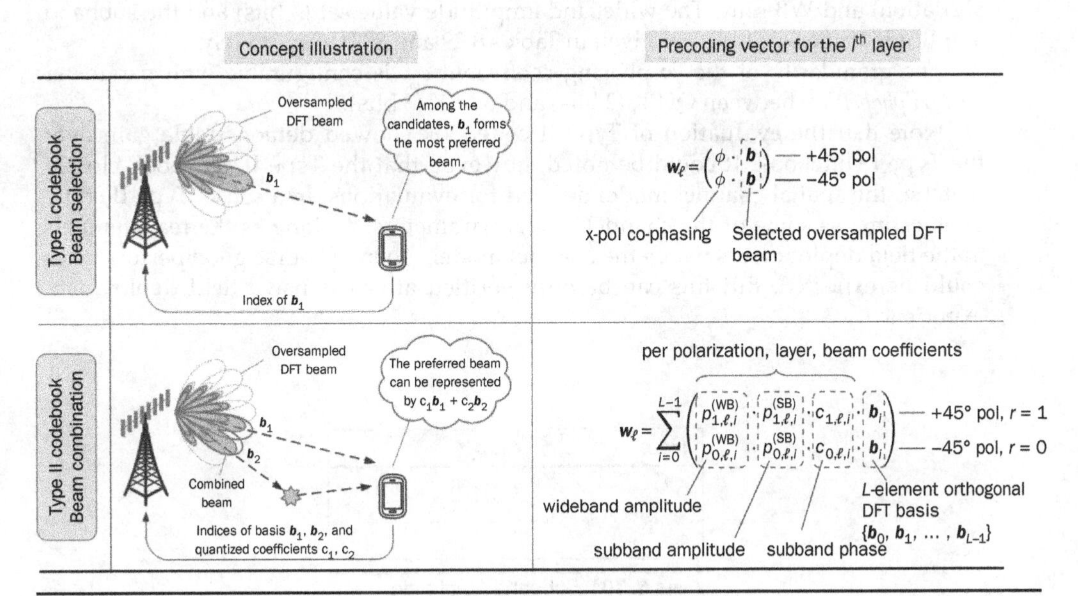

Figure 8.34 Comparison between the Type I and Type II codebook concepts.

Index	$p_{r,l,i}^{(WB)}$
0	0
1	$\sqrt{1/64}$
2	$\sqrt{1/32}$
3	$\sqrt{1/16}$
4	$\sqrt{1/8}$
5	$\sqrt{1/4}$
6	$\sqrt{1/2}$
7	1

TABLE 8.29 Wideband Amplitude Scaling Value Set

- $p_{r,l,i}^{(SB)}$ is the subband (SB) beam amplitude scaling factor for beam i on polarization r and layer l.
- $c_{r,l,i}$ is the beam combining phase coefficient for beam i on polarization r and layer l.

The beam selection for b_{k_1,k_2} is wideband only.

The amplitude scaling mode is configurable between WB+SB (with unequal bit allocation) and WB-only. The wideband amplitude value set (3 bits) and the subband amplitude value set (1 bit) are given in Tables 8.29 and 8.30, respectively.

The granularity of the co-phasing coefficient $c_{r,l,i}$ is configurable with parameter *phaseAlphabetSize* between QPSK (2 bits) and 8-PSK (3 bits).

Note that the evaluation of Type II codebook showed demonstrable gain over the Type I codebook. It could be noted, however, that the Type II codebook closely matches the spatial channel model defined for evaluations. In a sense, Type II codebook estimates some of the channel model parameters. As long as the real channels in the field deployments match the channel models, then of course good performance could be expected. But this can be only verified after extensive field deployment experience.

Index	$p_{r,l,i}^{(SB)}$
0	$\sqrt{1/2}$
1	1

TABLE 8.30 Subband Amplitude Scaling Value Set

8.17.3.4 Type II Codebook with Port Selection

The Type II codebook with port selection is an extension of the Type II codebook that uses selection among beamformed CSI-RS ports as the basis for linear combination. Similar to the Type II codebook without port selection, each layer can use a linear combination of beam directions. The Type II codebook with port selection supports $P_{\text{CSI-RS}} = \{4, 8, 12, 16, 24, 32\}$ Tx ports and supports rank up to 2 in Release 15, and rank up to 4 in Release 16. The first $\frac{P_{\text{CSI-RS}}}{2}$ Tx ports (i.e., first half) are assumed to be of a first polariza tion and the second half is assumed to be of a second polarization. The number of ports selected per polarization for the linear combination is L, with L being configurable, $L = \{2, 3, 4\}$. The port selection chooses L consecutive ports in the first half, starting from port index n, i.e., it selects port indices $\{n, n+1, \ldots, n+L-1\}$. The same selection is repeated in the second half starting from $\frac{P_{\text{CSI-RS}}}{2} + n$. This gives a total of $2 \cdot L$ ports (beam directions) selected over the two polarizations. The selection step size d is separately configured with higher layer parameter *portSelectionSamplingSize*, where $d \in \{1, 2, 3, 4\}$ and $d \leq min\left(\frac{P_{\text{CSI-RS}}}{2}, L\right)$. The selection step size means that while the first port index that could be chosen as n (i.e., as the first selected port index) is 3000, the next port index that could be chosen as the first selected port index is $3000 + d$, and the next port index that can be chosen as the first selected port index is $3000 + 2 \cdot d$ and so on. The port selec tion result is indicated with $m \in \left\{0, 1, \ldots, \left\lceil \frac{P_{\text{CSI-RS}}}{2d} \right\rceil - 1\right\}$, where m is the entry number in the sequence $\{3000, 3000 + d, 3000 + 2 \cdot d, \ldots\}$ that corresponds to the first selected port index. As mentioned before, the other selected port indices are consecutive after the first. The reporting of m is wideband and it uses $\left\lceil \log_2\left(\frac{P_{\text{CSI-RS}}}{2d}\right)\right\rceil$ bits.

The granularity of the co-phasing coefficient $c_{r,l,i}$ is configurable with parameter *phaseAlphabetSize* between QPSK (2 bits) and 8-PSK (3 bits).

The amplitude scaling modes (WB-only and WB+SB) and the corresponding amplitude quantization value sets are the same as for the Type II codebook without port selection given in Tables 8.29 and 8.30 in Sec. 8.17.3.3.

Further details of the codebook definitions are omitted for brevity, they can be found in specification Secs. 5.2.2.2.1 through 5.2.2.2.4 of Ref. [1].

8.17.4 CSI Processing Requirements

CSI processing is one of the most demanding tasks the UE must perform. In order to get the best system performance while managing UE complexity, a new CSI processing framework was adopted in NR [1]. The concept of CSI processing unit (CPU) was introduced. The UE is able to handle up to a certain number of simultaneous CPUs. That number, denoted as N_{CPU}, is reported by the UE as part of its capability record. There is a running count, L, of occupied CPUs, representing the processing units that are in use by ongoing CSI reports. At any given time, the $N_{CPU} - L$ unoccupied CPUs can be used to adding more CSI reports. Once there are no more unoccupied CPUs available, the UE will not process more CSI. The UE is still required to send CSI report even in this case, but for the CSI requests that are over the limit, the UE is allowed to send outdated reports. The minimum value the UE can declare as N_{CPU} is 5, which is across all configurable component carriers.

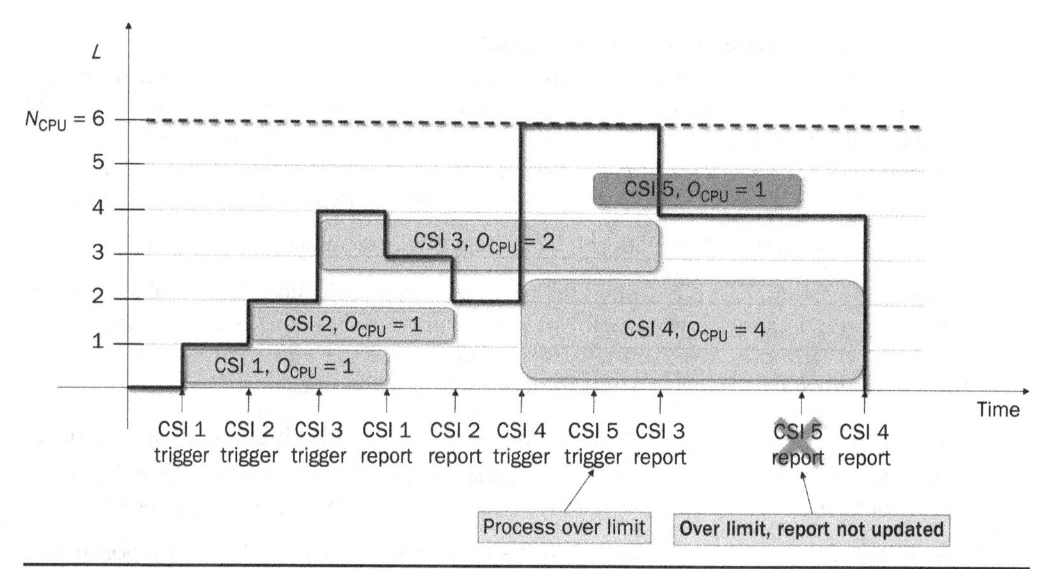

FIGURE 8.35 CSI processing capability example (N_{CPU} = 6).

Any time a CSI calculation starts, the count L is incremented by O_{CPU}, where O_{CPU} is the load designation of the new CSI process, and at any time a CSI calculation ends, the count L is decremented by O_{CPU}, where O_{CPU} is the load designation of the completed CSI process. This operation is depicted in Fig. 8.35.

A detailed set of definitions was developed for when a CSI process of a certain type starts and when it ends from the perspective of CPU count. Also, the designated load O_{CPU} was defined for each CSI type. These will be described next.

- For an aperiodic CSI report, the CPU becomes occupied at the end of the last symbol of the PDCCH carrying the CSI trigger and the CPU is released at the end of the last symbol of the PUSCH or PUCCH carrying the report.
- For the first report in a sequence of semi-persistent CSI reports on PUSCH, the CPU becomes occupied at the end of the last symbol of the PDCCH activating the CSI process and the CPU is released at the end of the last symbol of the PUSCH carrying the first report.
- For periodic and semi-persistent CSI reports, except for the first report in a sequence of semi-persistent CSI reports on PUSCH, the CPU becomes occupied at the latest CSI measurement resource (CSI-RS, CSI-IM or SSB) that is usable for the report and the CPU is released at the end of the last symbol of the PUCCH or PUSCH carrying the report. The latest such CSI resource is formally defined as the latest that is not later than the so-called CSI reference resource. The timing of the CSI reference resource is separately defined. If multiple CSI resources are used for a given report and they do not occur at the same time, then the earliest of those counts.

It is an interesting feature of NR that there are processes that require CSI processing similar to a CSI report but there is no actual reporting involved. One example is the case where the UE has to transmit precoded SRS in response to a DL CSI-RS, which is described in Sec. 10.12. Another example is the P-3 process mentioned in Sec. 8.18. In order to determine the end point of the CPU occupancy in these cases, the chosen solution was to define a "virtual report." The virtual report is assumed to be sent at the earliest time that complies with the UE's processing timeline capability, which will be discussed later in this section. Of course, no virtual report is in fact constructed or sent.

TRS is another case where NZP CSI-RS is processed but no report is sent. The TRS case, however, is not considered to require CSI processing; therefore, $O_{CPU} = 0$ and no start or end time is defined for CPU occupancy for TRS.

As it was mentioned, for periodic report (and for semi-persistent report, except the first report on PUSCH), the CPU is considered occupied only from the latest CSI resource usable in the CSI calculation, which is assumed to be the CSI resource used in the CSI calculation. CPU is not the only quantity describing the UE's processing load though. There is also a separate definition of the UE capability in terms of simultaneously active CSI-RS resources. The UE reports, for example, the following capabilities:

- Maximum number of simultaneously active NZP CSI-RS resources per component carrier (*maxNumberSimultaneousNZP-CSI-RS-PerCC*)
- Maximum total number of ports in all simultaneously active NZP CSI-RS resources per component carrier (*NumberPortsSimultaneousNZP-CSI-RS-PerCC*)
- Maximum number of simultaneously active NZP CSI-RS resources across all component carriers (*maxNumberSimultaneousNZP-CSI-RS-ActBWP-AllCC*)
- Maximum total number of ports in all simultaneously active NZP CSI-RS resources across all component carriers (*totalNumberPortsSimultaneousNZP-CSI-RS-ActBWP-AllCC*)

When determining the number of resources simultaneously active, a different rule than CPU occupancy applies. A CSI resource is considered active for the purposes of determining the number of simultaneously active resources according to the following:

- For an aperiodic CSI resource, the CSI resource and the ports within the resource become active at the end of the last symbol of the PDCCH carrying the CSI trigger and the CSI resource becomes inactive at the end of the last symbol of the PUSCH carrying the report.
- For a periodic CSI resource, the CSI resource and the ports within the resource become active at the time of the configuration of the CSI resource, and the CSI resource becomes inactive at the time of the release of the CSI resource configuration.
- For a semi-persistent CSI resource, the CSI resource and the ports within the resource become active at the time of the activation of the CSI resource, and the CSI resource becomes inactive at the time of the deactivation of the CSI resource.

As it can be seen, for periodic and semi-persistent CSI reports, the CSI measurement resource is considered active much longer than the time of CPU occupancy. The reason for this differentiation is that the count of the simultaneously active CSI resources is a measure of memory use, while the CPU is the measure of processing load. These metrics are clearly different. It is expected that the UE typically reports larger numbers for *maxNumberSimultaneousNZP-CSI-RS-ActBWP-AllCC* than for N_{CPU}.

The designated load, O_{CPU}, of a CSI process is defined as follows:

- For TRS, there is no report, $O_{CPU} = 0$, and there is no start or end time.
- For beam management related L1-RSRP and L1-SINR report, $O_{CPU} = 1$.
- For cases with CSI calculation but without CSI report (examples were given earlier), $O_{CPU} = 1$.
- For CSI report (with reported quantity being one of "cri-RI-PMI-CQI", "cri-RI-i1", "cri-RI-i1-CQI", "cri-RI-CQI", or "cri-RI-LI-PMI-CQI"), O_{CPU} equals the number of CSI-RS resources in the CSI-RS resource set used for the report. The reason for this is that with multiple CSI-RS resources in the CSI-RS resource set, the UE must calculate CSI for each resource before selecting the best one for feedback.
- For "fast" CSI report, $O_{CPU} = N_{CPU}$, is applicable to the case where there are no other concurrent reports, i.e., $L = 0$ at the start of CPU occupancy.

The fast CSI report mentioned above is a special case, which will be also mentioned in Sec. 8.17.4.1. It allows requesting a single CSI report with restrictions to be reported faster than other CSI report types or multi-CSI report.

8.17.4.1 CSI Processing Time Requirements

In NR, an effort was made to allow for faster turn-around of CSI reports than in LTE. This was made somewhat difficult by the introduction of new, more complicated reporting modes and codebook types in NR at the same time.

The time allowed for CSI processing is expressed with terms Z and Z', which can be described based on Ref. [1] as follows:

- Z is measured from the end of the PDCCH triggering the CSI report to the beginning of the PUSCH carrying the CSI report.
- Z' is measured from the end of the last CSI resource used for the CSI report to the beginning of the PUSCH carrying the CSI report.

For both the Z and Z' values, the effects of UL timing advance are included. Note that $Z > Z'$ typically holds because the CSI-RS triggering is causal. The definitions of Z and Z' are also illustrated in Fig. 8.36.

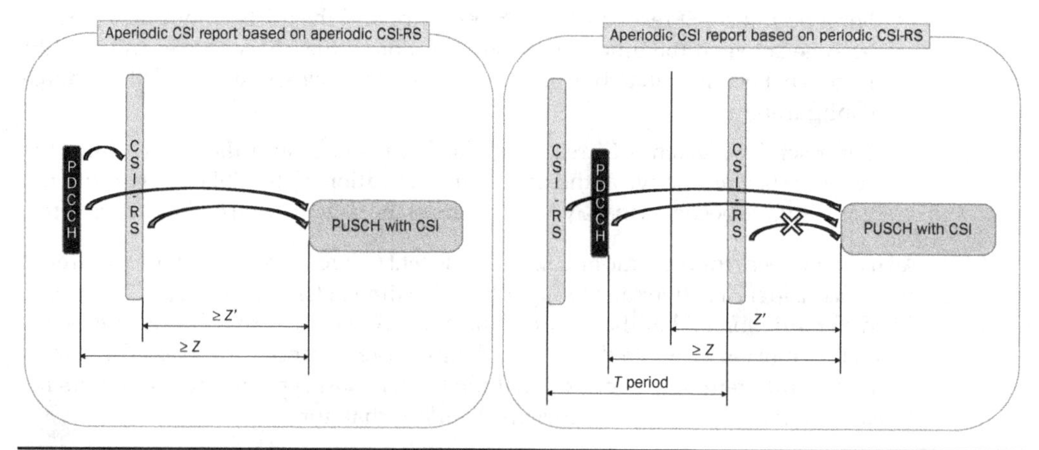

FIGURE 8.36 Processing time requirement Z and Z'.

The following CSI-RS processing requirements were defined for NR in Release 15:

- For the so-called fast CSI report, the processing requirements are defined according to Table 8.31. Fast CSI report is defined with the following attributes:
  - No other concurrent report, i.e., $L = 0$ at the start of occupancy
  - CSI report only without UL data and without HARQ ACK feedback in PUSCH
  - Single CSI-RS resource without CRI, with up to four ports
  - Wideband only report
  - Either
    - Type I single-panel codebook, or
    - No PMI report, i.e., *reportQuantity* = 'cri-RI-CQI' (CRI is not reported because of single CSI-RS resource.)
- For low complexity CSI report, the processing requirements are defined according to Table 8.32. Low complexity CSI report is defined with the following attributes:
  - There may be other concurrent reports, i.e., $L \geq 0$ at the start of occupancy
  - Single CSI-RS resource without CRI, with up to four ports
  - Wideband only report
  - Either
    - Type I single-panel codebook, or
    - No PMI report, i.e., *reportQuantity* = 'cri-RI-CQI' (CRI is not reported because of single CSI-RS resource.)

	Z (symbols)		Z (μs)	
SCS (kHz)	Z	Z′	Z	Z′
15	10	8	713.5	570.8
30	13	11	463.8	392.4
60	25	21	446	374.6
120	43	36	383.5	321.1

TABLE 8.31 CSI Computation Delay Requirement for Fast CSI Report

SCS (kHz)	Z (symbols)		Z (μs)	
	Z	Z′	Z	Z′
15	22	16	1569.8	1141.7
30	33	30	1177.3	1070.3
60	44	42	784.9	749.2
120	97	85	865.2	758.1

TABLE 8.32 CSI Computation Delay Requirement for Low Complexity CSI Report

SCS (kHz)	Z (symbols)		Z (μs)	
	Z	Z′	Z	Z′
15	40	37	2854.2	2640.1
30	72	69	2568.8	2461.7
60	141	140	2515.2	2497.4
120	152	140	1355.7	1248.7

Table 8.33 CSI Computation Delay Requirement for High-Complexity CSI Report

SCS (kHz)	Z (symbols)	
	Z	Z′
15	22	X_1
30	33	X_2
60	min (44, X_3 + KB_1)	X_3
120	min (97, X_4 + KB_2)	X_4

Table 8.34 CSI Computation Delay Requirement for Beam-management Related CSI Report

- For high-complexity CSI report, the processing requirements are defined according to Table 8.33. High-complexity CSI report includes all cases not already included in the low-complexity CSI and fast CSI categories above and excluding beam management–related CSI report.
- For beam management–related CSI report, the processing requirements are defined according to Table 8.34, where X_1, X_2, X_3, X_4 are UE reported values that are chosen from the following sets: $X_1 = \{2, 4, 8\}$, $X_2 = \{4, 8, 14, 28\}$, $X_3 = \{8, 14, 28\}$, $X_4 = \{14, 28, 56\}$, and KB_1, KB_2 are the reported *beamSwitchTiming* capability, which will be described in Sec. 8.18, and which are chosen from the following set: $KB_1 = \{14, 28, 48\}$, $KB_2 = \{14, 28, 48\}$.

The difference between fast CSI report and low-complexity CSI report is that for the former neither concurrent CSI calculation nor multiplexing with data or HARQ ACK is required, while for the latter these relaxations do not apply.

We can note that the processing time for high-complexity CSI report is comparable to the LTE CSI processing time, which was 2300 μs or 3300 μs when timing advance is included. However, the NR processing load for the same processing time is still larger due to the larger bandwidth and more complex codebook types compared to LTE.

8.18 Beam Management for the PDSCH

For the support of multi-beam operation, three "processes" P-1, P-2, and P-3 are supported, which are described as follows:

- P-1: Tx beam selection is used to enable UE measurement on different Tx beams to support selection of gNB Tx beams and UE Rx beam(s).
    - For the gNB, P-1 typically includes a Tx beam sweep from a set of different beams.

- For the UE, P-1 typically includes signal strength (L1-RSRP) measurement and Rx beam sweep from a set of different beams.
- P-2: Tx beam refinement is used to enable UE measurement on different Tx beams from a possibly smaller set of beams compared to P-1.
  - CSI-RS with different Tx beams are sent.
  - Typically, the UE measures the Tx beams without Rx beam sweep.
  - The UE reports the index (CRI) and signal strength (L1-RSPR) for the strongest beam(s).
- P-3: Rx beam refinement is used to enable UE measurement on the same gNB Tx beam to change UE Rx beam.
  - CSI-RS with common Tx beam are sent in quick succession.
  - The UE selects an Rx beam but does not report any measurement results to the gNB.

In Fig. 8.37, the P-1, P-2, and P-3 beam management processes are illustrated.

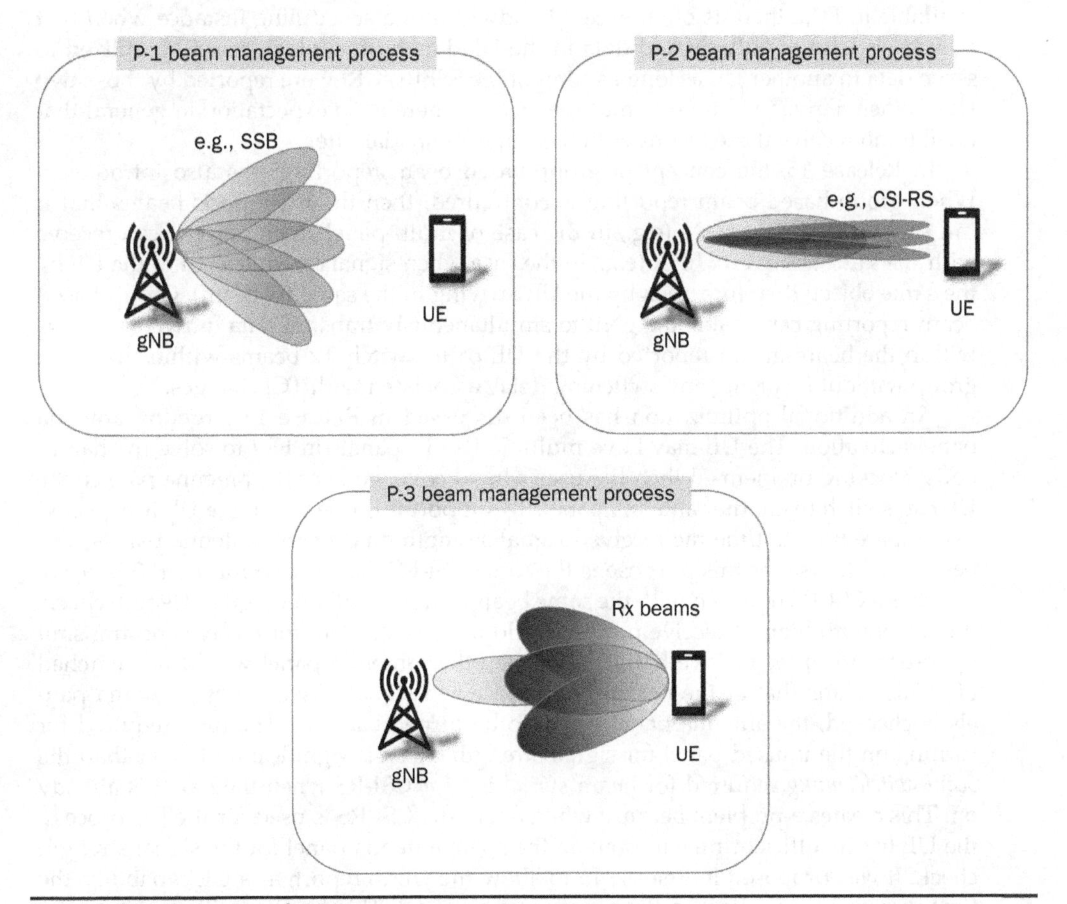

FIGURE 8.37 Illustration of the beam management process types.

Note that the Rx beam sweep component of P-1 can be performed as a background process by the UE. Due to the analog beamforming restriction, the Rx beam sweep could interrupt the PDSCH reception though; therefore, it is beneficial if the gNB configures designated signals in the form of CSI-RS repeated with the same beamforming to initiate the P-3 process to support Rx beam sweep by the UE.

For beam reporting purposes, the UE reports the index of the best beams in the form of CSI-RS resource indicator (CRI) or SS/PBCH block resource index (SSBRI). The UE also reports signal strength measurements in the form of L1-RSRP for the indicated beams. When multiple beams are reported, then the absolute RSRP is reported for the strongest beam (with highest L1-RSRP) and differential RSRP is reported for the other beams. This serves to save signaling overhead. The absolute RSRP is a 7-bit value in the range [−140, −44] dBm with 1 dB resolution. The differential RSRP uses only 4 bits. The reason for reporting multiple beams is to enhance the gNBs frequency multiplexing or SDM multiplexing capability. As it was mentioned in Sec. 2.4, the gNB is expected to be able to transmit with only one spatial beam at a time. If there are, say, three different beams observed by the UE with comparable RSRP, and the UE reports all three, then the chances are three times higher that this UE can be served with data at the same time when the gNB has to send data to some other UEs. Because of the wide bandwidth available in FR2, there is often excess bandwidth in a scheduling instance when that scheduled UE has only limited data in the DL data buffer, which could be utilized to serve data to another UE as long as compatible SSBRI/CRI were reported by those two UEs. When a given UE reports multiple beams, there is no expectation in general that the UE can receive these beams with the same Rx spatial filter.

In Release 15, the concept of group-based beam reporting was also introduced. When group-based beam reporting is configured, then the UE reports beams that it can receive simultaneously (e.g., in the case of multi-panel receiver) or it can receive with the same spatial Rx filter (e.g., in the case when signals are reflected to the UE by the same object; therefore, seen by the UE arriving in the same Rx beam). Group-based beam reporting can enable the gNB to simultaneously transmit data in two Tx beams within the beam group reported by the UE or to switch Tx beams within the beam group without incurring any switching delay associated with TCI changes.

An additional optimization has been discussed in Release 15 targeting antenna panel activation. The UE may have multiple Rx/Tx panels in FR2 to solve the hand/body blocking problems. When the user's hand covers one of the antenna panels, the UE can switch to another antenna panel. To support this operation, the UE is expected to compare time to time the received signal strength on different antenna panels. The best signal to use for this purpose is the same CSI-RS that is used for the P-3 process, which is a CSI-RS repeated with the same beamforming. After the signal strength check, the panel with highest receive power would be selected both for receive and transmit operation. In order to save battery life, the other antenna panel would be switched off. This means that at the next time the received signal power across antenna panels is checked, the antenna panel needs to be turned back on. The time required for turning on the unused panel for signal strength check is significantly longer than the *beamSwitchTiming* required for beam switching for CSI-RS when the panel is already on. This creates a problem because when aperiodic CSI-RS is used for the P-3 process, the UE has insufficient time to prepare the spare antenna panel for the signal strength check. It was proposed in Release 15 to allow the UE to report, as a UE capability, the time required for turning on the spare antenna panel. This functionality was not fully defined in Release 15, but it was completed in Release 16. When the UE reports *beamSwitchTiming* = 224 symbols (2 ms) or 336 symbols (3 ms), then this longer time duration is used as an offset between triggering a CSI-RS used for the P-3 process and the CSI-RS transmission. This allows the UE to turn on the additional antenna panel. For the other applications of *beamSwitchTiming*, e.g., timing between aperiodic CSI-RS trigger and the aperiodic CSI-RS for the P-2 process, 48-symbol offset is used for the UE that reported *beamSwitchTiming* = 224 or 336. Note that once the UE turns on the unused antenna panel, switching Rx beams across different antenna panels does not take longer time than switching Rx beams within a single panel. The indicated longer time is only needed for powering on an unused panel.

8.19 Signal Quasi Co-Location

In order to support multi-beam operation and potentially multi-TRP operation, the concept of quasi co-location (QCL) is used in the NR DL [1]. The NR QCL definition can be viewed as an evolution of the LTE definition introduced for CoMP in Release 11.

When two signals are "QCL-ed," more precisely, when a first signal is the QCL source for a second signal, it means that an estimation of certain channel parameters based on observing the first signal can be used in processing the second signal.

For example, a signal transmitted from a given transmission point is QCL-ed with any other signal transmitted from the same transmission point with respect to all channel properties except possibly for spatial (beamforming) parameters. But when different signals are transmitted from different transmission points, or different beams are used, then not all channel parameters are transitive between the different signals. Depending on the network architecture and the transmission scheme, the set of shared parameters can be different. Accordingly, four different QCL types were defined: 'QCL-TypeA', 'QCL-TypeB', 'QCL-TypeC' and 'QCL-TypeD'. The list of shared parameters between the first and second signal is listed below for each QCL type:

- QCL-TypeA: {Doppler shift, Doppler spread, average delay, delay spread}
- QCL-TypeB: {Doppler shift, Doppler spread}
- QCL-TypeC: {Doppler shift, average delay}
- QCL-TypeD: {Spatial Rx parameter}

The QCL parameters can be summarized as follows:

- Doppler shift is the parameter pertaining to frequency tracking, serves as a correction for transmit and receive oscillator frequency errors and possibly frequency offsets due to linear motion.
- Doppler spread is a parameter pertaining to channel time variability estimation, can be used to set the type and length of the time-domain filtering used in the processing of the received signal.
- Average delay is the parameter pertaining to time tracking, serves as a correction for receiver timing errors and propagation delay variations and in general is used for adjusting receiver timing.
- Delay spread is a parameter pertaining to time-domain dispersion caused by multipath in the propagation channel, causing variability in the channel in the

frequency domain. The estimated delay spread can be used to set the type and bandwidth of the frequency-domain filtering used in processing the received signal.

- Spatial Rx parameter pertains to analog beamforming in FR2. The spatial Rx parameter is used in setting the analog beamforming weights for the individual antenna elements of the receiver.

For DL QCL operation, the UE is provided with a QCL source for each signal to be processed, except for the initial SSB signal. The UE must acquire the initial SSB signal without any QCL assistance in both SA and NSA modes. In the case of PDCCH and PDSCH, a QCL source is given (either explicitly or implicitly) for the DM-RS to be used in the channel estimation, which is then also used in the decoding of control and data, respectively.

For any signal, one or two other signals can be configured as QCL source. When two signals are used as QCL source, they are of different QCL-types. For example, the first QCL source can be of QCL-TypeA and the second QCL source can be of QCL-TypeD. The reason for allowing this QCL configuration is that different signal designs are optimal for different QCL-Types. For example, multiple time-domain observation with reasonable time separation is needed for Doppler shift or Doppler spread estimation, while large frequency spread is required for accurate average delay and delay spread estimation. In the following we give a general description of QCL sources and their most typical applications:

- SSB: The SSB can be used for any QCL-type but due to its confinement in relatively short burst in time and narrow bandwidth in frequency, the precision of SSB-based Doppler spread and delay spread estimation can be poor. Because of this, SSB is often used as QCL-TypeC source, while another signal, e.g., CSI-RS is used for Doppler spread and delay spread estimation. SSB is often used in the P-1 beam management process.
- TRS: The specification section in Ref. [1] about QCL identifies TRS as "CSI-RS resource in a *NZP-CSI-RS-ResourceSet* configured with higher layer parameter *trs-Info*". TRS can be used for all QCL-types due to its spread in time and frequency. Although TRS can be used as QCL-TypeD source, it is not optimum for this purpose because its large overhead makes resource utilization poor if TRS had to be sent in a beam-swept fashion to support beam measurement and selection. TRS is the best signal to use for Doppler shift and Doppler spread and is often used as QCL-TypeB source for this reason. When multiple transmission points (TRPs) or panels use a common frequency reference, then a single SFN TRS can be used across multiple TRPs or panels, at least in FR1, to serve as reference for Doppler shift and Doppler spread estimation, while other lower overhead signals can be transmitted individually from each TRP or panel to estimate other channel parameters.
- CSI-RS for beam management: The specification section in Ref. [1] about QCL identifies CSI-RS for beam management as "CSI-RS resource in a *NZP-CSI-RS-ResourceSet* configured with higher layer parameter *repetition*". CSI-RS for beam management can be used as QCL-TypeD source. CSI-RS for beam management can be used in the P-2 process when a sequence of CSI-RS resources is sent with Tx beam sweep. In this case, the configured *repetition* is set to "off." It can also

be used in the P-3 process when the same CSI-RS is repeated with a fixed Tx beam. In this case, the configured *repetition* is set to "on." CSI-RS for beam management is not well-suited for Doppler shift or Doppler spread estimation due to the fact that it provides a single time observation from each CSI-RS resource. The use for Doppler shift or Doppler spread estimation is not precluded but these parameters are often just "inherited" from the signal that is the QCL source for the CSI-RS for beam management.

- CSI-RS for CSI: The specification section in Ref. [1] about QCL identifies CSI-RS for CSI with the somewhat cumbersome term: "CSI-RS resource in a *NZP-CSI-RS-ResourceSet* configured without higher layer parameter *trs-Info* and without higher layer parameter *repetition*". CSI-RS for CSI can be used as QCL-TypeD source. It is not well-suited for Doppler shift or Doppler spread estimation due to the fact that it provides a single time observation from each CSI-RS resource. The use for Doppler shift or Doppler spread estimation is not precluded but these parameters are often just "inherited" from the signal that is the QCL source for the CSI-RS for CSI.

The QCL configuration is defined via a three-level process:

- The UE is provided via RRC configuration with up to 128 transmission configuration indication (TCI) states. The maximum number of configurable TCI states can be lower if limited by lower UE capability. Each TCI state defines a different QCL source for the PDSCH DM-RS.
- A set of up to eight TCI states gets activated with a MAC CE. Any one of the RRC configured TCI states can be part of the set, as long as the limit eight on the maximum set size is not exceeded. Separately, MAC CE also activates a QCL source for each of the PDCCH CORESETs monitored by the UE.
- For a PDSCH granted by DCI format 1_1, a three-bit field in the grant selects one of the TCI states from the set of up to eight active TCI states. The QCL source defined by this TCI state will be used for the PDSCH.

A special complication is presented by the fact that applying the spatial Rx parameters cannot be done retroactively, as it was already mentioned in Sec. 2.4. If the UE needs to buffer PDSCH while it is decoding the TCI information conveyed in the DL grant, then it cannot use the correct TCI state for the buffered PDSCH since the information has not been decoded yet. Even after the TCI state information is decoded, it takes certain time to apply the indicated beamforming coefficients. The total time between the PDCCH carrying the TCI information and the application of the indicated beamforming coefficients is denoted as *timeDurationForQCL* and it is measured from the end of the PDCCH to the beginning of the PDSCH. Note that the *timeDurationForQCL* time threshold only applies in FR2 when QCL-TypeD is configured. The value of *timeDurationForQCL* depends on the UE's reported capability, and the UE can indicate 7, 14, or 28 symbols for 60 kHz SCS and 14 or 28 symbols for 120 kHz SCS as the required time separation between the PDCCH and PDSCH. Note that this time separation does not represent a reception gap. During the 7, 14, or 28 symbol period, the UE can continue to receive signals, in fact it can receive the granted PDSCH as well. But during this reception earlier than *timeDurationForQCL* after the PDCCH, the UE will use a default QCL instead of the QCL indicated by the TCI. For the case of same carrier scheduling (self-scheduling), the default QCL is the QCL source of the PDCCH CORESET with the lowest CORESET ID in the latest monitoring occasion before the PDSCH. For the case of cross-carrier scheduling, the default QCL is that of the lowest indexed TCI state active for the PDSCH. As a general rule, the UE uses this default QCL while the UE buffers OFDM symbols prior to reaching the *timeDurationForQCL* time threshold. There are exceptions, however, when there is a conflict between the default QCL determined by the above rules and the QCL assumption needed for some other signal that the UE would be monitoring without the PDSCH. For example, if the UE would be monitoring a search space in a PDCCH CORESET not with the lowest CORESET ID in the first symbol of the PDSCH, then the default QCL for the PDSCH is based on that overlapping CORESET.

DCI format 1_0 does not include TCI indication. When the PDSCH is granted by DCI format 1_0, then the QCL source for the PDSCH is determined to be the same as the QCL source of the CORESET containing the PDCCH granting the PDSCH. The minimum application time *timeDurationForQCL* applies in this case as well. This is because there can be multiple PDCCH decoding attempts with different QCL sources and until

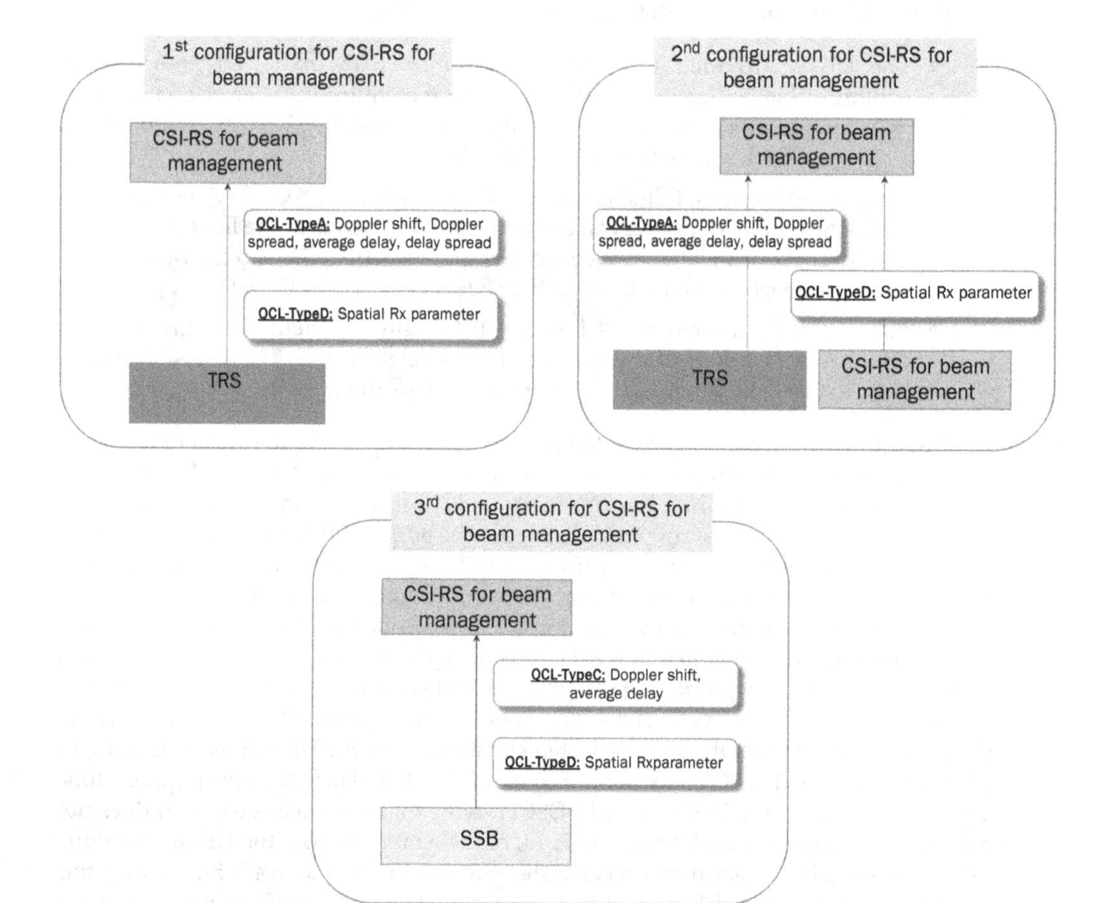

FIGURE 8.38 TCI state configuration options for CSI-RS for beam management.

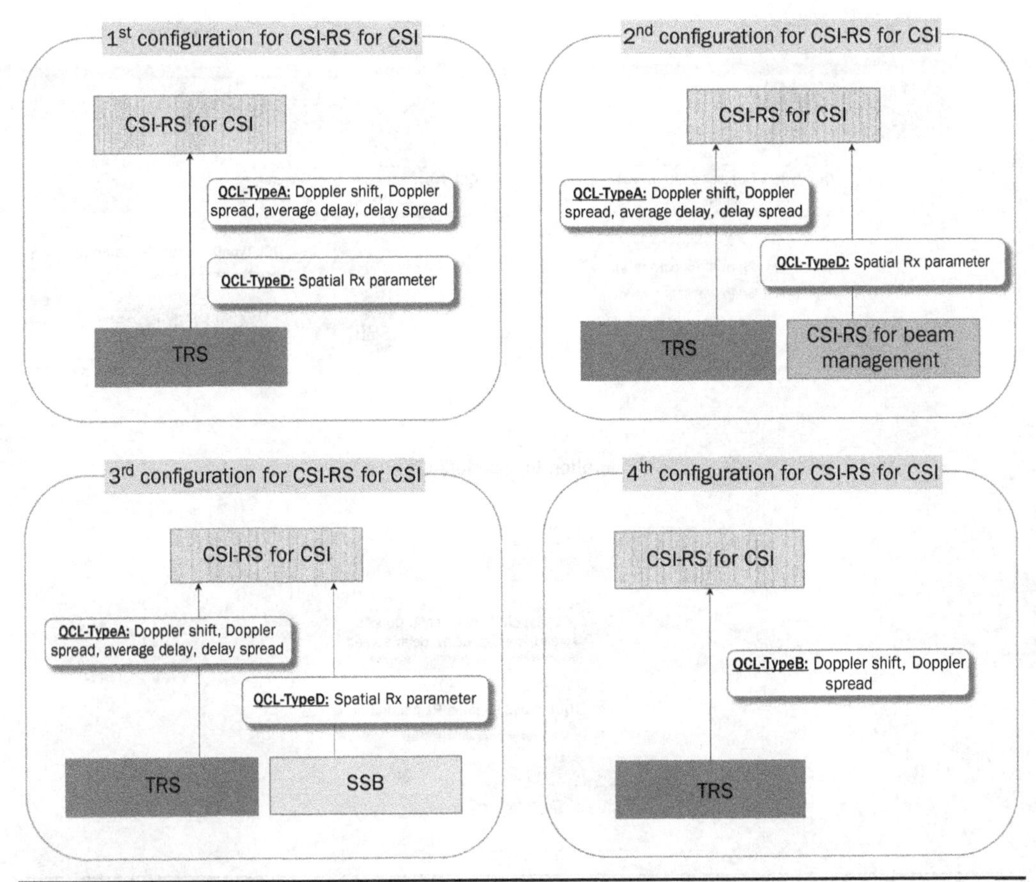

FIGURE 8.39 TCI state configuration options for CSI-RS for CSI.

the UE decodes the PDCCH it does not know which PDCCH carried the grant for the PDSCH, and therefore the UE does not know which analog PDCCH Rx beam should be applied to the PDSCH reception. The specification requires the UE to use the default QCL in this case as well. Note that the procedure mentioned above for DCI format 1_0 also applies to DCI format 1_1 whenever TCI bits are not configured.

In Figs. 8.38 through 8.42, we give a summary of all possible TCI state configuration cases defined.

The QCL source of some of the signals in Figs. 8.38 through 8.42 can also have a QCL source of its own. Assume for example that signal 1 is a QCL source for signal 2 and signal 2 is QCL source for signal 3. In these cases, the UE cannot always assume that signal 1 is a QCL source for signal 3. However, as it was mentioned earlier, there are cases where signal 2 is not an efficient QCL source for a given parameter. For example, CSI-RS is not well-suited for Doppler shift or Doppler spread estimation. So, if signal 2 is a CSI-RS, other than TRS, then Doppler shift or Doppler spread estimation can be "passed through" directly from signal 1 to signal 3, without observing it on signal 2.

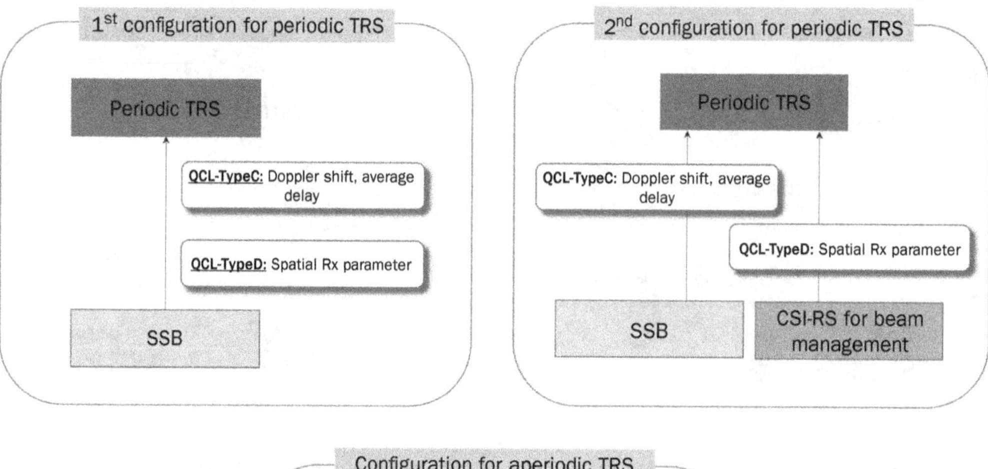

Figure 8.40 TCI state configuration options for TRS.

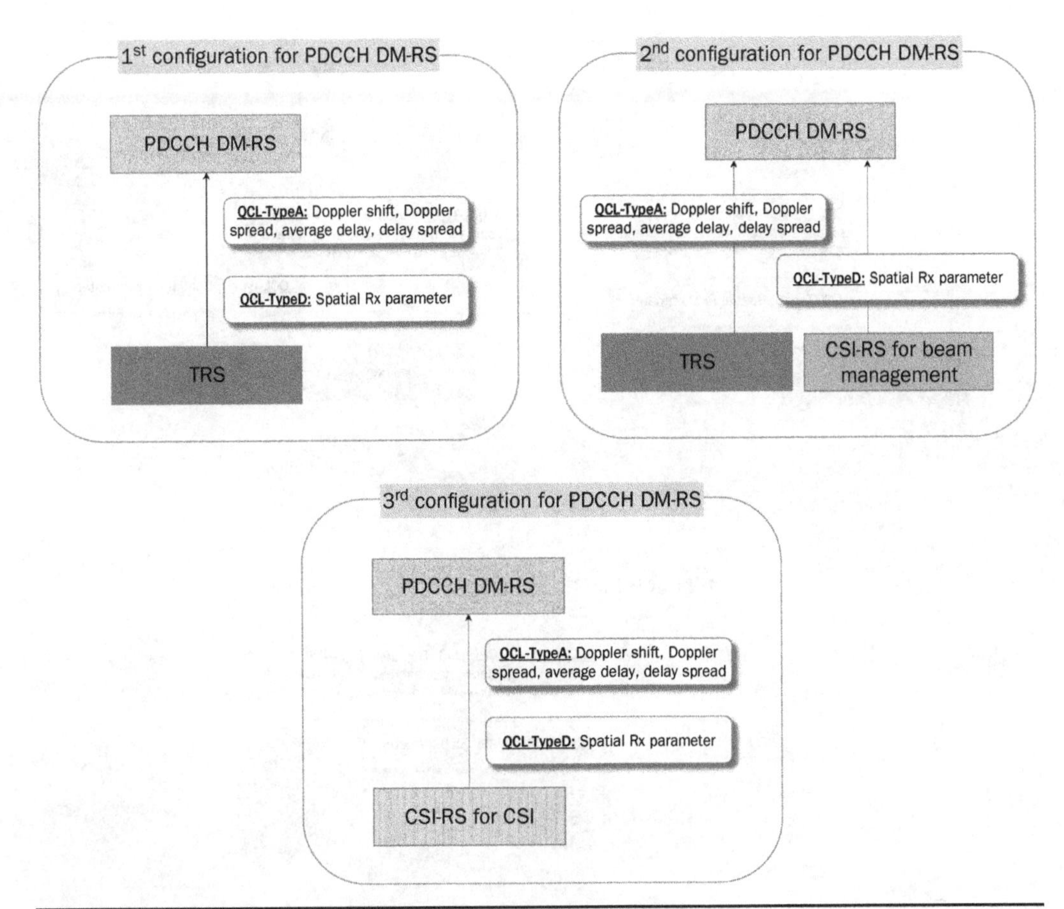

FIGURE 8.41 TCI state configuration options for PDCCH DM-RS.

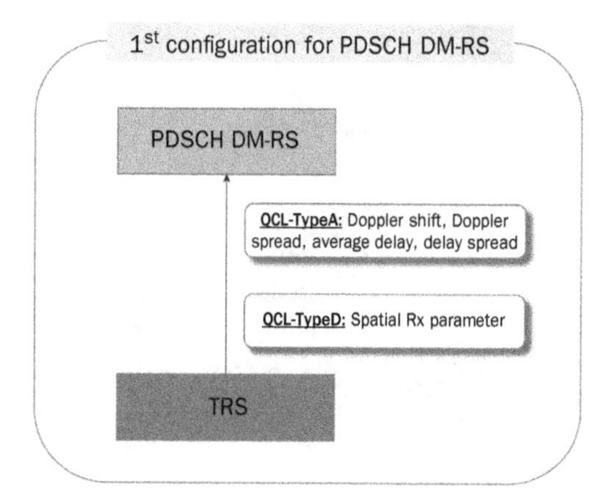

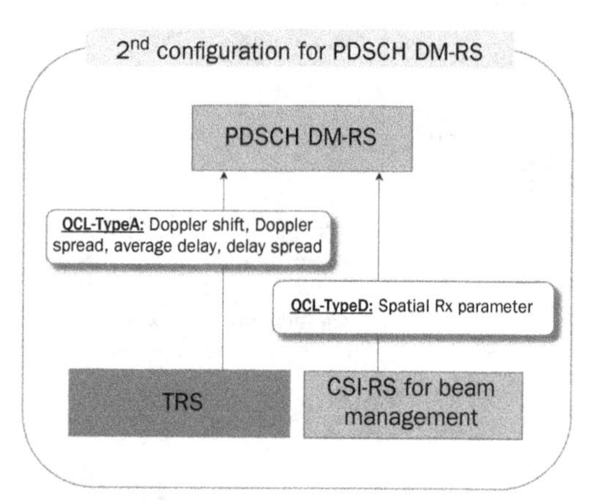

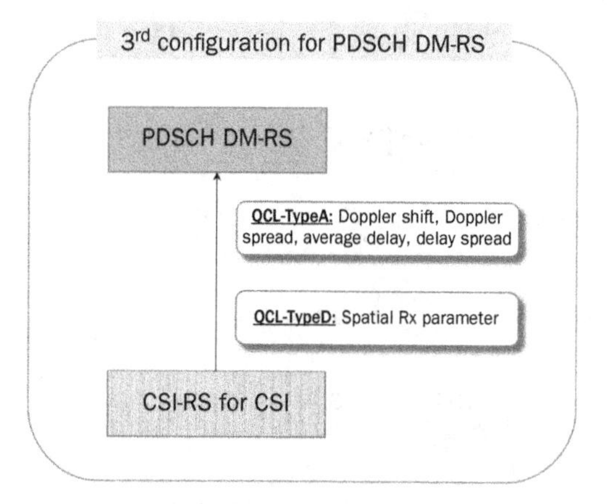

Figure 8.42 TCI state configuration options for PDSCH DM-RS.

References

[1] 3GPP, "3rd Generation Partnership Project; Technical Specification Group Radio Access Network; NR; physical layer procedures for data (Release 15)," 38.214, v15.8.0, December 2020.

[2] Tom Richardson and Shrinivas Kudekar, "Design of low-density parity check codes for 5G new radio," IEEE Communications Magazine, vol. 56, no. 3, pp. 28–34, March 2018.

[3] Ido Tal and Alexander Vardy, "List decoding of polar codes," IEEE Transactions on Information Theory, vol. 61, no. 5, pp. 2213–2226, May 2015.

[4] Alexios Balatsoukas-Stimming, Pascal Giard and Andreas Burg, "Comparison of polar decoders with existing low-density parity-check and turbo decoders," 2017 IEEE Wireless Communications and Networking Conference Workshops (WCNCW), 19–22 March 2017.

[5] 3GPP, "3rd Generation Partnership Project; Technical Specification Group Radio Access Network; NR; multiplexing and channel coding (Release 15)," 38.212, v15.8.0, December 2020.
[6] 3GPP, "3rd Generation Partnership Project; Technical Specification Group Radio Access Network; NR; Radio Resource Control (RRC) protocol specification (Release 15)," 38.331, v15.8.0, December 2020.
[7] 3GPP, "3rd Generation Partnership Project; Technical Specification Group Radio Access Network; NR; physical channels and modulation (Release 15)," 38.211, v15.8.0, December 2020.
[8] 3GPP, "3rd Generation Partnership Project; Technical Specification Group Radio Access Network; NR services provided by the physical layer (Release 15)," 38.202, v15.6.0, December 2020.
[9] 3GPP, "3rd Generation Partnership Project; Technical Specification Group Radio Access Network; NR; user equipment (UE) radio access capabilities (Release 15)," 38.306, v15.8.0, December 2020.
[10] 3GPP, "3rd Generation Partnership Project; Technical Specification Group Radio Access Network; "User Equipment (UE) radio transmission and reception; Part 1: Range 1 Standalone (Release 16)," 38.101-1, v16.2.0, December 2020.
[11] 3GPP, "3rd Generation Partnership Project; Technical Specification Group Radio Access Network; User Equipment (UE) radio transmission and reception; Part 2: Range 2 Standalone (Release 16)," 38.101-2, v16.2.0, December 2020.
[12] 3GPP, "3rd Generation Partnership Project; Technical Specification Group Radio Access Network; NR; physical layer procedures for control (Release 15)," 38.213, v15.8.0, December 2020.

CHAPTER 9

Uplink Control Operation

Generally, Uplink control is motivated by either Downlink transmissions, or Uplink transmissions, or a combination thereof. For the Downlink HARQ operation, HARQ has to be transmitted in the Uplink by the UE. A UE may have to provide a HARQ response for one or more Downlink transmissions, especially in a TDD system and/or when the UE is configured with multiple carriers. Channel state information (CSI) feedback by a UE is necessary to enable Downlink rate adaption for efficient operation (see Chap. 8). CSI may consist of different types, e.g., a rank indicator (RI), a precoding matrix indicator (PMI), a channel quality indicator (CQI), etc. A UE may have to provide CSI feedback for multiple CCs in one Uplink transmission as well. Due to the *primary-secondary* nature of a gNB and a UE in terms of system resource management, scheduling request (SR) is an indispensable component to enable efficient and low-latency Uplink scheduling when Uplink data arrives at the UE side. To enable Uplink data rate adaptation, Uplink sounding reference signal (SRS) is crucial. SRS can also be used for Downlink rate adaptation as well, especially when there is channel reciprocity.

Similar to Downlink control design, efficient Uplink control design is critical for wireless communications as well. Uplink control should provide sufficient information for Downlink and Uplink resource management. At the same time, the performance of Uplink control, including both link-level performance and system-level performance, should be satisfactory, especially for ultra-reliable communications.

In this chapter, we will first introduce the Uplink control design in 4G LTE, followed by the detailed Uplink control design for 5G NR.

9.1 Uplink Control in 4G LTE

In 4G LTE, the following Uplink control information (UCI) types are supported:

- HARQ,
- CSI, and
- SR.

UCI is intended to be transmitted using a physical Uplink control channel (PUCCH). To provide good frequency diversity for the transmission, intra-subframe frequency hopping is supported for PUCCH, as illustrated in Fig. 9.1. In particular, the intra-subframe frequency hopping is also known as *mirror-hopping*, as illustrated in Fig. 9.2. To be more specific, if a PUCCH transmission uses an RB $m \geq 0$ in the first slot, the RB for the PUCCH in the second slot is mirror-hopped to $N_{RB}^{UL} - m - 1$, where N_{RB}^{UL} is the

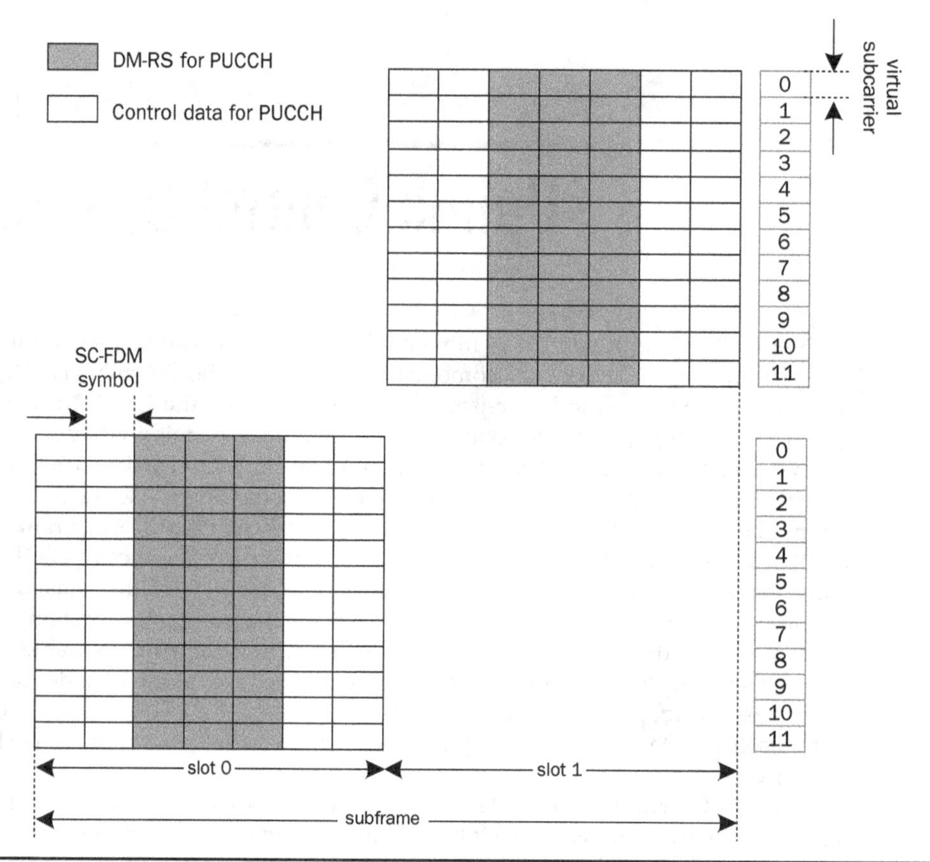

Figure 9.1 Illustration of the intra-subframe hopping of 4G LTE PUCCH (applicable to PUCCH formats 1/1a/1b).

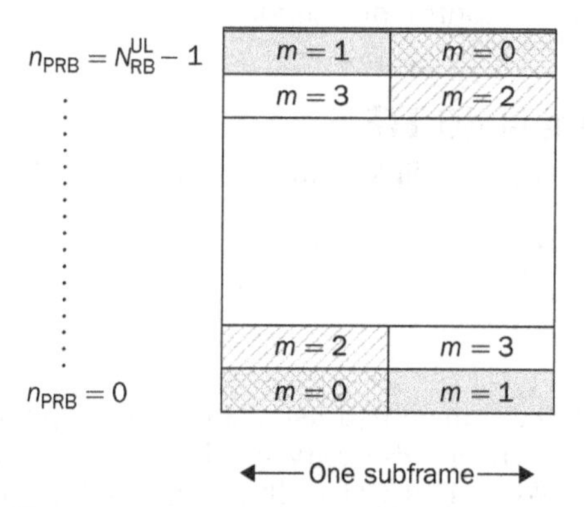

Figure 9.2 Illustration of the mirror-hopping for 4G LTE PUCCH.

PUCCH formats	Description
1, 1a, 1b	For 1-bit SR (ON_OFF keying), 1-bit HARQ (BPSK), and 2-bit HARQ (QPSK), respectively. One RB transmission, multiplexing up to 36 PUCCH transmissions in one RB.
2, 2a, 2b	For CSI feedback (up to 11 bits), possibly in combination with HARQ (1-bit or 2-bit). One RB transmission, multiplexing up to 12 PUCCH transmissions in one RB.
3	For multi-bit HARQ, SR, and CSI, up to 22 bits. One RB transmission, multiplexing up to 5 PUCCH transmissions in one RB.
4	For multi-bit HARQ, SR, and CSI, >22 bits. One or more RB transmission, multiplexing only one PUCCH per RB.
5	For multi-bit HARQ, SR, and CSI, >22 bits. One RB transmission, multiplexing up to 2 PUCCH transmissions in one RB.

TABLE 9.1 PUCCH Formats in 4G LTE

total Uplink bandwidth (in units of RBs). Although the 4G LTE specifications provide the necessary flexibility for an eNB to manage a PUCCH transmission by a UE in any RBs within the Uplink bandwidth, a typical PUCCH transmission occupies the two Uplink bandwidth edges. In this case, the mirror-hopping provides the maximum possible frequency diversity. Moreover, it minimizes the resulting resource fragmentation for Uplink data transmissions, especially considering the fact that 4G LTE has a single-carrier waveform in the Uplink (which requires a frequency-contiguous resource allocation for PUSCH).

Table 9.1 summarizes the PUCCH formats supported in 4G LTE. Note that there are also short PUCCH formats (1/1a/1b/3/4) introduced in Release 15 [1], for the purpose of low-latency communications using a short transmit time interval (sTTI), as discussed in Sec. 13.1.

HARQ payload size is a variable and can be more than 1 bit. In TDD, multiple DL transmissions in different DL subframes may map to a single UL subframe for HARQ feedback. A UE configured with CA or DC may have multiple PDSCH transmissions on different CCs requiring HARQ feedback in a same UL subframe. Instead of providing HARQ feedback *individually* on a per CC basis, it is more efficient to *jointly* encode and transmit HARQ feedback for the multiple PDSCH transmissions. This also reflects a case of configuring a UE with multiple DL CCs but with only a single UL CC, which is a very typical configuration or a capability for a UE considering the fact that in most cases, the traffic for a UE is DL heavy. In 4G LTE, in the first release of CA (Release 10), a UE is configured with only one CC (called the primary cell or Pcell) to possibly transmit PUCCH, while all other CCs are called secondary cells (or Scell). Note that different UEs may be configured with different CCs as the Pcell. In later releases, for additional flexibility (particularly the possibility to offload PUCCH from the Pcell to another Scell), a UE can be configured with a primary Scell (PScell) to additionally transmit PUCCH, forming two cell groups (the primary cell group or PCG, and the secondary cell group or SCG) for the UE. All other Scells are either in the PCG or the SCG (but not both). Figure 9.3 illustrates different CA cases with respect to PUCCH transmissions. In Case 1, there is a single cell group (PCG) with an asymmetric CA configuration (4 DL CCs vs. 1 UL CC), where PUCCH can only possibly be transmitted on the single UL CC (which

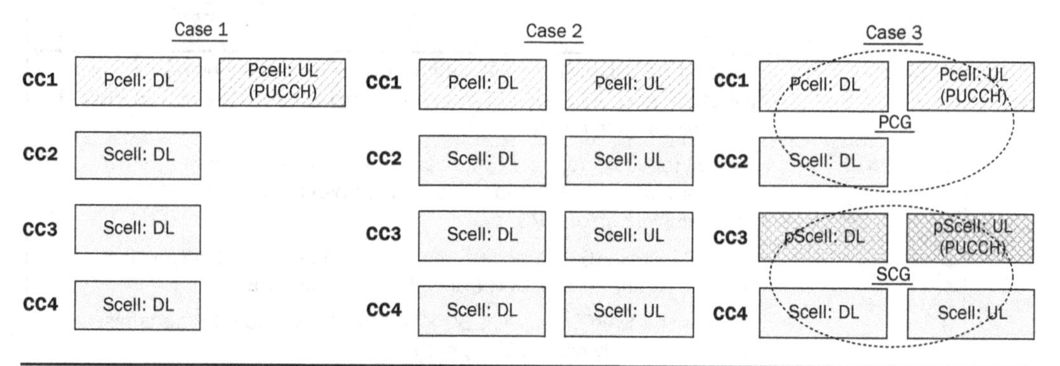

FIGURE 9.3 Illustration of different CA cases with respect to PUCCH transmissions. Case 1: asymmetric CA, a single cell group; Case 2: symmetric CA, a single cell group; Case 3: asymmetric CA, two cell groups.

is also the Pcell). In Case 2, it is also a single cell group but with a symmetric CA configuration, where CC1 is configured as the Pcell to carry PUCCH for the UE. In Case 3, a UE is configured with two cell groups (PCG and SCG), where PUCCH can be separately and simultaneously transmitted on the Pcell and on the PScell. In the example, CC2 belongs to the PCG while CC4 belongs to the SCG.

The PUCCH on the Pcell or PScell may also carry periodic CSI (P-CSI) reports for one or more DL CCs. Due to the limited PUCCH capacity, there are cases when some UCI has to be dropped in order to fit into a PUCCH transmission. In this case, HARQ and SR have a higher priority than P-CSI so that one or more CC's P-CSI reports may have to be dropped first.

Resource determination for a PUCCH transmission depends on the PUCCH format. For PUCCH formats 1a/1b, generally a UE determines a PUCCH resource based on the starting CCE of the corresponding PDCCH, along with other semi-statically configured parameters. This is also called *implicit* PUCCH resource allocation. On the other hand, for PUCCH formats 1/2/2a/2b, a UE is explicitly and semi-statically configured with a PUCCH resource, also known as *explicit* PUCCH resource allocation. For some cases (PUCCH formats 3/4/5 and PUCCH formats 1a/1b for DL SPS activation and release), a UE may be configured multiple PUCCH resources (e.g., up to 4) and is indicated by a DCI to dynamically choose one to use, possibly in combination with other semi-statically configured parameters. This provides a good tradeoff between DL control overhead and the resulting PUCCH resource efficiency.

In order to maintain a single carrier waveform, when there is a PUSCH transmission, UCI has to be multiplexed onto PUSCH. This is also known as *"UCI piggybacking on PUSCH"*. In order to maintain satisfactory UCI performance when piggybacking, a respective semi-static parameter is configured for each UCI type for the UE, e.g., denoted by, $\beta_{offset}^{HARQ-ACK}$, β_{offset}^{RI}, β_{offset}^{CQI} etc. These parameters are used, along with the parameters associated with PUSCH (e.g., MCS, the number of transmission layers, etc.), to determine the number of REs within the allocated PUSCH REs that should be used for the respective UCI transmission. In order to ensure the best performance possible for HARQ on PUSCH, it is placed right next to the DM-RS symbols for PUSCH, as illustrated in Fig. 7.4. Since RI in general has a higher performance requirement than CQI/PMI, it is separately mapped onto PUSCH symbols next to HARQ symbols. CQI/PMI is mapped together starting from the beginning of PUSCH in a time-first-frequency-second manner. Note that CQI/PMI/RI rate matches around data REs for PUSCH, while HARQ punctures data REs for PUSCH. The reason for such a distinction is because

- The presence of CQI/PMI/RI on PUSCH is either semi-statically configured (for periodic CSI feedback) or triggered as part of the same UL grant scheduling the PUSCH (for aperiodic CSI feedback). As a result, practically there is no misalignment between an eNB and a UE regarding whether or not CQI/PMI/RI is piggybacked on the PUSCH or not.
- However, the presence of HARQ on PUSCH depends on the scheduling of PDSCH. Especially when a PDSCH is dynamically scheduled by a DCI, there is a possibility that a UE may miss the DCI and thus the UE may not transmit the corresponding HARQ. The miss-detection of the DCI thus creates a misalignment between an eNB and a UE regarding whether HARQ is piggybacked on PUSCH or not. Puncturing (vs. rate matching) makes it easier for an eNB to perform UL data decoding separately from HARQ decoding, since the resource mapping function for UL data onto PUSCH is not dependent on HARQ.

If there is a need to avoid collision with SRS, the PUSCH REs also rate match around the last symbol of the subframe, as also illustrated in Fig. 9.4.

Table 9.2 provides a summary of the coding schemes for UCI on PUSCH.

Note also that 4G LTE supports a feature called simultaneous transmission of PUCCH and PUSCH, where UCI can still be transmitted on PUCCH along with a PUSCH transmission by a UE. However, although such a feature provides more flexibility in the UCI transmission and protection, it results in a non-single-carrier waveform where a substantial maximum power reduction (MPR) may be necessary due to implementation constraints and regulatory requirements. More discussion can be found in Sec. 9.5.

Power control for LTE PUCCH consists of two parts: open-loop power control and closed-loop power control. Open-loop power control is realized via path loss estimation,

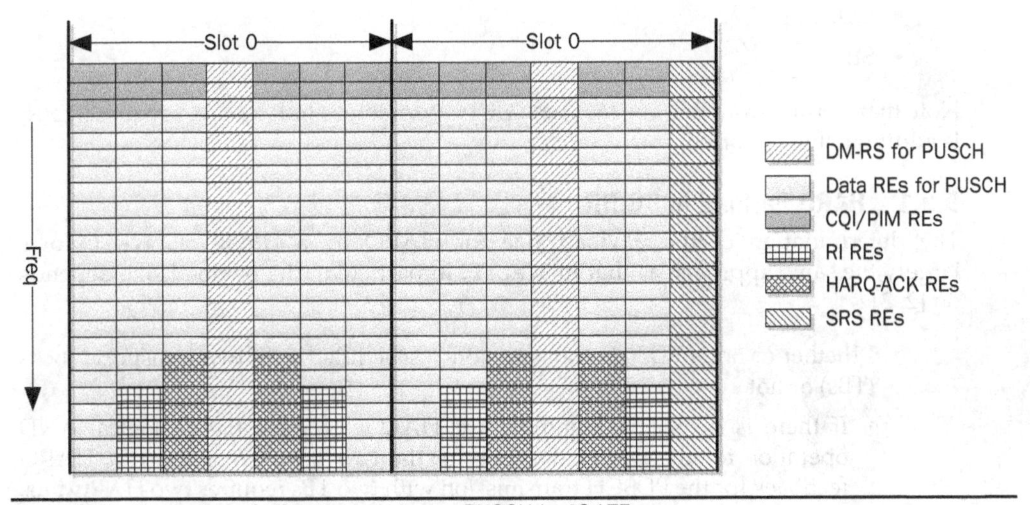

FIGURE 9.4 Illustration of UCI piggybacking on PUSCH in 4G LTE.

UCI type	Number of bits	Coding scheme
HARQ or RI	1	Repetition encoding
	2	Simplex encoding
	3–11	Reed-Muller (RM) encoding
	12–22	Dual RM encoding
	>22	8-bit CRC, TBCC encoding
CQI/PMI	Up to 11	RM encoding
	>11	8-bit CRC, TBCC encoding

Table 9.2 Summary of Coding Schemes for UCI Piggybacked on PUSCH in 4G LTE

which is to be combined with a set of parameters configured for a UE. Note also that there can be a semi-static power control adjustment for different PUCCH formats or different PUCCH payload sizes. Closed-loop power control is done via dynamic transmit power commands (TPC) issued by an eNB to a UE, where the TPC may be conveyed by an information field in a UE-specific DL scheduling DCI, or by an information field in a group-common DCI with the CRC scrambled by TPC-PUCCH-RNTI (see Chap. 7). The TPC can be a relative power control command so that the UE can accumulate the TPCs, or an absolute power control command so that the UE can apply it individually. The combination of open-loop power control and closed-loop power control drives the power setting for a PUCCH transmission so that a desirable performance target can be achieved.

9.2 UCI Types and Payload Sizes in 5G NR

There is a lot of similarity between 4G LTE UCI and 5G NR UCI. In particular, the same set of Uplink control information (UCI) types are also supported in 5G NR:

- HARQ,
- CSI, and
- SR.

Note that starting from Release 16, a new UCI type, namely, link recovery request (LRR) is additionally supported.

9.2.1 HARQ Payload in 5G NR

The determination of the payload size for HARQ by a UE in 5G NR follows largely the same approach as that in 4G LTE. In particular, the payload size depends on [2, 3]

- Whether or not a PDSCH transmission is scheduled with two transport blocks (TBs) or not.
    - If there is no spatial bundling for HARQ feedback (i.e., a logical AND operation for the two HARQ bits for the two TBs) across the TBs, HARQ feedback for the PDSCH transmission with two TBs requires two HARQ bits, one for each TB.

- A number of DL transmissions associated with a PUCCH transmission, in time, in frequency, or a combination thereof:
  - In the time domain,
    - For 4G LTE, in a TDD system, the availability of UL resources for a PUCCH transmission is limited to a subset of subframes on a carrier. Consequently, a PUCCH transmission may need to provide HARQ feedback for one or more DL transmissions in different time instances.
    - For 5G NR, the flexibility in managing the slot structure and the dynamic indication of HARQ timing on a carrier (see Chaps. 5 and 7) implies that the payload size for HARQ in a PUCCH transmission heavily depends on a gNB's resource management and scheduling decisions.
  - In the frequency domain,
    - A UE can be configured with more than one DL carrier as part of CA or DC operation, where one or more DL carrier(s) is associated with one UL carrier (i.e., the Pcell or PScell) for a PUCCH transmission within a PUCCH group. As discussed earlier, there are up to two PUCCH groups that can be configured for a UE (PCG and SCG). In 4G LTE, up to 32 DL CCs (each up to 20 MHz) can be configured in a PUCCH group, while in 5G NR, up to 16 DL CCs (each up to a much larger carrier bandwidth, see Sec. 5.4) can be configured in a PUCCH group.
- A configured HARQ feedback type (semi-static vs. dynamic)
  - For a semi-static HARQ codebook (also known as a type-1 HARQ codebook in 5G NR), the HARQ codebook is determined based only on higher-layer configured parameters for a UE.
  - For a dynamic HARQ-codebook (also known as a type-2 HARQ codebook in 5G NR), the HARQ codebook is determined based on the dynamic scheduling via one or more counters called Downlink assignment index or DAI (see Sec. 7.5), where a counter of a total number of PDSCH transmissions is used to determine the total HARQ payload size, and a counter of an accumulative number of PDSCH transmissions is used to determine how to construct the HARQ codebook. The accumulative counter is necessary for a UE to determine where to position a HARQ bit for a PDSCH transmission in the HARQ payload of a total size indicated by the total counter.

The need for two types of HARQ codebooks (semi-static vs. dynamic) in both 4G LTE and 5G NR is primarily driven by the following design considerations:

- Whether or not there is a tight coordination among CCs in a PUCCH group as part of a CA or a DC operation.
  - If the backhaul connection among the two or more CCs in a PUCCH group is not ideal, e.g., with a large latency or with a limited backhaul capacity, a semi-static HARQ codebook is more preferable. This helps minimize the dependency of scheduling decisions among the two or more CCs.
- Whether or not there is good predictability for an actual transmission of a scheduled transmission.

- It is possible that a scheduled PDSCH may not be transmitted. For example, let us consider cross-carrier scheduling for an unlicensed carrier, where a PDCCH transmitted in a licensed carrier schedules a PDSCH transmitted in an unlicensed carrier. Whether or not the PDSCH transmission may actually occur depends on the result of a clear channel assessment. If the assessment of a clear channel fails, the PDSCH cannot be transmitted.
- The uncertainty in the actual PDSCH transmissions makes it difficult to enable the dynamic HARQ codebook operation since the setting of the counter for the total number of PDSCH transmissions or the setting of the counter for the accumulative number of PDSCH transmissions in a DCI may have to wait until the knowledge of whether a correspondingly scheduled PDSCH transmission actually occurs or not.

- A general tradeoff between DL and UL overhead.
  - The usage of the semi-static HARQ codebook generally results in a larger HARQ codebook size, since the HARQ codebook size does not take into account the actual scheduling decisions. This implies that there is more UL overhead, which may not be desirable especially considering the fact that, generally speaking, UL link budget and UL capacity are more likely to be a bottleneck compared with those in DL from the system perspective. However, note that the dynamic HARQ codebook operation suffers from the additional DL overhead due to the constant presence of DAI bits in a DCI when the dynamic HARQ codebook is configured for a UE.

Note that in Release 16, as motivated by the support of 5G NR in unlicensed or shared spectrum, a new HARQ codebook type, namely, type 3, was introduced [3]. It serves as a one-shot HARQ feedback for all configured HARQ processes, which is necessary considering the fact that the HARQ feedback transmission by a UE in unlicensed/shared spectrum is opportunistic (subject to the clear channel assessment). The type 3 HARQ feedback thus provides a gNB an overall picture regarding how the past Downlink transmissions for each of the configured HARQ processes look like.

On the other hand, compared with 4G LTE, 5G NR has *new factors* impacting the HARQ codebook size.

One particular new factor is the introduction of a codebook group (CBG)–based HARQ feedback [4, Secs. 5.1.7 and 6.1.5]. The primary motivation for such a feature is for more efficient HARQ operation especially when different symbols or different frequency subbands for a PDSCH transmission may experience different interference conditions or different prioritization operations including possibly being pre-empted (see Secs. 13.5 and 13.6). More specifically, the interference variations across different symbols/subbands may come from different intra-cell MU-MIMO operations for the same link direction, different inter-cell operations for the same link direction, or different amount of cross-link interference from the same cell (e.g., due to a full-duplex operation in a TDD system) or a different cell. The different prioritization operations come from the need for 5G NR to support different types of UEs in a cell and the different service needs for a given UE. As one example, an eMBB PDSCH transmission scheduled for a UE may be pre-empted by another URLLC PDSCH transmission, where the pre-emption may impact only a subset of the symbols/subbands originally allocated for the eMBB PDSCH (see Sec. 13.5). Therefore, *the decoding performance for a PDSCH may be different in different symbols or subbands.*

For a PDSCH transmission, it is noted that the encoding is performed using one or more code blocks (CBs). It is known that low density parity check (LDPC) encoding for PDSCH may be based on one of the two base graphs (BGs) [5]. Each base graph has a respective maximum information block length K_{max}, where K_{max} = 8448 bits for BG1 and K_{max} = 3840 bits for BG2 (see more details in Sec. 8.1). Note that in 4G LTE, the maximum code block size is 6144 bits for turbo encoding. Each CB is attached with a CRC for error detection. The CBs are mapped in the assigned PDSCH resource in a frequency-first-time-second manner. In other words, when there are multiple CBs for a transport block, a CB k is mapped in the same or an earlier symbol than CB k+1. Such a frequency-first-time-second mapping makes it possible for a UE to decode PDSCH on a per CB basis in a streamline manner (i.e., CB k can be decoded first before CB k+1).

The number of CBs (denoted by C) for a given transport block (whose size is denoted by B bits) is thus determined by B and K_{max}, as shown below:

If $B \leq K_{max}$

$$C = 1,$$

otherwise

$$C = \lceil B / (K_{\text{max}} - L \rceil.$$

where L = 24 is the length of the CRC attached to each CB.

The maximum number of CBGs (denoted by N) on a carrier is configurable for a UE by the parameter *maxCodeBlockGroupsPerTransportBlock*, which can take a value of 2, 4, 6, or 8. The actual number of CBGs is determined by

$$M = \min(N,C).$$

The mapping of a set of CBs to a CBG is done in a sequential manner. In order to achieve uniform performance across CBGs, the number of CBs per CBG across CBGs in a TB is largely the same, differing at most by one CB. When the number of CBs in a TB is not divisible by M, the latter CBGs have one less CB compared with the earlier ones.

Let us use an example to illustrate the usage of CBG. Assuming that a TB has six CBs, and that the UE is configured with a maximum of four CBGs (thus $M = N = 4$). The association of the CBs with CBGs is as follows:

- CBG0: CB0 and CB1,
- CBG1: CB2 and CB3,
- CBG2: CB4, and
- CBG3: CB5.

Assuming a total of nine symbols for the transmission, the mapping of the CBs and the CBGs are illustrated by Fig. 9.5. Note that the CBGs are not necessarily aligned with symbol boundaries – in this example, while CBG0 and CBG1 are aligned with symbol boundaries, CBG2 and CBG3 are not. It is possible to design the mapping of CBGs such that it is always aligned with symbol boundaries, but it may cause uneven CB or CBG sizes and the resulting complexity can be quite high. Also note that the number of CBs per CBG is not necessarily equal – in this example, we have {2, 2, 1, 1} CBs for CBGs {0, 1, 2, 3}, respectively.

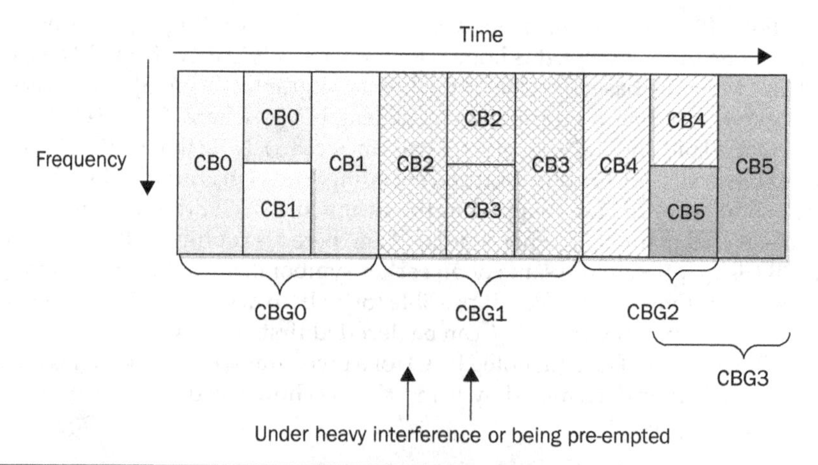

FIGURE 9.5 Illustration of the usage of CBG for 5G NR PDSCH.

In this example, assume that the 4th and the 5th symbols are under heavy interference or pre-empted (e.g., the symbols are being used by another URLLC transmission). As a result, while CBG0, CBG2, and CBG3 can be successfully decoded by the UE, CBG1 may be not correctly decoded. The usage of CBG makes it possible for the gNB to re-transmit a *part* of the TB, i.e., by re-transmitting CBG1 only, for successful decoding of the entire TB. Clearly, better Downlink link efficiency can be accomplished. This of course comes at the expense of the increased Downlink control overhead (for indicating the RBGs) and the increased HARQ feedback overhead in the Uplink. Instead of a single bit HARQ feedback for the TB, the UE in this example is required to provide a 4-bit (one bit for each CBG) HARQ feedback for the TB.

Another new factor is the flexibility in the HARQ feedback timing for a PDSCH transmission. In 4G LTE, the HARQ feedback timing is pre-determined. For FDD, the timing is fixed at 4 ms, i.e., a PDSCH transmission in subframe n is expected to have the corresponding HARQ feedback in subframe $n+4$. This thus creates a fixed one-to-one mapping between PDSCH and the corresponding HARQ. For TDD, due to the configuration of subframes into Downlink and Uplink subframes and hence the unavailability of UL in some subframes (see Sec. 5.1), the HARQ feedback timing can no longer be fixed at 4 ms. Instead, the HARQ timing is pre-determined depending on the Downlink/Uplink subframe configuration, the combinations of the carriers in CA or DC, and the conditions that

- The HARQ feedback delay is no less than 4 ms,
- Each PDSCH transmission in a subframe is expected to have a corresponding HARQ feedback in one Uplink subframe only, and
- The HARQ feedback payloads across different UL subframes are substantially equalized.

The last sub-bullet is necessary in order to ensure good performance for HARQ feedback, since the link-level performance particularly the Uplink coverage for HARQ feedback is primarily determined by the required maximum HARQ feedback payload size over all Uplink subframes.

In 5G LTE, the HARQ timing is no longer pre-determined. Instead, the timing relationships can be dynamically indicated via a DCI to a UE, including:

- The timing relationship between a DL grant and the corresponding PDSCH transmission, which is denoted as *K*0 in unit of slots; and
- The timing relationship between a PDSCH transmission and the corresponding HARQ feedback, which is denoted as *K*1 in units of slots.

The introduction of *K*0 provides the flexibility in gNB scheduling, while all UEs are required to support *K*0 = 0, so that the same-slot scheduling can be performed. Note also that the timing relationship between a UL grant and the corresponding PUSCH transmission can be dynamically indicated as well (denoted as *K*2 in units of slots).

The indication for *K*0 is via an information field in DCI called *Time domain resource assignment* (see Sec. 7.5). The parameter can be up to 4 bits, providing up to 16 different values, with a value range of 0 to 32 slots. Note that the parameter also provides other information related to PDSCH resource assignment, such as the mapping type of PDSCH (type A or type B, see Chap. 8), the starting symbol, and the number of symbols for PDSCH.

The indicator for *K*1 is via an information field in DCI called *PDSCH-to-HARQ_feedback timing indicator* (see Sec. 7.5). The parameter can be up to 3 bits, providing up to eight different values, with a value range of 0 to 15 slots.

Therefore, the HARQ feedback payload size for a UE is further dependent on factors such as the configured maximum value of *K*1, the dynamically managed slot structure (see Sec. 5.7) and the dynamics in PDSCH scheduling for a UE.

9.2.2 CSI Payload in 5G NR

The determination of the CSI payload size in 5G NR is similar to that of 4G LTE. In general, the CSI payload depends on

- CSI type, e.g., RI, CQI, PMI, CRI (CSI-RS resource indicator), LI ("strongest" layer indicator), layer 1 RSRP, etc.,
- A number of carriers for which CSI reporting is necessary, and
- A set of semi-static configurations, e.g., a number of CSI-RS ports, a CSI codebook type (type I or type II, single-panel or multi-panel, etc.), a CSI reporting subband size, etc.

The details on CSI feedback can be found in Chap. 8. The CSI payload size can range from a single bit to hundreds of bits or even higher.

9.2.3 SR Payload in 5G NR

In 4G LTE, a UE is configured with a *single* SR configuration, where a single bit SR is supported as part of the UCI. A positive SR indicates that there is a non-empty UL buffer, while a negative SR indicates otherwise.

In 5G NR, a UE can be configured with $K \geq 1$ SR configurations, where K can be up to 8. These different SR configurations make it possible for a UE to indicate a non-empty UL buffer associated with a particular QoS labeling. In order to fully support all *K* SR configurations, a full-*K* bitmap would be required, which is deemed to be very expensive in the UL operation. In addition, it may not be necessary to report the SR need for

UL data with a lower priority. This is because, after receiving a SR, a gNB typically responds with a UL schedule after which the UE can utilize the scheduled PUSCH to provide additional and more detailed UL buffer status information. Therefore, some compromised handling for SR over PUCCH is necessary.

To that end, 5G NR supports that a SR transmission should be able to explicitly indicate *one and which one* of the K configurations for which the request is associated with. The one SR configuration to be indicated can be a SR configuration of a highest priority, with details up to UE implementation. Note that in Release 16, a one-bit priority indication *schedulingRequestPriority* (low vs. high priority) [3] can be configured for a SR configuration for a UE to help SR prioritization management. To be more specific, given a set of K SR configurations, $\lceil \log_2(K+1) \rceil$ bits are necessary to indicate

- All K SR configurations have a negative SR, or
- The lowest *schedulingRequestResourceId* with a positive SR is 1, or
- The lowest *schedulingRequestResourceId* with a positive SR is 2, or
- ...
- The lowest *schedulingRequestResourceId* with a positive SR is K.

When $K = 8$, a total of 4 bits becomes necessary.

9.2.4 Link Recovery Request

In Release 16, to further improve beam failure recovery (BFR) operation for Scells, a new UCI type was introduced, namely, link recover request (LRR). This is illustrated in Fig. 9.6, where the steps for an improved BFR operation may include:

- Step 1: A UE detects that the conditions of failed DL control beams are met for a Scell on FR2 [3].
- Step 2: The UE sends LRR via PUCCH on a cell with PUCCH configured for LRR, e.g., Pcell on FR1.
- Step 3: The gNB schedules a UL grant, e.g., on Pcell, for the UE to report the failed Scell.
- Step 4: The UE reports the failed Scell, potentially along with an identified new candidate beam.
- Step 5: The gNB transmits a BFR response, acknowledging the reception of the UE report.

The management of the LRR transmission on PUCCH is similar to SR. In particular, the configuration is treated as if it were one of the multiple SR configurations, with its own unique resource ID *schedulingRequestIDForBFR* [3]. Note that a UE can be configured with up to one BFR on PUCCH for any BWP within a PUCCH group. Given that a UE has only one active BWP at any time instance, there is up to one LRR transmission per PUCCH group. When the LRR transmission collides with other PUCCH without a positive SR, the LRR is handled as if it were a SR transmission [3, Sec. 9.2]. When the LRR transmission collides with a positive SR, its transmission priority is handled in the same way as the positive SR.

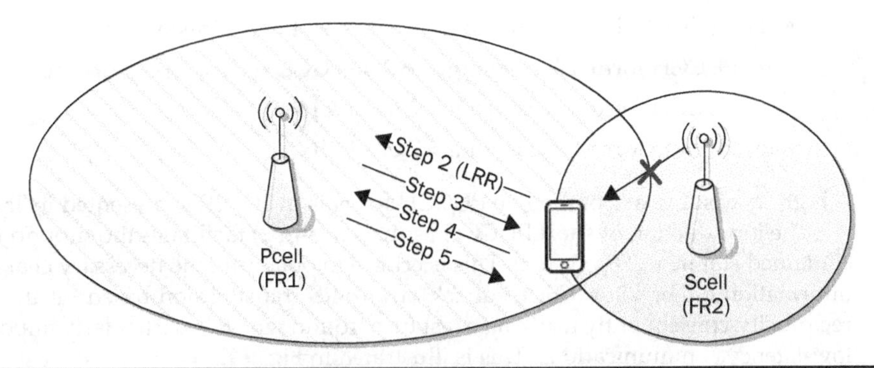

FIGURE 9.6 Illustration of LRR for beam failure recovery for a Scell.

9.3 PUCCH Formats in 5G NR

The design of PUCCH formats needs to take the following factors into account:

- *Variable UCI payload sizes.* As discussed in Sec. 9.2, some of the variations depend on semi-static configurations, while other variations may depend on dynamic scheduling, resource availability, etc. The PUCCH capacity thus needs to accommodate various UCI payload sizes.
- *Flexible slot structures.* The structure of a slot depends on both semi-static configurations and dynamic indication (see Sec. 5.7). The PUCCH transmission is thus necessary to be able to adapt to different slot structures.
- *Multiplexing capability of multiple PUCCH transmissions.* While it is necessary to ensure that a PUCCH transmission from a UE is optimized under a set of conditions (e.g., a given payload size, a particular slot structure, a certain channel condition, etc.), it is also important to provide efficient PUCCH transmissions from the system perspective. In particular, one measure of the efficiency is the so-called multiplexing capability of PUCCH in an RB. A larger multiplexing capability implies that a gNB can accommodate more PUCCH transmissions from different UEs using the same RB, thus leading to less UL overhead directly consumed by PUCCH and less UL resource fragmentation that can be potentially caused by PUCCH.
- *Coverage need.* It is critical to make sure that a PUCCH transmission by a UE can be reliably received by a gNB even when the UE is under a reasonably undesirable channel condition. The payload size on the PUCCH may be 1 bit or more.

In the following, we will discuss the details of 5G NR PUCCH formats.

9.3.1 General

5G NR supports five different PUCCH formats:

- Short PUCCH formats (1 to 2 symbols, a single slot):
  - PUCCH format 0, carrying 1 to 2 bits UCI
  - PUCCH format 2, carrying > 2 bits UCI

- Long PUCCH (4 to 14 symbols, a single slot or multi-slot):
  - PUCCH format 1, carrying 1 to 2 bits UCI
  - PUCCH format 3, carrying > 2 bits UCI
  - PUCCH format 4, carrying > 2 bits UCI

A high-level summary of the five PUCCH formats in 5G NR is presented in Table 9.3.

The introduction of short PUCCH formats (0 and 2) facilitates the support of a self-contained slot in a TDD system. This is critical in obtaining the necessary channel state information either via a CSI feedback or an SRS transmission based on the channel reciprocity, consequently realizing a fast turnaround which is particularly important for low-latency communications. This is illustrated in Fig. 9.7.

On the other hand, support of long PUCCH formats is necessary to carry a wider range of UCI payload sizes and to guarantee sufficient link coverage. This is realized by a fully configurable time duration within a slot (anywhere between 4 and 14 symbols) and across slots, and by the flexibility in configurability in the frequency domain (up to 16 RBs). PUCCH formats 3 and 4 provide increased coverage by utilizing the DFT-s-OFDM–based waveform, along with potentially repeated transmissions over multiple slots. Note that for PUCCH format 3, it is prohibited to use 7, 11, 13, or 14 RBs for easier implementation, following the same RB restriction for DFT-s-OFDM–based PUSCH transmissions as in LTE. In other words, for implementation simplicity, the number of RBs for a UL transmission has to be integer multiples of 2, 3, or 5 RBs.

Parameter	PUCCH format 0	PUCCH format 1	PUCCH format 2	PUCCH format 3	PUCCH format 4
# UCI bits	1 or 2	1 or 2	>2	>2	>2
Waveform	CGS	CGS	OFDM	DFT-s-OFDM	DFT-s-OFDM
# symbols	1 or 2	4 to 14	1 or 2	4 to 14	4 to 14
Starting symbol	0 to 13	0 to 10	0 to 13	0 to 10	0 to 10
# slots	1	1, 2, 4, 8	1	1, 2, 4, 8	1, 2, 4, 8
# RBs	1	1	1 to 16	1-6, 8-10, 12, 15, 16	1
RB location	Any	Any	Any	Any	Any
Frequency hopping	Yes (for 2-symbol only)	Yes	Yes (for 2-symbol only)	Yes	Yes

Table 9.3 A High-Level Summary of the five PUCCH Formats in 5G NR

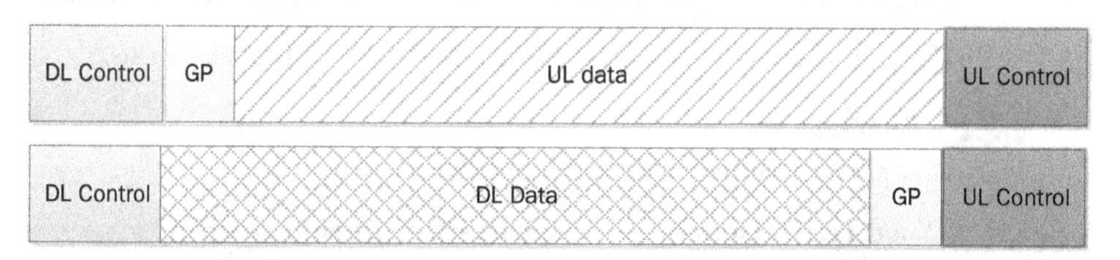

Figure 9.7 Illustration of short PUCCH formats for self-contained TDD slots: a UL-heavy slot (top) and a DL-heavy slot (bottom).

Frequency hopping for PUCCH is supported and can be enabled or disabled by a UE-specific signaling. If enabled, the resource for a PUCCH transmission hops only one time, and can start from a different RB in the second hop. As discussed in Sec. 9.1, in 4G LTE, frequency hopping for PUCCH is pre-determined, following a manner called *mirror-hopping*. The mirror-hopping makes it desirable to locate PUCCH on the two bandwidth edges of a cell's channel, in order to maximize the frequency diversity gain and to minimize the UL resource fragmentation. For 5G NR, such a pre-determined mirror hopping scheme is no longer desirable. This is primarily due to the fact that a UE's channel bandwidth is no longer necessarily the same as that of a gNB's (see Sec. 5.5). Indeed, a UE may be configured with one or more BWPs, where some BWPs may be narrower than a gNB's channel bandwidth and can be in any location of the gNB's channel bandwidth. As a result, a pre-determined hopping scheme may no longer provide the *most preferable* solution for all possible BWP configurations. Instead, the hopping for 5G NR PUCCH is done in a *fully flexible* manner – once the frequency hopping is enabled for a PUCCH transmission, the starting RB for the PUCCH in the second hop is separately configured by a parameter called *secondHopPRB* [3].

For PUCCH formats 0 and 2, frequency hopping is only possible when the PUCCH transmission has two symbols. In this case, each hop of the PUCCH transmission has one symbol. For PUCCH formats 1, 3, and 4 occupying N symbols in a slot, the symbols for the first hop are the first $\lfloor N/2 \rfloor$ symbols, while the symbols for the second hop are the remaining $\lceil N/2 \rceil$ symbols. Such fixed splitting of the set of PUCCH symbols under frequency hopping is for simplicity.

9.3.2 Details for PUCCH Format 0

PUCCH format 0 can be used to carry up to a 2-bit HARQ and a 1-bit SR, via a sequence selection scheme. In this case, the SR transmission does not distinguish the association of the SR with a SR configuration if two or more SR configurations are configured. The sequence selection scheme results in a low peak-to-average power ratio (PAPR) for advanced link coverage. The Length-12 base sequence, to be mapped to a single RB of 12 tones, is generated based on a set of 30 computer-generated sequences (CGS), which is represented by $e^{j\phi(n)\pi/4}$, $0 \leq n \leq 11$, where $\varphi(n)$ is shown in Table 9.4 [6, Table 5.2.2.2-2]. Note that these sequences are also used for other purposes, e.g., DM-RS for PUCCH formats 1, 3, and 4, DM-RS for DFT-s-OFDM based PUSCH, etc.

For PUCCH format 0, each sequence can have 12 different cyclic shifts, implying a multiplexing capability of up to 12 different selections or transmissions in an RB. For a given PUCCH by a UE, the selection of a cyclic shift for a base sequence provides the capability of conveying different information, where the selected cyclic shift depends on a UE-specifically configured initial cyclic shift, denoted by $C_{initial}$, and the combinations of the UCI. The primary design consideration in the selection is for good detection performance by the gNB, by maximizing the minimal gap between different cyclic shifts in order to differentiate various UCI combinations. This is summarized in Table 9.5. As can be seen, when there is a single bit HARQ, the gap is 6; when there are two bits, the minimal gap is 3, while the maximum possible gap of 6 is used for the two completely opposite values [e.g., (NAK, NAK) vs. (ACK, ACK)].

In order to randomize intra-cell and inter-cell interference for improved performance, the base sequence and its cyclic shift have a pre-determined hopping pattern, which is a function of a cell ID, a slot index, a symbol index, etc. Such a function is inherited from that of 4G LTE PUCCH.

Sequence index	$\varphi(0),\ldots,\varphi(11)$											
0	−3	1	−3	−3	−3	3	−3	−1	1	1	1	−3
1	−3	3	1	−3	1	3	−1	−1	1	3	3	3
2	−3	3	3	1	−3	3	−1	1	3	−3	3	−3
3	−3	−3	−1	3	3	3	−3	3	−3	1	−1	−3
4	−3	−1	−1	1	3	1	1	−1	1	−1	−3	1
5	−3	−3	3	1	−3	−3	−3	−1	3	−1	1	3
6	1	−1	3	−1	−1	−1	−3	−1	1	1	1	−3
7	−1	−3	3	−1	−3	−3	−3	−1	1	−1	1	−3
8	−3	−1	3	1	−3	−1	−3	3	1	3	3	1
9	−3	−1	−1	−3	−3	−1	−3	3	1	3	−1	−3
10	−3	3	−3	3	3	−3	−1	−1	3	3	1	−3
11	−3	−1	−3	−1	−1	−3	3	3	−1	−1	1	−3
12	−3	−1	3	−3	−3	−1	−3	1	−1	−3	3	3
13	−3	1	−1	−1	3	3	−3	−1	−1	−3	−1	−3
14	1	3	−3	1	3	3	3	1	−1	1	−1	3
15	−3	1	3	−1	−1	−3	−3	−1	−1	3	1	−3
16	−1	−1	−1	−1	1	−3	−1	3	3	−1	−3	1
17	−1	1	1	−1	1	3	3	−1	−1	−3	1	−3
18	−3	1	3	3	−1	−1	−3	3	3	−3	3	−3
19	−3	−3	3	−3	−1	3	3	3	−1	−3	1	−3
20	3	1	3	1	3	−3	−1	1	3	1	−1	−3
21	−3	3	1	3	−3	1	1	1	1	3	−3	3
22	−3	3	3	3	−1	−3	−3	−1	−3	1	3	−3
23	3	−1	−3	3	−3	−1	3	3	3	−3	−1	−3
24	−3	−1	1	−3	1	3	3	3	−1	−3	3	3
25	−3	3	1	−1	3	3	−3	1	−1	1	−1	1
26	−1	1	3	−3	1	−1	1	−1	−1	−3	1	−1
27	−3	−3	3	3	3	−3	−1	1	−3	3	1	−3
28	1	−1	3	1	1	−1	−1	−1	1	3	−3	1
29	−3	3	−3	3	−3	−3	3	−1	−1	1	3	−3

Table 9.4 The Length-12 CGS in 5G NR

9.3.3 Details for PUCCH Format 1

PUCCH format 1 can be used to carry up to a 2-bit HARQ and a 1-bit SR, via a multiplication of a BPSK (1-bit) or a QPSK (2-bit) modulation symbol and a length-12 sequence within each OFDM symbol. The DM-RS symbols and the UCI data symbols are arranged in an alternating manner, starting from DM-RS, within the assigned PUCCH symbols (≥4). This is illustrated in Fig. 9.8. Such an alternating pattern implies roughly a 50%

UCI combination	# of Bits	Value	Cyclic shift
SR only	1-bit	Positive	$C_{initial}$
		Negative	No transmission (ON-OFF Keying)
HARQ only	1-bit	NAK	$C_{initial}$
		ACK	$(C_{initial} + 6) \bmod 12$
	2-bit	[NAK, NAK]	$C_{initial}$
		[NAK, ACK]	$(C_{initial} + 3) \bmod 12$
		[ACK, ACK]	$(C_{initial} + 6) \bmod 12$
		[ACK, NAK]	$(C_{initial} + 9) \bmod 12$
HARQ + Positive SR	1-bit	NAK	$(C_{initial} + 3) \bmod 12$
		ACK	$(C_{initial} + 9) \bmod 12$
	2-bit	[NAK, NAK]	$(C_{initial} + 1) \bmod 12$
		[NAK, ACK]	$(C_{initial} + 4) \bmod 12$
		[ACK, ACK]	$(C_{initial} + 7) \bmod 12$
		[ACK, NAK]	$(C_{initial} + 10) \bmod 12$

TABLE 9.5 Sequence Selection for PUCCH Format 0 in 5G NR

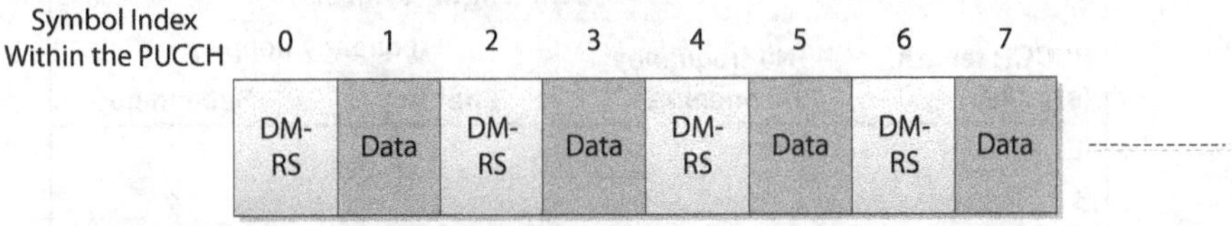

Symbol Index Within the PUCCH 0 1 2 3 4 5 6 7

DM-RS Data DM-RS Data DM-RS Data DM-RS Data

FIGURE 9.8 DM-RS and UCI data arrangement for PUCCH format 1.

DM-RS overhead for PUCCH format 1, which is close to an optimal DM-RS overhead ratio when the UCI payload is 1 or 2 bits in the intended unfavorable channel conditions. Within each OFDM symbol for PUCCH format 1, the length-12 sequence is based on the same sequence as that of PUCCH format 0.

A time-domain orthogonal code cover (OCC) is applied within the set of symbols for DM-RS (or UCI data) for 5G NR PUCCH format 1. This is similar to 4G LTE PUCCH formats 1/1a/1b. The usage of the OCC can provide an additional multiplexing capability for PUCCH format 1 such that within a same RB, more PUCCH transmissions can be accommodated. For simplicity, the length of the OCC is implicitly derived and is the same as the number of DM-RS (or UCI data) symbols. Since the length of PUCCH format 1 ranges from 4 to 14 symbols, the length of DM-RS (or UCI data) symbols is thus in between 2 and 7. Depending on whether or not the UE is configured with frequency hopping for PUCCH format 1, we have

- If no frequency hopping, the OCC lengths range from 2 to 7, and
- If frequency hopping is enabled, the OCC lengths range from 1 to 4.

The OCC length for DM-RS and UCI data for PUCCH format 1 are summarized in Tables 9.6 and 9.7, respectively.

PUCCH length, (symbols)	OCC length (symbols)		
	No frequency hopping	Frequency hopping	
		First hop	Second hop
4	2	1	1
5	2	1	1
6	3	1	2
7	3	1	2
8	4	2	2
9	4	2	2
10	5	2	3
11	5	2	3
12	6	3	3
13	6	3	3
14	7	3	4

Table 9.6 The OCC Length of DM-RS for PUCCH Format 1

PUCCH length, (symbols)	OCC length (symbols)		
	No frequency hopping	Frequency hopping	
		First hop	Second hop
4	2	1	1
5	2	1	1
6	3	1	2
7	3	1	2
8	4	2	2
9	4	2	2
10	5	2	3
11	5	2	3
12	6	3	3
13	6	3	3
14	7	3	4

Table 9.7 The OCC Length of UCI Data for PUCCH Format 1

For PUCCH format 1, the sequences for OCC lengths of 2 to 5 reuse the same sequences from 4G LTE (DFT-based), i.e.,

- Length 2:
  - [1, 1]
  - [1, −1]
- Length 3:
  - [1, 1, 1]

- $\left[1, e^{\frac{j2\pi}{3}}, e^{\frac{j4\pi}{3}}\right]$
- $\left[1, e^{\frac{j4\pi}{3}}, e^{\frac{j2\pi}{3}}\right]$

- Length 4:
  - [1, 1, 1, 1]
  - [1, −1, 1, −1]
  - [1, 1, −1, −1]
  - [1, −1, −1, 1]
- Length 5:
  - [1, 1, 1, 1, 1]
  - $\left[1, e^{\frac{j2\pi}{5}}, e^{\frac{j4\pi}{5}}, e^{\frac{j6\pi}{5}}, e^{\frac{j8\pi}{5}}\right]$
  - $\left[1, e^{\frac{j4\pi}{5}}, e^{\frac{j8\pi}{5}}, e^{\frac{j2\pi}{5}}, e^{\frac{j6\pi}{5}}\right]$
  - $\left[1, e^{\frac{j6\pi}{5}}, e^{\frac{j2\pi}{5}}, e^{\frac{j8\pi}{5}}, e^{\frac{j4\pi}{5}}\right]$
  - $\left[1, e^{\frac{j8\pi}{5}}, e^{\frac{j6\pi}{5}}, e^{\frac{j4\pi}{5}}, e^{\frac{j2\pi}{5}}\right]$

For lengths 6 and 7, the OCC sequences are also DFT-based, namely:

- Length 6:
  - [1, 1, 1, 1, 1, 1]
  - $\left[1, e^{\frac{j\pi}{3}}, e^{\frac{j2\pi}{3}}, -1, e^{\frac{j4\pi}{3}}, e^{\frac{j5\pi}{3}}\right]$
  - $\left[1, e^{\frac{j2\pi}{3}}, e^{\frac{j4\pi}{3}}, 1, e^{\frac{j2\pi}{3}}, e^{\frac{j4\pi}{3}}\right]$
  - [1, −1,1, −1,1, −1]
  - $\left[1, e^{\frac{j4\pi}{3}}, e^{\frac{j2\pi}{3}}, 1, e^{\frac{j4\pi}{3}}, e^{\frac{j2\pi}{3}}\right]$
  - $\left[1, e^{\frac{j5\pi}{3}}, e^{\frac{j4\pi}{3}}, -1, e^{\frac{j2\pi}{3}}, e^{\frac{j\pi}{3}}\right]$
- Length 7
  - [1, 1, 1, 1, 1, 1, 1]
  - $\left[1, e^{\frac{j2\pi}{7}}, e^{\frac{j4\pi}{7}}, e^{\frac{j6\pi}{7}}, e^{\frac{j8\pi}{7}}, e^{\frac{j10\pi}{7}}, e^{\frac{j12\pi}{7}}\right]$
  - $\left[1, e^{\frac{j4\pi}{7}}, e^{\frac{j8\pi}{7}}, e^{\frac{j12\pi}{7}}, e^{\frac{j2\pi}{7}}, e^{\frac{j6\pi}{7}}, e^{\frac{j10\pi}{7}}\right]$
  - $\left[1, e^{\frac{j6\pi}{7}}, e^{\frac{j12\pi}{7}}, e^{\frac{j4\pi}{7}}, e^{\frac{j10\pi}{7}}, e^{\frac{j2\pi}{7}}, e^{\frac{j8\pi}{7}}\right]$

- $\left[1, e^{\frac{j8\pi}{7}}, e^{\frac{j2\pi}{7}}, e^{\frac{j10\pi}{7}}, e^{\frac{j4\pi}{7}}, e^{\frac{j12\pi}{7}}, e^{\frac{j6\pi}{7}}\right]$
- $\left[1, e^{\frac{j10\pi}{7}}, e^{\frac{j6\pi}{7}}, e^{\frac{j2\pi}{7}}, e^{\frac{j12\pi}{7}}, e^{\frac{j8\pi}{7}}, e^{\frac{j4\pi}{7}}\right]$
- $\left[1, e^{\frac{j12\pi}{7}}, e^{\frac{j10\pi}{7}}, e^{\frac{j8\pi}{7}}, e^{\frac{j6\pi}{7}}, e^{\frac{j4\pi}{7}}, e^{\frac{j2\pi}{7}}\right]$

When the UCI contains SR only, ON-OFF keying is used for PUCCH format 1. When there is a positive SR and HARQ, HARQ is transmitted using the resource originally allocated to SR. This is the same as the operation in 4G LTE for PUCCH format 1/1a/1b.

9.3.4 Details for PUCCH Format 2

PUCCH format 2 can be used to carry more than 2-bit UCI (using the QPSK modulation), including HARQ, SR, and CSI, where the UCI bits are jointly coded and mapped into the allocated PUCCH resource over 1 or 2 symbols. As discussed in Sec. 9.2, the number of bits for SR is given by $\lceil \log_2(1+K) \rceil$, where $K \geq 1$ is the number of SR configurations.

PUCCH format 2 follows a DM-RS–based coherent transmission scheme, where DM-RS and UCI data are FDM based on a pre-defined pattern. Different from PUCCH format 1, which only supports 1 or 2 bits and thus is optimized for a coverage limited operation region, PUCCH format 2 can carry a large amount of UCI data and hence is intended to operate in relatively more favorable channel conditions. As a result, the channel estimation is less challenging, and a smaller DM-RS overhead ratio is thus more desirable. Indeed, PUCCH format 2 has a fixed DM-RS overhead ratio of 1/3, as shown Fig. 9.9.

9.3.5 Details for PUCCH Format 3 and Format 4

Both PUCCH format 3 and PUCCH format 4 support >2 (bits) UCI (using the QPSK or the $\pi/2$-BPSK modulation, configurable for a UE), including HARQ, SR, and CSI, where the UCI bits are jointly coded and mapped into the allocated PUCCH resource over four or more symbols in a slot. Again, the number of bits for SR is given by $\lceil \log_2(1+K) \rceil$, where $K \geq 1$ is the number of SR configurations.

Similar to PUCCH format 1, DM-RS and UCI data for PUCCH format 3 or 4 are partitioned in a TDM manner in units of symbols. However, since the intended operating region for PUCCH formats 3 or 4 can vary, in order to optimize the PUCCH performance, the DM-RS overhead is a function of several factors such as the PUCCH length (in symbols), whether or not frequency hopping is enabled, and an additional configuration parameter *additionalDMRS* indicating whether there is additional DM-RS or not (for better channel estimation at the expense of additional DM-RS overhead). The DM-RS symbols within the set of symbols allocated for PUCCH format 3 and PUCCH format 4 are summarized in Table 9.8.

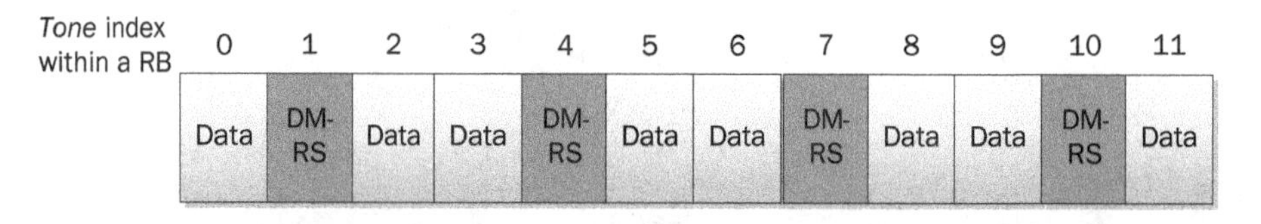

FIGURE 9.9 DM-RS and UCI data arrangement for PUCCH format 2 within an RB.

PUCCH length (symbols)	DM-RS symbol position within a PUCCH span			
	No additional DM-RS		Additional DM-RS	
	No hopping	Hopping	No hopping	Hopping
4	1	0, 2	1	0, 2
5	0, 3		0, 3	
6	1, 4		1, 4	
7	1, 4		1, 4	
8	1, 5		1, 5	
9	1, 6		1, 6	
10	2, 7		1, 3, 6, 8	
11	2, 7		1, 3, 6, 9	
12	2, 8		1, 4, 7, 10	
13	2, 9		1, 4, 7, 11	
14	3, 10		1, 5, 8, 12	

Table 9.8 DM-RS Symbol Locations Within a PUCCH Format 3 or 4 Transmission

Up to 16 RBs can be allocated to a PUCCH format 3 transmission (similar to PUCCH format 2), where the two or more RBs for a multi-RB PUCCH transmission are always contiguous in frequency. The number of RBs allocated for the PUCCH format 3 transmission depends on the corresponding resource configuration, the UCI payload, and a configured maximum coding rate. The maximum coding rate provides a necessary knob for a gNB to ensure good PUCCH detection performance.

Similar to a DFT-s-OFDM-waveform–based PUSCH transmission, the modulated symbols for UCI on PUCCH format 3 and PUCCH format 4 are DFT-precoded to form a DFT-s-OFDM-waveform–based PUCCH transmission, which is more link-efficient.

Similar to PUCCH format 5 in 4G LTE, PUCCH format 4 in 5G NR supports multiplexing of multiple PUCCH transmissions in the same RB via a pre-DFT OCC, with multiplexing capacity of 2 or 4 within one PRB. The OCCs are given by

- For multiplexing capacity of 2:
    - [+1 +1 +1 +1 +1 +1 +1 +1 +1 +1 +1 +1]
    - [+1 +1 +1 +1 +1 +1 −1 −1 −1 −1 −1 −1]
- For multiplexing capacity of 4:
    - [+1 +1 +1 +1 +1 +1 +1 +1 +1 +1 +1 +1]
    - [+1 +1 +1 −j −j −j −1 −1 −1 +j +j +j]
    - [+1 +1 +1 −1 −1 −1 +1 +1 +1 −1 −1 −1]
    - [+1 +1 +1 +j +j +j −1 −1 −1 −j −j −j]

9.3.6 Multi-Slot PUCCH

For long PUCCH formats (1, 3, 4), a PUCCH transmission can be configured to repeat over multiple slots (up to 8 slots). Multi-slot PUCCH transmissions provide improved cell coverage, especially for HARQ. This is necessary especially when the UCI payload size is large (see Sec. 9.2).

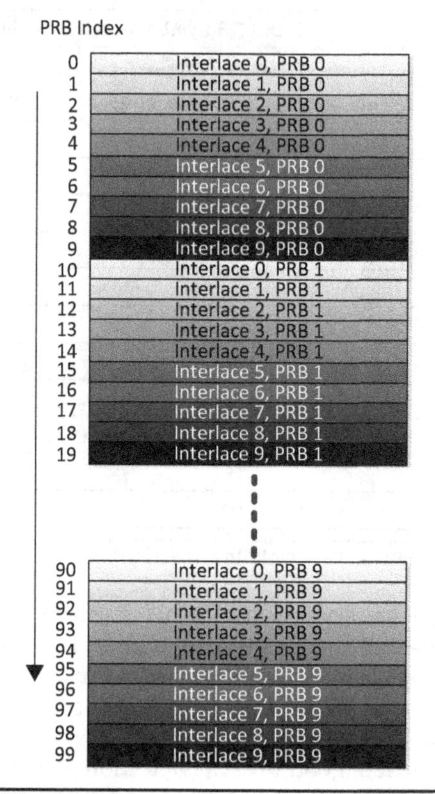

FIGURE 9.10 Illustration of the interlaced structure for PUCCH, (M, N) = (10, 10).

When a PUCCH is transmitted over multiple slots, the same information is repeated over the set of slots. In other words, each slot is *self-decodable.* For additional performance enhancement, inter-slot hopping can be supported by a configuration parameter, where different frequency resources can be used in different slots for increased frequency diversity.

9.3.7 Interlaced PUCCH

In Release 16, for NR operation in unlicensed or shared spectrum, in order to maximize Uplink link transmit power in observation of the regulatory requirements (particularly in terms of occupied channel bandwidth, see Chap. 12), interlaced PUCCH transmissions are supported based on the existing PUCCH formats 0, 1, 2, and 3. The interlaced structure has a uniform pattern in the frequency domain, where the number of interlaces (M) and the number of PRBs per interlace (N) are configurable and SCS-dependent. As an example, we have

- For 15 kHz SCS: (M, N) = (10, 10, or 11), and
- For 30 kHz SCS: (M, N) = (5, 10, or 11).

Figure 9.10 provides an illustration of the interlaced structure for PUCCH, where (M, N) = (10, 10). It can be seen that the 10 PRBs for each interlace (0 to 9) are evenly distributed, occupying a total bandwidth of 91 RBs or 16.38 MHz (for 15 kHz SCS). This is in comparison with the case of 10 frequency-consecutive PRBs, corresponding to a bandwidth of 1.8 MHz only.

9.4 PUCCH Resource Determination in 5G NR

In a slot, a UE may transmit up to two PUCCH transmissions mapped to different sets of symbols in the slot. If there are two PUCCH transmissions in a slot by a UE, one of the PUCCH transmissions has to be associated with a short PUCCH format (0 or 1). For simplicity of HARQ codebook management, a UE does not expect to transmit more than one PUCCH with HARQ information in a slot. Note that these restrictions are relaxed in Release 16, as discussed in Chap. 13.

When there is no HARQ in the UCI, the corresponding UCI (SR or CSI) is based on parameters from semi-static configurations. As a result, the corresponding UCI payload can be determined semi-statically. In this case, it is sufficient to determine a PUCCH resource for transmitting the UCI in a semi-static manner as well.

However, HARQ is driven mostly by the scheduling need for a UE. The payload size can be dynamic, as discussed in Sec. 9.2. Therefore, with HARQ as part of the UCI, it is necessary to *dynamically* determine a PUCCH resource for the PUCCH transmission, so that the PUCCH resource can be flexibly and appropriately chosen to accommodate the UCI need. Ideally, the chosen PUCCH resource should be such that

- The PUCCH transmission has sufficient amount of resource (frequency and time) for satisfactory link-level performance,
- The PUCCH transmission is efficient from the link performance perspective. In other words, it should not consume excessive amount of resource, and
- Whenever possible, it can be multiplexed with those used by other PUCCH transmissions, in order to minimize the overall PUCCH resource utilization and the resource fragmentation in the Uplink from the system perspective.

To that end, the PUCCH resource is determined using a three-step procedure when the UCI has the HARQ payload, as illustrated in Fig. 9.11.

In the first step, a UE first selects a PUCCH resource set from one or more (up to 4) configured PUCCH resource sets. Each resource set may contain one or more PUCCH resources:

- PUCCH resource set 0: up to 32 resources
- PUCCH resource set 1: up to 8 resources
- PUCCH resource set 2: up to 8 resources
- PUCCH resource set 3: up to 8 resources

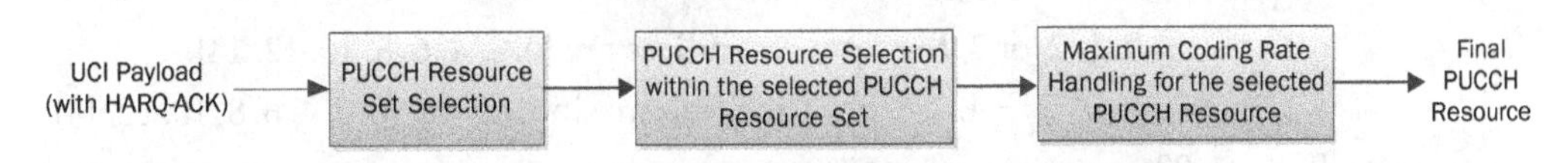

Figure 9.11 Illustration of the three-step PUCCH resource selection procedure.

The selection of a PUCCH resource set is based on a comparison between the UCI payload size O_{UCI} with a threshold associated with each PUCCH resource set. The threshold for PUCCH resource set 0 is fixed at 2 (bits), which implies that PUCCH resource set 0 can be used for 1-bit or 2-bit O_{UCI}, but O_{UCI} >2 has to use a PUCCH resource set with a higher index. Note that the case of 2-bit HARQ with a positive SR is considered as $O_{UCI} = 2$.

PUCCH resource sets 1, 2, or 3 may each be separately configured with a threshold (up to 1706 bits, a limit chosen for the coding chain to ensure good performance). If the threshold parameter for a PUCCH resource set (1, 2, or 3) is not configured, the threshold is assumed to be 1706 (bits), which implies the PUCCH resource set can support up to 1706 bits. The number 1706 is the maximum possible bits for UCI on PUCCH. A UE with O_{UCI} >2 would then sequentially compare O_{UCI} and the thresholds for PUCCH format set 1, 2, and 3, respectively, and determine the appropriate PUCCH resource set for a PUCCH transmission.

Now let us take a look at the second step. For PUCCH resource set 0, the selection of a PUCCH resource within the set is determined based on an explicit indication and under some condition, an implicit parameter. The explicit indication is done via a 3-bit information field in DCI called *PUCCH resource indicator*, or PRI (see Sec. 7.5), which can select one of the eight configured resources. To further improve the PUCCH resource selection flexibility without additional DCI overhead (i.e., still keeping the same 3-bit PRI), implicit resource selection can be additionally used when there are >8 resources configured in PUCCH resource set 0. To be more specific, the selected PUCCH resource index r_{PUCCH} among the R_{PUCCH} >8 resources in PUCCH resource set 0 is given by [3, Sec. 9.2.3]

$$r_{\mathrm{PUCCH}} = \begin{cases} \left\lfloor \dfrac{n_{\mathrm{CCE},p} \cdot \lceil R_{\mathrm{PUCCH}}/8 \rceil}{N_{\mathrm{CCE},p}} \right\rfloor + \Delta_{\mathrm{PRI}} \cdot \left\lceil \dfrac{R_{\mathrm{PUCCH}}}{8} \right\rceil & \text{if } \Delta_{\mathrm{PRI}} < R_{\mathrm{PUCCH}} \bmod 8 \\ \left\lfloor \dfrac{n_{\mathrm{CCE},p} \cdot \lfloor R_{\mathrm{PUCCH}}/8 \rfloor}{N_{\mathrm{CCE},p}} \right\rfloor + \Delta_{\mathrm{PRI}} \cdot \left\lfloor \dfrac{R_{\mathrm{PUCCH}}}{8} \right\rfloor + R_{\mathrm{PUCCH}} \bmod 8 & \text{if } \Delta_{\mathrm{PRI}} \geq R_{\mathrm{PUCCH}} \bmod 8 \end{cases} \tag{9.1}$$

where $N_{CCE,p}$ is a number of CCEs in CORESET p of the PDCCH reception, $n_{CCE,p}$ is the index of the first CCE for the PDCCH reception, and Δ_{PRI} is a value (from 0 to 7) of the PRI. To better understand the role that Δ_{PRI} plays in selecting a PUCCH resource, let us take a look at one specific example. Assuming $N_{CCE,p} = 8$, the respectively selected resource indices for $\Delta_{PRI} = \{0, 1, 2, 3, 4, 5, 6, 7\}$ among the resources configured for PUCCH resource set 0 are

- $R_{PUCCH} = 12$
  - If $n_{CCE,p} = 0, 1, 2$, or 3, the resource indices are {0, 2, 4, 6, 8, 9, 10, 11}.
  - Otherwise ($n_{CCE,p} = 4, 5, 6$, or 7), the resource indices are {1, 3, 5, 7, 8, 9, 10, 11}.
- $R_{PUCCH} = 16$
  - If $n_{CCE,p} = 0, 1, 2$, or 3, the resource indices are {0, 2, 4, 6, 8, 10, 12, 14}.
  - Otherwise ($n_{CCE,p} = 4, 5, 6$, or 7), the resource indices are {0, 2, 4, 6, 8, 10, 12, 14}.
- $R_{PUCCH} = 32$

- If $n_{CCE,p} = 0$ or 1, the resource indices are {0, 4, 8, 12, 16, 20, 24, 28}.
- If $n_{CCE,p} = 2$ or 3, the resource indices are {1, 5, 9, 13, 17, 21, 25, 29}.
- If $n_{CCE,p} = 4$ or 5, the resource indices are {2, 6, 10, 14, 18, 22, 26, 30}.
- If $n_{CCE,p} = 6$ or 7, the resource indices are {3, 7, 11, 15, 19, 23, 27, 31}.

As can be seen, the first CCE of the PDCCH reception ($n_{CCE,p}$) is used as another input to select a set of eight PUCCH resources from the configured $R_{PUCCH} > 8$ resources, while the 3-bit PRI in DCI can explicitly indicate which PUCCH resource from the selected set of eight resources is to be used for a PUCCH transmission. The combination of the first CCE and the PRI makes it possible to select a PUCCH resource from $R_{PUCCH} > 8$ resources while not increasing the DCI overhead.

It is worth noting that the implicit PUCCH resource selection mechanism using the first CCE is only applicable to PUCCH resource set 0, when the UCI payload (with HARQ) is no more than 2 bits. Such an implicit mechanism is also supported in 4G LTE. Indeed, it was a primary mechanism for the PUCCH resource determination in the first release of 4G LTE.

For PUCCH resource sets 1, 2, and 3, each set can have up to eight PUCCH resources. The selection of a PUCCH resource in the respective PUCCH resource set is explicitly indicated by the 3-bit PRI.

In the final step (the 3rd step), the selected resource may be further adjusted. Note that this step is applicable for PUCCH formats 2, 3, and 4 only, and only when the selected PUCCH resource has 2 or more RBs. Why? This is because the selected PUCCH resource may have excessive amount of resource and thus the PUCCH transmission may not be very efficient. To better manage the link efficiency for PUCCH, a maximum coding rate ($r_{max,\,UCI}$) can be configured for a UE, with the following possible coding rate values:

$$\{0.08, 0.15, 0.25, 0.35, 0.45, 0.60, 0.80\}$$

An effective coding rate is computed based on the total UCI payload potentially with a CRC overhead (denoted by O'_{UCI}, where the details of the CRC for UCI are to be discussed later), the modulation order (denoted by Q_m, where the modulation can be QPSK or $\pi/2$-BPSK), and the total amount of resources (excluding DM-RS REs) in the selected PUCCH. The total amount of resources is based on the resources on a per RB basis (denoted $N_{RE,\,per\,RB}$) and the total number of RBs ($N_{RB,\,Selected} \geq 2$), i.e., $N_{RB,\,Selected} \times N_{RE,\,per\,RB}$. A minimum number of RBs ($N_{RB,\,min}$) can be computed such that with $N_{RB,\,min}$, the resulting effective coding rate is ideally no more than $r_{max,\,UCI}$. That is

$$O'_{UCI}/(N_{RB,\,min} \times Q_m \times N_{RE,\,per\,RB}) \leq r_{max,\,UCI} \tag{9.2}$$

The final number of RBs for the PUCCH transmission is thus $\min(N_{RB,\,Selected}, N_{RB,\,min})$. When $N_{RB,\,min} \leq N_{RB,\,Selected}$, the RBs used for the PUCCH transmission are the lowest contiguous RBs in the $N_{RB,\,Selected}$ RBs of the selected PUCCH resource.

9.5 UCI on PUSCH in 5G NR

It is possible that when UCI is due and when a PUCCH resource is configured or selected for a PUCCH transmission, the UE is also configured with or scheduled for a PUSCH transmission, either on the same carrier or on a different carrier. Note that from the

UE perspective, PUCCH may be transmitted only on the Pcell or PScell (see Chap. 7). PUSCH transmissions, however, can be on any cell or any UL carrier configured for the UE. In such a case, one question is

How to transmit UCI in case there are also one or more PUSCH transmission(s)? Generally, there are two design alternatives:

- Alternative 1: Keeping the UCI on PUCCH, which is to be transmitted in parallel with the PUSCH transmission(s), or
- Alternative 2: Dropping PUCCH, and piggybacking the UCI onto PUSCH.

The first alternative is also generally known as *simultaneous PUCCH and PUSCH transmission*. It comes with two flavors:

- Case 1.1: PUCCH and PUSCH are on a same carrier, or
- Case 1.2: PUCCH and PUSCH are on different carriers.

Case 1.1 clearly results in a non-single-carrier waveform transmission. Depending on the frequency locations, resource allocation sizes, transmission power, etc., of the PUCCH and the PUSCH, a large power backoff (e.g., 10dB) may be necessary in order to ensure satisfaction of performance and regulatory requirements. Case 1.2 may or may not request a large power backoff, depending on whether or not the implementation is based on a single power amplifier (PA) or multiple PAs. For instance, if the two Uplink carriers are of a same frequency band, a likely implementation for the two different carriers is based on a single PA. On the other hand, if the two carriers are of different frequency bands, using two PAs each for one of the two carriers is a more likely implementation. Using two PAs may not require a large power backoff since one PA can be dedicated to the PUCCH transmission while the other to the PUSCH transmission. The heavy dependency on many parameters for power backoff makes it very difficult to specify performance requirements in case when simultaneous transmission of PUCCH and PUSCH is supported.

The benefits of having the simultaneous PUCCH and PUSCH transmission are the following. First, it is easier to manage UCI performance if it is carried on PUCCH. The selection of a PUCCH resource for UCI transmission, as discussed in Sec. 9.4, is carefully designed. The design considerations include both the link-level efficiency and the system-level efficiency. Although the PUSCH resource can be dynamically managed if it is scheduled by a DCI, the target performance for HARQ-based UL data on PUSCH is naturally different from that for the HARQ-less UCI. This is particularly true if the UL data and the UCI data are of different QoS types (e.g., eMBB vs. URLLC). Although some dynamic adjustments are possible (to be discussed later), it is still less straightforward comparing with the direct management of resources for PUCCH to carry the UCI.

Moreover, it is likely that a gNB may manage the UL frequency resource in a way that a certain subband(s) may be primarily used for the control region in one cell and across the cells, while other subband(s) can be primarily used for the UL data operation. These different regions may have different interference-over-thermal (IoT) operation points. As an example, the control subband(s) may be subject to a lower IoT especially considering that UCI is HARQ-less and typically associated with a higher performance target. On the other hand, the UL data subband(s) may tolerate a relatively higher IoT. Piggybacking UCI on PUSCH (instead of transmitting on PUCCH) may thus move the UCI to a different IoT operation region.

In light of the different performance requirements for UL data and UCI data, and the different performance needs for different UCI types, another possibility is to have PUCCH convey a part of the UCI (e.g., only HARQ and SR), while to have PUSCH carry the remaining part of UCI (e.g., CSI only). By doing so, a good compromise may be achieved.

Managing UCI by possibly using simultaneous PUCCH and PUSCH transmission can be semi-static or dynamic. A UE may be semi-statically configured regarding whether the UCI is transmitted on PUCCH or piggybacked on PUSCH. This semi-static management of UCI is supported in 4G LTE. Another way is to support a dynamic selection of UCI on PUCCH versus UCI on PUSCH. Such dynamic switching makes it possible to adjust the transmission scheme for UCI in response to the dynamics such as the transmission parameters and the QoS needs. However, it certainly complicates the UL operation especially at the UE side.

Despite the clear benefits of *simultaneous PUCCH and PUSCH transmission,* it is NOT supported in 5G NR in Release 15 and in Release 16, for both case 1.1 and case 1.2. Indeed, it was originally planned for the first 5G NR release, but was de-prioritized and removed later due to the overwhelming load and the time pressure to complete the first 5G release. It can be argued that supporting case 1.2 is more natural since multiple PUSCHs on different carriers can be simultaneously transmitted by a UE anyway. It can also be argued that since this feature is supported in 4G LTE, it should be supported in 5G NR as well. At the same time, it should be noted that although the feature is supported in 4G LTE, it never gets commercial deployments so far.

Therefore, in 5G NR, when UCI is due for transmission and there is at least one PUSCH transmission, the UE would always drop PUCCH and piggyback the UCI on one PUSCH, regardless of whether the PUSCH transmission is a DFT-s-OFDM–based or a CP-OFDM–based waveform.

The general philosophy of mapping UCI on PUSCH in 5G NR is similar to that of 4G LTE. The UCI is classified into different types, each type being separately coded, modulated, and mapped so that each type can be managed according to its respective performance target. The classified UCI types on PUSCH include

- HARQ,
- CSI type 1, and
- CSI type 2.

Note that SR is not piggybacked on PUSCH, since it may be replaced by a buffer status report (BSR) which contains more detailed information about the UE's Uplink buffer status.

For each classified type of UCI, a high-layer parameter is used by a UE to determine the amount of resource within PUSCH to be dedicated for the UCI. This parameter is known as a β parameter. The value range of β is typically very large, making it possible to re-purpose from a minimal amount resource to a large amount of resource within PUSCH to the UCI transmission, via different values of β. The detailed computation of the amount of REs for a UCI type on PUSCH can be found in Ref. [7, Sec. 6.3.2.4]. Besides β, the amount of REs is further dependent on the UCI payload size (including potentially a CRC overhead) and the spectral efficiency of PUSCH. Note that in order to prevent an excessive resource usage for UCI (hence insufficient left-over REs for UL data), an upper bound for the total amount of resources allocated to UCI can be explicitly controlled by a gNB via a higher-layer parameter specifically configured for a UE.

Note that even within one classified UCI type, the corresponding payload range can be very large. It is possible that a same UCI type may require different operation handlings, resulted from distinct operation conditions. Therefore, further refinement of the UCI handling on PUSCH within one classified UCI type is also necessary. To that end, we have

- HARQ
    - Case 1: when the payload is 1 or 2 bits, HARQ *punctures* the PUSCH REs originally scheduled for UL data, and the β parameter is configured via a dedicated parameter *betaOffsetACK-Index1*. This approach is similar to that in 4G LTE. The reason for such a special treatment is to reduce any potential misalignment and detection complexity between a gNB and a UE due to, e.g., a potential miss-detection of a PDCCH scheduling PDSCH. In this case, regardless of whether there is 0, 1, or 2 bits HARQ, the resource mapping for UL data on PUSCH remains the same (irrespective of HARQ). This was also discussed earlier in Sec. 9.1 for 4G LTE.
    - Case 2: when the payload is between 3 and 11 bits, HARQ rate-matches around the UL data REs on PUSCH, and the β parameter is configured via a dedicated parameter *betaOffsetACK-Index2*.
    - Case 3: when the payload is >11 bits, HARQ rate-matches around the UL data REs on PUSCH, and the β parameter is configured via a dedicated parameter *betaOffsetACK-Index3*.
- CSI Part 1
    - Case 1: when the payload is up to 11 bits, CSI part 1 rate-matches around the UL data REs on PUSCH, and the β parameter is configured via a dedicated parameter *betaOffsetCSI-Part1-Index1*.
    - Case 2: when the payload is >11 bits, CSI part 1 rate-matches around the UL data REs on PUSCH, and the β parameter is configured via a dedicated parameter *betaOffsetCSI-Part1-Index2*.
- CSI Part 2
    - Case 1: when the payload is up to 11 bits, CSI part 2 rate-matches around the UL data REs on PUSCH, and the β parameter is configured via a dedicated parameter *betaOffsetCSI-Part2-Index1*.
    - Case 2: when the payload is >11 bits, CSI part 2 rate-matches around the UL data REs on PUSCH, and the β parameter is configured via a dedicated parameter *betaOffsetCSI-Part2-Index2*.

In order to adapt to various UL data scheduling needs, dynamic operation conditions for UCI on PUSCH, and potentially different QoS requirements for UCI, the β parameter (although semi-statically configured) is not semi-statically indicated to a UE. Instead, a UE can be configured with up to four β values (separately for different UCI types and different cases within a UCI type), where one of the up to four semi-statically configured values can be dynamically indicated to the UE via an information field in DCI.

In general, when mapping the UCI onto PUSCH, the top priority is given to HARQ, followed by CSI type 1, and lastly CSI type 2. The prioritization is reflected by the proximity to DM-RS symbols when placing UCI onto PUSCH symbols (closer to DM-RS symbols typically implies better performance) and the possibility of being over-written by other UCI types. More specifically, the following rules are defined:

- If HARQ is 1 or 2 bits, a number of REs are reserved for HARQ (for puncturing) assuming a 2-bit HARQ payload, where the UL data REs in the reserved REs can be punctured by either 1 or 2 bits of the actual HARQ payload. Thus, there may be some left-over reserved REs (if zero or 1-bit actual HARQ payload), which can still be used for the UL data transmission.
- For HARQ with a payload >2 bits, CSI part I and CSI part II, the rate-matching is done by performing HARQ first (starting from the REs on the first available non-DMRS symbol right after the first DMRS symbol(s)), followed by CSI part I and finally CSI part II.
- CSI part I is not mapped to the reserved REs intended for HARQ. This is to protect CSI part I from being punctured by HARQ.
  - Both CSI part II and UL data can map to the reserved REs (thus subject to being punctured).

UCI mapping on PUSCH follows a frequency-first-time-second manner, starting from the lowest available tone of the lowest available symbol. This makes it possible for a gNB to decode UCI earlier. In a UL symbol, in order to further maximize the frequency diversity gain for UCI on PUSCH, the REs for UCIs are *frequency-distributed* when possible. To be more specific, the distance d between two adjacent REs in a symbol on which UCI is mapped is determined by

- If the number of to-be-mapped modulation symbols for the UCI is larger than or equal to the number of available REs in the UL symbol, d =1,
- Otherwise, d is defined as floor (number available REs in the UL symbol / the number of to-be-mapped modulation symbols for the UCI).

When $d > 1$ for a UL symbol, the mapping of the UCI modulation symbols in the symbol becomes frequency distributed.

All types of UCI are mapped to all layers of the PUSCH transmission, following the same PUSCH modulation order. This is slightly different from 4G LTE, where CSI is only mapped to the layers of the TB with a higher MCS. However, it is noted that the primary reason for such a difference is due to the fact that for a UL MIMO transmission, two TBs may be used in 4G LTE while a single TB is used for up to a four-layer UL MIMO transmission in 5G NR.

In case when frequency hopping for PUSCH is enabled, the UCI modulation symbols are partitioned into two parts of roughly the same size, by using a floor(.) operation for the first hop and a ceiling(.) operation for the second hop. The partitioned two parts are then mapped to the two hops, respectively. By an approximately equal partition, the negative impact on UL data due to UCI piggybacking on PUSCH can be roughly equally distributed between the first hop and the second hop.

9.6 Channel Coding for UCI

Table 9.9 provides a summary of the coding schemes for UCI, including the related CRC overhead. The coding schemes for UCI in 5G NR is similar to those in 4G LTE, except for the case when the UCI payload size is >11 bits, where polar coding is used in 5G NR in contrast to TBCC in 4G LTE.

The detailed polar coding scheme for UCI is illustrated in Fig. 9.12. The primary differences between polar coding for UCI and polar coding for DCI are the absence of the payload interleaver, the different CRC lengths, the additional utilization of polar codes of lengths 32, 64, and 1024 bit to accommodate the smaller and larger UCI payloads, and the introduction of a channel interleaver. Rate-matching (including the rate-matching interleaver) remains the same as in DCI.

For a UCI payload size K within a range of 12 to 19 bits, inclusive for both, 3 parity-check bits and a 6-bit CRC are generated [7]. For $K > 19$, an 11-bit CRC is generated without any other parity-check bits. Segmentation is applied when $K \geq 360$ and $M \geq 1088$, where M is the total number of coded bits for the UCI payload. In this case, there are two segments of equal sizes, each individually appended with an 11-bit CRC and applied with a channel interleaver. The coding chain ensures that the two segments have the same payload size and the same number of coded bits to facilitate decoding and to avoid cases where the segments would have used different polar code lengths.

# Info bits K	1	2	[3, 11]	[12, 19]	> 19
Coding scheme	Repetition	Simplex	Reed-Muller	PC-CA Polar	CA-Polar
CRC	N/A	N/A	N/A	Length 6	Length 11

Table 9.9 The Coding Schemes and CRC for UCI in 5G NR

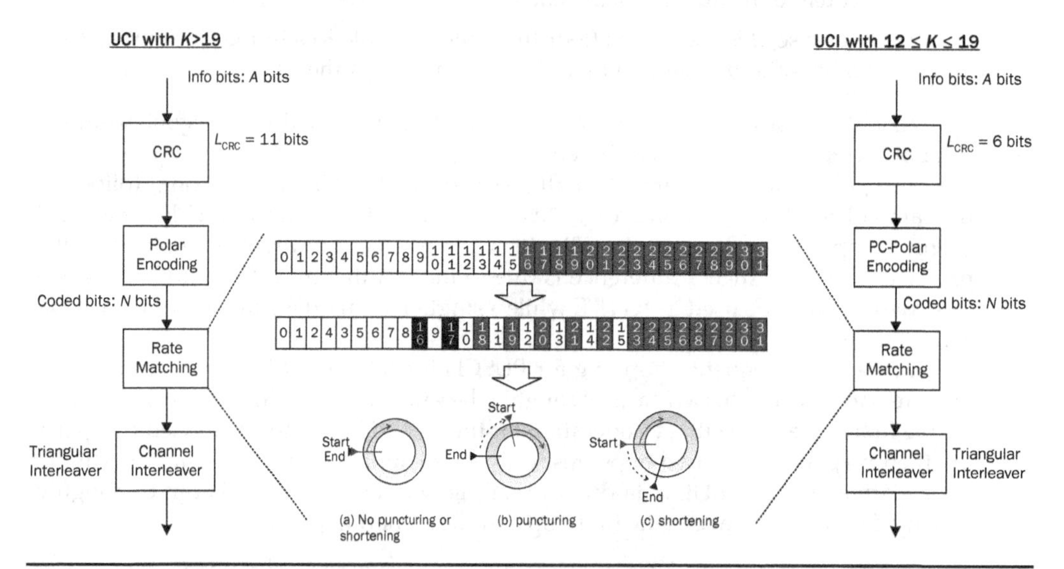

Figure 9.12 Illustration of polar coding for UCI in 5G NR.

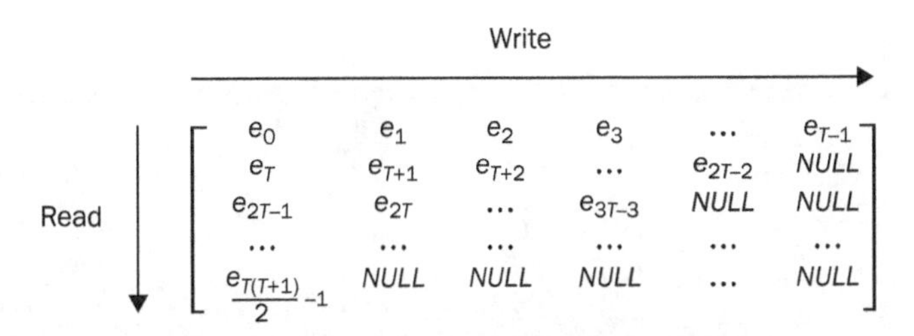

FIGURE 9.13 Illustration of the triangular channel interleaver for UCI in 5G NR.

A channel interleaver after rate-matching is necessary for UCI to maintain good performance when the modulation order is higher than QPSK. The regular structure of polar codes also necessitates the use of an interleaver with a non-uniform stride compared to the row-column rectangular interleavers with regular strides by other channels (e.g., Sec. 7.3). The triangular interleaver applied in the UCI coding chain provides performance almost identical to random interleaving [8], but with significantly lower implementation and description complexity. Figure 9.13 illustrates the operation of the triangular interleaver. Bits are written row-wise into the structure, where the size of each row is one less than the previous row, and are read column-wise. NULL elements are skipped when reading bits. The dimension, T, of the interleaver structure is chosen to ensure that $T(T+1)/2 \geq E$, where E is the number of bits after rate-matching. If $E < T(T+1)/2$, NULL elements are used to fill the structure.

References

[1] 3GPP, "3rd Generation Partnership Project; Technical Specification Group Radio Access Network; Evolved Universal Terrestrial Radio Access (E-UTRA); physical channels and modulation (Release 16)," 36.211, v16.1.0, March 2020.

[2] 3GPP, "3rd Generation Partnership Project; Technical Specification Group Radio Access Network; Evolved Universal Terrestrial Radio Access (E-UTRA); physical layer procedures (Release 16)," 36.213, v16.0.0, January 2020.

[3] 3GPP, "3rd Generation Partnership Project; Technical Specification Group Radio Access Network; NR; physical layer procedures for control (Release 16)," 38.213, v16.0.0, January 2020.

[4] Nokia, Samsung, Huawei, ZTE, MediaTek, Qualcomm, LG, Ericsson, Intel, CATT, "WF on LDPC parity check matrices," R1-1711982, 3GPP TSG RAN1 NR AH #2, June 2017.

[5] 3GPP, "3rd Generation Partnership Project; Technical Specification Group Radio Access Network; NR; physical layer procedures for data (Release 16)," 38.214, v16.0.0, January 2020.

[6] 3GPP, "3rd Generation Partnership Project; Technical Specification Group Radio Access Network; NR; physical channels and modulation (Release 15)," 38.211, v15.8.0, January 2020.

[7] 3GPP, "3rd Generation Partnership Project; Technical Specification Group Radio Access Network; NR; multiplexing and channel coding (Release 16)," 38.212, v16.0.0, January 2020.

[8] Qualcomm Inc., "Design and evaluation of interleaver for polar codes," R1-1711222, 3GPP TSG RAN1 NR AH #2, June 2017.

CHAPTER 10
Uplink Data Operation

Two different transmission schemes are supported for PUSCH [1]:

- Codebook-based transmission, and
- Non-codebook-based transmission

The UE is configured with codebook-based transmission when the RRC parameter *txConfig* is set to 'codebook' and the UE is configured with non-codebook-based transmission when the RRC parameter *txConfig* is set to 'nonCodebook'.

The codebook-based and non-codebook-based schemes are referring to UL MIMO. A special case is PUSCH transmission based on a single antenna port. This is of course a non-MIMO scheme, so in that sense it should be neither codebook-based nor non-codebook-based. However, somewhat confusingly, it is not possible to not select one of codebook-based scheme or non-codebook-based scheme once parameter *txConfig* is received by the UE. It was decided that PUSCH transmission based on a single antenna port is categorized as codebook-based transmission. It is distinguished from the MIMO variant of the codebook-based transmission by the number of configured SRS ports. When single-port SRS is configured per SRS resource, it is UL transmission based on single port (non-MIMO), and when multi-port SRS is configured, it is codebook-based MIMO UL transmission. This also means that when RRC parameter *txConfig* is set to 'codebook', at least one SRS resource must also be configured.

When RRC parameter *txConfig* is not configured, for example, during initial access before RRC configuration, the UE only supports PUSCH transmission based on a single antenna port. In this case, i.e., when RRC parameter *txConfig* is not configured, the UE only processes UL grants carried in DCI format 0_0, not in DCI format 0_1. This is because the lack of *txConfig* would make the DCI format 0_1 size undetermined.

10.1 UL MCS and TBS Determination

The UL MCS and TBS determination follows the procedure as for the DL, described in Sec. 8.4. When the UL waveform is CP-OFDM, the exact same procedure applies as in the DL (with replacing PDSCH with PUSCH and replacing DL grant with UL grant), which we do not repeat here.

The only significant difference between DL and UL is for the case when the UL waveform is DFT-S-OFDM. For the DFT-S-OFDM waveform, the lowest modulation order can be either QPSK or Pi/2-BPSK. Pi/2-BPSK is a new modulation order introduced in NR, which, when used together with a transparent UE implementation dependent transmit signal filtering, can significantly reduce the transmit signal's peak-to-average power ratio (PAPR) compared to quadrature phase shift keying (QPSK) or other modulation orders. Unlike in the UL, the transmitter in the DL typically operates with multiplexing multiple channels and signals in the frequency domain within the transmitter, so the PAPR benefit would be anyhow lost in the DL, and therefore the DL supports neither DFT-S-OFDM nor Pi/2-BPSK. In order to accommodate this change in the UL, when the UL waveform is DFT-S-OFDM, MCS table 1 (CP-OFDM, 64QAM)

MCS index I_{MCS}	Modulation order Q_m	Target code rate R x 1024	Spectral efficiency
0	q	240 / q	0.2344
1	q	314 / q	0.3066
2	2	193	0.3770
3	2	251	0.4902
4	2	308	0.6016
5	2	379	0.7402
6	2	449	0.8770
7	2	526	1.0273
8	2	602	1.1758
9	2	679	1.3262
10	4	340	1.3281
11	4	378	1.4766
12	4	434	1.6953
13	4	490	1.9141
14	4	553	2.1602
15	4	616	2.4063
16	4	658	2.5703
17	6	466	2.7305
18	6	517	3.0293
19	6	567	3.3223
20	6	616	3.6094
21	6	666	3.9023
22	6	719	4.2129
23	6	772	4.5234
24	6	822	4.8164
25	6	873	5.1152
26	6	910	5.3320
27	6	948	5.5547
28	q	reserved	
29	2	reserved	
30	4	reserved	
31	6	reserved	

Table 10.1 MCS Table 4 for DFT-S-OFDM PUSCH (64QAM)

is replaced with MCS table 4 (DFT-S-OFDM, 64QAM) and MCS table 3 (CP-OFDM, Low Spectral Efficiency) is replaced with MCS table 5 (DFT-S-OFDM, low spectral efficiency). MCS tables 4 and 5 are shown in Tables 10.1 and 10.2, respectively, based on Ref. [1].

MCS index I_{MCS}	Modulation order Q_m	Target code rate R x 1024	Spectral efficiency
0	q	60 / q	0.0586
1	q	80 / q	0.0781
2	q	100 / q	0.0977
3	q	128 / q	0.1250
4	q	156 / q	0.1523
5	q	198 / q	0.1934
6	2	120	0.2344
7	2	157	0.3066
8	2	193	0.3770
9	2	251	0.4902
10	2	308	0.6016
11	2	379	0.7402
12	2	449	0.8770
13	2	526	1.0273
14	2	602	1.1758
15	2	679	1.3262
16	4	378	1.4766
17	4	434	1.6953
18	4	490	1.9141
19	4	553	2.1602
20	4	616	2.4063
21	4	658	2.5703
22	4	699	2.7305
23	4	772	3.0156
24	6	567	3.3223
25	6	616	3.6094
26	6	666	3.9023
27	6	772	4.5234
28	q	reserved	
29	2	reserved	
30	4	reserved	
31	6	reserved	

TABLE 10.2 64QAM MCS Table 5 for DFT-S-OFDM PUSCH (Low Spectral Efficiency)

As it can be seen in Tables 10.1 and 10.2, the modulation order for the lowest entries has been changed from QPSK to being switchable between QPSK and Pi/2-BPSK by setting the value of q. When Pi/2-BPSK is enabled by RRC configuration, $q = 1$, and in this case all the switchable entries are Pi/2-BPSK. When Pi/2-BPSK is disabled by RRC configuration, $q = 2$, and in this case all the switchable entries are QPSK. There is no PDCCH-based dynamic switching between the QPSK and Pi/2-BPSK switchable entries. But, of course, PDCCH-based selection of the modulation order is still possible by selecting the different MCS index ranges.

For the low spectral efficiency table (MCS table 5), the switch point is at rate $R = 0.2$ for Pi/2-BPSK. Since channel coding rate matching for both LDPC base graphs 1 and 2 just repeats bits for rate $R \leq 0.2$ (actually, for base graph 1, channel bits are repeated already for $R \leq 0.33$), this ensures that there is no coding gain reduction by switching to Pi/2-BPSK. In the case of the 64QAM table (MCS table 4), the same selection method would not work, since there are no entries with $R \leq 0.2$. For this reason, the switch point in MCS table 4 is at $R = 0.33$ for Pi/2-BPSK. This ensures that at least with base graph 1, which can be applicable to $I_{MCS} = 1$, there is no coding gain loss with Pi/2-BPSK because the lowest code rate for base graph 1 without bit repetition is $R = 0.33$. When MCS table 4 is applied to LDPC base graph 2, some moderate coding gain reduction can occur with $I_{MCS} = 1$ and very minor coding gain reduction with $I_{MCS} = 0$.

Note that MCS table 2 (256QAM) given in Sec. 8.4 can be used with both CP-OFDM and DFT-S-OFDM but MCS table 2 does not support Pi/2-BPSK in either case.

10.2 UL Resource Allocation in the Time Domain

Similar to the DL resource allocation, the UL resource allocation in the time domain can refer to two mapping types: mapping type A and mapping type B [1]. Mapping type A is also known as slot-based scheduling and mapping type B is also known as mini-slot-based scheduling. Their properties are similar to that described for the DL. The main difference between the two UL mapping types is that PUSCH with mapping type A always starts at the beginning of the slot, while PUSCH mapping type B can start in any symbol of the slot, as long as it starts and ends in the same slot. Straddling slot boundaries is disallowed for both mapping types in Release 15.

Another difference between the two UL mapping types is that for PUSCH mapping type A, the DM-RS symbols are more or less fixed relative to the slot, while for PUSCH mapping type B, the DM-RS symbols slide together with the PUSCH.

The main differences compared to DL (which was described in Sec. 8.5) are the following:

- PUSCH with mapping type A always starts in the first symbol of the slot, while PDSCH with mapping type A can starts in one of the first four symbols of the slot. In the FDD UL, there is no control region at the beginning of the slot potentially delaying the PUSCH start, which is one of the reasons for this difference, even though the same statement is not necessarily true for TDD in general.
- PUSCH with mapping type B can be of any length, even a single symbol, while PDSCH with mapping type B can only be of length 2, 4, or 7. The reason for this difference is that it is less burdensome for the UE to implement a larger variety of slot formats in the UL than in the DL, since there is no channel estimation

PUSCH mapping type	Normal cyclic prefix			Extended cyclic prefix		
	S	**L**	**S + L**	**S**	**L**	**S + L**
Type A	0	{4, ..., 14}	{4, ..., 14}	0	{4, ..., 12}	{4, ..., 12}
Type B	{0, ..., 13}	{1, ..., 14}	{1, ..., 14}	{0, ..., 11}	{1, ..., 12}	{1, ..., 12}

Table 10.3 Start and Length of PUSCH

involved in the UL processing by the UE. As to the gNB implementation burden, the gNB can anyhow configure the UE to use the subset of slot formats that the gNB supports.

The allowed start S and length L in terms of symbols for PUSCH mapping type A and B are given in Table 10.3 based on Ref. [1].

The UL grant defines the time domain allocation as a combination of two parameters: K_2 and *SLIV*. The first parameter, K_2, gives a slot index offset between the slot in which the UL grant is received and the slot in which the PUSCH is transmitted. The second parameter, the so-called start and length indicator value (*SLIV*) gives the exact combination of start S and length L values, for which the allowed values were described in Table 10.3.

The UL grant actually does not include K_2 and *SLIV* explicitly but rather it includes a single time domain resource assignment field value m that provides a row index $m + 1$ to an allocation table. The allocation table defines a specific K_2, *SLIV* and PUSCH mapping type (A or B) combination for each row indexed by $m + 1$. The entries are configurable with RRC parameter *pusch-TimeDomainAllocationList*, which itself has the following format [2].

```
PUSCH-TimeDomainResourceAllocation ::= SEQUENCE {
    k2                                  INTEGER(0..32) OPTIONAL, -- Need S
    mappingType                         ENUMERATED {typeA, typeB},
    startSymbolAndLength                INTEGER (0..127)
}
```

The maximum UL slot offset is 32 slots.

SLIV itself is a 7-bit value that is mapped to a pair of start S and length L in terms of symbols according to the following equation, which is a definition identical to the DL *SLIV*.

$$\text{if } (L - 1) \leq 7 \text{ then } SLIV = 14 \cdot (L - 1) + S \quad (10.1)$$

$$\text{else } SLIV = 14 \cdot (14 - L + 1) + (14 - 1 - S), \text{ where } 0 < L \leq 14 - S \quad (10.2)$$

If there were no restrictions, the set of all possible start S and length L combinations would be $(14 \cdot 13)/2 = 91$, which is a subset of what can be expressed in 7-bits. Therefore, some of the *SLIV* values do not give valid start S and length L combinations.

In lack of RRC configuration, or before RRC configuration during initial access, one of a set of default tables is used. An example default table for normal CP is given in Table 10.4, based on Ref. [1].

Note that in the default tables, such as in the example Table 10.4, S and L are explicitly included, instead of being jointly signaled via *SLIV*.

Row index	PUSCH mapping type	K_2	S	L
1	Type A	j	0	14
2	Type A	j	0	12
3	Type A	j	0	10
4	Type B	j	2	10
5	Type B	j	4	10
6	Type B	j	4	8
7	Type B	j	4	6
8	Type A	$j + 1$	0	14
9	Type A	$j + 1$	0	12
10	Type A	$j + 1$	0	10
11	Type A	$j + 2$	0	14
12	Type A	$j + 2$	0	12
13	Type A	$j + 2$	0	10
14	Type B	j	8	6
15	Type A	$j + 3$	0	14
16	Type A	$j + 3$	0	10

TABLE 10.4 Example Default PUSCH Time-Domain Resource Allocation A for Normal CP

PUSCH SCS (kHz)	j
15	1
30	1
60	2
120	3

TABLE 10.5 Definition of Minimum UL Scheduling Slot Delay Value j

The K_2 value itself is a variable, the gNB can choose j, $j + 1$, $j + 2$, or $j + 3$, where j is the nominal minimum UL scheduling delay. Table 10.5 shows the defined values of j as a function of the PUSCH SCS. The values of j were determined based on selecting the largest values of allowed PUSCH preparation times as per UE capability for each SCS and rounding them up to the next integer number of slots.

As it can be seen from Table 10.5, the UL default time-domain resource allocation does not allow same slot scheduling. The non-default configurable resource allocation usable after RRC configuration does allow same slot scheduling, subject to meeting the UE minimum processing time requirements. The UL default time-domain resource allocation is notably different from the DL case, because in the DL, only same-slot scheduling is used in the default tables. The reason for this difference is that in preparation for a UL transmission after the UL grant reception, the UE has to fetch data, form MAC PDUs, and encode the data, which takes longer time, while in the DL, the UE just has to launch the PDSCH decoding task after decoding the DL grant. As it was mentioned in

PUSCH SCS (kHz)	Δ
15	2
30	3
60	4
120	6

Table 10.6 Definition of Additional RAR Slot Delay Value Δ

Sec. 8.5, even negative grant-to-data offsets are supported in the DL (requiring increased buffering). This would be clearly impossible to support in the UL.

For the special case of random access response (RAR) transmission of PUSCH, an additional Δ number of slots of delay is allowed for the UE. This is because the RAR grant parameters are obtained by decoding a PDSCH in Message 2, which takes longer than just decoding PDCCH. The values of Δ are also defined as a function of the PUSCH SCS. Table 10.6 shows the specified Δ values, based on Ref. [1].

The discussion so far was about the PUSCH time-domain resource allocation for the cases when PUSCH is transmitted with a transport block (data) included. When CSI-only PUSCH is transmitted without a transport block, a very similar procedure is followed. The CSI-only UL grant also includes the time-domain resource assignment field value m that provides a row index $m + 1$ to the allocation table. But instead of one allocation table, two tables are used. The first allocation table is *pusch-TimeDomainAllocationList*, which is the same as used for PUSCH with data. The values of *SLIV* and PUSCH mapping type (A or B) are read from the row indexed by $m + 1$ in this table. But the value of K_2 in this table is ignored. Instead, K_2 is read from another table denoted as *reportSlotOffsetList*. The entries of this table are also RRC configurable. The number of rows (entries) in *reportSlotOffsetList* must be the same as the number of rows in *TimeDomainAllocationList*. Different CSI reports included in one PUSCH may have different *reportSlotOffsetList* resulting in different K_2 values for the same index $m + 1$. In such cases, the conflict is resolved by taking the maximum value over those corresponding to the activated reports as K_2. Having a separate table for the case of CSI-only PUSCH can enable same slot CSI feedback scheduling, even when the processing time for data transport blocks would not allow same slot PUSCH scheduling for data.

When the PUSCH includes both a transport block and CSI, then the time-domain resource allocation procedure is the same as it would be with transport block only without CSI. When CSI is present, the gNB scheduler must select parameters that meet both the data processing timeline requirements and CSI processing timeline requirements as applicable to the given PUSCH.

10.2.1 PUSCH Slot Repetition

Another aspect of the UL time-domain resource allocation is PUSCH slot repetition, also called PUSCH slot aggregation or PUSCH slot bundling. The motivation for introducing slot bundling is improving UL coverage and link budget. The same TB is repeated in the bundled slots. PUSCH slot bundling is enabled when the UE is configured with *pusch-AggregationFactor*, in which case the same symbol allocation is applied across the *pusch-AggregationFactor* consecutive slots with *pusch-AggregationFactor* set to 2, 4 or 8, in response to a single received UL grant [1].

The details of PUSCH slot repetition are the same as PDSCH slot repetition described in Sec. 8.5 and will be omitted here. We only mention the differences compared to the DL.

PUSCH slot repetition can be dynamically switched off with using DCI format 0_0, since slot repetition is only used with DCI format 0_1. Unlike in the DL, PUSCH slot repetition can also be semi-persistently switched on/off for configured grant (CG) PUSCH. The NDI field is used for this, setting *NDI* = '1' enables PUSCH slot repetition.

Unlike in the DL for PDSCH, inter-slot frequency hopping can be configured in the UL for PUSCH. This will be also discussed in Sec. 10.3.

When PUSCH slot repetition is used, the PUSCH is limited to single-layer transmission. This is motivated by the fact that PUSCH slot repetition is used primarily as a tool to improve UL link budget, where spatial multiplexing would not be relevant.

Note that it is not a requirement for the UE to keep phase coherence across the slots in PUSCH slot repetition. Therefore, it is not expected that the gNB performs joint channel estimation and/or joint decoding across the repeated slots. Instead, the baseline assumption is that the gNB decodes the repeated slots as if they were separate HARQ retransmissions.

A further complication with PUSCH slot repetition, similar to PDSCH slot repetition in the DL, is that some of the repeated slots may not be available. This can occur to the PUSCH, for example, due to the presence of a DL slot in TDD. In these cases, the interrupted slots are omitted by the UE. The remaining slots are neither postponed nor cancelled.

10.3 UL Resource Allocation in the Frequency Domain

In the frequency domain of NR UL, similar to the DL, two resource allocation types, type 0 and type 1 are used [1]. Type 1 is the default, e.g., used with DCI format 0_0. The details of the frequency allocation scheme are the same as for the DL, so instead of detailed description, we refer to Sec. 8.6, and only describe the differences between UL and DL here.

Unlike the DL, two waveforms are used in the UL: CP-OFDM, when transform precoding is disabled, and DFT-S-OFDM when transform precoding is enabled. The details on these have been described in Chap. 5.

Only resource allocation type 1 is allowed for DFT-S-OFDM. Both resource allocation type 0 and 1 can be used for CP-OFDM. The reason is that resource allocation type 0 allows non-contiguous allocations, i.e., the frequency-domain allocation can consist of clusters of PRBs disjoint from each other. The DFT-S-OFDM waveform is fundamentally not designed to work with non-contiguous allocations in the sense that the main benefit of DFT-S-OFDM waveform, namely PAPR reduction, is not preserved once the allocation is non-contiguous. We note that the DFT-S-OFDM PAPR degradation, i.e., PAPR increase, in response to non-contiguous allocation is somewhat gradual. For example, if we remove one PRB from an otherwise contiguous 100-PRB allocation, the PAPR degradation is slight. This is because the ratio of gap versus the total allocation is small. Nevertheless, there was no work on defining allowable gap ratios for DFT-S-OFDM, and instead non-contiguous allocation for DFT-S-OFDM was simply disallowed.

Non-contiguous allocation is also problematic for CP-OFDM but for a different reason. There is no PAPR degradation for CP-OFDM as a result of non-contiguous allocation. But there is a clustering of transmit emissions that can lead to failing to meet emission requirements. The spectral shape of emissions is a function of the shape of the

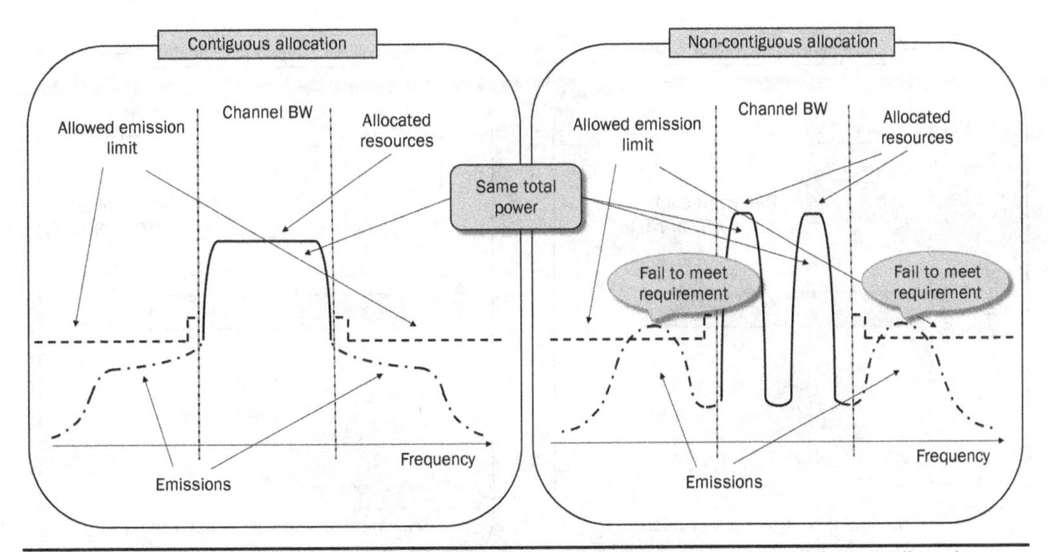

FIGURE 10.1 Illustration of comparison of emissions with contiguous and non-contiguous allocations.

fundamental transmit signal spectrum. In general terms, the more concentrated into isolated clusters the transmit signal gets, the more uneven the emission in the frequency domain becomes. An illustration of this effect is shown in Fig. 10.1. Note that signal levels are not to scale in Fig. 10.1.

The same effect holds both for CP-OFDM and DFT-S-OFDM waveforms; however, in the case of DFT-S-OFDM waveform, the PAPR degradation effect is more dominant and for that reason non-contiguous frequency-domain allocation for DFT-S-OFDM signal remains to be disallowed. On the other hand, after significant debate, non-contiguous allocation with some limitations has been decided to be allowed for CP-OFDM in Release 15. This scheme is called "almost contiguous allocation." Some of the motivation was that when some narrow subband needs to be reserved for some other uses, for example, for eMTC or NB-IoT UL, and the subband is in the middle of the channel, then allowing only contiguous allocation would limit the possible maximum frequency allocation to about one half of the total available. This is illustrated in Fig. 10.2.

Of course, the need for non-contiguous allocation could be avoided by placing the reserved subband at the edge of the channel, so the motivation was not universally accepted. Nevertheless, to increase the frequency configuration flexibility, almost contiguous allocation was introduced [3] in Release 15.

Almost contiguous allocation has the following attributes:

- $\dfrac{N_{RB\_gap}}{N_{RB\_alloc} + N_{RB\_gap}} \leq 0.25$
- $N_{RB\_alloc} + N_{RB\_gap} > 106$ RBs for 15 kHz SCS, $N_{RB\_alloc} + N_{RB\_gap} > 51$ RBs for 30 kHz SCS, and $N_{RB\_alloc} + N_{RB\_gap} > 24$ RBs for 60 kHz SCS

where N_{RB_alloc} is the total number of allocated RBs and N_{RB_gap} is the total number of unallocated RBs. Note that unallocated RBs at the edges of the allocation are ignored both for N_{RB_alloc} and N_{RB_gap}. Therefore, unallocated RBs are those that are situated in

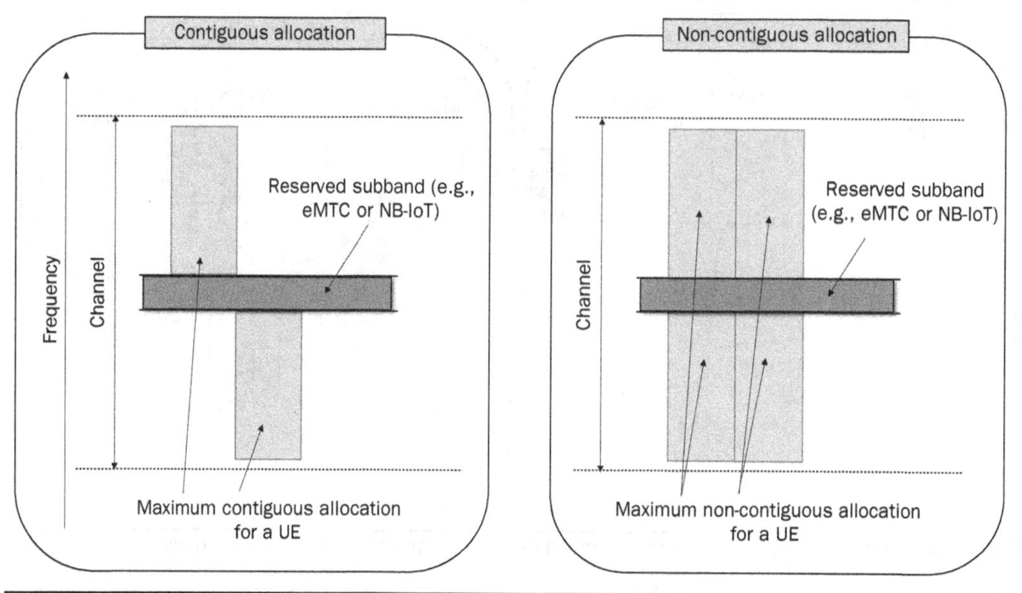

Figure 10.2 Maximum possible contiguous and non-contiguous resource allocation with reserved narrow band at the channel center.

frequency gaps, where each gap is bounded by allocated RBs on both sides. Note also that in general, there is no limitation on the number of gaps or on the number of allocated clusters (which is one more than the number of gaps), or on the arrangement of allocated clusters other than the allocation unit granularity expressed as the resource block group (RBG) size. The concept of RBG and RBG size has been described in Sec. 8.6.

For almost contiguous allocations, allowed maximum power reduction was introduced, expressed in dB as $10 \times \log_{10}\left(\frac{N_{\mathrm{RB+alloc}} + N_{\mathrm{RB\_gap}}}{N_{\mathrm{RB+alloc}}}\right)$ and rounded upwards to the closest 0.5 dB. The principle of deriving this limit is that when the UE applies the maximum power reduction with non-contiguous allocation, it can ensure that power spectral density (PSD) is never higher than what it would be with contiguous allocation and without the maximum power reduction. This principle is illustrated in Fig. 10.3.

For any almost contiguous allocation, no matter how small the gap is, the allowed maximum power reduction is at least 0.5 dB. Because of the following limitation: $\frac{N_{\mathrm{RB\_gap}}}{N_{\mathrm{RB\_alloc}} + N_{\mathrm{RB\_gap}}} \leq 0.25$, the only allowed maximum power reduction values in fact are 0.5 dB, 1 dB, and 1.5 dB.

In Release 15 and Release 16, almost contiguous allocation is only allowed in the FR1 UL, it is not allowed in the FR2 UL. The reason for this is that there is no need to reserve bandwidth for eMTC or NB-IoT in FR2 and, in general, the analog beamforming restriction makes UL FDM multiplexing of different UEs less likely in FR2 than in FR1.

The applicable resource allocation type (type 0 or type 1) can be dynamically indicated in the UL grant, when both CP-OFDM waveform and dynamic switching are

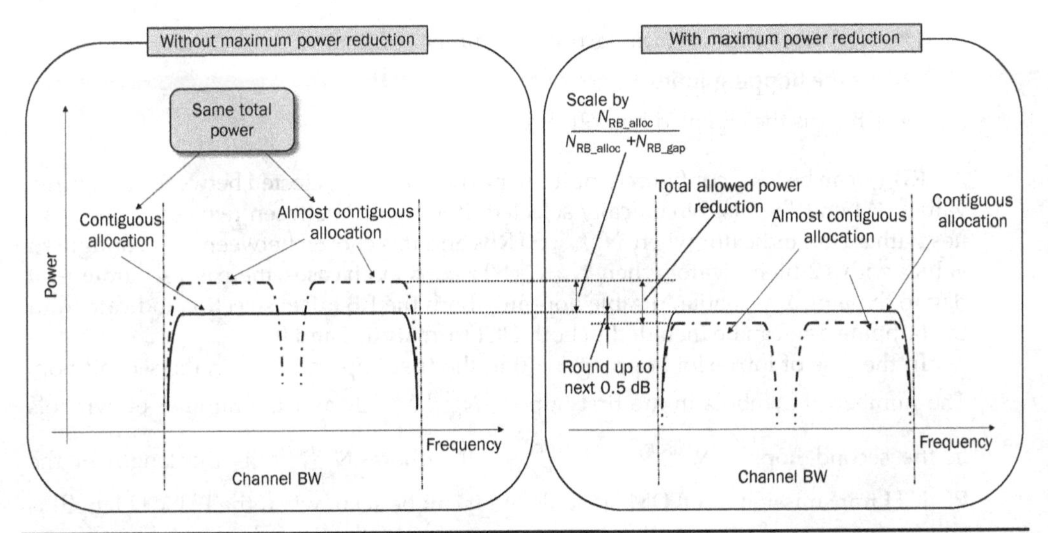

Figure 10.3 Illustration of PSD levels at maximum power with contiguous and almost contiguous allocation.

configured by RRC. When dynamic switching is configured, one bit is included in DCI format 0_1 to indicate the resource allocation type applicable to the granted PUSCH. Otherwise, the resource allocation type itself is RRC configured for CP-OFDM. This allows the gNB to choose between maximum scheduling flexibility and DCI payload minimization. When dynamic switching is configured, the maximum of the type 0 and type 1 bit-width is used in the DCI format 0_1 to make the payload size invariant, and an additional bit is included in DCI format 0_1 as the switching flag.

Frequency hopping is supported for PUSCH transmissions when configured by parameter *frequencyHopping* and frequency domain resource allocation type 1 is used [1]. Frequency hopping is not supported when resource allocation type 0 is used since type 0 can already allocate non-contiguous RBs, even though the frequency diversity it can provide is bounded by the 25% limit on the allowed maximum gap mentioned earlier.

The following two frequency hopping modes are supported:

- Intra-slot frequency hopping, applicable to both single-slot and multi-slot PUSCH transmission
- Inter-slot frequency hopping, applicable only to multi-slot PUSCH transmission

The frequency hopping scheme is relatively simple, and can be described by the following equation for both the intra-slot and inter-slot cases:

$$RB = \begin{cases} RB_{start} & i = 0 \\ (RB_{start} + RB_{offset}) \bmod N_{BWP}^{size} & i = 1 \end{cases} \quad (10.3)$$

where

- RB is the first RB of the PUSCH transmission,
- RB_{start} is the first RB from the *RIV*,

- N_{BWP}^{size} is the bandwidth part size in number of RBs,
- i is the hopping index,
- RB_{offset} is the signaled hop offset value.

RB_{offset} can be RRC configured, or it can be dynamically selected between configured values. When RB_{offset} is dynamically selected, it is chosen between two configured values with a 1-bit indicator when $N_{BWP}^{size} < 50$ RBs and it is chosen between four configured values with a 2-bit indicator when $N_{BWP}^{size} \geq 50$ RBs. In both cases, there is a separate 1-bit flag to dynamically enable/disable hopping. Both the RB offset selection indicator and the hopping flag can be included in both DCI formats 0_0 and 0_1.

In the case of intra-slot hopping, $i = 0$ in the first hop and $i = 1$ in the second hop. The number of symbols in the first hop is $\lfloor N_{symb}^{PUSCH,s} / 2 \rfloor$ and the number of symbols in the second hop is $N_{symb}^{PUSCH,s} - \lfloor N_{symb}^{PUSCH,s} / 2 \rfloor$, where $N_{symb}^{PUSCH,s}$ is the length of the PUSCH transmission in OFDM symbols. As it can be seen, when the PUSCH length is an even number of symbols, the two hops are equal, otherwise the second hop is one symbol longer than the first.

In the case of inter-slot hopping, hopping index i alternates between 0 and 1 for consecutive repeated PUSCH slots. For the first PUSCH slot in the sequence of repeated slots, $i = 0$ if the first slot is an even numbered slot within the 10-ms radio frame, and $i = 1$ if it is an odd numbered slot in the frame. This designation helps avoiding pattern collisions between different UEs when they start their PUSCH transmissions in different slots. Of course, this assumes that the different UEs have been assigned compatible RB_{offset} values and their RB allocations were also coordinated.

Note that the combination of the resource allocation and hopping parameters can easily lead to situations with invalid PUSCH resource allocation in the frequency domain. It is the base station scheduler's responsibility to avoid those cases.

10.4 UL Rate Matching

There was a significant debate about whether a rate matching scheme similar to that in the DL, as described in Sec. 8.7, should be introduced in the UL as well. The stated motivation for introducing it was mostly forward compatibility.

The first consideration was that when looking at the collection of symbols allocated to a given PUSCH transmission, the set of frequency-domain resources after rate matching should be identical across all symbols, otherwise there would be power fluctuation and as a result phase continuity would not be maintained by the UE. In order to make the frequency-domain resources constant, the rate matching pattern would have to be either a "strip of RBs" in frequency, or a "strip of symbols" in time.

If it is a strip of RBs, then frequency-domain resource allocation type 0 can already assign the needed RBs without rate matching. Although there is a limit of 25% maximum gap size for "almost contiguous" UL frequency-domain allocation, this was not viewed as overly limiting for forward compatibility–related resource reservation. At the same time, making the resource allocation non-contiguous by rate matching for DFT-S-OFDM PUSCH waveform was not viewed acceptable due to the PAPR increase it would result in.

If it is a strip of symbols then if it is in the middle of the time-domain resource allocation, then it would cut the PUSCH into two non-contiguous time segments, which would again create phase discontinuity. If it is either at the beginning or at the end of the time-domain allocation, then there is no discontinuity created, but again simply changing the time domain resource allocation by adjusting the *SLIV* can already achieve the same result without rate matching.

Due to these considerations, UL rate matching was not adopted in Release 15.

10.5 UL HARQ Operation

As it was mentioned before, the UL HARQ timing is dynamically configurable in NR.

For the UL, a maximum of 16 HARQ processes per UL carrier is supported by the UE [1]. Unlike in the DL, the actual maximum number of processes used for the UL is not configurable by RRC. The reason for this is that there is no soft buffer operation in the UE for the UL. The UE is assumed to perform full encoding of the UL TB for every retransmission.

The UE is not required to transmit PUSCHs overlapped in time in the same carrier, even if they are not overlapped in frequency.

There were various provisions introduced to exclude so called out-of-order HARQ in the UL [1]. These included the following restrictions:

- For a given HARQ process ID in a carrier, the UE is not required to process a UL grant for a retransmission or new transmission of PUSCH with the same HARQ process ID carrying data until after the UE completed the transmission of the previous PUSCH of the same HARQ process ID carrying data. Note that for a PUSCH carrying CSI only, the same restriction does not apply.
- For two different HARQ process IDs on the same carrier, the PUSCH and the corresponding UL grant cannot be in reverse order. That is, the UE cannot be required to transmit a first PUSCH before a second PUSCH, if the UL grant for the first PUSCH was received later than the UL grant for the second PUSCH. For deciding the relative timing of the two UL grants, the end of the last symbol of the PDCCH carrying the grant is counted. Note that receiving the UL grants at the same time for the two PUSCHs is allowed, and it is mandatory for the UE to be able to process the two UL grants received at the same time in NR TDD. This feature was introduced in support of TDD configurations with more UL than DL slots (so called "UL-heavy" configurations).

Some of these restrictions may be relaxed in a future release of the specification, for example, in order to enhance eMBB-URLLC multiplexing in a UE in a given carrier.

All the above restrictions apply to the HARQ processes associated with a given carrier, and there is no additional restriction across HARQ processes occurring on different carriers.

Similar restrictions apply to cases where there is potential conflict between a configured grant (CG) PUSCH transmission and a dynamically granted (DG) PUSCH transmission. Like in LTE, a dynamic grant can overwrite a configured grant. That is, if the CG PUSCH and DG PUSCH overlap in time, it is not considered an error case but rather the UE is required to drop the CG PUSCH and replace it with the DG PUSCH. This is required irrespective of whether the CG PUSCH and DG PUSCH have the same HARQ

ID or not. Obviously, it would be burdensome to the UE implementation, if the UE was required to replace a CG PUSCH transmission that the UE has already started transmitting or that the UE has already prepared to transmit. To avoid this burden, a timeline requirement was defined as follows:

- A UE is not required to handle the case where a CG PUSCH occasion overlaps in time on the same carrier, irrespective of having or not having any frequency domain overlap, with a DG PUSCH scheduled by a PDCCH ending in symbol i, if i is not at least N_2 symbols before the start of the first symbol of the CG PUSCH occasion. The value N_2 in symbols is determined according to the UE processing capability.

The above applies irrespective whether the UE intended to use the CG PUSCH occasion or not, i.e., whether or not it had applicable data to transmit. Since the timeline condition is in effect a requirement on the gNB, and the gNB cannot know whether the UE has data to transmit, differentiating between the UE having data or not would not have been helpful.

Note that the term "UE is not required to handle" above doesn't intend to mean that the UE should drop or the UE is allowed to drop the DG grant. Rather, it means that the scenario is classified an "error case," and the UE behavior in this scenario is completely up to UE implementation. Since the gNB has no way to predict the UE behavior, it is expected that the gNB would take effort to avoid this scenario altogether.

The next requirement is addressing a version of out-of-order HARQ operation for a CG PUSCH and DG PUSCH with the same HARQ ID on the same carrier. The requirement is best described with introducing the concept of a "virtual grant" for the CG PUSCH, which is a hypothetical UL grant assumed to be generated before the CG PUSCH at the latest possible time that still complies with the UE's processing time capability. In reality, no such virtual grant is sent, received, or generated for the CG PUSCH. The purpose of the concept is to define a deadline for certain events to take place. With this virtual grant concept, the requirement can be described as saying that the time between the end of the virtual grant and the start of the corresponding CG PUSCH occasion serves as an exclusion period for any dynamic grant for PUSCH with the same HARQ process ID as the HARQ process ID of the CG PUSCH. More formally, the requirement can be expressed as follows:

- For a given HARQ process ID in a carrier, the UE is not required to handle the case where a UL grant is received for a PUSCH with the same HARQ process ID as the HARQ process ID of a CG PUSCH occasion, irrespective of having or not having any time-domain overlap between the DG PUSCH and CG PUSCH occasion, if the PDCCH carrying the UL grant ends in symbol i and i is not at least symbols before the start of the first symbol of the CG PUSCH occasion. The value (in symbols) is determined according to the UE processing capability.

The above applies irrespective of the UE intended to use the CG PUSCH occasion or not, i.e., whether it had applicable data to transmit or not.

10.6 UL Soft Buffer Management

NR introduced the possibility of limited buffer rate matching (LBRM) in the UL [4], in addition to LBRM in the DL. Note that LTE only had LBRM in the DL, so having LBRM in the UL is a new feature in NR.

One difference between the NR DL LBRM and UL LBRM is that LBRM is always enabled in the DL, while it can be enabled/disabled with UE-specific configuration in the UL. Another difference is that while the support of DL LBRM is mandatory for the UE, UL LBRM is optional. The default assumption for the UL is that LBRM is disabled and this is also the assumption before initial UE capability information exchange. When UL LBRM is enabled, it applies to all UL HARQ processes. The reason for the differentiation between DL and UL LBRM configurability is that the gNB receiver complexity reduction is seen much less essential compared to the UE receiver complexity reduction. It was debated in fact whether UL LBRM is necessary at all. Note that when the gNB receiver would require LBRM, but the UE does not support UL LBRM, operation is still possible with some performance loss. In this case, the gNB can adjust the granted MCS and rv_{id}.

10.7 UL Data Rate Capability

As it was mentioned in Sec. 8.9, it has been decided that NR will not introduce the concept of UE categories for the time being. Instead, a standardized formula was provided, by which the UE's UL data rate capability is computed [5]. All input elements to the formula can be extracted from the UE reported capabilities. The performance requirements mandate that the UE can achieve this calculated maximum data rate in a sustained manner, at least under ideal channel conditions.

The formula introduced for calculating the maximum data rate in Ref. [5] is the following:

$$\text{UL data rate (Mbps)} = 10^{-6}\sum_{j=1}^{J}\left(v_{\text{Layers}}^{BW(j)} \cdot Q_m^{(j)} \cdot f^{(j)} \cdot R_{\max} \cdot \frac{N_{PRB}^{BW(j),\mu} \cdot 12}{T_S^{\mu}} \cdot (1 - OH^{(j)})\right) \quad (10.4)$$

where

- J is the number of aggregated UL component carriers in a band or band combination.
- $R_{\max} = 948/1024$ is the nominal maximum code rate.
- For the jth component carrier:
  - $v_{\text{Layers}}^{BW(j)}$ is the maximum number of supported layers given by maximum of higher. layer parameters *maxNumberMIMO-LayersCB-PUSCH* and *maxNumberMIMO-LayersNonCB-PUSCH*.
  - $Q_m^{(j)}$ is the maximum supported modulation order given by higher layer parameter *supportedModulationOrderUL*, which can take the following values: $Q_m^{(j)} = \{1, 2, 4, 6, 8\}$.
  - $f^{(j)}$ is a UE selected scaling factor given by higher layer parameter *scalingFactor*, which can take the values $f^{(j)} = \{1, 0.8, 0.75, 0.4\}$.

- μ is the SCS index ($\mu = 0$ for 15 kHz, $\mu = 1$ for 30 kHz, $\mu = 2$ for 60 kHz, $\mu = 3$ for 120 kHz).
- T_S^{μ} is the average OFDM symbol duration in a subframe for numerology μ, i.e., $T_S^{\mu} = \frac{10^{-3}}{14 \cdot 2^{\mu}}$ for normal cyclic prefix.
- $N_{PRB}^{BW(j),\mu}$ is the maximum RB allocation in bandwidth $BW^{(j)}$ with numerology μ, where $BW^{(j)}$ is the UE supported maximum bandwidth in the given band or band combination in Ref. [3] or [6].
- $OH^{(j)}$ is the nominal overhead and takes the following values: $OH^{(j)} = 0.08$ for UL in FR1 and $OH^{(j)} = 0.10$ for UL in FR2.

In order to determine the UE's UL peak data rate capability, the data rate is calculated in every band combination and feature combination that the UE indicated to support, and the maximum among all those calculated values is selected.

As an example, assume that the UE signals the following capabilities:

- $J = 1$, single carrier only
- $v_{\text{Layers}}^{BW(0)} = 1$, maximum supported rank is one
- $Q_m^{(j)} = 8$, reference modulation order is 256QAM
- $f^{(j)} = 1$, scaling factor of 1
- $\mu = 1$, 30 kHz SCS support
- $N_{PRB}^{BW(0),1} = 273$, 100 MHz channel support in FR1 and 273 RB maximum transmission bandwidth configuration based on Ref. [3] as mentioned in Sec. 2.4.1

With these, the maximum supported data rate will be calculated as

- UL data rate = 625 Mbps

It is important to note that $Q_m^{(j)}$ is not the modulation order the UE supports. It is mandatory for the UE to support 64QAM in all bands in the UL. Yet, the UE can signal $Q_m^{(j)} = 1$ (Pi/2-BPSK) or $Q_m^{(j)} = 2$ (QPSK), for example. This is because $Q_m^{(j)}$ represents only a scaling factor to be used in the maximum data rate formula, independent of the supported modulation order. It is interesting to note that in the UL, the UE can also indicate support of $Q_m^{(j)} = 1$ (Pi/2-BPSK), even if the UE indicated no support of this modulation scheme. There is also another scaling factor, $f^{(j)}$, serving a similar purpose. In the data rate calculation, the product of the two applies. Both $Q_m^{(j)}$ and $f^{(j)}$ were introduced with the purpose of allowing UE UL data processing complexity reduction for the cases of maximum aggregated bandwidth.

Similar to the DL, although $Q_m^{(j)} \cdot f^{(j)}$ is set independent for each band, their effect is aggregated across carriers. Also, similar limitations as in the DL apply to this data rate aggregation, including no transfer of unused data rate from any carrier to a carrier configured with processing time capability 2. Processing time capability 2 is discussed in Sec. 10.8.

10.8 Processing Time for UL Data

In LTE, the nominal preparation time for PUSCH transmission granted by PDCCH was 4 ms minus the UL timing advance. One of the main advancements made in NR was reduced latency, which required reducing the UL processing time relative to LTE. In this context, the UL processing time is measured from the end of the PDCCH granting the PUSCH to the beginning of the PUSCH. During this time, the UE has to decode the PDCCH and parse the DCI contents, fetch UL data, form MAC PDUs, encode the data, and generate the UL waveform.

Similar to the DL, two UE capability classes were introduced [1]: UE UL processing time capability 1 and UE UL processing time capability 2. The UE can be capability 1 in the DL and capability 2 in the UL, and vice versa.

The PUSCH preparation time requirement denoted with $T_{proc,2}$ was defined in Ref. [1] as

$$T_{proc,2} = max\ (N_2 + d_{2,1}, d_{2,2}) \tag{10.5}$$

where

- $d_{2,1}$ is a term dependent on whether the first OFDM symbol of the PUSCH allocation consists of DM-RS only or also includes data. If the first PUSCH symbol contains no data, then $d_{2,1} = 0$, otherwise $d_{2,1} = 1$. The effect of this term appears as one extra symbol being added until the beginning of the first data symbol in the PUSCH. However, in effect, the term can be also described as a one-symbol PDCCH-to-PDSCH delay reduction attained for the case of front-loaded DM-RS without data in the first symbol. The rationale for this was that preparing the PUSCH DM-RS does not need to be delayed by the time needed for data preparation.
- $d_{2,2}$ is a term dependent on whether the PDCCH granting the PUSCH also switches the active BWP. If the scheduling DCI triggered a switch of BWP, $d_{2,2}$ equals to the allowed BWP switching time, otherwise $d_{2,2} = 0$.
- N_2 is based on the SCS and is given in Tables 10.7 and 10.8 for UE processing time capability 1 and 2, respectively. The SCS corresponds to either the PDCCH or PUSCH, with details described later in this section.

	PUSCH preparation time	
SCS (kHz)	**N_2 (symbols)**	**μs**
15	10	713.5
30	12	428.1
60	23	410.3
120	36	321.1

TABLE 10.7 PUSCH Preparation Time for PUSCH Processing Time Capability 1

	PUSCH preparation time	
SCS (kHz)	**N_2 (symbols)**	**μs**
15	5	356.8
30	5.5	196.2
60	11 (for FR1)	196.2 (for FR1)

TABLE 10.8 PUSCH Preparation Time for PUSCH Processing Time Capability 2

Note that the Eq. (10.5) is not precise because it omits certain scaling factors, but it suffices for the current description.

The advanced UE processing capability 2 was not defined for FR2 in Release 15. It was seen that the required beam switching delays could possibly negate the benefits of reduced FR2 processing times.

The values for N_2 were agreed after significant debate. Even though N_2 increases in terms of number of symbols for increasing SCS, it actually decreases in terms of time. On the one hand, an argument can be made that when going to higher SCS, the UL waveform generation, PDCCH decoding and other aspects have to be sped up anyway with a factor corresponding to the SCS ratio, which would result in a constant processing delay in terms of number of symbols. On the other hand, it can also be argued that many of the data preparation and channel coding aspects are not dependent on SCS; therefore, due to these, the number of symbols should be scaled up with a factor corresponding to the SCS ratio. The actual N_2 selection represented a middle way between these two viewpoints.

The N_2 values are not significantly different between 30 kHz SCS and 60 kHz SCS. There was limited interest in aggressive optimizations for FR1 60 kHz, which was seen having less commercial relevance than 30 kHz.

Overall, it can be said that the PUSCH processing time, which is at most 713.5 μs, is indeed a significant improvement upon LTE, which had 3110 μs as the minimum PUSCH processing time. It can be also noted that with processing time capability 2, NR enables a so-called self-contained slot operation, wherein the reception of the UL grant, preparing the UL data and transmission of the PUSCH by the UE all occur within the same slot.

Note that the $d_{2,2}$ term in the Eq. (10.5) for $T_{proc,2}$ is actually not in units of symbols, so the units of the different terms in the formula have to be reconciled. We omit the details of this here for the sake of brevity.

The actual time between PDCCH and PUSCH is affected by the UL timing advance. The limits given by $T_{proc,2}$ must be applied after including the effects of timing advance and any relevant DL-to-DL or UL-to-UL timing differences.

As it was noted before, some part of the PUSCH preparation time like PDCCH decoding actually depends on the DL SCS, not the UL SCS. Also, it is possible to have processing time capability 2 in the DL carrying the PDCCH and processing time capability 1 in the UL carrying the PUSCH, or vice versa. To cover these cases, the general rule was adopted that in the determination of $T_{proc,2}$, a first value is determined based on the relevant DL parameters (DL SCS and processing time capability of the PDCCH carrier), a second value is determined based on the relevant UL parameters (UL SCS and processing time capability of the PUSCH carrier), and $T_{proc,2}$ is finally determined by selecting between the first and second value the one that gives the larger delay in absolute terms.

10.9 PUSCH DM-RS

In LTE, the DL and UL DM-RS designs had little or no commonality. In NR, a different approach was taken. The NR DM-RS designs of the DL and UL were made very similar. As a matter of fact, the PDSCH DM-RS and PUSCH DM-RS for CP-OFDM waveform are identical. Using the same design makes it possible to enhance demodulation performance and possibly enhance multiplexing capability for cases of interference from a disparate duplex direction in dynamic TDD. For example, when the UE receives

PDSCH DM-RS, it can be made orthogonal to another UE's PUSCH DM-RS transmission in a neighboring cell. Even the PUSCH DM-RS for DFT-S-OFDM waveform can be orthogonalized with PDSCH DM-RS type 1 because the frequency-domain patterns of the two are the same.

The main difference between the DM-RS in the DL and UL is the use of different DM-RS sequence for the DFT-S-OFDM waveform in the latter. For the PUSCH with DFT-S-OFDM waveform, computer generated sequence (CGS) were also introduced for PUSCH with one-RB or two-RB allocation in the frequency domain. In Release 16, an additional waveform type, Pi/2-BPSK was also introduced for the PUSCH DM-RS.

10.9.1 PUSCH DM-RS Frequency-Domain and Code-Domain Patterns

The frequency-domain and code-domain patterns of the PUSCH DM-RS for CP-OFDM waveform are identical to the PDSCH DM-RS described in Sec. 8.12.1 and it is not repeated here.

On the other hand, for the PUSCH DM-RS for DFT-S-OFDM waveform, a low PAPR design was needed, which has an impact also on the required frequency-domain DM-RS pattern. Low PAPR sequences can be relatively easily constructed when the frequency-domain pattern is equidistant, meaning that for any RE of a DM-RS port, the frequency distance to the closest RE of the same DM-RS port is the same on both sides. This is true for DM-RS type 1 but untrue for DM-RS type 2. For DM-RS type 2, the distance between REs of a given DM-RS port is one subcarrier on one side and five subcarriers on the other side, which are clearly unequal. For this reason, only DM-RS type 1 can be used for DFT-S-OFDM PUSCH.

The application of a frequency-domain OCC code, as the one defined for DM-RS type 1, might appear to be in conflict with achieving the desired low PAPR property of the DM-RS for DFT-S-OFDM waveform. However, there is no such conflict in reality because the one-dimensional frequency-domain OCC pattern has an effect equivalent to a time-domain cyclic shift for the DM-RS for the DFT-S-OFDM waveform, and a time-domain shift does not alter PAPR. Applying the [+ –] pattern is the same as a quarter-symbol cyclic time shift of the OFDM symbol. It is one quarter, instead of one half, because of the comb-2 arrangement of DM-RS type 1 applicable to the DFT-S-OFDM DM-RS. This can be seen with considering that for the comb-2 pattern the [+ –] code extends to a [... + – + – ...] series that is stretched in time by the comb factor 2.

In order to maintain the low PAPR for the PUSCH DM-RS for the DFT-S-OFDM waveform, the DM-RS in the DM-RS symbols cannot be multiplexed with data. Therefore, CDM groups with data applies to CP-OFDM DM-RS only.

10.9.2 PUSCH DM-RS Time-Domain Patterns

The time-domain patterns of the DM-RS for PUSCH without intra-slot frequency hopping are the same as those of the PDSCH DM-RS described in Sec. 8.12.2. This holds for both the CP-OFDM and DFT-S-OFDM waveforms without intra-slot frequency hopping [7]. For the PUSCH with intra-slot frequency hopping, some modifications in the time-domain patterns were needed, which will be described later in this section. Using the same sequence generation makes it possible to enhance multiplexing capability for cases of interference from a disparate duplex direction in dynamic TDD.

l_d in symbols	DM-RS positions $\bar{l}$							
	PUSCH mapping type A				PUSCH mapping type B			
	dmrs-AdditionalPosition				*dmrs-AdditionalPosition*			
	pos0	*pos1*	*pos2*	*pos3*	*pos0*	*pos1*	*pos2*	*pos3*
<4	-	-	-	-	l_0	l_0	l_0	l_0
4	l_0	l_0	l_0	l_0	l_0	l_0	l_0	l_0
5	l_0	l_0	l_0	l_0	l_0	l_0, 4	l_0, 4	l_0, 4
6	l_0	l_0	l_0	l_0	l_0	l_0, 4	l_0, 4	l_0, 4
7	l_0	l_0	l_0	l_0	l_0	l_0, 4	l_0, 4	l_0, 4
8	l_0	l_0, 7	l_0, 7	l_0, 7	l_0	l_0, 6	l_0, 3, 6	l_0, 3, 6
9	l_0	l_0, 7	l_0, 7	l_0, 7	l_0	l_0, 6	l_0, 3, 6	l_0, 3, 6
10	l_0	l_0, 9	l_0, 6, 9	l_0, 6, 9	l_0	l_0, 8	l_0, 4, 8	l_0, 3, 6, 9
11	l_0	l_0, 9	l_0, 6, 9	l_0, 6, 9	l_0	l_0, 8	l_0, 4, 8	l_0, 3, 6, 9
12	l_0	l_0, 9	l_0, 6, 9	l_0, 5, 8, 11	l_0	l_0, 10	l_0, 5, 10	l_0, 3, 6, 9
13	l_0	l_0, 11	l_0, 7, 11	l_0, 5, 8, 11	l_0	l_0, 10	l_0, 5, 10	l_0, 3, 6, 9
14	l_0	l_0, 11	l_0, 7, 11	l_0, 5, 8, 11	l_0	l_0, 10	l_0, 5, 10	l_0, 3, 6, 9

TABLE 10.9 PUSCH DM-RS Positions $\bar{l}$ Within a Slot for Single-Symbol DM-RS and Intra-Slot Frequency Hopping Disabled

The symbol locations are summarized in Table 10.9 from Ref. [7].

The definitions of $\bar{l}$, l, l_d, l_0 for the PUSCH are the same as for the PDSCH. Note that l_0 for PUSCH mapping type A is configurable, similar to the PDSCH mapping type A case. For the PDSCH this was motivated by moving the DM-RS to after the PDCCH occupying the first 2 or 3 symbols of the slot. The same does not apply to the PUSCH because PUSCH mapping type A always starts in the first symbol of the slot. Nevertheless, by allowing the PUSCH DM-RS location to match that of the PDSCH, it was made possible to enhance multiplexing capability for cases of interference from a disparate duplex direction in dynamic TDD.

When PUSCH with intra-slot frequency hopping is used, the DM-RS pattern is different compared to the case with no frequency hopping [7]. The DM-RS symbol allocation was modified for this case in order to provide approximately even channel observation opportunity in the first and second hops. In Figs. 10.4, 10.5, and 10.6, we give example illustration for PUSCH DM-RS symbol positions in the time domain for the case of PUSCH with intra-slot frequency hopping. Figure 10.4 shows DM-RS time-domain patterns with PUSCH mapping type A with no additional DM-RS position. The illustration on the left side is for the case of $l_0 = 2$, while the illustration of the right side is for the case of $l_0 = 3$. Figure 10.5 shows DM-RS time-domain patterns with PUSCH mapping type A with one additional DM-RS position. The illustration on the left side is for the case of $l_0 = 2$, while the illustration of the right side is for the case of $l_0 = 3$. Figure 10.6 shows DM-RS time-domain patterns with PUSCH mapping type B. The illustration on the left side is for the case of no additional DM-RS symbol position, while the illustration of the right side is for the case of one additional DM-RS symbol position.

As it can be noted in Fig. 10.5, the number of DM-RS symbols is unequal across the two hops in some cases for PUSCH mapping type A with intra-slot frequency hopping.

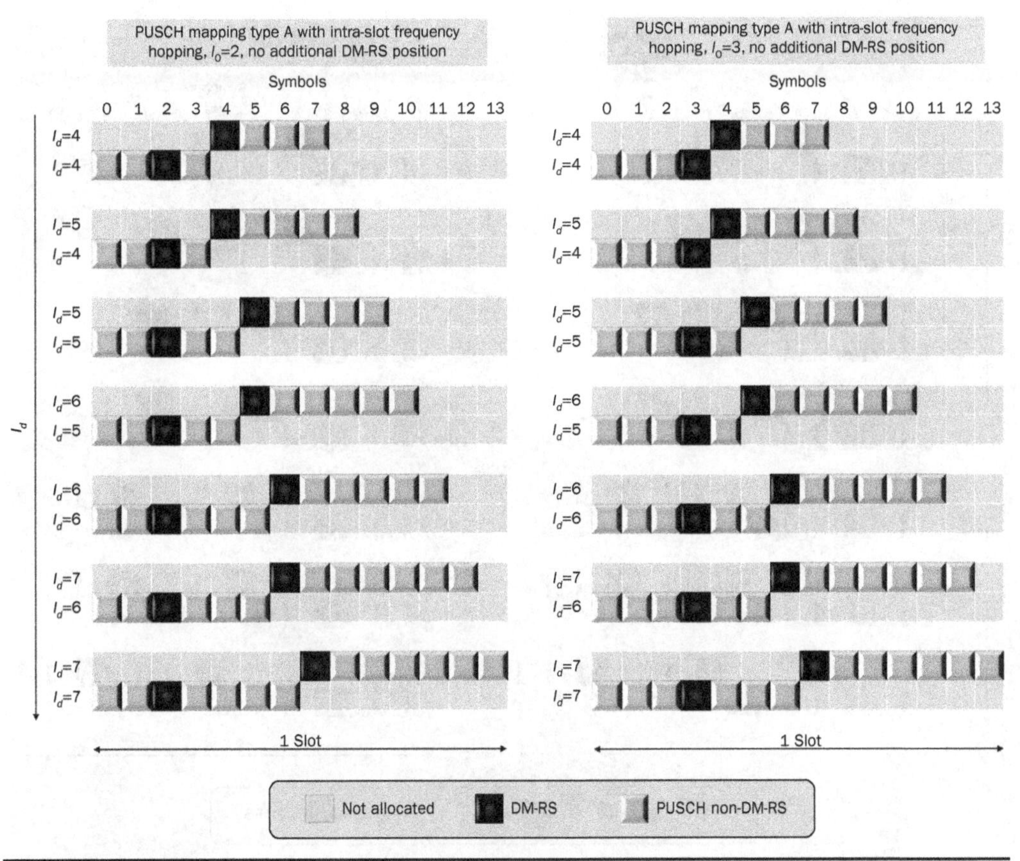

FIGURE 10.4 Time-domain mapping for single-symbol DM-RS with no additional DM-RS positions for PUSCH mapping type A with intra-slot frequency hopping.

The DM-RS symbol allocation for PUSCH with intra-slot frequency hopping is also summarized in Table 10.10, based on Ref. [7].

Note that in Table 10.10, as well as in Fig.10.4 through Fig. 10.6, the length l_d is counted within each hop individually. Also, the symbol numbers in Table 10.10 are within each hop, even for PUSCH mapping type A.

Double-symbol DM-RS is not supported for PUSCH with intra-slot frequency hopping.

10.9.3 PUSCH DM-RS Sequence for CP-OFDM

The PUSCH DM-RS sequences define the DM-RS waveform type, which can be CP-OFDM or DFT-S-OFDM. The definition of the scrambling sequence for CP-OFDM PUSCH is the same as for the PDSCH described in Sec. 8.12.3. Using the same sequence generation makes it possible to enhance multiplexing capability for cases of interference from a disparate duplex direction in dynamic TDD.

The Release 16 enhancement to the scrambling ID determination that was introduced in order to mitigate the multi-port DM-RS PAPR increase, as described in the

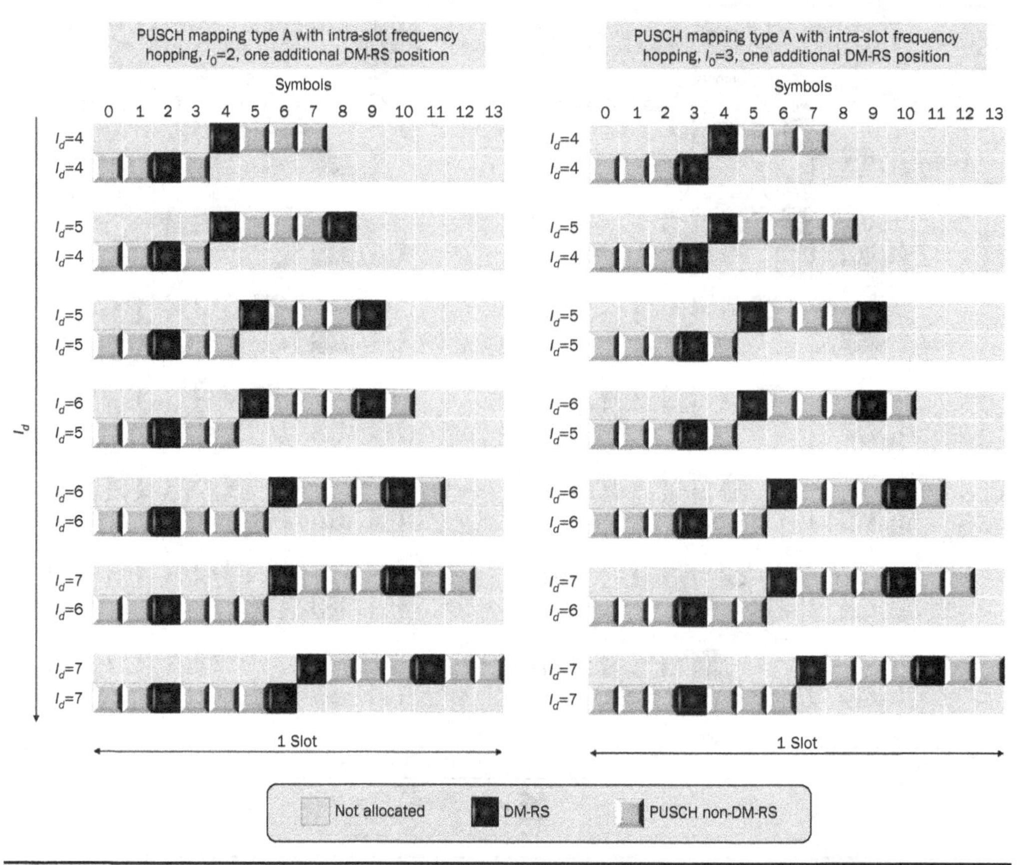

FIGURE 10.5 Time-domain mapping for single-symbol DM-RS with one additional DM-RS positions for PUSCH mapping type A with intra-slot frequency hopping.

in Sec. 8.12.3, equally applies to the DM-RS for CP-OFDM PUSCH. As a matter of fact, managing the DM-RS PAPR is even more important for the PUSCH than for the PDSCH because of the stricter limitations on transmitter complexity and power consumption for the former. Therefore, in Release 16, the scrambling ID n_{SCID} was changed between the first and second CDM groups of the DM-RS for CP-OFDM waveform for the reasons described in Sec. 8.12.3. Since the n_{SCID} is binary, this can be described as simply flipping the n_{SCID} value from 0 to 1 or from 1 to 0. With this change, the scrambling initialization is

$$c_{init}(\lambda) = \left(2^{17}\left(N_{symb}^{slot} n_{s,f}^{\mu} + l + 1\right)\left(2N_{ID}^{n'_{SCID}(\lambda)} + 1\right) + 2N_{ID}^{n'_{SCID}(\lambda)} + n'_{SCID}(\lambda) + 2^{17}\left\lfloor\frac{\lambda}{2}\right\rfloor\right) \bmod 2^{31} \tag{10.6}$$

where λ is the CDM group index, which can take on value {0, 1, 2}, and $n'_{SCID}(0) = n_{SCID}, n'_{SCID}(1) = 1 - n_{SCID}, n'_{SCID}(2) = n_{SCID}$. As in the case of the PDSCH, using the n_{SCID} values this way preserves the opportunity to MU-MIMO multiplex Release 15 and Release 16 UEs with each other.

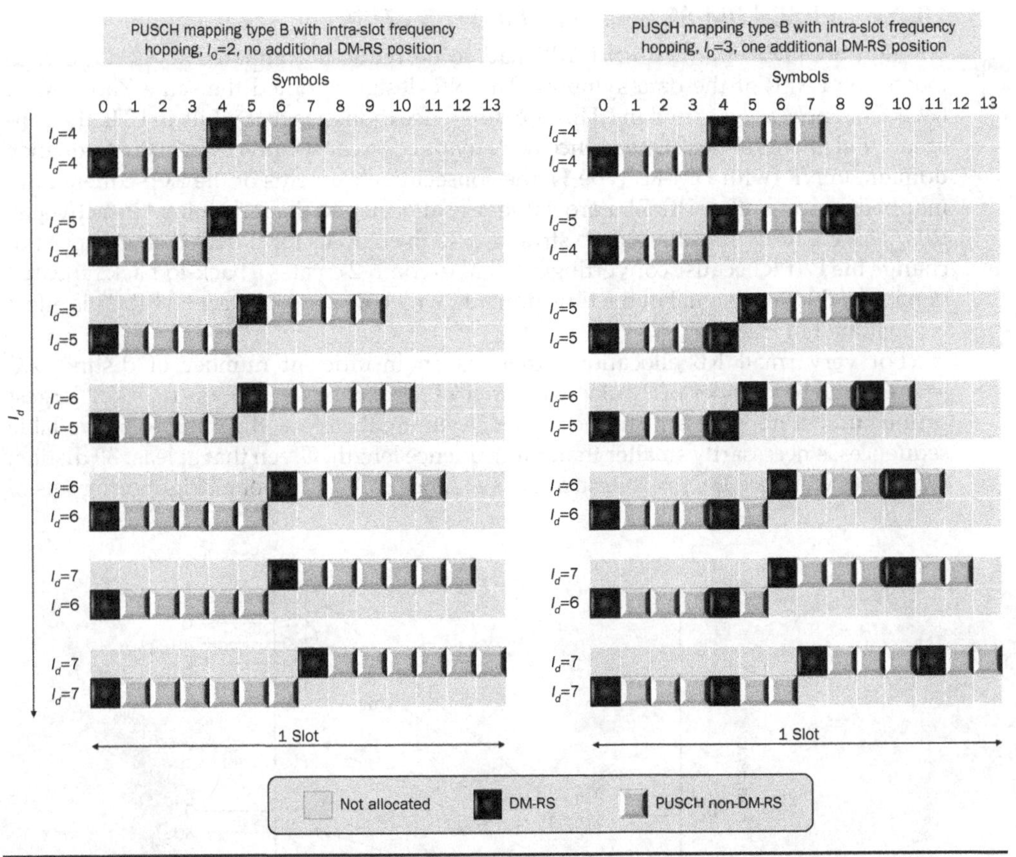

Figure 10.6 Time-domain mapping for single-symbol DM-RS with zero and one additional DM-RS positions for PUSCH mapping type B with intra-slot frequency hopping.

l_d in symbols	DM-RS positions $\bar{l}$											
	PUSCH mapping type A								PUSCH mapping type B			
	l_0 = 2				l_0 = 3				l_0 = 0			
	dmrs-AdditionalPosition				*dmrs-AdditionalPosition*				*dmrs-AdditionalPosition*			
	pos0		*pos1*		*pos0*		*pos1*		*pos0*		*pos1*	
	1st hop	2nd hop	1st hop	2nd hop	1st hop	2nd hop	1st hop	2nd hop	1st hop	2nd hop	1st hop	2nd hop
≤3	-	-	-	-	-	-	-	-	0	0	0	0
4	2	0	2	0	3	0	3	0	0	0	0	0
5, 6	2	0	2	0, 4	3	0	3	0, 4	0	0	0, 4	0, 4
7	2	0	2, 6	0, 4	3	0	3	0, 4	0	0	0, 4	0, 4

Table 10.10 PUSCH DM-RS Positions $\bar{l}$ Within a Slot for Single-Symbol DM-RS with PUSCH Intra-Slot Frequency Hopping

10.9.4 PUSCH DM-RS Sequence for DFT-S-OFDM

For DFT-S-OFDM, the DM-RS PAPR had to be reduced compared to CP-OFDM to match the PAPR of the data symbols. The NR design adopted the same Zadoff-Chu (ZC) sequences as was used in LTE. One difference though is that while in LTE, the consecutive elements of the ZC sequence were mapped to consecutive REs in the frequency domain, in NR (with DM-RS type 1), the consecutive elements of the ZC sequence are mapped to every other RE [7]. Zero values are allocated to the remaining "in-between" REs. This creates a so-called comb structure, with comb factor 2. This change does not change the PAPR because converting a signal to comb-2 creates a back-to-back repeated version of the same signal in the time domain, which does not impact PAPR. This effect is illustrated in Fig. 10.7.

For very small RB allocations, there is an insufficient number of distinct ZC sequences. The number of available distinct ZC sequences is one less than the largest prime number not greater than the sequence length; therefore, the number of available sequences is necessarily smaller than the sequence length. Given that at least 30 distinct sequences were needed in order to avoid sequence collision among neighboring sets of

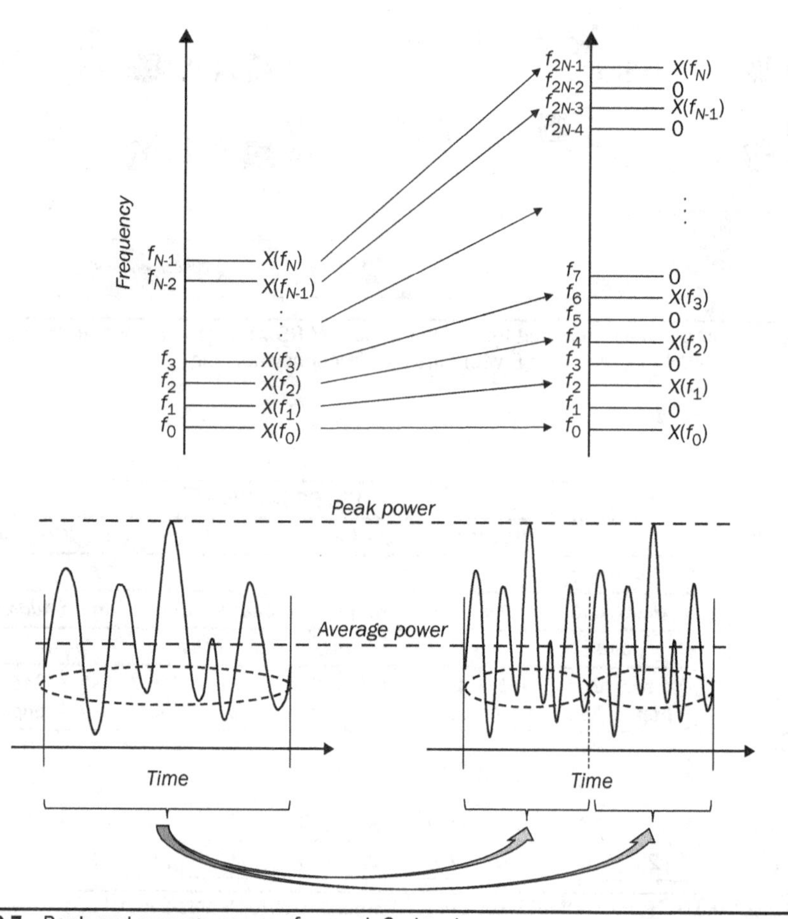

FIGURE 10.7 Peak and average power of a comb-2 signal.

30 cells, ZC sequences were not suitable for smaller than 3-RB allocations. Because of this reason, computer generated sequences (CGS) were used in LTE. CGS is simply a set of frequency domain QPSK sequences that were found with computer search and have low PAPR and cross-correlation. Note that they also need to have zero out-of-phase auto-correlation but this was not subject of the computer search because perfect autocorrelation is already ensured by the construction method: a sequence that has flat power spectrum, such as a frequency domain QPSK sequence, always has perfect auto-correlation in the time domain.

Because of the comb-2 structure of DM-RS type 1 used for DFT-S-OFDM, the sequence length for an RB allocation in NR is one half to that compared in LTE. In LTE, the only CGS length needed was 12 and 24, corresponding to LTE one-RB and two-RB allocations. In Release 15 NR, additionally CGS sequence length 6 and 18 had to be added for one-RB and three-RB allocations. The CGS sequences are given as elements of tables in Ref. [7] and not repeated here.

Although the DM-RS sequences introduced for DFT-S-OFDM in Release 15 were able to match the PAPR of DFT-S-OFDM data with QPSK, 16QAM, 64QAM, or 256QAM modulation, they have still higher PAPR than DFT-S-OFDM data with filtered Pi/2-BPSK modulation. In order to correct this, new DM-RS sequences were introduced specifically for Pi/2-BPSK modulation in Release 16. These sequences match the PAPR of Pi/2-BPSK simply by using the same signal construction as Pi/2-BPSK data. The DM-RS is constructed with a binary Gold sequence as an input to the Pi/2-BPSK modulation. This construction guarantees low PAPR but doesn't guarantee flat power spectrum; therefore, doesn't guarantee perfect auto-correlation in the time domain. Nevertheless, the evaluations showed that the DM-RS auto-correlation performance was adequate in the SNR regimes of interest. The new DM-RS sequences introduced in Release 16 for Pi/2-BPSK have 1.6 … 2 dB lower PAPR than ZC sequences.

Similar to the case of ZC sequences, the number of distinct Pi/2-BPSK sequences available for very small RB allocations proved to be insufficient. For these cases, new CGS sequences based on time-domain 8-PSK construction were developed and adopted in Release 16.

10.9.5 PUSCH DM-RS Power Allocation

It was a desirable property for the DM-RS design that the Tx power in OFDM symbols containing DM-RS should be the same as in other PUSCH OFDM symbols containing only data. This is even more important in the UL than in the DL due to the fact that symbol-to-symbol power variation can cause phase discontinuity in the UE transmitter, for similar reasons as in the case of non-coherent MIMO UE capability described in Sec. 10.12.

Equal power across symbols would not hold without special handling because the number of REs occupied by DM-RS in the OFDM symbol containing DM-RS varies based on the number of co-scheduled PUSCH spatial layers. To solve this, the same scaling of the energy per resource element (EPRE) was introduced [1] in the UL as was mentioned for the DL in Sec. 8.12.4. The scaling factor for ERPE is defined as

$$\frac{\text{DM-RS EPRE}}{\text{Data EPRE}}(\text{dB}) = 10 \cdot \log_{10}(\text{Number of CDM groups without data}) \tag{10.7}$$

Note that when the scheduler uses MU-MIMO multiplexing and CDM groups are used for DM-RS of other UEs, those will be counted as CDM groups without data.

Number of DM-RS CDM groups without data	DM-RS type 1	DM-RS type 2
1	0 dB	0 dB
2	−3 dB (10 · $\log_{10}(2)$)	−3 dB (10 · $\log_{10}(2)$)
3	-	−4.77 dB (10 · $\log_{10}(3)$)

TABLE 10.11 The Ratio of PUSCH EPRE to DM-RS EPRE

The scaling factor values are summarized in Table 10.11.

In the case of DM-RS for PUSCH with DFT-S-OFDM waveform, only DM-RS type 1 applies; therefore, the only possible scaling factors are 0 dB and 3 dB.

Applying the power scaling given in Table 10.11 ensures that the Tx power in OFDM symbols containing DM-RS is the same as in other OFDM symbols that contain data only without DM-RS.

10.10 UL Phase Tracking Reference Signal

The UL phase tracking reference signal (PT-RS) was introduced for similar reasons as the DL PT-RS to help mitigate phase noise. The UL PT-RS enables the gNB receiver to estimate and remove the phase noise present in the UE's transmission. Because phase noise is more prevalent at higher frequencies, the main target was introduction in FR2; however, the use in FR1 is not precluded.

The UL PT-RS for CP-OFDM waveform, when single UL PT-RS port is used, is largely similar to the DL PT-RS, but there are significant differences when UL PT-RS for DFT-S-OFDM waveform is used or when two PT-RS ports are used in the UL. These will be described in Secs. 10.10.1 and 10.10.2.

10.10.1 UL PT-RS for CP-OFDM Waveform

Unlike the DL where only one PT-RS port is supported in Release 15, up to two PT-RS ports can be configured in the UL [1]. The UE indicates whether one or two port PT-RS should be used for UL MIMO. The choice depends on the UE transmitter architecture, namely depends on whether common oscillator signal with common jitter is applied for up-conversion in the different Tx chains or not. This is illustrated in Fig. 10.8.

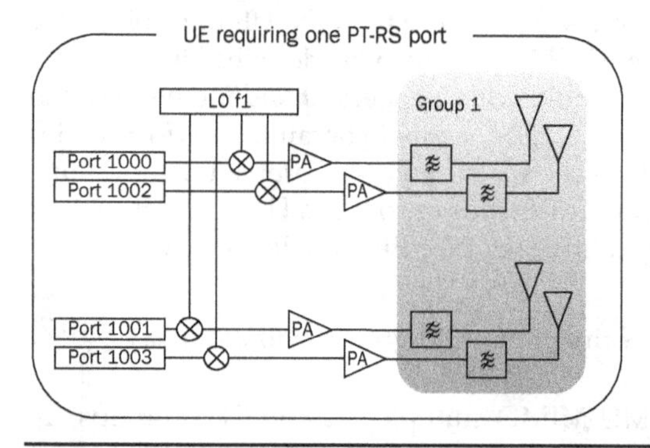

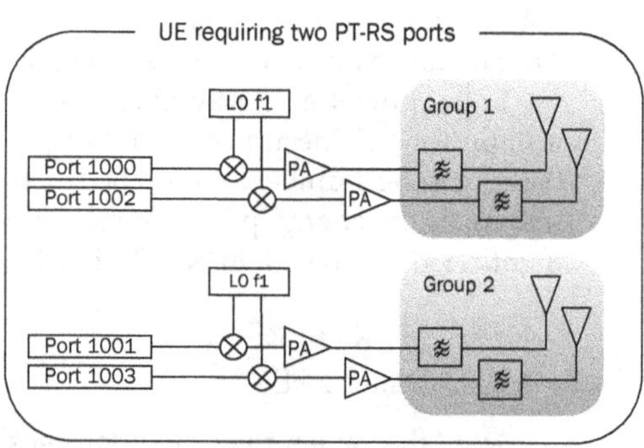

FIGURE 10.8 Example UE architecture requiring one and two PT-RS ports.

Note that when we say two oscillators, we mean two independent phase noise sources. This could occur because of two actually independent oscillators or, more commonly, because the distribution of the signal of a single oscillator incurs independent jitter.

If the UE supports coherent UL MIMO, then it can maintain phase coherence across antenna ports, so it is assumed that common oscillator with common jitter is used. Because of this, only single-port UL PT-RS is used for coherent UL MIMO.

When the UE is configured to use PT-RS in the UL, it is also configured with the maximum number of PT-RS ports that can be used, which can be one or two. The selection of the actual number of ports and the DM-RS port to PT-RS port association is signaled in the UL grant. The main reason to allow for this flexibility is to enable associating PT-RS with the layer with the highest SNR. Since the gNB selects the UL precoding in the case of codebook-based UL transmission or selects the SRS ports in the case of non-codebook-based UL transmission, the gNB can estimate the expected SNR on each layer before sending the UL grant.

When there is a single PT-RS port, a 2-bit "*PTRS-DMRS association*" value in DCI format 0_1 can select any one of the up to four scheduled DM-RS ports for PT-RS association. When there are two PT-RS ports, the association becomes a little more complicated because in selecting the DM-RS port to PT-RS association, the restriction should be observed that one layer from each "oscillator group" needs to be selected, where the groups are organized based on which "oscillator" the port is connected to. For example, if four layers are used, and the UE had recommended using two-port PT-RS based on its Tx architecture, it is beneficial if each of the two PT-RS ports transmitted represents the phase noise of each of the two oscillators, even if this means not picking the two MIMO layers with the highest SNR.

Note that when we say two oscillators, we mean two independent phase noise sources. This could occur because of two actually independent oscillators or, more commonly, because the distribution of the signal of a single oscillator incurs independent jitter.

In any case, only up two PT-RS ports are supported, so when the UE operates with four Tx ports, it must make sure that the phase noise is common within a pair of ports, where the four Tx ports are organized into two pairs. When two PT-RS ports are configured, the MSB of the 2-bit "*PTRS-DMRS association*" value in DCI format 0_1 selects between the first and second DM-RS port within the first group and the LSB selects between the first and second DM-RS port within the second group.

The "oscillator groups" are formed by the UE; however, in order to have a common understanding between the UE and gNB, the following assumptions are made:

- For codebook-based coherent MIMO, all ports are assumed to be connected to a common oscillator. A single PT-RS port is used that can be associated with any of the DM-RS ports allocated for transmission.
- For codebook-based partially coherent MIMO, when the UE recommended using two PT-RS ports then one PT-RS port is used for each of two coherent precoding groups. The coherent precoding groups are formed by the layers that utilize co-phasing coefficients according to the TPMI definition. The first coherent precoding group is formed by PUSCH port 1000 and 1002 and is associated with PT-RS port 0. The second coherent precoding group is formed

by PUSCH antenna port 1001 and 1003 and is associated with PT-RS port 1. The PT-RS port can be associated with any of the DM-RS ports within the group.

- For codebook-based non-coherent MIMO with two ports, when the UE recommended using two PT-RS ports then one PT-RS port is used for each DM-RS port.
- For codebook-based non-coherent MIMO with four ports, when the UE recommended using two PT-RS ports then one PT-RS port is used for each of two groups. The first group is formed by PUSCH port 1000 and 1002 and is associated with PT-RS port 0, and the second group is formed by PUSCH antenna port 1001 and 1003 and is associated with PT-RS port 1. The PT-RS port can be associated with any of the DM-RS ports within the group.
- For non-codebook-based MIMO, each SRS resource is configured with a PT-RS port association. The association maps the SRS resource to either PT-RS port 0 or PT-RS port 1. The SRI field in DCI 1_0 selects SRS port(s), which then also selects the corresponding PT-RS port(s) to be used. If all selected SRS ports are associated with the same PT-RS port then one PT-RS port is used in the transmission, otherwise two PT-RS ports are used.

The density of the UL PT-RS for CP-OFDM waveform both in time and frequency is controlled with the same method as in the DL. The density is a function of other PUSCH allocation parameters [1] with a configurable dependency similar to the DL. The configuration is part of the *PTRS-UplinkConfig* record. One difference between the DL and UL is that when PT-RS is configured but thresholds ptrs-MCS_i and N_{RBi} are not configured, then in the UL, PT-RS is always included with $L_{PT-RS} = 1$ (every OFDM symbol), $K_{PT-RS} = 2$ (every other RB), while in the DL, PT-RS is not included for QPSK and when the number of allocated RBs is less than 3. The reason for this difference is that the UL PT-RS in general might be used for channel tracking by the gNB even in the cases of low SNR or small bandwidth allocation.

The power control of UL PT-RS for CP-OFDM waveform is different compared to the DL. Recycling the power of muted MIMO layers to boost the PT-RS power in the non-muted layer is not supported. This is because the power boost could lead to a symbol-to-symbol power variation across OFDM symbols for the physical antenna ports participating in the PT-RS transmission. This in turn could lead to undesirable loss of phase coherence between symbols, impacting demodulation performance. The same was not a concern for the DL because the gNB transmitter is not expected to vary gain state settings as a function of transmit power. But for the UE transmitter, symbol-to-symbol power variations are to be avoided in general on each physical antenna port.

In the cases when two PT-RS ports are used, a 3 dB PT-RS power boost is still possible without creating a symbol-to-symbol power variation because in this case the power of one muted RE in a layer can be recycled for the non-muted PT-RS RE within the same layer. Therefore, the total power of the layer within the OFDM symbol remains unchanged. This power boost can be turned on or off by configuration parameter *ptrs-Power* included in the *PTRS-UplinkConfig* record. Similar to the DL, there are four possible power offset values configurable by parameter *ptrs-Power* but only two are specified, the other two are reserved for future use [1]. The two specified values indicate whether 3 dB power boost is applied or not, which only applies to the two PT-RS port cases. Note that when the power boost is turned off, there is still a slight symbol-to-symbol power variation introduced. The same is true for the layers that have no associated PT-RS port but are muted for the PT-RS associated with other layers. This power variation was deemed acceptable because the actual power drop is quite small, only 0.2 dB when $K_{PT-RS} = 2$ and only 0.1 dB when $K_{PT-RS} = 4$, and most importantly because the power is only reduced, not increased.

10.10.2 UL PT-RS for DFT-S-OFDM Waveform

In the case of DFT-S-OFDM waveform, there is always a single MIMO layer and single DM-RS port. If UL PT-RS is configured, it is always associated with the single DM-RS port.

The main characteristics of the UL PT-RS for DFT-S-OFDM waveform are that the PT-RS symbols are inserted in the modulation symbol stream before the DFT operation and that instead of QPSK modulation, Pi/2-BPSK modulation is used [7]. The reason for using Pi/2-BPSK modulation is that the PAPR of the PT-RS segment should not be higher than the PAPR of data. Of course, the same modulation for PT-RS as for data could have been used to achieve this goal; however, this would have resulted in variable PT-RS waveform. In order to simplify the design, a common PT-RS waveform was chosen, which then had to be the waveform with lowest PAPR, which is Pi/2-BPSK.

For the PT-RS with DFT-S-OFDM waveform, there is no adjustable frequency-domain density or configurable frequency location, since the DFT spreading distributes the PT-RS across all REs within the allocated RBs.

The density of the PT-RS in the time domain is configurable and comprises two constituents [1]. The first constituent is a configurable density in terms of occupied OFDM symbols. The PT-RS density is once per every L_{PT-RS} OFDM symbols, with $L_{PT-RS} = 1$ or 2. Unlike in CP-OFDM, $L_{PT-RS} = 4$ is disallowed for DFT-S-OFDM, i.e., the minimum density of PT-RS is higher for DFT-S-OFDM. L_{PT-RS} is directly configurable with parameter *timeDensity*, which is part of the record *PTRS-UplinkConfig*. Note that unlike in CP-OFDM, there are no configurable MCS thresholds for the PT-RS time-density selection.

The other constituent is a configurable pattern within the OFDM symbol. The PT-RS is inserted before the DFT in the form of sample groups. There are either 2 or 4 samples per sample group and there are 2, 4, or 8 sample groups that are distributed over time in the OFDM symbol in a relatively even fashion. The pattern selection is dependent on the number of allocated RBs, N_{RB}, as shown in Table 10.12, based on Ref. [1].

Scheduled bandwidth	Number of PT-RS groups	Number of samples per PT-RS group
$N_{RB} < N_{RB0}$	PT-RS is not present	PT-RS is not present
$N_{RB0} \le N_{RB} < N_{RB1}$	2	2
$N_{RB1} \le N_{RB} < N_{RB2}$	2	4
$N_{RB2} \le N_{RB} < N_{RB3}$	4	2
$N_{RB3} \le N_{RB} < N_{RB4}$	4	4
$N_{RB4} \le N_{RB}$	8	4

TABLE 10.12 PT-RS Group Pattern as a Function of Scheduled Bandwidth

The thresholds are configurable, by configuring five N_{RBi} parameters with $i = 0, 1, \ldots, 4$. The purpose of this configurability is to optimize overhead. As N_{RB} increases, the number of pre-DFT samples is increasing; therefore, if the number of pre-DFT samples for PT-RS was constant, it would lead to a diminishing proportion of PT-RS samples. It is reasonable to increase somewhat the number of pre-DFT samples for PT-RS as N_{RB} increases. At the same time, the number of PT-RS samples doesn't need to be increased linearly with N_{RB} because increased intra-symbol time resolution for the phase noise observation is not needed, and furthermore, the PUSCH power increases with increasing N_{RB} anyway, so the PT-RS power would not be diminishing even if the number pre-DFT samples for PT-RS was kept constant.

The distribution of the PT-RS samples prior to DFT is defined in Table 10.13 based on Ref. [7].

M_{sc}^{PUSCH} in Table 10.13 is the number of REs in the allocated RBs, with $M_{sc}^{PUSCH} = 12 \cdot N_{RB}$.

The distribution of the PT-RS samples is illustrated in Fig. 10.9 with taking $N_{RB} = 6$ ($M_{sc}^{PUSCH} = 72$) as example.

For PT-RS scrambling, a BPSK sequence is generated by the same Gold sequence generator as for CP-OFDM DM-RS and where the sequence generator is initialized with the same c_{init} formula as for DM-RS but using the UE specific parameter *nPUSCH-Identity* as the scrambling ID.

Since with DFT-S-OFDM there is always a single layer, there is no layer muting and hence there is no PT-RS power boosting. The PUSCH-to-PT-RS power offset is

Number of groups	Number of samples per group	Index of PT-RS samples in OFDM symbol prior to DFT operation
2	2	$s\lfloor M_{sc}^{PUSCH}/4\rfloor + k - 1$ where $s = 1,3$ and $k = 0,1$
2	4	$sM_{sc}^{PUSCH} + k$ where $\begin{cases} s = 0 \text{ and } k = 0,1,2,3 \\ s = 1 \text{ and } k = -4,-3,-2,-1 \end{cases}$
4	2	$\lfloor sM_{sc}^{PUSCH} / 8\rfloor + k - 1$ where s = 1, 3, 5, 7 and k = 0, 1
4	4	$sM_{sc}^{PUSCH} / 4 + n + k$ where $\begin{cases} s = 0 \text{ and } k = 0,1,2,3 & n = 0 \\ s = 1,2 \text{ and } k = -2,-1,0,1 & n = \lfloor M_{sc}^{PUSCH} / 8\rfloor \\ s = 4 \text{ and } k = -4,-3,-2,-1 & n = 0 \end{cases}$
8	4	$\lfloor sM_{sc}^{PUSCH} / 8\rfloor + n + k$ where $\begin{cases} s = 0 \text{ and } k = 0,1,2,3 & n = 0 \\ s = 1,2,3,4,5,6 \text{ and } k = -2,-1,0,1 & n = \lfloor M_{sc}^{PUSCH} / 16\rfloor \\ s = 8 \text{ and } k = -4,-3,-2,-1 & n = 0 \end{cases}$

TABLE 10.13 PT-RS Sample Group Patterns

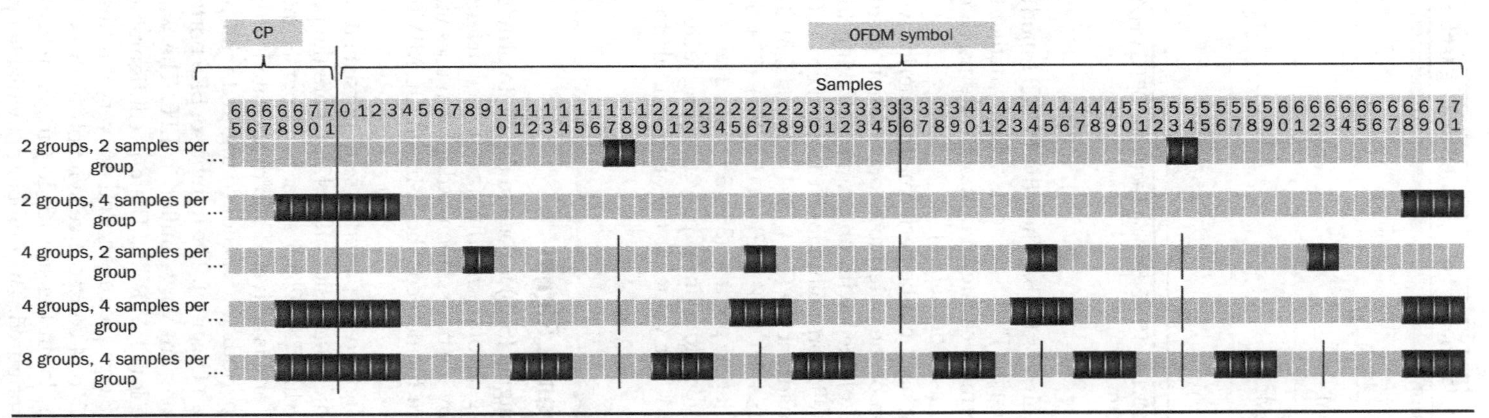

FIGURE 10.9 Distribution of PT-RS sample groups in an OFDM symbol.

Scheduled modulation	PT-RS scaling factor
Pi/2-BPSK	1
QPSK	1
16QAM	$3/\sqrt{5}$
64QAM	$7/\sqrt{21}$
256QAM	$15/\sqrt{85}$

Table 10.14 PT-RS Scaling Factor for DFT-S-OFDM Waveform

not configurable and is always 0 dB. When the PUSCH modulation order is 16QAM, 64QAM, or 256QAM, then the PT-RS power is actually higher than the average PUSCH power because the outermost constellation points are used for PT-RS. This scaling is captured in Table 10.14, based on Ref. [1].

Note that the scaling need not be performed numerically, since it simply captures the effect of the selection of the outermost constellation points. The derivation of the scaling factors can be explained with an example. Taking the 16QAM constellation, assuming that the constellation point closest to the zero is (1, 1), the four constellation points in the first quadrant are (1, 1), (3, 1), (1, 3), (3, 3). The average power of the constellation points is $(1^2 + 1^2) + (3^2 + 1^2) + (1^2 + 3^2) + (3^2 + 3^2) = 40$, while the power of the outermost constellation point is $(3^2 + 3^2) = 18$. The ratio of these two values is $18/40 = 9/5$, which converts to a linear power scaling factor of $\sqrt{9/5} = 3/\sqrt{5}$, same as in Table 10.14 for 16QAM. The value for the other modulation orders can be similarly derived.

10.11 Sounding Reference Signal

The sounding reference signal (SRS) is used in the UL to provide a reference for UL frequency and time tracking, a source for timing advance determination, power control, and channel estimation by the base station. Additional uses include UL beam management, precoded reference for non-codebook-based UL MIMO, and reciprocity-based DL channel estimation.

The SRS can be configured as periodic, semi-persistent, or aperiodic. Periodic is configured with a periodicity and slot offset [2], where the periodicity can be {1, 2, 5, 10, 20, 40, 80, 160, 320, 640, 1280, 2560} slots. Semi-persistent SRS can be configured with similar periodicities and is activated and deactivated with a MAC control element (MAC CE) command. Aperiodic SRS is not configured with periodicity, it can be triggered dynamically with an UL grant. The number of SRS ports can be {1, 2, 4}.

The SRS configurations are very similar to LTE. The waveform is ZC sequence based in order to enable low PAPR [7]. There is configurable comb factor of 2 or 4, which means that the SRS occupies every 2nd or every 4th REs in the allocated RBs.

The SRS occupies 1, 2, or 4 consecutive symbols in the slot, which must be within the last 6 symbols of the slot in Release 15. To further increase multiplexing capability, cyclic shifts can be configured. The number of available cyclic shifts is 12 for comb factor 4 and it is 8 for comb factor 2. When more than one antenna SRS ports are transmitted in the same OFDM symbol, they are on the same comb offset and use equally separated cyclic shifts, where the separation depends on the number of ports. The minimum cyclic shift time separation between antenna ports multiplexed in the same OFDM symbol is given in Table 10.15.

Note that even though the addressable number of cyclic shifts was increased from 8 to 12 for the higher comb factor, the number of practically available cyclic shifts may be reduced, since the comb factor creates repeated copies of shorter base sequences within the symbol, and resolvable cyclic shifts must be confined within those shorter segments. For example, the minimum cyclic shift separation for 4 ports with comb factor 4 is 1/16 = 6.25% of the symbol duration, which is already slightly less than the CP duration (CP duration is 144/2048 = 7.03% of the symbol duration).

There are many bandwidth configuration options for SRS [7]. The bandwidth of an individual SRS transmission is a multiple of 4 RBs. The 4 RB boundaries are counted from reference point A to enable multiplexing SRS of UEs that have different notion of communication channel bandwidth. The maximum SRS bandwidth is 272 RBs. Many of the options with bandwidth 96 RBs or less are the same as in LTE, in order to enhance the multiplexing capability between NR and LTE UEs in a shared LTE-NR UL carrier.

The SRS can be configured with or without frequency hopping. When frequency hopping is configured, it can be intra-slot, where hopping is between the repeated symbols in a slot, or inter-slot where hopping is across the periodic SRS transmission opportunities.

In Table 10.16, some of the SRS configuration parameters are summarized.

SRS resources can be RRC configured [2] with usage {'codebook', 'nonCodebook', 'beamManagement', 'antennaSwitching'}. The configured usage determines the detailed process of the SRS transmission. The first three uses, namely 'codebook', 'nonCodebook' and 'beamManagement' will be described in Sec. 10.12.

Number of SRS ports	Minimum separation as fraction of OFDM symbol duration	
	Comb factor 2	Comb factor 4
1	N/A	N/A
2	1/4 symbol	1/8 symbol
4	1/8 symbol	1/16 symbol

Table 10.15 Minimum Time-Domain Cyclic Shift Separation Between SRS Ports

Parameter	Value
Number of ports	1, 2, 4
Comb factor	2, 4
Number of repeated symbols in slot	1, 2, 4
Periodicity in number of slots	1, 2, 5, 10, 20, 40, 80, 160, 320, 640, 1280, 2560

Table 10.16 SRS Parameters

The use 'antennaSwitching', also called SRS switching, is similar to the LTE SRS switching feature. It is used to provide reciprocity-based channel estimation in the DL when the number of Rx ports in the UE is larger than the number of Tx ports. The SRS transmitted in the Tx ports used for PUSCH, when observed by the gNB, can already provide reciprocity-based channel estimation to the corresponding Rx ports. In order to provide channel estimation to the remaining Rx ports, the UE that has SRS switching capability can transmit SRS from those ports as well. This will create an interruption in the normal UL operation though because if the UE was able to transmit SRS on the other ports simultaneously (with normal UL operation), then the UE would be capable to operate on the same number of Tx ports as the number of Rx ports to begin with.

The UE capability regarding SRS switching is indicated with the parameter *supportedSRS-TxPortSwitch* [5], which can take on the following values in Release 15: {1T2R, 2T4R, 1T4R, 1T4R + 2T4R, 1T=1R, 2T=2R, 4T=4R}. The UE can report only one value from the set for each supported band in a CA band combination. The value x in xTyR represents the number of SRS ports in the SRS resource configured for SRS switching the UE can transmit, while value y in xTyR represents the total number of Rx ports the UE is capable of sounding.

When $x = y$, there is no actual switching involved, the xT = xR capabilities indicate that the UE has the same number of Rx and Tx ports and the same ports are used for Rx and Tx. Note that even with $x = y$, it is not obviously true in all cases that the UE always uses the same physical ports for Rx and Tx; therefore, the explicit reporting of xT = xR capability is meaningful.

When $x < y$, there is actual switching involved. The UE can transmit an x-port SRS resource in any OFDM symbol configured for SRS switching. To sound all y Rx ports, y/x OFDM symbols are dedicated to SRS switching. The UE is allowed a gap to perform the switching, with a one-symbol gap for SCS of 15, 30, and 60 kHz, and a two-symbol gap in the case of 120 kHz. The UE is not required to transmit any signal in the gap.

In Release 16, further SRS switching capability options were introduced in order to indicate capability to perform partial sounding. Partial sounding means that SRS switching is used over a set of ports that is greater than the number of Tx ports but less than the number of Rx ports. For example, the UE can be configured with 1 Tx port and 4 Rx ports. In this case, configuring 1T2R antenna switching represents partial sounding. The additional Release 16 SRS capability options are the following:

- (1T = 1R) + (1T2R)
- (1T = 1R) + (1T2R) + (1T4R)
- (1T = 1R) + (1T2R) + (2T = 2R) + (2T4R)
- (1T = 1R) + (2T = 2R)
- (1T = 1R) + (2T = 2R) + (4T = 4R)
- (1T = 1R) + 1T2R) + (2T = 2R) + (1T4R) + (2T4R)

The UE can report one of the rows. Once the row is selected, the UE must support every element in the row. Note that the Release 16 UE can choose to report none of the above but instead report one of the Release 15 values from the set {1T2R, 2T4R, 1T4R, 1T4R + 2T4R, 1T = 1R, 2T = 2R, 4T = 4R}. It can be observed that each of the possible Release 15 values is included as a subset in at least one Release 16 capability option.

10.12 UL MIMO Scheme

Two different MIMO schemes are supported for the PUSCH [1]:

- Codebook-based transmission, and
- Non-codebook-based transmission

The UE is configured with codebook-based transmission when the RRC parameter *txConfig* is set to 'codebook' and the UE is configured with non-codebook-based transmission when the RRC parameter *txConfig* is set to 'nonCodebook'.

For both UL MIMO schemes, every UL channel transmission has the UL DM-RS precoded the same way as data. In the case of non-codebook-based UL MIMO, the manner in which the PUSCH is precoded is not specified and the precoding itself is transparent to the gNB. In the case of codebook-based UL MIMO, the precoding follows the TPMI included in the UL grant; hence, the precoding is non-transparent to the gNB in this case.

In the UL, up to rank 4 spatial multiplexing is supported with CP-OFDM waveform. The MIMO layers are mapped to a single codeword [1], which means that channel bits from a code block are distributed evenly across the layers. The modulation order and code rate on all layers is the same. The SNR on each layer can be different, but the single-codeword arrangement ensures that the code blocks distributed across the layers will experience the same average SNR and therefore using the same code rate is appropriate. The benefits of the single-codeword mapping include more stable rate prediction.

The layer mapping cases are shown in Fig. 10.10.

Only rank 1, i.e., no spatial multiplexing is supported with the DFT-S-OFDM waveform.

There were a number of UL transmit diversity schemes considered for NR. Both transparent and non-transparent schemes were studied. Similar to the DL, only transparent transmit diversity scheme was adopted, which can be, for example, small delay cyclic delay diversity (SD-CDD). SD-CDD means that multiple transmit ports transmit the same signal but with a small time delay difference relative to each other. Although the diversity provided by SD-CDD is not the best, especially when small number of RBs are allocated for data transmission, this is compensated by the fact that SD-CDD requires only a single DM-RS port (per layer), not two, which reduces overhead. Another advantage of SD-CDD is simpler interference estimation. When interfered by PUSCH, the interference can always be accurately estimated by observing the DM-RS of the interferer.

10.12.1 Codebook-Based Transmission

For codebook-based transmission, the precoder is selected by the gNB based on observing prior SRS transmissions. If the UE supports up to 2-layer MIMO, then SRS resources

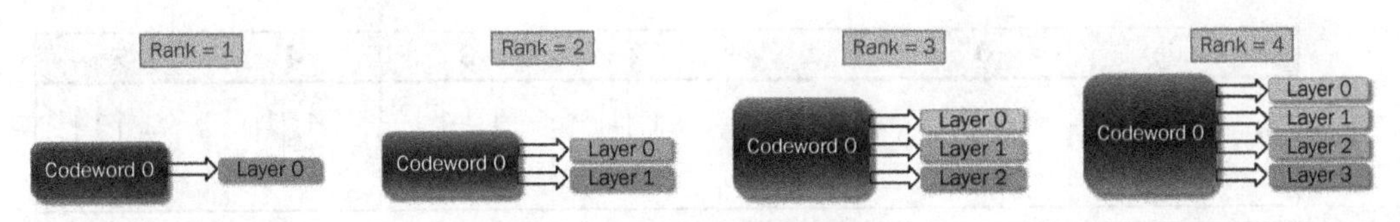

FIGURE 10.10 Illustration of the codeword to layer mapping cases in the UL.

with two ports in every SRS resource can be configured. If the UE supports up to 4-layer MIMO, then SRS resources with four ports in every SRS resource can be configured. The gNB selects a precoder from the specified set of precoders and indicates the selected precoder in the transmit precoder matrix indicator (TPMI) field in the UL grant. The TPMI defines precoding that is to be performed on top of the ports used for SRS transmission. Therefore, the TPMI defines a linear relationship between SRS ports and DM-RS ports. In order to apply the precoding vector, the UE should be able to replicate the same phase difference between physical ports used for the PUSCH transmission as was observed between the SRS ports. Not all UEs are capable to perform this though. The UEs indicate their capability for codebook-based MIMO as one of the following:

- Coherent MIMO capable
- Partial coherent MIMO capable
- Non-coherent MIMO capable

The coherent MIMO capable UE can replicate the same phase offset between antennas for PUSCH transmissions as for the prior SRS transmissions. The specified precoding for coherent MIMO includes co-phasing coefficients between any antenna port pair. As it can be seen in Table 10.17, the co-phasing coefficient can be $\{1, j, -1\ -j\}$, which gives the set of precoders for TPMI = 2, 3, ..., 5 for single-layer 2 Tx MIMO [7].

Partial coherent MIMO applies only to 4 Tx. In the case of partial coherent MIMO, the UE can keep phase coherence within the first coherent precoding group formed by PUSCH port 1000 and 1002. The UE can also keep phase coherence within the second coherent precoding group formed by PUSCH antenna port 1001 and 1003. But the UE cannot keep phase coherence across the two groups.

Non-coherent MIMO can apply to both 2 Tx and 4 Tx. In the case of non-coherent MIMO, the UE is not able to keep phase coherence between any pair of antenna ports. One of the reasons phase coherence is not kept is that when power changes happen, the RF hardware applies various switches to optimize performance and to minimize power consumption. An example with a two-stage PA is shown in Fig. 10.11. Note that these types of PAs are not commonly used today but other techniques, like envelope tracking, can create similar effects. The two-stage PA has two power states: a low-power (or low-gain) state and a high-power (or high-gain) state. In the high-power state, both PA stages are utilized, while in the low-power state, the second stage is bypassed and only the first stage is used.

In the example shown in Fig. 10.11, the phase offset between the two antennas in the low-power state is $\Delta\Phi_L = (\varphi_{1,1} - \varphi_{2,1})$ and in the high-power state it is $\Delta\Phi_H = (\varphi_{1,1} - \varphi_{2,1}) + (\varphi_{1,2} - \varphi_{2,2})$, where $\varphi_{1,1}, \varphi_{1,2}, \varphi_{2,1}, \varphi_{2,2}$ are phase changes introduced by each of the PA stages in the two PAs. Because of the different RB allocations for SRS and PUSCH, it is often possible that they are transmitted in different power

TPMI index	0	1	2	3	4	5
W	$\frac{1}{\sqrt{2}}\begin{bmatrix}1\\0\end{bmatrix}$	$\frac{1}{\sqrt{2}}\begin{bmatrix}0\\1\end{bmatrix}$	$\frac{1}{\sqrt{2}}\begin{bmatrix}1\\1\end{bmatrix}$	$\frac{1}{\sqrt{2}}\begin{bmatrix}1\\-1\end{bmatrix}$	$\frac{1}{\sqrt{2}}\begin{bmatrix}1\\j\end{bmatrix}$	$\frac{1}{\sqrt{2}}\begin{bmatrix}1\\-j\end{bmatrix}$

TABLE 10.17 Precoding Matrix W for Single-Layer Transmission Using Two Antenna Ports

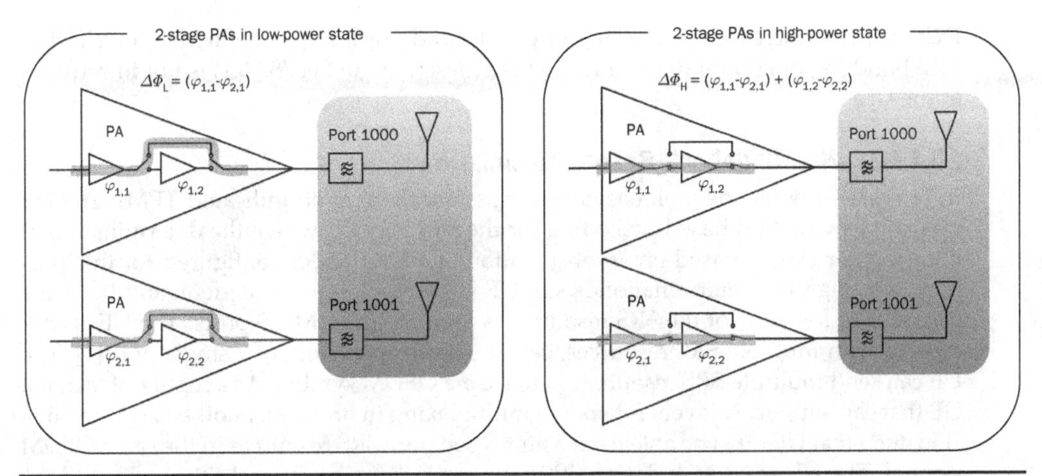

FIGURE 10.11 Example two-stage PAs in low-power and high-power states.

states. For example, a wide band SRS may be transmitted in the high-power state and a narrow band PUSCH may be transmitted in the low-power state. In this case, there will be a $\Delta\Phi_H - \Delta\Phi_L = (\varphi_{1,2} - \varphi_{2,2})$ phase difference between the SRS and PUSCH antenna phase offsets. Since typically $\varphi_{1,2} \neq \varphi_{2,2}$, the antenna phase offset for SRS and PUSCH will be different and hence phase continuity is not kept.

Because of the lack of control of phase offsets between the antenna ports, no co-phasing terms are defined for the TPMI for non-coherent MIMO. Taking the 2 Tx rank 1 transmission as an example, only TPMI = 0 and TPMI = 1 in Table 10.17 are used for non-coherent MIMO in Release 15. As it can be seen, this means that with rank 1, the non-coherent MIMO capable UE will use only a single antenna port at a time. For UEs that have two PAs, each providing half of the total power required, this means that the non-coherent MIMO UE cannot reach the required maximum power with rank 1 transmission. For example, a 2 Tx power class 3 UE can utilize two 20 dBm PAs to reach the required 23 dBm total power. If this UE is non-coherent MIMO capable, then the UE can only reach 20 dBm total power with rank 1 transmission. This shortfall is corrected in Release 16, where TPMI = 2 can also be used for non-coherent MIMO capable UEs, together with applying a phase randomization scheme, such as SD-CDD, thereby allowing these UEs to reach 23 dBm total power with rank 1 transmission.

An added feature introduced in NR is the ability to configure more than one SRS resource [1]. In the case of multiple SRS resources, the UL grants include an SRS resource indicator (SRI) field, and in the case of codebook-based transmission, the PUSCH precoding indicated by the TPMI is to be performed based on the SRS resource selected by SRI, i.e., the precoding is applied on transmit ports that are equivalent to the SRS ports in the selected SRS resource. This feature can be used, for example, for UL beam sweep. The SRI field comprises $\log_2(N_{SRS})$ bits, where N_{SRS} is the number of configured SRS resources. In the case of codebook-based transmission, all configured SRS resources must have the same number of SRS ports.

There was a debate about whether to introduce subband-based precoding for coherent, non-coherent, or partial coherent MIMO in the UL for the CP-OFDM waveform. It was decided not to introduce this scheme due to its limited gain and relatively large

DCI overhead increase. As a result, only wideband precoding is defined in the UL. The PRB bundling operation defined for the DL (described in Sec. 8.16.1) is not introduced in the UL.

10.12.2 Non-Codebook-Based Transmission

In the case of non-codebook-based transmission, there is no indicated TPMI. The UE uses implementation-based precoding for the SRS already, where the determination of the precoder can be based on an observation of a DL CSI-RS configured for this purpose, although the determination is still UE implementation dependent, and the same precoder as for some of the SRS resources is used for the DM-RS ports. The UE is configured with multiple SRS resources, each resource consisting of a single port [1]. The UE can send multiple SRS resources in the same OFDM symbol. As a matter of fact, if a UE that can support N layers for spatial multiplexing in non-codebook-based transmission, then that UE must be able to transmit N separate SRS resources in the same OFDM symbol. The UL grant includes an SRS resource indicator (SRI) field that selects SRS k resources, each of which will be mapped to a spatial layer of the PUSCH with a one-to-one mapping. In turn, the layers are mapped to PUSCH ports in the range of port 1000,..., 1003. The SRI field jointly codes the selection of k and the $\binom{N_{\text{SRS}}}{k}$ possible choices of the SRS resources, in which N_{SRS} is the number of SRS resources. The number of bits of the SRI field is thus determined as $\left\lceil \log_2\left(\sum_{k=1}^{M}\binom{N_{\text{SRS}}}{k}\right)\right\rceil$, where M is the maximum number of configured SRS resources that can be selected, which is limited by the lesser of the number of SRS resources and the maximum configured number of MIMO layers [4].

There is no precoding performed on top of the selected set of SRS resources, or saying it differently, the precoding is the identity. Of course, this does not mean that there is no precoding at all. The precoding was performed already as part of the SRS transmission. We note that the functionality of non-codebook-based transmission has a lot of similarity with non-coherent codebook-based MIMO when each SRS resource is chosen to correspond to a physical Tx antenna port in the UE.

10.13 Beam Management for the PUSCH

The beam management in the UL is somewhat different from the beam management in the DL. The UL beam management relies heavily on the feature called beam correspondence, where the UE selects the Tx spatial domain filter (i.e., analog beamforming and panel selection for Tx) for the UL that is the same as the Rx spatial domain filter (i.e., analog beamforming and panel selection for Rx) for the DL. Due to this, at least for UEs supporting full beam correspondence (which is a mandatory Release 15 feature), there are no beam management processes and/or QCL relationship types defined in the UL beyond those already used in the DL. However, for UEs with only partial beam correspondence, beam management processes in the UL, in addition to those already used in the DL, can be still used to utilize beam sweep based on multiple configured SRS resources. Also, instead of QCL relationships in the DL, spatial relations are defined. The spatial relations define a mapping between a source signal, which can be either a DL or UL signal, and a target UL signal/channel. The Tx spatial domain filter (i.e., analog beamforming and panel selection) for PUSCH, and in certain cases for PUCCH, is set to be the same as for one or multiple of the SRS resources, and the requested UE Tx beamforming is indicated from the gNB to the UE with setting the SRI [1]. The Tx spatial domain filter for the SRS is primarily derived based on beam correspondence to a selected DL signal at least for UEs supporting full beam correspondence.

Note that beam correspondence represents an even higher level of requirements for the UE than coherent MIMO described earlier. For beam correspondence, the UE must manage not only phase differences of PAs driving the Tx antenna elements but also the phase difference of low noise amplifiers (LNAs) connected to the Rx antenna elements employed in the receiver. This requires close integration of the relevant RF components. Nevertheless, in order to avoid the high cost in network resources that would have been incurred by accommodating random beam sweep–based UL beam management for all UEs served by a gNB, beam correspondence–based operation was made mandatory for Release 15 UEs.

There is a reason why fully autonomous UL beam selection by the UE, solely based on beam correspondence, even with perfect beam correspondence, would not work well. The DL Tx beam and DL Rx beam form a DL beam pair. The corresponding UL Rx and UL Tx beams form an UL beam pair. This is illustrated in Fig. 10.12. If the UE autonomously selected the best UL Tx beam for transmission, it may be out of sync with the gNB's UL Rx beam selection. For example, if the UE selected UL Tx beam 1 as the best beam, while the gNB prepared to receive the PUSCH with UL Rx beam 2, then this would cause a beam mismatch.

In order to support synchronized beam selection at the gNB and the UE for the PUSCH, a multi-step process was introduced, which is described in Fig. 10.13.

The beam management allows multiple SRS resources to be used. The gNB configures spatial relations for each SRS resource, with configuring an associated DL reference signal for each SRS resource to be used by the UE as the source for beam correspondence. At the first reception of the SRS, the gNB may use an initial Rx beamforming that corresponds to the Tx beamforming that the gNB had been using to transmit the DL signal. Subsequently, the gNB can refine Rx beams for each SRS resource. This beam management is based on SRS transmissions. When it comes to PUSCH transmissions,

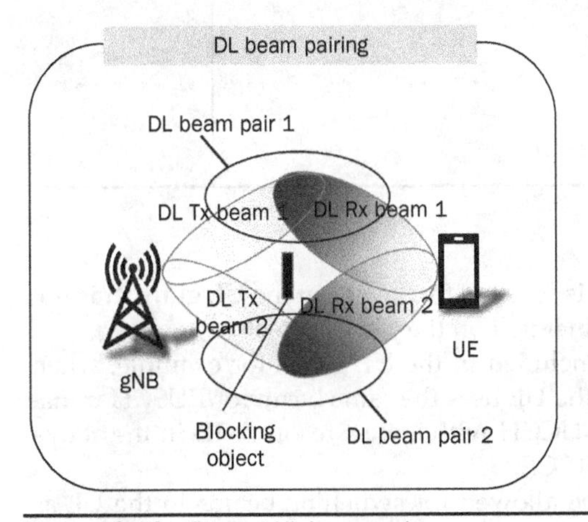

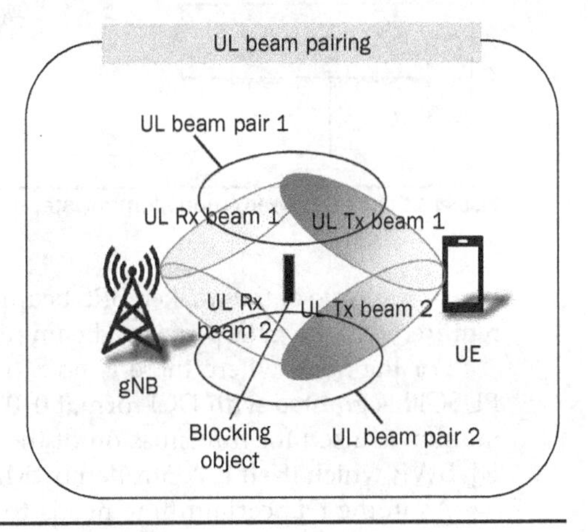

FIGURE 10.12 DL and UL beam pairs.

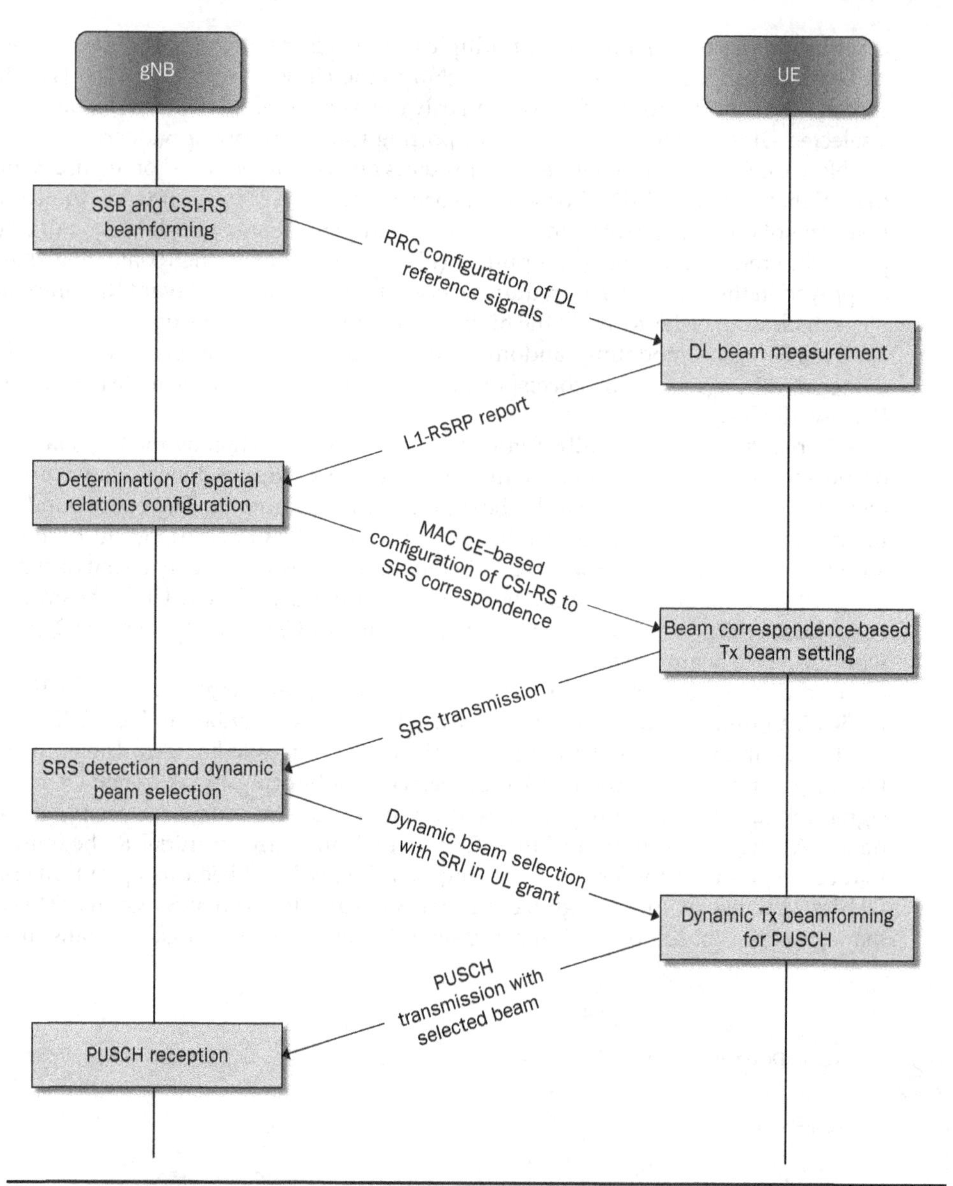

FIGURE 10.13 UL beam management steps.

one or multiple of the tracked SRS beams is selected by the SRI in the UL grant. It is not required that the gNB performs beam refinement on the PUSCH itself.

For the cases where there is no SRI included in the UL grant, for example, when PUSCH is granted with DCI format 0_0, the UE uses the same beam for PUSCH transmission as used for transmission of the PUCCH with lowest resource ID in the active UL BWP, which itself is controlled by MAC CE.

As in the DL, certain time needs to be allowed for switching beams in the UL as well. The UE needs time to decode the UL grant containing the SRI and even after the

SRI information is decoded, it takes certain time to apply the indicated beamforming coefficients. The total time between the PDCCH carrying the SRI and the application of the indicated beamforming coefficients, as measured from the end of the PDCCH to the beginning of the PUSCH, could be reasonably assumed to be similar to the *timeDurationForQCL* used in the DL. The largest reportable value for *timeDurationForQCL* was 28 symbols [5]. We can note that the smallest number of symbols allowed for PUSCH preparation for 120 kHz SCS is 36, as discussed in Sec. 10.8; therefore, the time allowed for SRI-based switching can be absorbed in the UE's PUSCH preparation time, and no specific beam switch time was needed to be defined for the UL.

In the case of aperiodic SRS that is associated with an aperiodic CSI-RS, the UE is allowed 42 symbols between the reception of the NZP CSI-RS resource and the transmission of the SRS. This allows the UE to process the received NZP CSI-RS, similar to that for CSI calculation. The aperiodic SRS can be triggered to be transmitted at an earlier time but in this case, the UE is not required to update the SRS precoding information based on the aperiodic NZP CSI-RS.

10.14 UL Power Control

The principles of UL power control applicable in a single carrier are similar to LTE. Open-loop and closed-loop power control is applied [8]. The UL power control equation for PUSCH applicable to a single carrier is given in Fig. 10.14.

Similar to LTE, fractional power control can be applied with setting $a_{b,f,c}(j) < 1$, which results in scaling the open-loop power control so that it doesn't fully compensate for the estimated pathloss. This allows increasing power for cell center UEs that are expected to cause less interference to other cells even with the increased power.

The new features of NR power control compared to LTE are listed below:

- Accommodating different SCS by scaling bandwidth in the power control equation in Fig. 10.14
- Up to two separate power control loops introduced
- Configurable reference signal selection as pathloss measurement source introduced
- Accommodating variable grant to PUSCH scheduling delay
- Dependency on BWP

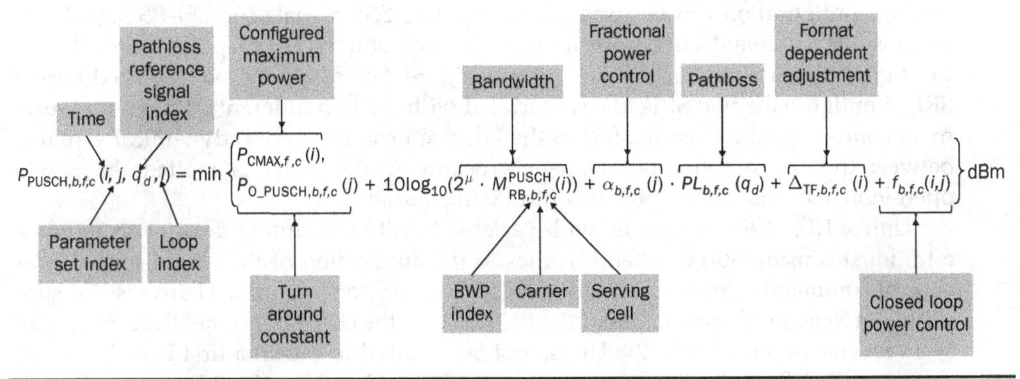

FIGURE 10.14 Single-carrier UL power control equation for PUSCH.

Adopting the power control to different SCS is achieved with scaling the number of allocated RBs by 2^{μ}, where $\mu = 0, 1, 2, 3$ for SCS 15, 30, 60, and 120 kHz, respectively. This is because the bandwidth of one RB is scaled by the same factor.

NR supports two independent power control loops when configured [8]. The switch between the two loops is accomplished by setting the SRS resource indicator (SRI) in the UL grant. When more than one SRS resource is configured, each SRS resource can be associated with either the first or the second power control loop. The SRI setting in the UL grant has two effects:

- The SRI selects the power control loop (first or second) whose accumulated closed-loop value gets updated by adding the transmit power control (TPC) value read in the current UL grant
- The SRI selects the power control loop (first or second) whose accumulated closed-loop value is to be used in the power control equation in Fig. 10.14 for the granted PUSCH transmission.

The main purpose of introducing two power control loops was to enable the UE communicating with two transmission/reception points or communicating with a base station using two panels or multiple beams in a dynamic time-switched manner. In these cases, the SRI in the UL grant allows dynamic selection of the reception point/panel/beam. As the reception point/panel/beam for the PUSCH keeps getting switched back and forth, the proper receive power at the gNB can be maintained by adjusting the transmission power independently for each reception point/panel/beam. In order to allow power control monitoring even if PUSCH is not transmitted, two separate SRS resources are configured, where each SRS resource has a different target reception point/panel/beam.

Configurable DL reference signal as pathloss measurement source was introduced to enable multiple independent open loops for power control [8]. The number of pathloss measurement sources is not limited to two. For example, each BWP can be configured with a different pathloss measurement source. Dynamic switching between open loops is accomplished by setting the SRI in the UL grant. Each SRI value can be preconfigured with an associated pathloss reference source. The SRI setting in the UL grant activates the corresponding open loop in the power control equation in Fig. 10.14.

In the setting where the UE is communicating with two transmission/reception points or communicating with a base station using two panels or two beams, the following configuration can be used. Configure two SSB signals (or CSI-RS signals) as pathloss measurement source, one for each transmission/reception point/panel/beam. Configure two SRS resources linked to two different closed loops and to two different SRIs. Configure the two SRIs to be associated with the two different pathloss measurement source signals. Then the SRI in the UL grant can be used to dynamically switch between the two reception points/panels/beams. As the switch is applied, the correct open-loop and closed-loop power control will be used.

Unlike LTE, NR can operate with variable HARQ scheduling delay. This requires additional considerations when it comes to the application of the closed-loop power control commands. As it was mentioned in Sec. 10.5, out-of-order HARQ is not supported in Release 15, meaning that the PUSCH and the corresponding UL grant cannot be in reverse order. That is, the UE cannot be required to transit a first PUSCH before a second PUSCH if the UL grant for the first PUSCH was received later than the UL

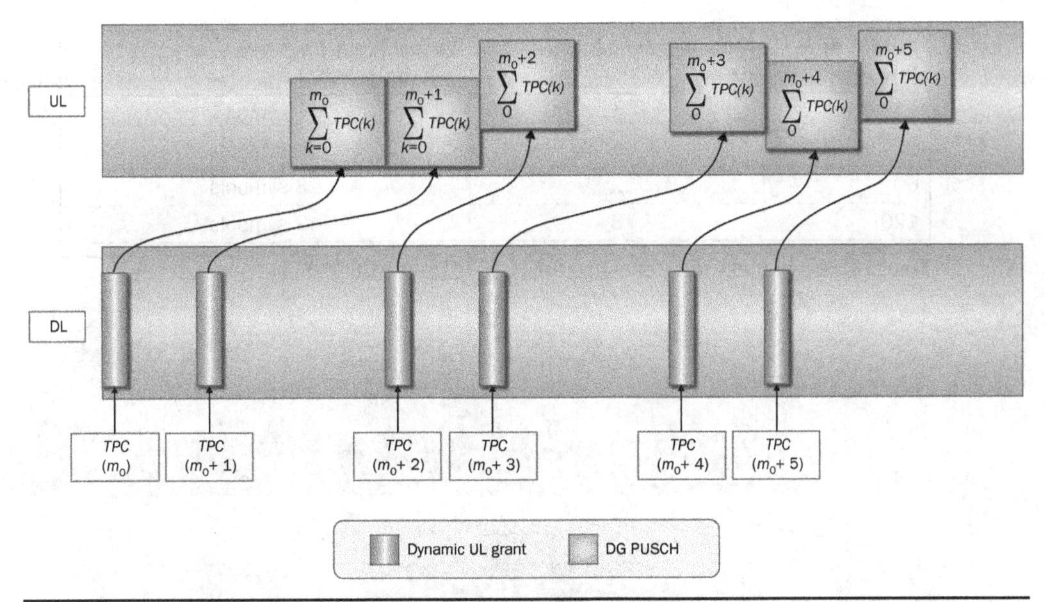

FIGURE 10.15 Power control accumulation with in-order HARQ PUSCH.

grant for the second PUSCH. This makes the application of TPC at least for dynamic grant–based PUSCH straightforward. The TPC commands can be applied in the order the grants are received, or in the order of PUSCH transmissions; the power control operation is identical in both cases. An example is shown in Fig. 10.15.

In addition to PUSCH with dynamic grants (DG PUSCH), PUSCH transmissions with configured grant (CG PUSCH) can be present. CG PUSCH transmissions do not have a dynamic grant, so it becomes ambiguous which power control command, i.e., accumulation up to what point, should apply to them.

In addition to TPC command in dynamic grants, TPC command in group power control messages can be present. Group power control messages are DCI format 2_2 with CRC scrambled by a TPC-PUSCH-RNTI. Since these do not have scheduling time associated with them, there is ambiguity in their time of applicability.

Both the above-mentioned ambiguities were resolved by the following two additional definitions:

- A PUSCH transmission with a configured grant (CG PUSCH) has a corresponding "virtual grant" defined. The virtual grant is $K_{PUSCH}(i)$ symbols before the CG PUSCH transmission, where $K_{PUSCH}(i)$ is the number of symbols corresponding to the minimum UL scheduling slot delay in the default time-domain resource allocation tables. The default time-domain allocation tables were described in Sec. 10.2 and is repeated in Table 10.18.
- The group power control message is inserted in the sequence of the received dynamic UL grants and self-generated CG PUSCH virtual grants in the order they were received or generated.

After the above definitions are adopted, the power control accumulation operation can be performed in the order of time sequence of received power control commands.

PUSCH SCS (kHz)	Minimum scheduling slot delay	Virtual grant time advance $K_{PUSCH}(i)$ before CG PUSCH
15	1	14 symbols
30	1	14 symbols
60	2	28 symbols
120	3	42 symbols

Table 10.18 Virtual Grant Time Advance $K_{PUSCH}(i)$ Before CG PUSCH

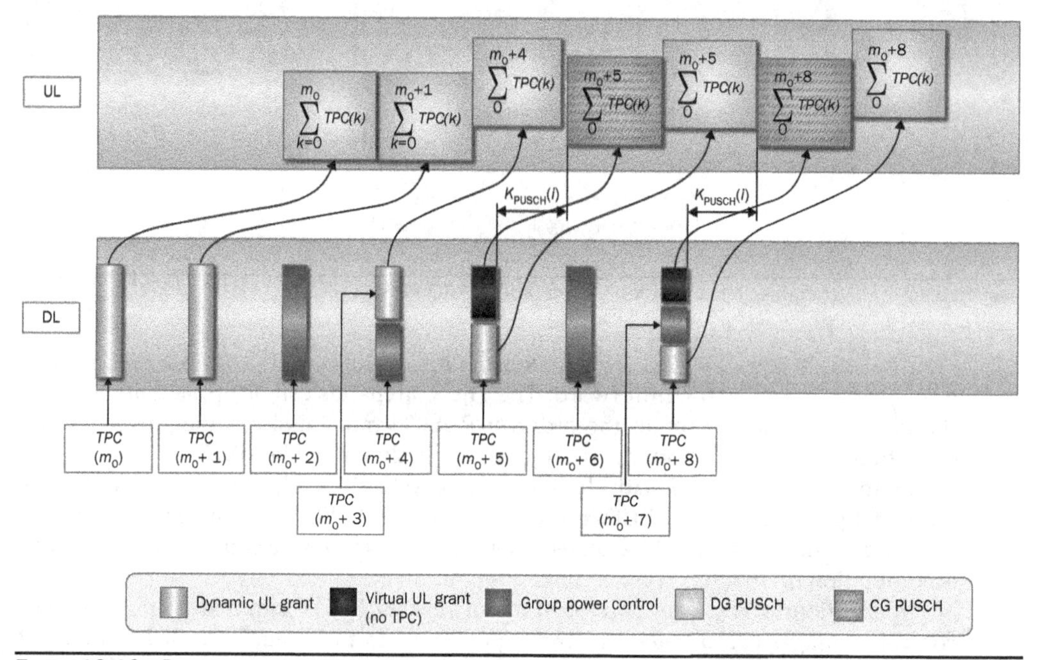

Figure 10.16 Power control accumulation with group power control and CG PUSCH.

The DG PUSCH uses all TPC received, including unicast TPC and group TPC up to the time of its grant. Similarly, the CG PUSCH uses all TPC received, including unicast TPC and group TPC up to the time of its virtual grant. This operation is illustrated in Fig. 10.16.

The power control parameter sets are independently configured for each bandwidth part (BWP). At the time of BWP switch, the power control parameters are also switched.

10.14.1 Power Scaling in Carrier Aggregation

The operation with carrier aggregation (CA) is defined similarly to the single carrier operation with the additional functionality that the power of certain transmissions needs to be scaled down, in order to make sure the total permitted power across carriers is not exceeded. Such power scaling is only applied within a given frequency range, i.e., within FR1 or FR2 because up to now, joint power limits across FR1 and FR2 have not been specified.

Unlike in LTE, NR allows the time-domain resource allocation for UL transmissions to be variable. This can give rise to frequent partial overlap situations across the aggregated UL carriers. When a new transmission on a CC starts in the middle of another transmission in another CC, then the power of the ongoing transmission needs to be re-scaled in some cases. In order to solve this, the NR UL power control equations according to Fig. 10.14 are evaluated on a per symbol basis, not on a channel transmission occasion basis as in LTE. This means that the transmission power of a given channel is evaluated in every symbol and it may change during the given channel transmission [8]. An example for this is shown in Fig. 10.17.

When the transmission power of a channel is changed during its transmission, then it is not guaranteed that signal phase continuity is maintained, examples of which are shown in Fig. 10.17. Phase continuity in the UE transmitter is lost for similar reasons as in the case of non-coherent MIMO UE capability described in Sec. 10.12. Losing phase continuity is of course detrimental to demodulation performance at the gNB (except in certain special cases, such as power change at a frequency-hop boundary). It is expected that the schedulers of the aggregated cells will coordinate with each other within the gNB to avoid the cases leading to phase discontinuity, or at least indicate internally in the gNB when phase continuity cannot be counted on. In order to enhance the chances of phase continuity, the UE is allowed not to increase power during a transmission if it would lead to phase discontinuity. This is illustrated in Fig. 10.18. The UE is allowed to use Option 2 shown in Fig. 10.18, which avoids phase discontinuity.

Somewhat different considerations apply to intra-band contiguous UL CA. When the UE is configured with intra-band contiguous UL CA, then the signal of two or more UL CCs will be transmitted by the same RF chain. In this case, whenever partial overlap occurs, there is a power change for the common PA, even without any power scaling. An example case is shown in Fig. 10.19. Whenever a PA power change happens, phase continuity may not be maintained. Note that due to this limitation, it is expected that in the intra-band contiguous UL CA case, the schedulers coordinate so that either no overlap or full overlap is maintained across carriers, thereby precluding partial overlap.

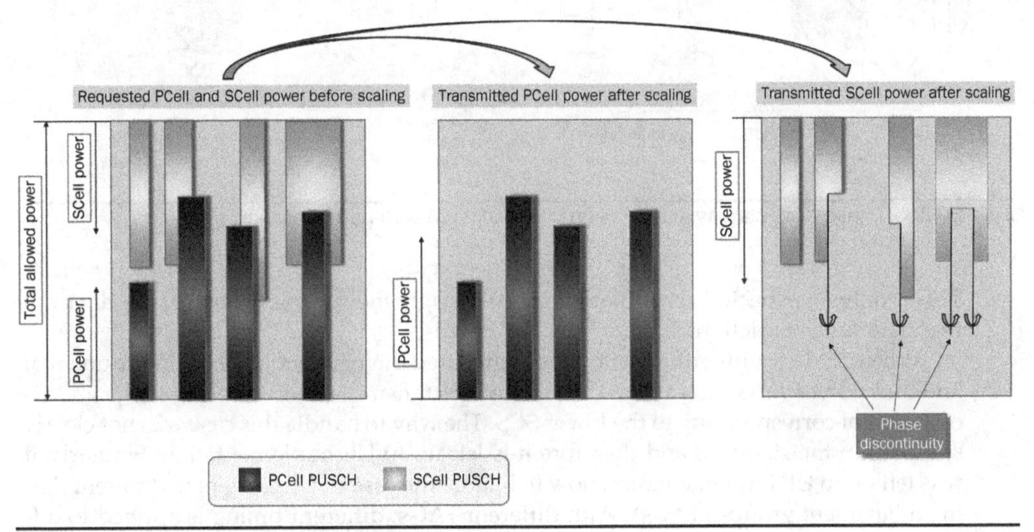

FIGURE 10.17 Power scaling in UL carrier aggregation.

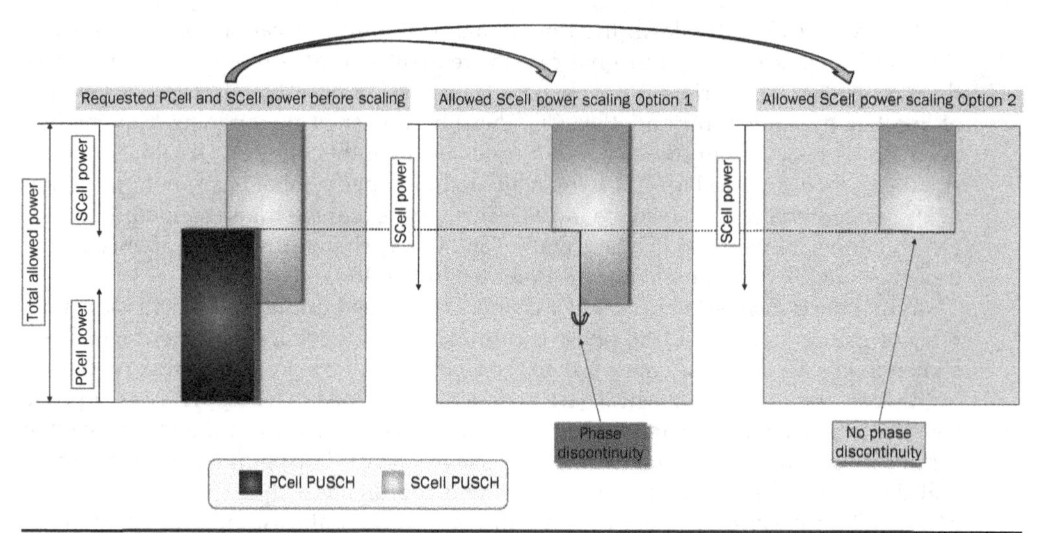

FIGURE 10.18 Allowed SCell power scaling options.

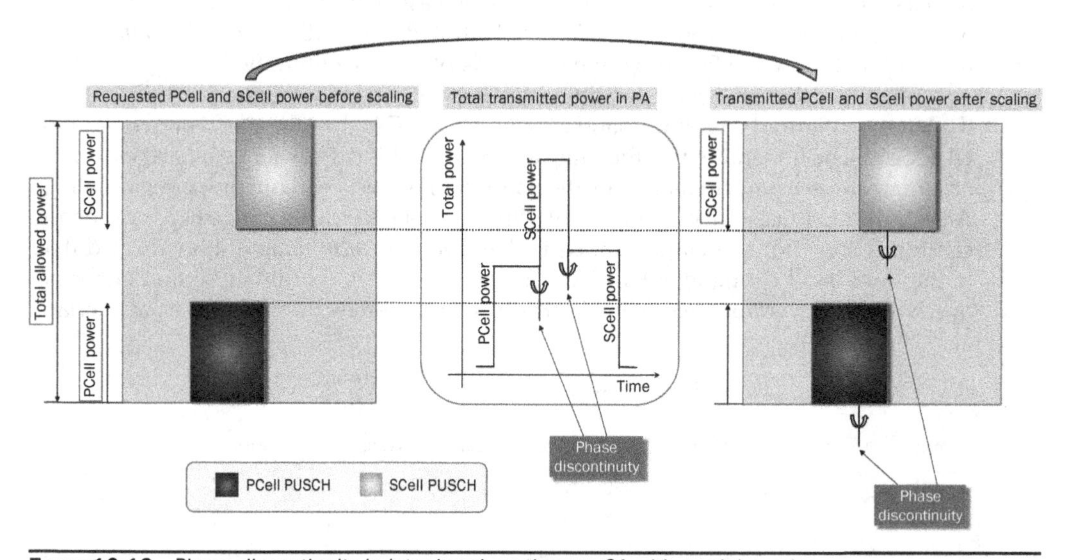

FIGURE 10.19 Phase discontinuity in intra-band contiguous CA with partial overlap.

This is only an expected scheduler behavior though; the specification actually does not mandate such restrictions.

When carriers with different SCS are aggregated then the symbol lengths are unequal and thus power variations due to the higher SCS transmission could occur in the middle of a symbol corresponding to the lower SCS. The way to handle this case was not clearly specified in the standard and therefore it is left up to UE implementation. Similarly, it was left up to UE implementation how to handle the case of UL carriers in different timing adjustment groups (TAGs). With different TAGs, different timing is applied to different UL carriers, so the power transitions and symbol boundaries may not be aligned.

As a UE implementation based option, the misalignment can be absorbed within the allowed power transient periods around the power change, similar to LTE.

When scaling the transmission power due to power limitation in CA, certain priority order was established. In general, PCell transmissions are prioritized over SCell, and control information transmissions are prioritized over data transmissions. The list of priority order was defined [8] as follows, with (1) being the highest priority and (5) being the lowest:

1. PRACH transmission on the PCell
2. PUCCH transmission with HARQ-ACK information and/or SR or PUSCH transmission with HARQ-ACK information
3. PUCCH transmission with CSI or PUSCH transmission with CSI
4. PUSCH transmission without HARQ-ACK information or CSI
5. SRS transmission, with aperiodic SRS having higher priority than semi-persistent and/or periodic SRS, or PRACH transmission on a serving cell other than the PCell

When scaling needs to be applied across channels at the same priority level, then PCell has higher priority level than SCell. When scaling needs to be applied across channels at the same priority level within SCells, then applying equal scaling (in dB) to all such channels can be assumed, although this is not explicitly specified.

We again note that all the scaling rules are applied within the given frequency range (FR1 or FR2), there is no power control scaling interaction between FR1 and FR2.

10.14.2 Power Scaling in NR-NR Dual Connectivity

As it was mentioned, the maximum power-related power scaling has a risk of introducing phase discontinuities in the UL transmission, and in order to mitigate the impact, it is expected that there is close coordination of the scheduling among aggregated cells. While this works in CA, it is not expected that the same would work in dual connectivity (DC). In DC, the serving cells for a given UE are not expected to be able to perform dynamic coordination. So, when one cell issues a UL grant, that cell will not know whether and where phase discontinuity might occur in the received signal due to power scaling. This was seen unacceptable in DC. Although DC was not expected to be part of Release 15, it was added as a late feature. Since there was insufficient time to develop appropriate power control specifications, the Release 15 version of DC was limited to FR1-FR2 DC only, where the two cell groups (CGs) are fully confined in FR1 and FR2, respectively. Since there is no interaction between FR1 and FR2 power control, this restriction ensured that the lack of inter-CG coordination had no impact on power control or phase continuity. In Release 16, the remaining cases, including FR1-FR1 DC and FR2-FR2 DC were added. For this, additional power scaling methods had to be developed. The following four power scaling methods were introduced:

- Semi-static power sharing without duplex direction check
- Semi-static power sharing with duplex direction check
- Dynamic power sharing without look-ahead
- Dynamic power sharing with look-ahead

The specification work for the above is still ongoing at the time of writing, so the following description may require future updates.

A common feature of all four DC power control variants is that within cell groups, the CA power control rules were retained. In DC, there are two cell groups (CGs), the master cell group (MCG) and secondary cell group (SCG). For the description we assume that within the UE, there are separate processing blocks, sometimes called tiles, dedicated to processing the MCG and SCG, respectively. This is, of course, just a conceptual description. It is not actually specified whether the UE indeed have separate processing blocks or not. It may well be also possible that the UE has two processing blocks in the UL but a common processing block in the DL, or vice versa, or common processing block in both the DL and UL, and so on. Using the conceptual description of processing blocks, the DC power control can be described with saying that there is only a one-time unidirectional information exchange for any transmission occasion, where the requested total MCG power is communicated from the MCG block to the SCG block. This is illustrated in Fig. 10.20.

Next, we give a brief description of the four power control schemes.

In the semi-static power sharing method, a configurable power limit is specified that is independently settable for the MCG and SCG. The linear sum of the semi-static thresholds must be less than the total power allowed. When a CG includes both FR1 and FR2 cells, a separate power limit is specified for the CG in FR1 and FR2 and the linear sum of the semi-static thresholds for MCG and SCG must be less than the total power allowed within the frequency range. This is because of the independent power control across FR1 and FR2.

For the case of semi-static power sharing without duplex direction check, the semi-static thresholds must be complied with at all time. This case is illustrated in Fig. 10.21.

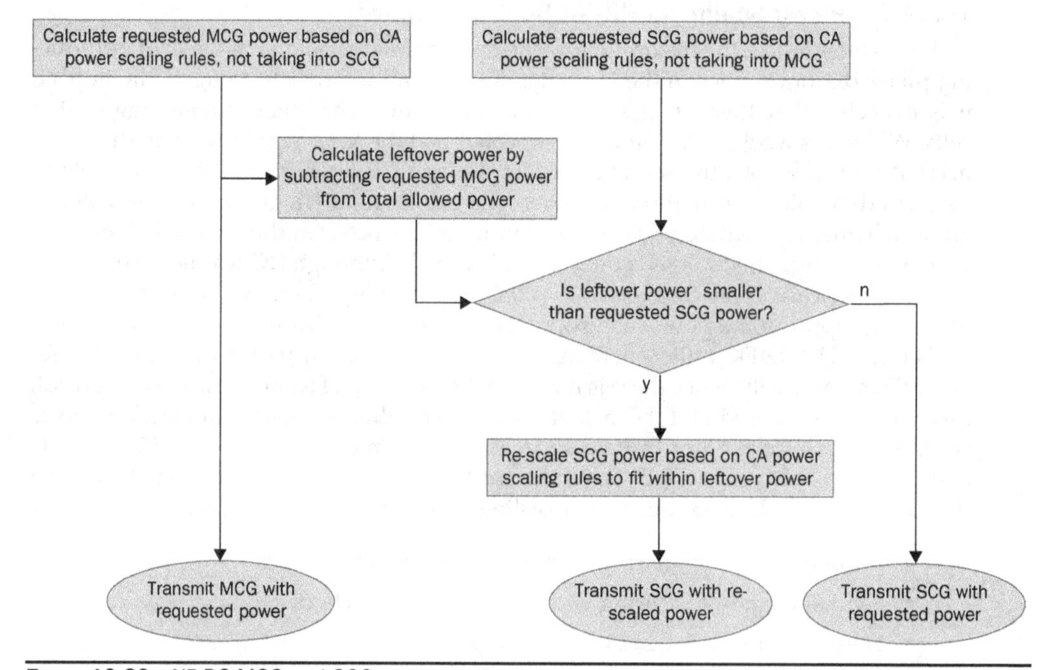

FIGURE 10.20 NR-DC MCG and SCG power scaling process.

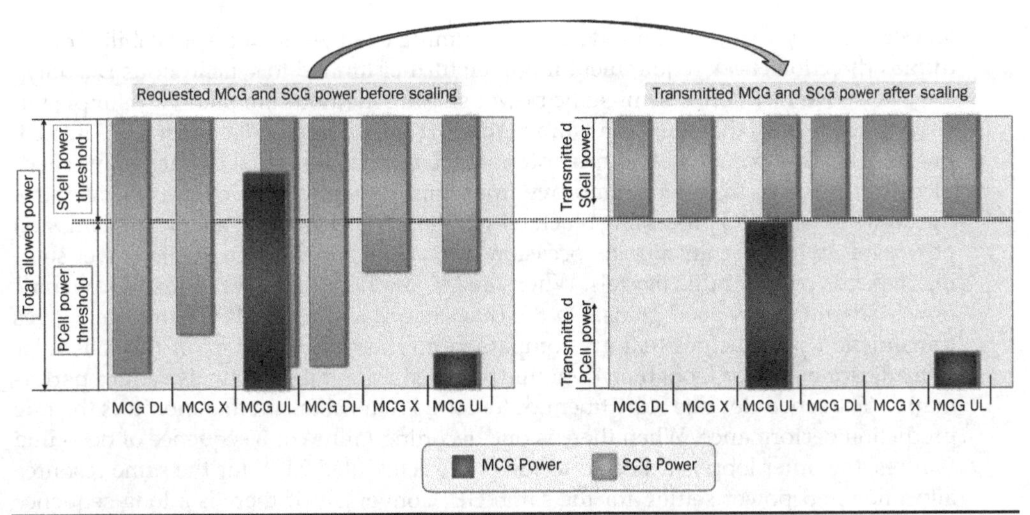

FIGURE 10.21 NR-DC semi-static power sharing without duplex direction check.

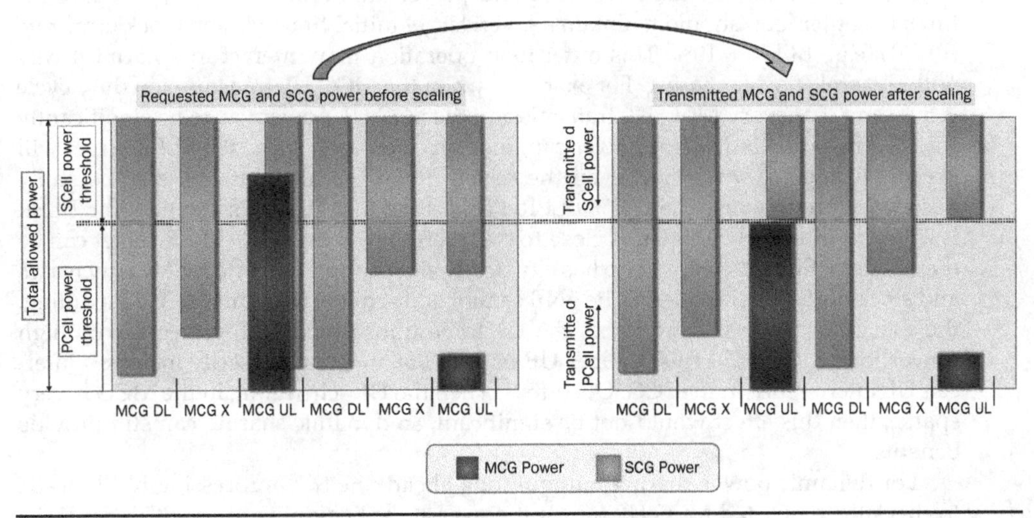

FIGURE 10.22 NR-DC semi-static power sharing with duplex direction check.

Note that instead of a semi-static threshold for a CG, a reserved power for the other CG could be introduced. The two ways of defining thresholds are not identical but there is no essential difference in their principles of operation.

The case of semi-static power sharing with duplex direction check is a modification of the previous method, where the UE is required to ignore the semi-static threshold for a CG when the UL time period in the CG doesn't overlap, even in part, with any potential UL transmission of the other CG. This method is illustrated in Fig. 10.22. It is not expected that the UE would be required to evaluate this duplex direction check dynamically, therefore the determination of the duplex direction of the other CG should be based on semi-static DL–UL configuration only, not based on dynamic SFI or dynamic grants. Of course, even with semi-static DL–UL configuration only, the occurrence of an overlap may vary based on dynamic UL timing changes, so the applicability of the duplex direction check requirement might be further limited to synchronous DC only.

The above mentioned semi-static power sharing methods are relatively simple but have the drawback that they don't allow utilizing full power available for a CG in a UL period when the other CG also has potential UL but has no actual dynamically scheduled transmission. In order to improve upon this, dynamic power sharing was introduced where the SCG processing block within the UE checks the available transmission power for each SCG transmission occasion. The actual benefit of dynamic power sharing depends on the traffic models. When the MCG scheduling is very sparse, dynamic power sharing gives good gains. On the other hand, when the MCG duty cycle gives transmission periodicities that are comparable to the "forget factor" of the scheduler rate adjustment outer loop, then dynamic power sharing might actually reduce performance. The scheduler rate adjustment outer loop is a mechanism that monitors the rate prediction performance. When there is one decoding failure or a sequence of decoding failures, the outer loop is expected to lower the scheduled MCS for the same resource allocation and power setting for the same UE. Conversely, if there is a long sequence of successfully decoded transport blocks, the outer loop is expected to increase the scheduled MCS for the same resource and power allocation for the same UE. Over time, the outer loop should maintain a given target initial transmission block error rate (BLER), e.g., BLER = 10%. This outer-loop operation may interact in a harmful way with dynamic power sharing. For example, when the MCG UL transmission duty cycle is 50% and if at every MCG UL transmission the SCG UL power has to be significantly scaled down due to the joint power cap, then an outer loop with 10% BLER target will keep reducing the MCS until even the scaled down UL transmission is successfully decoded with no worse than 20% BLER. (This is because if 50% of the time the BLER is 20% and in the other 50% it is close to 0%, then the average 10% BLER target can be maintained.) Since the outer loop has a typically slow adaptation rate for MCS increases and since the SCG scheduler in the gNB cannot adapt to the dynamic MCG scheduling, the dynamic power sharing forces the UL to be transmitted with unnecessarily high power for a low MCS. This reduces UE battery life and unnecessarily increases intercell UL interference in the SCG. Of course, when the UL scheduling in the MCG is very sparse, then this effect would not be significant, so dynamic sharing can still provide benefits.

For dynamic power sharing without look ahead, the SCG processing block needs to check the requested MCG UL transmit power at the beginning of any SCG transmission. This is expected to be done for each new UL channel transmission. For example, if a PUSCH transmission starts in an SCG cell at symbol n and another PUSCH transmission in a different SCG cell starts at symbol $n+1$, then the UE needs to check the requested MCG transmit power at both symbol n and $n+1$. However, the ongoing PUSCH transmission power that started at symbol n is not expected to be rescaled. The check at symbol $n+1$ should impact only the second PUSCH starting in symbol $n+1$. The operation with dynamic power sharing without look-ahead is illustrated in Fig. 10.23.

For dynamic power sharing with look-ahead the operation is similar to that without look-ahead, but here the UE has to check the requested MCG power for the full duration of the SCG channel transmission and take the maximum. Therefore, the SCG power control accommodates the requested MCG power even if an MCG transmission starts in the middle of the SCG transmission. The operation with dynamic power sharing with look-ahead is illustrated in Fig. 10.24.

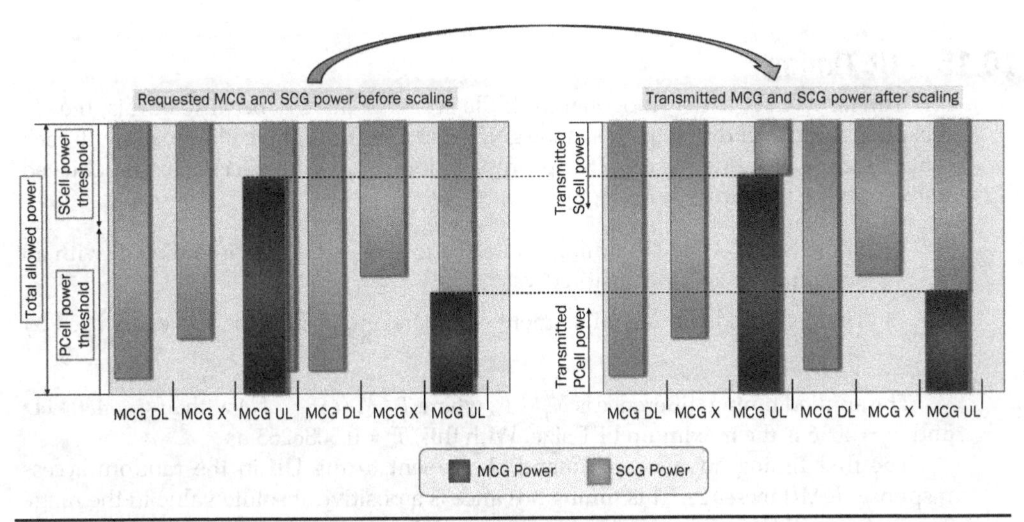

FIGURE 10.23 NR-DC dynamic power sharing without look-ahead.

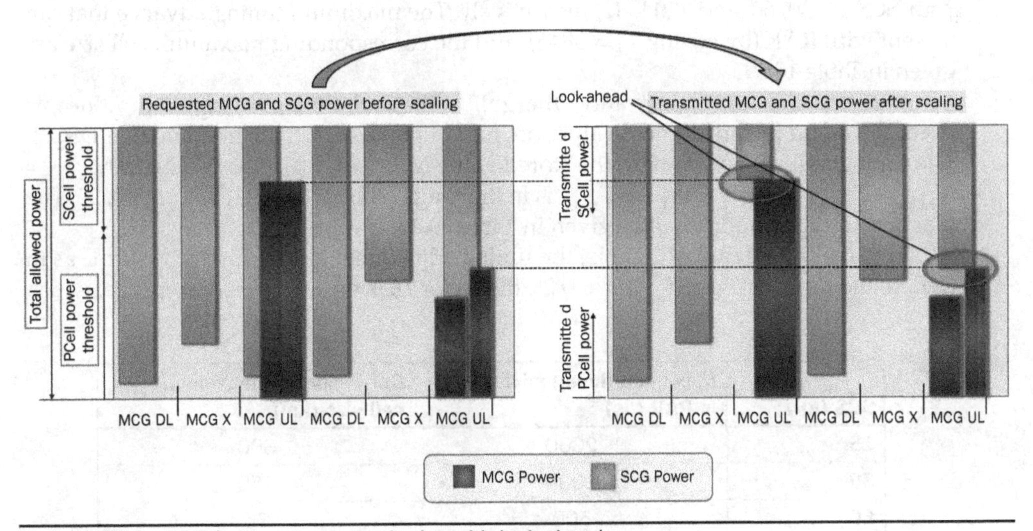

FIGURE 10.24 NR-DC dynamic power sharing with look-ahead.

For dynamic power sharing, both with and without look-ahead, the following additional considerations apply. Since every control decoding requires some time, and calculating and applying the SCG transmit power also take some time, a processing time–related deadline needs to be defined, after which the UE will not be able to accommodate any grant requesting additional MCG power during an ongoing SCG transmission. Note that at least in the case of dynamic power sharing with look-ahead, this restriction can be satisfied with "slowing down" the MCG scheduling, i.e., where all MCG grants are cross-slot grants with large enough K_2 value (K_2 was described in Sec. 10.2). In this case, satisfying the restriction by the scheduler will not require any information exchange between the MCG scheduler and SCG scheduler in the base stations.

10.15 UL Timing

NR, similar to LTE, uses orthogonal multiple access in the UL. Because of this, the UL reception timings of different UEs at the gNB need to be brought to alignment with each other by compensating for round-trip propagation delay differences. This is achieved with applying two procedures [8, 9]:

- The UE tracks the DL timing and advances the UL relative to the DL with an amount derived from a stored value N_{TA}.
- The gNB sends timing adjustment commands to update the N_{TA} value used by the UE.

The applied timing difference is $N_{TA} \cdot T_c$, where $T_c = 1/(\Delta f_{max} \cdot N_f)$ with $\Delta f_{max} = 480$ kHz and $N_f = 4096$ is the maximum FFT size. With this, $T_c = 0.5086263$ ns.

The first timing advance command, T_A, is sent to the UE in the random access response (RAR) message. This timing advance is a positive absolute value in the range $T_A = 0, 1, 2, \ldots, 3846$, advancing the UL timing relative to the DL. The value of T_A is converted to time offset as $N_{TA} = T_A \cdot 16 \cdot 64/2^{\mu}$, where μ is the SCS index with $\mu = 0, 1, 2, 3$ for SCS 15, 30, 60, and 120 kHz, respectively. The maximum timing advance that can be sent with RAR (by setting $T_A = 3846$), and the corresponding maximum cell size are given in Table 10.19.

After the first timing advance, the gNB will send timing adjustment values as needed. The timing adjustment values are part of the MAC CE included in the PDSCH. The timing adjustment changes the stored value N_{TA_old} to N_{TA_new} according to $N_{TA\_new} = N_{TA\_old} + T_{Adjust} \cdot 16 \cdot 64/2^{\mu}$ where T_{Adjust} is in the range $-31, -30, \ldots, -1, 0, 1, \ldots, 31, 32$. The maximum timing adjustment is given in Table 10.20.

Note that in the case of 15 kHz, the timing adjustment granularity is the same as in LTE. In the formula $N_{TA} = T_A \cdot 16 \cdot 64/2^{\mu}$, the factor 16 is the adjustment step size, which

SCS (kHz)	Maximum timing advance in RAR (μs)	Corresponding maximum cell size (km)
15	2000	300
30	1000	150
60	500	75
120	250	37.5

Table 10.19 Maximum Timing Advance and Corresponding Cell Size for Each SCS

SCS (kHz)	Maximum timing adjustment (μs)
15	16.667
30	8.333
60	4.167
120	2.083

Table 10.20 Maximum Timing Adjustment for Each SCS

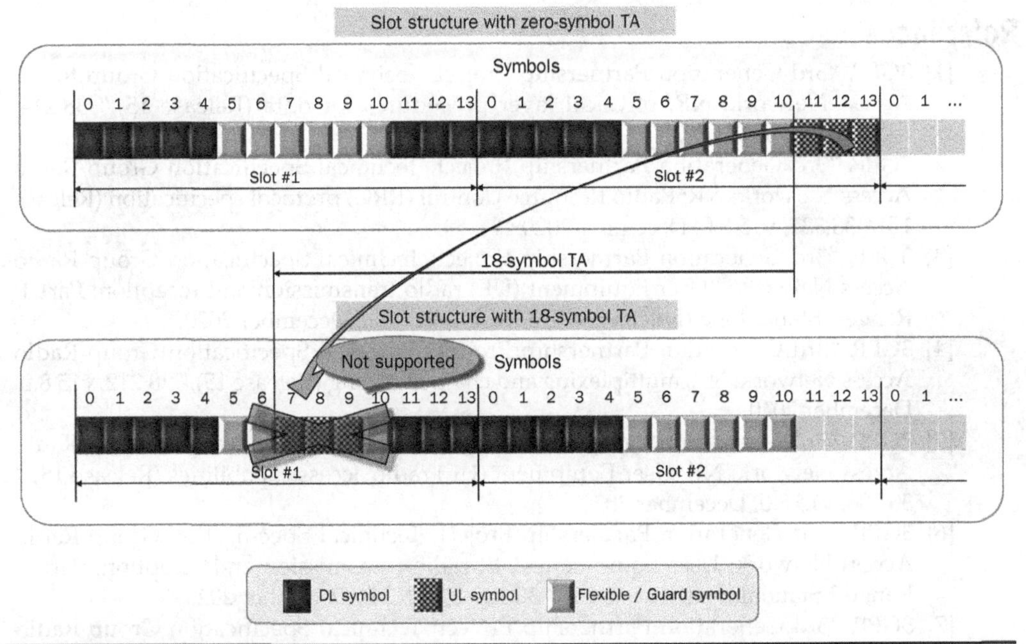

FIGURE 10.25 Invalid slot format with large TA.

is the same as in LTE, and the factor 64 is the conversion factor between T_c and T_s, which are the sample rates for NR and LTE, respectively. For SCS larger than 15 kHz, the timing adjustment step size is reduced by 2^{μ}. This is because the CP length is also reduced by the same factor 2^{μ}.

When multiple serving cells are in the same timing adjustment group (TAG), then their timing is simultaneously changed with the same timing adjustment command. When there are active BWPs with different SCS across cells within a TAG, then their timing adjustment step size and range are according to the largest SCS. This is motivated by the need to achieve sufficient accuracy in the highest SCS carrier; however, this can also put undue burden on the UE implementation by requiring increased granularity in the smaller SCS carriers. In order to mitigate this, the requirement was relaxed by allowing that the timing adjustment applied to the UL carrier with lower SCS may be rounded to align with the timing advance granularity corresponding to its own SCS.

As it was described in Table 10.19, the maximum supportable cell size in NR is 300 km, which is a significant increase corresponding to the LTE maximum supportable cell size of 100 km. This very large cell size creates some artifacts as shown below in Fig. 10.25.

In general, the baseline support in Release 15 is for slot formats for which whenever a slot contains both DL and UL symbols, the DL symbols cannot follow the UL symbols within the slot. In Fig. 10.25, a non-compliant slot format would be created by the large TA from the UE's perspective. Even though the slot formats are defined with zero TA assumption, the restrictions on supported slot formats must be satisfied even after applying the actual non-zero TA value.

References

[1] 3GPP, "3rd Generation Partnership Project; Technical Specification Group Radio Access Network; NR; physical layer procedures for data (Release 15)," 38.214, v15.8.0, December 2020.

[2] 3GPP, "3rd Generation Partnership Project; Technical Specification Group Radio Access Network; NR; Radio Resource Control (RRC) protocol specification (Release 15)," 38.331, v15.8.0, December 2020.

[3] 3GPP, "3rd Generation Partnership Project; Technical Specification Group Radio Access Network; "User Equipment (UE) radio transmission and reception; Part 1: Range 1 Standalone (Release 16)," 38.101-1, v16.2.0, December 2020.

[4] 3GPP, "3rd Generation Partnership Project; Technical Specification Group Radio Access Network; NR; multiplexing and channel coding (Release 15)," 38.212, v15.8.0, December 2020.

[5] 3GPP, "3rd Generation Partnership Project; Technical Specification Group Radio Access Network; NR; User Equipment (UE) radio access capabilities (Release 15)," 38.306, v15.8.0, December 2020.

[6] 3GPP, "3rd Generation Partnership Project; Technical Specification Group Radio Access Network; User Equipment (UE) radio transmission and reception; Part 2: Range 2 Standalone (Release 16)," 38.101-2, v16.2.0, December 2020.

[7] 3GPP, "3rd Generation Partnership Project; Technical Specification Group Radio Access Network; NR; physical channels and modulation (Release 15)," 38.211, v15.8.0, December 2020.

[8] 3GPP, "3rd Generation Partnership Project; Technical Specification Group Radio Access Network; NR; physical layer procedures for control (Release 15)," 38.213, v15.8.0, December 2020.

[9] 3GPP, "3rd Generation Partnership Project; Technical Specification Group Radio Access Network; requirements for support of radio resource management (Release 15)," 38.133, v15.8.0, December 2020.

CHAPTER 11
Coexistence of 4G and 5G

LTE-NR coexistence can occur in two different scenarios:

- LTE and NR operating in the same band but in different channels, e.g., adjacent channels
- LTE and NR operating in the same band and in the same channel

In the following sections, we discuss both scenarios.

11.1 Adjacent Channel Coexistence

When LTE and NR operate in the same band but in different channels, their coexistence is similar to LTE-LTE coexistence due to the similarities between the LTE and NR general waveform design. In the case of LTE TDD, the LTE-LTE coexistence usually requires different channels in the same band to be operated with the same DL/UL configuration. This is to ensure that there is no interference across opposing duplex directions, i.e., the UL receiver of one eNB is not interfered by the DL transmitter of another eNB in an adjacent channel and the DL receiver of one UE is not interfered by the UL transmitter of another nearby UE in an adjacent channel. When it comes to LTE-NR TDD coexistence in adjacent channels in the same band, similar considerations apply. This necessitates that there should be a way to configure NR with duplex directions matching LTE. Since NR provides much more flexibility for DL/UL configuration than LTE, finding an NR DL/UL configuration matching LTE wouldn't seem difficult. There is, however, a particular issue that creates some complications. The principles of DL/UL configuration are somewhat different between LTE and NR:

- In LTE, as the DL-to-UL ratio increases, the UL-DL switch point gets moved earlier in the frame, while the DL-UL switch point stays fixed [1].
- In NR, as the DL-to-UL ratio increases, the DL-UL switch point gets moved later in the frame, while the UL-DL switch point stays fixed. This is because in NR, the DL/UL configuration is defined as a number of DL slots followed by slot(s) including guard period followed by a number of UL slots [2].

This difference in DL/UL configuration principles is illustrated in Fig. 11.1, where cases with 5-ms switch periodicity are shown for both LTE and NR. For LTE, 15 kHz SCS is assumed, while for NR, 30 kHz SCS.

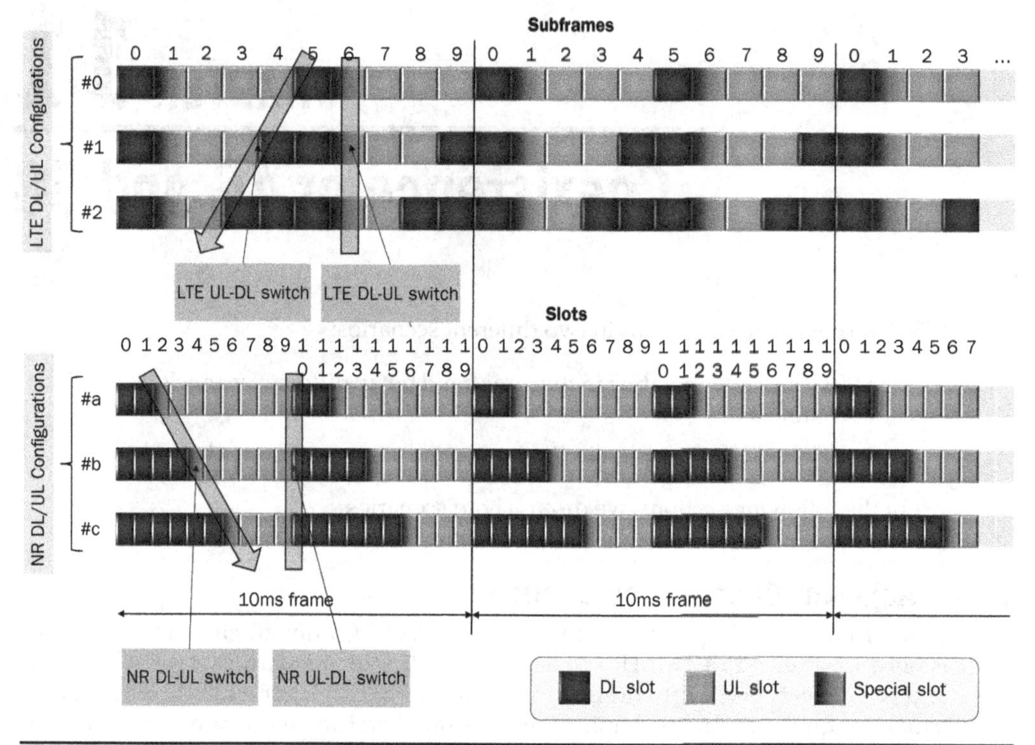

FIGURE 11.1 DL-UL and UL-DL switch points in example LTE and NR TDD configurations with 5-ms switch periodicities.

In Fig. 11.1, DL slot means that all symbols in the slot are designated as DL, UL slot means that all symbols in the slot are designated as UL, and special slot means one of the following three options:

- DL symbol(s) followed by guard symbol(s), followed by UL symbols
- DL symbol(s) followed by guard symbol(s)
- Guard symbol(s) followed by UL symbols

If NR TDD is to be deployed in a band where there is also LTE TDD deployed, then first an NR DL/UL configuration is chosen whose DL-to-UL ratio is the same as that of LTE. For example, if LTE uses configuration #1 in Fig. 11.1, then NR will use configuration #b. However, with this choice there will be a conflict in duplex directions in certain slots as shown in Fig. 11.2.

In order to solve the duplex direction misalignment, two solutions are supported in Release 15. The first solution is to introduce a slot offset between LTE and NR system time. Although offsetting system time in independent carriers is transparent to the standard, at least for EN-DC, the slot offset does impact the UE operation. The solution with slot offset is illustrated in Fig. 11.3, where there is a −1 ms offset between LTE and NR system time modulo 5 ms.

Offsetting system time does create some inconvenience in network management. For this reason, a second solution was also introduced in Release 15, which allows for defining a dual TDD configuration in NR [2]. This is illustrated in Fig. 11.4.

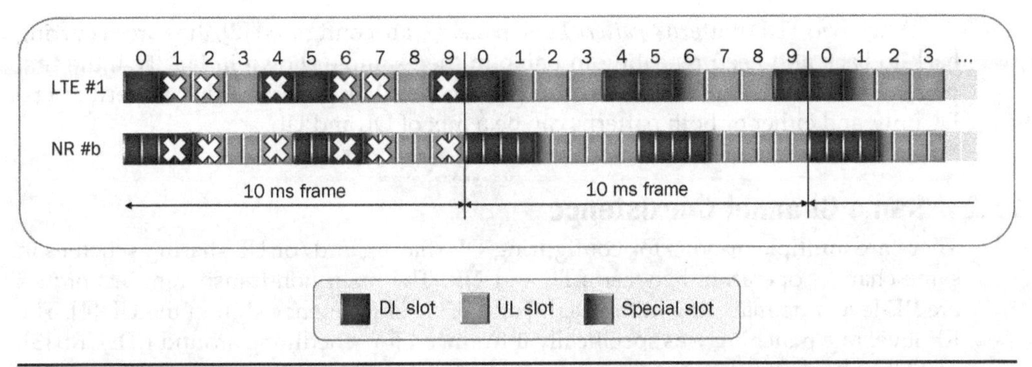

FIGURE 11.2 Conflict in duplex directions between LTE and NR with matching DL-to-UL ratios.

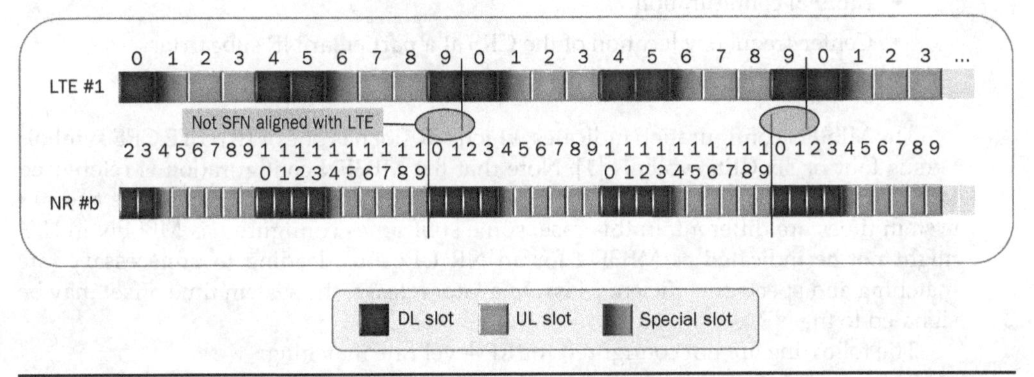

FIGURE 11.3 Solving the duplex direction conflict between LTE and NR with introducing a slot offset.

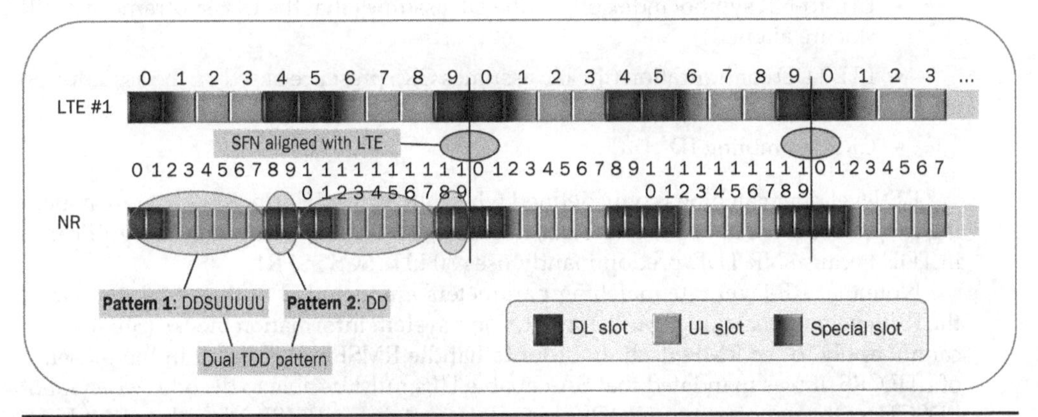

FIGURE 11.4 Solving the duplex direction conflict between LTE and NR with introducing dual TDD configuration in NR.

When two TDD patterns, *pattern1* and *pattern2*, are configured [2], they are occurring back-to-back and keep repeating in pairs. In the example shown in Fig. 11.4, the first pattern is DDSUUUUU and the second pattern is DD. In general, either pattern can be DL only, and either or both patterns can be a mix of DL and UL.

11.2 Same Channel Coexistence

There are multiple options for configuring DL sharing and/or UL sharing schemes in same channel operation between LTE and NR. The main non-transparent techniques are RE-level rate matching in the DL [3] and 7.5 kHz frequency shift in the UL [4]. The RE-level rate matching was specifically introduced for scheduling around LTE CRS [3]. The UE is being informed of the following CRS parameters:

- Number of CRS ports
- MBSFN configuration
- Center-frequency location of the CRS at a particular NR subcarrier
- Subcarrier offset (*vshift*) of the CRS pattern

The MBSFN configuration indicates which NR slot has one or two LTE CRS symbols versus four or six CRS symbols [1]. Note that the MBSFN configuration is referenced to NR system time. This leads to some suboptimality when LTE system time and NR system times are different. In this case, some subframes configured as MBSFN in LTE might not be indicated as MBSFN for an NR UE; thus, leading to unnecessary rate matching and spectrum efficiency loss. In a later release, the system time offset may be signaled to the NR UE.

The following are not configured for RE-level rate matching:

- LTE-to-NR system time offset (the UE assumes the same system time modulo 40 ms)
- LTE-to-NR symbol index offset (the UE assumes that the LTE subframe and NR slot are aligned)
- LTE TDD configuration (the UE assumes CRS presence based on the assumption that every NR DL symbol is DL in LTE)
- CRS scrambling ID

RE-level rate matching is only defined when NR SCS is 15 kHz. RE-level rate matching is applicable to both FDD and TDD; however, it has much less practical significance in TDD because NR TDD predominantly uses 30 kHz SCS in FR1.

Note that RE-level rate matching parameters are included in the RMSI; therefore, the RE-level rate matching could apply to later system information blocks (SIBs), but it cannot apply to the RMSI itself. In order to handle RMSI transmission in the presence of LTE CRS, it was mandated that SA-capable UEs must be able to decode two-symbol PDSCH with mapping type B (DL mini-slot) containing RMSI. With this, RMSI can be scheduled in the two-symbol gap between CRS symbols. After this solution was adopted, the same solution was also applied to the other system information (OSI, SIBs other than RMSI) as well. That is, it was mandated that SA-capable UEs must be able to decode two-symbol PDSCH mini-slots containing OSI, and OSI also does not assume RE-level rate matching.

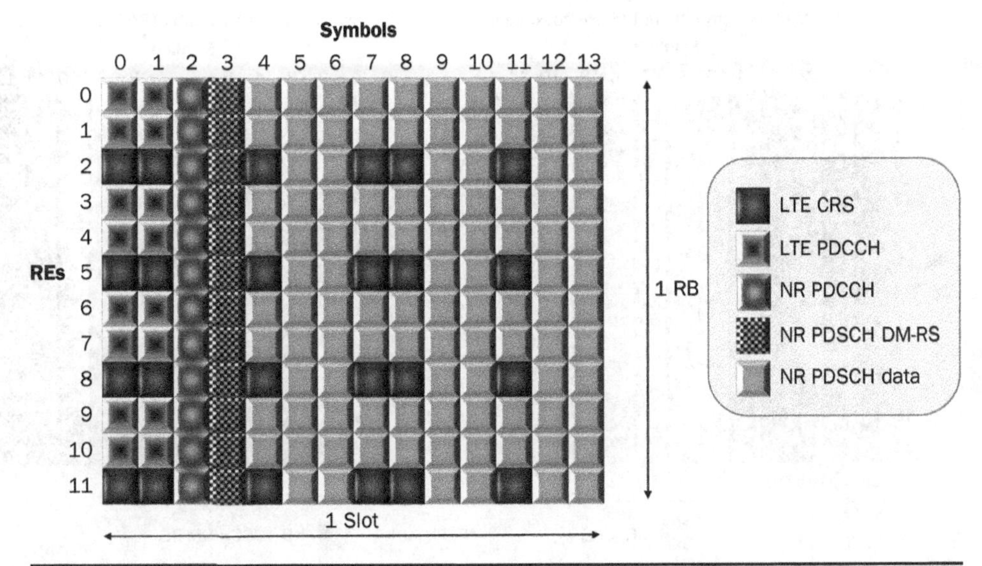

FIGURE 11.5 Example DSS DL subframe/slot in a non-MBSFN subframe.

It has been left unspecified in detail what the UE behavior is when OSI is scheduled overlapping with CRS. The UE does not assume any rate matching in these cases, but it may or may not set the LLRs derived in the overlapping REs to zero.

In Fig. 11.5, we give an example of a DSS DL subframe/slot with LTE/NR coexistence in a non-MBSFN subframe.

In Fig. 11.6, we show the special accommodation introduced as a modified NR DM-RS pattern for two-symbol DM-RS for PDSCH mapping type A [5]. Without this accommodation, the additional DM-RS symbol would collide with the LTE CRS, as shown in the left half of Fig. 11.6. In order to avoid this, the additional DM-RS symbol is shifted to the right [5], as shown in the right half of Fig. 11.6.

Note that once RE-level rate matching is configured, the shift is applied, irrespective of whether or not an actual collision without the shift would occur. For example, the shift applies irrespective of whether it is an MBSFN or non-MBSFN subframe. The main reason for this is simplicity, but also to make sure that it is possible to configure all UEs to use the same DM-RS symbols to facilitate their MU-MIMO multiplexing. As a counterexample, to illustrate what could happen if the DM-RS symbol shift was made dependent on actual collision with CRS, consider the case of a 40 MHz NR channel that coexists with a single 20 MHz LTE channel. In this case, a 40 MHz UE with full bandwidth PDSCH allocation would have the DM-RS symbol shifted due to the collision with CRS. Another UE that is scheduled within the 20 MHz that does not contain LTE CRS would not have the DM-RS symbol shifted. Scheduling both these UEs at the same time would require MU-MIMO multiplexing but this would be made more difficult if their DM-RS symbols are not fully aligned. Making the shift invariable helps avoiding this situation.

Rate matching around other LTE signals, such as PSS/SSS/PBCH, must be accomplished by using either RB-symbol-level rate matching or gNB scheduling restriction.

Another limitation is that the rate matching only applies to the data part of PDSCH. Any other NR channel or signal, such as SSB, PDCCH, CSI-RS, PDSCH DM-RS, must not overlap with any LTE channels or signals.

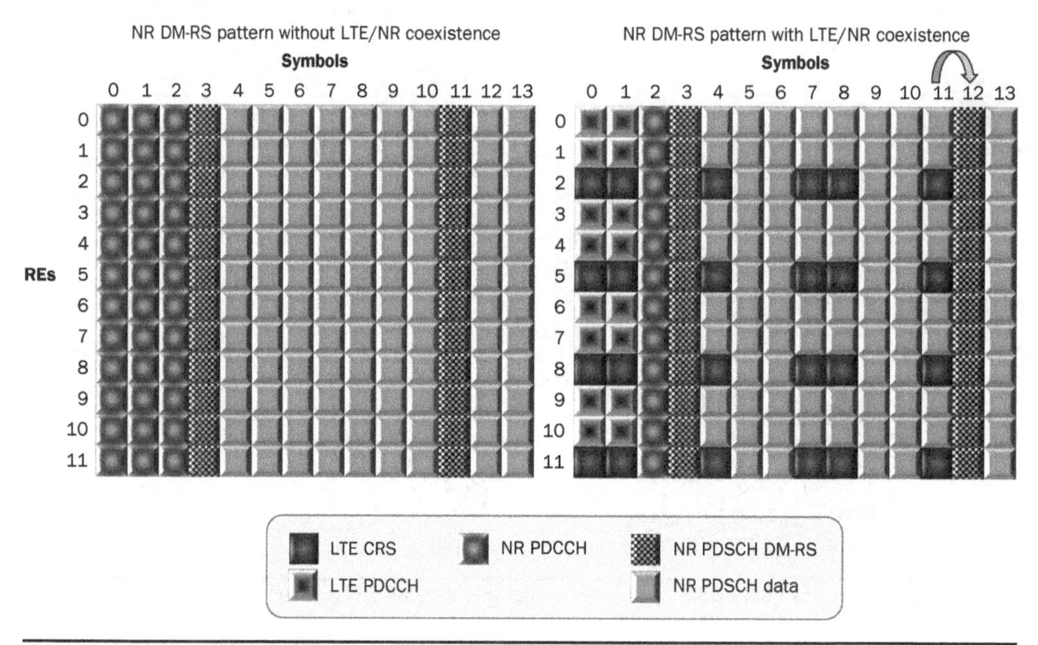

FIGURE 11.6 NR DM-RS symbol shift with DSS.

Figure 11.7 shows two examples of how NR SSB can be accommodated in DSS. The first option shown in the left half of Fig. 11.7 utilizes 30 kHz NR SSB SCS, even though the NR data SCS is 15 kHz. In order to enable this option, both 15 kHz and 30 kHz SSB option is defined in a few bands (n5 and n41). The drawback of this option is the gNB and UE implementation complexity and potentially reduced SSB link budget (depending on available SSB transmission power boost). The second option is shown in the right half of Fig. 11.7. This option uses LTE MBSFN subframes for the NR SSB. The drawback of this option is that in these subframes many LTE UEs, namely UEs that do not support LTE Transmission Mode 9 or 10, cannot be scheduled. However, this only impacts one out of twenty subframes, and even in these subframes, other NR signals or channels can be scheduled to offset the loss. Overall, the second, MBSFN-based option is expected to be used more widely.

In the UL, there are no rate matching patterns, neither RB-symbol level nor RE-level defined. The gNB is expected to schedule LTE and NR UEs with non-overlapping resources in a transparent manner to the UEs. Since it is mandatory for the NR UE to support PUSCH with mapping type B (UL mini-slot), it was seen that there is sufficient scheduling flexibility for the gNB to accomplish this.

As mentioned before, the primary non-transparent mechanism for UL sharing is the 7.5 kHz UL frequency shift [4]. The following is defined for all SUL bands, for the uplink of all FDD bands and for Band n90.

$$F_{REF,shift} = F_{REF} + \Delta_{shift}, \qquad \Delta_{shift} = 0 \text{ kHz or } 7.5 \text{ kHz}, \qquad (11.1)$$

where Δ_{shift} is signaled by the network in higher layer parameter *frequencyShift7p5khz*.

The frequency arrangement with and without 7.5 kHz shift is shown in Fig. 11.8.

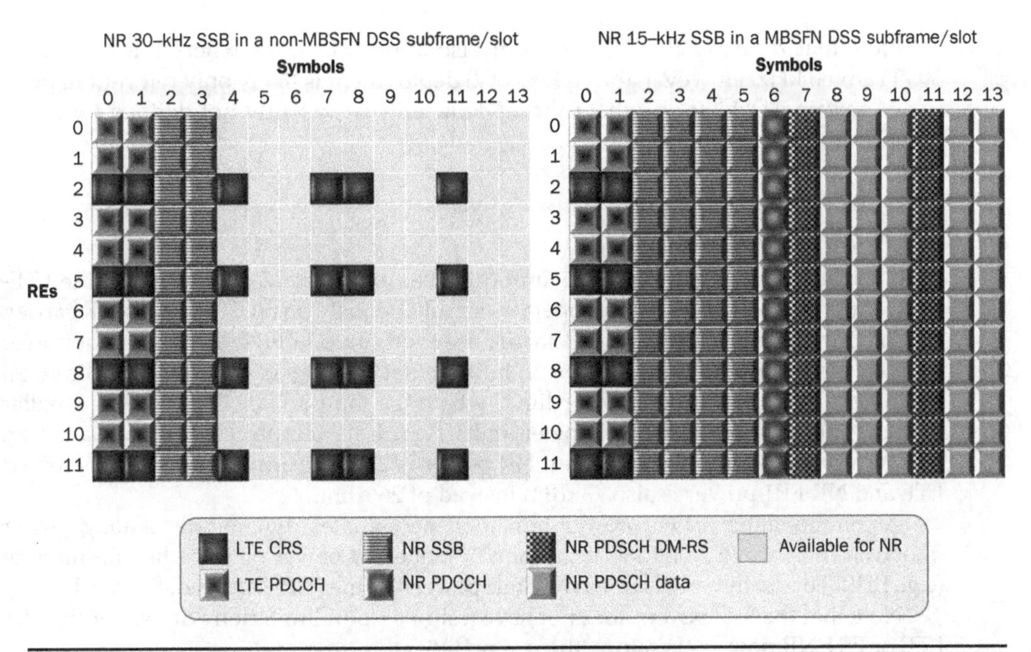

FIGURE 11.7 NR SSB multiplexing in non-MBSFN and MBSFN DSS subframes.

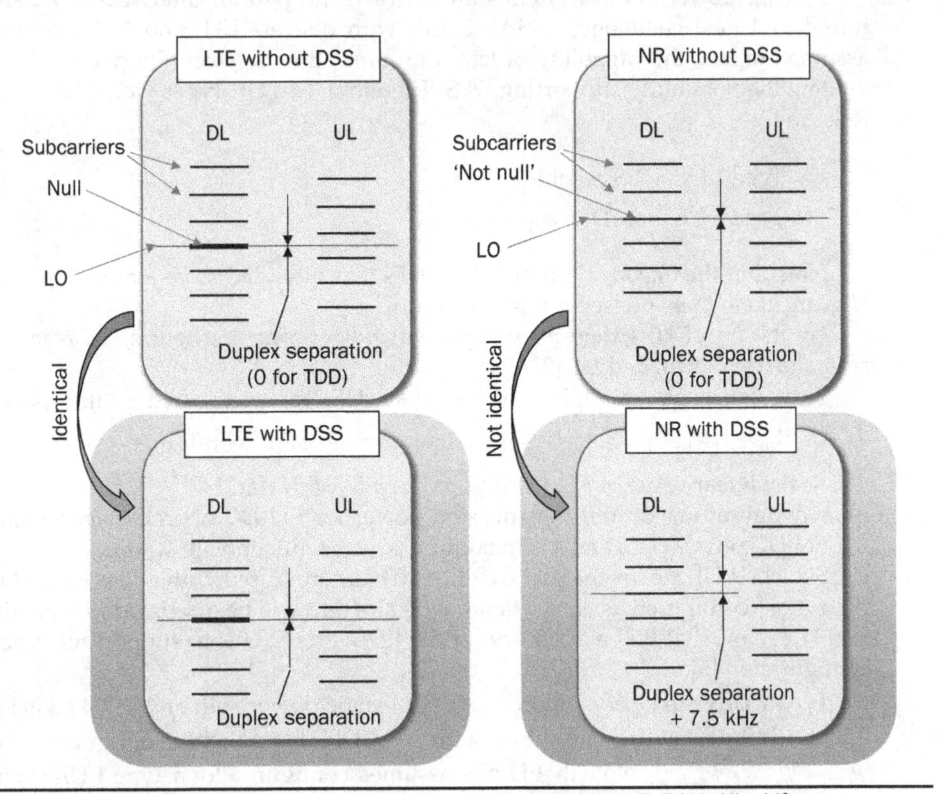

FIGURE 11.8 LTE and NR frequency arrangement with and without 7.5 kHz UL shift.

Note that formally the 7.5 kHz shift also applies to cases where the NR SCS is 30 kHz or 60 kHz. However, the 30 kHz SCS deployment is likely only relevant in practice to certain TDD bands, where the 7.5 kHz shift is currently not defined (with the exception of band n90).

11.3 EN-DC Power Control

As it has been discussed, NSA was introduced as an essential component of the LTE-to-NR migration. In NSA, the UE operates simultaneously on an LTE and an NR carrier. When both the LTE carrier and NR carrier are configured in FR1, there is a joint power limit on the combined transmit power. In more detail, there is an LTE power limit, an NR power limit and a joint power limit, where the joint power limit may be smaller than the sum of the LTE and NR power limits. A typical example is when the maximum LTE power is 23 dBm, the maximum NR power is 23 dBm, and the allowed combined LTE and NR FR1 power is also 23 dBm instead of 26 dBm.

Accommodating a joint power limitation necessitates dynamically scaling power in one air-interface (e.g., NR) as a function of requested power on the other air-interface (e.g., LTE). This feature was named dynamic power sharing (DPS) and is defined in Ref. [2].

Note that the FR2 power does not have a similar joint limitation with either the FR1 LTE or FR1 NR power; therefore, there is no FR2 DPS.

Having power control interaction between the two air-interfaces (LTE and NR) introduced new challenges. NSA devices with discrete LTE and NR modems were assumed to lack the capability of fast communication between the two modems, preventing the possibility supporting DPS. Therefore, two UE Types were identified [6] in Release 15:

- Type 1 UE: DPS capable
- Type 2 UE: non-DPS capable

Note that the introduction of Type 2 UEs was intended to be temporary, and these UEs are likely to be phased out in Release 16.

For the Type 2 UEs, there are two primary power control schemes, semi-static power split and TDM switched UL [2].

Semi-static power split means that UE is configured with parameters $\hat{P}_{\text{LTE}} + \hat{P}_{\text{NR}} \leq \hat{P}_{\text{Total}}^{\text{EN-DC}}$, where $\hat{P}_{\text{LTE}}$ is the linear value of P_{LTE} (configured as *p-MaxEUTRA*), $\hat{P}_{\text{NR}}$ is the linear value of P_{NR} (configured as *p-NR-FR1*), and $\hat{P}_{\text{Total}}^{\text{EN-DC}}$ is the linear value of a configured maximum transmission power for EN-DC operation (represented by the EN-DC power class and the maximum power reduction allowance).

Switched UL means that the LTE FDD UE is configured with a reference TDD configuration (configured as *tdm-PatternConfig-r15*) and will be discussed in more detail in Sec. 11.4. Note that it is mandatory for the FDD Type 2 UEs to support reference TDD configuration [6].

Type 1 UEs can be configured with semi-static power split and TDM switched UL, but in addition, can be configured with DPS. DPS is implicitly configured whenever $\hat{P}_{\text{LTE}} + \hat{P}_{\text{NR}} > \hat{P}_{\text{Total}}^{\text{EN-DC}}$. Note that DPS is assumed configured for a Type 1 UE even when reference TDD configuration for LTE is configured, because Type 1 UEs, unlike Type 2

UEs, cannot assume that NR UL is not scheduled at the same time when LTE UL is also scheduled under the reference TDD configuration.

There was substantial debate about how to define DPS operation itself. There were differing views on whether in the cases of exceeding the joint power limit, LTE power or rather NR power should be scaled down. Both approaches have benefits and drawbacks. One special consideration is the different processing timelines. LTE processing time is slower than NR, so for an overlapping LTE and NR transmission, the LTE UL grant is received and processed earlier than the typical reception time of the NR grant. This makes accommodating the power limit by scaling down LTE power difficult. Although a possible way to circumvent this would be to "slow down NR," i.e., to send NR UL grants according to the slower LTE timeline, this solution was not adopted. In the end, DPS was defined as NR power scaling only. Note that at least in the case of EN-DC, this approach can be further justified by considering that LTE is the master cell group (MCG) in EN-DC; therefore, its power allocation should be protected.

As a further allowance, a new threshold parameter X_{SCALE} was introduced. If NR power is scaled down by more than X_{SCALE} as a result of DPS, the UE is allowed to drop NR transmission altogether [2]. The rationale for this was that when the power scaling is large, the gNB is likely not able to decode the UL transmission; therefore, not dropping but performing the transmission would just add interference to the system and waste UE battery power.

11.4 Switched EN-DC UL

There were various concerns raised regarding NSA operation, as already mentioned in Sec. 2.1. One concern is simply that the UE may not be capable to operate two ULs at the same time, even if they are in different bands. This, however, did not prove convincing enough in itself to introduce switched UL solutions. Switched UL here means that the UE operates on two carriers but on each carrier performs discontinuous operation so that the active periods on the two carriers are complementary. This allows the UE to switch between the two carriers so that active transmission happens only on one carrier at a time [2]. Note that configuring the UE with reference TDD configuration for LTE (by *tdm-PatternConfig-r15*), as mentioned in Sec. 11.3, is one form of switched UL operation. As a matter of fact, operating with reference TDD configuration for LTE is called Case 1 TDM operation. The other case, when the TDM operation is achieved in a transparent manner by coordinated gNB scheduling without configuring any TDM pattern, is called Case 2 TDM operation. In the remaining part of this section, we will discuss Case 1 TDM operation only.

Switched UL was introduced [2] in order to solve the following problems:

- Potential intermodulation issues or reverse intermodulation issues created when the simultaneous transmission on two carriers gets mixed via some nonlinearity.
- Link budget limitation due to necessary power scaling when two simultaneous transmissions have to meet a joint power limit.

Note that the reason on link budget limitation is debatable, since the UL discontinuous transmission itself creates equal or worse link budget limitation than the power scaling.

When both UL carriers are NR, no special accommodation was introduced for switching because it was seen that the various NR scheduling features, such as PUSCH mapping types A and B, flexible cross-slot scheduling, and flexible HARQ timing, already give enough flexibility to the gNB to schedule NR in a switched manner in a transparent fashion. Therefore, any switched UL operation among NR carriers could be seen as a version of Case 2 TDM operation.

When one UL carrier is LTE and the other UL carrier is NR then, due to the fixed HARQ timing of LTE, any UL subframe made unavailable to LTE would create a corresponding DL hole due to the missing HARQ ACK feedback in the UL. To mitigate the resulting DL losses, a new LTE HARQ scheme was introduced. The new HARQ scheme is called Case 1 HARQ timing. The basic attributes of Case 1 HARQ timing are the following:

- The LTE UL transmissions, including both dynamically scheduled and semi-statically configured transmissions, are limited to a subset of the subframes configured by *tdm-PatternConfig-r15*.
  - Note that in Release 16, this limitation is relaxed, so that LTE transmission can also take place outside of the restricted subframes.
- The LTE UL HARQ is 10 ms based (as opposed to the nominal 8-ms–based UL HARQ).
- The LTE DL HARQ is the same as a corresponding case of LTE FDD-TDD CA with TDD PCell.

When Case 1 HARQ timing is configured, an additional HARQ subframe offset can be also configured. The benefit of this is that UL of all UEs configured with Case 1 HARQ timing can be evenly distributed over all subframes of the frame, as opposed to making them congested in the same subframe subset.

In Release 15, Case 1 HARQ timing was only defined for LTE FDD. In Release 16, this was extended also to LTE TDD. In addition, in Release 16 a similar method was introduced also for mitigating UL harmonic issues. This case is actually not aimed at solving possible problems created by LTE and NR UL transmitting simultaneously. The reason for using Case 1 HARQ timing for this case is that an LTE UL transmission prevents NR DL reception by the UE due to harmonic interference; therefore, it needs to be possible to perform UL transmission only in a subset of the subframes, but without sacrificing LTE DL throughput. Without using Case 1 HARQ timing, any UL subframe unavailable for LTE transmission would create a corresponding DL hole in LTE due to the missing HARQ ACK feedback in the UL.

References

[1] 3GPP, "3rd Generation Partnership Project; Technical Specification Group Radio Access Network; Evolved Universal Terrestrial Radio Access (E-UTRA); physical channels and modulation (Release 15)," 36.211, v15.8.1, December 2020.

[2] 3GPP, "3rd Generation Partnership Project; Technical Specification Group Radio Access Network; NR; physical layer procedures for control (Release 15)," 38.213, v15.8.0, December 2020.

[3] 3GPP, "3rd Generation Partnership Project; Technical Specification Group Radio Access Network; NR; physical layer procedures for data (Release 15)," 38.214, v15.8.0, December 2020.
[4] 3GPP, "3rd Generation Partnership Project; Technical Specification Group Radio Access Network; "User Equipment (UE) radio transmission and reception; Part 1: Range 1 Standalone (Release 16)," 38.101-1, v16.2.0, December 2020.
[5] 3GPP, "3rd Generation Partnership Project; Technical Specification Group Radio Access Network; NR; physical channels and modulation (Release 15)," 38.211, v15.8.0, December 2020.
[6] 3GPP, "3rd Generation Partnership Project; Technical Specification Group Radio Access Network; NR; User Equipment (UE) radio access capabilities (Release 15)," 38.306, v15.8.0, December 2020.

CHAPTER 12

5G in Unlicensed and Shared Spectrum

In this chapter we present some background information on how unlicensed operation was first defined in 3GPP. Then, we will present the main characteristics of NR access to unlicensed spectrum which is being defined as part of Release 16.

12.1 Unlicensed Operation in LTE

In an attempt to evolve LTE and take it to the next level in terms of aiming at Gbps LTE, Qualcomm and Ericsson submitted at the 3GPP RAN Plenary #63 meeting of December 2013 the first paper proposing to take LTE into unlicensed spectrum [1]. This initial paper had two proposals:

1. Define LTE as supplemental Downlink in the 5.725 to 5.850 GHz band in the USA for which regulations allowed to take LTE without modifications and deploy it in that band. A RAN4-led work item was proposed for this proposal [2].
2. Study the necessary modifications to LTE to enable operation in unlicensed spectrum. A RAN1-led study item was proposed for this proposal [3].

Far from a *warm welcome* to the idea, it took three-quarters until the study item on *Licensed-Assisted Access using LTE* was approved in September 2014 [4] after endless discussions, education, and even a workshop to overcome the initial reluctance. This study lasted until June 2015 at which point a work item for the same was approved [5]. The technical report with all the findings identified during the study phase can be found at Ref. [6].

As a result, 3GPP defined LTE operation in unlicensed spectrum in Release 13 after heavy debates in 3GPP revolving around two main areas:

1. The value of licensed spectrum and its possible dilution if operation on unlicensed spectrum was enabled. Clearly, operation in licensed spectrum offers guarantees that are only possible in this type of spectrum enabling mobile network operators (MNOs) the offering of services with guaranteed quality of service (QoS) including guaranteed bit rate (GBR). Also, transmit power limits set forth in unlicensed spectrum are typically much lower than those allowed in licensed spectrum – making blanket coverage (outdoors, indoors, and outdoor-to-indoor) purely based on unlicensed spectrum unrealistic.

2. Fairness and coexistence issues with the other wireless system in the same band, that is, Wi-Fi in all its 5 GHz flavors, e.g., 11a, 11n, 11ac, 11ax. This issue became quite heated in 3GPP and was extensively discussed.

Irrespectively, licensed spectrum is, on one side, scarce and, on the other, may not be available for the given operating entity with an interest to offer wireless operation in its own private premises (e.g., factory, office, enterprise campus, etc.). At that point, access to unlicensed spectrum becomes important for traditional MNOs, as well as verticals, e.g., with the desire to move from wireline to wireless connectivity.

There are multiple unlicensed bands defined in various regions, namely, below 1, 2.4, 5, 6, 37 (United States only), 60 GHz. 3GPP targeted, initially, all the efforts on the 5 GHz band, which was the band with larger available bandwidth at the time. Figure 12.1 shows the 5 GHz band definition from FCC. Other regions may deviate from the precise definition but 5,150 to 5,925 MHz constitutes the frequency range of interest. Note that the 5 GHz band in 3GPP is denoted band 46 (5,150–5,925 MHz).

A newer band for LAA operation was later defined as band 49 (3,550–3,700 MHz) in what the FCC defined as the citizens broadband radio service (CBRS) band to experiment the concept of shared spectrum.

Access to unlicensed spectrum is governed by regional regulations which set the access rules enabling the deployment of possibly different technologies and focusing on enabling coexistence among those potentially different technologies and also between nodes within the same technology.

Therefore, defining LTE access in unlicensed spectrum triggered discussions on coexistence with the other technologies in the 5 GHz band, namely 802.11 Wi-Fi technologies.

While the LTE access to unlicensed spectrum was defined in Release 13 and evolved over two more releases, namely, Release 14 and Release 15, all of them shared the same restriction. LTE access to unlicensed spectrum requires the presence of a licensed anchor carrier or primary cell (Pcell) and access to unlicensed spectrum takes place by resorting to carrier aggregation (CA) whereby the unlicensed carriers are always secondary cells (Scells). This is what 3GPP designated as licensed assisted access (LAA).

The first release of LAA (Release 13) enables unlicensed operation in the Downlink (DL) only. The UL control channels to enable DL operation, namely, feedback channels for HARQ-ACK and channel state information (CSI) are all carried in the licensed UL Pcell carrier. Therefore, the medium access to unlicensed spectrum is done exclusively by eNBs. All UE transmissions take place in licensed spectrum for both data and control information.

A new frame structure, namely, frame structure type 3, was introduced in the 3GPP specifications, specifically for unlicensed operation. Relegating the unlicensed operation to a new and different frame structure precluded an automatic applicability of

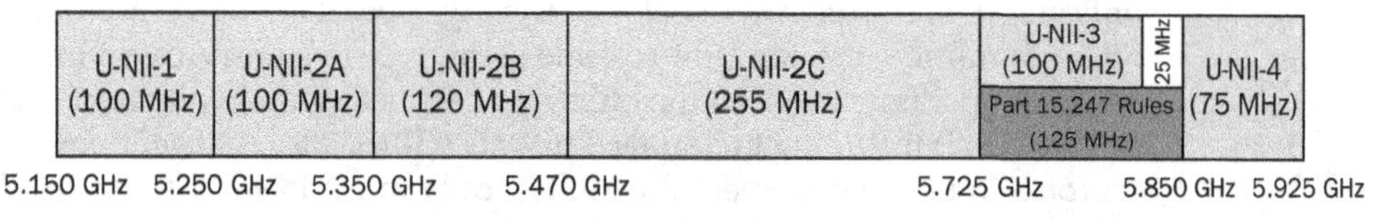

FIGURE 12.1 Summary of existing and proposed new FCC part 15 rules for 5-GHz unlicensed spectrum usage.

general LTE aspects to unlicensed operation, since an explicit applicability to frame structure type 3 would have to be called out in the objectives for the various projects. Indeed, unlike all the other radio specifications, the specification of the medium access rules for unlicensed operation is not in the 36.xxx or 38.xxx series for LTE and NR, respectively, but in the 3GPP Technical Specification 37.213 "Physical layer procedures for shared spectrum channel access" [7].

While at first sight the first release taking LTE to unlicensed spectrum may appear quite limiting, Release 13 LAA enabled the first Gbps LTE implementations marking an important milestone for cellular wireless operation.

LAA Release 14 (aka eLAA) added unlicensed operation in the Uplink (UL). The physical layer structure of the UL of LTE encompasses the following physical layer channels and signals:

- **Physical random access channel** (PRACH), enabling contention-based and contention-free random access to target cells and facilitating the computation, by the receiving eNB, of the required timing advance and transmit power level at the UE.
- **Physical Uplink shared channel** (PUSCH), UL data channel.
- **Physical Uplink control channel** (PUCCH), UL control channel for HARQ-ACK, as well as CSI feedback.
- **Sounding reference signals** (SRS), enabling sounding of the UL channel facilitating UL link adaptation or reciprocity-based DL link adaptation.

Out of all the above physical layer channels and signals, only PUSCH and SRS were within the scope of the original Release 14 work item description [8]. The definition of PUCCH and PRACH transmissions for unlicensed operation was conditional to its necessity and, after multiple meetings of deliberations, there was no consensus to support either of them for unlicensed operation. PRACH was not deemed necessary assuming small timing difference between the Pcell with the licensed carrier and the, possibly non-co-located, Scell with the unlicensed carriers. Also, the transmission of UL control information in support of the corresponding unlicensed DL carriers was determined to be sufficiently enabled by PUCCH transmission on the licensed carrier Pcell.

Effectively, the CA framework under which LTE LAA was built upon forced many constraints to expanding unlicensed operation for LTE beyond licensed assisted.

Indeed, the work item for LAA Release 15 [9] was approved at the same 3GPP RAN Plenary meeting in March 2017 as the study item on NR unlicensed [10]. At the time, all companies with interest on cellular unlicensed operation put all the energy on the NR-U study item which got approved, with a very broad scope, including stand-alone unlicensed operation. In turn, the Release 15 work item on LTE LAA was approved with some limited scope, including improvements to autonomous UL transmissions not relying on UL grants from the network, and an increase on the possible starting and ending points for the DL and UL transmissions.

12.2 Overview

The first release of NR, i.e., Release 15, was defined for operation in licensed spectrum only. As discussed in earlier chapters, two frequency regimes were defined for sub-7-GHz and mmWave operation; however, both considered licensed bands only. Nonetheless, from the early stages of the NR study, applicability and forward compatibility toward unlicensed operation were considered making the extension of NR to unlicensed operation relatively straight forward in Release 16.

A study item on NR access to unlicensed spectrum (NR-U) was approved at the 3GPP RAN Plenary meeting of March 2017 [10] coinciding with the start of Release 15 and the conversion of the NR study to work item; hence, NR entering the normative phase. This 3GPP RAN Plenary meeting was quite eventful as it also approved the plan to accelerate the standardization of the non-standalone (NSA) option of 5G, specifically architecture Option 3 (see Chap. 3) pulling it in by 6 months from the original plan [11]. With this new schedule, the stage-3 freeze for NR Option 3 was aimed at December 2017 breaking the ground for what became a very busy year for 3GPP. Indeed, a small number of study items for NR were approved in March 2017 constituting the first batch of studies after the jumbo study on NR which had just occupied the entire Release 14.

Because of the tremendous amount of work around the specification of the first release of NR and despite the early approval of the NR-U study item, discussions at the RAN working groups (WGs) would not start until the first Quarter of 2018. From that point, the study phase would take place during the entire 2018 and the technical report capturing all the decisions and recommendations can be found at Ref. [12].

It is worth noting that the NR-U study initially had scope for unlicensed bands in FR1 and FR2. The initial scope even had provisions for looking into unlicensed bands above 52.6 GHz, i.e., beyond the upper range of FR2, to the extent that the waveform design principles remained unchanged with respect to below 52.6 GHz.

Prioritization of the work during the study phase led to focusing the study on the 5 GHz and 6 GHz bands only. Unlicensed bands below 5 GHz were considered either too loaded (e.g., 2.4 GHz) or too narrow for NR operation (e.g., sub-1 GHz). An unlicensed band in FR2 available in the United States, namely 37 GHz, did not gather sufficiently broad interest and was also dropped from the study. The work on unlicensed bands above 52.6 GHz, most notably around the 60 GHz band, was also dropped due to the outstanding issue on which waveform to assume for operation in this frequency regime – a critical aspect to consider for looking into operation in this band. The Release 17 package approved at the RAN Plenary of December 2019 includes a sequential tandem study/work item for the extension of NR up to 71 GHz [13, 14] assuming the same waveform as below 52.6 GHz with possibly new (larger) subcarrier spacing being used. More details about this can be found in Chap. 19.

The deployment scenarios covered by the NR-U study [10] included a license-assisted model, as it had been assumed for LTE, as well as, a stand-alone option without the involvement of a licensed carrier. This fact, as discussed in more details later in this chapter, is very relevant and constituted the first time that 3GPP looked into the possibility to enable a fully unlicensed operation not relying on a licensed anchor carrier.

The NR-U study would lead to a Release 16 work item approved at the 3GPP RAN Plenary meeting of December 2018 [15] with a pretty broad scope, particularly compared to the unlicensed access defined from Release 13 through Release 15 for LTE. It was at that same 3GPP RAN Plenary meeting where the definition of the frequency range for NR FR1 in 3GPP was redefined from an initial range from 450 MHz to 6 GHz, which led to the often used term of "sub-6 GHz" or simply "sub-6," to a newly expanded range from 410 MHz to 7.125 GHz to cover, within this frequency range, the 6 GHz band being actively developed for unlicensed access by ETSI and FCC.

Indeed, 3GPP is closely following regulators' work to define the so-called 6 GHz band for possible deployment of LAA and NR-U. A technical report summarizing the regulations and aiming at recommending band plans is underway and is available at Ref. [16].

12.2.1 How NR-U Came About

Possibly, there is no single reason why the NR-U study item in Ref. [10] came about, rather a combination of reasons made it possible and, probably more importantly, the timing was right. MNOs had witnessed the standardization of LAA from Release 13. Indeed, some of them enjoyed commercial success of the feature vis-à-vis their own Wi-Fi offerings. MNOs had also witnessed the surge in 3GPP of new players coming from verticals not traditionally served by 3GPP before. MulteFire, in addition, had been created with the potential to take over full ownership in the development of cellular-based access to unlicensed spectrum and, hence, potentially diluting the reach of 3GPP.

On the technical side, in parallel, 3GPP had just completed the study phase of NR. A number of architecture options for the deployment of 5G had been considered with one flavor, Option 3, gaining traction for early deployments based on non-standalone NR anchored to an LTE carrier connected to 4G's EPC (see Fig. 12.2 and Chap. 3 for more information). Indeed, as mentioned earlier, it was in this very same 3GPP RAN Plenary meeting when 3GPP decided to accelerate the standardization of Option 3 targeting completion by the end 2017 [11]. The 3GPP schedule was set to complete its standalone Option 2 connected to the 5G core six months later, i.e., by June 2018 with allegedly no PHY layer impact.

The development of NR-U hinged on the same paradigm. Defining NR-U operation assisted by an LTE licensed carrier would lead to a definition of NR-U as a PScell of a dual connectivity scenario where the Pcell is an LTE carrier, very much the same as for Option 3 in Fig. 12.2.

Going through this route, one would need to define all the physical layer channels and signals for this dual-connectivity based LAA operation to materialize, exactly in the same way as the development of Option 3 for regular NR operation in licensed spectrum. Once all the physical layers and signals are there, removing the (licensed) anchor does not pose any issues at the physical layer (allegedly, there was no physical layer impact to go from Option 3 to Option 2) and we only need to define procedures, most notably,

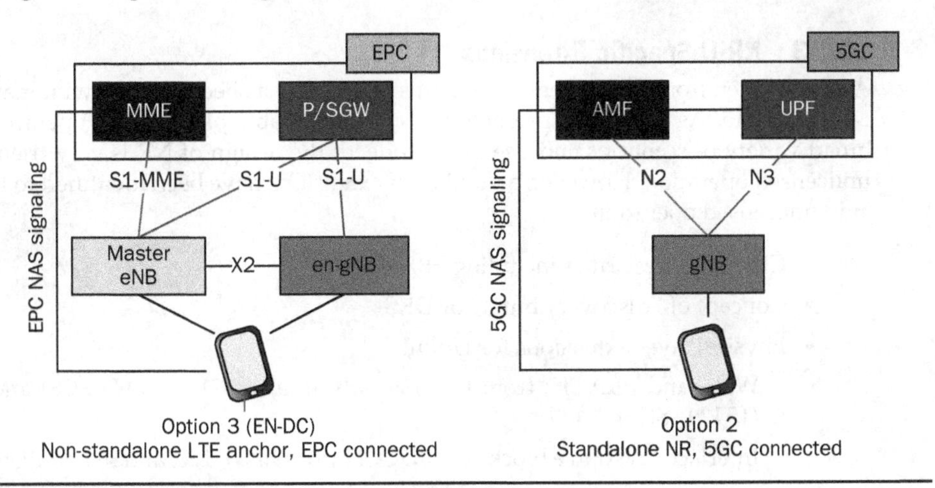

FIGURE 12.2 5G architecture Option 3 and Option 2.

Idle mode and mobility procedures, to be able to remove the dependency to the anchor (licensed) carrier. This is, in summary, the rationale behind the specification of the first release of NR-U including non-standalone, i.e., LAA as well as standalone options.

12.2.2 NR-U Scenarios

As an outcome of the SI on NR-U [12], the following scenarios were deemed of interest and were captured in the corresponding subsequent WID [15]:

- **Scenario A**: Carrier aggregation between licensed band NR (PCell) and NR-U (SCell).
  - NR-U SCell may have both DL and UL, or DL-only.
  - In this scenario, NR PCell is connected to 5G-CN.
- **Scenario B**: Dual connectivity between licensed band LTE (PCell) and NR-U (PSCell).
  - In this scenario, LTE PCell connected to EPC as higher priority than PCell connected to 5G-CN.
- **Scenario C**: Standalone NR-U.
  - In this scenario, NR-U is connected to 5G-CN.
- **Scenario D**: A standalone NR cell in unlicensed band and UL in licensed band (single cell architecture).
  - In this scenario, NR-U is connected to 5G-CN.
- **Scenario E**: Dual connectivity between licensed band NR and NR-U.
  - In this scenario, PCell is connected to 5G-CN.

As discussed earlier, unlike in LTE where a new frame structure, namely frame structure type 3, was created for LAA, unlicensed operation for NR did not resort to any new frame structure. Indeed, the entire NR operation does not have the concept of different frame structures. Instead, a single flexible frame structure exists.

12.2.3 NR-U Specific Extensions

As can be seen from the presentation in this book, NR has been defined with many use cases in mind. As a result, it presents a very configurable platform for operation in a broad variety of scenarios and use cases. Indeed, the design of NR is very friendly to unlicensed operation. However, a number of extensions have been identified to further aide unlicensed operation:

- Channel access rules including FBE support
- Concept of "discovery burst" or DRS
- Physical layer extensions for Uplink
  - Wideband PRACH: from L = 139 only to also 571 (30 kHz SCS) and 1151 (15 kHz SCS)
  - Interlaced resource blocks for PUCCH (formats 0, 1, 2, and 3) and PUSCH

- Increased scheduling flexibility
  - Dynamic PDCCH monitoring switching including PDSCH transmissions for type B scheduling beyond 2, 4, and 7
  - HARQ enhancements
  - Multi-PUSCH grants
  - Enhanced configured grant UL

These extensions will be discussed in further details in subsequent sections.

12.3 Channel Access

A key aspect for the operation in unlicensed spectrum is the definition of the channel access. This item would trigger, initially and still today, many discussions on coexistence fairness with deployed technologies, most notably Wi-Fi.

At the highest level, regional regulators, e.g., ETSI in Europe, FCC in the United States, set the rules for accessing the given spectrum in a technology agnostic way and facilitating coexistence among the candidate technologies.

Specific standards development organizations (SDOs), e.g., IEEE for Wi-Fi, 3GPP for LTE and NR, will define the access rules abiding by the aforementioned regulations. Typically, the SDO rules are stricter and more demanding that those set by the regulations. Beyond this point, fairness in the coexistence of different technologies becomes a complex matter not easily measurable.

12.3.1 ETSI Rules

ETSI regulations for operation in the 5 GHz band are provided in Ref. [17]. Note that the access rules for the 6 GHz band are being actively developed in 2020. This harmonized standard covers all the aspects of operation in the 5 GHz band, including the band's nominal center frequencies, nominal channel bandwidth and occupied channel bandwidth, RF output power limits, transmit power control (TPC) and power density requirements, transmitter unwanted emissions limits, receiver spurious emissions requirements, dynamic frequency selection (DFS), adaptivity (channel access mechanism), receiver blocking requirements, user access restrictions, and geo-location capability. For the purpose of this discussion, we focus on the channel access mechanism discussed in Sec. 4.2.7 of Ref. [17].

Importantly, the standard defines two types of *adaptive equipment*:

- Frame-based equipment (FBE)
- Load-based equipment (LBE)

The access rules for both types of equipment are well differentiated. FBE access rules are presented in Sec. 4.2.7.3.1 [17] and LBE access rules are presented in Sec. 4.2.7.3.2. There is an additional Sec. 4.2.7.3.3 on short control signaling transmissions applicable to both.

Wi-Fi and LAA access rules follow the access rules for LBE. NR-U channel access can follow LBE or FBE access rules.

12.3.1.1 LBE Access Rules

The LBE access rules have the following components:

- **Medium energy sensing**: based on *clear channel assessment* (CCA) measurements which are energy measurements over a candidate transmission channel on *observation slots* 9 μs long. There is an associated *energy detection* (ED) *threshold level* (TL) reflecting whether the channel is deemed to be free or occupied. Equipment shall consider a channel to be occupied as energy is detected at a level greater than the TL. The ED threshold, assuming a 0 dBi receive antenna, is specified to be dependent of the equipment's maximum transmit power (P_H) as follows:
  - For $P_H \leq 13$ dBm: TL = –75 dBm/MHz (i.e., –62 dBm / 20 MHz)
  - For 13 dBm < P_H < 23 dBm: TL = –85 dBm/MHz + (23 dBm – PH)
  - For $P_H \geq 23$ dBm: TL = –85 dBm/MHz (i.e., –72 dBm/20MHz)

Note that for equipment supporting only IEEE 802.11a, 802.11n, and/or 802.11ac, TL is independent of P_H and is set to be TL = –75 dBm/MHz, i.e., –62 dBm/20MHz.

- ***Priority class***: associated with a particular type of traffic and enabling transmission prioritization within a node according to different priorities or QoS requirements. There are four priority classes defined. An initiating device can have up to four *channel access engines* with their respective priority class, with only one *channel access engine* for each priority class implemented. The priority class allows prioritization of access by specifying different ***priority counter***, *p*, values per priority class. Higher priority classes get lower priority counter values than lower priority classes.
- ***Channel occupancy time*** (COT): duration of transmission burst. There are limits to the maximum COT duration for the various priority classes. Higher priority classes get shorter maximum COT duration than lower priority classes.
- ***Contention Window*** (CW): countdown counter governing the so-called *backoff procedure* for getting access to the channel (when the counter becomes negative). There is a minimum value, CW_{min}, a maximum value, CW_{max}, and a computed value CW. The CW_{min} and CW_{max} values are also utilized to prioritize access. There are four different set of values defined, one for each priority class.

There is an *initialization phase*, a *prioritization period*, and a *backoff procedure* defined.

Initiating Device

Initialization phase:

- CW is set to CW_{min}, defined for the corresponding priority class.
- A random number, *q* (backoff counter), uniformly distributed between 0 and CW is drawn. This *q* value counts 9 μs long *observation slots*.

Prioritization period:

- Priority counter, *p*, set to the priority value p_0 of the corresponding priority class.
- Wait for 16 μs.

- Upon CCA satisfying the ED threshold over *observation slot* countdown. If ED threshold not satisfied, re-start the prioritization period upon CCA satisfying the ED threshold.
- When priority counter, *p*, becomes 0, start *backoff procedure*.

Backoff procedure:

- Set the backoff counter, *q*, to a random number uniformly distributed between 0 and CW.
- Upon CCA satisfying the ED threshold over *observation slot* countdown. If ED threshold not satisfied, re-start the *prioritization period*.
- When backoff counter, *q*, becomes negative, if *channel access engine* is ready for transmission, go to *internal collision handling*. Otherwise, keep counting down according to the previous bullet.

Internal collision handling stage:
If only one *channel access engine* is in this stage, go to the *start of the transmission* stage. Otherwise, i.e., internal collision event, the *channel access engine* with the highest priority goes to *start of the transmission*, while all other *channel access engines* go to the *CW adaptation in case of collision* stage.

Start of the transmission stage:
COT transmission takes place with the following conditions:

- *Channel access engine* can have multiple transmissions without additional CCAs provided the gaps between transmissions do not exceed 16 μs. For gaps between 16 μs and 25 μs, transmissions may continue provided that CCA is satisfying the ED threshold.
- *Channel access engine* may issue grants to a responding device.
- Simultaneous transmissions of multiple priority classes are allowed provided that the overall COT duration is within the allowed limits of the priority class of the channel access engine.

Collision detection phase stage:
Upon COT completion and when at least one transmission that started at the beginning of the COT was deemed to be successful, go back to the *initialization phase*. Otherwise, go to the *CW adaptation in case of collision* stage.

CW adaptation in case of collision stage:

- If the initiating device wishes to re-transmit, the contention window is adjusted as follows: CW = ((CW + 1) x m – 1) saturated to the CW_{max} value ($m \geq 2$). Go back to second step of *initialization phase*.
- Otherwise, go back to the *prioritization period*.

Responding Device When a grant is received, the responding device can start transmission without CCA provided that the gap for its transmission is less than 16 μs from the last transmission of the initiating device. For larger gaps, CCA on the operating channel is required during a single 9 μs observation slot within 25 μs period immediately preceding the granted transmission time.

The responding device may perform transmissions for the remaining of the COT. The responding device may have multiple transmissions provided that the gap in between such transmissions does not exceed 16 μs.

12.3.1.2 FBE Access Rules Components

Frame-based equipment is equipment where the transmit/receive structure has a periodic timing with a periodicity equal to the *fixed frame period.* The *fixed frame period* is in the range of 1 to 10 ms. Transmissions by the initiating device can only start at the beginning of the *fixed frame period.* Equipment can change the *fixed frame period* no more than once every 200 ms.

Figure 12.3 illustrates the timeline associated with FBE operation.

The *initiating device* starts assessing the channel availability by performing a CCA right before transmissions at the start of the *fixed frame period.* This is denoted by "Cat 2 LBT here" in Fig. 12.3 which will be expanded in next section. The CCA consists of an energy measurement over one 9 μs *observation slot* and a comparison with an *energy detection threshold level* with the same values, as a function of the device max transmit power, P_H, as for LBE.

Grants can be issued to one or more *responding devices* to transmit within the COT. Note that the COT cannot be greater than 95% of the *fixed frame period* so that the *idle period* following the COT in the corresponding *fixed frame period* is at least 5% of the *fixed frame period.* The minimum duration of the *idle period* is 100 μs.

The rules for the *responding device* are the same as for LBE.

12.3.2 Wi-Fi Medium Access Rules

Wi-Fi medium access follows LBE rules. It is worthwhile mentioning that Wi-Fi *clear channel assessment* examines energy detection (as discussed thus far) but also performs virtual carrier sensing to detect Wi-Fi waveform in the form of Wi-Fi preambles. This procedure is also referred as preamble detection (PD) procedure. The thresholds for PD are more stringent than those for ED, namely, –82 dBm / 20 MHz versus –62 dBm / 20 MHz. Indeed, –82 dBm is considered to be the sensitivity level for the reception of a Wi-Fi packet and, hence, the detection of its preamble. As a result, if a Wi-Fi preamble is detected, Wi-Fi initiating devices attempt decoding the Wi-Fi packet and will refrain from attempting transmission and CCA will not clear.

12.3.3 NR-U Medium Access Rules

The NR-U TR [12] defines the following types of *listen before talk* (LBT):

The channel access schemes for NR-based access for unlicensed spectrum can be classified into the following categories:

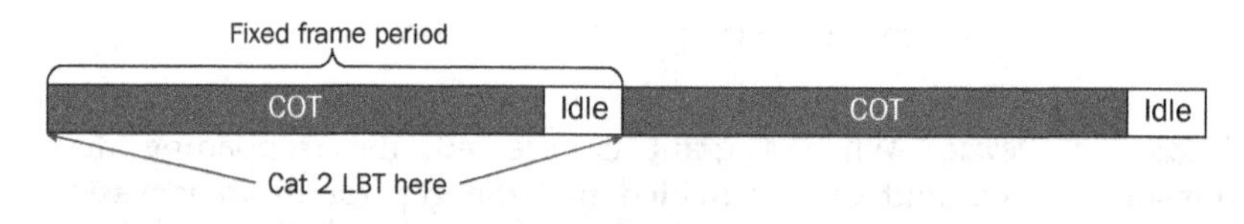

Figure 12.3 FBE timeline example.

Category 1: Immediate transmission after a short switching gap

- This is used for a transmitter to immediately transmit after a switching gap inside a COT.
- The switching gap from reception to transmission is to accommodate the transceiver turnaround time and is no longer than 16 μs.

Category 2: LBT without random backoff

- The duration of time that the channel is sensed to be idle before the transmitting entity transmits is deterministic.

Category 3: LBT with random backoff with a contention window of fixed size

- The LBT procedure has the following procedure as one of its components. The transmitting entity draws a random number N within a contention window. The size of the contention window is specified by the minimum and maximum value of N. The size of the contention window is fixed. The random number N is used in the LBT procedure to determine the duration of time that the channel is sensed to be idle before the transmitting entity transmits on the channel.

Category 4: LBT with random backoff with a contention window of variable size

- The LBT procedure has the following as one of its components. The transmitting entity draws a random number N within a contention window. The size of contention window is specified by the minimum and maximum value of N. The transmitting entity can vary the size of the contention window when drawing the random number N. The random number N is used in the LBT procedure to determine the duration of time that the channel is sensed to be idle before the transmitting entity transmits on the channel.

For different transmissions in a COT and different channels/signals to be transmitted, different categories of channel access schemes can be used.

NR-U rules for channel access are specified in Ref. [7]. Note that the same specification covers LTE LAA and NR-U. Basically, the channel access rules from ETSI are followed for LBE and FBE.

LBE In general, NR-U resorts to Cat 4 LBT to contend for COT, and use Cat 2 for inside COT. There are a few exceptions, most notably, a Cat 2 LBT is used for a discovery burst (see Sec. 12.4) transmission when it does not carry unicast data and has certain scheduling restrictions, namely, a duration no longer than 1 ms and a duty cycle <5%.

Channel access LBT mechanisms defined for NR-U along with their designation in Ref. [7]:

- Cat 4 LBT with a contention window (Type 1)
- Cat 2 LBT with 25 μs gap (Type 2A)
- Cat 2 LBT with 16 μs gap (Type 2B)

- Cat 1 LBT with no more than 16 μs gap without channel sensing (Type 2C)
  - A transmission burst length limit of 0.584 ms is applied when using this.

FBE This channel access mechanism has been discussed in Sec. 12.3.1. As discussed, Wi-Fi does not have a channel access mechanism attempting to exploit this possibility provided by the regulations. In the context of NR, this mechanism is of particular interest for industrial IoT use cases using unlicensed spectrum. In controlled environments, FBE will be able to guarantee QoS and URLLC traffic on unlicensed spectrum becomes feasible.

The channel access relies on Cat 2 LBT for contending for the channel at a fixed time grid. There is no Cat 4 LBT and, hence, there is no uncertainty in channel access time.

The FBE mode of operation will be announced in SIB1 along with the fixed frame period configuration (see Fig. 12.3). This configuration can also be signaled to a UE with UE-specific RRC signaling in the case of FBE Scell use case.

In Release 16, only the gNB contends for the channel. The UE transmissions within a fixed frame period can occur if DL signals/channels from the gNB, e.g., PDCCH, SSB, PBCH, SIBs, GC-PDCCH, within the fixed frame period are detected.

12.3.3.1 UL to DL COT Sharing

As discussed earlier, UE initiated COTs will undergo Cat 4 LBT. The goal is for the gNB to be able to exploit the fact that the UE gained channel access and allow a UL to DL COT sharing. This would apply to both scheduled UL transmissions and configured grant (CG) UL transmissions.

A new energy detection (ED) threshold for UL to DL COT sharing can be configured (to increase fairness in the case when the gNB has higher transmit power than UE).

For CG-UL, the COT sharing information is included in CG-UCI.

12.3.3.2 Channel Access for Wider Bandwidth Operation

NR is designed for wide bandwidth operation, see Chaps. 2 and 5. On the other hand, Wi-Fi is designed for operation in 20 MHz channels. Also, at least the 5 GHz regulations are written for 20 MHz channels [17] and regulations govern access to the entire channel, i.e., all or nothing type approach. Regulations for 6 GHz are not available at the time of publishing this work but chances are that they will be similar to those for 5 GHz.

As a result, NR-U operation needs to adapt to operation in 20 MHz basic channel units. We call these basic channel access units *LBT bandwidth*. The available resource blocks (RBs) in each *LBT bandwidth* is called *RB set*. The *RB set* is derived, separately for DL and UL, from the intra-band guardband signaling. Figure 12.4 shows an illustration of the RB sets and the intra-band guardbands.

Note that it is possible for the guardband to be of 0 size if the transmitting node, i.e., gNB or UE, performs an "all or nothing" transmission. The "all or nothing" approach would entail the transmitting node to transmit only when it clears channel access on all the underlying channels. As long as one channel would not pass channel access, it would not transmit anything. This radical approach would enable the transmission without guardbands because the transmission would not be affected by the independent clearing of channels.

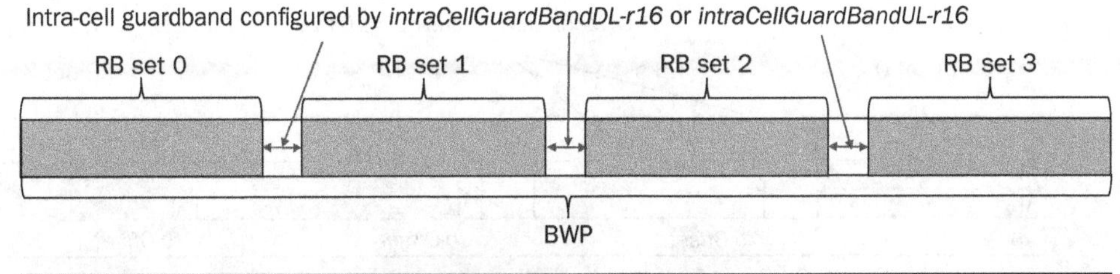

Figure 12.4 Illustration of RB sets and intra-band guardband.

12.3.3.3 COT Structure Indication by DCI 2_0

When the gNB contends for access in 20 MHz channel units, the gNB will provide the UE with information on time- and frequency-domain span of the current channel occupancy. This indication is done expanding the role of Release 15 DCI format 2_0 (SFI):

- Frequency-domain COT: bitmap to indicate the available LBT bandwidths (valid until the end of the corresponding channel occupancy)
- Time-domain COT: duration bit-field per serving cell indicating remaining length from the beginning of the slot where the information is received. The interpretation of the field is configurable by RRC:
  - If this field is not present (by configuration), the UE should use SFI indication to determine end-of-COT (if SFI is available), i.e., the UE may assume that the duration of the COT is the same as the duration for which SFI is provided in DCI format 2_0.
  - When a UE receives a COT duration indication with a given symbol being within the COT duration, the UE is not expected to receive a subsequent COT duration indication that suggests that symbol not being within the COT duration.

12.3.3.4 CORESET Enhancements

While the legacy CORESET definitions and operation are fully supported for NR-U, due to the channel access in units of 20 MHz channels, it is desirable to confine each PDCCH in one 20 MHz subband and allow the UE to independently decode PDCCH in each 20 MHz.

A search space set configuration is introduced allowing multiple frequency-domain associations. Also, a new way to define CORESET frequency-domain resources with *rb-offset* is introduced. In this sense, a search space set can be associated with multiple "images" of a CORESET confined within one LBT bandwidth and with the additional CORESET "images" being indicated by the parameter *freqMonitorLocations-r16*. Figure 12.5 illustrates this operation with an example.

12.3.3.5 Cyclic Extension of the First OFDM Symbol

For some channel access types, the UE needs to maintain a strict gap before transmission, e.g., 25 μs, 16 μs. For those cases, the CP extension is used to control the gap.

The CP extension is transmitted before the first UL symbol, using a long CP of the symbol. For PUCCH, the length of the CP extension is DCI controlled and is LBT type dependent. For PUSCH, similarly, the length of the CP extension is DCI controlled and is LBT type and channel access priority class (CAPC) dependent. The CP extension value also depends on the current timing advance of the UE.

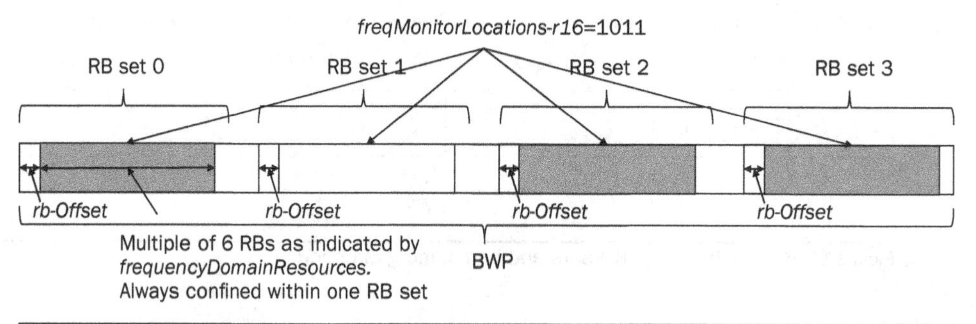

FIGURE 12.5 Illustration of CORESET configuration for multiple RB sets.

12.3.4 Discussion of NR-U versus Wi-Fi Access

NR-U and Wi-Fi are fundamentally two distinct wireless systems. Probably, the biggest difference between them affecting their stance vis-à-vis medium access is the fact that Wi-Fi access is fully asynchronous, down to the 9 μs *observation slot* granularity. While NR-U access could also be considered asynchronous, as contention can start at any time, transmissions start following the underlying synchronous slot and mini-slot structure.

There are pros and cons of both designs. Async Wi-Fi design is more agile from the medium access point of view. Transmissions take place packet by packet in an either point-to-point communication link or in point-to-multipoint communication links exploiting multi-user MIMO (MU-MIMO) techniques. It requires a preamble preceding every packet (the so-called STF), enabling time-domain preamble detection.

NR-U, and LTE for that matter, relies on synchronization signals for the terminal devices to acquire synchronization. This *initial acquisition* process is carried out via time-domain correlation processing at the receiving node. Once timing is acquired, all the receiver processing relies on frequency domain with an underlying time tracking for the placement of the FFT at the beginning of each OFDM symbol. Synchronizing within a cell enables simultaneous transmissions to different users on the DL, as well as simultaneous transmissions from different users on the UL. Synchronization across cells further enables interference control, particularly useful for higher power transmissions typical in licensed spectrum deployments.

While LAA had some limitations, relaxed over time, on the possible starting and ending positions for transmissions within a slot. NR has been designed with a way more flexible time-domain allocation capability.

For the DL of NR, PDSCH scheduling Type B (see Chap. 8), also known as mini-slot scheduling, enables the possibility to define mini-slots of two OFDM symbols duration with possibility to be transmitted across the slot in seven different positions. On the other hand, the starting and ending points for the PUSCH transmissions in the UL can be fully indicated in the corresponding grant.

12.4 Discovery Burst

The concept of discovery burst is introduced in the 3GPP specifications motivated by operation in unlicensed or shared spectrum. The discovery burst mainly consists of the SSB transmission. As we have seen in Chap. 6, the SSB (also known as SS/PBCH block) transmission consists of the transmission of the sync signals (PSS and SSS) and the physical broadcast channel (PBCH) carrying the MIB. For the frequencies of interest for unlicensed operation in Release 16, i.e., 5 GHz and 6 GHz bands, the SS/PBCH block transmission can consist of up to eight unique SSB indices or beam directions for its transmission.

The discovery burst, however, can also contain Type0-PDCCH (also known as CORESET#0, which is the CORESET for SIB1), SIB1 (also known as RMSI), CSI-RS, Paging messages, and other broadcast/unicast signals. Note that depending on the contents of the discovery burst, the channel access rule differs.

In absence of unicast data and provided that the discovery burst duty cycle is ≤5% and the total duration is less than 1 ms, a so-called Cat 2 LBT, consisting of 25 μs energy sensing prior to transmission, can be applied. Otherwise, Cat 4 LBT is to be applied to the discovery burst transmission.

We have seen in Chap. 6 that SSB transmission in FR1 can use 15 kHz or 30 kHz SCS. Also that the UE assumes a single SSB numerology for a given frequency band with a few exceptions. For NR-U operation in 5 GHz and 6 GHz bands, UEs will assume SSB transmissions with 30 kHz SCS. Note that SSB with 15 kHz SCS can be configured by higher layers. No change of numerology is assumed between SS/PBCH block and CORESET#0 (allowed for regular NR operation).

CORESET#0 for NR-U operation is assumed to have a bandwidth configuration of 48 RBs for 30 kHz SCS and 96 RBs for 15 kHz SCS; and a time span of one or two OFDM symbols (only).

SSBs are a critical piece of NR as they are needed for initial access and for RLM/RRM (see Chap. 6). Under NR-U operation, SSB transmission is subject to listen before talk (LBT) and, therefore, SSBs transmission is not guaranteed. In order to increase the chances for its transmission, measures have been taken in the NR-U design.

The actual extension from regular NR operation is quite straight forward and easy to see from Figs. 12.6 and 12.7 and how they relate to their counterparts in Chap. 6 for regular NR operation. Regular NR operation defines the maximum number of SSB indices for transmission within a frequency band, L_{max}. In all cases, the L_{max} SSB transmissions span no more than a 5 ms half-frame. For FR1 bands above 3 GHz, $L_{max} = 8$.

Looking into the 15-kHz-SCS– and 30-kHz–SCS–based SSB transmission and extending it to candidates in a 5 ms span yields what is shown in Figs. 12.6 and 12.7. Looking now into the candidate SSB transmissions which one could have within those 5 ms, one can easily find that for 15 kHz SSB numerology, we can have 10 candidate SSB positions (Fig. 12.6). Similarly, for 30 kHz SSB numerology, we can have 20 candidate SSB positions (Fig. 12.7).

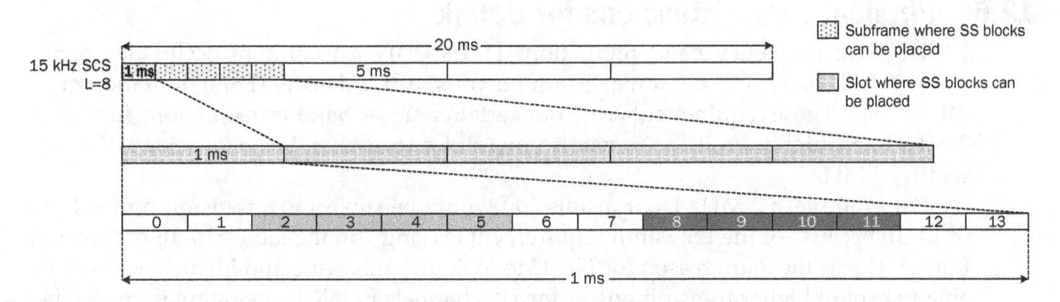

FIGURE 12.6 Illustration of extended SSB with 10 candidate SSB positions (15 kHz SCS).

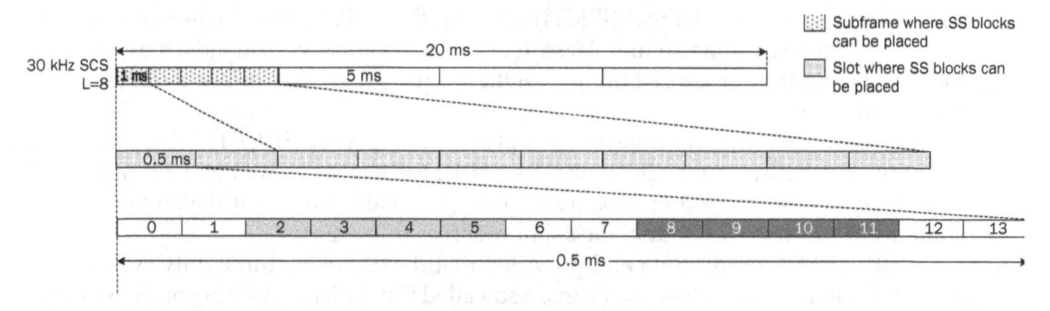

FIGURE 12.7 Illustration of extended SSB with 20 candidate SSB positions (30 kHz SCS).

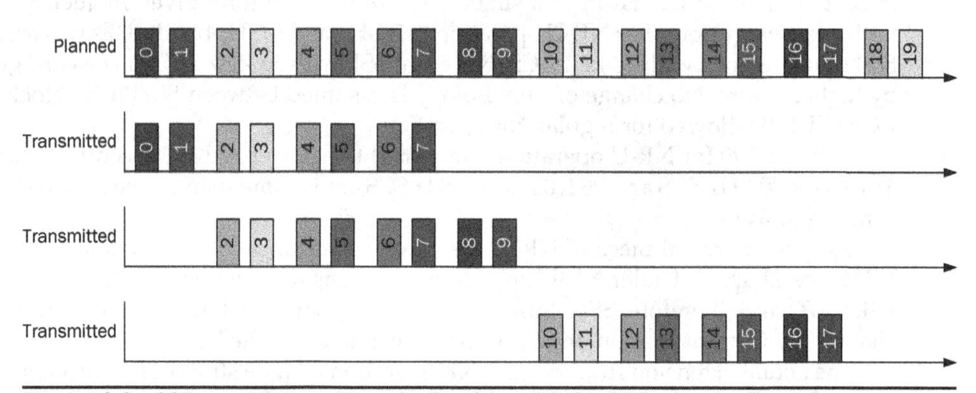

FIGURE 12.8 SSB candidate positions for Q = 8 and 30 kHz SCS.

In NR-U, the concept of QCL relation factor, Q, is introduced such that the SSB positions Q positions apart are QCL'd. If the gNB failed to transmit the SSB at position x, it has another chance to send it at position x+Q as illustrated in Fig. 12.8. Q can take the following values {1,2,4,8} and its value (2-bits) is included in MIB or UE specifically configured for RLM/RRM.

12.5 Physical Layer Extensions for Uplink

ETSI 5 GHz frequency band regulations [17] specify a minimum occupancy bandwidth (OCB) of 2 MHz and a maximum power spectral density (PSD) limitation of 10 dBm/MHz. These requirements rule out certain narrow band transmissions (less than 2 MHz) and highly limit the transmit power of narrow band transmissions of bandwidth ≥2 MHz.

For example, a 2 MHz UL transmission would be subject to a transmit power limit of 13 dB because of the PSD limit requirement leaving "on the table" 10 dB of transmit power. This is the main reason for NR-U to look into this issue and identify ways to be able to exploit higher transmit power for UL channels in NR-U operating in the 5 GHz band while leveraging to the maximum extent the NR design.

12.5.1 Wideband PRACH

As we have seen in Chap. 6, NR has defined a large number of RACH preamble formats. RACH preamble formats 0, 1, 2, and 3 do not apply to NR-U. These are the long sequence-based (L = 839) RACH preamble formats, which are, indeed, narrow band (format 0, 1, and 2 are 1.08 MHz wide while format 3 is 4.32 MHz wide, see Chap. 6) and of long duration which is not anticipated to be required for NR-U operation.

For this reason, a new set of PRACH preamble formats has been introduced in NR for the purpose of NR-U. These formats consist of taking the short sequence-based (L_{RA} = 139) RACH preamble formats, namely, formats A1, A2, A3, B1, B2, B3, B4, C0, and C2, and apply a larger length, L_{RA} = 1151 for 15 kHz and with L_{RA} = 571 for 30 kHz. Note that old and new formats are applicable for NR-U operation. SIB1 will inform which formats to use.

Figure 12.9 illustrates the formats for L_{RA} = 1151 and 15 kHz SCS. Their time span is the same as that for shorter sequence (L_{RA} = 139). However, the transmission bandwidth has increased from 2.16 to 17.28 MHz. Note that these figures also apply to L_{RA} = 571 and 30 kHz SCS considering the time-ticks to be half of what is shown in Fig. 12.9 as OFDM symbols for 30 kHz SCS are half than those for 15 kHz SCS.

As we have done for the Release 15 RACH preambles, Tables 12.1 and 12.2 summarize the high-level parameters for these new RACH preamble formats.

12.5.2 Interlaced Resource Blocks for PUCCH and PUSCH

We already discussed the ETSI requirements for operation in the 5 GHz frequency band [17] in terms of max PSD limit (10 dBm/MHz) and minimum OCB (2 MHz). In addition, there is a requirement on OCB of 80% of the 20 MHz channel bandwidth.

Because of that requirement, NR-U looks into ways to increase the transmit bandwidth of the UL transmissions. We have already discussed PRACH preambles in the previous section. We look now into PUCCH and PUSCH.

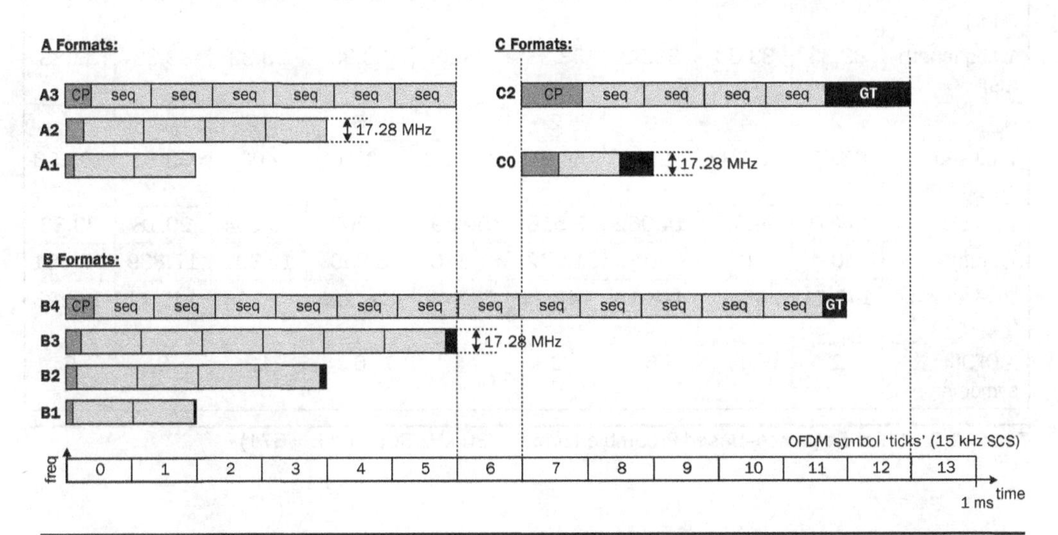

Figure 12.9 Illustration of new wider-bandwidth PRACH formats (15 kHz SCS and L = 1151).

Concept	A1	A2	A3	B1	B2	B3	B4	C0	C2
L	1151	1151	1151	1151	1151	1151	1151	1151	1151
N	1152	1152	1152	1152	1152	1152	1152	1152	1152
SCS (kHz)	15	15	15	15	15	15	15	15	15
Bandwidth (MHz)	17.28	17.28	17.28	17.28	17.28	17.28	17.28	17.28	17.28
1 seq length (μs)	66.67	66.67	66.67	66.67	66.67	66.67	66.67	66.67	66.67
N_{SEQ}	2	4	6	2	4	6	12	1	4
Total seq length (μs)	133.33	266.67	400	133.33	266.67	400	800	66.67	266.67
T_{CP} (μs)	9.375	18.75	28.125	7.031	11.719	16.406	30.469	40.36	66.67
T_{GT} (μs)	0	0	0	2.344	7.031	11.719	25.781	35.677	94.922
Total length (μs)	142.71	285.42	428.13	142.71	285.42	428.125	826.25	142.71	428.26
# OFDM symbols	2	4	6	2	4	6	12	2	6

TABLE 12.1 Short-Sequence–Based Preamble Formats (15 kHz SCS and L = 1151)

Concept	A1	A2	A3	B1	B2	B3	B4	C0	C2
L	571	571	571	571	571	571	571	571	571
N	576	576	576	576	576	576	576	576	576
SCS (kHz)	30	30	30	30	30	30	30	30	30
Bandwidth (MHz)	17.28	17.28	17.28	17.28	17.28	17.28	17.28	17.28	17.28
1 seq length (μs)	33.33	33.33	33.33	33.33	33.33	33.33	33.33	33.33	33.33
N_{SEQ}	2	4	6	2	4	6	12	1	4
Total seq length (μs)	66.67	133.33	200	66.67	133.33	200	400	33.33	133.33
T_{CP} (μs)	4.6875	9.375	14.062	3.516	5.859	8.203	15.234	20.18	33.33
T_{GT} (μs)	0	0	0	1.172	3.516	5.859	12.891	17.839	47.461
Total length (μs)	142.71	285.42	428.13	142.71	285.42	428.125	826.25	142.71	428.26
# OFDM symbols	2	4	6	2	4	6	12	2	6

TABLE 12.2 Short-Sequence–Based Preamble Formats (30 kHz SCS and L = 571)

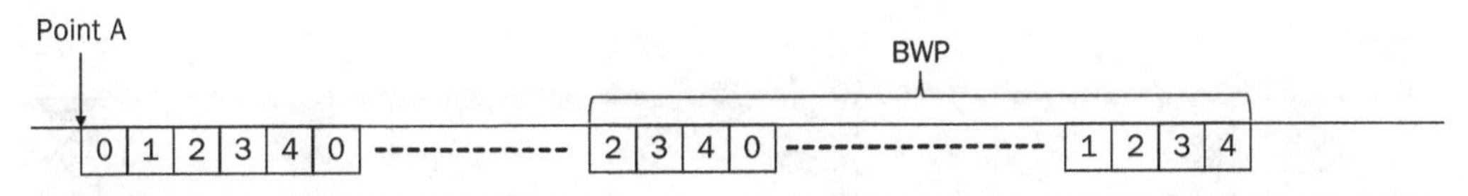

FIGURE 12.10 Illustration of five interlaces for UL transmission (30 kHz).

Similar to LTE-LAA, a PRB block interlaced waveform is introduced for PUCCH/PUSCH in NR-U. Point A (see Chap. 2) is the reference for the interlace definition.

- For 15 kHz SCS, M = 10 interlaces are defined for all bandwidths.
- For 30 kHz SCS, M = 5 interlaces are defined for all bandwidths.

Figure 12.10 illustrates the five interlaces for 30 kHz SCS and how they relate to Point A.

PUSCH Interlaced PUSCH is introduced for both DFT-s waveform and CP-OFDM waveform. For DFT-s version, if the allocated number of RBs does not satisfy $2^a 3^b 5^c$, the ending RBs can be dropped.

For resource allocation, interlace assignment and RB set assignment are required.

- X bits for interlace assignment (same as LTE-LAA design)
  - For 30 kHz SCS, X = 5 (full bitmap for all possible interlace combinations)
  - For 15 kHz SCS, X = 6 to indicate start interlace index and number of contiguous interlace indices (RIV). Remaining up to 9 RIV values to indicate specific pre-defined interlace combinations
- Y bits for RB set assignment (for DCI 0_1)
  - RB set assignment is RIV format for starting and ending RB sets (always contiguous).
  - When two adjacent RB sets are assigned, the guardband in-between is also assigned.

Figure 12.11 shows an illustration of interlaced PUSCH allocation across two RB sets.

The legacy waveform is still supported (for regions without OCB requirement and **where** full power transmit power **may not be** needed). The interlaced waveform for common/dedicated PUCCH/PUSCH is separately configured, but the UE will not expect the configuration to be different in a given cell **and there is no** dynamic switching of waveform.

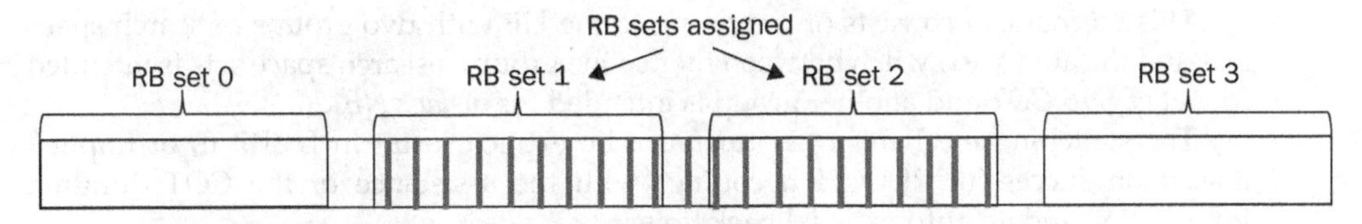

FIGURE 12.11 Interlaced PUSCH allocation example.

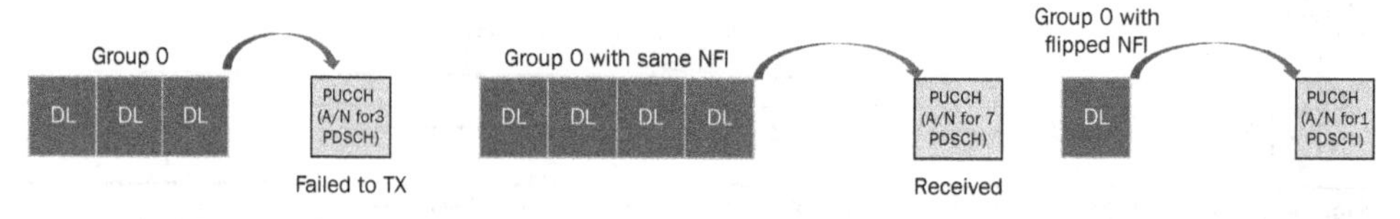

Figure 12.12 Example of enhanced dynamic ACK/NACK codebook.

PUCCH NR-U has also introduced the interlaced flavor of PUCCH formats 0, 1, 2, and 3. Note that these extensions only apply within one 20 MHz channel. Therefore, one interlace corresponds to 10 or 11 RBs.

- PUCCH formats 0 and 1 are extended to 1 interlace from 1 RB.
    - Cyclic shift ramping across PRBs for better PAPR
- PUCCH formats 2 and 3 are extended to 1 or 2 interlaces from 1 to 16 RBs in Release 15.
    - Interlaced PUCCH format 2 further supports frequency-domain OCC (1/2/4) to support user multiplexing when only one interlace is configured.
    - Interlaced PUCCH format 3 further supports pre-DFT OCC (1/2/4) to support user multiplexing when only one interlace is configured.

Chapter 9 has more details for these new PUCCH formats.

12.6 Increased Scheduling Flexibility

In this section we review the enhancements introduced by NR-U to improve the scheduling flexibility.

12.6.1 Dynamic PDCCH Monitoring Switching

NR is inherently a slot-based system. Data grants are transmitted in PDCCH using a corresponding CORESET with its own determined time and frequency location. As discussed in Chap. 8, NR has defined two PDSCH mapping types, namely, PDSCH mapping type A or slot-based scheduling, and PDSCH mapping type B or mini-slot–based scheduling.

Clearly, the UE complexity requirement for mini-slot–based scheduling is greater than for slot-based scheduling. From the NR-U viewpoint, having access to more "entry points" in the slot for scheduling is of great value, particularly as it relates to gaining channel access.

Once the channel access has been gained, however, the incentive to continue using mini-slot–based scheduling greatly diminishes. With this realization, NR-U has introduced a dynamic PDCCH monitoring switching mechanism.

This mechanism consists of providing to the UE with two groups of search space sets and the ability to switch between them. One group of search space sets is intended for *out of gNB COT* and another group is intended for *inside gNB COT*.

The switching mechanism, in turn, can be explicit (a bit in DCI 2_0) or implicit (based on successful PDCCH decoding) with the assistance of the COT duration information and an automatic fallback timer.

Indeed, in order to support a more flexible mini-slot operation at the beginning of the gNB COT, all the lengths between 2 symbols and 13 symbols for PDSCH mapping type B are introduced in Release 16. Note that Release 15 only allowed for mini-slot durations of 2, 4, or 7 OFDM symbols.

12.6.2 HARQ Enhancements

12.6.2.1 Non-Numerical K1

PDSCH transmissions from the network are acknowledged by the UE. The timing relationship between the DL assignment with subsequent DL data transmission and the UE's acknowledgment is governed by the parameter K1.

In NR-U operation, the problem that the HARQ ACK transmission time could be out of the COT and that there is uncertainty on the ability to transmit outside the COT arises. For this case, the devised solution is to introduce a non-numerical K1 as a "wildcard."

If a DL assignment indicates this non-numerical K1, the UE will hold on the HARQ ACK report until the gNB provides a normal (numerical) K1 in a later DL assignment, e.g., at the next gNB COT. Note that the fallback DCI does not support signaling a non-numerical value of K1.

12.6.2.2 Enhanced Dynamic ACK/NACK Codebook

Release 15 dynamic codebook is enhanced to support retransmission of HARQ ACK previously transmitted or failed to transmit due to LBT failure.

A PDSCH group concept is introduced where HARQ ACK retransmission triggering is in the unit of PDSCH groups, i.e., two groups are enabled when enhanced dynamic codebook is configured. An explicit group index is included in the DCI scheduling the PDSCH.

A new ACK-feedback group indicator (NFI) for each PDSCH group is also introduced. This indicator operates as a toggle bit like the new data indicator (NDI).

The size of a group can increase when more PDSCH of the same group are granted, until NFI for the group flips. The group size can increase even after the ACK is transmitted. T-DAI and C-DAI are per group and count until the NFI flips. An example of this operation is depicted in Fig. 12.12.

12.6.2.3 One-Shot HARQ Feedback

This is another way to provide a mechanism to trigger A/N retransmission enabling the gNB the possibility to request feedback of a HARQ-ACK codebook containing all DL HARQ processes (one-shot feedback) for all component carriers (CCs) configured for a UE in the PUCCH group.

One-shot feedback can be configured with semi-static codebook, (non-enhanced) dynamic HARQ codebook, and enhanced dynamic codebook. When configured, a triggering bit will be included in DCI 1_1. There is also possibility for DCI 1_1 to be used exclusively to trigger the one-shot HARQ feedback, without an associated PDSCH.

In the ACK/NACK report, NDI for each HARQ process can be configured to be included, and CBG level ACK/NACK report can be configured if a CC is configured with CBG level ACK/NACK.

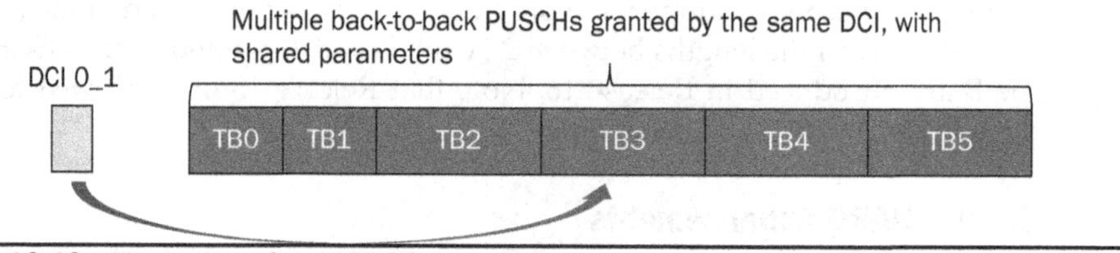

Figure 12.13 Illustration of multi-PUSCH grant.

12.6.3 Multi-PUSCH Grants

This feature was introduced in LTE-eLAA to support the use case of having a long UL burst in a COT, while reducing the control overhead.

The goal is to schedule PUSCH over multiple consecutive slots/mini-slots with a single DCI. Figure 12.13 shows an illustration of this mechanism. Note that each PUSCH will have its own separate transport block (TB) and each TB is mapped to one slot or one mini-slot.

Different PUSCHs can have different lengths, but they are contiguous in time domain. Multi-PUSCH transmissions are granted with DCI 0_1 with modifications that support both single PUSCH and multi-PUSCH scheduling. For the multi-PUSCH scheduling, most parameters are shared across individual PUSCHs, except HARQ process ID, redundancy version, NDI, and time-domain resource allocation.

Scheduling multiple PUSCHs is not supported with the UL fallback DCI, i.e., DCI 0_0.

12.6.4 Configured Grant Enhancements

The following improvements to UL CG transmissions have been adopted:

- More flexible time-domain resource allocation
  - Number of allocated slots
  - Number of consecutive PUSCHs in a slot for mini-slot support
- CG-UCI extended to carry HARQ process ID, redundancy version, NDI, COT sharing information to provide UE with more flexibility
- CG-UL resource allowed to be used for retransmission
- DL feedback information (DFI) DCI added to provide HARQ A/N for PUSCH

References

[1] Qualcomm, Ericsson, "Introducing LTE in unlicensed spectrum," RP-131635, 3GPP RAN#62, December 2013.

[2] Verizon, "New band for LTE deployment as supplemental Downlink in unlicensed 5.8 GHz in USA," RP-131680, 3GPP RAN#62, December 2013.

[3] Ericsson, Qualcomm, "Study on LTE evolution for unlicensed spectrum deployments," RP-131788, 3GPP RAN#62, December 2013.

[4] Ericsson, Qualcomm, Huawei, Alcatel-Lucent, "Study on licensed-assisted access using LTE," RP-141664, 3GPP RAN#65, September 2014.

[5] Ericsson, Huawei, Qualcomm, Alcatel-Lucent, "New work item on licensed-assisted access to unlicensed spectrum," RP-151045, 3GPP RAN#68, June 2015.

[6] 3GPP, "Feasibility study on licensed-assisted access to unlicensed spectrum," 3GPP TR 36.889, V13.0.0, July 2015.

[7] 3GPP, "3rd Generation Partnership Project; Technical Specification Group Radio Access Network; physical layer procedures for shared spectrum channel access (Release 16)," 3GPP TS 37.213, V16.0.0, January 2020.

[8] Ericsson, Huawei, "New work item on enhanced LAA for LTE," RP-152272, 3GPP RAN#70, December 2015.

[9] Nokia, Ericsson, Intel, Qualcomm, "New work item on enhancements to LTE operation in unlicensed spectrum," RP-1700848, 3GPP RAN#75, March 2017.

[10] Qualcomm, "New SID on NR-based access to unlicensed spectrum," RP-170828, 3GPP RAN#75, March 2017.

[11] Alcatel-Lucent Shanghai-Bell, Alibaba, Apple, AT&T, British Telecom, Broadcom, CATT China Telecom, China Unicom, Cisco, CMCC, Convida Wireless, Deutsche Telekom, DOCOMO, Ericsson, Etisalat, Fujitsu, Huawei, Intel, Interdigital, KDDI, KT, LG Electronics, LGU+, MediaTek, NEC, Nokia, Ooredoo, OPPO, Qualcomm, Samsung, Sierra Wireless, SK Telecom, Sony, Sprint, Swisscom, TCL, Telecom Italia, Telefonica, TeliaSonera, Telstra, Tmobile USA, Verizon, vivo, Vodafone, Xiaomi, ZTE, "Way forward on the overall 5G-NR eMBB workplan," RP-170741, 3GPP RAN#75, March 2017.

[12] 3GPP, "Study on NR-based access to unlicensed spectrum," 3GPP TR 38.889, V16.0.0, December 2018.

[13] Intel Corporation, "New SID: Study on supporting NR from 52.6 GHz to 71 GHz," RP-193258, 3GPP RAN#86, December 2019.

[14] Qualcomm, "New WID on extending current NR operation to 71 GHz," RP-193229, 3GPP RAN#86, December 2019.

[15] Qualcomm, "New WID on NR-based access to unlicensed spectrum," RP-182878, 3GPP RAN#82, December 2018.

[16] 3GPP, "Feasibility study on 6 GHz for LTE and NR in licensed and unlicensed operations (Release 16)," 3GPP TR 37.890, V0.7.9, January 2020.

[17] ETSI, "5 GHz RLAN; Harmonised standard covering the essential requirements of article 3.2 of Directive 2014/53/EU," ETSI EN 301 893, v2.1.1, May 2017.

CHAPTER 13

Vertical Expansion: URLLC

As discussed earlier, at the start of the standardization, 5G is designed to accommodate diverse services, a wide range of spectrum and various deployment scenarios. Ultra reliable low latency communications (URLLC) is one pillar of the three types of services offered by 5G. By its name, URLLC is optimized for ultra-low latency and/or ultra-low reliability, which represents traffic characteristics of many applications, e.g., Refs. [1–8]:

- Consumer-oriented, such as gaming, XR
- Factory-oriented, such as factory automation
- Transportation-oriented, such as remote driving
- Power-distribution, such as grid fault and outage management, differential protection
- Health-oriented, such as remote surgery
- Tactile internet

At the same time, not all URLLC applications have exactly the same detailed performance requirements. Indeed, the end-to-end latency can be from as low as one millisecond to as high as tens of milliseconds [9, 10]. For example, motion control in factory automation, tele-surgery, and autonomous driving may require almost near-real-time connectivity (e.g., 1 ms), while XR services may tolerate latency around 10 ms. There are also different levels of reliability requirements, e.g., from 99.9% to 99.9999%. In addition, throughput requirements have a large range, which can be represented by a payload size from tens of bytes (e.g., for motion control in factory automation) to thousands of bytes (e.g., remote driving, XR, etc.).

The stringent requirements for URLLC services pose extreme challenges for wireless networks, especially considering the time-varying channel and interference conditions, the need to optimize for both the link-level efficiency and the system-level efficiency, and the necessity to coexist with other types of services such as eMBB and mMTC. In particular, the following key questions need to be addressed in order to support URLLC:

- What are the use cases and deployment scenarios that URLLC design should be focused on?
- How to schedule resources for URLLC? In a dynamic manner via a DCI, a semi-static manner via a configured grant (CG), or an activation/release–based semi-persistent scheduling (SPS)?

- How to optimize the link-level efficiency for URLLC? This may include factors such as HARQ operation (synchronous vs. asynchronous, a number of re-transmissions, etc.), processing time requirements, channel feedback, and interference handling.
- How to share Downlink or Uplink resources among different services (e.g., URLLC, eMBB, and mMTC)? What if there is an ongoing low-priority transmission occupying resources which could become necessary for scheduling a higher-priority URLLC packet?
- Should the UE be able to support different service types concurrently, particularly when their scheduling and/or the HARQ feedback can be out of order (e.g., a PDSCH scheduled at a later time may require a HARQ feedback earlier than a HARQ feedback for an earlier scheduled PDSCH)?
- How to improve availability of URLLC services by a cell, particularly in terms of coverage?
- How to handle time-sensitive applications?

In the subsequent sections, we will provide a detailed description for the URLLC design in 5G NR addressing the above key questions. Before that, we will first briefly introduce similar efforts in 4G LTE.

13.1 A Brief History of 3GPP Standardization Related to URLLC

Since the beginning of the first 4G LTE release, tremendous efforts have been spent in 3GPP to further evolve the 4G LTE standardization, primarily targeting enhancing throughputs from both the cell's and the UE's perspectives. For instance, these additional features include advanced SU/MU-MIMO, carrier aggregation, dual-connectivity, coordinated multi-point transmission (CoMP), enhanced inter-cell interference coordination/cancellation (eICIC), higher-order modulation schemes (e.g., 256QAM, 1024QAM), etc. In these enhancements, the minimum time-schedule unit for a transport block (TB), also known as the transmit time interval (TTI), remains at 1 ms, in combination with a minimum HARQ round-trip-time (RTT) of 8 ms. For a packet transmitted over the air, this results in a minimum of 9 ms latency for just one HARQ re-transmission where a minimum of 3 ms processing time is budgeted for a UE or an eNB, as illustrated in Fig. 13.1. Very little efforts have been devoted to reducing over-the-air latency using 4G LTE until Release 15 [11], where a study item was approved dedicated to studying how to achieve lower latency in 4G LTE particularly using a short TTI (sTTI). The study item was followed by a work item (WI), which was completed in

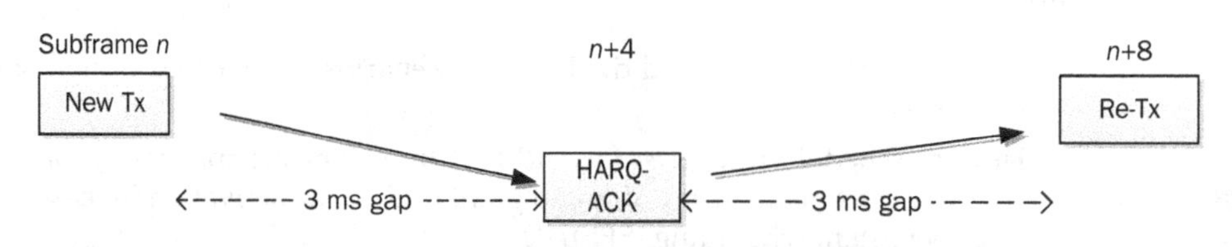

FIGURE 13.1 Illustration of a minimum of 8 ms HARQ RTT in 4G LTE.

December 2017 as part of the LTE Release 15. A related work item for high reliability low latency communications (HRLLC) [12] was also completed in LTE Release 15.

This feature on the low latency enhancements in LTE Release 15 has the following two sets of requirements:

- Set 1: a block error rate (BLER) of 1×10^{-5} within 1 ms for a payload size of 32 bytes, and
- Set 2: a BLER of 1×10^{-4} within 10 ms for a payload size of 32 bytes

To satisfy these requirements, two different approaches were adopted:

- Shortened processing time (sPT), and
- sTTI.

The idea of sPT is to re-use the existing 1 ms TTI, but introducing a reduced processing time for a UE and an eNB. More specifically, instead of a minimum of 3 ms processing delay assumed for a UE or an eNB, a new minimum of 2 ms processing delay is assumed for a sPT-capable UE or eNB. This effectively brings down the over-the-air latency from 9 to 7 ms for one HARQ re-transmission as illustrated in Fig. 13.2.

The reduced processing delay inevitably increases the processing requirement for a UE or an eNB. It should be noted that the 3 ms or 2 ms gap as illustrated in Figs. 13.1 and 13.2 is intended to accommodate not only the processing time, but also the round-trip propagation delay between the two nodes, as shown in Fig. 13.3. As can be seen, although there is a 3 ms gap between subframe n and subframe $n + 4$ at the eNB side, there is only a $3 - 2T_{prop}$ ms gap, i.e., the effective processing time available at a UE, between subframe n and subframe $n + 4$ at the UE side, where T_{prop} is the propagation delay between an eNB and a UE. In LTE, an eNB can have a coverage area as far as 100 km, which implies a maximum of $2T_{prop} = 667$ µs round-trip propagation delay. Thus, the effective minimum processing time budget for a UE with an 8 ms HARQ

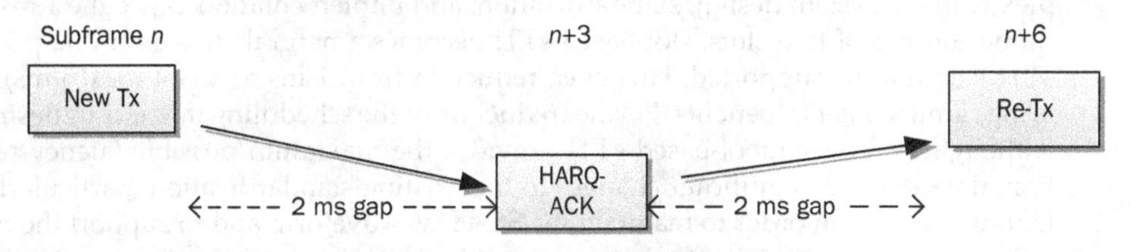

FIGURE 13.2 Illustration of a minimum of 6 ms HARQ RTT for sPT in 4G LTE.

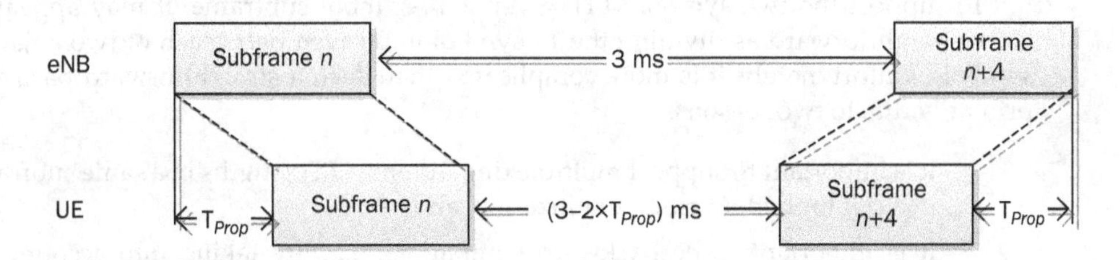

FIGURE 13.3 Illustration of the effective processing time at the UE due to the effect of the round-trip propagation delay.

RTT is 3 ms – 667 μs = 2.333 ms. If the same maximum round-trip propagation delay is assumed for the sPT case, a UE would only have a minimum 2 ms – 667 μs = 1.333 ms processing time budget. This is deemed to be a challenging requirement for a 4G LTE UE. To alleviate the challenge, a maximum round-trip delay of $2T_{prop}$ = 200 μs is assumed for the sPT-capable UEs, which corresponds to a maximum distance of 30 km from the serving eNB.

The idea of sTTI is much more revolutionary. Reference [13] provides a nice summary of the enabling techniques, and Ref. [14] presents some link-level evaluations.

The sTTI operation no longer relies on the 1 ms "subframe" (or "TTI") unit for scheduling. Instead, it zooms into the subframe and utilizes a small set of symbols in the subframe as the minimum scheduling unit. In designing the sTTI, the foremost criterion is backward compatibility, which implies that:

- The LTE numerology is re-used without any change,
- A cell can smoothly integrate the 1 ms TTI-based UEs and the sTTI-based UEs in the subframe, and
- A legacy UE can still access an eNB supporting sTTI without any impact.

In addition, it is desirable that:

- The legacy procedure for the initial access can be re-used without any change in order to minimize the specification impact. As a result, sTTI operation is only limited to RRC_CONNECTED UEs, and
- A new sTTI-capable UE can be scheduled by a cell with the legacy TTI and the new sTTI dynamically.

Note that there are 14 symbols in a 1 ms subframe with NCP. Naturally, one may choose the minimum scheduling unit as *any* number of symbols, i.e., {1, 2, 3, 4, ..., 14} symbols. However, such full flexibility is not necessary. It creates challenges and complexity in the system design, standardization, and implementation. Since the 1 ms subframe consists of two slots, slot-based sTTI becomes a natural choice for one possible sTTI length to be supported. However, reduction from 1 ms to 1 slot (or 0.5 ms) only brings limited latency benefits. Further reduction of the scheduling unit is thus desirable. Although the one-symbol-based sTTI provides the maximum possible latency reduction, it necessitates significant changes to the existing standardization, particularly for Uplink channels in order to maintain the SC-FDM waveform and to support the possibility of frequency hopping for better performance. Consequently, a two-symbol-based sTTI becomes appealing and eventually is specified as another sTTI length.

To support the two-symbol sTTI, given a 14-symbol subframe, it may appear to be as straightforward as dividing the 14-symbol into seven pairs each of two adjacent symbols. Unfortunately, it is more complicated than such a straightforward partition, primarily due to two reasons:

- It is important to support multiplexing different TTI lengths in a same subframe in a cell for better resource utilization, and
- It is important to be backward compatible, e.g., by taking into account the legacy control region in Downlink, the legacy CSI-RS patterns in Downlink, the SRS symbols in Uplink, etc.

Considering the need to coexist with the 1 ms TTI and the 1 slot sTTI, it is necessary to organize the two-symbol sTTI structure based on one *slot* instead of one *subframe*. A seven-symbol slot has thus to be partitioned into three two-symbol based sTTIs, where one of them consists of three symbols. In Downlink, due to the fact that the legacy control region may have either 1, or 2, or 3 symbols (for reasonably large bandwidths), the structure for the two-symbol sTTI in a subframe naturally depends on the legacy Downlink control region size as illustrated in Fig. 13.4. As can be seen, while the second slot follows a fixed {2, 2, 3} (symbols) structure, the first slot has a structure of {3, 2, 2}, {2, 3, 2}, or {3, 2, 2} for 1, 2, or 3 legacy Downlink control symbols, respectively. It is noted that for simplicity, short PDSCH (sPDSCH) is not supported in the Downlink legacy control region. This means that sTTI0 shown in Fig. 13.4 for case 2 and case 3 cannot be used for the sTTI data operation.

In Uplink, the two-symbol sTTI structure is shown in Fig. 13.5. Note that the last sTTI in a subframe has three symbols, with the last symbol flexibly managed to facilitate the coexistence with the legacy SRS symbol.

In order to come up with a reasonable processing requirement, the maximum round-trip propagation delay is similarly specified for the two-symbol sTTI. To be more specific, for a sTTI transmission at sTTI n, if the minimum HARQ RTT is 8 sTTIs or at sTTI $n + 8$, the maximum tolerable round-trip propagation delay is 67 μs, corresponding to a maximum site distance of 10 km. If the minimum HARQ RTT is relaxed to 12 sTTIs or at sTTI $n + 12$, the maximum tolerable round-trip propagation delay is 167 μs,

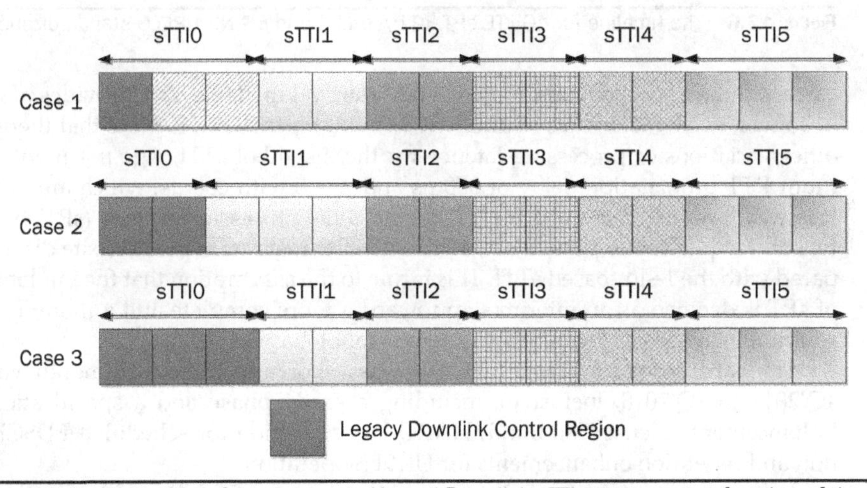

FIGURE 13.4 Illustration of the two-symbol-based Downlink sTTI structure as a function of the legacy Downlink control region size in 4G LTE.

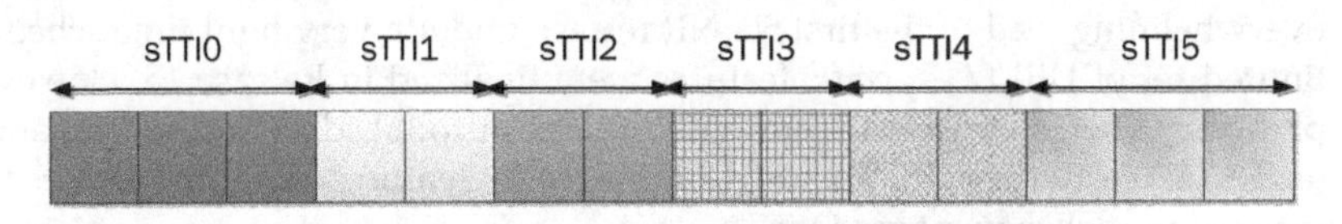

FIGURE 13.5 Illustration of the two-symbol-based Uplink sTTI structure in 4G LTE.

TTI length	Processing latency	Min HARQ RTT (ms)	Max RTT propagation delay (μs)	Max site distance (km)	Min UE processing time (ms)
1 ms	n + 4	8	667	100	2.33
	n + 3	6	200	30	1.80
1 slot	n + 4	4	310	46.5	1.19
2 symbol	n + 4	1.33 (on average)	67	10	0.36
	n + 6	2	167	25	0.62

Table 13.1 Summary of Some Key Characteristics of Different TTI Lengths in 4G LTE

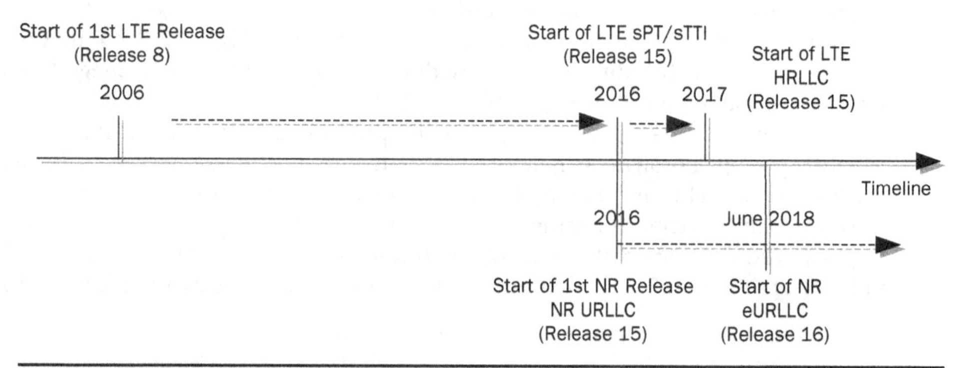

Figure 13.6 The timeline for 4G LTE sPT/sTTI/HRLLC and 5G NR URLLC standardization.

corresponding to a maximum site distance of 25 km. Table 13.1 provides a summary of some key characteristics of different TTI lengths in 4G LTE. Note that there are also other variations of processing latency for the 2-symbol sTTI, e.g., $n + 6$ with a maximum RTT propagation delay of 350 μs and $n + 8$ with 450 μs, which are not shown. It is worth noting that the 1 ms TTI with a 3 ms processing latency (sPT) has a more relaxing requirement, particularly with a smaller maximum possible site distance compared with the 1-slot-based sTTI. This is due to the assumption that the implementation of sPT is deemed as an interim step toward a more complete and a more demanding solution using either the one-slot or two-symbol–based sTTI.

Standardization for LTE HRLLC [12] was a short and intensive 9-month work (from 4Q'2017 to 2Q'2018, inclusive), including a study phase and a specification phase. Enhancements focused on blind/HARQ-less repetition for scheduled PDSCH operation and repetition enhancements for UL SPS operation.

Different from 4G LTE, URLLC considerations for 5G NR standardization started from the first NR release in 2016, as illustrated in Fig. 13.6. Indeed, the start of NR URLLC work almost concurred with the LTE sPT/sTTI/HRLLC work. Due to the overwhelming load of the first 5G NR release under a very tight time schedule, only a limited set of URLLC-specific features were finalized in Release 15. However, a comprehensive set of features were additionally introduced to further enhance URLLC (eURLLC) in Release 16. These URLCC-specific features for both releases will be the focus of the subsequent sections.

13.2 Use Cases and Deployment Scenarios for 5G NR URLLC

TR 38.802 [9] contains a first set of findings for URLLC from the physical layer perspective as part of the study on 5G NR technology, which was a preface for the first NR release. In particular, it contains separate sections dedicated to simulation assumptions related to URLLC evaluations for both the link-level simulations and the system-level simulations. These simulation assumptions contain some key parameters such as deployment scenarios, carrier frequencies, channel bandwidths, gNB antenna configurations, UE antenna configurations, packet sizes, delay bounds, etc. In particular, the following aspects are identified [9, Sec. 11.1]:

- User plane latency,
- Reliability, and
- URLLC capacity and URLLC/eMBB multiplexing capacity.

Both urban macro [10, Sec. 6.1.4] and indoor hotspot [10, Sec. 6.1.1] scenarios were considered, with the former intended to focus on large cells and continuous coverage, while the latter on small coverage per site or TRP (transmission and reception point) and high user throughput or user density in buildings. A set of requirements on the packet sizes (32, 50, and 200 bytes) in combination with a 1 ms delay bound were also included.

Dedicated study to enhance URLLC was carried out later in Release 16 and summarized in TR 38.824 [10]. One objective of the study is to establish the baseline performance achievable with NR Release 15, which contains URLLC-specific considerations (e.g., 10^{-5} BLER-based CSI feedback, etc.) and design considerations applicable to both eMBB and URLLC (e.g., mini-slots, UE processing timelines, etc.). Based on the baseline performance, additional enhancements for URLLC were investigated and introduced in Release 16. To that end, the following use cases are identified:

- **Electrical power distribution** [16]
- **Factory automation** [16]
- **Transport industry** including remote driving [17] and intelligent transport system 15]
- **Augmented reality** (AR) [15] and **virtual reality** (VR) [18]

It is worth emphasizing that the above use cases are the ones for URLLC design optimization. These use cases already represent a wide range of scenarios and requirements. Naturally, the enabling techniques introduced in Release 16 based on these use cases can be readily applied to many other URLLC use cases as well.

Although TR 22.804 [16] contains many use cases for electrical power distribution, the following use cases are explicitly considered for Release 16 URLLC:

- Power distribution grid fault and outage management focusing on distributed automated switching for isolation and service restoration for over-head lines:
    - The reliability requirement is 99.9999% with a delay target <5 ms.

- Differential protection in distribution network of smart grids:
  - The target peer-to-peer transfer interval is 0.8 ms, with a packet size of 250 bytes. The target latency is less than the peer-to-peer transfer interval, with jittering less than half of the peer-to-peer transfer interval.
  - The end-to-end latency is expected to be less than 15 ms.

For factory automation, motion control was the focus for Release 16 URLLC [16]. Motion control is among the most challenging and demanding closed-loop control applications in industry. Examples can be found in printing machines, machine tools, or packaging machines, where the motion control system is in charge of controlling moving and/or rotating parts of machines in a cyclic and deterministic manner. A 5G-based wireless communication facilitates handling of movements/rotations of components leading to less abrasion, maintenance effort, and costs. The target transfer interval is 2 ms, with a packet size of 20 bytes. The moving speed can be up 20 m/s, while the service area can be as large as 50 m × 10 m × 10 m with up to 100 UEs. The end-to-end latency is expected to be less than the target transfer interval, with a reliability level in the range of 99.9999% to 99.999999%.

For remote driving, as part of the transport industry use cases [17], the information exchange between a UE supporting V2X application and a V2X application server should accommodate an absolute speed of up to 250 km/h. The maximum end-to-end delay is expected to be around 5 ms (with a 3 ms air interface latency), with a reliability level of 99.999%. The data rates can be up to 25 Mbps and 1 Mbps for Downlink and Uplink, respectively.

Intelligent transport systems (ITS) [18] comprises traffic control centers and road-side infrastructure (e.g., traffic light controllers, roadside units, traffic monitoring, etc.), which are wirelessly connected to provide management and control of transportation. Road-side infrastructure is deployed alongside streets in urban areas and alongside major roads and highways every 1 to 2 km. ITS is deemed to increase travel safety, minimize environmental impact, improve traffic management, and maximize the benefits of transportation to both commercial users and the general public. The maximum end-to-end delay is expected to around 30 ms, although a 10 ms end-to-end delay (with a 7 ms air interface latency) was adopted in the 5G NR URLLC evaluations. A reliability level of 99.999% is assumed when the reliability is related to a given system entity while a reliability level of 99.9999% is assumed when the reliability is related to communication service availability related to the service interfaces. The payload ranges from small (e.g., ≤256 bytes) to large (e.g., >1000 bytes). Based on the assumption that all connected applications within the service volume require the user-experienced data rate of 10 Mbps along with a connection density of 1000/km^2, the traffic density requirement can thus be as high as 10 Gbps/ km^2.

eXtended reality (XR) applications are one of the most important and popular edge applications offered by 5G NR. XR is an umbrella term for various types of realities including real-and-virtual combined environments and human–machine interactions generated by computer technology and wearables. Examples of XR include AR/VR, mixed reality (MR), cloud gaming, etc. TR 26.928 [19] contains the result of the study for XR use cases in 3GPP, including

- **Streaming,** where the typical media streaming experience can be enhanced with the capability of 6 degrees of freedom (6DoF) within a scene.

- **Gaming,** where multi-player VR games allow remote people to play and interact in the same game space. A game spectator can take two possible views: a player's view and a spectator's view.
- **Real-time 3D communication,** where video chats are captured using 3D models of people's heads, which can be rotated by the receiving party. Multi-party VR conferences support the blended representation of the participants into a single 360-degree video with a pre-recorded office background.
- **Industrial Services,** e.g., an AR-guided assistant at a remote location for augmented instructions/collaboration. The remote assistant can see in real time, through the local person's AR glasses, the local environment.

Some parameters for AR/XR were identified in Release 16 as follows:

- AR [16]: an end-to-end delay of less than 10 ms with a reliability level of 99.9%.
- XR [18]: a motion-to-photon latency, defined as between the physical movement of a user's head and the updated picture in the VR headset, in the range of 7 to 15 ms while maintaining the required user data rate of 1 Gbps and a motion-to-sound delay (defined as the latency between the physical movement of a user's head and the updated sound waves from a head mounted speaker reaching their ears) of <20 ms.

More recently, a new study item on XR performance characterization was approved as in Ref. [3]. Similar to the voice performance characterization in 4G LTE, XR performance characterization in 5G NR is necessary to fully understand how the existing 5G NR standardization supports it as one of the most popular services. The focusing areas include:

- Traffic models for applications of interest (e.g., viewport-dependent streaming, viewport rendering with time warp in device, XR distributed computing, XR conversational, cloud game, etc.), including different upper layer assumptions, e.g., rendering latency, codec compression capability,
- Evaluation methodology to assess XR and cloud gaming performance along with identification of key performance metrics (KPIs) of interest for the relevant deployment scenarios, and
- Performance evaluations toward characterization of identified KPIs.

Depending on the outcome of the performance characterization, a follow-up work item is possible if there are areas identified for potential improvements, in terms of system-level capacity, power consumption, link-level performance, etc.

13.3 Resource Management for URLLC

One key aspect in supporting URLLC services is the issue of the coexistence of URLLC with other types of services, such as eMBB and mMTC, in the same cell. Note that the coexistence can be viewed from both the cell's perspective and the UE's perspective. From the cell's perspective, there may exist one or more UEs with different service needs, thus requiring resource scheduling for different service types. From the UE's perspective, there are cases when a UE may be involved in URLLC-related applications (e.g., cloud gaming) in parallel with other conventional mobile broadband tasks (e.g., a video download).

Addressing services with different QoS requirements is not a new task for cellular standardization. In the era of 3G standardization, there were already requirements of supporting voice and data simultaneously in a cell. Voice type of traffic is delay sensitive (e.g., an end-to-end delay in the order of tens of milliseconds), and quality sensitive (e.g., no more than 1% packet dropping rate.) At the same time, there is always a desire to support as many as voice users as possible with good reachability (e.g., to serve voice traffic for ≥99% of the users in the cell.) At the same time, data traffic in the 3G era was deemed to be insensitive to latency; thus, can be treated with a relatively lower priority.

However, it is noted that handling of voice versus data in 3G, although challenging, is relatively less complicated compared with 5G. This is because the latter has to handle a much more flexible and complicated system, such as a flexible slot structure, various scheduling and HARQ timings, durations of transmissions, ultra-high reliability (a 10^{-4} or lower BLER), ultra-low latency (e.g., 0.5 ms), discrepancy in the cell bandwidth and the UE-specific bandwidth (see Chap. 5), dynamically scheduled versus grant-free (or configured grant) transmissions, etc.

To support both eMBB and URLLC in a cell, generally there are two ways:

- *Semi-static resource sharing*, where the resource for eMBB and the resource for URLLC in a cell are non-overlapping. The partitioning of the resources is semi-static, which can be in the form of FDM, TDM, or a combination thereof. This is shown in Fig. 13.7 for the FDM-based semi-static resource sharing.
- *Dynamic resource sharing*, where the resource for eMBB and the resource for URLLC in a cell are at least partially overlapped. For the overlapped resource, the usage of the resource for eMBB or URLLC is determined by the actual scheduling need, which can change dynamically. This is shown in Fig. 13.8.

It is obvious that semi-static resource sharing is simpler for implementation. In the resource region exclusively set aside for a particular traffic type (either eMBB or URLLC), it is easier for a gNB to schedule UEs of the same traffic type without the complexity of prioritizing and/or multiplexing different traffic types. Under the same

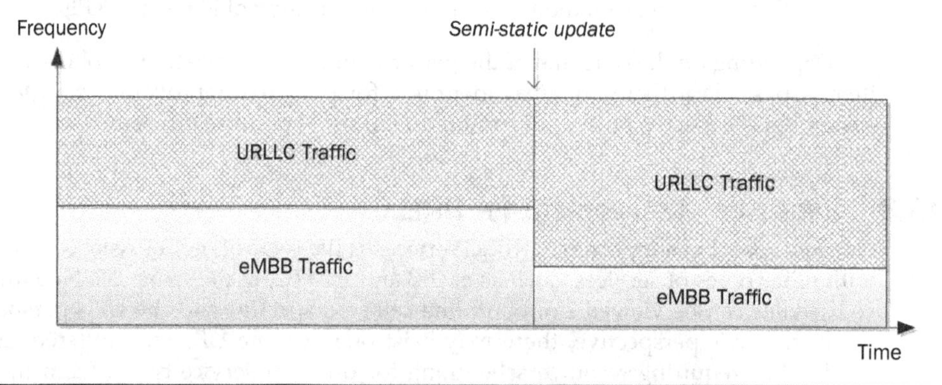

FIGURE 13.7 Semi-static resource sharing for eMBB and URLLC in a cell.

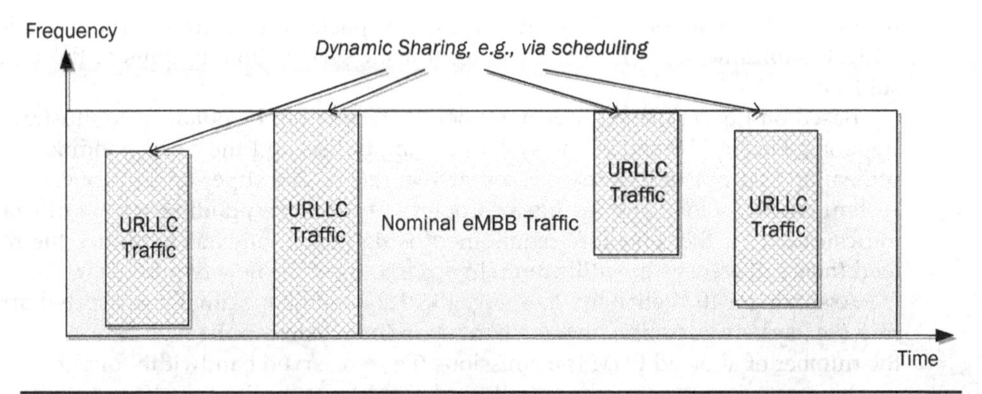

Figure 13.8 Dynamic resource sharing for eMBB and URLLC in a cell.

traffic type, there is no strong need to revert a previous scheduling decision and replace it with a new and more urgent scheduling need. However, the shortcoming is also clear – due to its semi-static nature of resource partitioning and the burstiness of packet arrivals, the separate resources are more likely to be under-utilized causing undesirable system resource wastage. This is the consequence of less statistical multiplexing or trunking efficiency. To address the need for high reliability and low latency, it is desirable to fully utilize all available resources in both the frequency domain and the time domain. This is not only important from the link-efficiency perspective, but also from the system-efficiency perspective. A wider bandwidth (or a longer time duration) provides improved frequency-diversity gain (or time-diversity gain), which is crucial particularly when the required reliability level is high (e.g., 99.999%). More resources also make it possible to complete transmission of a packet *earlier*, more friendly for the much-needed *low* latency.

A simple yet effective theoretical analysis of the tradeoff among URLLC capacity, system resource utilization and reserved system bandwidth (in terms of the number parallel transmissions in a time instance) is presented in Ref. [20]. A *M/D/m/m* queueing model is assumed, where *M* represents that the arrival process is Poisson, *D* indicates a hard transmission deadline (assuming a single transmission without HARQ), the first *m* denotes the maximum number of concurrent transmissions, while the second *m* indicates the dropping of a packet when the deadline cannot be met (no queuing delay). Under the *M/D/m/m* queueing model, the loss of system reliability is the *Erlang-B* formula [21]:

$$p_{loss} = \frac{(\lambda / \mu)^m / m!}{\sum_{n=0}^{m} (\lambda / \mu)^n / n!} \quad (13.1),$$

while the corresponding system utilization, defined as the time-average proportion of the allocated resources, is given by

$$\sum_{k=1}^{m} \left(\frac{k}{m}\right) \frac{(\lambda / \mu)^k / k!}{\sum_{n=0}^{m} (\lambda / \mu)^n / n!} \quad (13.2)$$

where λ is the Poisson arrival rate in units of packets for a time duration called a URLLC subframe, and $(1/\mu)$ is the deterministic service time in units of the URLLC subframe.

Based on Eqs. (13.1) and (13.2), numerical results can be obtained to illustrate the tradeoffs. Figure 13.9 shows the system reliability loss and the corresponding system utilization as a function of the Poisson arrival rate λ. The super-linear property of the system reliability loss, and the linear property of the corresponding system utilization indicates that as the reliability requirement is tightened, one has to reduce the traffic load (hence, lower system utilization) in order to meet the new requirement.

Assuming a 10^{-5} reliability loss, Fig. 13.10 shows the maximally supported arrival rate (i.e., system capacity) and the corresponding system utilization as a function of the number of allowed FDM transmissions (i.e., a reserved bandwidth for URLLC). As can be seen, when the number of allowed FDM transmissions or the reserved bandwidth for URLLC increases, the system capacity increases super-linearly, indicating the increased efficiency in the supported URLLC traffic per unit of bandwidth.

System-level simulations are also shown in Ref. [20] assuming an FDD system with a 60 kHz SCS, a 10^{-5} reliability level, a 2×2 SU-MIMO, a 32-byte URLLC payload, and a 1732-m inter-site distance. Other simulation parameters can be found in Ref. [20]. Table 13.2 summarizes the resulting system resource utilization (%) as a function of the delay budget (0.5, 0.75, or 1 ms). It can be observed that as the reserved bandwidth increases, the resource utilization increases super-linearly. It is also not surprising that when the delay budget increases, utilization of system resource also increases. These left-over resources can be readily utilized by eMBB UEs. For example, the system resource utilization is only 32.4% when the reserved bandwidth is 10 MHz and the latency is 0.75 ms.

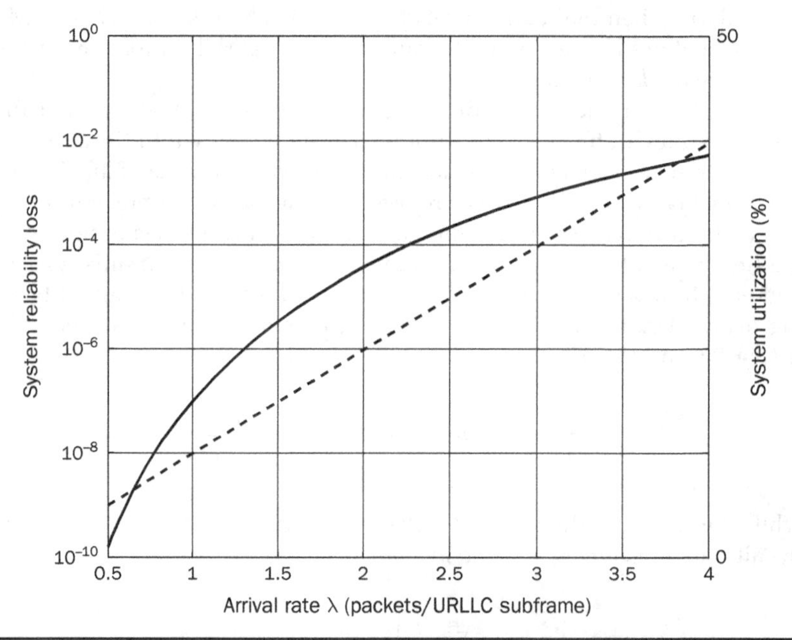

FIGURE 13.9 Loss of system reliability and the system utilization as a function of Poisson arrival rate in the assumed $M/D/m/m$ queueing model with $m = 10$, $\mu = 1$ [20].

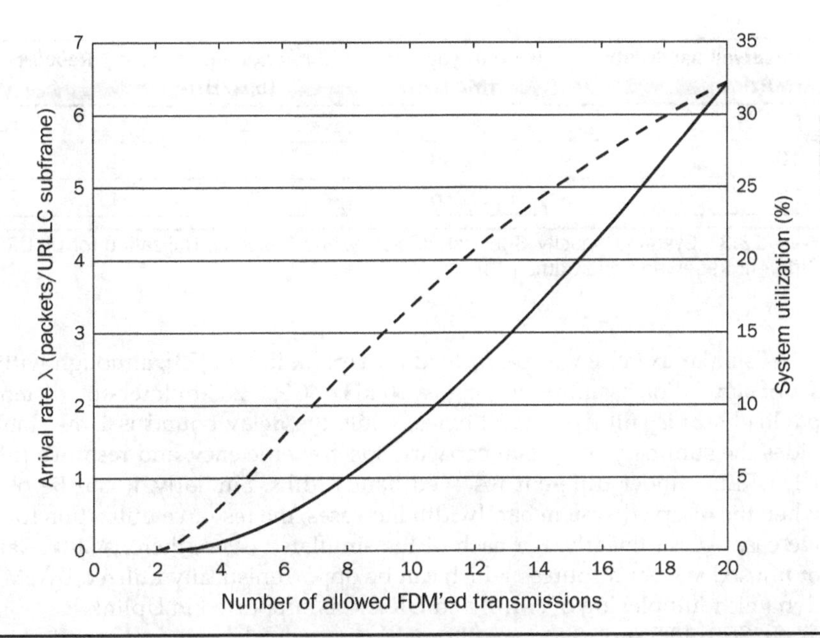

FIGURE 13.10 Maximally supportable Poisson arrival rate and the associated system utilization subject to a 10^{-5} reliability requirement in the assumed *M/D/m/m* queueing model with $\mu = 1$ [20].

Delay budget (milliseconds)	Reserved bandwidth (MHz)	Resource utilization (%)
0.5	5	0
	10	2.3
	20	43.3
0.75	5	0
	10	32.4
	20	64.0
1.0	5	9
	10	50.9
	20	68.6

TABLE 13.2 System Resource Utilization as a Function of the Delay Budget in Downlink [20]

Assuming a 4.18 bps/Hz spectral efficiency under a full-buffer traffic model with eMBB traffic [22], the un-used 67.6% of the system resource can translate into

$$4.18\ bps/Hz * 10\ MHz * 67.6\% = 28.26\ Mbps,$$

if the 67.6% of the system resource can be opportunistically used by eMBB UEs. This indicates the strong need for dynamic multiplexing between eMBB traffic and URLLC traffic in Downlink.

Reserved bandwidth (MHz)	System capacity (Mbps)	Spectral efficiency (bps/Hz)	Resource utilization (%)
5	4.51	0.9	51.5
10	12.39	1.24	65.3
20	28.16	1.41	74.4

TABLE 13.3 System Capacity, Spectral Efficiency, and Resource Utilization for UL URLLC under Different Reserved Bandwidths [23]

A similar exercise was performed for Uplink in Ref. [23], although with a different set of simulation assumptions, e.g., a 30 kHz SCS, a 200-m inter-site distance, etc. The payload size is still fixed at 32 bytes, while the delay bound is 1 ms. Table 13.3 provides the summary of system capacity, spectral efficiency, and resource utilization for UL URLLC under different reserved bandwidths. Similarly, it can be observed that when the reserved system bandwidth increases, the resource utilization for UL URLLC increases super-linearly. For each of the simulated cases, there is still a large amount of unused system resources which can be opportunistically utilized by eMBB traffic if dynamic multiplexing of eMBB and URLLC is supported in Uplink.

In Secs. 13.5 and 13.6, we will provide more details on how to multiplex eMBB and URLLC in Downlink (which is specified in Release 15) and in Uplink (which is specified in Release 16).

13.4 Optimizing Link Efficiency for URLLC

The requirements for URLLC are two aspects: latency and reliability. While it is relatively easier to satisfy one of the two aspects, it is very challenging to optimize for both aspects, while keeping the standardization efforts and eventually the implementation efforts at a reasonable level.

13.4.1 Latency and Processing Time for URLLC

As discussed earlier, both sPT and sTTI are specified in 4G LTE for low latency by utilizing a reduced processing time and/or a reduced TTI length. These two aspects are utilized by NR URLLC as well. Moreover, 5G NR has flexible tone spacings, i.e., 15, 30, 60, and 120 kHz. Under the same number of symbols for a TTI, a larger tone spacing results in a shorter time duration, which is not an option in 4G LTE (as it has a fixed 15 kHz tone spacing.) The reduced processing time and the reduced transmission time make it possible to have more HARQ transmission opportunities within a delay budget; thereby, improving the link-level efficiency and consequently, the system-level efficiency. To illustrate the system-level benefits, Fig. 13.11 presents the evaluation results comparing URLLC system capacity for three combinations of SCS, TTI length, and HARQ RTT (processing time). Key evaluation assumptions include a 20 MHz bandwidth, a −3dB geometry with a bursty traffic for a UE and a 99.99% reliability level. A similar study can also be found in Ref. [24]. As can be seen, significant system capacity increase can be achieved via a tighter processing time, a shorter TTI length, and a larger SCS.

It should be clear that the increase of link-efficiency and system-efficiency comes at the cost of incurring larger control overhead, primarily due to more frequent scheduling

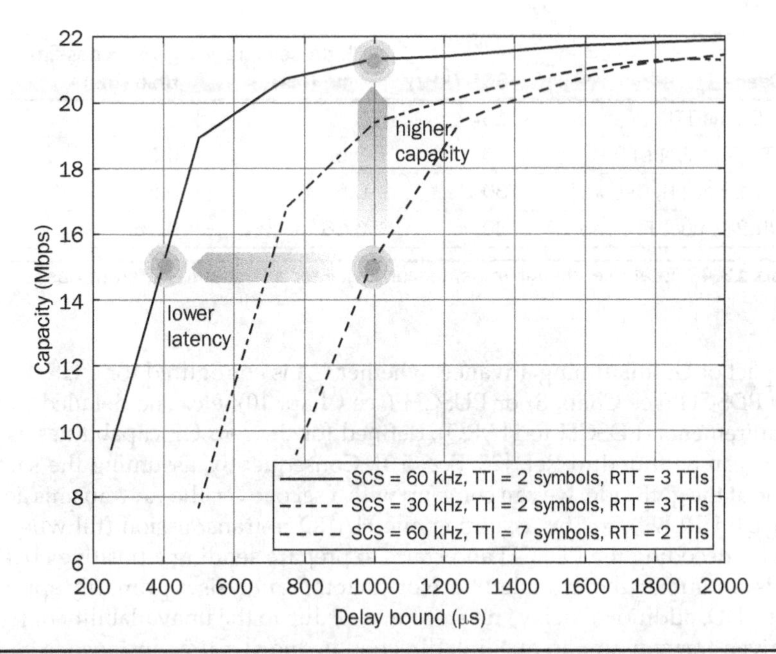

FIGURE 13.11 System capacity under different HARQ RTT timelines and OTA latency requirements.

in time; however, such overhead increase can be alleviated by allowing for grant-free transmissions. A grant-free transmission is a transmission for which the transmission parameters can be completely or partially configured by a gNB (instead of being dynamically indicated via a DCI), potentially with the assistance of a dynamic activation/release via a DCI (see Sec. 7.5).

Moreover, although a higher SCS is a straightforward way to reduce the duration of an OFDM symbol, and hence the TTI length for a packet transmission, the number of available tones is inevitably decreased for a given system bandwidth. As a result, a gNB can only accommodate a fewer number of UEs at a given time instance, which may increase queuing delays for some UEs. Moreover, a larger SCS also has a reduced CP length; hence, it is likely to experience increased sensitivity to channel impairments. Therefore, although the 5G NR standard specifications provide a variety of ways in reducing the scheduling TTI (e.g., via mini-slots, a higher SCS, etc.), these factors need to be carefully managed in the actual implementation and deployments to achieve a reasonable trade-off.

Considering the need to support different latency requirements, two UE processing capabilities are introduced in NR (also known as UE Capability #1 and UE Capability #2). The processing times by the two UE capabilities may be commonly viewed as applicable for eMBB and URLLC, respectively. However, there is no such an explicit labeling or restriction. Indeed, even for some eMBB services, it is desirable to have the possibility of realizing low latency for improved user experiences.

The standardization process in finalizing the processing times was very involved. This is not surprising as it is heavily dependent on the implementation considerations. To that end, many factors had been considered during the 5G NR standardization, e.g.,

Cases	SCS (kHz)	UE processing time (ms)	gNB processing time (ms)	Total delay (ms)
LTE 1ms TTI	15	3	3	~12
LTE 2-symbol sTTI	15	0.5	0.5	~1.9
NR 1-slot TTI, Cap #1	30	0.5	0.5	~2.7
NR 2-symbol TTI, Cap #2	30	0.16	0.16	~0.7

TABLE 13.4 Total Over-the-Air Transmission Delay for a Packet in Different Cases

impact of Uplink timing advance, whether CA is configured for a UE, DM-RS patterns for PDSCH (see Chap. 8) or PUSCH (see Chap. 10), etc. The detailed processing time requirements (PDSCH to HARQ), defined for the two UE capabilities as a function of SCS, can be found in Ref. [25, Sec. 5.3]. Consequently, assuming the same processing time at the gNB side, we can come up with a set of over-the-air transmission total delays for DL URLLC, e.g., by assuming one HARQ re-transmission (till when the UE completes decoding the PDSCH and starts to prepare sending a positive HARQ feedback). This is summarized in Table 13.4. Note that this analysis is primarily applicable to FDD. For TDD, additional delays may occur, e.g., due to the unavailability of resources when a given transmission in one link direction is due at a time instance (e.g., the resources are of a different link direction). As can be seen, while the introduction of sTTI in LTE brings down the total over-the-air delay by more than six times, with UE processing capability #2, a 30 kHz SCS, and a two-symbol TTI, 5G NR can further reduce the latency by roughly 2.7 times. Indeed, the total over-the-air latency is <1 ms, making it possible to satisfy the stringent over-the-air delay requirements for some low latency URLLC services as discussed earlier. It should be noted that in Table 13.4, it is assumed there is no restriction of a single HARQ feedback opportunity per slot as in Release 15 (see Sec. 13.4.3 for more details). Further reduction of the over-the-air delay is also possible by NR URLLC, e.g., by using a 60 kHz SCS or a 120 kHz SCS.

In order to mitigate the challenges in supporting URLLC-related processing, some relaxation of PDSCH and PUSCH processing is necessary. To that end, the URLLC PDSCH and URLLC PUSCH are restricted to:

- A single codeword (or transport block) only, and
- No support of CBG (code-block-group) (see Sec. 9.2 for more details on CBG).

13.4.2 PDCCH for URLLC

Naturally, the design of PDCCH for ULRRC can also be classified into two aspects: reliability and latency.

It is not sufficient to ensure a 99.999% or better reliability for data channels only. Although it is possible to use SPS or configured grants to minimize the usage of DCI, it is still necessary to ensure sufficient reliability for PDCCH. This is because a PDCCH based PDSCH and PUSCH operation is inevitable (even for the SPS-based operation, see Sec. 7.5) and is generally a dominant scheduling scheme, especially for bursty traffic.

To understand whether there is a need to further enhance PDCCH dedicated to URLLC, it is necessary to perform extensive evaluations to check how the existing PDCCH design performs under various scenarios. This was done as part of the Release 16 URLLC

Cases	PDCCH BLER	PDCCH payload (excluding CRC)	
		40 bits	24 bits
Carrier frequency 4 GHz, 4Tx/4Rx	10^{-6}	−7.2	−7.9
	10^{-5}	−7.7	−8.4
Carrier frequency 700 MHz, 2Tx/2Rx	10^{-6}	−3.8	−5.3

TABLE 13.5 Minimum Required SINR (dB) to Achieve a Target BLER for PDCCH under Various Cases for Urban Macro Deployments

study item, where both urban macro and indoor hotspot scenarios were considered. The detailed results can be found in Table 6.1.1-1 of Ref. [10] capturing the evaluation results for each of the evaluated scenarios. The deployment scenario of the indoor hotspot case is less of a concern with respect to the PDCCH reliability, since the channel conditions are more benign than that of the urban macro case. Table 13.5 shows the performance results for some selected urban macro cases by *averaging* over the results from multiple sources (when there are five or more sources), with key simulation assumptions such as a TDL-C 300 ns channel model, a 30 kHz SCS, an aggregation level 16, and a mobility speed of 3 km/h. It is noted that due to the natural differences of the simulation results from various sources and the different numbers of sources for various cases (as can be observed in Table 6.1.1-1 of Ref. [10]), the averaging may be arguably not that desirable, especially in terms of the resulting absolute numbers. However, by doing so, it is still valuable to see how well PDCCH reliability may stand in general. As can be observed, the minimum required SINR for PDCCH to achieve a target BLER of 10^{-5} or 10^{-6} depends on the evaluated cases and the DCI size. In particular, when the DCI size decreases from 40 to 24 bits, the minimum required SINR drops, with a value in the range of 0.7 to 1.5 dB. This is generally in line with the linear decrease of the PDCCH payload size; a payload reduction from 40 bits (hence, 64 bits in total after including the 24-bit CRC) to 24 bits (hence, 48 bits in total including the CRC) represents a 25% drop, or 1.2 dB.

Although the minimum required SINR for URLLC PDCCH generally meets the requirements for different use cases, it is still desirable to consider reducing the DCI size for the benefit of increased PDCCH overhead efficiency (e.g., targeting ~1 dB savings). This is the motivation for introducing a new DCI size, primarily aiming for a smaller size, for URLLC in Release 16. Obviously, the introduction of a new DCI size for URLLC does not come without any cost. Indeed, new standardization efforts have to be spent in terms of re-defining the set of information fields in the new DCI, and the additional complexity (e.g., a total number of PDCCH blind decodes, a total limit of PDCCH channel estimation, etc.) for standardization and implementation. As discussed in Sec. 7.5, the new DCI formats are called DCI format 0_2 (for Uplink) and 1_2 (for Downlink). Note that the information fields in the new DCI formats can be separately configured from those in DCI formats 0_1/1_1 and as a result, the size of the new DCI formats 0_2 (or 1_2) is not necessarily smaller than that of DCI formats 0_1 (or 1_1). Such an arrangement is primarily for flexibility.

Another option of improving PDCCH reliability is to utilize more resources, e.g., by repetitions in the time domain. However, such an approach is not preferable especially when considering the tight delay requirement for URLLC. Specifically, a repetition in time for PDCCH would imply a longer transmission duration; thereby, increasing the latency for the overall URLLC operation.

Monitoring PDCCH for URLLC operations can be challenging and power-consuming for a UE. A per-slot PDCCH monitoring frequency is a natural choice if scheduling of DL and UL data is intended to be on a per-slot basis for a UE. For an extremely tight latency budget, per-slot scheduling may be insufficient to meet the latency requirement especially when the SCS is small (e.g., 15 kHz). In addition, the per-slot scheduling may be inefficient since it may not provide a reasonable number of HARQ re-transmission opportunities for a delay-constrained TB. More frequent PDCCH monitoring opportunities are thus necessary to ensure efficient and effective HARQ operations for URLLC. To that end, the monitoring opportunities *zoom* into within a slot, and are specified in units of symbols, including (see also Sec. 7.4):

- Every two symbols,
- Every four symbols, or
- Every seven symbols.

This is similar to the sTTI operation as in 4G LTE. It should be clear that the increase of PDCCH monitoring frequency would entail more power consumption, and incur additional complexity in both the standardization and the implementation efforts.

13.4.3 HARQ Feedback Enhancements for URLLC

Similarly, there are primarily two aspects for HARQ feedback for URLLC: reliability and latency.

Due to the different performance targets for eMBB and URLLC, it is crucial to be able to differentiate and prioritize packets of different QoS requirements (see Chap. 3) as appropriate and as necessary. This is particularly important when different services are multiplexed in a set of overlapped resources in a dynamic manner (see Sec. 13.3). The handling of PDSCH and PUSCH transmissions with different priorities is to be discussed in Secs. 13.5 and 13.6, respectively.

Is it necessary to differentiate and/or prioritize HARQ feedback for different service types? Since HARQ operation is an integral part of efficient over-the-air PDSCH or PUSCH transmissions, the general answer is yes. However, as in many other cases, this would come at a cost. Such a cost prevented it from happening in Release 15. In other words, in Release 15, the HARQ feedback for eMBB and URLLC is treated the same way—the same HARQ codebook construction (including the Downlink assignment index (DAI) counting), coding, resource mapping and transmission. It is impossible for a UE to differentiate HARQ for PDSCH transmissions for eMBB versus URLLC, and consequently, it is impossible for the UE to perform any special handling for HARQ for URLLC PDSCH transmissions. Therefore, depending on the scheduling needs, there are cases when HARQ for URLLC PDSCH may have a risk in satisfying the target performance metrics especially when there are a large number of HARQ bits for eMBB PDSCH transmissions and/or when the UE is power limited.

The enhancement was introduced in Release 16, where up to two HARQ codebooks can be indicated to a UE. The differentiation of different HARQ priorities (e.g., one for eMBB and the other for URLLC) is done as follows:

- A 1-bit priority indicator in DCI formats 1_1 and 1_2 (see Sec. 7.5) for PDSCH assignments, and
- Two sets of DAI indicators in DCI formats 0_1 and 0_2 (see Sec. 7.5) for PUSCH assignments.

The priority indicator in a DCI helps identify the index of one of the two HARQ codebooks that the corresponding PDSCH transmission is expected to be associated with, where the DAI counters in the DCI provide the necessary total number HARQ bits for the PDSCH transmissions (see Sec. 9.2) for a UE. For UL grants (DCI formats 0_1 and 0_2), each of the two sets of DAI indicators indicate a respective total number of HARQ bits for the two HARQ codebooks for a UE.

Correspondingly, a UE can construct and manage two separate HARQ codebooks, one for high-priority PDSCH transmissions (e.g., URLLC), and the other for low-priority PDSCH transmissions (e.g., eMBB). This makes it possible for a UE to either handle them separately, or jointly. A simple way is to prioritize the high-priority HARQ codebook over the low-priority one by completely dropping the low-priority HARQ feedback. Although simple, it may still be desirable if:

- The performance targets for the high-priority HARQ feedback and the low-priority HARQ feedback are drastically different, and/or,
- The HARQ payload size for the low-priority one is so large that satisfaction of the performance target for the high-priority HARQ feedback is at risk, and/or,
- The UE becomes power limited if carrying both the low-priority and the high-priority HARQ codebooks.

This approach was indeed adopted in Release 16. Although not specified yet, it is also possible to multiplex the two codebooks together potentially under certain constraints. As an example, the HARQ payload size for the low-priority HARQ feedback may be reduced before multiplexing with the high-priority one, e.g., by HARQ feedback compression or a simple HARQ bundling (i.e., a logical AND operation among the HARQ bits). The multiplexing operation of the two HARQ codebooks may be introduced in Release 17.

The latency of HARQ feedback depends on the UE processing capability (see Sec. 13.4.1) and the frequency of possible PUCCH transmission opportunities. In Release 15, for simplicity, PUCCH for HARQ can be as frequent as up to one slot. In other words, regardless of the PUCCH format, i.e., short or long, a UE is not expected to transmit more than one PUCCH for HARQ within one slot (see Sec. 9.3). It is noted, however, that it is still possible to have up to two PUCCH transmissions in one slot by a UE in Release 15. In this case, only one of them can carry HARQ, while the other one has to be used for some other UCI types (e.g., CSI). The restriction of up to one HARQ feedback per-slot is driven by the tight time schedule for Release 15 and the amount of work to be done otherwise. To be specific, if there are two or more HARQ feedback transmission opportunities in a slot by a UE, defining the association of a PDSCH transmission with the two or more feedback opportunities for both the semi-static HARQ codebook and the dynamic semi-static HARQ codebook can be complicated (see Sec. 9.2).

On the other hand, the once per-slot HARQ feedback restriction inevitably limits how fast and how frequently a HARQ re-transmission can occur. Similar to the discussion earlier with respect to the frequency of PDCCH monitoring, the frequency of HARQ feedback opportunities impacts the URLLC operation efficiency and effectiveness, subject to the additional standardization and implementation efforts.

Such a restriction was lifted in Release 16. To that end, a Release 16 URLLC UE may have up to seven HARQ feedback transmissions in a slot. The maximum number of seven is connected with the maximum frequency of PDCCH monitoring and the maximum number of TBs that can be scheduled within each slot. Note that the introduction of more frequent HARQ feedbacks is particularly beneficial when a UE is configured with CA where the CCs are of different numerologies. As an example, a UE may be configured with two CCs, one of a 15 kHz SCS and the other of a 30 kHz SCS, where the 15 kHz CC is the primary CC (see Sec. 9.1). Note that the slot duration is 1 ms for the 15 kHz SCS and 0.5 ms for the 30 kHz SCS. Since the 15 kHz CC may carry the PUCCH transmission for both CCs, a configuration of two PUCCH transmissions in a 15 kHz based slot (thus, one PUCCH every 0.5 ms) makes it possible to provide HARQ feedback for the 30 kHz CC on a per slot (0.5 ms for the 30 kHz) basis.

13.4.4 CSI/MCS for URLLC

Channel adaption is one key enabler for efficient data transmissions in wireless networks (see Chap. 8). CSI and MCS are two integral components to enable data rate adaptation. Although CSI stands for *channel state information*, it includes a set of necessary operations such as configuration or triggering reference signals for channel measurement and/or interference measurement, configuring or triggering different ways of CSI feedback (periodic, aperiodic, semi-persistent), evaluation of the *best* or the *most* preferable CSI, resources for CSI feedback transmission, etc. Chapter 8 contains more details of various aspects of CSI management.

Before the support of URLLC, it is typically assumed that under a certain set of time and frequency resources, and a set of assumed operation conditions (e.g., a PDSCH transmission scheme), the CSI should be such that the initial success rate after one transmission (i.e., without any HARQ re-transmission) is 90%, or, equivalently, a 10% initial BLER for one PDSCH transmission. The 10% initial BLER is derived based on extensive evaluations assuming a full-buffer traffic model and a certain number of total HARQ re-transmissions to maximize the link efficiency. However, while link efficiency is optimized with a roughly 10% initial BLER, latency is not.

Under a tight latency requirement for URLLC, there are cases where it is preferable or even necessary from the latency budget perspective to have a very limited number of HARQ re-transmissions, e.g., only one re-transmission, or even no possibility for HARQ re-transmissions (i.e., HARQ-less transmissions). The typical assumption of a 10% initial BLER for CSI feedback and MCS indication is thus not only sub-optimal, but even risky for meeting the requirements of highly delay-sensitive applications. Therefore, it is necessary to consider a different operation point for CSI feedback by assuming an initial BLER target tailored for the intended applications.

As discussed in Sec. 13.2, there are various use cases for URLLC with different performance requirements. It is thus sensible to consider more than one initial BLER target in deriving the CSI for URLLC such that a good tradeoff between the link efficiency and the latency can be achieved. To that end, extensive study was carried out in Release 15, which ended up with the following possible options for the initial BLER targets to derive the CSI for URLLC:

- Option A: (10^{-1}, 10^{-4}),
- Option B: (10^{-1}, 10^{-5}),
- Option C: (10^{-3}, 10^{-5}), and
- Option D: (10^{-2}, 10^{-4}).

These different options offer different tradeoffs in terms of the link-level efficiency, feasibility for a UE to produce an accurate CSI estimation, a gap between the SNR targets in achieving the two respective BLER targets, achievable latency targets, and the associated UE complexity. For instance, a combination of (10^{-2}, 10^{-4}) (option D) may result in a relatively narrower gap between the set of required SNR operation points for 10^{-2} and the set for 10^{-4}. It may bring relatively more impact on UE complexity as well considering the need to support the 10^{-1} initial BLER target for eMBB services (thus, up to 3 BLER targets for CSI per UE). On the other hand, it may achieve a good tradeoff between efficiency and latency for many URLLC use cases since the BLER targets are more or less situated in the middle of the set of URLLC performance targets. Considering the existing 10^{-1} initial BLER target for eMBB and the possibility of achieving a 10^{-5} BLER for URLLC with a single transmission (without HARQ), option B (10^{-1}, 10^{-5}) was eventually adopted in Release 15.

For the 10^{-1} initial BLER target, URLLC and eMBB share the same CSI and MCS tables. For the 10^{-5} initial BLER target, a new CSI table and a corresponding new MCS table have to be developed. The detailed CQI tables for both the 10^{-1} initial BLER target and the 10^{-5} initial BLER target can be found in Sec. 5.2.2.1 of Ref. [26]. In coming up with a 4-bit CQI table, typically a lowest possible spectral efficiency and a maximum possible spectral efficiency are assumed, while the CQI entries in between are such that the corresponding SNR operation points are roughly equally spaced. Based on the 4-bit CQI table, a 5-bit MCS table can be derived, which is designed to be consistent with the spectral efficiency entries in the 4-bit CQI table, with an additional interpolation to obtain a roughly twice number of entries. The MCS tables can be found in Sec.5.1.3 of Ref. [26].

13.5 Downlink Resource Sharing for Distinct Service Types

In Sec. 13.3, two types of resource sharing for eMBB and URLLC were discussed. In case of dynamic resource sharing (which is more efficient than semi-static resource sharing), eMBB and URLLC share a set of resources via scheduling decisions. Due to different latency and reliability requirements, it is possible that a gNB may decide to schedule a more urgent/higher priority URLLC packet mapped to resources at least partially overlapped with those being occupied by an ongoing earlier scheduled low-priority packet transmission (e.g., for eMBB). Although it is possible to do superposition coding in the overlapped resources, it is more practical to have the later scheduled URLLC packet *overwrite* the modulation symbols of the earlier scheduled packet in the overlapped resources. Such a behavior is also known as *pre-emption* or *puncturing*.

When the resources for a lower priority packet for a UE are preempted by the transmission for a higher priority packet in the Downlink, one key question is:

should the low-priority UE know whether there is preemption, and if so, how?

To answer this question, it is important to know the performance difference between the case without the preemption knowledge and the case with the preemption knowledge at the UE side. If a UE has no knowledge of the preemption, the UE would still assume that the preempted resources contain the packet information and proceed with decoding of the Downlink packet accordingly. However, due to the preemption, the impacted resources would only contribute to noise and interference, which surely causes degradation of the decoding performance. On the other hand, with the assistance of the preemption knowledge, the UE can utilize this information in decoding the packet, e.g., by nulling out the log-likelihood ratios (LLRs) corresponding to the impacted resources, and consequently, the impact of the preemption can be reduced. The performance gap between the two cases depends on many factors, e.g., the MCS of the low-priority transmission, the amount of resources being punctured, etc.

Reference [27] provides the evaluation results investigating the impact of the knowledge of preemption. Different MCS values are considered, where more detailed simulation assumptions can be found in Ref. [27]. It is noted that Turbo coding was assumed in the evaluations, although it is expected that a similar trend would hold if LDPC coding were used. Figure 13.12 shows the low MCS case, where there is only one CB for a transmit-diversity (TxDiv) scheme–based PDSCH. The PDSCH has a time span of an entire slot, except symbols 0 and 1. Resources in symbols 2 and 3 for the PDSCH are punctured. It can be observed that without the knowledge of the preemption, the impact of puncturing is significant—at a 10% BLER, the degradation is roughly 5 dB. However, with the knowledge of the preemption, the performance degradation drops to slightly over 1 dB. This demonstrates the significant benefit for a UE to know the preemption information.

When the MCS increases (from medium to high MCS ranges), the TB size increases where one transport block may contain multiple CBs. Figure 13.13 illustrates the impact of the preemption as a function of carrier to interference/noise ratio (CINR), where the increase of CINR implies the increase of MCS and hence the increase of the number of

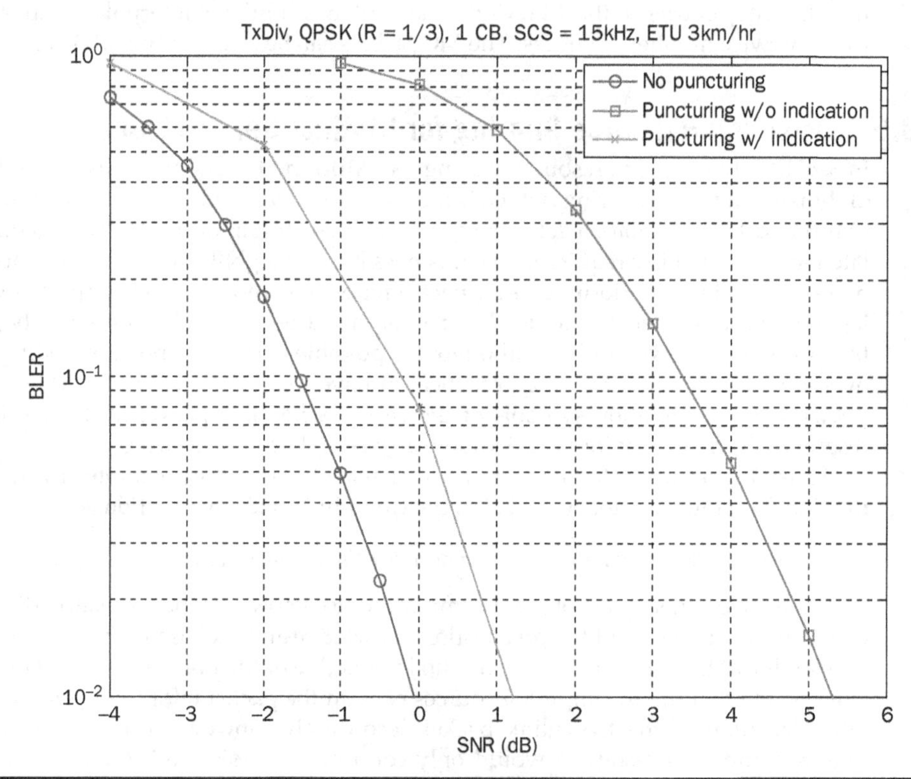

Figure 13.12 Impact of preemption, with vs. without the knowledge at the UE side, a low MCS [27].

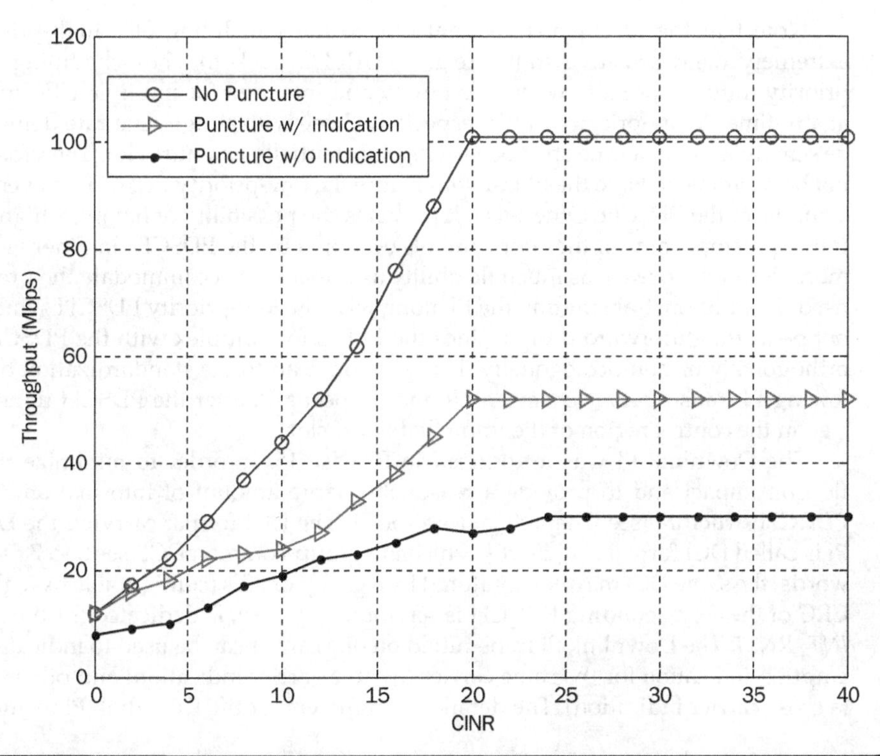

FIGURE 13.13 Impact of preemption, with vs. without the knowledge at the UE side, as a function of the carrier to interference/noise ratio (CINR) [27].

CBs for the TB. It can be observed that in general, when the MCS increases, the impact of the preemption, with or without the knowledge at the UE side, becomes more pronounced. This is partly due to the assumption of TB-based HARQ feedback, while the puncturing of symbols 2 and 3 would primarily cause a decoding failure only for a subset of CBs, it would eventually cause a decoding failure for the entire TB; hence, it would trigger a re-transmission of the entire packet. This of course can be alleviated by using a CBG-based HARQ feedback scheme [27] (see also Sec. 9.2). However, the performance benefit of the knowledge of the preemption is still significant, where the performance gain increases with the CINR.

Given the significant performance benefit, it is thus necessary to indicate to a UE the preemption information. In theory, since there is always a time gap between a PDSCH transmission and the corresponding HARQ transmission, the Downlink *preemption indication* (PI) for a PDSCH transmission can be transmitted to the UE:

- Before the PDSCH transmission, or
- During the PDSCH transmission, or,
- After the PDSCH transmission.

In addition, the PI may be transmitted in multiple time instances for a PDSCH transmission, e.g., every other symbol in the set of symbols for the PDSCH. This would help increase the reliability of the indication. So, how should the PI be transmitted?

Note that the PI is generally sent due to the scheduling of a high-priority and extremely delay-sensitive traffic (e.g., a URLLC packet.) The scheduling of high-priority traffic, especially when it is bursty and intended for multiple UEs, may occur at any time. A low-priority traffic, especially those with a high data rate demand, may be scheduled with a time span occupying substantially an entire slot. Therefore, it may not be desirable to have the PI transmit before the low-priority PDSCH, or even in early symbols of the PDSCH. Otherwise, it prevents the possibility of having a high-priority traffic preempt some of the later symbols occupied by the PDSCH. In other words, it is more desirable to leave as much flexibility as possible to accommodate the preemption need. In addition, transmitting the PI along with the low-priority PDSCH symbols may not be a straightforward design, since the PI has to multiplex with the PDSCH, either orthogonally or non-orthogonally. These factors lead to the standardization of the following PI transmission scheme: a UE may expect a PI after the PDSCH transmission, e.g., in the control region of the immediate next slot.

The Downlink PI was introduced in Release 15. In order to minimize the specification impact and to provide a reasonably large amount of information, the same PDCCH structure (see Chap. 7) is used for PI. The DCI format carrying the Downlink PI is called DCI format 2-1 [25, 28], which is a group-common DCI (see Sec. 7.5). In other words, the same DCI may be monitored by a group of UEs (configurable by a gNB). The CRC of the corresponding PDCCH is scrambled by a RNTI dedicated for the PI, called *INT_RNTI*. The Downlink PI transmitted on one carrier can be used to indicate the preemption indication for the same carrier (a same-carrier indication) and other carrier(s) (a cross-carrier indication). The detailed parameters for the Downlink PI include

- A set of serving cells where the Downlink PI is applicable.
  - Each cell has a specific bit-position in the DCI, where each cell has a 14-bit indication. Note that two or more cells may share the same bit-position.
- An information payload size for DCI format 2_1.
  - Excluding the 24-bit CRC, up to 126 bits (equivalently, indication for up to 9 unique PI patterns per DCI)
- An indication granularity for time-frequency resources.
- A monitoring periodicity for the PDCCH denoted by T_{INT}.
  - In units of slots, chosen from {1, 2, 4} slots

The applicability of the Downlink PI is in terms of a frequency bandwidth (denoted by B_{INT}) and a set of symbols (denoted by N_{INT}). The applicable frequency bandwidth B_{INT} is fixed at the entire bandwidth of the active BWP of the given serving cell. The applicable set of symbols N_{INT} depend on the numerology of the serving cell carrying the PDCCH, the numerology of the intended serving cell, the monitoring periodicity T_{INT}, and the number of symbols per slot [25, Sec. 11.2]. A one-bit indication for the time-frequency resource granularity determines which one of the following two modes should be applied for the intended serving cell:

- Mode 1: each bit of the 14-bit bitmap is applicable to the entire B_{INT} on a per symbol basis for a set of 14 symbols, and
- Mode 2: each bit of the 14-bit bitmap is applicable to one half of B_{INT} on a per symbol basis for a set of 7 symbols.

Note that the one-bit indication is the same for all the serving cells configured in the same PDCCH for the Downlink PI. Depending on the configured monitoring periodicity, each bit may be applicable to multiple symbols (e.g., a periodicity of two or four slots).

To illustrate how the Downlink PI works, consider two serving cells configured for a UE (e.g., as part of a CA operation). The first serving cell has a SCS of 15 kHz and the second serving cell has a SCS of 30 kHz. The first cell carries the PDCCH indicating the PI for both the first cell and the second cell. This is shown in Fig. 13.14. A periodicity of one slot is assumed. As can be seen, the DL PI indication is sent in a slot subsequent to the set of symbols where the DL PI applies. For the same carrier indication, the example in the figure illustrates that the 14-bit bitmap for serving cell 1 is applicable to each of the 14 symbols in the preceding slot, where symbols 6, 7, 10, and 11 are indicated being punctured for the entire band of the active BWP. In the same PDCCH, the DL PI also indicates for serving cell 2, covering a set of 28 symbols. The 14-bit bitmap is applicable on each set of a span of 14 contiguous symbols, and the same DL PI pattern is repeated over the two sets of a 14-symbol span. In each of the 14-symbol span, symbols 2, 3, 4, and 9 are indicated as being punctured for the entire bandwidth of the active BWP.

Figure 13.15 shows the second mode of the indication granularity, where the 14-bit bitmap is applicable in a 7 × 2 manner, where 7 refers to a set of seven symbols and 2 refers to the two halves of B_{INT}. The 14-bit bitmap is applicable in a frequency-first-time-second manner. That is, a first bit indicates the first half of B_{INT} in a first group of symbols, the second bit indicates the second half of B_{INT} in a first group of symbols, the third bit indicates the first half of B_{INT} in a second group of symbols, and so on. In this example, in the first half of B_{INT}, symbols 2, 3, and 5 are punctured. In the second half

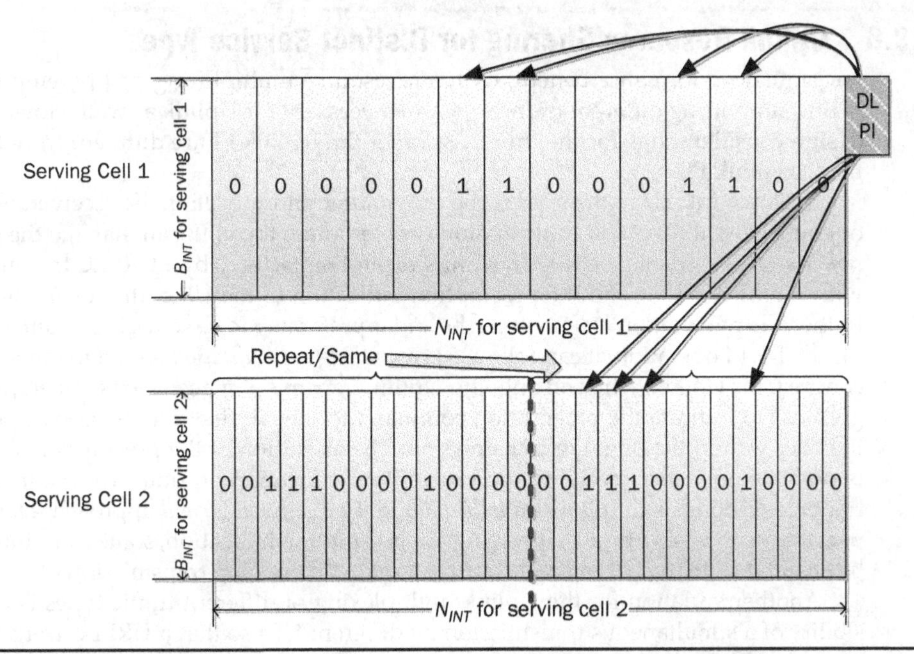

FIGURE 13.14 Illustration of the DL PI, including the same-carrier and cross-carrier indication, with mode 1 for the indication granularity.

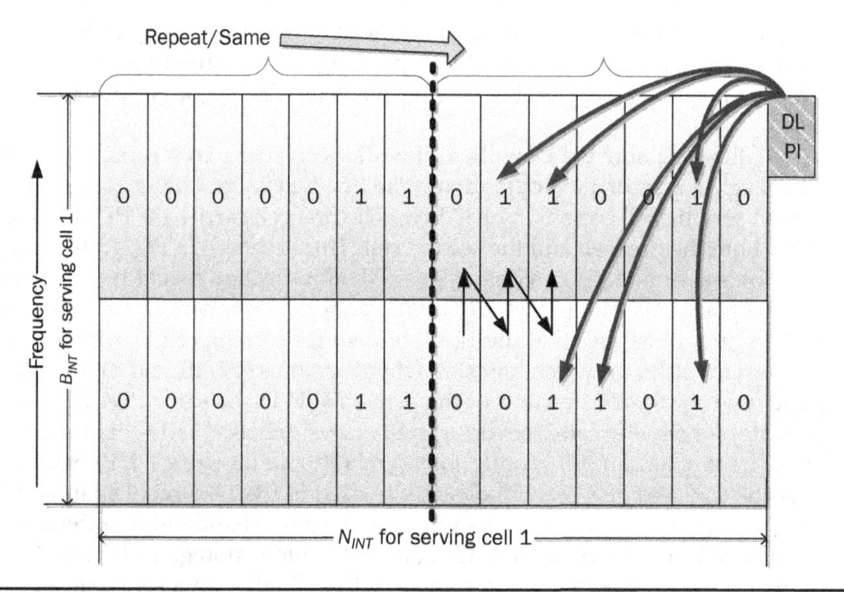

FIGURE 13.15 Illustration of the DL PI, including the same-carrier and cross-carrier indication, with mode 2 for the indication granularity.

of B_{INT}, symbols 1, 2, and 6 are punctured. The same DL PI pattern is also repeated over the two sets each of a seven-symbol span.

13.6 Uplink Resource Sharing for Distinct Service Types

For similar performance benefit, dynamic resource multiplexing and preemption targeting supporting different traffic types are necessary for Uplink as well. However, the design considerations for the transmission of the Uplink PI are different from that for the Downlink PI.

In Downlink, since the gNB is the transmitter and the UE is the receiver, when an ongoing Downlink eMBB transmission is preempted, the gNB can manage the *preemption first* (by stopping an eMBB transmission and replacing it by a URLLC transmission) without the need to send the preemption indication to the UE earlier or immediately. Rather, the preemption *indication* can be and is preferably to be sent *later*. In other words, for the DL PI operation, it can follow a *preemption-first-indication-second* manner. On the contrary, for Uplink, when an ongoing Uplink transmission needs to be preempted, the gNB has to indicate the preemption command to the UE first and as soon as possible, in order to stop the ongoing transmission. The execution of the preemption at the UE is possible only *after* receiving the preemption indication. In other words, in Uplink, the preemption has to follow an *indication-first-preemption-second* approach. Due to the necessary processing time of receiving the preemption indication, some lead time has to be ensured so that a UE can react in time to stop an ongoing transmission.

Another variation for the Uplink multiplexing of different traffic types is the possibility of a simultaneous transmission by different UEs so that a URLLC transmission by a first UE does not necessarily preempt an eMBB transmission by a second UE, even if the two transmissions have overlapped resources. Compared with the preemption operation, the approach of continuing an eMBB transmission even with the presence of a URLLC transmission has the benefit of minimizing the service interruption for eMBB. The shortcomings are clear – the URLLC traffic has to experience additional intra-cell interference from the one or more concurrent eMBB transmission(s) in the overlapped resources. In order to mitigate the negative impact to URLLC, a *soft preemption* mechanism can be taken, e.g., by some power control. In particular, depending on the need, a gNB can indicate to a URLLC UE to temporarily increase the transmit power in order to mitigate the expected additional interference.

Both the hard Uplink preemption indication and the soft Uplink preemption indication are supported in Release 16. It can also be termed as a priority indication (with the abbreviation of UL PI) or a cancellation indication (UL CI). Reference [29] provides analysis and comparison for both the hard and soft preemption indications.

Figure 13.16 provides an illustration of the UL PI operation. After sending a PDCCH scheduling an eMBB PUSCH, a more urgent scheduling need may arise for a URLLC traffic. The gNB would then transmit a PDCCH to schedule a URLLC PUSCH using a much shorter UL scheduling time. As a result, the later scheduled URLLC PUSCH overlaps with the earlier scheduled eMBB PUSCH. To inform the UE(s) about the updated scheduling, an UL PI can be sent in the same time instance of the PDCCH scheduling the URLLC PUSCH, or at a later point as long as there is sufficient time for the UL PI to take effect. For the hard UL preemption, the UL PI is intended for the eMBB UEs. This can be sent as a group-common indication, similar to the DL PI indication. For the soft UL preemption, the indication is intended for the URLLC UEs. In this case, the UL PI does not have to be a separate transmission. Rather, for simplicity and better Downlink control overhead efficiency, the power boosting command may be embedded in the DCI (e.g., indicating one of a plurality of configured open-loop power control parameters) scheduling the URLLC PUSCH.

Upon reception of the UL PI, particularly the hard UL preemption indication, the eMBB UE needs to take some actions. In general, there are two types of actions:

- *Action 1*: the eMBB PUSCH stops the transmission starting from an indicated symbol, and does NOT resume after that, and
- *Action 2*: the eMBB PUSCH stops the transmission starting from an indicated symbol, and may resume after that.

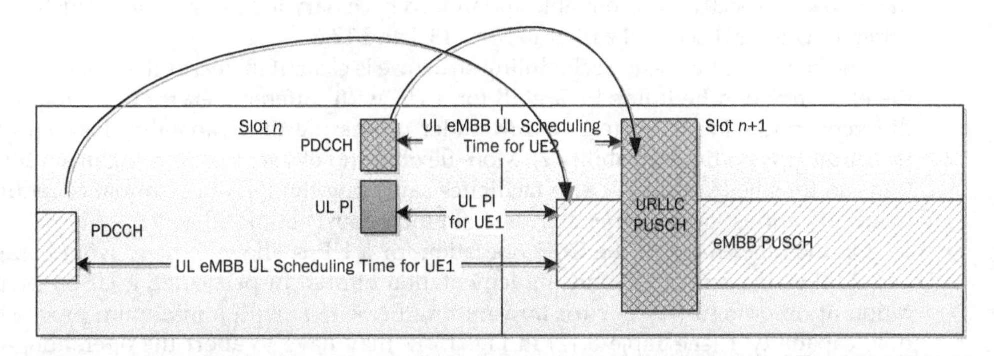

Figure 13.16 Illustration of the UL PI/CI operation.

Action 2 is preferable from the UL eMBB performance perspective, since the resumed transmission may increase the likelihood of successful decoding of the PUSCH. However, due to the preemption operation, the eMBB UE has to transition from an active PUSCH transmission to a silent period without the PUSCH transmission. To resume the eMBB PUSCH transmission, the UE then has to transition from the silence duration back to an active transmission. However, these transitions impose some challenge at the UE side in terms of maintaining UL phase continuity. Without the UL phase continuity, the resume of the UL transmission may not be warranted for two reasons:

- The resumed UL transmission portion may not be coherently combined with the portion before the preemption, and
- More importantly, depending on the DM-RS pattern and where the preemption takes place, the resumed portion may not contain any DM-RS symbols. Losing phase contiguity implies that the resumed portion may not utilize the DM-RS before the preemption for coherent decoding. Therefore, the resumed portion may become practically useless.

Although it is possible that some UEs may be capable of maintaining the phase continuity during the above transitions in implementation, for simplicity, only Action 1 is supported in Release 16.

13.7 Handling Distinct Services at the UE

A UE may be configured to handle different service types such as eMBB and URLLC at the same time, either in the Downlink, or the Uplink, or both.

For a given UE, due to the dynamic indication of the Downlink scheduling timing in the DL scheduling DCI, it is possible that a later PDCCH may schedule an PDSCH (or PUSCH) with a transmission time earlier than another PDSCH (or PUSCH) scheduled by an earlier PDCCH. Due to the dynamic indication of the DL HARQ timing, it is also possible that a later PDSCH may have a HARQ feedback time earlier than another HARQ feedback corresponding to an earlier PDSCH. However, such a nested operation (also known as an out-of-order or OoO operation) is not allowed in Release 15. Figure 13.17 presents an illustration of the prohibited OoO operation for DL HARQ and UL scheduling. Note that such restriction is from the UE's perspective, not from the gNB's perspective. It is possible and indeed necessary to have a nested structure for different UEs, as discussed earlier in Secs. 13.5 and 13.6.

The benefit of the nested scheduling structure is clear. It increases the flexibility and the efficiency of scheduling by a gNB for a UE with different service types requiring different processing times (e.g., eMBB under processing time capability 1 vs. URLLC under processing time capability 2). More flexible and efficient system resource utilization can thus be expected. It also facilitates satisfying the target performance requirements of various service types activated simultaneously at the UE side.

On the other hand, if the OoO operation for a UE is allowed, there would surely be additional specification and implementation efforts. In particular, a UE in preparation of decoding *PDSCH 1* (or transmitting *PUSCH 1*) with a minimum processing time capability 1 (see Table 8.12) in Fig. 13.17 may have to abort the operation, and replace it with a new preparation of decoding *PDSCH 2* (or transmitting *PUSCH 2*)

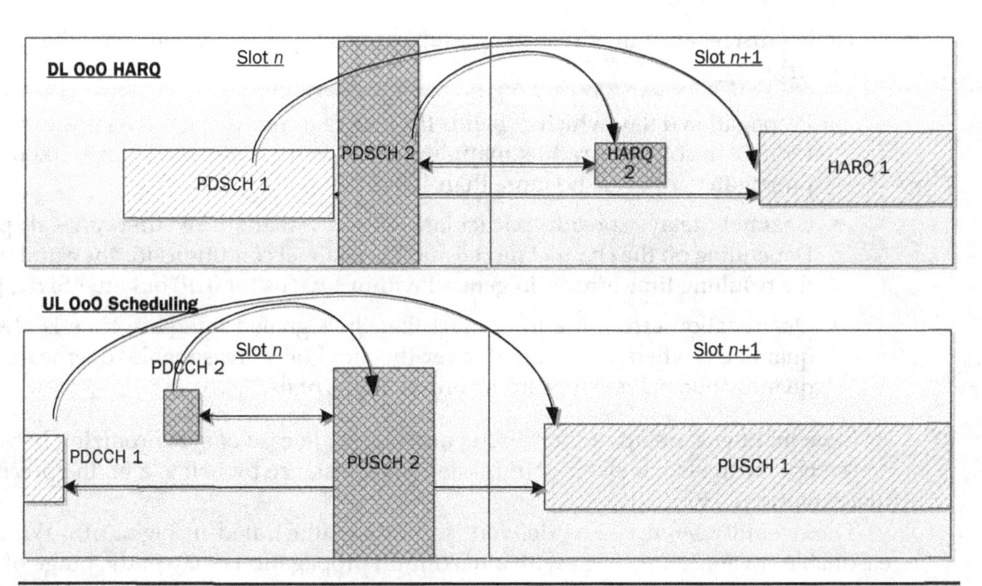

Figure 13.17 Illustration of DL OoO HARQ and UL OoO scheduling, not supported from the UE's perspective.

with a minimum processing time capability 2 (see Table 8.13). Such a switch may not be straightforward, heavily depending on the capability and implementation details at the UE side. Therefore, from the specification perspective, it is difficult to define a single condition under which the nested structure can be handled by a UE.

Intensive discussion on how to support the OoO operation was carried out during Release 16 standardization. Due to the direct connection with various possible implementation approaches, a large number of implementation-dependent conditions or alternatives were proposed and discussed [30], making it impossible to converge. Eventually, this led to the removal of the support of the OoO operation in Release 16.

13.8 Other Related Aspects

Two additional aspects related to URLLC were also specified in Release 16, one is the time-sensitive communications (TSC) as part of industrial internet of things (IIoT) [31], and the support of the multiple transmit/receive points (multi-TRP) operation [32].

As discussed earlier in Sec. 13.2, factory automation is one of the targeted use cases for URLLC. Factory automation is further helped by IIoT, which utilizes internet-connection to connect machinery and provides an advanced analytics platform processing the collected data. As part of IIoT, certain operations are highly time-sensitive forming a so-called TSC. Reference [33] provides a precise definition of TSC:

- *A communication service that supports deterministic communication and/or isochronous communication with high reliability and availability. It is about providing packet transport with bounds on latency, loss, packet delay variation (jitter), and reliability, where end systems and relay/transmit nodes can be strictly synchronized.*

Timing discrepancy in wireless communications is primarily driven by the following factors:

- Propagation delay, which depends the coverage area of a gNB. As one example, for a 57 m coverage radius mapping to a coverage area of 100 m × 100 m, the propagation delay is no more than 190 ns.
- Channel delay spread, particularly the estimation of first arrival path. Depending on the channel model and the channel conditions for the estimation, the resulting time error is in general within ±100 ns for 0 dB or better SNRs [34].
- Quantization error, due to the fact that the signaled reference time is always quantized when transmitted over-the-air. For a reasonable overhead, the quantization error can be in the order of tens of ns.

Note that there are also other discrepancies, e.g., in case of synchronizing between a reference master clock and a gNB master clock instance by using, e.g., the precision time protocol (PTP).

The overall reference time delivery for TSC is illustrated in Fig. 13.18. Within a reasonable coverage area (e.g., with a maximum propagation delay in the range of low hundreds of ns), an over-all maximum timing accuracy of less than 1 μs is feasible. Enhancements on supporting a large coverage area is also possible by using some propagation delay compensation enhancements, which is expected to be specified in Release 17 [35].

Due to the extreme high reliability requirement by URLLC, careful resource management particularly with respect to interference management is critical. Interference may come from the same cell or one or more neighboring cells, the same frequency or adjacent frequency, and the same direction or different link directions. Multi-TRP is a

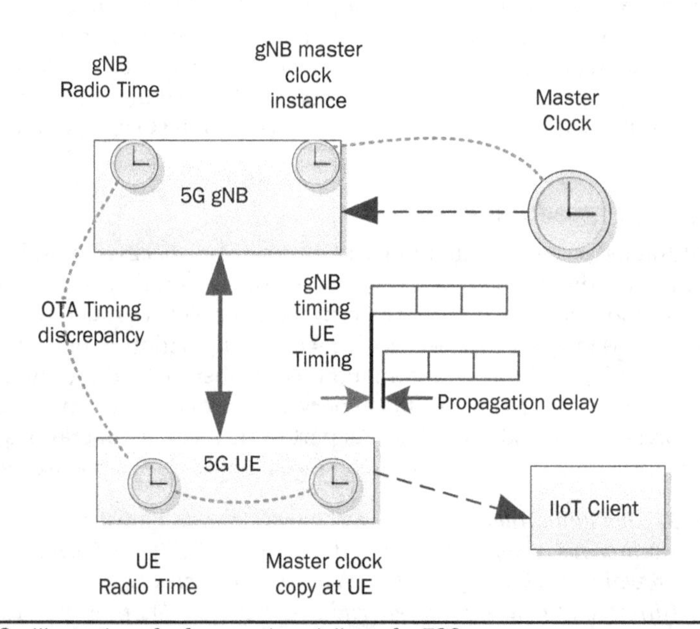

FIGURE 13.18 Illustration of reference time delivery for TSC.

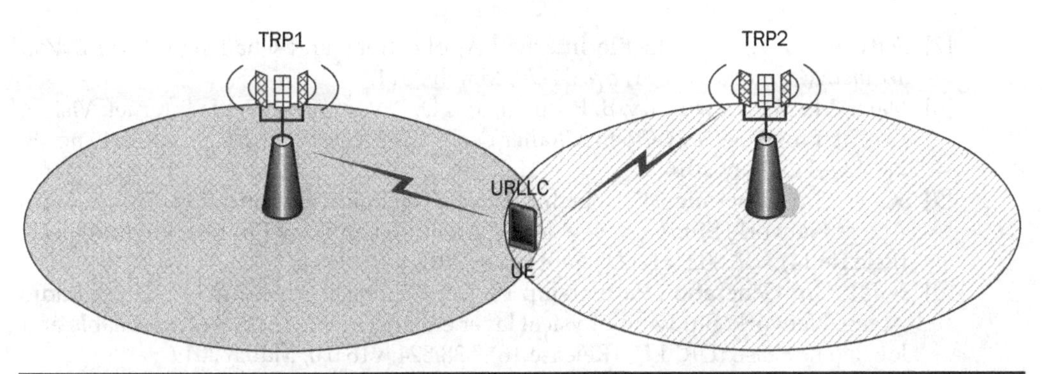

Figure 13.19 Illustration of the multi-TRP operation for URLLC.

technique to reduce inter-cell interference or even to boost the signal power through inter-cell coordination as illustrated in Fig. 13.19. This technique was also carefully studied and specified in 4G LTE, also known as the Coordinated Multi-Point (CoMP) operation [36]. Depending on the details of inter-cell coordination, the multi-TRP operation can be coherent or non-coherent. The techniques may include joint transmission, coordinated beamforming, dynamic TRP switching, etc. There can be a single or multiple Downlink control channels scheduling multiple data transmissions via the multiple TRPs. The TRPs can be co-located or physically separated. However, for simplicity of standardization and implementation, in Release 16, a UE can assume the TRPs have the same physical cell ID. Further enhancements to multi-TRP are expected to be done in Release 17 [37].

References

[1] Elbamby, M. S., C. Perfecto, M. Bennis and K. Doppler, "Toward Low-Latency and Ultra-Reliable Virtual Reality," in *IEEE Network*, vol. 32, no. 2, pp. 78–84, March–April 2018.

[2] Bastug, E., M. Bennis, M. Medard, and M. Debbah, "Toward Interconnected Virtual Reality: Opportunities, Challenges, and Enablers," in *IEEE Commun. Mag.*, vol. 55, no. 6, pp. 110–117, June 2017.

[3] Qualcomm, "New SID on XR Evaluations for NR." RP-193241, 3GPP TSG RAN#86, December 2019.

[4] Wollschlaeger, M., T. Sauter, and J. Jasperneite, "The future of Industrial Communication: Automation Networks in the era of the Internet of Things and Industry 4.0," *IEEE Ind. Electron. Mag.*, vol. 11, no. 1, pp. 17–27, March 2017.

[5] Schulz, P., M. Matthe, H. Klessig, M. Simsek, G. Fettweis, J. Ansari., S. A. Ashraf, et al., "Latency critical IoT applications in 5G: Perspective on the design of radio interface and network architecture," *IEEE Commun. Mag.* vol. 55, no. 2, pp. 70–78, February 2017.

[6] Samii, S., and H. Zinner, "Level 5 by Layer 2: Time-sensitive Networking for Autonomous Vehicles," *IEEE Commun. Stand. Mag.*, vol. 2, no. 2, pp. 62–68, June 2018.

[7] Fettweis, G. P., "The Tactile Internet: Applications and Challenges," *IEEE Veh. Technol. Mag.*, vol. 9, no. 1, pp. 64–70, March 2014.

[8] Maier, M., M. Chowdhury, B. P. Rimal, and D. P. Van, "The Tactile Internet: Vision, Recent Progress, and Open Challenges," *IEEE Commun. Mag.*, vol. 54, no. 5, pp. 138–145, May 2016.

[9] 3GPP, "3rd Generation Partnership Project; Technical Specification Group Radio Access Network; Study on New Radio Access Technology Physical Layer Aspects (Release 14)," 38.802, v14.2.0, September 2017.

[10] 3GPP, "3rd Generation Partnership Project; Technical Specification Group Radio Access Network; Study on physical layer enhancements for NR ultra-reliable and low latency case (URLLC) (Release 16)," 38.824, v16.0.0, March 2019.

[11] Ericsson, Huawei, "New SI Proposal: Study on latency reduction techniques for LTE," RP-150465, 3GPP TSG RAN#67, March 2015.

[12] Ericsson, Nokia, Alcatel-Lucent Shanghai Bell, "New Work item on Ultra Reliable Low Latency Communication for LTE." RP-170796, 3GPP TSG RAN#75, March 2017.

[13] Damnjanovic, A., W. Chen, S. Patel, Y. Xue, K. Hosseini, and J. Montojo, "Techniques for Enabling Low Latency Operation in LTE Networks," *2016 IEEE Globecom Workshops (GC Wkshps)*, Washington, D.C., 2016, pp. 1–7.

[14] Hosseini, K., S. Patel, A. Damnjanovic, W. Chen, and J. Montojo, "Link-Level Analysis of Low Latency Operation in LTE Networks," *2016 IEEE Global Communications Conference (GLOBECOM)*, Washington, D.C., 2016, pp. 1–6.

[15] 3GPP, "3rd Generation Partnership Project; Technical Specification Group Radio Access Network; Study on Scenarios and Requirements for Next Generation Access Technologies (Release 14)," 38.913, v14.3.0, August 2017.

[16] 3GPP, "3rd Generation Partnership Project; Technical Specification Group Services and System Aspects; Study on Communication for Automation in Vertical Domains (Release 16)," 22.804, v16.2.0, December 2018.

[17] 3GPP, "3rd Generation Partnership Project; Technical Specification Group Services and System Aspects; Enhancement of 3GPP support for V2X scenarios; Stage 1 (Release 16)," 22.186, v16.2.0, June 2019.

[18] 3GPP, "3rd Generation Partnership Project; Technical Specification Group Services and System Aspects; Service requirements for the 5G system; Stage 1 (Release 16)," 22.261, v16.10.0, December 2019.

[19] 3GPP, "3rd Generation Partnership Project; Technical Specification Group Services and System Aspects; Extended Reality (XR) in 5G (Release 16)," 26.928, v1.3.0, February 2020.

[20] Qualcomm Inc., "URLLC system capacity and URLLC/eMBB multiplexing efficiency analysis," R1-166395, 3GPP TSG RAN1#86, August 2016

[21] Bertsekas, D., and R. Gallager, "Data Networks," 2nd ed., Prentice Hall, 1992

[22] Qualcomm Inc., "Updated Sub6 DL Full-buffer KPI evaluation for eMBB," R1-166390, 3GPP TSG RAN1#86, August 2016.

[23] Qualcomm Inc., "UL URLLC capacity study and URLLC/eMBB dynamic multiplexing design principle," R1-1708634, 3GPP TSG RAN1#89, May 2017.

[24] Li, C., K. Hosseini, S. B. Lee, J. Jiang, W. Chen, G. Horn et al., "5G-Based Systems Design for Tactile Internet," in *Proceedings of the IEEE*, vol. 107, no. 2, pp. 307–324, February 2019.

[25] 3GPP, "3rd Generation Partnership Project; Technical Specification Group Radio Access Network; NR; Physical layer procedures for control (Release 16)," 38.213, v16.0.0, January 2020.
[26] 3GPP, "3rd Generation Partnership Project; Technical Specification Group Radio Access Network; NR; Physical layer procedures for data (Release 16)," 38.214, v16.0.0, January 2020.
[27] Qualcomm Inc., "URLLC DL pre-emption and UL suspension indication channel design," R1-1713452, 3GPP TSG RAN1#90, August 2017.
[28] 3GPP, "3rd Generation Partnership Project; Technical Specification Group Radio Access Network; NR; Multiplexing and channel coding (Release 16)," 38.212, v16.0.0, January 2020.
[29] Yang, W., C.-P. Li, A. Fakoorian, K, Hosseini, and W. Chen, "Dynamic URLLC and eMBB Multiplexing Design in 5G New Radio," *IEEE Consumer Communications & Networking Conference*, Las Vegas, NV, USA, January 2020.
[30] Qualcomm Inc., "Summary #4 of Enhancements to Scheduling/HARQ," R1-1911708, 3GPP TSG RAN1#98b, October 2019.
[31] Nokia, Nokia Shanghai Bell, "New WID: Support of NR Industrial Internet of Things (IoT)," RP-190728, 3GPP TSG RANP#83, March 2019.
[32] Samsung, "Revised WID: Enhancements on MIMO for NR," RP-182067, 3GPP TSG RANP#81, September 2018.
[33] 3GPP, "3rd Generation Partnership Project; Technical Specification Group Services and System Aspects; Study on enhancement of 5G System (5GS) for vertical and Local Area Network (LAN) services (Release 16)," 23.734, v16.2.0, June 2019.
[34] Qualcomm Inc., "Discussion on Timing Requirements for Industrial IoT," R1-1813389, 3GPP TSG RAN1#95, November 2018.
[35] Nokia, Nokia Shanghai Bell, "New WID on enhanced Industrial Internet of Things (IoT) and URLLC support." RP-193233, 3GPP TSG RANP#86, December 2019.
[36] Samsung, "Coordinated Multi-Point Operation for LTE," RP-111365, 3GPP TSG RANP#53, September 2011.
[37] Samsung, "New WID: Further Enhancements on MIMO for NR," RP-193133, 3GPP TSG RANP#86, December 2019.

CHAPTER 14

Vertical Expansion: MTC

As one of the three pillars of 5G targeted services, mMTC is one integral part of 5G design considerations from the very beginning. Similar to URLLC, mMTC is another vertical expansion from the traditional eMBB service. As discussed in Chap. 5, the numerology and slot structure in 5G NR are designed to be fully compatible with supporting various services from both the gNB's perspective and the UE's perspective.

Discussion on supporting MTC started as early as in3GPP Release 12 [1, 2]. It was motivated by maximizing the investment that operators already have on LTE deployments, so that a single LTE node can not only support the traditional mobile broadband services, but also support low cost/complexity MTC devices at the same time. This makes it possible for operators to migrate low-end MTC devices from GSM/GPRS networks to LTE networks over time.

There are various use cases associated with MTC. The key characteristics of MTC services are summarized into the following three different design goals:

- Low cost/complexity,
- Low power consumption, and
- Extreme coverage.

In the sequel, we will start with a brief history of MTC in 3GPP, followed by the key use cases and the standardization efforts achieving the three different design goals. We will also briefly discuss the integration of MTC into 5G NR and some future trends.

14.1 A Brief History of MTC in 3GPP

There are many types of MTC devices. Reference [3] provides some examples such as metering, road security, and consumer electronic devices.

Metering applications can be related to power, gas, water, heating, grid control, industrial or electrical metering, etc. This type of MTC services make the corresponding applications smarter, in terms of monitoring, information collection (such as energy consumption, billing), efficiency improvement, meter alerts, etc. The corresponding traffic may be classified into three categories [4]:

- Command-response traffic (triggered reporting) such as energization status message, with a payload of ~20 bytes for commands (Downlink) and a payload of ~100 bytes for responses (Uplink), in combination with a 10-second round-trip latency. Such a message may have the frequency of occurring on a daily to monthly basis.

- Exception reporting such as meter alerts (tamper, fire, etc.), where the report sent on the Uplink can be ~100 bytes with a latency of 3 to 5 seconds. The frequency of occurring may also be in the order of daily to monthly.
- Periodic reports such as power (kilowatts), volume (e.g., in units of cubic meter or m^3), where the report may be ~100 bytes (Uplink) and not sensitive to latency (e.g., tolerance of 1 hour) with the frequency occurring of daily to monthly.

For road security, it can be related to an in-vehicle emergency call service, where the service may provide information such as the vehicle's location. Such information helps bring faster responses in handling the emergency such as collisions. Other possible use cases may also include ticketing, intelligent traffic management, congestion avoidance, and fleet management.

There are a wide range of consumer electronic devices, which become an integral part of many people's daily life. For instance, these devices may include digital cameras, eBook readers, personal computers, smart wearables, GPS automotive navigation systems, smart homes (e.g., smart TV and smart microwave), etc. These devices can provide functions beyond the basic communication need, such as entertainment with or without an internet connection, improved office productivity.

There are surely other use cases for MTC, e.g., industry automation-related applications, health-related monitoring devices, etc. It is obvious that different use cases may translate into different sets of requirements in terms of cost, complexity, power consumption, coverage, etc. Chapter 13 covered requirements related to URLLC.

It is crucial for 3GPP to address the need of MTC in an integrated manner with the existing and future evolved networks, while reusing the existing framework of standardization whenever possible. Indeed, the start of MTC discussion in 3GPP was as early as in Release 12 in 2011 [1]. Even today, the standardization evolution still continues in a nonstop manner. Figure 14.1 provides an illustration of a brief history of MTC-related standardization in 3GPP.

MTC began with a study item with detailed justification and objectives listed in Ref. [1]. In particular, it intends to achieve low cost/complexity with data rates and power consumption no worse than the existing GSM/GPRS-based MTC devices, while aiming for higher spectral efficiency and coexistence with the existing LTE devices on a same carrier. To conduct cost, complexity, and performance evaluations, the following is assumed as a reference LTE modem to compare with an MTC modem [4]:

- System bandwidth: 20 MHz,
- A Category-1 LTE UE [5] (about 10 Mbps in DL and 5 Mbps in UL),

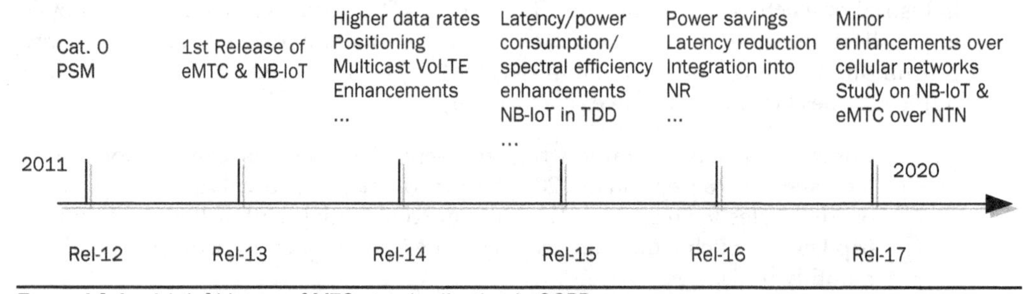

FIGURE 14.1 A brief history of MTC standardization in 3GPP.

- Single RAT,
- Single band,
- TDD/full duplex FDD, and
- Direct DL and UL wide-area-network access from MTC devices to an eNB.

To analyze the cost and complexity, two main factors were considered: RF components and processing. The study identified the percentage cost of each of the two parts, and, for each cost reduction technique, the relative percentage of cost reduction to that of the reference LTE modem. The key findings are summarized in Table 5.3.1 of Ref. [4]. From the findings, the following are listed as the main drivers for cost savings:

- Reduction of the maximum bandwidth,
- Single receive RF chain,
- Reduction of the peak rate,
- Reduction of the transmit power,
- Half duplex (HD) operation, and
- Reduction of supported Downlink transmission modes.
    - Note that in LTE, there are multiple DL transmission modes defined, where a UE can be configured with one of the modes in a semi-static manner. Each mode is associated with some transmission scheme(s) (e.g., a single-layer transmission, a closed-loop MIMO transmission, an open-loop MIMO transmission).

However, the above techniques may also have implications on coverage and power consumption. As one example, a reduction of transmit power helps power saving, but shrinks the link coverage for a UE.

One particular goal for MTC is coverage enhancement which was later added as in Ref. [2] by updating [1], where it states:

- *A 20 dB improvement in coverage in comparison to defined LTE cell coverage footprint engineered for "normal LTE UEs" should be targeted for low-cost MTC UEs, using very low rate traffic with relaxed latency (e.g., size of the order of 100 bytes/message in UL and 20 bytes/message in DL, and allowing latency of up to 10 seconds for DL and up to 1 hour in uplink, i.e., not voice).*

To that end, various DL and UL channels and signals were extensively evaluated particularly in terms of the so-called maximum coupling loss (MCL) [4]. This is summarized in Table 14.1, and the detailed evaluation assumptions can be found in Ref. [4].

Physical channel name	PUCCH Format 1A	PRACH	PUSCH	PDSCH	PBCH	Sync channel	PDCCH DCI Format 1A
MCL (FDD)	147.2	141.7	140.7	145.4	149.0	149.3	146.1
MCL (TDD)	149.4	146.7	147.4	148.1	149.0	149.3	146.9

TABLE 14.1 Summary of MCL for Different DL and UL Channels/Signals (unit: dB)

Note that an eNB is assumed to be equipped with 2 Tx and 2 Rx in FDD systems, and with 8 Tx and 8 Rx in TDD systems. It can be noticed that the maximum gap among different channels/signals in terms of MCL (maximum MCL vs. minimum MCL) is 8.6 dB for FDD and 2.7 dB for TDD. If a 20 dB coverage enhancement is targeted, all channels have to be enhanced to achieve the improved coverage.

In Release 12, a new UE category was introduced for MTC, called UE Category 0. It has the following key properties for cost/complexity reduction:

- A 1 Mbps peak rate for unicast data,
- A max transport block size (TBS) of 1000 bits for unicast and 2216 bits for broadcast data,
- A 1-Rx antenna, and
- A HD operation with a 1 ms switching time (also known as a type B HD operation).

In addition, for the purpose of power savings, a new feature called power saving mode (PSM) has been introduced. The PSM feature allows a UE to go from the idle state to a "deep sleep" state, as illustrated in Fig. 14.2. During the "deep sleep" state, the UE has minimal activity in order for the maximum power saving benefit. The UE cannot be paged in this state, which is not friendly for non-scheduled mobile terminated (MT) data applications. However, it is very efficient for mobile originated (MO) data applications and scheduled MT data applications.

Note that MTC UEs of Category 0 do not have any reduction in channel bandwidth (i.e., still mandated to support a maximum of 20 MHz) or any enhancement in increasing the coverage, including recovering the coverage loss due to 1 Rx antenna. Indeed, as the first release of MTC-related standardization in 4G LTE, Category 0 is a simplified way of supporting MTC, with minimal specification and implementation impacts.

Major standardization efforts dedicated to MTC started from Release 13. The standardization resulted in two branches of MTC support, namely:

- Enhanced MTC (eMTC), and
- Narrow-band internet of things (NB-IoT).

Table 14.2 provides a comparison of Release 12 Category 0 MTC, Release 13 eMTC and NB-IoT, where

- Extended I-DRX means extended DRX for IDLE state, with a periodicity up to 43.69 minutes, and
- Extended C-DRX means extended DRX for CONNECTED state, with a periodicity up to 10.24 seconds.

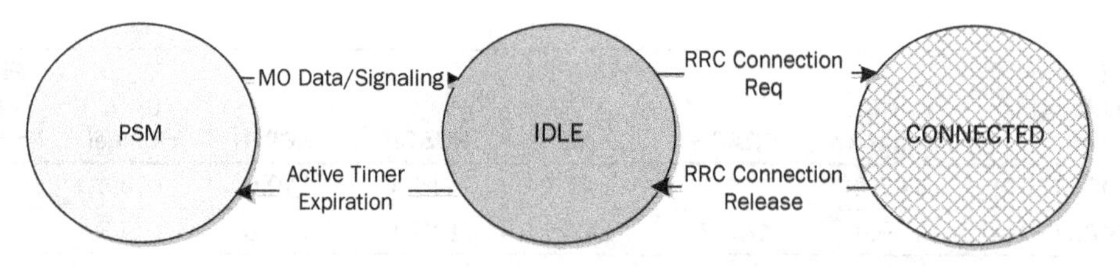

Figure 14.2 Illustration of the PSM mode introduced in Release 12.

	Release 12 Category 0	Release 13 eMTC	Release 13 NB-IoT
Deployment	In-band LTE	In-band LTE	In-band, guardband LTE, standalone
Coverage (MCL)	140.7 dB	155.7 dB	164 dB
Downlink	OFDMA, 15 kHz tone spacing, Turbo code, 64 QAM, 1 Rx	OFDMA, 15 kHz tone spacing, Turbo code, 16 QAM, 1 Rx	OFDMA, 15 kHz tone spacing, TBCC, QPSK, 1 Rx
Uplink	SC-FDMA, 15 kHz tone spacing, Turbo code, 16 QAM	SC-FDMA, 15 kHz tone spacing, Turbo code, 16 QAM	Single tone, 15 kHz and 3.75 kHz spacing, SC-FDMA, 15 kHz tone spacing, Turbo code, QPSK
Bandwidth	20 MHz	1.08 MHz	180 kHz
Peak rate (DL/UL)	1 Mbps DL and UL	1 Mbps DL and UL	DL: ~32 kbps(in-band), 34 kbps (standalone) UL: ~66 kbps multitone, ~17 kbps single tone
Duplexing	FD, HD (type B), FDD/TDD	FD, HD (type B), FDD/TDD	HD only (type B), FDD only
Mobility	Full support	Full support for normal/ small coverage, limited for large coverage	No mobility support in connected mode
Power saving	PSM	PSM, extended I-DRX & C-DRX	PSM, extended I-DRX & C-DRX
Power class	23 dBm	23 dBm, 20 dBm	23 dBm, 20dBm

Table 14.2 A Comparison of Release 12 Category 0, Release 13 eMTC, and Release 13 NB-IoT

The standardization work in Release 13 for eMTC and NB-IoT builds up a solid foundation for continued and non-interrupted evolutions of MTC support in 3GPP afterwards, including Releases 14, 15, 16, and 17. The evolutions focus on further extension of various MTC types (e.g., higher data rates, support of VoLTE), new services (e.g., positioning, multicast), improved efficiency/latency/power consumption, applicability (e.g., support of NB-IoT in TDD), etc. One milestone is the integration of eMTC and NB-IoT into 5G NR systems, both from radio perspective and core network perspective (connectivity to 5GC), making them the *de-facto* mMTC solution in 5G NR. This not only avoids duplication of standardization work in 3GPP, but also avoids potential fragmentation of commercial deployments thanks to a single continuous standardization track for MTC shared by 4G LTE and 5G NR in 3GPP.

14.2 Key Technical Enablers for eMTC

As mentioned earlier, one primary goal of supporting eMTC is to deploy the service in any LTE spectrum or other spectrum re-farmed to LTE, and to coexist with other LTE services within a same bandwidth. Ideally, the additional support of eMTC can be deployed by an operator using the existing base stations with software update only.

Specific to eMTC, there are a set of very ambitious goals:

- Long battery life: **5 to 10 years** of operation (depending on the traffic pattern and coverage needs),
- Low device cost: comparable to that of GPRS/GSM devices (as mentioned earlier),
- Extended coverage: **>155.7 dB** MCL, and
- Variable data rates: up to **1 Mbps** depending on the coverage.

To that end, the key technical enablers include

- Narrow-band operation (up to 1.08 MHz or 6 RBs) by reusing the same RB structure in LTE,
- HD operation,
- De-featuring or feature restriction,
- Coverage enhancement techniques (repetition, power boosting, frequency hopping, etc.),
- Power saving techniques,
- Efficiency enhancements, and
- Latency reduction.

We will provide the details of each technical enabler in the sequel.

14.2.1 Narrow-Band Operation

With a reduced bandwidth, the cost of RF and baseband components in a UE can be decreased. To minimize the device cost for eMTC, the bandwidth is preferably as small as possible. However, as in many other design cases, there are also other related issues by reducing the bandwidth. The bandwidth has to be large enough to support the intended data rates for both Downlink and Uplink, including the throughput requirement for firmware/software updates. To ensure a smooth integration with the existing LTE networks, the bandwidth should be in units of RBs (or 180 kHz), since one RB is the basic resource unit in 4G LTE. In addition, it is highly undesirable to design a completely new initial access mechanism, although it may become necessary for extreme cost savings (as in the case of NB-IoT). This implies that the bandwidth for eMTC is ideally not narrower than the bandwidth of PSS/SSS/PBCH and PRACH, so that the same LTE channels/signals for initial access can be readily reused especially for eMTC UEs not in the need of extreme coverage.

Considering the above factors, it is not surprising that in Release 13, eMTC UEs are designed to have a bandwidth of 1.08 MHz or 6 RBs, the same bandwidth as PSS/SSS/PBCH/PRACH in 4G LTE. In Release 14, more flexible bandwidths are supported for eMTC devices addressing additional use cases or services such as VoIP and improved support of firmware/software upgrades. In particular, in DL, two additional bandwidths are specified: 24 RBs (~5 MHz) and 96 RBs (~20 MHz); while in UL, one additional bandwidth is supported: 24 RBs (~5 MHz). A UE can be configured with different combinations of DL and UL bandwidths depending on the actual service need, e.g., a 5 MHz DL in combination with a 1.08 MHz UL.

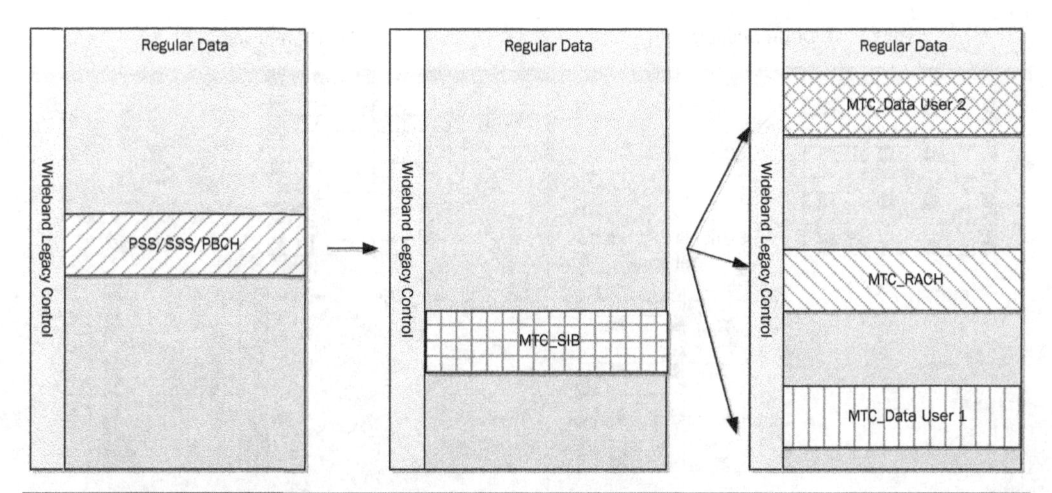

FIGURE 14.3 Illustration of the narrow-band operation for eMTC.

The 6-RB narrow-band requirement for eMTC makes it possible to completely reuse the legacy PSS/SSS/PBCH/PRACH channels, as shown in Fig. 14.3. However, due to the wide-band nature of the legacy LTE control channels (see Sec. 7.1), it is impossible for an eMTC UE to decode the legacy control channels. As a result, eMTC UEs are designed to skip decoding the legacy control channels. More specifically, eMTC UEs rely on a semi-static signaling to skip an indicated size of the legacy control region, and are required to monitor a new narrow-band control channel located in the DL data channel region, also called eMTC physical Downlink control channel (MPDCCH), for both DL and UL scheduling. Different from regular UEs in LTE which follow a synchronous UL HARQ procedure, eMTC UEs follow an asynchronous UL HARQ procedure (hence, PHICH-less), where the HARQ re-transmissions are completely driven by MPDCCH.

For load balancing across different narrow-bands in a wideband of a cell, it is necessary to redirect an eMTC from the central narrow-band for PSS/SSS/PBCH reception to a different narrow-band for SIB reception and possibly yet another different narrow-band for unicast services, as illustrated in Fig. 14.3. It is possible that different eMTC UEs may re-tune to different narrow-bands for unicast services, which may be different from the narrow-band used for the RACH procedure. Such a narrow-band re-tuning (or frequency hopping) requires some re-tuning time, depending on the implementation. From the specification perspective, a two-symbol re-tuning gap is assumed in Release 13, to a large extent corresponding to the most sensible implementation. This is illustrated in Fig. 14.4. Note that in this example, the switching of eMTC narrow-bands is from DL to DL, or from UL to UL (PUSCH to PUSCH). There are also other cases (e.g., PUCCH to PUSCH, PUSCH to PUCCH, PUCCH to PUCCH), where the location of the two symbols for re-tuning depends on the priorities of the channels in the respective combinations (e.g., in the case of PUCCH vs. PUSCH, the two-symbol gap "penalty" is always located on the PUSCH side so that the prioritized PUCCH transmission is not impacted).

The partitioning of a wideband into eMTC narrow-bands/subbands is pre-defined once the cell bandwidth is known (from PBCH decoding). All eMTC narrow-bands

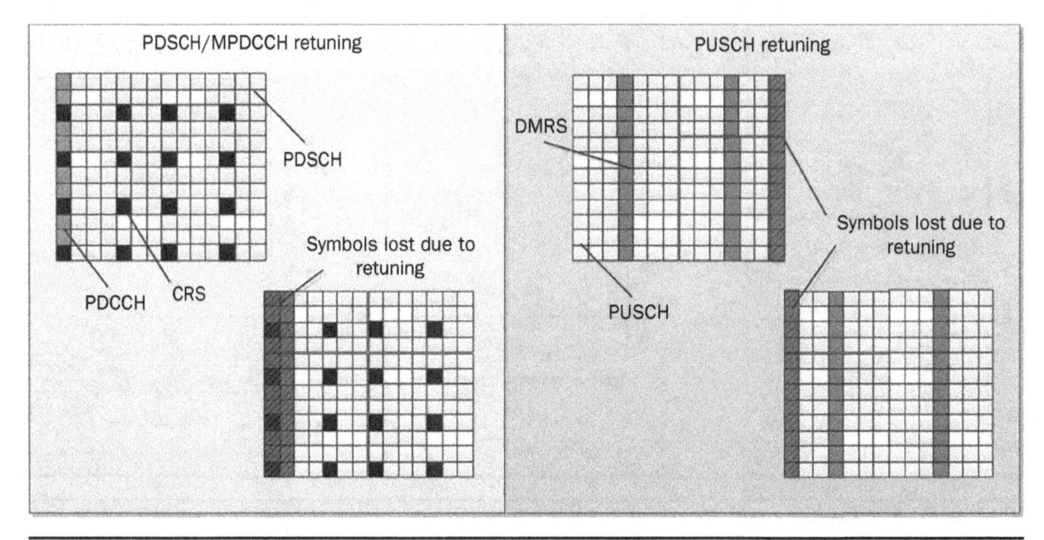

FIGURE 14.4 Illustration of a two-symbol switching gap between two eMTC narrow-bands.

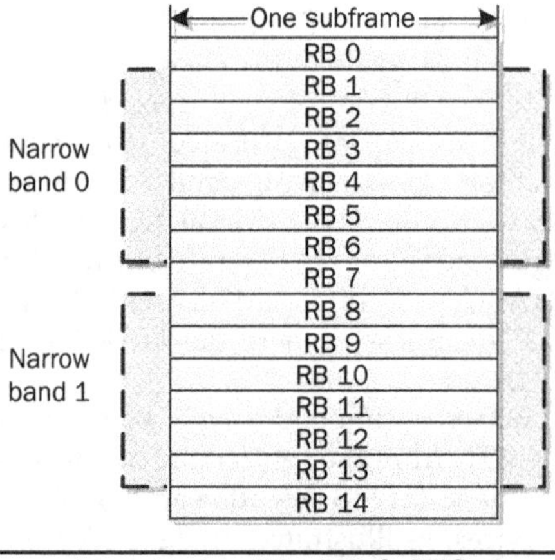

One subframe
RB 0
RB 1
RB 2
RB 3
RB 4
RB 5
RB 6
RB 7
RB 8
RB 9
RB 10
RB 11
RB 12
RB 13
RB 14
Narrow band 0
Narrow band 1

FIGURE 14.5 An exemplary pre-determined partitioning of eMTC narrow-bands from the cell channel bandwidth.

with a fixed size of 6 RBs are non-overlapping in the wideband of a cell. Note that the possible cell bandwidth configurations in LTE are {6, 15, 25, 50, 75, 100} RBs, corresponding to a total cell channel bandwidth of {1.4, 3, 5, 10, 15, 20} MHz, respectively. Obviously, the cell bandwidths in RBs are not divisible by 6. In this case, the remainder of the RBs (after dividing 6) are evenly split at both ends of the cell channel bandwidth; if there is still an extra odd PRB, it is located in the center of the cell bandwidth. To illustrate (see Fig. 14.5), consider a cell of 15 RBs, with numbering from 0 to 14. The first eMTC narrow-band comprises of RBs 1 to 6, while the second narrow-band comprises of RBs 8 to 13. There is one RB in the center and one RB each on the bandwidth edge that are not part of the eMTC narrow-bands.

14.2.2 Half-Duplex Operation

For a regular FDD operation, a UE is expected to be capable of full-duplex (FD) so that it can operate on a Downlink frequency and on an Uplink frequency simultaneously. Such a FD operation requires a duplexer in UE implementation, which imposes extra cost. Replacing the relatively expensive duplexer by a cheaper switch leads to half-duplex (HD) operation and cost reduction of an eMTC device (also for a NB-IoT device, to be discussed later). In order to accommodate HD-FDD UEs, a base station (which may still be FD-capable since it may also serve some FD-FDD UEs) needs to take extra care in resource management and scheduling. The resource management should consider data and control traffic in both directions when making scheduling decisions for an eMTC UE. In addition, the base station has to consider the fact that a HD-FDD UE needs some time to switch from transmission (Tx) to reception (Rx) or vice versa.

It should be noted that a TDD UE is not required to transmit and receive at the same time and is thus expected to be HD. Therefore, the benefit of cost savings (i.e., HD-FDD vs. FD-HDD) may not be applicable to a TDD UE.

The requirement of the switching time (from Tx to Rx or vice versa) necessary for HD-FDD eMTC UEs is quite relaxed – assumed to be 1 ms. This switching time allows for UE implementations with a single-phase-locking-loop for both Uplink and Downlink. The 1-ms guard period is created by skipping the appropriate DL reception subframe as detailed below [6, Sec. 6.2.5]:

- Not receiving a Downlink subframe immediately preceding an Uplink subframe from the same UE, and
- Not receiving a Downlink subframe immediately following an Uplink subframe from the same UE.

In order to provide more flexibility for resource management at the base station, a UE can be indicated a set of DL subframes (via parameter *fdd-DownlinkOrTddSubframeBitmapBR*) and/or a set of UL subframes (*fdd-DownlinkOrTddSubframeBitmapBR* for TDD or *fdd-UplinkSubframeBitmapBR* for FDD) that are valid for the eMTC operation. The remaining subframes that are not part of the indicated sets are considered invalid or reserved from the eMTC UE's perspective. Consequently, the base station can allocate these subframes for other purposes.

14.2.3 De-featuring or Feature Restriction

To reduce the cost and complexity of eMTC devices, it is imperative to downsize the features developed for non-eMTC UEs. In addition, even if an eMTC device supports a feature, it is also necessary to see whether the feature can be simplified or restricted, so that additional cost/complexity reduction can be achieved.

The channels with de-featuring or restriction include DL control, DL data, UL control, and UL data.

For DL control, a new eMTC PDCCH (MPDCCH) is designed. An eMTC UE is required to monitor a reduced number of blind decodes (see also Sec. 7.1) – that is, the maximal number of blind decoding attempts decreases from 44 to 20.

For DL data, it has the following relaxations:

- Relaxed scheduling timing: For a non-eMTC UE, its PDCCH and the scheduled PDSCH are located in a same subframe. The scheduling timing offset for

PDSCH can be considered as "zero" milliseconds, i.e., a PDCCH in subframe n scheduling a PDSCH in the same subframe n. For an eMTC UE, the scheduling timing is extended to 2 ms, i.e., an MPDCCH in subframe n scheduling a PDSCH in subframe $n + 2$. The relaxed timing not only reduces processing requirements for a UE, but also circumvents its need to store the samples for a potential PDSCH reception in the same PDCCH subframe, both of which result in cost reduction for eMTC UEs. The relaxed scheduling timing is illustrated in Fig. 14.6, where both MPDCCH and PDSCH are repeated in multiple subframes for improved coverage (details later), and the timing is defined as the gap between the last subframe for MPDCCH and the first subframe for the corresponding PDSCH.

- TB size and modulation constraint: A maximum TB size of 1000 bits is supported with rank-1 only data transmissions (no SU-MIMO), and the modulations are constrained to be QPSK or 16QAM. Note that these constraints were relaxed later, e.g., in Release 15, the modulation order restriction is relaxed on DL by supporting 64QAM as well.
- Simplified CSI feedback: For eMTC UEs, while CSI measurement and feedback are still feasible and thus supported, the CSI measurement is based on CRS only (no UE-specific CSI-RS–based CSI feedback) and the set of CSI feedback modes are restricted. For eMTC UEs under extreme coverage (details later), CSI feedback is not supported.

For UL data, it has the following relaxations:

- TB size and modulation constraint: A maximum TB size of 1000 bits and rank-1 only transmission are supported (no SU-MIMO), and the modulations are constrained to be QPSK or 16QAM. Similarly, relaxations were introduced later, e.g., in Release 14, the maximum TBS increased to 2984 bits.
- No support of PHICH: Only MPDCCH-based HARQ feedback for PUSCH is supported.

For UL control, an eMTC UE is only required to support a limited set of PUCCH formats (see also Sec. 9.1), namely, PUCCH formats 1/1a/1b/2/2a/2b (i.e., no support of PUCCH formats 3/4/5), where PUCCH formats 1b and 2b are only supported for TDD [6].

14.2.4 Coverage Enhancements

To address the need for extreme coverage enhancement, it is important to make sure that the enabling techniques reuse the existing mechanisms as much as possible to

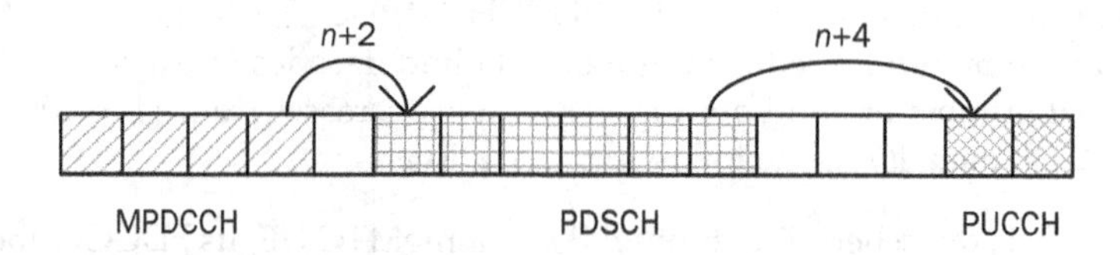

Figure 14.6 Illustration of a relaxed PDSCH scheduling timing for eMTC UEs.

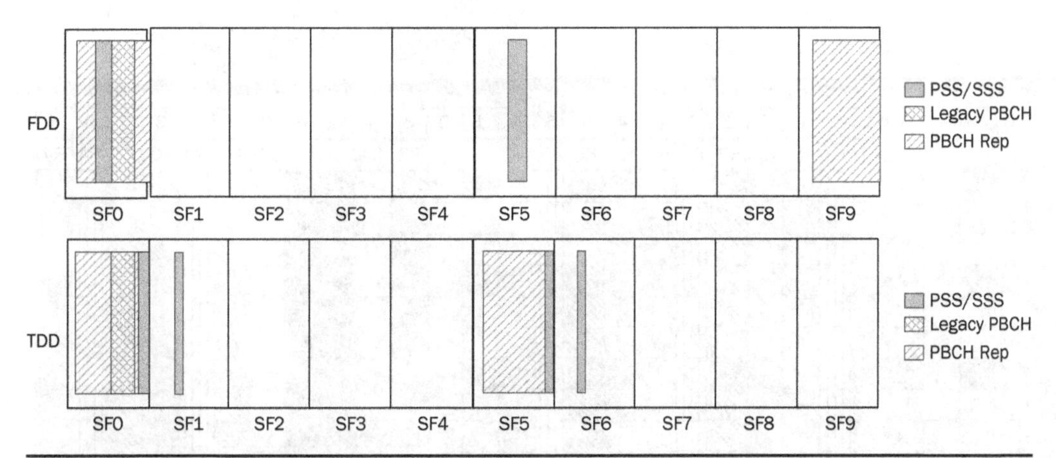

FIGURE 14.7 PBCH repetition for eMTC.

minimize standardization and implementation impact. The coverage need should be evaluated not only for unicast channels (DL control, DL data, UL control, and UL data), but also for channels/signals associated with the initial access procedure.

For the initial access, the legacy PSS/SSS/PBCH is reused for eMTC. To increase the coverage, the legacy PSS/SSS can be power-boosted to reduce the acquisition latency, while PBCH would require some repetitions. To minimize the impact on the legacy operations, the symbols and the subframes that may be used for PBCH are carefully chosen. Therefore, it is not surprising that the repetition pattern is dependent on the frame structure (FDD vs. TDD), as illustrated in Fig. 14.7, where a 10-ms subframes (SF) duration is shown.

Under the extreme coverage, it is crucial to facilitate improved frequency tracking before PBCH decoding for an eMTC UE. To that end, when the PBCH symbols are repeated in other symbol locations of a subframe, the PBCH repetitions need to follow the same CRS-dependent pattern of legacy PBCH, as illustrated in Fig. 14.8. Within the four legacy PBCH symbols (denoted as R0), only the first two symbols contain CRS REs. In repeating PBCH (first repetition R1, second repetition R2, third repetition R3, and fourth repetition R4), the same four-symbol PBCH pattern are exactly duplicated whenever possible for each repetition, including the CRS REs.

A new system information block (SIB) is designed for eMTC UEs, also known as an MTC-SIB or bandwidth-reduced SIB1 (SIB1-BR). To ensure decodability under the extreme coverage and improved efficiency, the MTC-SIB is not scheduled by a control channel. Rather, based on the cell ID, the subframe index derived from PSS/SSS, and the additional information indicated via re-interpreting some original spare bits in PBCH, limited scheduling information of MTC-SIB transmissions can be determined. The scheduling information includes a TB size (up to six different sizes), a repetition level (no repetition, 4, 8, or 16), and a time and frequency location. In particular, in order to improve the MTC-SIB performance, depending on the cell channel bandwidth (derived from PBCH), MTC-SIB transmission may be subject to frequency hopping. To be more specific, when the cell bandwidth is

- 12 to 50 RBs, MTC-SIB is frequency hopped between two narrow-bands.
- 51 to 110 RBs, MTC-SIB is frequency hopped between four narrow-bands.

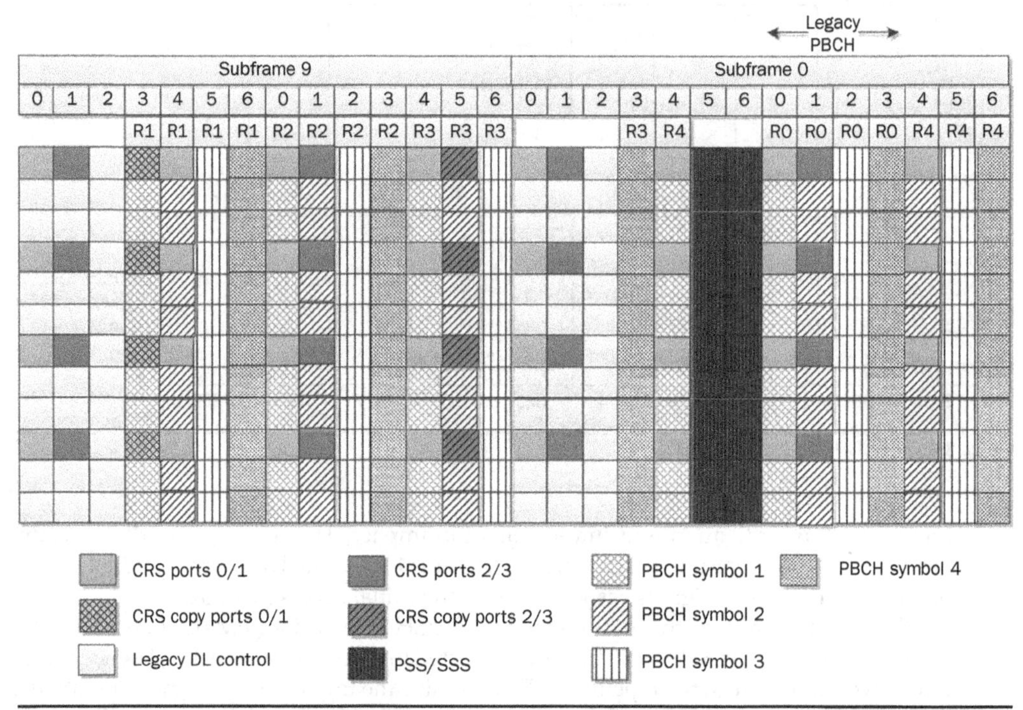

FIGURE 14.8 The detailed repetition pattern for PBCH for eMTC UEs, FDD, normal CP.

While the determination of the coverage area for PSS/SSS/PBCH/MTC-SIB depends on cell-specific deployment scenarios, the subsequent procedures for the initial access are more efficient if UE-specifically managed. The UE-specific configuration of the coverage for the four-step PRACH procedure makes it possible to not always repeat the PRACH to the maximum repetition level supported. Rather, depending on the DL measurements, a UE-specific coverage level can be determined and used for the four-step PRACH procedure, as illustrated in Fig. 14.9. In particular, based on the DL RSRP (reference signal received power) measurement [7], a repetition level can be determined out of four options, which is then one-to-one mapped to a PRACH resource set. Each PRACH resource set has a respectively configured maximum number of attempts for PRACH transmissions. Across different attempts, both power ramping and repetition level increase can be conditionally used by a UE. The subsequent procedures after PRACH transmissions (i.e., random access response reception, message 3 transmission, and message 4 reception) have the corresponding repetitions as well [8].

For unicast operation, it is important to make sure that the design is flexible enough to accommodate different deployments and UE-specific coverage requirements. To that end, two coverage extension (CE) modes are introduced for eMTC:

- CE Mode A: no or minimal repetitions, and
- CE Mode B: a large number of repetitions.

Specifically, MPDCCH can be repeated up to a repetition number of 256. The maximum aggregation level for MPDCCH is 24, i.e., the entire six RBs in a narrow-band (note

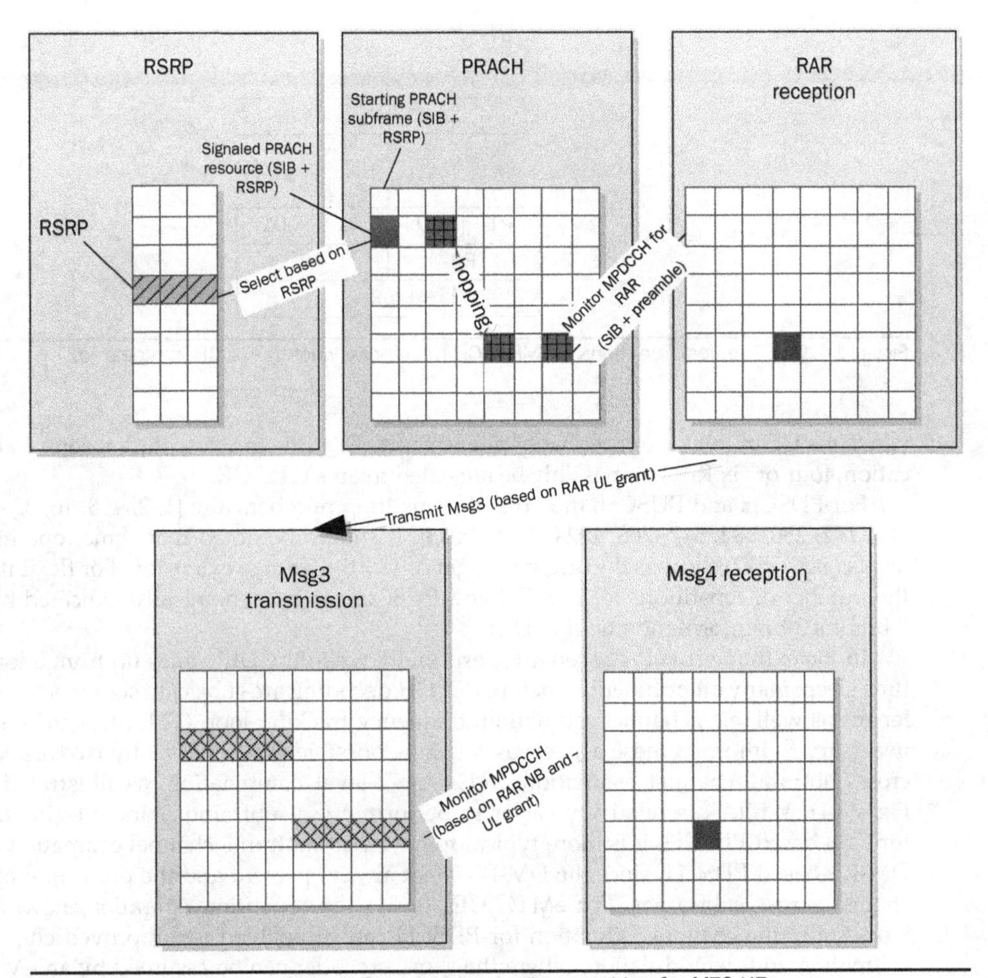

FIGURE 14.9 Illustration of the RACH procedure under repetition for MTC UEs.

that each RB contains up to four CCEs). For a better tradeoff between the MPDCCH scheduling flexibility and the MPDCCH monitoring complexity, two options are supported regarding how a repeated MPDCCH transmission is monitored by an eMTC UE, namely,

- Option 1: different repetitions start in a same subframe, intended for paging CSS, and
- Option 2: different repetitions can start in different subframes, intended for USS and other CSS.

This is illustrated in Fig. 14.10.

Two sets of eMTC DCI formats are defined corresponding to the two CE modes (A or B). For CE mode B under a deep coverage need, the resource allocation is designed to be less flexible for overhead savings but the restriction is tied with the link direction. In particular, for Uplink resource allocation, only one or two RBs can possibly be allocated

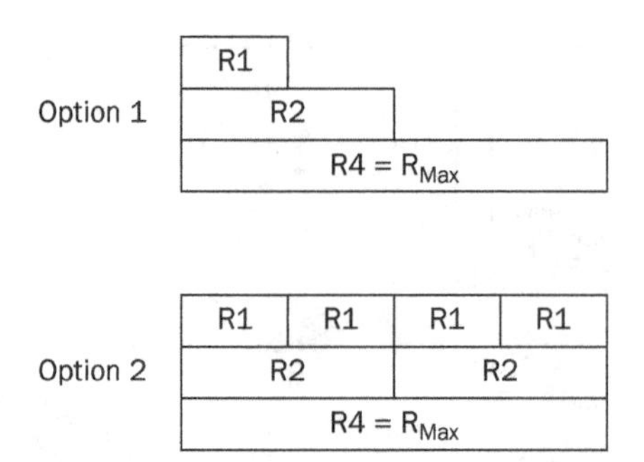

FIGURE 14.10 Two design options for MPDCCH monitoring when MPDCCH is repeated.

(since the UE in this case is probably power limited), while for Downlink resource allocation, four or six RBs can possibly be allocated to an eMTC UE.

For PDSCH and PUSCH, the possible repetition numbers are: {1, 2, 4, 8, 16, 32, 64, 128, 192, 256, 384, 512, 768, 1024, 1536, 2048}. It should be noted that some repetition levels (e.g., >512) exceeds the original target of 15 dB coverage extension. For flexibility, the number of repetitions for PDSCH and PUSCH can be dynamically indicated by a 2-bit or a 3-bit information field in DCI.

In the extreme coverage scenarios, especially when the UE wakes up from a long-time sleep, many other functions before PDSCH decoding are subject to significant challenges as well, e.g., channel estimation, frequency tracking loop (FTL). It is thus also necessary to improve these aspects as much as possible. This is done by two aspects: cross-subframe channel estimation and symbol-level combination, as illustrated in Fig. 14.11. While it is relatively easier to perform cross-subframe channel estimation for CRS-based PDSCH, it is more typical to have per-subframe channel estimation for DM-RS–based PDSCH, since the DM-RS is not always present and the precoding may change across subframes. For eMTC UEs under the repetition operation, however, cross-subframe channel estimation for PDSCH can be enabled for improved channel estimation and demodulation, where the same precoder can be assumed by an eMTC

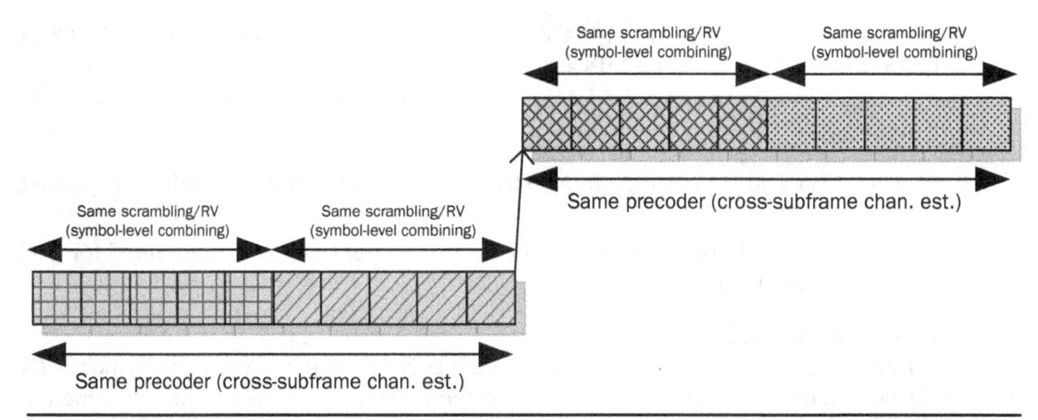

FIGURE 14.11 Cross-subframe channel estimation and symbol-level combining for improved PDSCH decoding for eMTC UEs.

UE within the same set of resources before frequency hopping (if any). In addition, to enable I/Q (real/imaginary) samples combining, the same redundancy version (RV) and scrambling sequence can be applied to DM-RS and PDSCH/PUSCH for a consecutive number of subframes, when the number Z is specified as:

- For CE Mode A: Z = 1.
- For CE Mode B:
  - For FDD: Z = 4.
  - For TDD PUSCH Z = 5, PDSCH Z = 10.

This symbol-level combining (same symbols across the Z subframes) makes it possible to perform frequency error estimation based on the differential operation applied to data symbols.

For PUCCH, repetition is supported as well. Specifically, the possible number of repetitions are

- For CE Mode A: {1, 2, 4, 8}, and
- For CE Mode B: {4, 8, 16, 32}.
  - Note that in Release 14, two additional values of {64, 128} were introduced.

In order to further improve the quality of channel estimation (at the expense of frequency diversity gain), slot-level hopping (within a subframe) for PUCCH can be disabled.

As mentioned earlier, CSI feedback is still supported for eMTC UEs under CE Mode A, but the support is restricted. The measurement for CSI feedback is only based on CRS, using the narrow-bands associated with MPDCCH. For subband CQI, the CQI is measured using a single narrow-band (of six RBs), while for wideband CQI, the CQI is averaged over the set of narrow-bands used for MPDCCH hopping. This is illustrated in Fig. 14.12, where subband CQI may be measured using one of the three narrow-bands (1, 2, or 3), while wideband CQI is measured by averaging the channel quality over narrow-bands 1, 2, and 3.

Note that while for CE Mode A, SRS is still supported, SRS is not supported for CE Mode B for simplicity.

14.2.5 Power Saving Techniques

Coverage extension is generally in conflict with the power saving need of an eMTC UE, due to reception and/or transmission over a prolonged time duration. Careful deployments are often crucial in minimizing the power consumption for eMTC UEs, e.g., by minimizing the repetition number for both DL and UL, especially via the usage of small

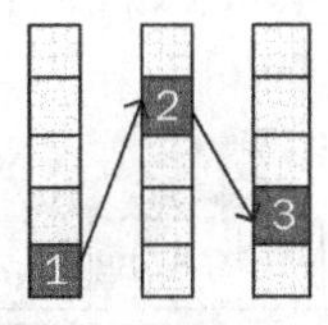

Figure 14.12 CSI measurement based on the MPDCCH narrow-bands, where each box represents one MPDCCH narrow-band in a set of subframes.

cells, relaying, decoupling a DL serving cell from an Uplink serving cell [9]. At the same time, it is also important to have the standardized mechanisms to improve power savings for eMTC UEs.

Besides the support of PSM, extended I-DRX, and C-DRX for power savings (see Sec. 14.1), there are additional technical enablers for power savings, including

- Wake-up signal (WUS),
- Pre-configured UL Resource (PUR), and
- Multi-TB Scheduling for DL and UL.

The support of WUS is intended for RRC_IDLE UEs, particularly related to paging. The WUS for a UE is transmitted only if there is paging for the UE. Therefore, by detecting a sequence-based WUS [6], an eMTC UE benefits from the reduction in unnecessary MPDCCH/PDSCH monitoring/detection in every DRX cycle, as illustrated in Fig. 14.13. The WUS sequence is transmitted in one or multiple subframes, where the maximum WUS duration as well as the gap between the end of a WUS duration and the associated PO is indicated in SIB. In a subframe, the WUS sequence is repeated in a pair of contiguous two PRBs, where the starting PRB within a six PRB bandwidth is indicated in SIB.

Assuming a 10% probability of paging PDCCH for eMTC IDLE UEs, it was observed in Ref. [10] that comparing with the case of no WUS, the power saving gain is

- ~70% in case of a DRX cycle of 2.56 seconds, and
- ~50% in case of an eDRX cycle of 20.56 seconds.

The detailed split of power consumption and the corresponding power saving are illustrated in Fig. 14.14, where it shows the significant gain from reduced PDCCH monitoring (note that the power consumption for PDSCH detection is not accounted for).

Instead of waking up all UEs of the same paging occasion (PO) by a WUS, a group-based WUS can be used to wake up only a group of UEs of the same PO for further power savings, as illustrated in Fig. 14.15. Besides the FDM- or TDM-based WUS grouping as shown in Fig. 14.15, a single-sequence CDM-based WUS grouping is possible as well (i.e., one sequence per WUS resource at a time). A total of up to 32 WUS groups, resulted from up to two FDM WUS resources, up to two TDM WUS resources, and up to eight single-sequence CDM groups per WUS resource, can be configured for eMTC UEs.

It is possible that some eMTC UEs may have a predictable traffic pattern with infrequent and periodic small UL payloads, e.g., in the case of smart meters. Relying on the legacy four-step PRACH procedure to transmit these UL packets incurs a relatively long delay and relatively large power consumption. To improve services for this type of traffic, a UE in the RRC_IDLE state can be enabled with a pre-configured UL resource

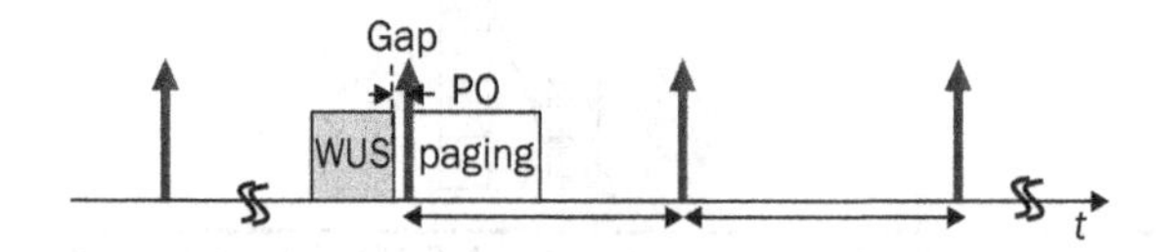

FIGURE 14.13 The wake-up signal (WUS) associated with paging for power savings for eMTC UEs.

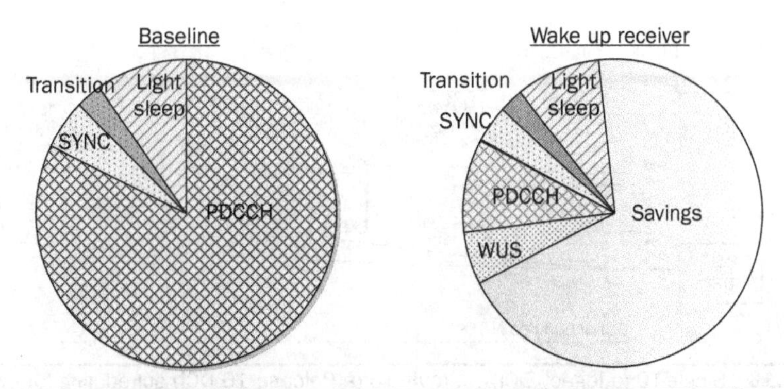

FIGURE 14.14 Detailed power consumption split and power savings with WUS for eMTC UEs.

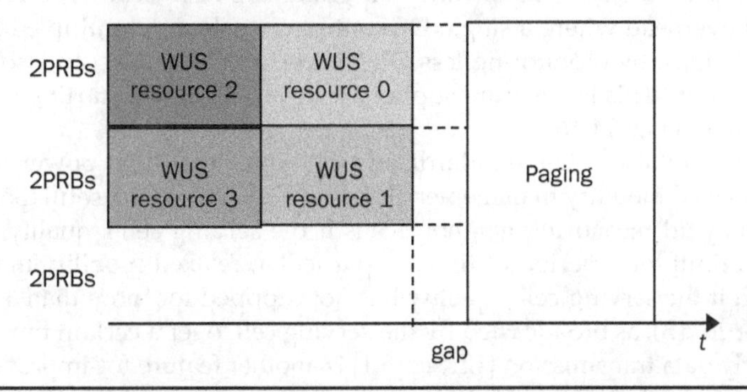

FIGURE 14.15 Group-based WUS (four different WUS resources) for eMTC UEs.

(PUR). With the PUR, if the UE has a valid UL timing advance (i.e., the UL is not out of synchronization), it can transmit UL data directly using the PUR. This feature can be very useful especially for static or semi-static eMTC UEs. Because in this case, the validity of UL timing advance typically holds for a long-time duration. Note that an eMTC UE may assist an eNB in configuring the PUR, either at subscription or at a previous RRC_CONNECTED state, by providing assistance information such as a periodicity, UL grant information, TA validation parameters, a preference of one-shot or infinite-shot transmissions. Both dedicated PUR and contention-free-shared PUR (up to two UEs) are supported.

It is typical that a single DL control channel schedules only one PDSCH or only one PUSCH. This provides the necessary flexibility and simplicity in managing resources in DL and UL, considering the fact that

- The channel and/or the interference conditions may change over different scheduling opportunities for a UE,
- The overall system resource availability may be different in different scheduling opportunities, and
- The complexity in system design, standardization, and implementation if supporting a multi-TB grant.

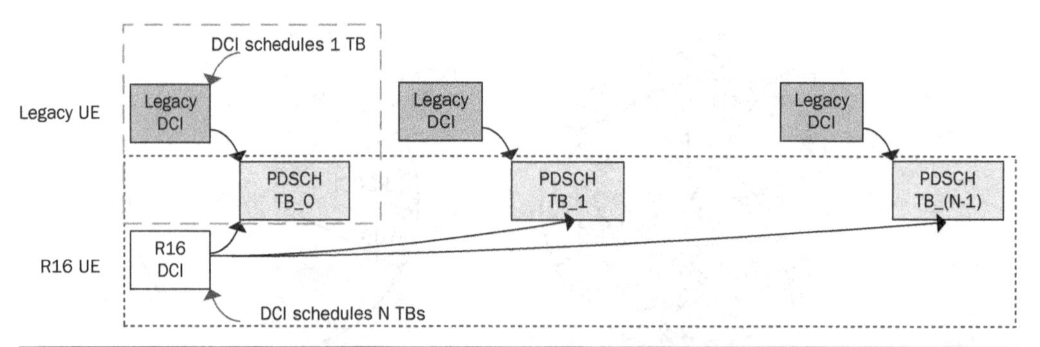

FIGURE 14.16 Single-TB (a legacy DCI) vs. multi-TB (a Release 16 DCI) scheduling for eMTC UEs.

However, the benefits of multi-TB grants are also clear. It certainly saves the DL control overhead where a single DL control can schedule multiple PDSCH or PUSCH transmissions. By monitoring less DL control, an eMTC UE can also enjoy less power consumption. This is a feature supported by an eMTC UE starting from Release 16, as illustrated in Fig. 14.16.

There are also other standardized techniques that help power savings for eMTC UEs. Relaxed mobility management [11, Sec. 5.2.4.12] is one useful tool, where an eMTC UE can avoid measuring neighbor cells if the serving cell's quality remains within a defined limit for a period of time. In particular, relaxed mobility management can be enabled if the serving cell's quality has not dropped for more than a threshold (e.g., 6, 9, 12, or 15 dB), as broadcasted by the serving cell, over a certain time.

Early data transmission [12, Sec. 5.1] is another feature for improved UE power savings, where a UE is allowed to transmit UL data in message 3 and DL data in message 4 as part of the four-step PRACH procedure. This feature is similar to PUR discussed earlier.

14.2.6 Efficiency Enhancements

Ever since the introduction of eMTC in Release 13, additional evolutions for further enhancing eMTC have been introduced in a non-stop manner. One aspect is related to efficiency enhancements. The enhancements include [13, 14, 15, 16]

- Higher data rates,
- Sub-PRB resource allocation for PUSCH,
- Early termination for PUSCH,
- Multicast,
    - Intended for UEs in the RRC_IDLE state. The multicast feature is also known as single-cell point-to-multipoint (SC-PTM), which enables more efficient DL transmissions when the service is intended for a group of eMTC UEs, e.g., firmware/software updates.
- VoLTE enhancements.
    - More efficient offering of voice services at low cost (half-duplex UEs).

To support higher data rates, several aspects have been considered, such as larger bandwidths and transport block sizes (as discussed earlier), HARQ-ACK bundling for half-duplex FDD (HD-FDD), more HARQ processes for full-duplex FDD (FD-FDD), a reduced number of retuning symbols (e.g., from 2 symbols to 1 or 0 symbol), 64QAM for PDSCH (in addition to QPSK and 16QAM, only for PDSCH without repetition), a flexible starting PRB (for better coexistence with non-eMTC UEs), etc. To ensure backward compatibility and smooth coexistence with eMTC UEs restricted with the legacy narrow eMTC bandwidth (six RBs), support of large bandwidths is built based on the same six-RB narrow-band partitioning. For Downlink, only the RBs belonging to the legacy narrow-bands can be allocated to a wide-bandwidth PDSCH. That is, as illustrated in Fig. 14.5, if there are any edge RBs and/or center RBs not part of the narrow-band definition, these RBs cannot be assigned to a wide-bandwidth PDSCH. However, due to the constraint of continuous resource allocation in UL, the RBs not part of the narrow-band definition may be assigned to a wide-bandwidth PUSCH. This is illustrated in Fig. 14.17.

The support of HARQ-ACK bundling for HD-FDD is motivated by the fact that any subframe used for the transmission of HARQ-ACK for a PDSCH transmission implies that the same resource cannot be used for any DL reception for the same UE. Instead of a one-to-one mapping between PDSCH and HARQ-ACK, a many-to-one mapping can be enabled to reduce the number of HARQ-ACK transmissions and, consequently, to create more opportunities for PDSCH transmissions (hence, higher PDSCH data rates). To that end, an eMTC UE can be indicated that two or more PDSCHs are mapped to the same PUCCH transmission instance. For FD-HDD, the increase of HARQ processes makes it possible for a UE to process more pending PDSCH transmissions.

One way to achieve more efficient UL operation for PUSCH for eMTC UEs is via a finer resource allocation granularity, potentially combined with modulation enhancements. In particular, three sub-PRB resource allocation schemes are additionally supported for eMTC:

- Six subcarriers with a 2 ms duration, QPSK,
- Three subcarriers with a 4 ms duration, QPSK, and
- Two subcarriers with an 8 ms duration, $\pi/2$-BPSK.

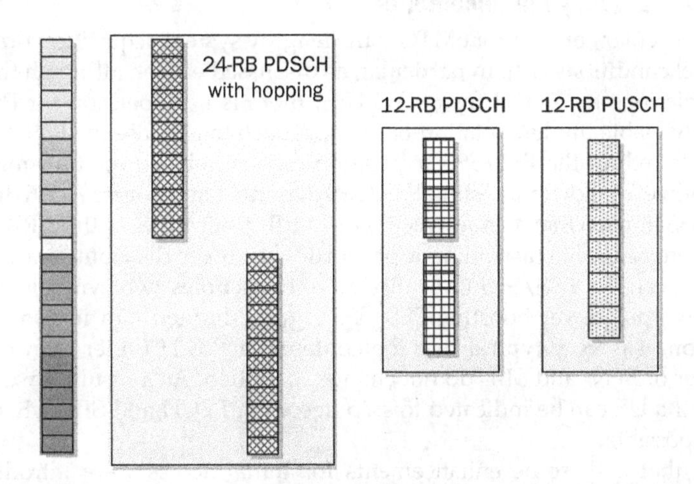

FIGURE 14.17 Illustration of resource allocation for wideband PDSCH and PUSCH transmissions for eMTC UEs.

For better coexistence among different resource allocation granularity schemes (2, 3, 6, and 12 subcarriers), the partitioning of the 12 tones in an RB is pre-determined and non-overlapping, resulting in a pair of 6 tones and a set of four 3-tone splits. That is, for a 12-tone RB indexed by {0, 1, 2, ..., 11}, we have the following sub-PRB resource allocation possibilities:

- For the 6-tone granularity: {0, 1, 2, 3, 4, 5} and {6, 7, 8, 9, 10, 11},
- For the 3-tone granularity: {0, 1, 2}, {3, 4, 5}, {6, 7, 8}, and {9, 10, 11}.

Obviously, it follows a *tree*-structure. For the 2-tone granularity, it is based on the 3-tone partitioning, but selecting two consecutive tones out of each 3-tone split. For improved inter-cell interference randomization, the selection of the 2-tone resource allocation is a function of cell ID (N_{ID}^{cell}), as given by

$$N_{ID}^{cell}\ mod\ 2 + (k, k + 1), \text{ where } k = 0, 3, 6, 9. \quad (14.1)$$

For simplicity, an ongoing PUSCH transmission with a nominal repetition level is expected to be transmitted during the entire duration, irrespective of whether the base station has successfully decoded the PUSCH or not. In reality, due to the non-ideal channel and interference estimation along with channel/interference variations during a PUSCH transmission, it is possible that the base station may already decode the PUSCH before the end of the nominal repetition. As a result, it is beneficial (e.g., for power saving, for more efficient system resource utilization, etc.) to terminate the transmission *earlier* than the intended duration, which is possible for full-duplex FDD and TDD eMTC UEs. The early-termination can be done either explicitly (by an indication in DCI), or implicitly (by using an MPDCCH scheduling a new transport block on PUSCH overlapping with the previous one). This is illustrated in Fig. 14.18. Note that, as expected, there is a time gap between the explicit or implicit indication of the early termination and the actual action at the UE side to terminate the PUSCH transmission.

14.2.7 Latency Enhancements

One particular concern for eMTC is the lengthy system acquisition time for UEs in bad channel conditions [17]. In particular, as discussed earlier, although there are repeated transmissions for PBCH for eMTC UEs, there is no repetition for PSS/SSS. Instead, the only viable implementation-based approach to increase the PSS/SSS coverage and hence to reduce the PSS/SSS acquisition time is to boost the transmit power of PSS/SSS. However, power boosting PSS/SSS does not come for free – a 6 dB power boost of PSS/SSS implies that an equivalence of 24 RBs (four times of the 6-RB for the PSS/SSS) is no longer usable, assuming a proportional power distribution over the cell bandwidth, even if a PSS/SSS transmission only occupies two symbols in a subframe. In other words, power boosting PSS/SSS is quite limited and in general inefficient. In addition, it is very typical that the contents of PBCH (other than the system frame number or SFN) and SIBs do not change that often. As a result, it would be very beneficial if a UE can be indicated to skip decoding PBCH and SIBs when applicable and when possible.

To that end, some enhancements for initial access were introduced, which are summarized in Table 14.3.

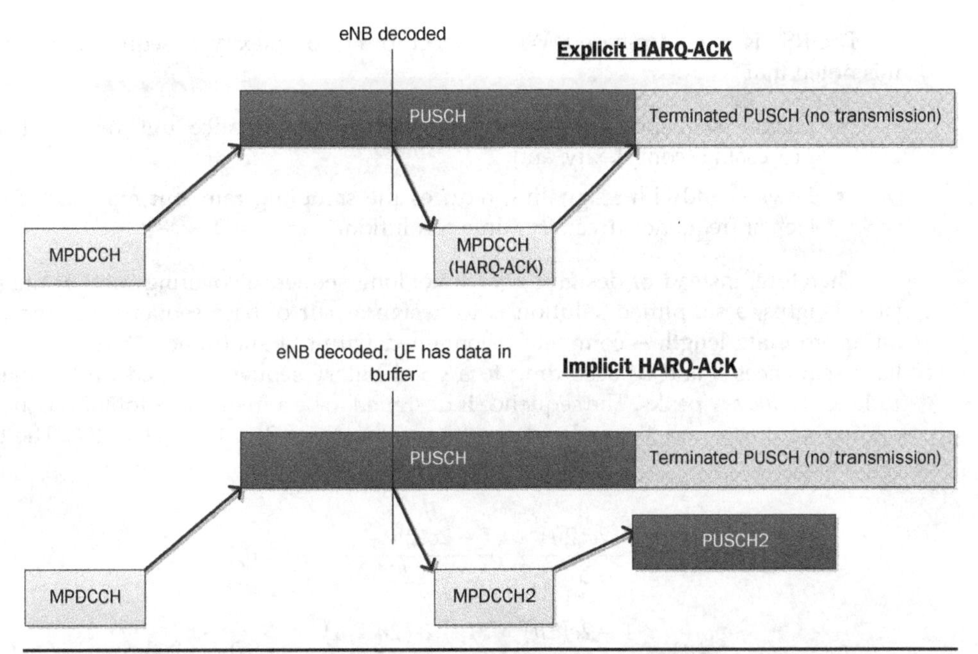

Figure 14.18 Early termination of PUSCH transmissions for eMTC UEs.

Features	Enhancements
Sync signals	Introduction of RSS (re-synchronization signal)
PBCH	Cross-TTI combining
SIBs	MIB Indication for SIB skipping

Table 14.3 Summary of the Enhancements for System Acquisition for eMTC UEs

To design the RSS, the following aspects were considered:

- It is not intended to be transmitted very often (minimal system overhead).
- Once transmitted, it can be detected in a short time (dense transmissions in a short time).
- It can provide an indication to skip PBCH decoding.
- It can be detected with reasonable implementation complexity (in both the frequency domain and the time domain).

To address the first and second bullets, the RSS is designed to be transmitted in a bursty manner over a long period of time. More specifically, the periodicity for the RSS can be configured to be 160, 320, 640, or 1280 ms, while within each transmission burst, the bursty length can be configured to be a consecutive 8, 16, 32, or 40 subframes (or milliseconds). Within each subframe, it occupies the entire 11 symbols after the maximum possible legacy control region size of 3 symbols (assuming a system bandwidth >10 RBs).

The RSS is sequence-based. With respect to the complexity of sequence detection, it is noted that

- Longer sequences generally have better performance but require higher processing/complexity, and
- Lower bandwidth signaling reduces the sampling rate, but may suffer from lack of frequency diversity/time resolution.

Therefore, instead of designing a single, long sequence covering each of the possible lengths, a simplified solution is to design a pair of base sequences S and S^* of an appropriate length – completely contained within a subframe – but arrange the base sequences S and S^* according to a good binary sequence to reduce the magnitude of secondary peaks. The sequence is designed to be a frequency-domain sequence, occupying 2 RBs and 11 symbols, resulting a length of $2 \times 12 \times 11 = 264$. The base sequences are

$$S(n) = \frac{1-2c(2n)}{\sqrt{2}} + j\frac{1-2c(2n+1)}{\sqrt{2}}, \quad n = 0,1,\ldots,263, \tag{14.2}$$

$$S^*(n) = \frac{1-2c(2n)}{\sqrt{2}} - j\frac{1-2c(2n+1)}{\sqrt{2}}, n = 0,1,\ldots,263, \tag{14.3}$$

where the sequences are initialized by $N_{ID}^{cell} + 2^9 \times \delta$, and where N_{ID}^{cell} is the cell ID and δ indicates when the PBCH contents are updated or not. To provide low sidelobes when performing cross-correlation–based detection of the RSS, the binary "cover code" $b(i)$ in arranging the base sequences, as in

$$d_i(n) = \frac{1-2c(2n)}{\sqrt{2}} + jb(i)\frac{1-2c(2n+1)}{\sqrt{2}}, \quad n = 0,1,\ldots,263, i = 0,1,\ldots,N_{RSS} \tag{14.4},$$

is designed as summarized in Table 14.4 [6, Sec. 6.11.3.1], where N_{RSS} is the RSS bursty length in milliseconds.

Reference [18] provides a comprehensive set of evaluation results comparing a baseline long PN sequence of a 40 ms duration (which is computationally prohibitive for implementation by a typical eMTC UE), the adopted S and S^* based RSS of 40 ms, and an S and $-S$ based scheme. It can be observed that the adopted S and S^* based RSS

N_{RSS}	$b(i)$
8	[1, 1, −1, 1, −1, −1, 1, 1]
16	[1, 1, −1, −1, 1, −1, 1, 1, 1, −1, −1, 1, 1, −1, 1, −1]
32	[−1, −1, 1, 1, −1, 1, 1, −1, 1, −1, −1, −1, 1, 1, 1, −1, −1, −1, 1, −1, 1, −1, 1, 1, −1, 1, 1, 1, −1, −1, 1, −1]
40	[1, −1, −1, 1, −1, −1, 1, 1, 1, −1, 1, −1, 1, 1, −1, −1, −1, 1, −1, −1, −1, 1, 1, 1, 1, −1, −1, −1, 1, −1, 1, 1, −1, −1, 1, −1, 1, −1, −1, 1]

Table 14.4 Binary Cover Codes for RSS for eMTC

performs almost identically as the baseline long sequence, while the *S* and –*S* based scheme suffers due to the phase coherence loss under a non-ideal carrier frequency offset.

Besides the possibility of skipping PBCH decoding via the RSS, an eMTC UE may also skip decoding of SIBs via an explicit indication in PBCH during a validity period of the SI.

Other enhancements that are beneficial for latency reduction include PUR and early data transmission, as discussed earlier for power savings in Sec. 14.2.5.

14.3 Key Technical Enablers for NB-IoT

The set of design goals for NB-IoT are very similar to those of eMTC, although the detailed numbers may be different, as summarized in Table 14.2. NB-IoT aims for the low-end MTC market, where there are a massive number of devices with even lower data rates, lower cost but still a long battery life. The targeted cell coverage is even further extended: a 164 dB MCL to be more specific. The deployment can also be more flexible: besides the same in-band deployment as eMTC, NB-IoT is also targeted to be deployed in a standalone carrier or in a guardband of an LTE carrier. A standalone NB-IoT deployment makes it more flexible, e.g., by migrating one or more GSM carriers to NB-IoT carriers. Using the guard-band of an LTE carrier provides improved system efficiency, although extra care has to be taken to ensure that all the regulatory and performance requirements are met.

14.3.1 NB-IoT versus eMTC

There are a lot of similarities in terms of the technical enablers for NB-IoT and eMTC in achieving low complexity/cost, a long battery life and the extended coverage. Herein, we focus on the key differences between the two MTC types.

One major difference between eMTC and NB-IoT is that although eMTC is designed completely based on the LTE numerology, channels, signals, and procedures with necessary extensions and updates, the extreme low-cost NB-IoT with a 164 dB MCL requires a clean-slate system design, starting from the tone spacing, sync channels, etc. Indeed, NB-IoT is based on the following two tone spacings:

- 15 kHz (DL and UL): same as eMTC, with 1, 3, 6, and 12 tones per assignment, and
- 3.75 kHz (UL only): a single-tone assignment in a 2-ms slot.

This results in a maximum bandwidth of 180 kHz (15 kHz/tone × 12 tones).

A new set of channels/signals are necessary to support NB-IoT, as summarized in Table 14.5, where the channels for NB-IoT start with a prefix "N." Note that besides the

Link	Channels/Signals
DL	NPSS/NSSS, NPBCH, NPDCCH, NPDSCH
UL	NPRACH, NPUSCH

Table 14.5 New Channels/Signals for NB-IoT

necessary half-duplex operation for the low-cost purpose, the modulation and coding scheme for NB-IoT is also simplified. Specifically,

- DL: QPSK, TBCC, single RV only
  - In comparison with up to 64QAM, Turbo coding and four different RVs for PDSCH for eMTC.
- UL: QPSK, π/2-BPSK and π/4-QPSK, Turbo coding with two RVs
  - In comparison with up to 16QAM, and four RVs in UL for eMTC.

Note that in Release 17, the maximum modulation order for NB-IoT is expected to be further extended to 16QAM for improved spectral efficiency [16].

14.3.2 Synchronization Signals for NB-IoT

In designing NPSS and NSSS, a bandwidth of 180 kHz is assumed, occupying one LTE RB. Since each RB is 180 kHz, it is not possible to use all the RBs in an LTE cell in order to align the channel raster (which is 100 kHz, the same for all the three deployment scenarios, i.e., in-band, guardband, and standalone) with the center of an RB. Instead, for simplicity and necessary flexibility, it is targeted that NPSS/NSSS is no more than 7.5 kHz from the channel raster. With this rule, the possible PRBs for NPSS/NSSS transmissions are summarized in Table 14.6. Note that in the table, it is assumed that the center six PRBs of LTE system bandwidth is not used for N-NPSS/NSSS transmissions.

The physical layer design of NPSS and NSSS is intended to be common for all deployment scenarios. To achieve that, the following factors were considered for all deployment scenarios:

- No assumption of the presence of CRS for NPSS and NSSS transmissions
  - This is primarily driven by the fact that there is no CRS in standalone or guardband deployments.
  - For in-band deployments, if CRS is present, NPSS/NSSS is punctured. Note that in the MBSFN region of MBSFN subframes [6], there is no CRS transmission.
- Exclusion of the first three symbols in a subframe for NPSS/NSSS transmissions
  - In order to avoid impact to the legacy control operation as in the in-band deployments.

To minimize impact to legacy operation and for better system efficiency, in the case of in-band deployments, the synchronization channel transmissions are preferably located in a particular 180 kHz narrow-band, also known as the anchor carrier, where

LTE system bandwidth	3 MHz	5 MHz	10 MHz	15 MHz	20 MHz
PRB indices for NPSS/NSSS transmissions	2, 12	2, 7, 17, 22	4, 9, 14, 19, 30, 35, 40, 45	2, 7, 12, 17, 22, 27, 32, 42, 47, 52, 57, 62, 67, 72	4, 9, 14, 19, 24, 29, 34, 39, 44, 55, 60, 65, 70, 75, 80, 85, 90, 95

TABLE 14.6 Possible PRBs for NPSS/NSSS Transmissions for NB-IoT UEs

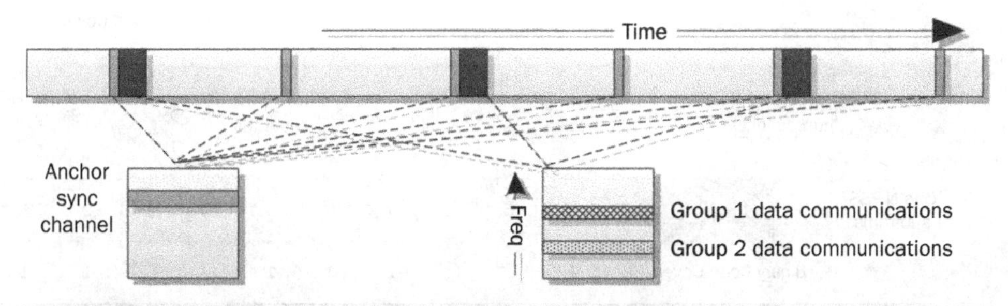

Figure 14.19 Sparse sync channel transmissions along with re-tuned NB-IoT data communications (to different 180 kHz narrow-bands).

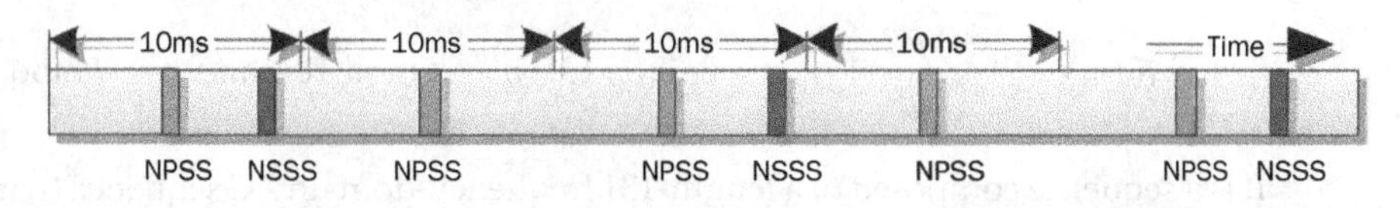

Figure 14.20 A 10-ms periodicity NPSS and a 20-ms periodicity NSSS for NB-IoT.

the transmissions are in a sparse manner in the time domain, as illustrated in Fig. 14.19. After the initial acquisition, NB-IoT UEs can move to a different 180 kHz carrier for data communications.

The synchronization channel for NB-IoT still consists a PSS and an SSS, where

- NPSS is for time and frequency synchronization only, and is transmitted on subframes 5 of every radio frame with a periodicity of 10 ms, and
- NSSS conveys the cell ID and is transmitted on subframes 9 of a radio frame, with a periodicity of 20 ms.
    - Note that in a later release (Release 15), NB-IoT is additionally supported in TDD, where the subframes for NSSS are subframes 0 of even frames.

This is illustrated in Fig. 14.20.

Within each subframe (subframes 5 for NPSS and subframes 9 for NSSS), the synchronization signals occupy the last 11 OFDM symbols, where the first three symbols are reserved for the legacy control. This leaves a total of 12 (tones/symbol) × 11 (symbols) = 132 REs.

For NPSS, the sequences are generated in the frequency domain, on a per symbol basis using a short length-11 Zadoff-Chu sequence, given by [6, Sec. 10.2.7.1.1]:

$$d_l(n) = S(l) \cdot e^{-j\frac{\pi un(n+1)}{11}}, \quad n = 0,1,\ldots,10, \tag{14.5}$$

with the root index $\mu = 5$ using a fixed zero cyclic shift. The length-11 binary cover code $s(l)$ = [1, 1, 1, 1, −1, −1, 1, 1, 1, −1, 1] provides a concatenation of the sequences in the 11 symbols (l = 3, 4, …, 13) in the time domain. In each symbol, the length-11 sequence is mapped to the first 11 tones of an RB. This is illustrated in Fig. 14.21.

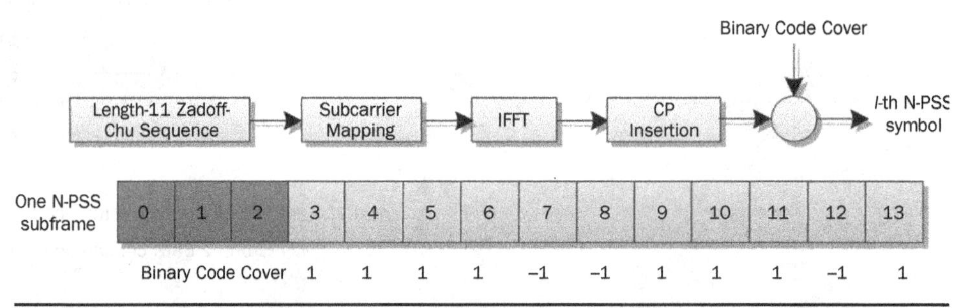

FIGURE 14.21 A block diagram for NPSS for NB-IoT.

For NSSS, the frequency-domain sequence is given by [6, Sec. 10.2.7.2.1]:

$$d(n) = b_q(m)e^{-j2\pi\theta_f n}e^{-j\frac{\pi u n'(n'+1)}{131}}, n = 0, 1, ..., 131, n' = n \bmod 131, \text{ and } m = n \bmod 128 \quad (14.6)$$

It is a sequence composed of a length-131 frequency-domain ZC sequence (with root indices $\mu = \{3, \ldots, 128\}$) and one of four possible binary scrambling sequences (extended from length-128 Hadamard sequences) $b_q(m)$, derived based on the NB-IoT cell ID N_{ID}^{Ncell} as in $q = \left\lfloor \frac{N_{ID}^{Ncell}}{126} \right\rfloor$. The combination of root indices and scrambling sequences conveys a total of 504 cell IDs. An 80 ms boundary is indicated by one of the four time-domain cyclic shifts {0, 33, 66, 99}, via $\theta_f = \frac{33}{132}(n_f / 2) \bmod 4$ where n_f is the system frame number. The sequences are mapped to the available 132 REs in a frequency-first-time-second manner.

One possible low-complexity implementation of initial acquisition using NPSS/NSSS is illustrated in Fig. 14.22 [19]. By using a 240 kHz sampling frequency and the auto-correlation of NPSS over adjacent symbols, coarse timing and fractional frequency offsets can be estimated. Additional refinement of the estimation can be performed using a reference local copy of NPSS with a sampling rate of 1.92 MHz, resulting a total of 137 samples (including the CP) in a symbol. The auto-correlation and cross-correlation properties for NPSS are shown in Figs. 14.23 and 14.24, respectively. More details on the possible low-complexity implementation and its properties can be found in Ref. [19].

14.3.3 System Information, Paging, and PRACH for NB-IoT

The master information block for NB-IoT (NMIB) has a payload of 34 bits and a 16-bit CRC, transmitted on NPBCH in subframes 0 and subframes 9 of every radio frame for FDD and TDD, respectively. For efficient acquisition especially under the extreme coverage, similar to NPSS/NSSS, NPBCH fully occupies the last 11 symbols of a subframe. Within each NPBCH subframe, there are 12 (tones/symbol) × 11 (symbols) = 132 REs. In order to ensure the coexistence with CRS and NB-IoT reference signals (NRS), REs that can be potentially used for CRS (up to four ports) and NRS (up to two ports) are excluded from NPBCH transmissions. In the last 11 symbols of a subframe, since there are up to 16 REs for CRS and 16 REs for NRS, it results in a total of 132 – 16 – 16 = 100 REs available for NPBCH per subframe. This is shown in Fig. 14.25.

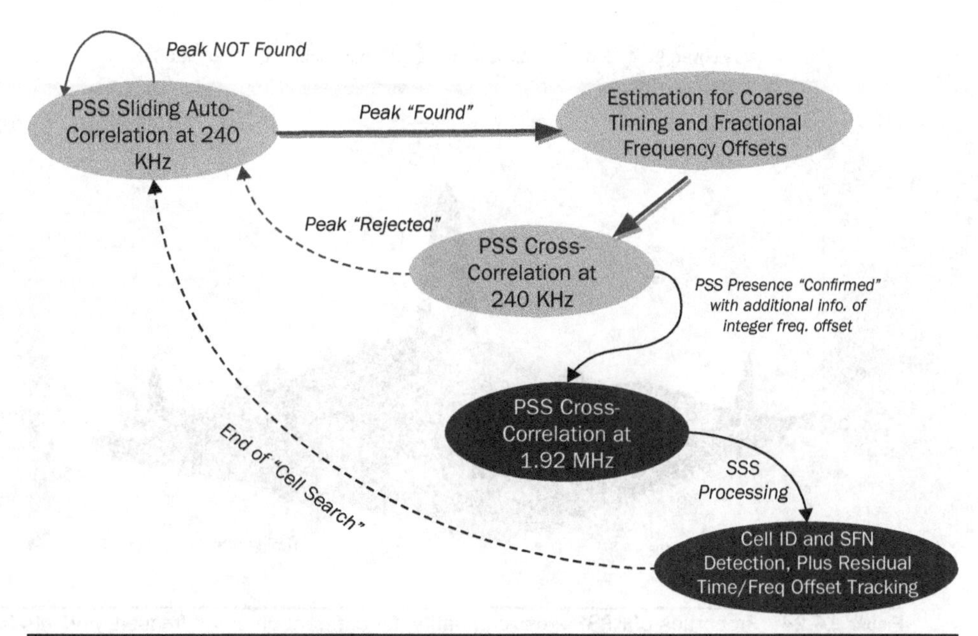

Figure 14.22 One possible implementation of the initial acquisition using NPSS/NSSS for NB-IoT UEs.

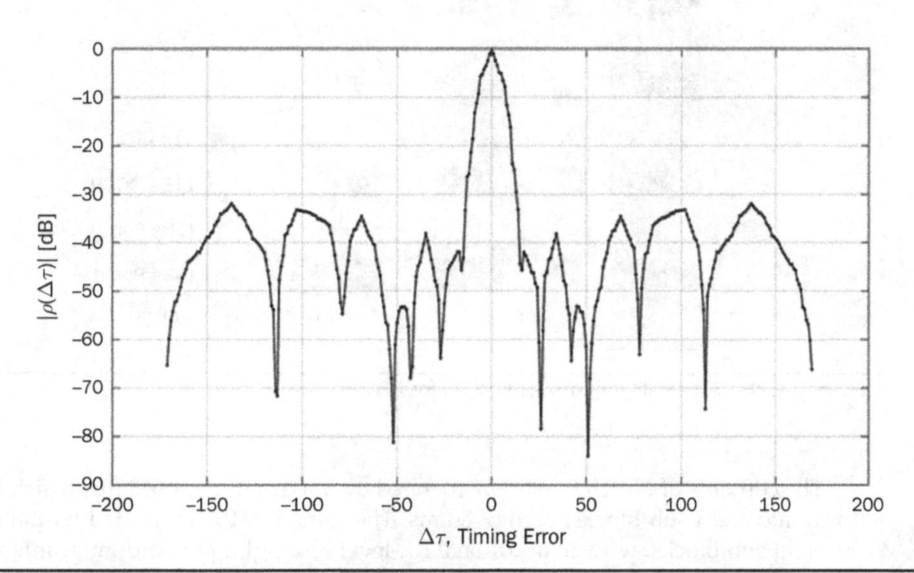

Figure 14.23 Auto-correlation properties of NPSS for NB-IoT (at a sampling frequency of 240 kHz).

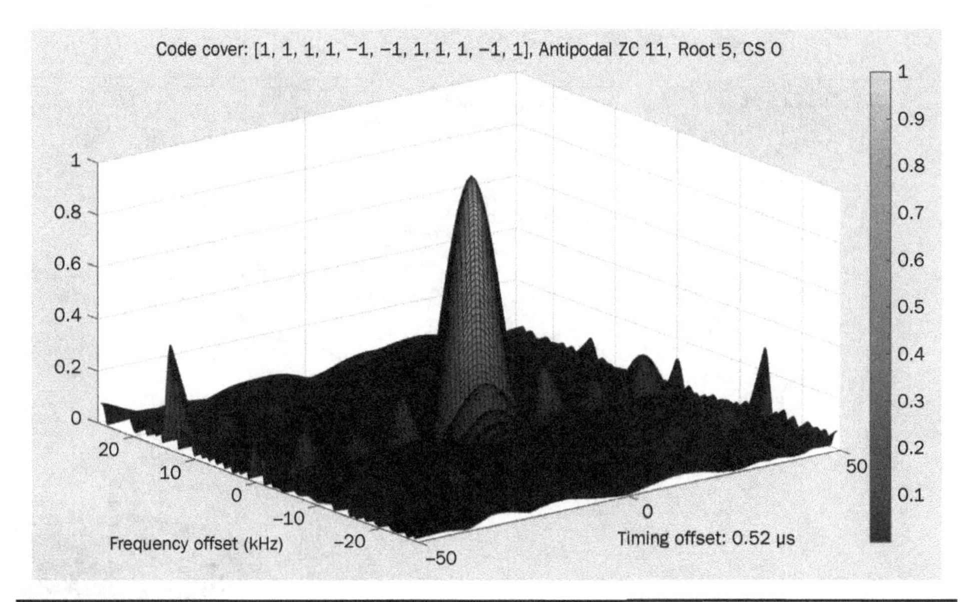

Figure 14.24 Properties of NPSS cross-correlation for different time and frequency offsets for NB-IoT.

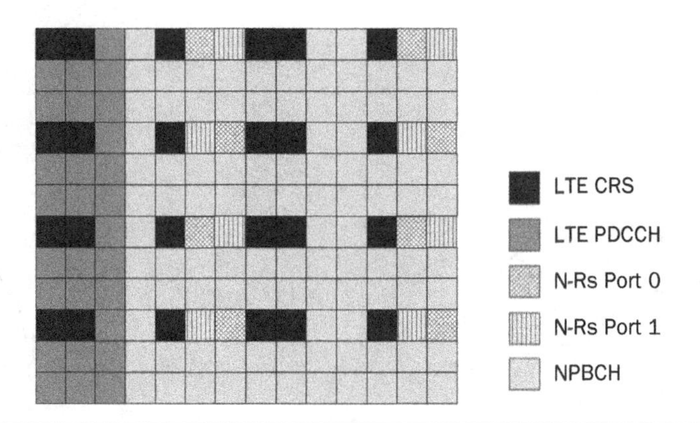

Figure 14.25 Availability of REs for NPBCH for NB-IoT.

The contents of NMIB remain unchanged over a duration of 640 ms, which is organized into eight sub-blocks, each of 80 ms. The same NMIB is repeated within each of the eight sub-blocks, with an additional RE-level scrambling to randomize interference across different cells. With the eight transmissions and QPSK, a total of $8 \times 100 \times 2$ (modulation order for QPSK) = 1600 bits can be used for NPBCH rate matching. Note that the coding scheme for NPBCH is TBCC.

The transmission of NSIB1 is similar to that of NMIB – the same information is repeated over eight subframes as a sub-block. More specifically, it is transmitted in one subframe (e.g., subframe 4) of every other frame in 16 continuous frames. The period of

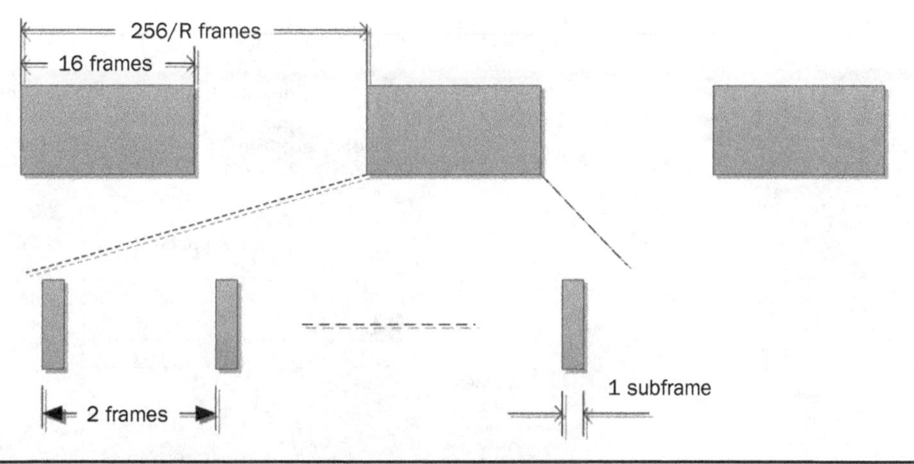

Figure 14.26 Illustration of NSIB1 transmissions for NB-IoT.

NSIB1 is 256 radio subframes, with a transport block size (TBS) and a repetition number R indicated by NMIB from four possible TB sizes and three possible repetition numbers R (4, 8, or 16). Figure 14.26 provides an illustration of NSIB1 transmissions for NB-IoT. NSIB1 also provides scheduling information of other NB-IoT SIBs, which also have repeated transmissions with a TBS up to 680 bits.

For paging, NB-IoT follows the same LTE mechanism in determining the paging subframes and paging occasions [20].

A PRACH transmission for NB-IoT (NPRACH) is subject to frequency hopping using a time unit of a *symbol group* and a frequency unit of a single subcarrier (3.75 kHz). The frequency hopping for an NPRACH transmission is limited to a bandwidth of 12 tones, which is necessary for NPRACH to be completely contained within the maximum bandwidth for low complexity. The hopping is composed of three different components:

- A single-subcarrier hopping between the 1st and the 2nd symbol groups, and between the 3rd and the 4th symbol groups.
- A six-subcarrier hopping between the 2nd and the 3rd symbol groups.
- A pseudo-random hopping between repetitions of a group of four symbol groups.

This is illustrated in Fig. 14.27. The detailed parameters depend on the NPRACH formats and the system frame structure (FDD or TDD), as specified in Ref. [6]. The parameters in the illustration in Fig. 14.27 are based on NPRACH format 1 and NPRACH format 2 for FDD, where the CP length is 66.6 μs and 266 μs, respectively. A large CP helps make sure that an NPRACH transmission can support large cell deployments where a large propagation delay is expected. Multiple hopping patterns for an NPRACH transmission helps timing estimation in terms of accuracy and potential ambiguity avoidance [21] because there is at least one large hopping value (e.g., 6 tones) to help improve accuracy and at least one small hopping value (e.g., 1 tone) to help resolve potential ambiguity. Detailed evaluation results can be found in Ref. [21].

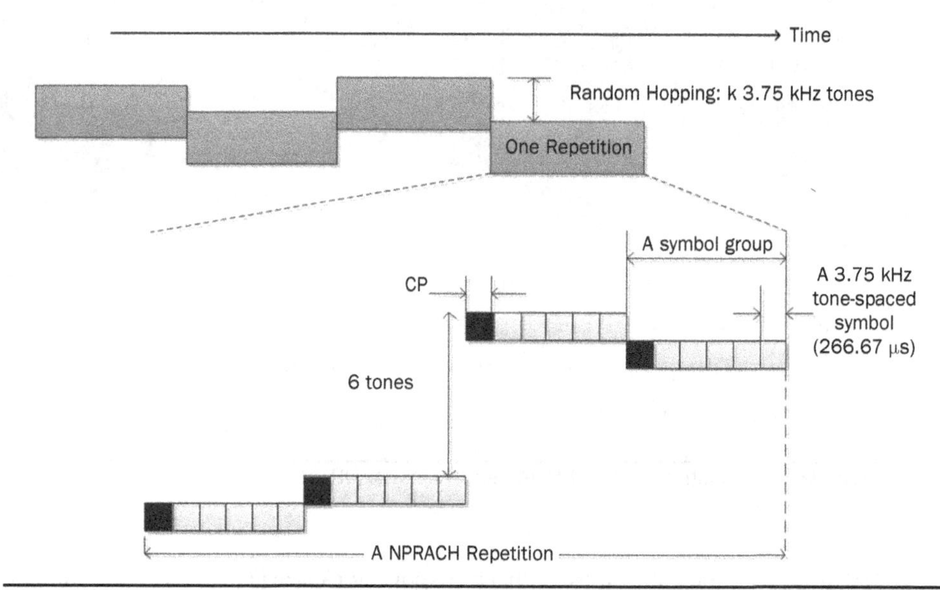

FIGURE 14.27 Frequency hopping for NPRACH for NB-IoT.

Resource allocation for NPRACH includes a set of parameters configurable by a base station, such as three possible coverage levels (i.e., repetition numbers, up to 128), a periodicity value (up to 2560 ms) and a starting time of a period, and frequency resource in terms of frequency locations and possible NPRACH subcarriers. Power ramping for NPRACH is supported for the lowest coverage level, but the maximum transmit power is assumed once an NPRACH transmission moves to the next level so that instead of using power ramp-ups, repetition-based ramp-ups are used to increase the chances of a successful NPRACH attempt.

14.3.4 Unicast Operation for NB-IoT

There is a lot of similarity between eMTC and NB-IoT in terms of unicast control and unicast data especially in addressing the coverage enhancement need. Therefore, we will not discuss in length herein, but rather emphasize some key properties specific to NB-IoT.

For NPDCCH (NB-IoT PDCCH) and NPDSCH (NB-IoT PDSCH), up to 2048 repetitions are supported for each channel.

For NPDCCH, one RB may be split into two control channel elements (CCEs), one is the bottom 6 tones and the other is the top 6 tones. Repetitions of NPDCCH are done with a granularity of subframes. For the NB-IoT UE's perspective, NPDCCH and NPDSCH are always located in different subframes, although it is possible that one subframe may carry NPDCCH and NPDSCH for different UEs over different NB-IoT carriers.

NPDSCH only has the 15 kHz tone spacing without the sub-RB based resource allocation. It has two transmission schemes: a single port and a two-port based SFBC scheme (see Fig. 14.25). For simplification, coding is based on TBCC (instead of Turbo) without the support of RV, while the maximum modulation is QPSK (although later, it is relaxed to additionally support 16QAM). The maximum TBS for NPDSCH is 680 bits (although it was later increased to 2536 bits in Release 14).

For DL, besides the newly defined DL NRS as shown in Fig. 14.25, there is a possibility to reuse CRS as well. It is noted that the cell ID conveyed by NSSS may be different from the LTE cell ID conveyed by PSS/SSS. However, when NPBCH indicates that the cell ID is the same for LTE and NB-IoT, LTE CRS can be used for NB-IoT, where CRS port 0 is mapped to NRS port 0 and CRS port 1 is mapped to NRS port 1.

Different from eMTC and NPDSCH, NPUSCH is able to operate not only with a 15 kHz tone spacing, but also with a 3.75 kHz tone spacing. One slot for the 3.75 kHz tone spacing has a duration of 2 ms and seven symbols (as in a slot for 15 kHz). The sampling frequency is assumed to be 1.92 MHz, such that the time unit T_s is defined as $1/(1.92 \times 10^6) \approx 0.52$ μs. The slot structure for the seven-symbol in a 2 ms duration for the 3.75 kHz tone spacing is shown in Fig. 14.28. Note that the guard period may help minimize or avoid overlapping with other channels/signals, e.g., LTE SRS (which is typically in the last symbol of a 1 ms subframe).

The selection of 3.75 kHz versus 15 kHz for NPUSCH for an NB-IoT UE is performed during the initial access, including whether or not to have multitone scheduling for NPUSCH based on 15 kHz (3, 6, or 12 tones). Note that there is only single-tone scheduling for 3.75 kHz. A base station can indicate NPRACH resources where some are intended for UEs with the multitone capability while others for UEs without such a capability. Upon selection of an NPRACH resource for initial access, an NB-IoT UE capable of multitone scheduling can then choose a corresponding NPRACH resource to indicate such a capability. In the RAR, an NB-IoT UE is indicated whether to use 3.75 kHz or 15 kHz for message 3 transmissions, which is also applied to subsequent NPUSCH transmissions by the UE. The selection of which tone spacing for NPUSCH for a UE can be done by an eNB based on the NPRACH transmissions – a successful NPRACH with a large number of repetitions from a UE generally implies that it is more preferable to use 3.75 kHz for the UE due to the expected large propagation delay and the extended coverage need.

The resource allocation for NPUSCH is based on a granularity of a resource unit (RU), which defines the mapping of NPUSCH to REs. Each RU spans $7N_{\text{slots}}^{\text{UL}}$ symbols and $N_{\text{sc}}^{\text{RU}}$ consecutive subcarriers, where the parameter $N_{\text{slots}}^{\text{UL}}$ is a function of the tone spacing and single-tone versus multitone scheduling [6, Table 10.1.2.3-1 and Table 10.1.2.3-2], and $N_{\text{sc}}^{\text{RU}} = 1$ for 3.75 kHz and 1, 3, 6, or 12 for 15 kHz. For instance, $N_{\text{slots}}^{\text{UL}}$ equals to 16 for the case of single-tone scheduling for both 3.75 kHz and 15 kHz,

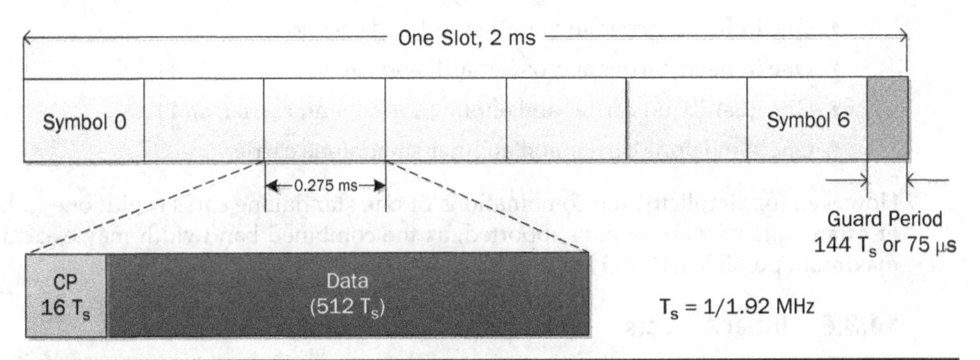

FIGURE 14.28 A 2 ms slot for the 3.75 kHz tone spacing: seven-symbol and a 75 μs guard period.

while $N_{\text{slots}}^{\text{UL}}$ = {8, 4, 2} for 15 kHz–based multitone scheduling using {3, 6, 12} tones, respectively. A UE is indicated by DCI a number of RUs (up to 10) and a repetition number (up to 128) in determining the resource allocated to NPUSCH. The maximum possible TBS for NPUSCH was 1000 bits in the first NB-IoT release (Release 13), although it was later increased to a larger number for higher data rates (e.g., 2536 bits in Release 14).

There is no dedicated PUCCH design for NB-IoT. Instead, the HARQ-ACK for DL transmissions are transmitted using the same NPUSCH design, where the HARQ-ACK payload is fixed at one bit. This is also called NPUSCH format 2 (while the NPUSCH for UL data transmissions is called NPUSCH format 1), where it is a single-tone transmission with $N_{\text{slots}}^{\text{UL}} = 4$ for both 3.75 kHz and 15 kHz. The simplest repetition coding is used for the transmission.

For reduced complexity, the scheduling and HARQ timings for NB-IoT are relaxed. For instance, the HARQ response time is relaxed from a typical 4 ms for non-MTC UEs to at least a 12 ms delay for NB-IoT UEs. Similarly, the UL scheduling timing (from the end of an NPDCCH to the start of an NPUSCH) is at least 8 ms. There is also an information field in DCI indicating a dynamic scheduling delay for NPDSCH and NPUSCH, which provides the necessary flexibility for a base station to manage the resources for NB-IoT services.

14.3.5 Multi-Carrier Operation for NB-IoT

As illustrated in Fig. 14.19, one important aspect for NB-IoT is the multi-carrier operation, where there is one anchor carrier and one or more non-anchor carriers for an NB-IoT UE. An anchor carrier for an NB-IoT UE is a carrier where the UE assumes that NPSS/NSSS/NPBCH/N-SIB are transmitted. Otherwise, a carrier serving an NB-IoT UE is a non-anchor carrier where unicast transmissions may occur. This ensures a good trade-off between UE complexity and a reasonable system load distribution. In order to better address load balance and different coverage needs of paging and NPRACH operations, a base station may have up to 15 non-anchor paging/NPRACH carriers, each associated with a respective coverage level. An NB-IoT UE may determine, in a controlled manner by a base station, a non-anchor carrier from a configured set of non-anchor carriers for paging/NPRACH. It should be noted that measurements for cell selection, reselection, and the coverage level determination are always performed based on the anchor carrier.

Possible multiple carriers for an NB-IoT UE may be a combination of

- One in-band carrier and another in-band carrier,
- One in-band carrier and one guardband carrier,
- One guardband carrier and another guardband carrier, and
- One standalone carrier and another standalone carrier.

However, for simplicity, the combinations of one standalone carrier with one in-band or guardband carrier are not supported, as the combined bandwidth may exceed the maximum possible LTE cell bandwidth of 20 MHz.

14.3.6 Other Aspects

In terms of power savings, efficiency enhancements, and latency reduction techniques, there is a lot of similarity between eMTC and NB-IoT. Indeed, in many cases in 3GPP standardization, a set of agreements made in one session (e.g., an eMTC session) may be easily applied with minimal or no modification to the other session (e.g., an NB-IoT session). Since the corresponding techniques were already discussed in Sec. 14.2, we will not repeat here.

One thing in particular for NB-IoT is the support of scheduling request (SR) in Release 15, where the SR can be transmitted using NPUSCH format 2 (by using a separate cover code other than the pure repetition for HARQ-ACK) or as a dedicated NPRACH resource. This helps reduce UL data latency when there is data arrival at an NB-IoT UE.

Additional NPRACH enhancements were also introduced in Release 15 in order to support large cell sizes (up to 120 km by using an 800 μs CP length for NPRACH) for NB-IoT.

14.4 Integration of eMTC and NB-IoT into 5G NR

Different from smart phones, MTC devices are generally not expected to be upgraded frequently (e.g., once a year) in commercial deployments. Instead, once deployed, MTC devices are expected to last for many years. In addition, it is not prudent to have separate MTC standardizations in 4G LTE and 5G NR, unless there are critical aspects that cannot be easily addressed. As discussed in Chap. 5, the standardization of 5G also provides a flexible framework making it possible to readily integrate eMTC/NB-IoT initiated in 4G LTE into the 5G family.

To integrate eMTC into 5G NR, one aspect to consider is the issue of RB alignment in DL. As analyzed in Ref. [22], while it is possible to perfectly align an eMTC subcarrier with an NR subcarrier (when the NR subcarrier has a 15 kHz subcarrier spacing), it is not possible to align an eMTC RB with an NR RB, due to the presence of a DC subcarrier in eMTC (but not in NR). This implies that a six-RB eMTC narrow-band may effectively require a seven-RB block in NR. In other words, due the RB-misalignment, 1 six-RB eMTC narrow-band may have at least one subcarrier not contained in the same six-RB NR narrow-band as other subcarriers in the eMTC narrow-band, as illustrated in Fig. 14.29. The at least one subcarrier is often called an outlying subcarrier. For an NR system bandwidth of an even number of RBs, the possible number of outlying subcarriers can be 1, 4, or 5; for an NR system bandwidth of an odd number of RBs, the possible number of outlying subcarriers can be 2, 3, or 6. Note that it is still possible to have a full six-RB eMTC narrow-band when the seven RBs in NR are within the system bandwidth of a cell. If an outlying subcarrier falls out of the NR system bandwidth (hence, within the NR guardband), additional specification impact may be necessary especially from the regulation and performance perspectives. To address the issue, we may consider:

- Puncturing REs at the outlying subcarrier(s) for an eMTC RB,
- Rate-matching around the outlying subcarrier(s) for an eMTC RB, or
- Exploitation of a portion of the NR guardband.

Due to the limited performance impact to eMTC operations, especially for the case of 1 or 2 outlying subcarriers, and more importantly being transparent to legacy eMTC UEs, the option of puncturing was finally adopted. In particular, an eMTC UE can be indicated to one of the four typical cases in terms of the number and the location of the punctured carrier(s) [8].

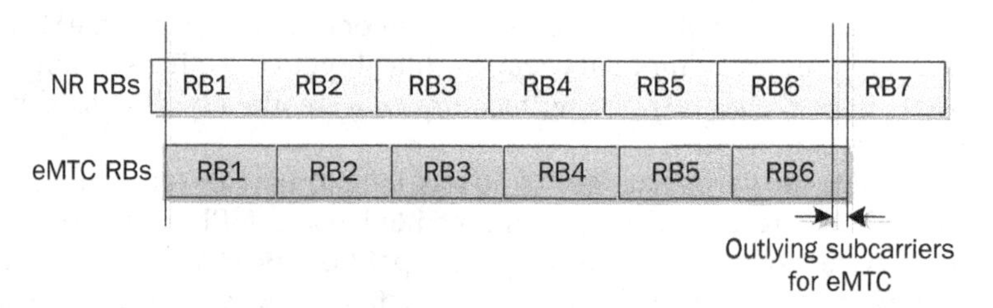

FIGURE 14.29 Outlying eMTC subcarriers when integrating into 5G.

In order to have smooth coexistence with other NR services, an eMTC/NB-IoT UE can be indicated that some resources are reserved (not available for eMTC/NB-IoT). The resource reservation needs to consider the following aspects:

- Granularity (frequency-domain and time-domain) of the resource reservation,
- Time scale of the resource reservation (semi-static or dynamic),
- Whether and how to support an eMTC/NB-IoT transmission in a portion of the subframe, and
- Whether an eMTC/NB-IoT transmission needs to be postponed or dropped when colliding with the reserved resources.

Considering that all other NR services have a granularity of an RB for frequency-domain resource management, an RB-level granularity for resource reservation is sufficient for eMTC/NB-IoT. However, in the time domain, although a subframe-level (or a slot-level) granularity is a natural choice, it is debatable whether a finer granularity is necessary, e.g., using a symbol-level granularity. Due to the potential impact on low-latency applications (e.g., URLLC), a symbol-level resource reservation granularity would be more flexible and more robust, hence adopted eventually.

Besides the semi-static resource reservation in a cell-specific manner, additional flexibility can be achieved via a dynamic UE-specific indication, e.g., by using an information field in DCI. The dynamic indication can over-write the semi-statically reserved resource so that it can be used for eMTC/NB-IoT services.

If an eMTC/NB-IoT transmission is overlapped with a reserved resource (unless it is over-written), for simplicity, the transmission is dropped in the overlapped resource.

14.5 Future Trends

There have been and will be new services continue driving the evolutions for MTC in 3GPP, including expansion into new fields.

New services are typically associated with new requirements, sometimes more than what the current eMTC and NB-IoT standardization can offer. One example is in Ref. [23], where it addresses applications such as pressure sensors, humidity sensors, thermometers, motion sensors, accelerometers and actuators in industrial environments. The requirements for these services can no longer be satisfied by what the currently standardized eMTC and NB-IoT can offer, while the existing eMBB and URLLC standardization is also deemed to be unnecessarily complicated and sophisticated for these services. Therefore, there is a strong need to further evolve the standardization to optimize for these new services particularly in terms of cost, complexity, power savings, and coverage (at least to recover the coverage loss due to the simplified devices, e.g., due to narrower bandwidths, fewer number of receive antennas, etc.).

While it is typical to offer MTC services over traditional terrestrial networks, there is a desire to provide the services over non-terrestrial networks as well. Recently, a study item was approved as in Ref. [24], which focuses on evaluating the minimum necessary specification changes for NB-IoT and eMTC so that MTC services can be accommodated by satellites. Such a study is necessary since it is very costly and sometimes impossible to deploy traditional terrestrial networks especially in remote areas, over the sea, etc. In order to provide MTC services in a global manner, i.e., anytime and anywhere on Earth, MTC over satellites becomes an important complementary tool. The services may include:

- Transportation (maritime, road, rail, air) & logistics,
- Solar, oil & gas harvesting,
- Utilities,
- Farming,
- Environment monitoring, and
- Mining.

The combinations of traditional terrestrial networks and non-terrestrial networks would then make it possible to serve MTC services in a ubiquitous manner with reasonable complexity and cost.

References

[1] Vodafone, "Proposed SID: Provision of low-cost MTC UEs based on LTE," RP-111112, 3GPP TSG RAN#53, September 2011.

[2] Vodafone, "Updated SID on: Provision of low-cost MTC UEs based on LTE," RP-121441, 3GPP TSG RAN#57, September 2012.

[3] 3GPP, "3rd Generation Partnership Project; Technical Specification Group Radio Access Network; Study on RAN Improvements for Machine-type Communications; (Release 11)," 37.868, v11.0.0, October 2011.

[4] 3GPP, "3rd Generation Partnership Project; Technical Specification Group Radio Access Network; Study on provision of low-cost Machine-Type Communications (MTC) User Equipments (UEs) based on LTE (Release 12)," 36.888, v12.0.0, June 2013.

[5] 3GPP, "3rd Generation Partnership Project; Technical Specification Group Radio Access Network; Evolved Universal Terrestrial Radio Access (E-UTRA); User Equipment (UE) radio access capabilities (Release 15)," 36.306, v15.7.0, January 2020.

[6] 3GPP, "3rd Generation Partnership Project; Technical Specification Group Radio Access Network; Evolved Universal Terrestrial Radio Access (E-UTRA); physical channels and modulation (Release 15)," 36.211, v15.8.1, January 2020.

[7] 3GPP, "3rd Generation Partnership Project; Technical Specification Group Radio Access Network; Evolved Universal Terrestrial Radio Access (E-UTRA); physical layer; Measurements (Release 15)," 36.214, v15.5.0, January 2020.

[8] 3GPP, "3rd Generation Partnership Project; Technical Specification Group Radio Access Network; Evolved Universal Terrestrial Radio Access (E-UTRA); physical layer procedures (Release 15)," 36.213, v15.8.0, January 2020.
[9] Qualcomm Inc., "Coverage enhancement techniques for MTC," R1-130589, 3GPP TSG RAN1#72, January 2013.
[10] Qualcomm Inc., "Efficient monitoring of DL control channels," R1-1718135, 3GPP TSG RAN1#90bis, October 2017.
[11] 3GPP, "3rd Generation Partnership Project; Technical Specification Group Radio Access Network; Evolved Universal Terrestrial Radio Access (E-UTRA); User Equipment (UE) procedures in idle mode (Release 15)," 36.304, v15.5.0, December 2019.
[12] 3GPP, "3rd Generation Partnership Project; Technical Specification Group Radio Access Network; Evolved Universal Terrestrial Radio Access (E-UTRA); Medium Access Control (MAC) protocol specification (Release 15)," 36.321, v15.8.0, December 2019.
[13] Ericsson, "New WI proposal on Further Enhanced MTC," RP-161321, 3GPP TSG RAN#72, June 2016.
[14] Ericsson, Qualcomm Inc., "New WID on Even further enhanced MTC for LTE," RP-170732, 3GPP TSG RAN#75, March 2017.
[15] Ericsson, "New WID on Release 16 MTC enhancements for LTE," RP-181450, 3GPP TSG RAN#80, June 2018.
[16] Huawei, HiSilicon, "New WID on Release 17 enhancements for NB-IoT and LTE-MTC," RP-193264, 3GPP TSG RAN#86, December 2019.
[17] 3GPP TSG RAN4, "LS to RAN1, RAN2 on FeMTC SI acquisition delay," R1-1701570, 3GPP TSG RAN1#88, February 2017.
[18] Qualcomm Inc., "Reduced system acquisition time," R1-1804912, 3GPP TSG RAN1#92bis, April 2018.
[19] Qualcomm Inc., "NB-PSS and NB-SSS Design," R1-161116, 3GPP TSG RAN1#84, February 2016.
[20] 3GPP, "3rd Generation Partnership Project; Technical Specification Group Radio Access Network; Evolved Universal Terrestrial Radio Access (E-UTRA) and Evolved Universal Terrestrial Radio Access Network (E-UTRAN); overall description; Stage 2 (Release 15)," 36.300, v15.8.0, January 2020.
[21] Qualcomm Inc., "Random Access Channel Design," R1-160883, 3GPP TSG RAN1#84, February 2016.
[22] Ericsson, "NR and LTE-M Coexistence," R1-1810188, 3GPP TSG RAN1#94bis, October 2018.
[23] Ericsson, "New SID on support of reduced capability NR devices," R1-193238, 3GPP TSG RAN#86, December 2019.
[24] MediaTek Inc., "New Study WID on NB-IoT/eMTC support for NTN," R1-193235, 3GPP TSG RAN#86, December 2019.

CHAPTER 15

5G Vertical Expansion: V2X

15.1 Overview

One important vertical which 3GPP cellular technologies expanded on starting in Release 14 is **automotive**. Indeed, in November 2014, Qualcomm [1] and LGE [2] had proposals in SA WG1 to study requirements for V2X communications. Subsequently, a study item for V2X requirements was agreed in Ref. [3] at the 3GPP SA WG1 meeting of February 2015. The first radio study on V2X was approved at the 3GPP RAN Plenary meeting of June 2015 [4]. Since then, 3GPP has worked uninterruptedly on V2X over the years and across multiple 3GPP releases (Release 14 through Release 17). This is illustrated in the 3GPP timeline with respect to work on V2X in Fig. 15.1.

The term V2X represents the following types of communication:

- **V2V**: vehicle-to-vehicle for communications among vehicles.
- **V2P**: vehicle-to-pedestrian for communications between a vehicle and a device carried by an individual (e.g., handheld terminal carried by a pedestrian, cyclist, driver, or passenger).
- **V2N**: vehicle-to-network for communications between a vehicle and network infrastructure.
- **V2I**: vehicle-to-infrastructure for communications between a vehicle and a roadside unit. A roadside unit (RSU) is a stationary infrastructure entity supporting V2X applications that can exchange messages with other entities supporting V2X applications.

In all V2X cases, importance is given to scenarios with and without cellular network coverage. The following figures from Ref. [5] illustrate different V2X scenarios of interest.

Figure 15.2 shows V2X communication scenarios via the sidelink or PC5 interface. Figure 15.3 shows V2X communication scenarios across the Uu interface. Finally, Fig. 15.4 shows V2X communication scenarios with a mix of sidelink and Uu interfaces.

With respect to frequency operation, two scenarios are considered, one where the given frequency carrier or carriers for V2X communications are dedicated to cellular-based V2X services (subject to regional regulation and operator policy), and another where the frequency carrier or carriers for V2X communications are licensed spectrum and are also used for regular cellular operation. The former spectrum case relies on the ITS spectrum which has been made available globally mainly in the 5.9 GHz frequency range. Active regulatory discussions are taking place to assign or divide the ITS spectrum across technologies, notably, DSRC and cellular-based (LTE and NR V2X).

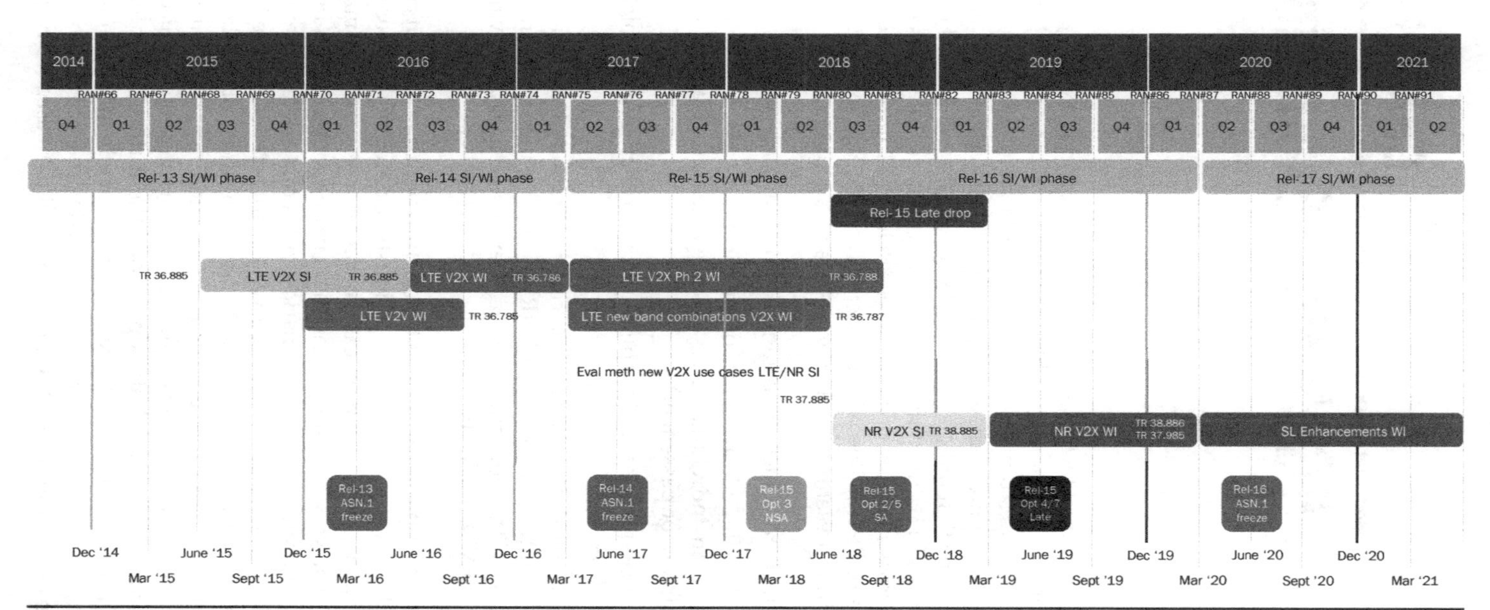

FIGURE 15.1 History of V2X in 3GPP.

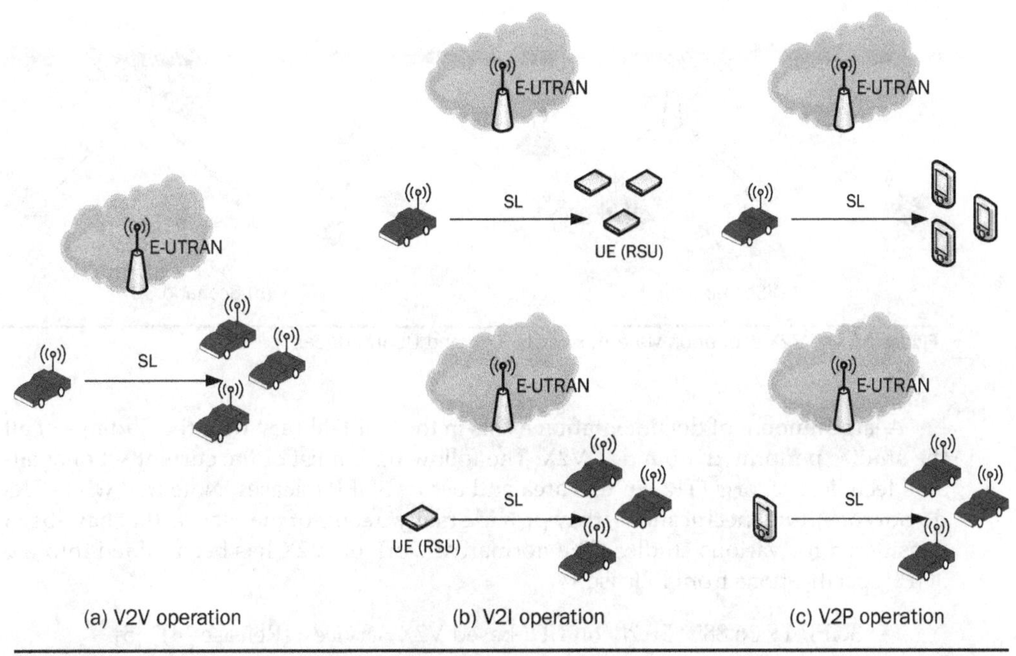

Figure 15.2 V2X scenarios via sidelink interface.

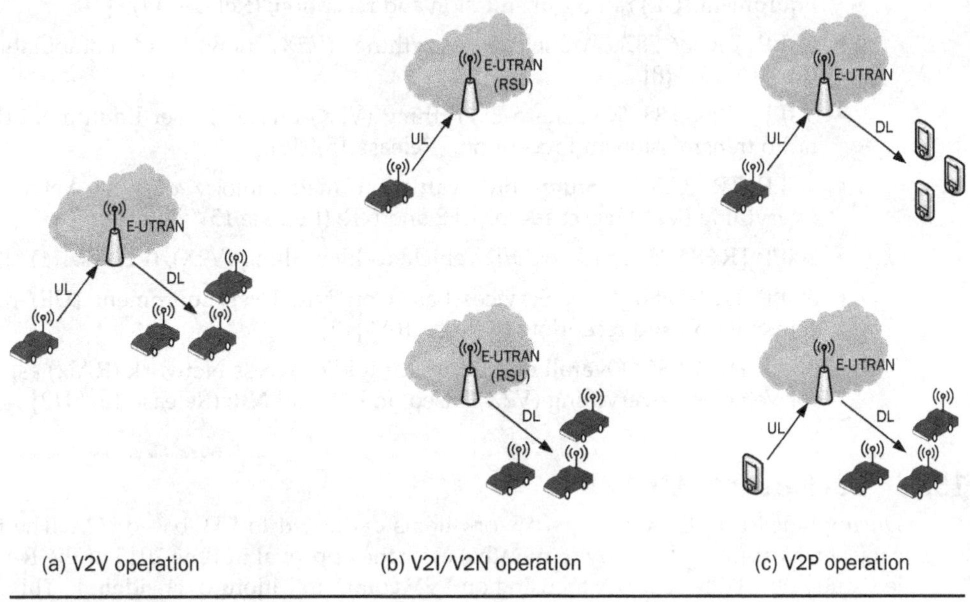

Figure 15.3 V2X scenarios via Uu interface.

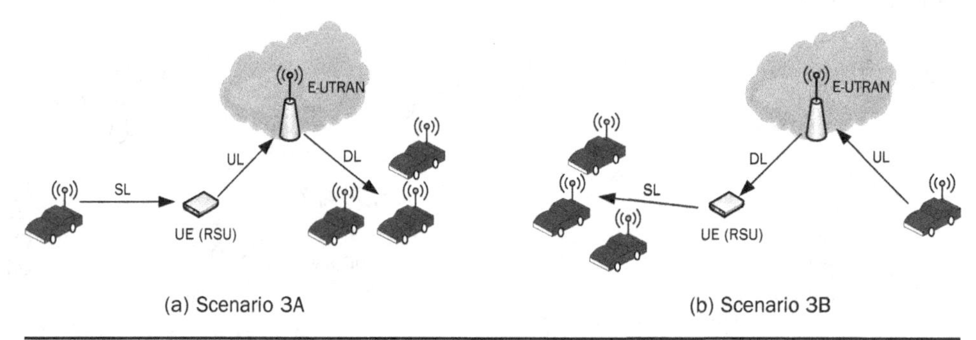

Figure 15.4 V2X scenarios via a mix of sidelink and Uu interfaces.

A large amount of documentation exists in the 3GPP library with the findings of all the studies performed around C-V2X. The following is a list of the current set of available technical reports (TRs) in this area and across 3GPP releases. Note that while TRs do not constitute specifications, they provide many details of the aspects that have been considered for various studies. The normative work on V2X has been folded into the 3GPP specifications from Release 14:

- 3GPP TR 36.885 "Study on LTE-based V2X Services; (Release 14)" [5].
- 3GPP TR 36.785 "Vehicle-to-Vehicle (V2V) services based on LTE sidelink; User Equipment (UE) radio transmission and reception; (Release 14)" [6].
- 3GPP TR 36.786 "Vehicle-to-Everything (V2X) services based on LTE; User Equipment (UE) radio transmission and reception; (Release 14)" [7].
- 3GPP TR 36.787 "Vehicle-to-Everything (V2X) new band combinations; (Release 15)" [8].
- 3GPP TR 36.788 "Vehicle-to-Everything (V2X) Phase 2; User Equipment (UE) radio transmission and reception; (Release 15)" [9].
- 3GPP TR 37.885 "Study on evaluation methodology of new Vehicle-to-Everything (V2X) use cases for LTE and NR; (Release 15)" [10].
- 3GPP TR 38.885 "Study on NR Vehicle-to-Everything (V2X); (Release 16)" [11].
- 3GPP TR 38.886 "V2X Services based on NR; User Equipment (UE) radio transmission and reception; (Release 16)" [12].
- 3GPP TR 37.985 "Overall description of Radio Access Network (RAN) aspects for Vehicle-to-everything (V2X) based on LTE and NR; (Release 16)" [13].

15.2 Background: LTE V2X

During Release 14 there were two work items dedicated to LTE-based V2X. The first project was approved shortly (6 months) after the approval in June 2015 of the RAN1-led study on LTE-based V2X focusing on V2V communications over sidelink. This had been the priority for the LTE-based V2X SI from its inception until December 2015.

The urgency to approve this project was the perceived importance for the auto industry to have a cellular-based solution for V2X. This cellular V2X or, in short, C-V2X solution would enter into direct competition with other technologies being developed for inter-car communications, most notably, DSRC based on IEEE 802.11p, and gaining favor in some regulatory bodies mainly due to the lack of an alternative technology.

In this section, we will cover the most salient features of LTE-based V2X without presenting a comparison with DSRC. The main goal is to provide some background to the later 3GPP work on NR-based V2X.

While C-V2X heavily relied on the sidelink design developed in 3GPP during Release 12 for the purpose of device-to-device or D2D communications, we will not dwell on it in the present work.

15.2.1 LTE V2X Traffic Characteristics

Early studies on the type of traffic generated for basic safety messages indicated that the traffic is of periodic nature and with a limited set of payload sizes. Also, the target recipient of this information was deemed to be every vehicle in the proximity. This realization led to a sidelink design for V2X based on broadcasting of messages.

The maximum latency associated with V2X messages was assumed to be typically 100 ms which governs the required periodicity of the sidelink transmissions. Multiple values are assumed for that, including 100 ms, 50 ms, and 20 ms.

15.2.2 Physical Channels

The Physical layer of LTE sidelink consists of two channels:

Physical sidelink control channel (PSCCH): used for carrying scheduling control information. This channel carries the sidelink control information (SCI).

Physical sidelink shared channel (PSSCH): used for carrying the sidelink data and shared among different nodes.

The V2X work focused on tailoring these two channels for vehicular communications. Compared to Rel-12 D2D, V2X communications entailed support of very high density of transmitters and support of very high relative speed.

One important aspect to consider for sidelink communications largely influencing its corresponding physical layer design is the so-called "half-duplex issue" whereby a node cannot transmit and receive at the same time. This limitation does not exist for frequency duplex division (FDD) communications where transmission and reception take place at different frequencies but is a limitation in current time division duplex (TDD) communications. Sidelink, indeed, is a form of TDD communication either on dedicated unpaired spectrum or on the UL resources of an FDD or TDD carrier.

15.2.3 Waveform and Coping with High Doppler

As stated earlier in this chapter, V2X work in 3GPP built upon the existing sidelink design which had just been standardized in 3GPP during Release 12 and Release 13 for the purpose of D2D communications. That design leveraged the UE transmissions on the UL for sidelink operation. As a direct result of that design choice, the entire physical layer sidelink design is based on the UL waveform of LTE, namely SC-FDMA, also denoted DFT-spread OFDMA.

While this single carrier waveform choice is optimized for power efficiency, i.e., it requires less power back-off, and, hence, improves the link budget for a given power amplifier, it imposes certain restrictions to maintain the single carrier property. First, data and pilot symbols are time-division multiplexed within a subframe.

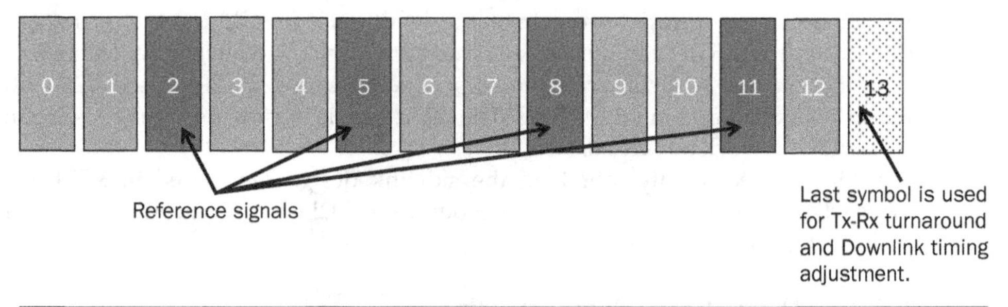

Figure 15.5 Slot structure for LTE V2X sidelink.

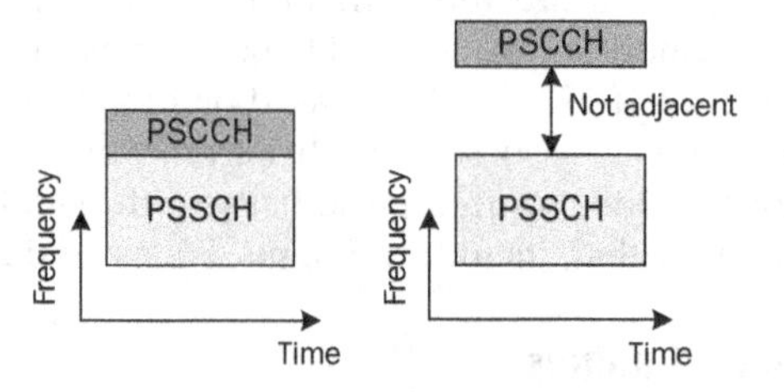

Figure 15.6 LTE PSCCH and PSSCH frequency arrangement possibilities.

V2X communications require operation at speeds of up to 250 km/h. The Doppler frequency associated with this speed is doubled when two cars are driving against each other. As stated earlier, ITS spectrum is around 5.9 GHz, which together with the 500 km/h relative speed leads to a Doppler frequency to cope with of 2.7 kHz.

This large Doppler frequency led to the need of increasing the density of pilots in the subframe, as shown in Fig. 15.5

Note that only normal cyclic prefix (CP) is supported. Intra-symbol frequency error estimation will be required to cope with the high Doppler shift and frequency error incurred from transmissions at high frequency (close to 6 GHz).

PSSCH and PSCCH are frequency division multiplexed in the same subframe as illustrated in Fig. 15.6. Separate DFTs and reference signals (two cluster SC-FDM) are assumed. Having control and data in the same subframe reduces the half duplex issue. Note that PSSCH and PSCCH may or may not be adjacent in frequency, depending on the resource pool configuration (resource pool discussed later in this section), as illustrated in Fig. 15.6.

Same open-loop power control parameters are used for both channels with a 3 dB power spectral density (PSD) boosting for PSCCH to avoid the control channel becoming the bottleneck for reliable communication.

The maximum number of transmissions is restricted to two, i.e., first transmission and retransmission. There is one PSCCH transmitted for each PSSCH (on the same subframe). Some more information on PSCCH and PSSCH follow.

15.2.3.1 LTE PSCCH

SCI format 1 for PSCCH limited to 48-bits and transmitted at coding rate 1/9. The SCI format 1 fields are summarized in Table 15.1.

There is no combining of PSCCH retransmission, as they will have different contents.

Field	Bit-width
Time gap between initial and final tx	4-bits
Retransmission index	1-bit
Frequency resource location	$[\log_2(N_{subschannel}(N_{subchannel} + 1)/2]$-bits
Data MCS	5-bits
PPPP (priority level discussed later)	3-bits
Resource reservation	4-bits
CRC	16-bits
Reserved bits	As need to achieve 48-bits

TABLE 15.1 LTE SCI Format 1 Fields

15.2.3.2 LTE PSSCH

Redundancy version (RV) ID sequence for HARQ transmissions hard coded to 0 and 2. The maximum distance between the initial transmission and the HARQ retransmission is 15 subframes.

Various physical layer parameters are a function of PSCCH CRC instead of "SA ID."

15.2.4 Resource Allocation, Scheduling (Considering In and Out of Coverage)

Resources for sidelink operation are arranged in so-called *resource pools*. The resource pools consist of a number of consecutive time-frequency resources where sidelink transmissions and receptions take place.

Resource pools are defined by offset, subframe bitmap, sub-channel and start RB index for PSCCH, number of sub-channels, sub-channel size, and, possibly, zone id, which is an optional feature discussed later.

Subframes with PSCCH/PSSCH resources indicated using a bitmap. Bitmaps repeat leading to a repeating pattern.

The possible number of sub-channels are in the set {1, 3, 5, 8, 10, 15, 20}.

- For adjacent case sub-channel sizes are {5, 6, 10, 15, 20, 25, 50, 75, 100} RBs.
- For non-adjacent case sub-channel sizes are {4, 5, 6, 8, 9, 10, 12, 15, 16, 18, 20, 30, 48, 72, 96} RBs.

The notion of sub-channel is introduced due to UE complexity–related issues. UE will select one or more sub-channels for PSSCH transmission subject to DFT size constraint, i.e., number of subcarriers in an assignment should satisfy $2^x.3^y.5^z$ for both, adjacent and non-adjacent multiplexing cases.

There is a one-to-one mapping between the number of sub-channels and the number of PSSCH resources.

The multiplexing of V2V transmissions with other signals and channels in the case where V2X transmissions take place in a frequency channel used for Uu communications is possible. In this case, the V2X transmissions are placed in resources corresponding to the UL of the Uu channel, i.e., on the UL frequency for FDD deployments and on

y \ x	0	1	2	0	1	2
0	Zone 0	Zone 1	Zone 2	Zone 0	Zone 1	Zone 2
1	Zone 3	Zone 4	Zone 5	Zone 3	Zone 4	Zone 5
2	Zone 6	Zone 7	Zone 8	Zone 6	Zone 7	Zone 8
0	Zone 0	Zone 1	Zone 2	Zone 0	Zone 1	Zone 2
1	Zone 3	Zone 4	Zone 5	Zone 3	Zone 4	Zone 5

FIGURE 15.7 Concept of LTE resource pool zones.

the UL subframes for TDD deployments. Note that a set of reserved subframes can be configured precluding sidelink transmissions on those.

Note that, if the notion of zones is used (optional feature), resource pools can be arranged with a geographical reuse, in so-called *zones*, to reduce interference from transmissions in different zones utilizing the same frequency channel for sidelink communications.

Each UE calculates the "zone id" based on its current longitude and latitude (GNSS assumed in the design of V2X). The number of zones (reuse ratio) and zone size are configured by the eNB. The zoning mechanism and parameters are the same for in-coverage (provided by eNB) and out-of-coverage (pre-configured). This is illustrated in Fig. 15.7.

15.2.5 Transmission Modes

While D2D communications created two transmission modes for operation on sidelink (mode 1 and mode 2), two transmission modes were introduced for V2X purpose:

- Mode 3: eNodeB scheduled resource selection
- Mode 4: UE autonomous resource selection

Mode 4 uses sensing with semi-persistent scheduling (SPS) transmission. Semi-persistent transmission exploits the semi-periodic traffic arrival pattern. The UE uses the past interference patterns to predict the future. Sensing consists of a combination of energy sensing, PSCCH decoding, and priority information.

The energy sensing operation ranks resources according to energy received and picks low-energy resources. By way of PSCCH decoding results, the goal is to avoid resources for which control is decoded and received energy is above a threshold. The priority information aims at avoiding resources that are being used for higher priority packet transmissions.

Note that in addition to the sidelink physical layer channels (PSCCH and PSSCH), extensions to the Downlink control information (DCI) carried by Uu's interface PDCCH are made to support mode 3 of V2X operation. In that sense, DCI format 5A is introduced for granting resources on the sidelink. After the reception of a DCI format 5A on the Uu interface, PSSCH/PSCCH transmission occurs on the first PSCCH/PSSCH configured subframes four subframes after the grant is received.

Given the importance of SPS transmissions for sidelink, the support of SPS for sidelink mode 3 and mode 4 is enhanced as follows:

- Multiple SPS can be activated simultaneously for V2X: up to eight SPS configurations per UE. All configured SPSs can be active at the same time. The legacy logical channel prioritization (LCP) procedure is not changed for V2X.
- V2X UE can send UE assistance information to eNB to enable adequate SPS configuration by the network.
    - The UE assistant information includes a set of preferred expected SPS interval, timing offset with respect subframe 0 of the SFN 0 (frame and subframe number).
    - UE assistance triggers are left to UE implementation.
    - The network should be able to configure UE assistance information.
    - DCI format 5A with SPS-V2V-RNTI is used for activation/deactivation
- In case of Uu SPS, DCI format 0 with SPS-V2V-RNTI is used for activation/deactivation

15.2.6 Synchronization

We cover the synchronization for NR in Sec. 15.3.7. Note that the same principles have been adopted for NR. Certain extensions for NR are still being discussed by 3GPP at the time of publishing this work.

15.2.7 Key Performance Indicators

V2X communications introduce the following important metrics:

- Channel busy ratio (CBR) measurement: For PSSCH, the fraction of sub-channels whose S-RSSI exceeds a (pre-) configured threshold. The measurement window $[n - 100, n - 1]$ for CBR measured at subframe n (physical subframe indexing).
- Channel occupancy ratio (CR) measurement: Fraction of sub-channels used for transmission in $[n - a, n - 1]$ and granted $[n, n + b]$ where a is positive and b is non-negative integer; $a + b + 1 = 1000$; $a >= 500$ and $n + b$ should not exceed the last transmission opportunity of the grant for the current transmission.

CBR is measured and CR is evaluated for each (re-)transmission.

- For a (re-)transmission in subframe $n + 4$, the CR is evaluated in subframe n.
- For a (re-)transmission in subframe $n + 4$, the CBR measured in subframe n is used.

15.2.7.1 Congestion Control

Congestion control is introduced so that UEs limit resource utilization (to a configured value) based on channel busy ratio (CBR).

15.2.7.2 V2P

Support for pedestrian-to-vehicle communications poses different challenges, most notably, UE power consumption from monitoring or transmitting on the sidelink.

Indeed, early on, it was realized that the power required for transmitting messages versus for monitoring them is substantially different.

In an attempt to viably support V2P communications, some enhancements were introduced to the V2X framework. Examples of that are the use of partial sensing for reduced power consumption, and the possibility for pedestrian UEs (P-UEs) to use random selection of transmit resources enabling the possibility of having P-UEs without RX chain for sidelink.

Cross-carrier scheduling is also supported enabling eNB in commercial carrier frequencies to schedule V2X in ITS carrier using sidelink mode 3.

The prioritization between Uu and V2X transmissions (in case of conflict) is based on the ProSe per packet priority (PPPP) prioritization. V2X is prioritized if the PPPP of a packet is above a configured threshold, otherwise UL is prioritized. Note that RACH and emergency call transmissions on Uu interface are always prioritized compared to sidelink.

15.2.8 UE Capabilities

Certain processing complexity limitations for sidelink operation are set forth in the specifications according to different possible UE capabilities. In summary, the maximum number of PSCCH decodings in a subframe (denoted by X) as well as the maximum number of RBs that a UE is expected to process for PSCCH and PSSCH (denoted by Y) are considered. Two UE capabilities are defined: (X = 10, Y = 100) and (X = 20, Y = 136).

The maximum number of sidelink transport block bits received within a TTI is set to 31,704 bits. Note that all those bits could come from a single transport block, i.e., the maximum number of bits in a sidelink transport block is 31,704 bits. The total sidelink soft buffer size is 737,280 bits.

15.2.9 V2X Enhancements in Release 15

The following enhancements were introduced for LTE V2X in Release 15:

- Carrier aggregation (up to eight PC5 carriers)
- Higher order modulation (64QAM)
- Latency reduction
- Radio resource pool sharing between UEs using mode 3 and UEs using mode 4.

15.2.10 Evaluation Methodology for New V2X Use Cases

The same 3GPP RAN Plenary meeting (RAN#75, March 2017) when the work item on LTE-based V2X Phase 2 was approved [14], a study on evaluation methodology of new V2X use cases for LTE and NR [15] was also approved with the goal to start bridging the roadmap of the cellular-based V2X from LTE to NR. A technical report for this study can be found in Ref. [10].

15.3 NR V2X

A Release 16 study item was approved in June 2018 [16]. Much discussion took place at the time to find a complementary role for NR V2X vis-à-vis LTE V2X to avoid market

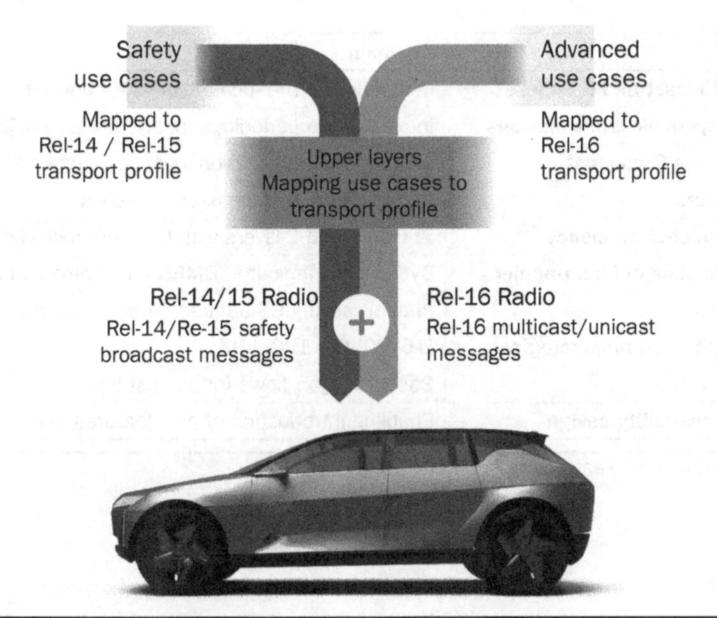

FIGURE 15.8 NR and LTE V2X complementarity.

confusion and having 3GPP defining two technologies for the same use case. As a result, it was broadly accepted that LTE V2X would be used for basic safety messages, while NR V2X would be designed for advanced use cases such as sensor sharing to assist autonomous driving and requiring higher data rates, higher reliability, and lower latency with an arbitrary traffic pattern in the sidelink. Figure 15.8 illustrates how LTE V2X and NR V2X complement each other.

Note that (pre)configuration provides application ID to RAT mapping. Also, the NR and LTE specifications provides support for

- NR Uu controlling LTE sidelink, and
- LTE Uu controlling NR sidelink.

The outcome of the SI on NR V2X is captured in the technical report in Ref. [11] and the subsequent work item was approved in Ref. [17].

15.3.1 Basics

Now that we have introduced the LTE V2X operation at high level, we examine NR V2X operation and see how NR V2X complements LTE V2X. The main new features that NR V2X supports when compared to LTE V2X are summarized in Table 15.2.

There are two feedback modes for Groupcast:

- Groupcast Option 1: Receiver UE transmits only NACK. This is the mode used for NACK-only distance-based feedback.
- Groupcast Option 2: Receiver UE transmits ACK or NACK.

Feature	Comments
Support of Unicast and Groupcast	In addition to Broadcast. HARQ for unicast and groupcast
Support of aperiodic transmissions	In addition to periodic, support "per packet" scheduling
Connection-less Groupcast	Enabled with application-aware, distance-based feedback
OFDM waveform	Unlike LTE V2X which uses SC-FDMA
Increased spectral efficiency	256QAM and 2-layers with CSI from receiver
Improved handling of high Doppler	Dynamically indicated DMRS time-domain density
Support of FR2	Phase tracking RS defined for sidelink operation
Support of different numerologies	{15, 30, 60, 120} kHz
Unified QoS model	Based on QoS flows for all cast types
Forward compatibility design	Enabling introduction of new features in the future

Table 15.2 Main New Features of NR V2X

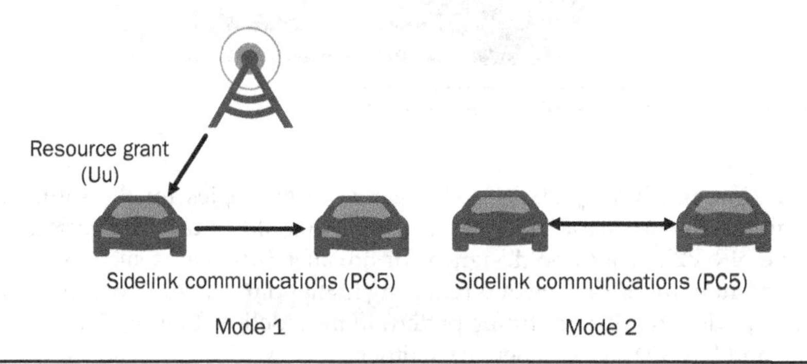

Figure 15.9 Illustration of NR V2X resource allocation modes.

Similar to LTE V2X, in NR V2X there are two resource allocation modes for the sidelink illustrated in Fig. 15.9:

- **Mode 1**: gNB allocates resources for sidelink (what in LTE V2X was mode 3)
- **Mode 2**: UEs autonomously select sidelink resources (what in LTE V2X was mode 4)

In either case, the signaling on sidelink is the same. In other words, from the receiver point of view, there is no difference between mode 1 and mode 2.

Communications on sidelink take place, as in LTE V2X, in transmission and reception *resource pools*. The resource allocation in time is one slot, the minimum resource allocation in frequency is a sub-channel. The sub-channel size can be (pre)configured to {10, 15, 20, 25, 50, 75, 100} PRBs. The resource allocation can be either pre-configured or gNB configured.

NR sidelink supports HARQ-based retransmission for unicast and groupcast. Note that in LTE, transmissions are broadcast and there is a limit of one re-transmission and there is no HARQ-ACK feedback defined.

On sidelink, the transmitting UE embeds control information facilitating the reception of the accompanying data transmission at the receiver. This control information is called sidelink control information (SCI). The SCI in NR is transmitted in two stages. While the first stage of SCI (SCI-1) is transmitted on a dedicated physical layer channel (PSCCH), the second stage (SCI-2) is embedded in the sidelink data channel (PSSCH). In the next section, we will cover the physical layer channels and signals introduced for NR sidelink operation.

For congestion control, NR sidelink supports channel busy ratio (CBR) as a metric of congestion. Sidelink RSSI is used to estimate CBR.

Congestion control can restrict the following transmission parameters: MCS indices and tables, number of sub-channels per transmission, number of re-transmissions, transmission power. Note that absolute UE speed can also restrict transmission parameters.

15.3.2 Physical Layer Structure

NR V2X defines the following physical layer channels and signals:

15.3.2.1 Physical Sidelink Channels

- Physical sidelink control channel (PSCCH) – carries SCI-1
- Physical sidelink shared channel (PSSCH) – carriers SCI-2 and sidelink data payload
- Physical sidelink feedback channel (PSFCH) – carries 1-bit HARQ-ACK feedback
- Physical sidelink broadcast channel (PSBCH) – equivalent of PBCH in sidelink

Note that the channel state information feedback for sidelink is supported for unicast traffic. The CSI is transmitted over the sidelink in the form of a MAC-CE with 1-bit for rank and 4-bits for CQI, i.e., no PMI information in Release 16.

15.3.2.2 Reference Signals

- Demodulation RS (DMRS) for PSCCH, PSSCH and PSBCH
- Phase tracking RS (PTRS) - for FR2 phase noise compensation
- Channel state information RS (CSI-RS) – enabling CSI measurement on sidelink
- Sidelink primary synchronization signal (S-PSS) – equivalent of PSS in sidelink
- Sidelink secondary synchronization signal (S-SSS) – equivalent of SSS in sidelink

15.3.2.3 Sidelink Slot Format

As in regular NR, sidelink slots consist of 14 OFDM symbols. Note that sidelink can be (pre)configured to occupy less than 14 symbols in a slot.

Depending on whether or not feedback is configured for a given slot, we have different slot formats. Figure 15.10 shows the slot format for the case without feedback resources configured. Figure 15.11, in turn, shows the slot format for the case with configured feedback resources.

In both cases, PSCCH and PSSCH are transmitted in the same slot. Note that, unlike LTE V2X, where PSCCH and PSSCH were purely FDM'd, in NR, PSCCH is transmitted

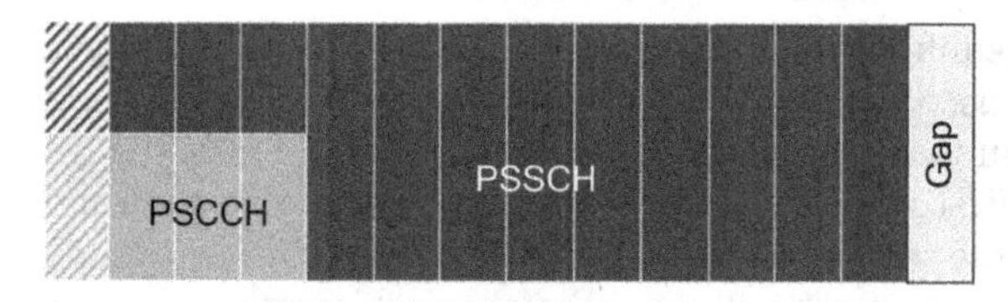

FIGURE 15.10 Illustration of slot format without feedback.

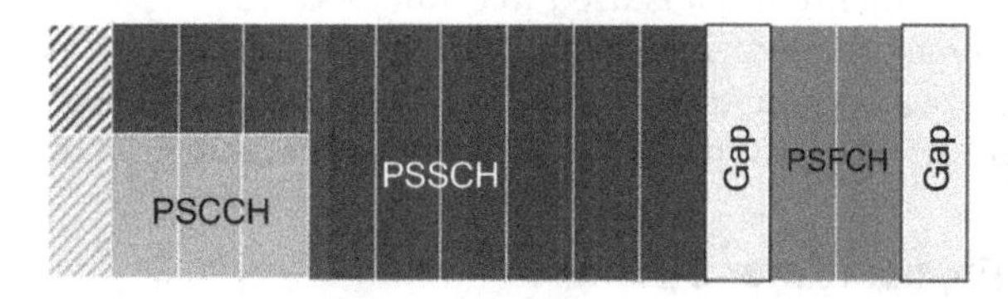

FIGURE 15.11 Illustration of slot format with feedback.

in the early part of the slot (three or four symbols including the repeated symbol) to enable pipelining control and data processing at the receiver.

Also, for both cases, the first symbol is repeated for AGC settling. Also, the last symbol of the slot is left as a gap so that the terminal has time to switch from transmitting to receiving.

For the case with feedback, another symbol gap is introduced before the PSFCH. In turn, the PSFCH symbol is also repeated for AGC settling, resources for PSFCH can be configured with a period of {0, 1, 2, 4} slots.

15.3.2.4 Sidelink Control Information

As discussed earlier, SCI is transmitted in two stages. The main motivation for this is to facilitate forward compatibility.

- First stage control (SCI-1) is transmitted on PSCCH and contains information for resource allocation and decoding second stage control.
- Second stage control (SCI-2) is transmitted on PSSCH and contains information for decoding data (SCH).

SCI-1 will be decodable by UEs in all releases, whereas new SCI-2 formats can be introduced in future releases. This ensures that new features can be introduced while avoiding resource collisions between releases. Note that, while the design of Uplink control information (UCI) in PUSCH was considered in the design of SCI-2, SCI-2 has its own characteristics. As will be discussed later, SCI-2 has separate scrambling than data, there is no interlacing of SCI-2 and data transmissions, and SCI-2 is transmitted at fixed QPSK modulation order with variable "beta" value to control the coding rate for transmission, irrespective of the modulation order of the data transmission.

Both SCI-1 and SCI-2 use the PDCCH polar code (see Chap. 7).

15.3.3 Resource Allocation Modes

Resource allocation is *reservation based* in NR sidelink. The resource allocation is in units of sub-channels in frequency domain and limited to one slot in time domain.

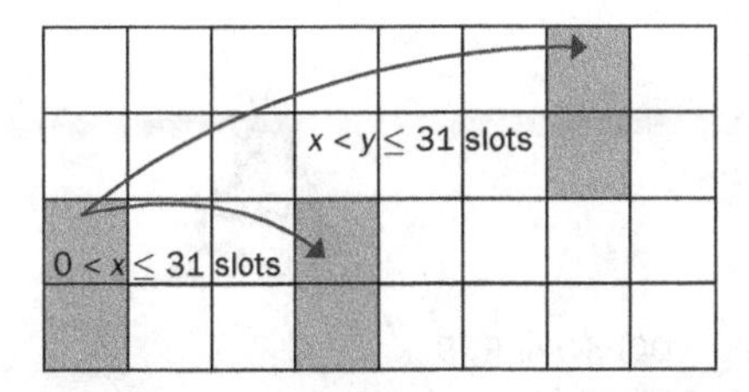

FIGURE 15.12 Illustration of resource reservation.

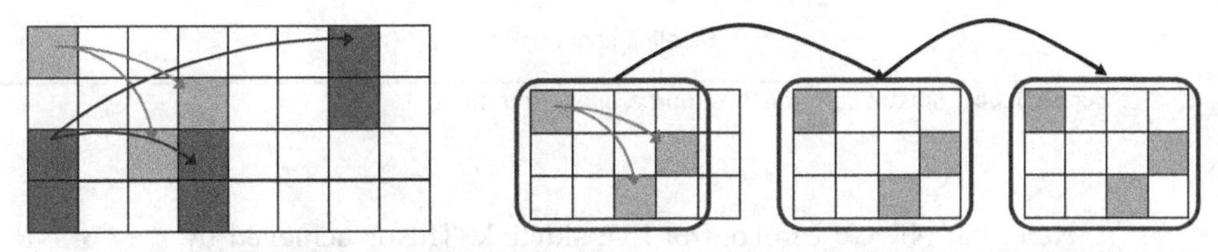

FIGURE 15.13 Aperiodic (left) and periodic (right) resource reservation.

Reservations are in a window of 32 logical slots, as illustrated in Fig. 15.12. A transmission reserves resources in the current slot and in up to two future slots (slot x and slot y in Fig. 15.12). Therefore, the maximum total number of reservations (Nmax) can be configured to be two or three. Reservation information is carried in SCI.

Aperiodic and periodic resource reservations are supported, as shown in Fig. 15.13. A period, with configurable values between 0 ms and 1000 ms, can be signaled in SCI. Periodic resource reservation and signaling can be disabled by (pre)configuration.

15.3.3.1 Mode 1

In resource allocation mode 1, the gNB manages the sidelink resources. Figure 15.14 illustrates, conceptually, resource allocation mode 1.

The sidelink data allocation can be done via dynamic grants (DG) for aperiodic traffic, or via configured grants (CG) of type 1 (RRC configured) or type 2 (DCI activated) for periodic traffic.

Over the Uu interface, the gNB provides the sidelink grant via DCI format 3-0 over PDCCH. This DCI can be a DG providing the allocation to use over sidelink or can activate a CG type 2 for sidelink. For DG, DCI format 3-0 indicates whether the allocation is for a retransmission.

In mode 1, the UE reports the buffer status report (BSR) to the gNB using a MAC-CE. The MCS selection is up to the UE, within limits set by the gNB. The UE can report sidelink feedback to the gNB over PUCCH resources assigned by the gNB.

For sidelink open-loop power control, a UE can be (pre)configured to use DL pathloss (between transmitting UE and gNB) only, sidelink pathloss (between transmitting UE and receiving UE) only, or both DL pathloss and sidelink pathloss. When both Downlink pathloss and sidelink pathloss are used, the minimum of the power values given by open-loop power control based on DL pathloss and the open-loop power control based on sidelink pathloss is taken.

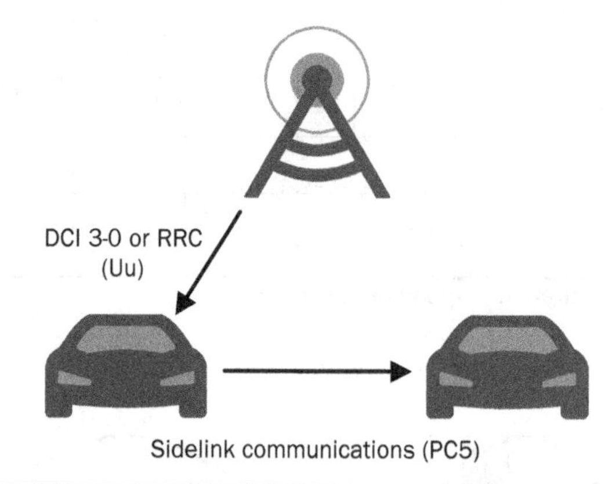

Figure 15.14 Illustration of NR resource allocation mode 1.

Note that NR Uu can control LTE sidelink. This is achieved by way of NR DCI format 3-1, which activates/deactivates the LTE sidelink and carries the LTE sidelink resource grant, i.e., LTE DCI format 5A.

15.3.3.2 Mode 2

Unlike mode 1 where the resource allocation is centralized at the gNB, for mode 2 the transmitting UE will have to autonomously select the resources for sidelink communication.

Resource selection in mode 2 consists of two steps:

1. Identification of candidate resources by *sensing* and *exclusion*
2. Candidate resource selection from the identified resources (random selection)

Reserved resources can be preempted by a higher priority reservation. At that point, the preempted UE can reselect resource if enabled by configuration. A selected, but not yet reserved, resource could be reserved by another UE. In that case, the resource selection procedure is triggered again.

Candidate resource identification (sensing):
A UE determines whether a resource is available or not by decoding SCI. The UE is measuring received power (RSRP) for reservations in decoded SCIs. The RSRP of the transmission associated with SCI reserving resources is projected onto the resource selection window, as illustrated in Fig. 15.15. RSRP is measured on PSCCH or PSSCH DMRS according to (pre)configuration. The length of the sensing window (where SCI is decoded) is (pre)configured. Each reservation has a priority indicated in SCI that is also tracked as part of the sensing.

Candidate resource identification (exclusion):
Resources associated with RSRP below a threshold are considered available. RSRP comparison threshold is (pre)configured per transmitter priority and receiver priority pair. If proportion of available resources in selection window is <20%, RSRP threshold is increased and process is repeated. Available resources in the selection window form the *candidate resource set*. The candidate resource set is reported to higher layers.

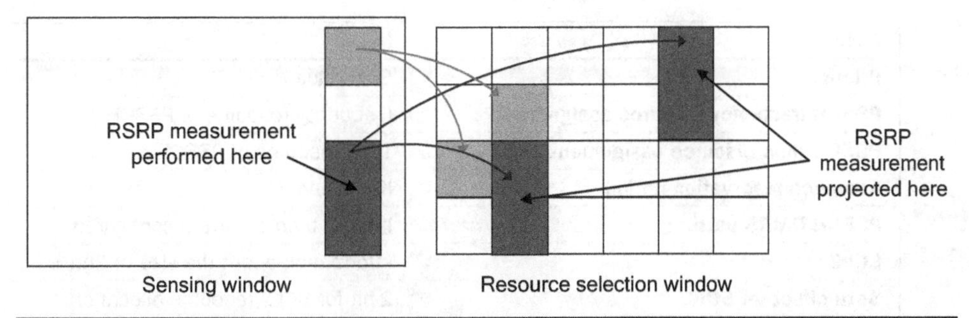

FIGURE 15.15 Illustration of mode 2 candidate resource identification (sensing).

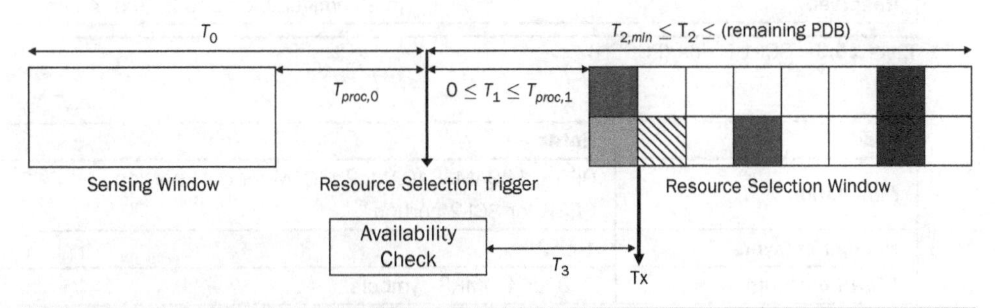

FIGURE 15.16 Resource selection timeline for mode 2.

Resources are selected such that all retransmissions for a packet must occur within its delay budget (PDB).

Resource selection timeline:
Figure 15.16 illustrates the resource selection timeline for mode 2. The following time durations are configured or pre-configured:

T_0: (pre-)configured to 100 ms or 1100 ms
$T_{2,min}$: (pre-)configured per priority $\{1, 5, 10, 20\}2^{\mu}$

Note that T_1 is up to UE implementation. A UE must check that the resource is still available T_3 before transmission.

15.3.4 Physical Sidelink Control Channel (PSCCH)

Time and frequency span of PSCCH is (pre)configured:

- Time span: two or three symbols from the beginning of the slot (in addition to the repeated symbol at the beginning of the slot)
- Frequency span: {10, 12, 15, 20, 25} PRBs and limited to be within a single sub-channel

DMRS is present in every PSCCH symbol and is placed on every 4th subcarrier (as in PDCCH). Note that a frequency-domain orthogonal cover is applied to the DMRS to

Field	Notes
Priority	QoS value
PSSCH frequency resource assignment	Frequency resource of PSSCH
PSSCH time resource assignment	Time resource of PSSCH
Resource reservation period	If enabled
PSSCH DMRS pattern	If more than 1 pattern configured
SCI-2 format	Information about the size of 2nd SCI
Beta offset for SCI-2	2-bit for SCI-2 resource allocation
Number of PSSCH DMRS ports	1 or 2
MCS	5-bits
Reserved	(pre)configurable up to 2, 3, or 4 bits

Table 15.3 SCI-1 Fields (PSCCH)

Field	Notes
Modulation	QPSK, 16QAM, 64QAM, 256QAM for data portion QPSK for SCI-2 portion
Number of layers	1 or 2
DMRS patterns	2, 3, or 4 DMRS symbols

Table 15.4 PSSCH Characteristics

reduce impact of colliding PSCCH transmissions from different UEs. The transmitter UE randomly selects from a set of predefined orthogonal covers.

Table 15.3 shows the fields of SCI-1 carried over PSCCH.

15.3.5 Physical Sidelink Shared Channel (PSSCH)

Table 15.4 shows the main characteristics of the PSCCH for NR and Fig. 15.17 illustrates the possible DMRS patterns for the slot duration of 14 symbols.

As discussed earlier, SCI-2 is transmitted as part of PSSCH starting from the first symbol with PSSCH DMRS and mapped to contiguous RBs. SCI-2 is scrambled separately from the data part. It uses QPSK modulation irrespective of the modulation of the data portion. There is no associated blind decoding to obtain the SCI-2, the format is indicated in SCI-1, as shown in Table 15.3. The number of REs for SCI-2 is also derived from SCI-1 information. When the data transmission is performed over two layers, SCI-2 modulation symbols are copied on both layers.

Table 15.5 shows the SCI-2 fields. The final details of SCI-2 were not available at the time of publishing. However, the intention is to have a set of fields constituting the general design followed by possibly specialized field depending on the situation. Note that SCI-2 design is meant to be forward compatible enabling modifications and extensions in future releases.

15.3.6 Physical Sidelink Feedback Channel (PSFCH)

Feedback resources are system wide (pre)configured with period N = {1, 2, 4} slots. As shown in Fig. 15.11, PSFCH occupies three OFDM symbols if configured. Two symbols

FIGURE 15.17 DMRS patterns for PSSCH.

Field	Notes
General design:	
HARQ process ID	Fields used to determine new tx or re-tx
NDI	
Source ID	
Destination ID	
CSI reporting trigger	Only applicable to unicast
For groupcast feedback Option 1:	
Zone ID	For groupcast feedback Option1 (NACK-only, distance-based feedback)
Max comm range for sending feedback	
...	

TABLE 15.5 SCI-2 Fields

for the actual duplicated transmission and one for gap. No other channels will be frequency multiplexed with PSFCH on those symbols because, due to AGC variations, those channels would be corrupted. Based on priority, a given UE decides to either receive or transmit PSFCH due to half duplex limitation. Note that a UE can perform N simultaneous PSFCH transmissions in a symbol for different UEs or groups. UE selects N PSFCH based on priority of the corresponding PSCCH/PSSCH.

The total number of available PSFCH resources is (pre)configured. PUCCH Format 0 on one RB carries HARQ-ACK information for a single PSSCH transmission (see Chap. 9).

As discussed earlier, PSFCH can be enabled for unicast and groupcast. For unicast, PSFCH carries 1 bit ACK/NACK. For groupcast, there are two PSFCH feedback modes as also discussed earlier:

- PSFCH for groupcast feedback mode Option 1: 1 bit NACK (only)
- PSFCH for groupcast feedback mode Option 2: 1 bit ACK/NACK

There is an implicit mapping between the actual PSSCH and its corresponding PSFCH resource. The mapping is based on the starting sub-channel of PSSCH, the slot containing PSSCH, the source ID, and the destination ID.

The number of available PSFCH resources must be equal to or greater than the number of UEs in groupcast option 2.

Distance-based feedback can be enabled for groupcast feedback option 1. In that case, a receiver UE within communication range sends NACK if failed PSSCH decoding. As seen in Table 15.5, SCI-2 communicates the maximum communication range as an index to the one of the following preconfigured set of values: {20, 50, 80, 100, 120, 150, 180, 200, 220, 250, 270, 300, 350, 370, 400, 420, 450, 480, 500, 550, 600, 700, 1000} meters.

The distance between the transmitter and the receiver is computed from the transmitter indicated zone (center of the zone used) and the receiver own location. The transmitting UE indicates its zone ID in SCI-2 and the receiving UE is assumed to know its own location and zone ID. Zones are square with dimensions (pre)configured from {5, 10, 20, 30, 40, 50} meters.

15.3.7 Synchronization

The synchronization mechanism is de-coupled from the communications. The receiving UEs communicating with a transmitting UE need not derive its time/frequency synchronization from the transmitting UE. The following sources of synchronization are available: GNSS, eNB/gNB, and SyncRef UEs (synchronization reference UEs). Figure 15.18 illustrates the various sources of synchronization for sidelink.

The procedure for signaling, identifying priority for one or more synchronization references, and selecting the synchronization reference from LTE [18] is reused for NR sidelink.

The synchronization source of choice is based on priorities in a set of predefined rules. Three predefined rules are possible with one being configured or pre-configured. Table 15.6 provides the set of synchronization predefined rules.

The synchronization priority of a detected S-SSB is determined from sidelink sync signal (SLSS) ID and coverage status indication in PSBCH. SLSS ID and coverage status indication together indicate synchronization priority of a detected S-SSB.

15.3.7.1 S-SSB Structure

Similar to the Uu interface (see Chap. 6), the sidelink SSB (S-SSB) block consists of S-PSS, S-SSS, and PS-BCH. Unlike the four OFDM symbols of the Uu SSB/PBCH block, S-SSB takes an entire slot as illustrated in Fig. 15.19 for normal CP. Table 15.7 summarizes the main characteristics of the S-SSB signals and channels.

PS-BCH has a payload of 56-bits with contents as shown in Table 15.8. Note that the 12-bits of TDD configuration indication provide the system-wide information, e.g., TDD UL-DL common configuration and/or potential sidelink slots.

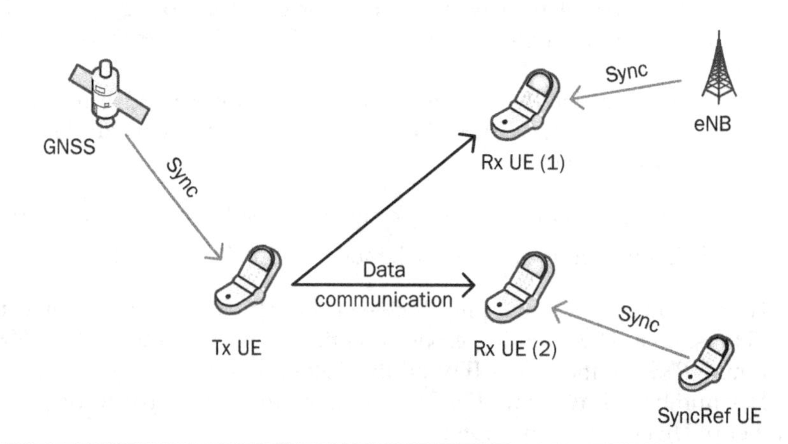

FIGURE 15.18 Illustration of the various sources of synchronization for sidelink.

GNSS-based sync Case 1	GNSS-based sync Case 2	gNB/eNB-based sync
P0: GNSS	**P0**: GNSS	**P0'**: gNB/eNB
P1: UE directly sync'd to GNSS	**P1**: UE directly sync'd to GNSS	**P1'**: UE directly sync'd to gNB/eNB
P2: UE indirectly sync'd to GNSS	**P2**: UE indirectly sync'd to GNSS	**P2'**: UE indirectly sync'd to gNB/eNB
P6: remaining UEs	**P3**: gNB/eNB	**P3'**: GNSS
	P4: UE directly sync'd to gNB/eNB	**P4'**: UE directly sync'd to GNSS
	P5: UE indirectly sync'd to gNB/eNB	**P5'**: UE indirectly sync'd to GNSS
	P6: remaining UEs	**P6'**: remaining UEs

Table 15.6 Synchronization Predefined Rules

PS-BCH	S-PSS	S-PSS	S-SSS	S-SSS	PS-BCH	PS-BCH	PS-BCH	PS-BCH	PS-BCH	PS-BCH	PS-BCH	PS-BCH	Gap

Figure 15.19 S-SSB structure for normal CP.

Channel	Description
S-PSS	Length 127 M-sequence and same generator/initial value as Uu PSS with cyclic shifts {22, 65}, repeated on 2 consecutive symbols
S-SSS	Length 127 Gold sequence and same generator/initial value/cyclic shifts as Uu SSS, repeated on 2 consecutive symbols
PS-BCH	11 PRBs and 9 or 7 OFDM symbols for normal CP or extended CP, respectively DMRS in every PS-BCH symbol with frequency density on every 4th subcarrier First symbol in slot can be used for AGC settling

Table 15.7 S-SSB–Related Information

Field	# bits
DFN	10
TDD config indication	12
Slot idx	7
In-coverage indicator	1
Reserved	2
CRC	24

Table 15.8 PS-BCH Payload

15.3.8 Coexistence with LTE Sidelink

FDM solutions for coexistence:

- Both intra-band and inter-band are considered feasible.
- Inter-band band solutions with static power assignment do not assume synchronization between LTE sidelink and NR sidelink.
- Intra-band and inter-band with dynamic power sharing solutions require synchronization between LTE sidelink and NR sidelink.

TDM solutions for coexistence:

- Long-term time-scale solutions where LTE and NR sidelink potential transmissions are statically determined are possible, but not preferred due to impact on latency, reliability, and data rate.
- Short time-scale coordination where LTE and NR sidelink transmissions are known to each RAT is possible and requires synchronization between the two RATs.

Prioritization between LTE sidelink and NR sidelink:

- LTE Rx/NR Rx coexistence is feasible: up to UE implementation how to manage overlap.
- LTE Tx/NR Tx overlap: If the packet priorities for both are known sufficiently in advance, the packet with higher priority is transmitted. Otherwise, it is up to UE implementation how to manage overlap.
- LTE Rx/NR Tx and LTE Tx/NR Rx overlap: If packet priorities for both are known sufficiently in advance, the packet with higher priority is transmitted/received. Otherwise, it is up to UE implementation how to manage overlap.

Sidelink SSB priority is (pre)configured for the UE and is used to manage overlap

References

[1] Qualcomm, "V2X Communication in 3GPP," S1-144374, 3GPP SA WG1#68, November 2014.
[2] LGE, "Introducing a New Set of Features for V2X over LTE," S1-144335, 3GPP SA WG1#68, November 2014.
[3] LGE, "Proposed study on LTE-based V2X," S1-150284, 3GPP SA WG1#69, February 2015.
[4] LG Electronics, CATT, Vodafone, Huawei, "New SI proposal: Feasibility Study on LTE-based V2X Services," RP-151109, 3GPP RAN#68, June 2015.
[5] 3GPP, "3rd Generation Partnership Project; Technical Specification Group Radio Access Network; Study on LTE-based V2X Services; (Release 14)," 3GPP TR 36.885 V14.0.0, July 2016.
[6] 3GPP, "3rd Generation Partnership Project; Technical Specification Group Radio Access Network; Vehicle to Vehicle (V2V) services based on LTE sidelink; User Equipment (UE) radio transmission and reception; (Release 14)," 3GPP TR 36.785 V14.0.0, October 2016.

[7] 3GPP, "3rd Generation Partnership Project; Technical Specification Group Radio Access Network; Vehicle-to-Everything (V2X) services based on LTE; User Equipment (UE) radio transmission and reception; (Release 14)," 3GPP TR 36.786 V14.0.0, March 2017.

[8] 3GPP, "3rd Generation Partnership Project; Technical Specification Group Radio Access Network; Vehicle-to-Everything (V2X) new band combinations; (Release 15)," 3GPP TR 36.787 V15.0.0, July 2018.

[9] 3GPP, "3rd Generation Partnership Project; Technical Specification Group Radio Access Network; Vehicle-to-Everything (V2X) Phase 2; User Equipment (UE) radio transmission and reception; (Release 15)," 3GPP TR 36.788 V15.0.0, July 2018.

[10] 3GPP, "3rd Generation Partnership Project; Technical Specification Group Radio Access Network; Study on evaluation methodology of new Vehicle-to-Everything (V2X) use cases for LTE and NR; (Release 15)." 3GPP TR 37.885 V15.3.0, June 2019.

[11] 3GPP, "3rd Generation Partnership Project; Technical Specification Group Radio Access Network; NR; Study on NR Vehicle-to-Everything (V2X) (Release 16)," 3GPP TR 38.885 V16.0.0, March 2019.

[12] 3GPP, "3rd Generation Partnership Project; Technical Specification Group Radio Access Network; V2X Services based on NR; User Equipment (UE) radio transmission and reception; (Release 16)," 3GPP TR 38.886 V0.5.0, February 2020.

[13] 3GPP, "3rd Generation Partnership Project; Technical Specification Group Radio Access Network; Overall description of Radio Access Network (RAN) aspects for Vehicle-to-everything (V2X) based on LTE and NR; (Release 16)," 3GPP TR 37.985 V1.1.0, February 2020.

[14] Huawei, CATT, LG Electronics, HiSilicon, China Unicom, "New WID on 3GPP V2X Phase 2," RP-170798, 3GPP TSG RAN#75, March 2017.

[15] LG Electronics, "New SI proposal: Study on evaluation methodology of new V2X use cases for LTE and NR," RP-170837, 3GPP TSG RAN#75, March 2017.

[16] Vodafone, "New SID: Study on NR V2X," RP-181480, 3GPP TSG RAN#80, June 2018.

[17] LG Electronics, Huawei, "New WID on 5G V2X with NR sidelink," RP-190766, 3GPP TSG RAN#83, March 2019.

[18] 3GPP, "3rd Generation Partnership Project; Technical Specification Group Radio Access Network; Evolved Universal Terrestrial Radio Access (E-UTRA); Radio Resource Control (RRC); Protocol specification (Release 15)," 3GPP TS 36.331, V15.8.0, January 2020.

CHAPTER 16

Vertical Expansion: Broadcast and Multicast

Cellular systems are built to support unicast communication wherein the network sends/receives data to/from individual UEs. But there are cases where identical content needs to be delivered to multiple devices, and in these cases, there is a promise of improved system efficiency by allowing multiple devices to receive the same transmission. In order to take advantage of this, broadcast delivery was designed for the following cases in LTE:

- System information delivery
- Emergency alert message delivery, such as commercial mobile alert system (CMAS) and earthquake and tsunami warning system (ETWS)
- Linear broadcast content delivery, multimedia broadcast multicast system (MBMS), which can be delivered via multicast broadcast single frequency network (MBSFN) or single cell point-to-multipoint (SC-PTM) mechanisms
- Push to talk (PTT) group communication services for commercial and public safety uses, delivered via single cell point-to-multipoint (SC-PTM) mechanism

Note that all the above functionalities were specified in LTE but only system information delivery and emergency alert message delivery have been widely deployed. In NR, so far only system information delivery and emergency alert message delivery have been specified.

The main difference between MBSFN and SC-PTM is in the assumption they make regarding the use of multicell single frequency network (SFN) transmission. MBSFN assumes the use of SFN, while SC-PTM does not.

SFN transmission means that multiple transmission points, e.g., base stations, transmit identical signals on the same frequency at the same time. The receiver sees a linear combination of these signals, which it can decode without attempting to distinguish individual transmissions from individual transmission points. It has been always seen an advantage of OFDM transmission schemes that the individual constituent transmissions of SFN add to the combined signal power, without increasing interference. In order to achieve this, the cyclic prefix (CP) utilized in the transmission has to be sufficiently long to not only cover the delay spread of the channel but also cover the worst case of the propagation delay differences from the different transmission points to any of the intended receivers. This means that a new CP length and a new subframe structure needed to be introduced in LTE [1]. The new subframe type was called MBSFN subframe. We'd like to point out the difference between eMBMS and MBSFN. eMBMS is the name of the broadcast service system, which includes all the necessary physical layer and higher layer components. MBSFN is the physical layer subframe structure used for broadcast services. MBSFN subframes can be used for other purposes in LTE as well, for example, for CRS overhead reduction in conjunction with unicast transmission modes 9 and 10.

As it was mentioned earlier, MBSFN increases the CP length. This in turn increases the overhead, where overhead is defined as the CP length divided by the sum of the CP length and the OFDM symbol length. In order to maintain the same level of efficiency, the OFDM symbol length should be increased proportionally by the same factor as the CP length increase. For example, if the CP length is doubled, the OFDM symbol length should also be doubled in order to maintain the same nominal efficiency. Increasing the OFDM symbol length changes the subcarrier spacing (SCS), which has fundamental impact on FFT size, FDM multiplexing capability, reference signal structure, resilience to channel Doppler, among other things.

The corresponding changes to the MBSFN structure were introduced in multiple steps in LTE, as summarized in Table 16.1.

An important consideration has been the maximum ratio of MBSFN subframes shown in Table 16.1. The values 60% and 80% indicate the cases that can only support mixed unicast/broadcast transmission. For these cases, four or two subframes are reserved for unicast in every frame of 10 ms. The value 97.5% indicates the cases that can also support dedicated broadcast-only mode. The 97.5% ratio is the result of designating one out of 40 subframes as cell acquisition subframe (CAS). This aspect will be discussed later in this section.

In Table 16.2, common assumptions for the various broadcast scenarios are listed. For the various scenarios, one of the following UE antenna options was assumed:

- Mobile handset (height 1.5 m)
- Rooftop antenna (height 10 m)
- Car-mounted antenna (height 1.5 m)

Target scenario name	CP length (μs)	OFDM symbol length (μs)	CP ratio (%)	SCS (kHz)	Maximum ratio of MBSFN subframes (%)	Release introduction
Low power, low tower	16.6675	66.667	20	15	60 or 80	9
	33.335	133.333	20	7.5	100	N/A (in Release 9 but incomplete)
Medium power, medium tower	200	800	20	1.25	80 or 97.5	14
High mobility	33.335	133.333	20	7.5	80 or 97.5	14
High power, high tower	300	2700	10	0.37	97.5	16
High mobility (250 km/h)	100	400	20	2.5	80 or 97.5	16

Table 16.1 Summary of LTE MBSFN Options

Target scenario name	Maximum ISD (km)	Base station power (dBm)	Base station antenna height (m)	Base station antenna gain (dBi)	Base station sectorization
Low power, low tower	15	46	35	15	3 sectors
Medium power, medium tower	50	60	100	10.5	omni
High power, high tower	125	70	300	13	omni

TABLE 16.2 Summary of Common Assumptions for Broadcast Scenarios

In Figs. 16.1 and 16.2, we give illustrations of the various MBSFN subframe structures.

For any terrestrial communication system, including broadcast, a mechanism is needed to support initial acquisition. For the cases supporting mixed unicast and broadcast, it is a natural choice to use the existing unicast initial acquisition signals. For the cases of broadcast only, there is a choice to be made between using the existing signal and designing a new initial acquisition signal. The benefit of using the existing signal is of course simplicity, while the drawback is that the existing initial acquisition signal will not benefit from SFN gains; therefore, its SNR will be lower. After extensive evaluations, it

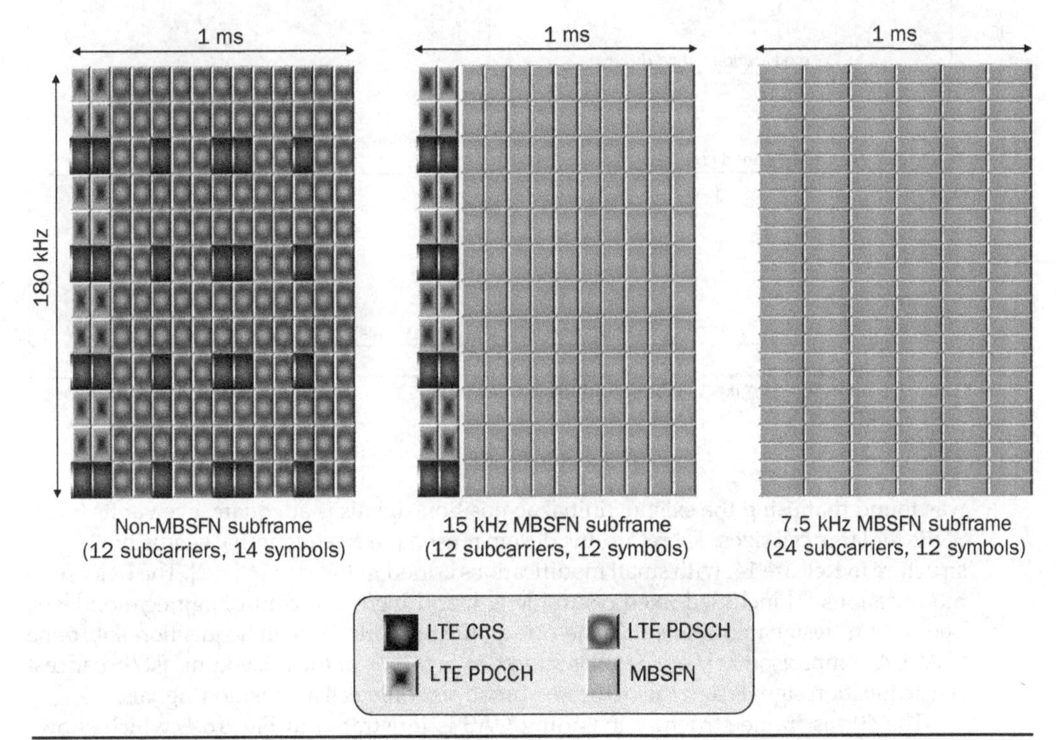

FIGURE 16.1 Non-MBSFN subframe and MBSFN subframe structures for 15 kHz and 7.5 kHz SCS.

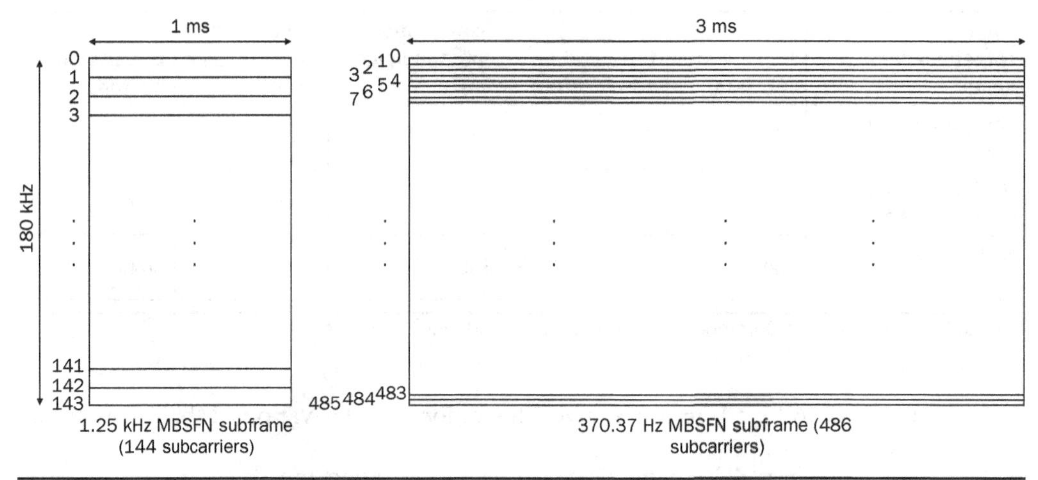

FIGURE 16.2 MBSFN subframe structures for 1.25 kHz and 370.37 Hz SCS.

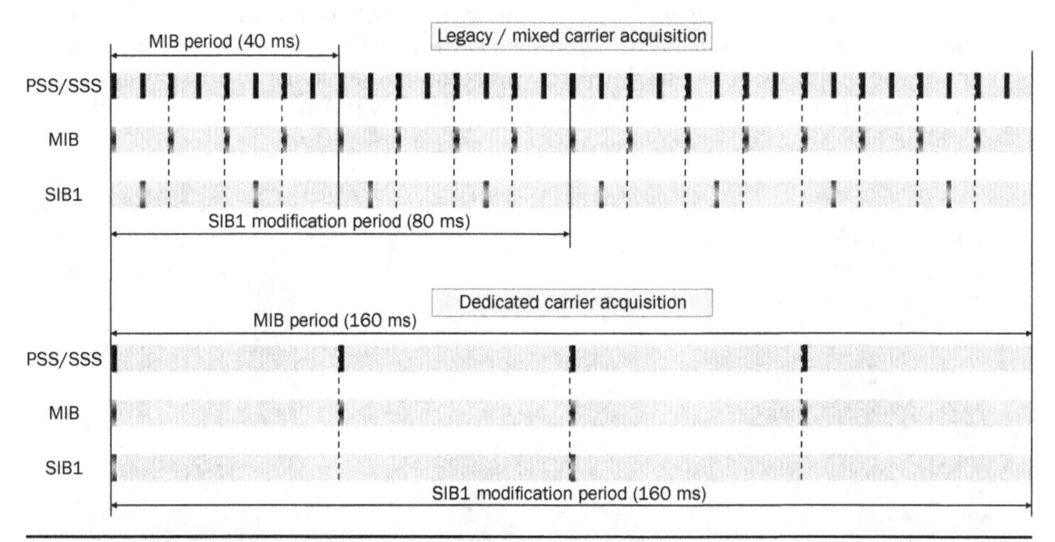

FIGURE 16.3 Comparison of mixed unicast/broadcast cell acquisition signals and dedicated broadcast cell acquisition signals.

was found that using the existing initial acquisition signals is adequate, even at its lower SNRs for large cell sizes. Therefore, the design reused the existing initial acquisition signal structure in Release 14, with small modifications added in Release 16 [1, 2]. The Release 16 modifications [1] included fixed control length and increased control aggregation level. The design designated one subframe out of 40 subframes as cell acquisition subframe (CAS). A comparison in more detail is shown in Fig. 16.3, of the mixed unicast/broadcast cell acquisition signals and the dedicated broadcast-only cell acquisition signals.

The 40 ms frame structure including CAS is illustrated in Fig. 16.4, which shows one CAS subframe of duration 1 ms followed by 13 MBSFN slots of duration 3 ms each

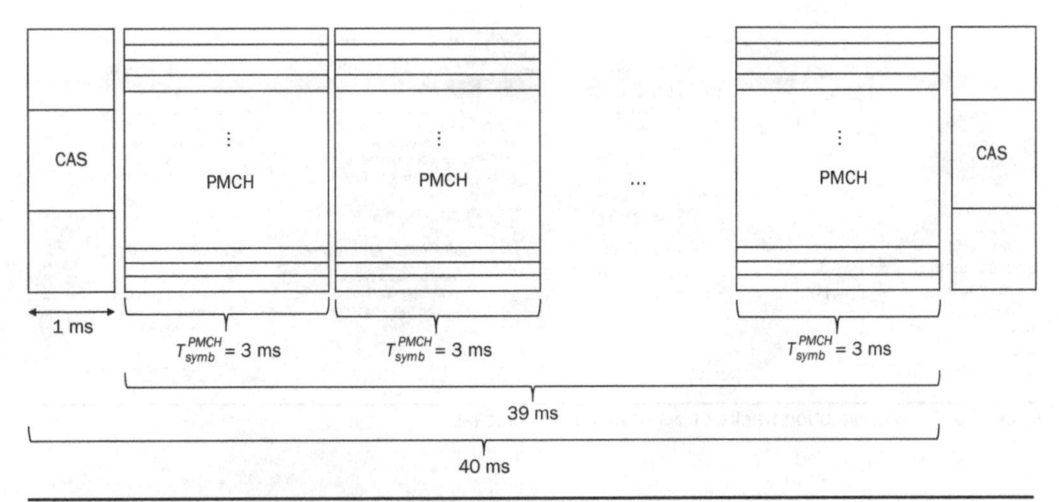

Figure 16.4 Frame structure for Release 16 MBSFN symbol duration of 3 ms.

(depicting the case of 370 Hz SCS). Note that each of the 3 ms MBSFN slots consists of a single OFDM symbol.

The currently implemented broadcast designs are DL only. This is because in normal use cases, different user devices do not have the same content to send to the network.

The currently specified broadcast/multicast systems do not use any feedback. As a consequence, there is no opportunity for retransmissions and, therefore, the packet losses due to temporal fade become a significant performance-limiting factor. An interesting solution to address this is the use of outer codes. Outer codes operate on the already decoded and CRC checked data, and outer codes are able to recover missing packets as long as the ratio of missing packets in a certain window is below a design limit. Outer codes are usually described as part of the application layer, without any physical layer impact. They are also agnostic to the underlying physical layer implementation; therefore, it is expected that the outer codes adopted for LTE may be reused for NR, if needed, in the future. One example of an applicable outer code is the Raptor code described in Ref. [3].

Another option to mitigate the impact of temporal fade is to introduce time-domain interleaving in the physical layer. When the UE operates in a broadcast-only reception mode, all the soft buffer that would be otherwise used for the HARQ operation can be made available to collect and jointly decode data in multiple MBSFN subframes. Time diversity is achieved when the transport blocks are distributed among those subframes. Proposals on this enhancement were made in Release 16 but not adopted. These enhancements may be considered in future releases.

There has been significant debate on the broadcast evolution path for NR. It has been decided that a two-track approach will be taken. It was seen that there is an urgent need to add mixed mode unicast/multicast operation support in Release 17. The mixed mode unicast/multicast system will be tightly integrated in the NR eMBB system and there will be maximum commonality between the two. The mixed mode unicast/multicast operation can address use cases such as push to talk (PTT) group communication services for commercial and public safety uses. The mixed mode unicast/multicast

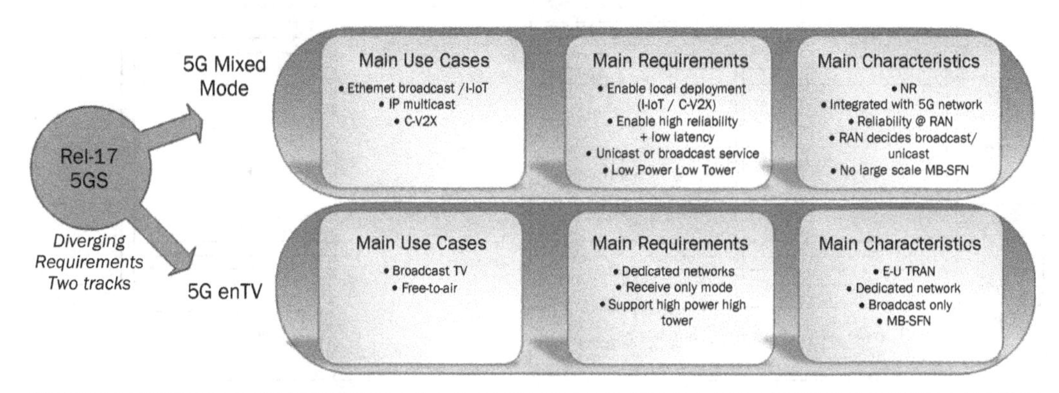

Figure 16.5 Two evolution tracks of 5G broadcast/multicast.

system will not support SFN transmission, other than the SFN implemented in a transparent manner.

The other track is for dedicated broadcast. The dedicated broadcast enhanced TV (enTV) evolution of LTE specified in Release 16 has already been designed to meet the 5G broadcast requirements [2]. Therefore, it was seen that there is no urgent need to add an NR-based version of dedicated broadcast. The main attributes of the two evolution tracks are shown in Fig. 16.5.

As it has been mentioned, the Release 17 multicast evolution work will be likely single cell–based, similar to the SC-PTM mode introduced in Release 13 LTE. While in a large multi-cell broadcast system, the benefits of any UE feedback are questionable, there may be potential benefits of feedback in an SC-PTM-like multicast system. Therefore, it is expected that some proposals on feedback for multicast will be discussed in Release 17.

If there was HARQ feedback, CSI feedback or closed-loop rate control introduced for multicast, then some version of SFN transmission in the UL may also be considered. For example, for HARQ feedback, the network does not necessarily need to know which UE received or not the multicast packet. The network only needs to know whether there is at least one UE that did not receive the packet or all UEs received it. For this, an SFN-based NACK only transmission may suffice. Similarly, for CSI feedback, the network does not necessarily need to know the CSI for each UE. Rather, it needs to know the lowest CQI among all UEs. Optimized feedback to address these requirements may be studied in Release 17 for NR.

References

[1] 3GPP, "3rd Generation Partnership Project; Technical Specification Group Radio Access Network; Evolved Universal Terrestrial Radio Access (E-UTRA); physical channels and modulation (Release 16)," 36.211, v16.0.0, December 2020.

[2] 3GPP, "3rd Generation Partnership Project; Technical Specification Group Radio Access Network; Overall description of LTE-based 5G broadcast (Release 16)," 36.976, v1.1.0, December 2020.

[3] Watson, M., T. Stockhammer, and M. Luby, "Raptor FEC Schemes for FECFRAME," IETF RFC6681, August 2012.

CHAPTER 17

Miscellaneous Topics for 5G

17.1 Overview

By now, the reader should have a fairly good idea of what the 5G platform offers. Release 15 set the foundation for 5G NR with a large configurability capability, as well as with a special emphasis put on forward compatibility. Forward compatibility enables the introduction of features in the future without affecting legacy operation.

After Release 15, 3GPP is evolving NR into improving its eMBB support and into enabling verticals. There is a good balance on both objectives and a common desire to mutually leverage so that, to the extent possible, implementations and economy of scales can be leveraged across use cases.

We have covered in dedicated chapters the expansion of 5G into new verticals for, namely, URLLC, V2X, and broadcast. We also had a dedicated chapter on NR for unlicensed spectrum which can be seen as a technology enabler to expand the reach of NR. All those features have been introduced in Release 16. Chapter 19 introduces the Release 17 projects and dwells on a few areas which will require multiple releases to get fully specified.

In this chapter, we introduce a few more areas that have been developed in 3GPP in the context of NR Release 16, namely:

- Interference management
- UE power savings
- NR positioning
- Two-step RACH
- Multi-RAT DC/CA enhancements
- Mobility enhancements
- Integrated access and backhaul (IAB)

17.2 Interference Management

In the context of interference management, the issues at hand are twofold:

1. gNB-to-gNB interference in a TDD deployment where the transmission of a faraway gNB gets into the receive window of another gNB, despite having the

same UL/DL TDD configuration, creating a potentially large interference over the first few symbols of an UL reception of the victim gNB.

2. UE-to-UE interference in a TDD deployment with different DL/UL partition in neighboring cells where the transmission of a first UE interferes with the reception of a second UE connected to a different gNB but, possibly, not far away from the first UE.

A single 3GPP project dealt with the normative work of two distinct interference issues. In this section, we look into these two issues and how 3GPP devised ways to alleviate them.

For the most part, the solution to the interference problems starts with the definition of a relevant measurement and corresponding reporting. The relevant measurement, in turns, requires a relevant signal to measure. Those two are, at high level, the "hooks" that 3GPP will put in place. How to use these "hooks" falls outside the scope of 3GPP and will be addressed by implementations.

17.2.1 Remote Interference Management

Networks configure and provision the guard period (GP) between DL transmissions and UL receptions of the TDD deployment in order to address the first issue reported in the previous section. The GP is provisioned to accommodate the propagation delay between gNBs which could cause noticeable interference.

The question is whether there could be some variability in the interference pattern due to, e.g., atmospheric ducting phenomenon, which could cause problems. The atmospheric ducting phenomenon, caused by lower densities at higher altitudes in the Earth's atmosphere, causes a reduced refractive index, causing the signals to bend back toward the Earth. In order to address this issue a study on remote interference management (RIM) was agreed for Release 16 [1].

This RIM study had the goal to characterize the inter-gNB interference from gNBs potentially faraway from each other operating NR in unpaired spectrum (TDD). This is a problem that has been reportedly experienced in LTE TDD deployments in China and for which proprietary solutions are in place.

The atmospheric ducting phenomenon causes the radio signals to travel a relatively long distance and the propagation delay goes beyond the GP discussed earlier. In this case, the DL signals of an aggressor gNB can travel a long distance and interfere with the UL signals of a victim gNB that is far away from the aggressor, as shown in Fig.17.1. The further the aggressor is to the victim, the more UL symbols of the victim will be impacted by the remote interference.

The goal was to look into standardized solutions to cope with the problem in NR networks yet to be deployed. The corresponding technical report capturing the main findings can be found in Ref. [2].

As discussed earlier, the procedure to identify that there is a remote interference situation starts with the identification of the interference at a victim gNB. Aggressor and victim gNBs can be grouped into semi-static sets, where each cell is assigned a set ID, and is configured with a RIM reference signal (RIM-RS) and the radio resources associated with the set ID. Each aggressor gNB can be configured with multiple set IDs and each victim gNB can be configured with multiple set IDs; however, each cell can have at most one victim set ID and one aggressor set ID. Consequently, each gNB can be an aggressor and a victim at the same time

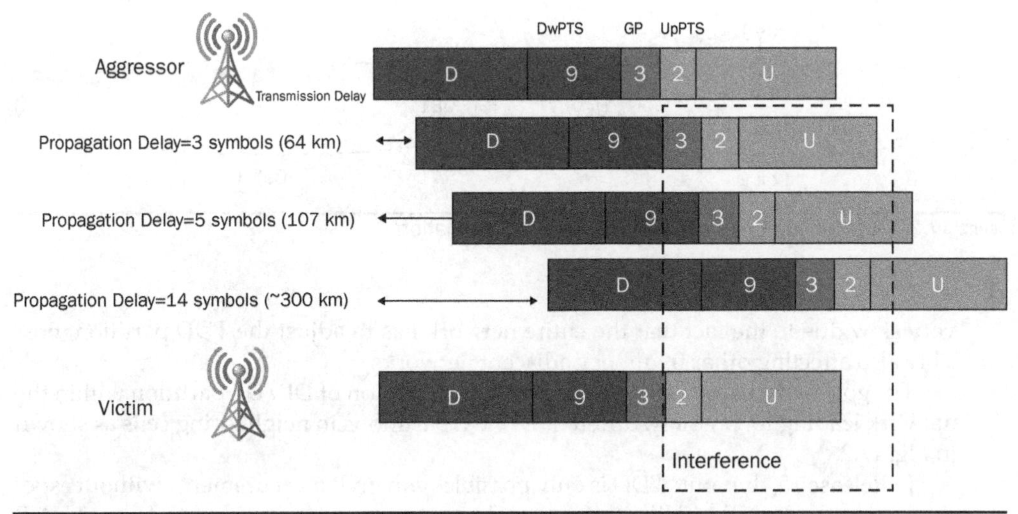

FIGURE 17.1 An illustration of how remote interference happens in TD-LTE network [2].

All gNBs in a victim set simultaneously transmit an identical RIM-RS carrying the victim set ID over the air. Upon reception of the RIM-RS from the victim set, aggressor gNBs undertake RIM measures, and either

1. Send back a RIM reference signal carrying the aggressor set ID (in the RIM wireless framework) or
2. Establish backhaul coordination toward the victim gNB set (in the RIM backhaul framework).

In the RIM wireless framework, the RIM-RS sent by the aggressor is able to provide information whether the atmospheric ducting phenomenon exists. The victim gNBs realize the atmospheric ducting phenomenon have ceased upon not receiving any reference signal sent from aggressors.

In the RIM backhaul framework, the RIM backhaul messages from aggressor to victim gNBs carry the indication about the detection or disappearance of RIM reference signal. Based on the indication from the backhaul message, the victim gNBs realize whether the atmospheric ducting and the consequent remote interference have ceased.

In both frameworks, upon realizing that the atmospheric ducting issue has disappeared, the victim gNBs stop transmitting the RIM reference signal.

17.2.2 Cross Link Interference

The UE-to-UE interference issue happens when a UE is allowed to transmit at a time when a nearby UE is receiving. This only happens in deployments where the DL/UL partition is flexible across the network which is, typically, denoted as dynamic TDD.

In typical TDD networks based on macro-cell deployments, the same DL/UL partition is assumed across the network. Adaptation to traffic conditions (if any) may be

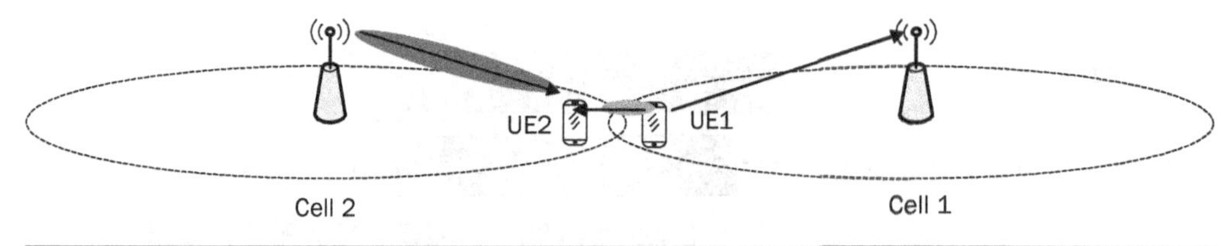

FIGURE 17.2 Illustration of cross link interference (CLI) situation.

very slow due to the fact that the entire network has to adjust the TDD partition, possibly also affecting other frequency adjacent networks.

The goal of dynamic TDD is to enable the adaptation of DL/UL partition within the network leading to, possibly, different DL/UL partitions in neighboring cells as shown in Fig. 17.2.

In Release 15, dynamic TDD is only possible with gNB measurements without especial provisions for UE-to-UE interference identification or measurements. Instead, gNB would typically use its own estimate of network load to determine an adequate UL/DL split not causing unwanted interference situations.

The gNB-to-gNB interference issue that may arise in dynamic TDD is similar to the one characterized by RIM. The UE-to-UE interference issue is addressed by the CLI project. Indeed, CLI has the goal to enable UE-to-UE interference measurements and reporting for improving the handling of dynamic TDD.

For CLI, some SRS resources may be configured for UE-to-UE measurements and corresponding reporting to the network. Indeed, two new UE measurements are defined, namely, SRS-RSRP and CLI-RSSI [3]. The UE reports can be used by the network to adapt the DL/UL partition for cases where no UE-to-UE interference is identified. The gNBs can coordinate via the exchange of their intended DL/UL configuration.

The TR on cross link interference (CLI) handling and remote interference management (RIM) for NR can be found in Ref. [4].

17.3 UE Power Savings

UE power consumption is one of the key performance indicators in today's wireless systems. Typically, there are many optimizations that can be done via implementation. We have seen, for example, how 4G LTE devices have been improving in power consumption over the years without necessarily seeing any specific 3GPP work to enable the savings.

As we have discussed in earlier chapters, 5G NR comes with a set of features which may not be best suited for a better power consumption than 4G. First off, 5G NR is expected to operate at substantially higher bandwidths than LTE. Operating at higher bandwidths implies running all the data converters at higher speeds and processing more bits which, in turn, consume more power. This phenomenon may even be more severe in mmWave systems with channel bandwidths of up to 400 MHz. Other aspects, such as reduction of "always on" signals (no CRS or the increase in the periodicity of sync channels) make it more difficult for the UE to find new cells at initial access or when moving around the network with a negative impact in power consumption.

One aspect that NR Release 15 brings, which may have an advantageous impact on UE power consumption, is the bandwidth part (BWP) concept. Similar to other UE power savings techniques, BWP requires the network to play a role in moving the UEs with little or no data activity to a narrowband BWP and, possibly, even reducing the periodicity of control monitoring. In addition, DRX procedures for NR are defined in Release 15 similar to those in LTE.

The objective of the Release 16 project on UE power savings was to specify promising techniques enabling power savings at the UE by exploiting adaptation in frequency, time, and antenna domains. Reference [5] captures the findings of the SI along with the model for estimating the power savings offered by different techniques.

Several techniques are specified to improve UE power consumption and we present and discuss them in subsequent subsections.

17.3.1 Wake-Up Signal

The study on power savings found that the introduction of a wake-up signal (WUS) preceding the DRX ON duration achieved good power savings. The WUS is transmitted in the form of a regular PDCCH and determines whether or not to wake-up a user sufficiently earlier than its DRX On duration to transition, in case of a positive indication, from a 1st stage wake-up (adequate for the reception of the wake-up signal) to a 2nd stage wake-up for processing of regular PDCCH and subsequent data. Figure 17.3 shows an illustration of how this WUS would be used along with the expected power consumption at each UE processing stage.

PDCCH-WUS delivers good power saving gain across different scenarios. Based on the results shown in Ref. [6], the gains are more visible for shorter DRX cycles and for lighter traffic which would lead to more "empty" DRX cycles.

Figure 17.4 illustrates the relative power savings [6] for PDCCH-WUS for a variety of traffic models, namely web browsing, instant message, and FTP, for various DRX configurations represented by {C-DRX cycle, Inactivity timer}: {40, 10}, {160,100}, and {320,80}.

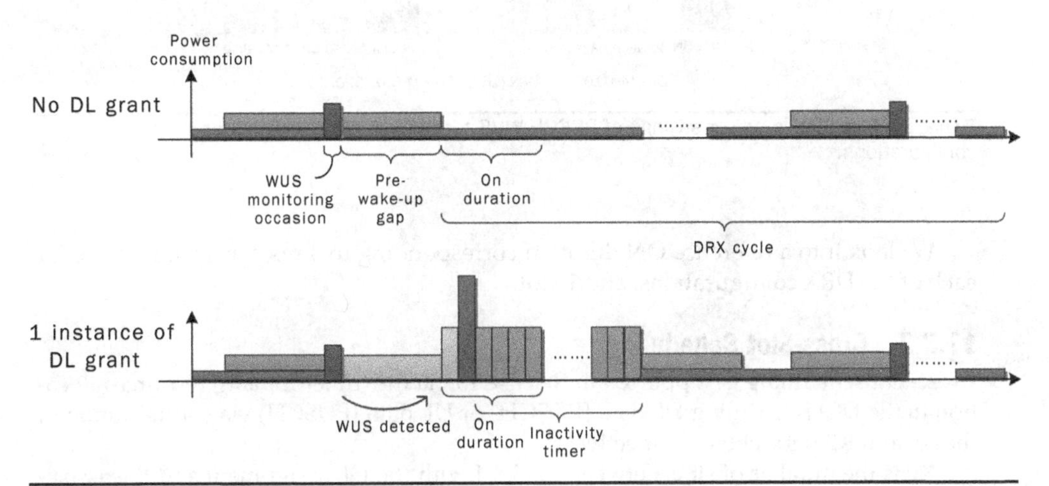

FIGURE 17.3 Illustration of wake-up signal.

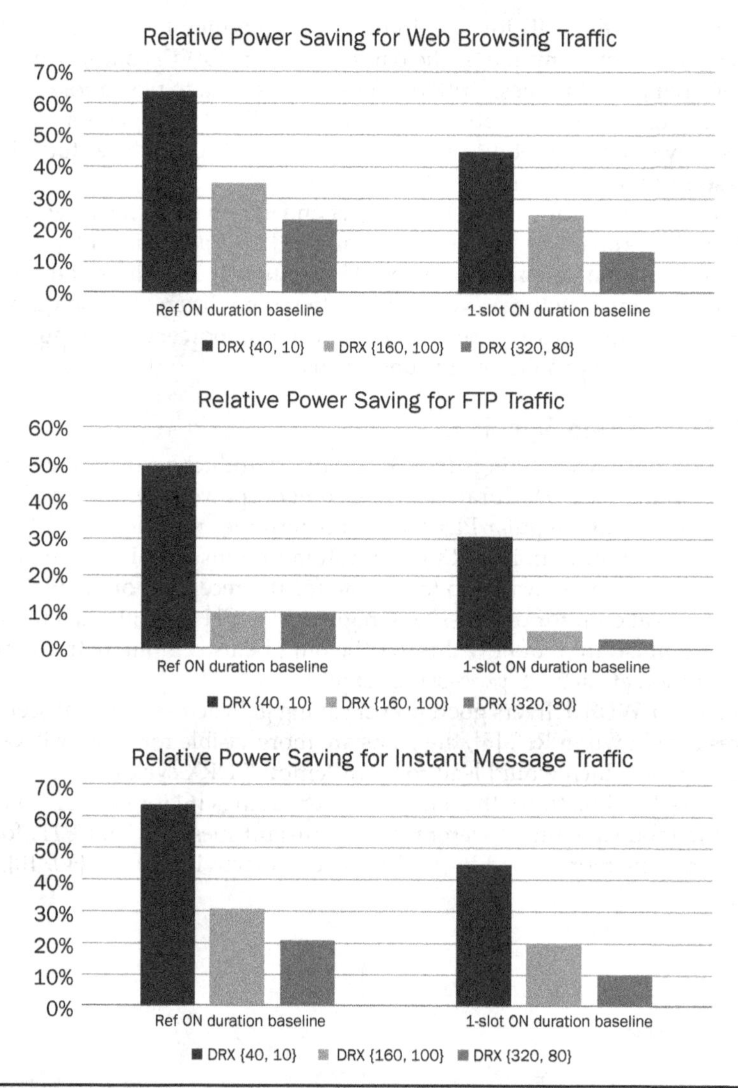

FIGURE 17.4 Relative power savings of PDCCH-WUS for various traffic models and DRX configurations.

We look into a reference ON duration corresponding to 4 ms, 8 ms, and 10 ms, for each of the DRX configurations, and 1-slot.

17.3.2 Cross-Slot Scheduling

Cross-slot scheduling is supported in Release 15 via the (time-domain) resource allocation in the DCI scheduling DL data (PDSCH) or UL data (PUSCH) via configuration of the K0 and K2 parameters, respectively.

K0 is the number of slots between the DCI with the DL assignment and the associated PDSCH data transmission.

K2 is the number of slots between the DCI with the UL grant and the associated PUSCH data transmission.

While K0 and K2 are parameters in the corresponding DCI and can be configured to large values, they cannot really be used for power savings purposes in Release 15. The reason is that the K0 and K2 values are not tied to the timing relationships used for aperiodic CSI-RS triggers and aperiodic SRS triggers that could come in the same DCIs. Having those aperiodic triggers shorter than the K0 and K2 values makes it impossible for the UE to process PDCCH at a lower clock rate because it has to be ready to react to the trigger sufficiently fast. As a result, UEs have to process all DCIs at the same nominal speed irrespective of the K0, K2 values. In addition, the K0 and K2 values in Release 15 are semi-statically configured for the given BWP making it difficult to quickly adapt to various traffic conditions.

This inefficiency of Release 15 is resolved in Release 16 introducing an explicit "minimum scheduling offset" which is RRC configured per BWP. Note that 1-bit in the DCI enables the dynamic switch between two preconfigured values.

17.3.3 Scell Dormancy

UE power consumption in carrier aggregation (CA) can be substantial. Activating and deactivating cells entail a fair amount of signaling and subsequent latency. To the point where by the time the activation has been completed, the data burst may have already been transmitted via the Pcell. Note that this has been identified as an issue in LTE networks where CA activation is certain that network implementations can take a considerable amount of time.

As a result, there is good incentive to be able to keep Scells activated and devise ways to introduce a *"dormancy"* sub-state where activity on the corresponding Scell is minimized to the maximum extent. This feature enables Scells to dynamically be switched between *"dormancy"* and *"non-dormancy"* sub-states which very much accelerates their availability for the transmission of arriving data bursts. For an illustration of the state machine associated with this operation, please refer to Fig. 7.14 in Chap. 7.

17.3.4 UE Adaptation to Maximum Number of MIMO Layers

UE power consumption increases with the number of MIMO layers for the simple reason that there are more bits to process. In Release 15, the number of MIMO layers is RRC configured per cell for DL transmissions. Release 16 introduces the support of per-DL-BWP configuration of maximum number of DL MIMO layers (*maxMIMO-layer* parameter in BWP configuration).

This configuration enables opportunities for power savings at the UE in a scenario where, for example, a UE is configured with two BWPs for different conditions. BWP1 as wideband with *maxMIMO-layer* = 4 suitable for high data rate and BWP2 as narrowband with *maxMIM-layer* = 2 suitable for low traffic and potentially UE power savings.

17.3.5 UE Assistance

The possibility for the UE to indicate its preference to be transitioned out of the RRC connected state is specified in Release 16. Despite the specification of this indication, there is no guaranteed commitment from the network to honor such request. As happens with other features, in order to improve UE power savings, network cooperation is necessary as it would have to take the UE reports and recommendations into account.

17.3.6 Relaxation of RRM Measurements

Relaxation of intra- and inter-frequency RRM measurements for neighboring cells for RRC_IDLE and RRC_Inactive are also considered in Release 16.

17.4 NR Positioning

A study on NR positioning techniques was carried at the beginning of Release 16 with corresponding 3GPP technical report captured in Ref. [7]. Regulatory as well as commercial use cases were considered of importance with the following target baseline requirements.

While the study and subsequent NR positioning items looked into RAT-dependent and RAT-independent techniques, we will focus presentation of RAT-dependent techniques which are the ones relevant to NR air interface. RAT-independent techniques are either GNSS-based or hybrid-based solutions outside the scope of this book.

For regulatory use cases, the following baseline requirements were set forth [7]:

- Horizontal positioning error ≤ 50 m for 80% of UEs
- Vertical positioning error < 5 m for 80% of UEs
- End-to-end latency and time to first fix (TTFF) < 30 seconds

NR positioning targets for commercial use cases with RAT-dependent solutions were agreed to be [7] following:

- Horizontal positioning error < 3m for 80% of UEs in indoor deployment scenarios
- Vertical positioning error < 3 m for 80% of UEs in indoor deployment scenarios
- Horizontal positioning error < 10 m for 80% of UEs in outdoor deployments scenarios
- Vertical positioning error < 3 m for 80% of UEs in outdoor deployment scenarios
- End-to-end latency < 1 s

In this section we will present the various RAT-dependent positioning techniques that NR supports at the concept level. Once the positioning methods are covered, we will present how NR has introduced reference signals and corresponding measurements to enable the various positioning techniques.

In addition, 3GPP has introduced the extensions of the LTE positioning protocol (LPP) for NR RAT-dependent positioning techniques.

17.4.1 Positioning Techniques

In this section we present conceptually the various positioning techniques for RAT-dependent solutions. We only cover the techniques which were identified during the study phase for specification at the work item phase.

The RAT-dependent techniques can be categorized as Downlink-based, Uplink-based, and Downlink- and Uplink-based solutions.

Downlink-based solutions:

- DL or observed time difference of arrival (DL-TDOA or OTDOA)
- DL angle of departure (DL-AoD) with beam sweeping

Uplink-based solutions:

- UL time difference of arrival (UL-TDOA)
- UL angle of arrival (UL-AoA)

Downlink- and Uplink-based solutions:

- Round-trip time (RTT) with one or more neighboring cells (multi-RTT)
- Enhanced cell ID (E-CID) based

Note that the E-CID method relies on RRM measurements already defined in Release 15.

Positioning methods have typically been network based and UE assisted via measurements and corresponding reports. This applies for NR to all the techniques mentioned above. In addition, NR has introduced UE-based methods for the DL-based solutions, namely, OTDOA and DL-AoD.

For UE-based methods, the actual positioning computation is performed at the UE without the need to report back to the network. For UE-based positioning to be possible, the network has to communicate to the UE the location of the various TRPs (for OTDOA) and the beam angles of each positioning signal resource (for DL-AoD).

UE-based positioning would be the first step to enable positioning in idle or inactive mode, when the UE is not actively connected to the network. In addition, it reduces the positioning latency and scales well with large number of UEs which are no longer required to send measurement reports (reduced UL overhead).

17.4.1.1 DL/UL-TDOA Method

The time difference of arrival (TDOA) is a well-known technique which has been used for LTE, UMTS, and CDMA systems before. OTDOA was introduced in Release 9 and UTDOA in Release 10 of the LTE specifications.

Conceptually, TDOA hinges on the fact that the time difference of arrival between two synchronized cells provides a positioning estimate along a hyperbola. Multiple TDOA measurements are used for triangulation, which requires hearability of multiple cells. Indeed, we need four or more cells, i.e., two pairs, to measure the corresponding TDOA at two pairs of cells and do the triangulation of the measurement.

Figure 17.5 pictorially shows this concept with the UE being able to observe and measure three cells.

For the DL, this technique requires the UE to be able to receive and accurately measure transmissions from multiple cells. In the UL, in turn, the UE transmissions are to be received and accurately measured at multiple cells.

For this technique to work, hearability beyond the serving cell becomes necessary. For this purpose, the network can configure muting patterns, which will enable interference-free reception from neighboring cells (assuming synchronized network) over some resources and, thus, will extend the coverage of the corresponding cell transmissions.

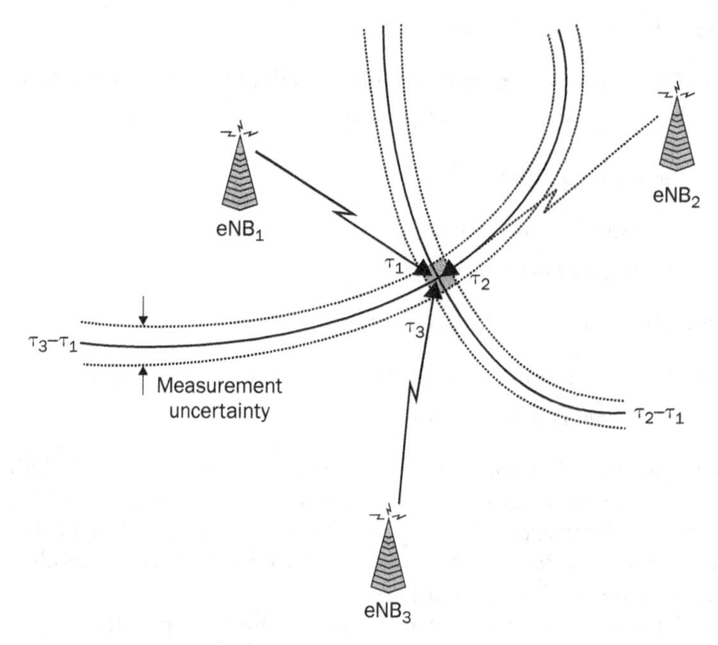

FIGURE 17.5 Illustration of TDOA concept.

The positioning measurement precision has a dependency on the propagation conditions, e.g., channel dispersion and availability or not of line of sight (LOS) component to the corresponding cell. In addition, the network synchronization error poses a floor to the positioning accuracy which can easily become the bottleneck for the positioning accuracy that NR is after. For example, GPS/GNSS synchronization accuracy is limited to 50 to 100 ns which equates to 15 to 30 m positioning accuracy, which may be sufficient for regulatory requirements but is insufficient for commercial requirements.

UL TDOA has a similar strict network synchronization requirement compared to OTDOA, with the additional challenge of limited Tx power of UEs impacting the corresponding coverage.

17.4.1.2 Multi-RTT Method

As discussed in Sec. 17.4.1.1, TDOA positioning methods have a dependency on the accuracy of the network synchronization. A typical network synchronization error of 50 to 100 ns equates to 20 to 30 m of positioning accuracy. This dependency on network synchronization motivates RTT-based methods.

Indeed, the time difference between receive and transmit at the UE and gNB provides a distance estimate between the two. Multiple RTT measurements enable triangulation, and the method requires being able to measure three or more cells, in contrast with the four or more cells required by TDOA methods.

Figure 17.6 illustrates the RTT concept with corresponding measurements from one UE to three cells. While this positioning method eliminates the dependency on network synchronization, it requires very accurate calibration of the transmit and receive front-end delays at the UE and gNB, so that timing measurements account only for the propagation delay over the air.

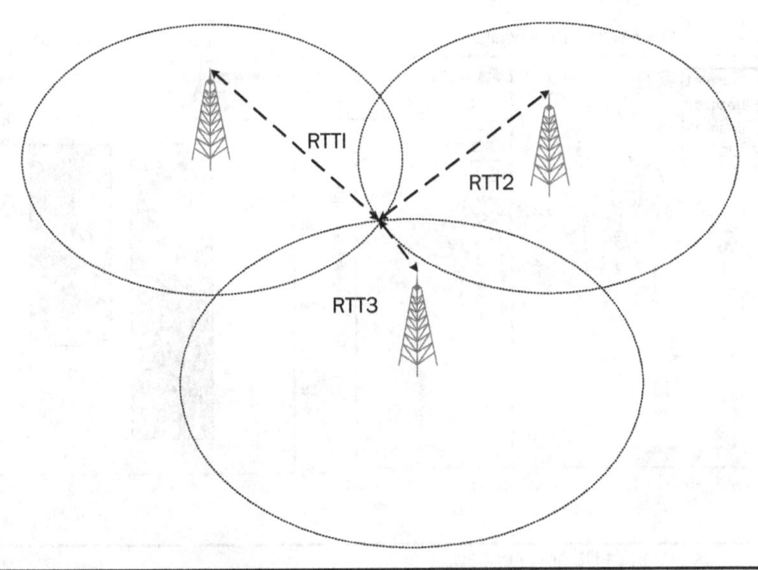

Figure 17.6 Illustration of RTT concept.

17.4.1.3 DL-AoD Method

In this method, the gNBs transmits a positioning signal in multiple directions on corresponding resources, the UE measures RSRP for each resource.

In **UE-assisted mode**, the measured RSRPs are reported through the positioning protocol to the location management function (LMF) where the corresponding AoDs are estimated and positioning computation is performed.

In **UE-based mode**, the UE uses the assistance data, including the TRP geographic locations, and the positioning signal beam information (e.g., beam azimuth, elevation) to compute its position.

The RSRP measurement vector can be considered as an "RF fingerprint" and AoD calculation may be performed using a pattern matching approach. Figure 17.7 illustrates this approach for an exemplary multi-beam transmission with three beam directions.

17.4.1.4 UL-AoA Method

The UL-AoA method is the UL counterpart of the DL-AoD positioning method discussed in the previous section.

An angle of arrival (AOA) estimate is made from base stations using a directional antenna such as a phased array to determine which direction a signal was transmitted from. When combined with a range estimate (e.g., using RTT), this can provide a location estimate even for a single-base station. UL SRS beam sweeping may not be needed in case of digital elements where the UL beam scan can be implemented with fine granularity through baseband processing.

Figure 17.8 illustrates the UL-AoA concept for multi-cell (left) and single-cell (right). The single-cell utilizes an RTT measurement to get the distance estimate to the given cell, and thus enables positioning computation with a single-cell measuring.

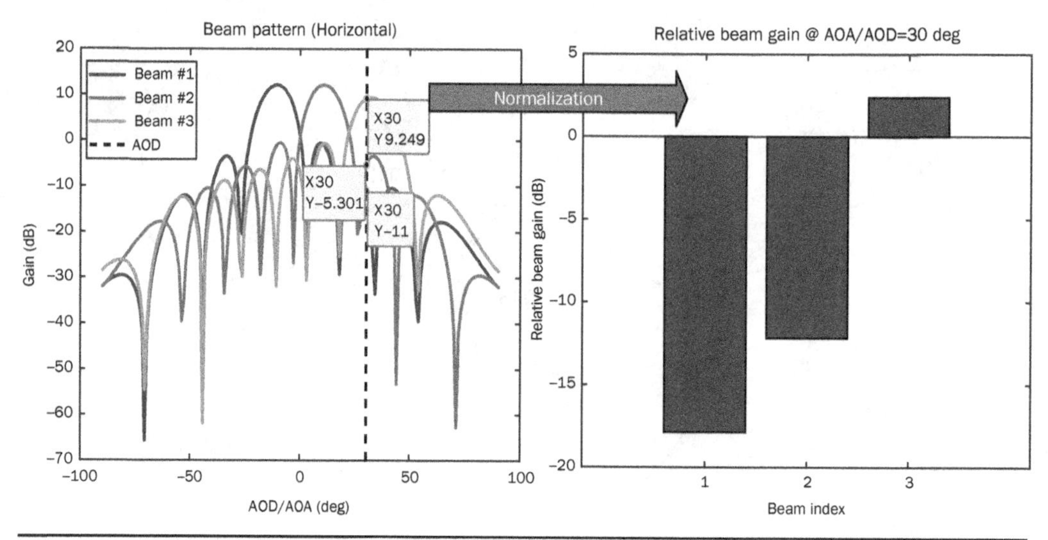

FIGURE 17.7 Illustration of DL-AoD concept.

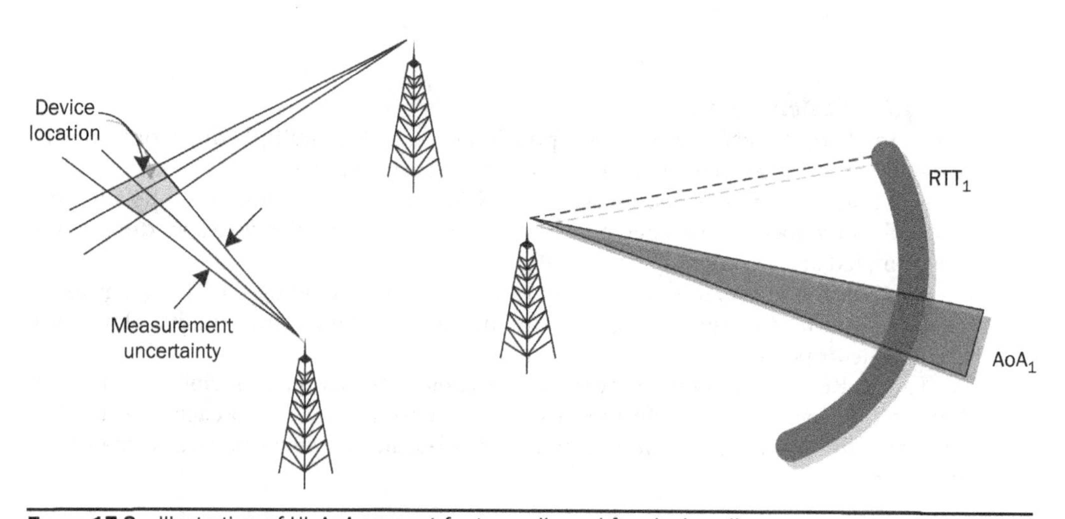

FIGURE 17.8 Illustration of UL-AoA concept for two cells and for single cell.

17.4.2 Measurements Defined for NR Positioning

Now that we have discussed the concepts behind the different positioning methods, the corresponding measurements easily follow. In summary, there are two types of measurements: timing measurements and receive power measurements at the UE and the gNB.

Timing measurements:

- Downlink relative signal time difference (DL RSTD), used for DL-TDOA
- Uplink relative time of arrival (UL RTOA), used for UL-TDOA

- UE Rx-Tx time difference and gNB Rx-Tx time difference, used for multi-RTT
- Time difference used for multi-RTT

Power measurements:

- DL PRS-RSRP, used for DL-AoD and E-CID
- UL SRS-RSRP and UL-AoA, used for UL-AoA
- SSB and CSI-RS RRM measurements (RSRP and RSRQ) for E-CID

All these measurements are defined in Ref. [3].

DL measurements, other than RRM ones, are based on a new reference signal introduced for positioning, called positioning reference signal (**PRS**).

UL measurements are based on periodic, semipersistent, and aperiodic transmission of **Release 15 SRS or SRS for positioning** for all the UL-based methods, i.e., UL TDOA and UL AoA. UL measurements for the multi-RTT method is based on **SRS for positioning**, newly introduced in Release 16 and presented in the next section.

17.4.3 Reference Signals for NR Positioning

17.4.3.1 Positioning Reference Signal

Positioning reference signals (PRS) are defined for NR positioning to enable UEs to detect and measure more neighbor cells. Several configurations are supported to enable a variety of deployments (indoor/outdoor, FR1/FR2).

A DL PRS resource spans within a slot 2, 4, 6, or 12 consecutive symbols with a fully frequency-domain staggered pattern. Note that after de-staggering, we get a full pilot density in frequency domain for all the patterns with various levels of processing gain at the receiver depending on the density prior to de-staggering and the number of OFDM symbols used for transmission.

The DL PRS resource can be configured in any high layer configured Downlink or Flexible symbol of a slot. Constant EPRE for all REs of a given DL PRS resource is assumed.

Table 17.1 summarizes all the possible DL PRS configurations where each number in curly bracket represents a relative subcarrier position on corresponding symbols to build the pattern.

Figure 17.9 illustrates all the PRS patterns except the comb-2, 4 symbol, i.e., {0,1,0,1} which would be a direct extension of the comb-2, 2 symbol transmitted over four consecutive symbols.

	2 symbols	4 symbols	6 symbols	12 symbols
Comb-2	{0,1}	{0,1,0,1}	{0,1,0,1,0,1}	{0,1,0,1,0,1,0,1,0,1,0,1}
Comb-4	N/A	{0,2,1,3}	N/A	{0,1,0,1,0,1,0,1,0,1,0,1}
Comb-6	N/A	N/A	{0,3,1,4,2,5}	{0,1,0,1,0,1,0,1,0,1,0,1}
Comb-12	N/A	N/A	N/A	{0,1,0,1,0,1,0,1,0,1,0,1}

Table 17.1 PRS Configurations

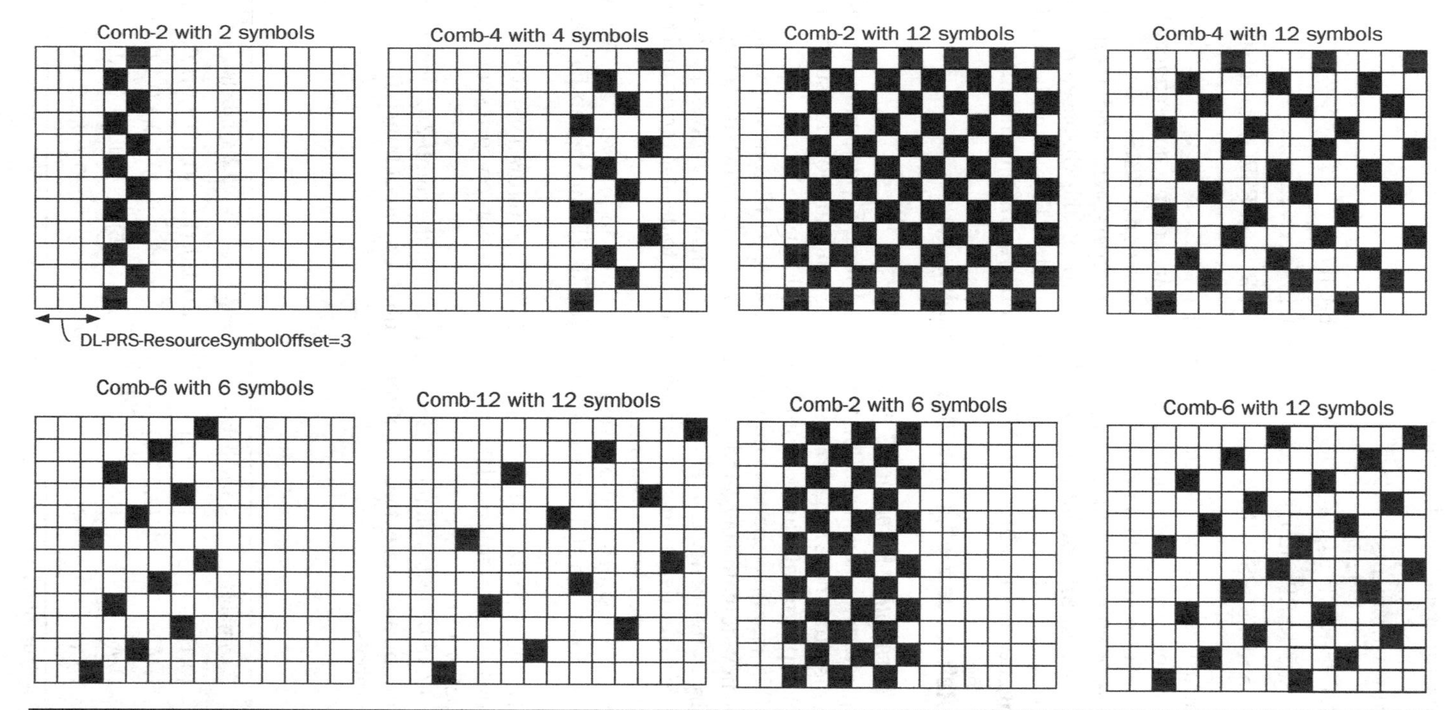

FIGURE 17.9 Illustration of DL PRS patterns.

	1 symbols	2 symbols	4 symbols	8 symbols	12 symbols
Comb-2	{0}	{0,1}	{0,1,0,1}	N/A	N/A
Comb-4	N/A	{0,2}	{0,2,1,3}	{0,2,1,3,0,2,1,3}	{0,2,1,3,0,2,1,3,0,2,1,3}
Comb-8	N/A	N/A	{0,4,2,6}	{0,4,2,6,1,5,3,7}	{0,4,2,6,1,5,3,7,0,4,2,6}

Table 17.2 SRS for Positioning Configurations

DL PRS transmission also supports resource repetition across multiple slots enabling beam sweeping and/or further increased processing gain at the receiver. The set of consecutive back-to-back slots carrying PRS transmission with a given periodicity cycle constitutes what we call a *PRS instance.*

As discussed earlier, the detectability of multiple TRPs is critical to improve the positioning accuracy. To that end, DL PRS supports muting configurations in two different ways, either within a PRS instance, across PRS instances, or a combination of both. Bitmaps are used to indicate the PRS resources to be muted.

Similar to regular DMRS, the DL PRS consists of QPSK symbols transmitted using CP-OFDM. The DL PRS sequence is generated using a Gold sequence generator.

17.4.3.2 Sounding Reference Signal for Positioning

Sounding reference signal (SRS) for positioning is the counterpart of the DL PRS for the Uplink. The transmission properties are similar to those of Release 15 SRS, i.e., same Zadoff-Chu sequences, transmitted within active BWP, periodic/semi-persistent/aperiodic, same timing advance as other UL transmissions, group or sequence group hopping supported.

On the other hand, there are some enhancements tailored for positioning-related measurements at the gNB. Specifically, path-loss reference can be a PRS or SSB from a serving or neighbor cell (open-loop power control only). A multi-symbol SRS resource is always staggered in frequency. No frequency hopping is supported and transmissions are with a single antenna port.

Table 17.2 summarizes all the possible SRS for positioning configurations where each number in curly bracket represents a relative subcarrier position on corresponding symbols to build the pattern.

Figure 17.10 illustrates all the SRS for positioning patterns.

17.5 Two-Step RACH

A study on non-orthogonal multiple access (NOMA) was carried during Release 15 and the earlier part of Release 16 with corresponding 3GPP technical report captured in Ref. [8]. Indeed, this was a continuation of one of the study areas during the Release 14 NR study item.

The difficulties, in 3GPP, in down-selecting a candidate scheme for NOMA, out of the more than 10 studied schemes, triggered a redirection of focus and led to the approval of a work item on two-step RACH.

We have discussed random access for NR in Chap. 6. In that chapter we focus on the Release 15 random access procedure based on four-step. This section presents the two-step counterpart introduced in Release 16.

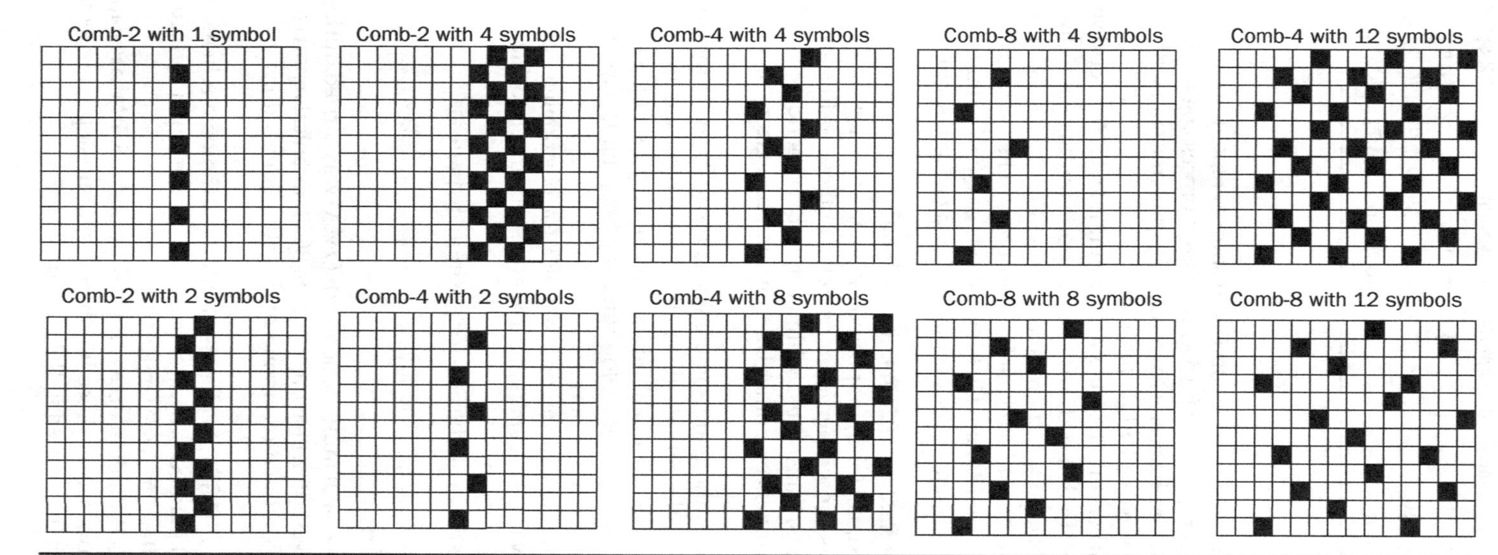

FIGURE 17.10 Illustration of SRS for positioning patterns.

Two-step RACH applies to contention based random access (CBRA) and to contention free random access (CFRA). The applicability to CFRA is limited to the handover case.

Figure 17.11 illustrates the four-step random access procedure for CBRA and CFRA. In contrast, Fig. 17.12 illustrates how two-step RACH modifies the procedures. Note that Message A of two-step RACH carries, at least, the contents of Message 3 in four-step RACH. In turn, Message B of two-step RACH carries at least the equivalent contents of Message 2 and Message 4.

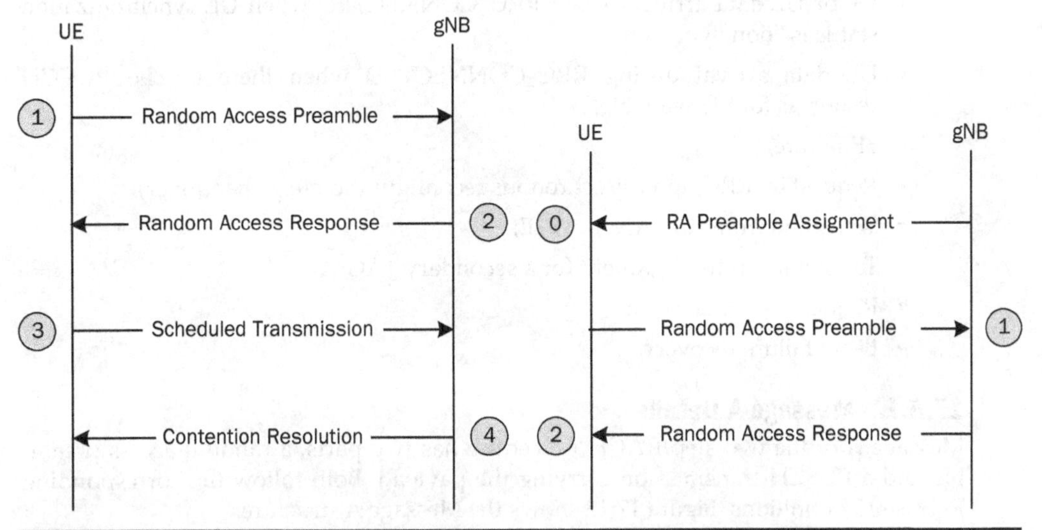

FIGURE 17.11 Four-step CBRA (left) and CFRA (right) random access procedure.

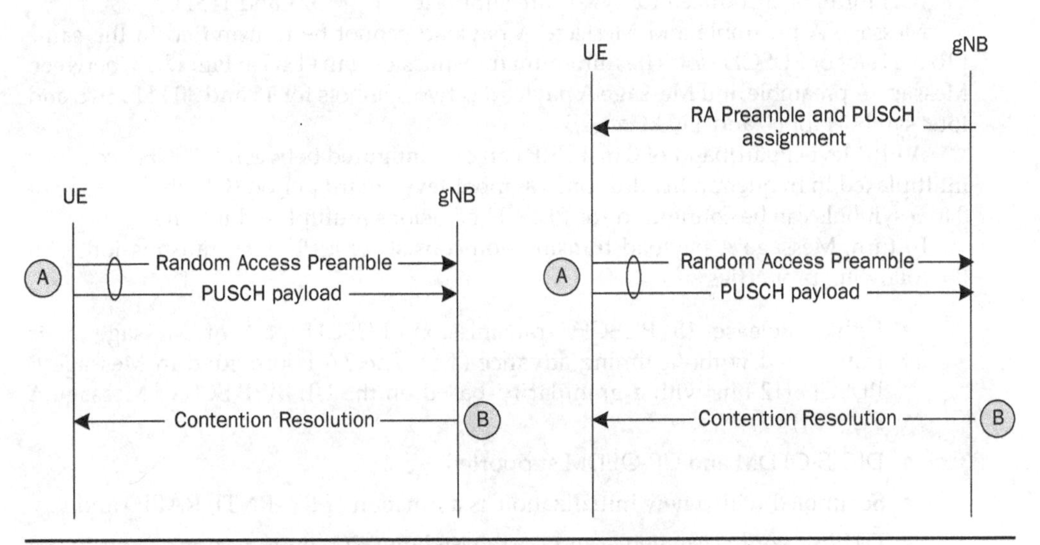

FIGURE 17.12 Two-step CBRA (left) and CFRA (right) random access procedure.

Two-step RACH applies to all the RRC states, namely, RRC idle, inactive, and connected. However, user plane (UP) data transmission in Message A is only supported for RRC Connected. Note that this restriction will be lifted in Release 17 with the project on small data transmission discussed in Chap. 19.

All CBRA triggers for Release 15 four-step RACH apply to two-step RACH but there are no new triggers defined. In summary, there are

- Initial access from RRC_IDLE;
- RRC connection reestablishment procedure;
- DL or UL data arrival during RRC_CONNECTED when UL synchronization status is "non-synchronized";
- UL data arrival during RRC_CONNECTED when there are no PUCCH resources for SR available;
- SR failure;
- Request by RRC upon synchronous reconfiguration (e.g., handover);
- Transition from RRC_INACTIVE;
- To establish time alignment for a secondary TAG;
- Request for other SI; and
- Beam failure recovery.

17.5.1 Message A Details

Message A of the two-step RACH procedure has two parts, a random-access preamble and a PUSCH transmission carrying the payload. Both follow the corresponding Release 15 definitions. Figure 17.13 shows the Message A structure.

All PRACH preamble formats in Release 15 for FR1 and FR2 (reference Chap. 6) are supported for the preamble part of Message A. In addition, the newly added longer PRACH formats introduced for NR-U are supported (L_{RA} = 571 and 1151).

Message A preamble and Message A payload cannot be transmitted in the same PRACH slot or PUSCH slot. The minimum transmission gap (TxG in Fig. 17.13) between Message A preamble and Message A payload is two symbols for 15 and 30 kHz SCS and four symbols for 60 and 120 kHz SCS.

A PRB-level guardband of 0 or 1 PRB can be configured between PUSCH occasions multiplexed in frequency. In addition, a symbol-level guard period (GP) in the range of 0 to 3 symbols can be configured for PUSCH occasions multiplexed in time.

In turn, Message A payload transmission consists of a PUSCH transmission with the following properties:

- Unlike Release 15 PUSCH transmission, PUSCH part of Message A is transmitted without timing advance (TA). The TA is provided in Message B PDSCH (12-bits with a granularity based on the UL BWP SCS of Message A PUSCH)
- DFT-S-OFDM and CP-OFDM supported
- Scrambled with a new initialization as a function of RA-RNTI, RAPID and n_{ID}.
- DMRS resource (sequence and port) based on RRC configuration
- RV = 0

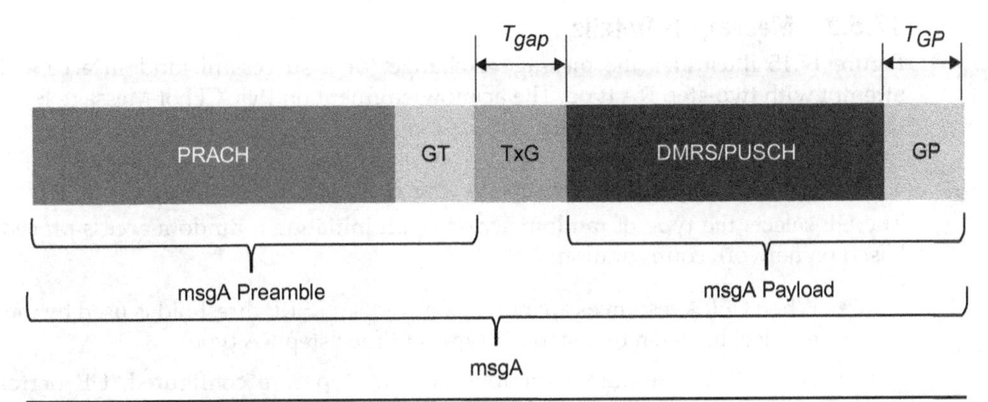

FIGURE 17.13 Message A structure.

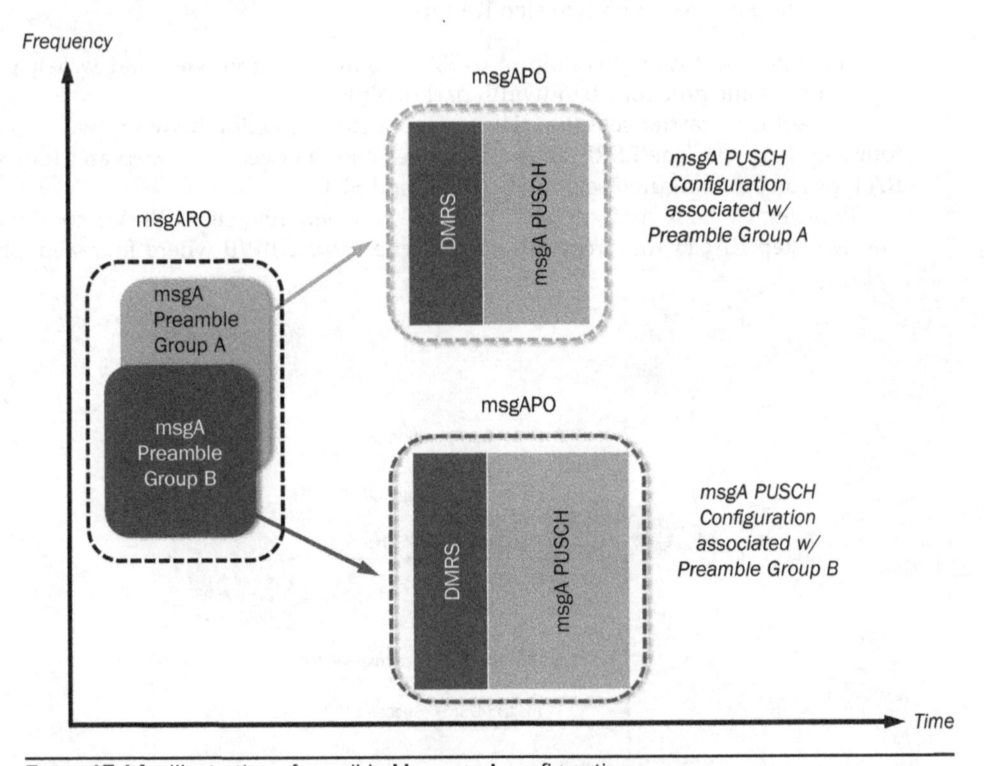

FIGURE 17.14 Illustration of possible Message A configuration.

The multiplexing of PRACH and DMRS/PUSCH is based on RRC configuration. UEs can share the same RACH occasion (RO), preamble group, DMRS resources, and PUSCH occasions (see Fig. 17.14 for an illustration). The RO for two-step RACH can be dedicated or shared with four-step RACH.

The power control of Message A is open-loop based on same power ramping counter used for the preamble and the payload parts. Power control parameters different from four-step RACH can be configured for two-step RACH on dedicated RO.

17.5.2 Message B Details

Figure 17.15 illustrates the message exchange for a successful random access (RA) attempt with two-step RA type. The acknowledgment on PUCCH of Message B reception is illustrated in the last message exchange.

17.5.3 Applicability of Two-Step RA and Fallback to Four-Step

The UE selects the type of random access upon initiating a random access procedure based on network configuration:

- When CFRA resources are not configured, an RSRP threshold is used by the UE to select between two-step RA type and four-step RA type.
- When CFRA resources for four-step RA type are configured, UE performs random access with four-step RA type.
- When CFRA resources for two-step RA type are configured, UE performs random access with two-step RA type.

The network does not configure CFRA resources for four-step and two-step RA types at the same time for a bandwidth part (BWP).

UE performs carrier selection (UL or SUL) before selecting between two-step and four-step RA type. The RSRP threshold for selecting between two-step and four-step RA type can be configured separately for UL and SUL.

Two-step CFRA is configured only on a BWP where two-step CBRA is configured. The two-step RACH resources can be configured on a BWP where four-step CBRA

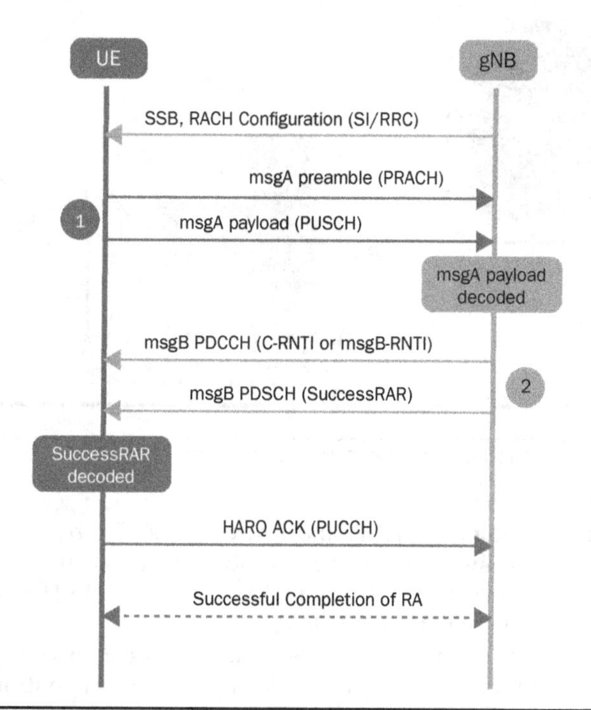

FIGURE 17.15 Illustration of successful RA attempt with two-step RA type.

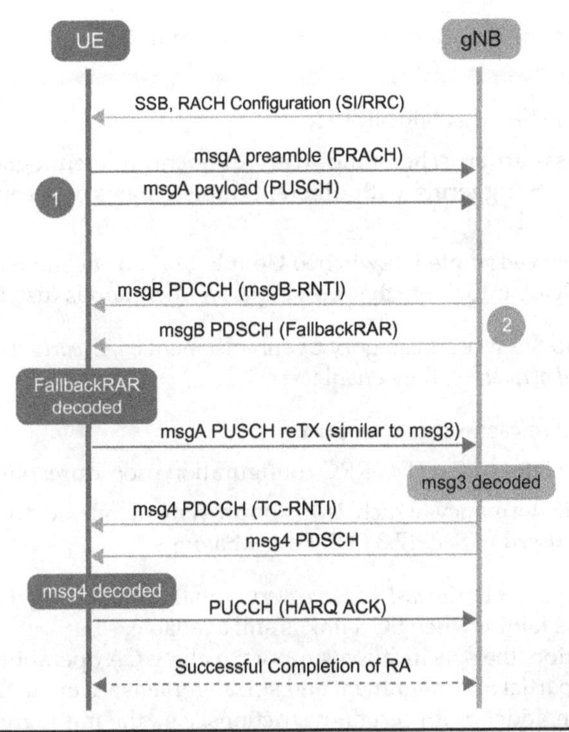

FIGURE 17.16 Illustration of fallback based on FallBackRAR.

resources are not configured. Two-step RACH resources can only be configured on SpCell*. The PDCCH triggered two-step CFRA is not supported in Release 16.

If fallback indication (i.e., fallbackRAR) is received in Message B, the UE performs Message 3 transmission and monitors contention resolution (Message 4), as illustrated in Fig. 17.16. If contention resolution is not successful after Message 3 (re)transmission(s), the UE goes back to Message A transmission.

If the random access procedure with two-step RA type is not completed after a number of Message A transmissions (*msgA-TransMax*), the UE can be configured to switch to CBRA with four-step RA type (Message 1 of four-step CBRA). If *msgA-TransMax* is not configured by network, UE is not allowed to switch to four-step CBRA.

Upon receiving fallback indication corresponding to random access preamble transmitted by UE, UE may stop monitoring PDCCH addressed to Message B RNTI.

17.6 Multi-RAT DC/CA Enhancements

A number of aspects related to EN-DC/NE-DC/NN-DC and other CA improvements were not completed as part of Release 15 and were packaged in Ref. [9] in a Multi-RAT DC/CA enhancement project led by the radio protocols working group (RAN2).

We can categorize the objectives for this project in three main areas:

1. Completion of Release 15 leftovers for DC and CA
2. Early measurements and fast DC/CA setup and activation
3. Fast RLF recovery

In relation to the first category, i.e., leftovers from Release 15, the following aspects are addressed:

- Asynchronous NR-NR DC.
- Cross-carrier scheduling with different numerologies including aperiodic CSI-RS triggering with different numerologies between CSI-RS and triggering PDCCH.
- Enhanced single TX switched Uplink solution beyond what is already supported in Release 15. Note that the single TX operation is discussed in Chap. 2.

In relation to the second category of enhancements, i.e., *early measurements and fast DC/CA setup and activation*, they enable

- Early measurement support.
- Low latency CA/DC RRC configuration procedure: blind SCell/SCG resume.
- Scell dormancy which is an improvement related to CA operation, already discussed in Sec. 17.3 "UE Power Savings."

Finally, in relation to the *fast RLF recovery*, enhancements target increased robustness in case of MCG failure when SCG link is still available.

In addition, the specification support to allow CA operation with unaligned frame boundary (partial SFN alignment and slot alignment are maintained) for NR inter-band CA has been added with certain restrictions, e.g., the misalignment should be limited to ±76800Ts and signaling support for slot offset if necessary. This CA operation is discussed in detail in Chap. 2.

17.7 Mobility Enhancements

One of the IMT-2020 requirements is 0 ms mobility interruption time [10]. The ability for NR to guarantee such requirement was studied during the latter part of Release 15. In order to fulfill such requirement in NR a Release 16 project was agreed for this and other robustness enhancements in the context of mobility [11].

The objectives of this project were twofold:

1. Mobility interruption reduction enhancements (with the goal to support 0 ms interruption)
2. Mobility robustness enhancements

In this section we will present, at high level, the techniques that have been introduced in NR to address the above objectives.

17.7.1 Mobility Interruption Reduction Enhancements

The solution to achieve a mobility interruption time of 0 ms as required for IMT-2020 [10] resorts to having a dual active protocol stack (DAPS) during handover from one source cell to a target cell. In other words, the terminal would maintain source connection during handover.

DAPS implies: dual physical layer, dual L2 stack (MAC, RLC), and single PDCP with multiple security keys. Therefore, from the network side, it requires simultaneous data Rx on both source gNB and target gNB. It also requires the support of new data transmission on either the source gNB or the target gNB link. In turn, the UE continues the DL user data reception from the source gNB until releasing the source cell and continues the UL user data transmission to the source gNB until successful random access procedure to the target gNB.

Figure 17.17 illustrates the call flow associated with a DAPS handover. As can be seen from the figure, there is a period of time where the UE maintains connection with the source and the target gNB very much resembling macro diversity and soft handover in 3G systems.

Release 16 support of DAPS handover comes with certain limitations. For example, there is no support for FR2 to FR2 DAPS handover offered in Release 16.

17.7.2 Mobility Robustness Enhancements

There are three identified areas to improve the mobility robustness. All of them ultimately relate to reducing the latency for a switch of serving cell or to quickly recover from a failure:

- Conditional Pcell handover
- Conditional PScell addition/change
- Fast failure recovery – borrowing some of the improvements introduced for LTE

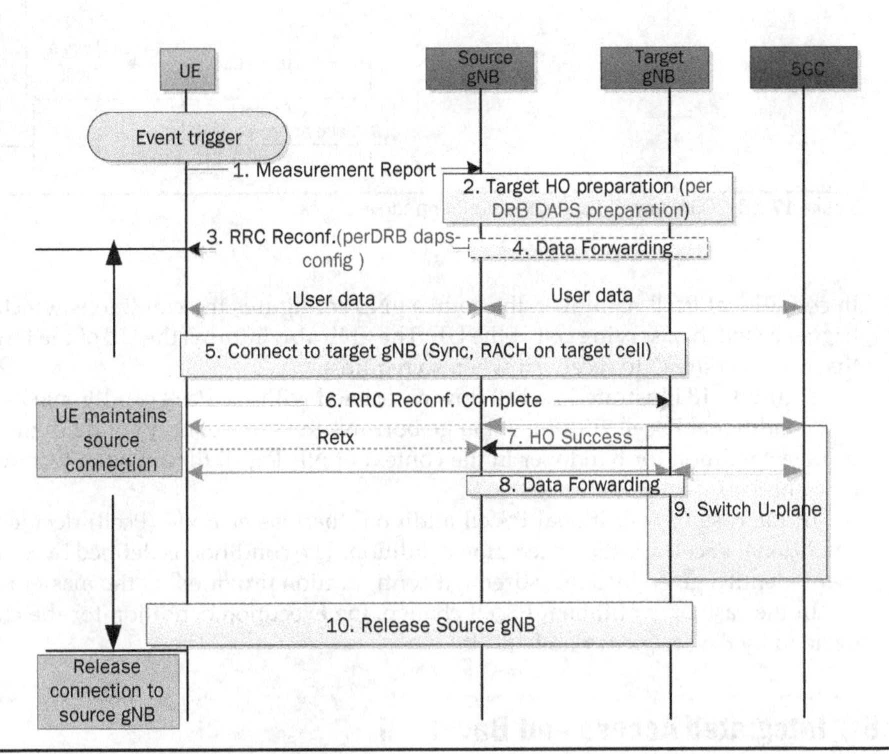

Figure 17.17 Call flow for DAPS handover.

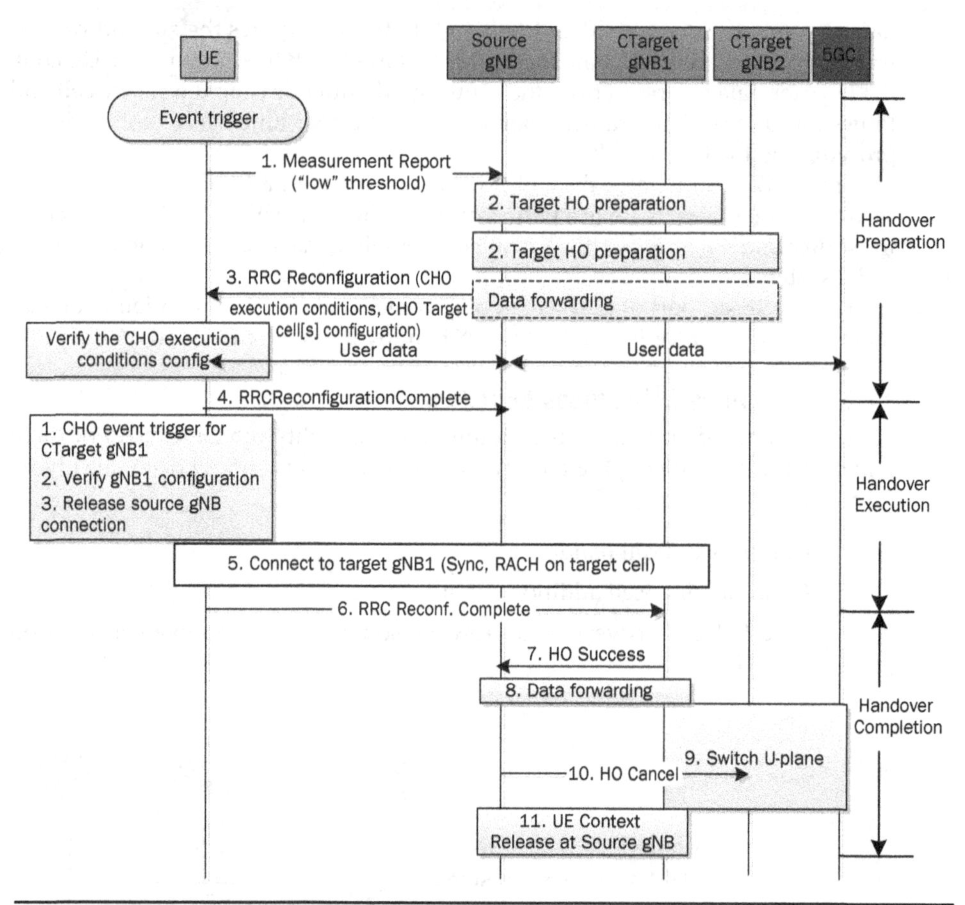

FIGURE 17.18 Call flow of Pcell conditional handover.

In conditional Pcell handover, the source gNB configures the conditions which would trigger a switch of serving cell at the UE. The gNB also informs the UE of the target cells that are "prepared" to receive it when so requires.

Figure 17.18 illustrates the call flow associated with the Pcell conditional handover.

Conditional PScell addition/change borrows the same concept of defining/configuring a condition for handover in the context of NR PScell for any architecture option (see Chap. 3).

In the case of conditional PScell addition, the master node (Pcell) decides on the conditional PScell addition execution condition. The condition is defined by a measurement identity, given by a measurement configuration provided by the master node.

In the case of conditional PScell change, the execution condition for the change is decided by the secondary node itself.

17.8 Integrated Access and Backhaul

Network densification is something that has been discussed in 3GPP for some time. Probably, it was in LTE-advanced days (Release 10) when companies started seriously looking into densifying networks with low-power nodes. The expectation was that low-power nodes would enable a much easier and more effective deployment than regular macro-cells. They would be used to offload traffic from the macro network which would be used mainly as a coverage layer of the network deployment.

The logistical issues for the deployment of small cells were underestimated. One of the key identified issues was the availability of a backhaul for that small cell. Indeed, it is not only the backhaul availability but also a sufficiently capable backhaul, which, in practice for the volume of data in today's networks, means fiber to the node.

While, the backhaul was not an issue for Home eNBs (HeNBs), as they had access to the cable, DSL, or fiber to the home, the backhaul was an issue for other types of small cells. Indeed, that justified the introduction of relays for LTE in Release 10. Back at the time, the relays adopted a Layer 3 architecture with full eNB capabilities. Interestingly, 3GPP never completed the definition of performance requirements for this new type of network nodes. LTE relays never saw the light of commercial deployments.

For NR, a study on integrated access and backhaul (IAB) was approved in March 2017 [12]. The study was carried during Release 15 and the earlier part of Release 16 with corresponding 3GPP technical report captured in Ref. [13]. IAB enables self-backhauling to remove the dependency of a wired backhaul. It is based on a Layer 2 architecture with end-to-end PDCP layer (from the donor IAB node to the UE for CP and UP).

The high-level objectives of the Release 16 work item are

- Support for FR1 and FR2 for both access and backhaul.
- In-band and out-of-band operation:
  - Out-of-band: e.g., access can be FR1 while backhaul is FR2.
  - The work item focused on in-band operation since this is more challenging case due to the half-duplex constraint and interference between access and backhaul links.
- Connectivity to EPC via EN-DC or to 5GC in SA mode.
- Multi-hop backhauling supported.
- Topology adaptation: Relays can change their backhaul attachment point during operation.
- Topological redundancy: Relays can have multiple backhaul paths to the wireline network.
- UE-bearer-specific QoS across the wireless backhaul.

And these are some of the key assumptions:

- Relays are assumed time-synchronized. IAB synchronization procedure required.
- Relays are assumed stationary, i.e., there is no optimization for physical relay mobility. This could be relaxed in future releases of IAB work.
- Release 15 UEs can connect to IAB, i.e., IAB node appears as a regular cell to the UE.

17.8.1 IAB Protocol and Architecture

IAB-related nodes are introduced in Fig. 17.19 [14] and we can define them as follows:

IAB-donor refers to a gNB that supports IAB. The IAB-donor consists of an IAB-donor-CU and one or more IAB-donor-DUs.

IAB-node refers to the wireless relay node. It holds an IAB-DU, which has gNB-DU functionality, and it holds an IAB-MT, which has UE functionality.

An illustration of the IAB topology can be found in Fig. 17.20. For more details on the generic RAN disaggregation options discussed for NR, please refer to Chap. 3.

17.8.1.1 IAB Topology

The IAB-nodes connected to an IAB-donor support a "directed acyclic graph" (DAG) topology:

- Each IAB-donor-DU and IAB-node can support multiple "child" nodes.
- Each IAB-node can have up to two "parent" nodes
- The IAB-node holds routing function on Layer 2.

A Release 16 IAB only supports NR-backhauling of NR-access traffic. Extensions to other RATs could be looked at in future releases.

A Release 15 UE can connect to the network via IAB and Release 15/16 UEs remain unaware of IAB operation.

The interfaces in the IAB architecture are illustrated in Fig. 17.21.

The backhaul user plane (UP) protocols are depicted in Fig. 17.22. Backhaul links hold RLC channel and a new *Backhaul Adaptation Protocol* (BAP) layer. RLC channels support UM and AM modes. BAP layer is used for routing across IAB-topology and carries IP layer.

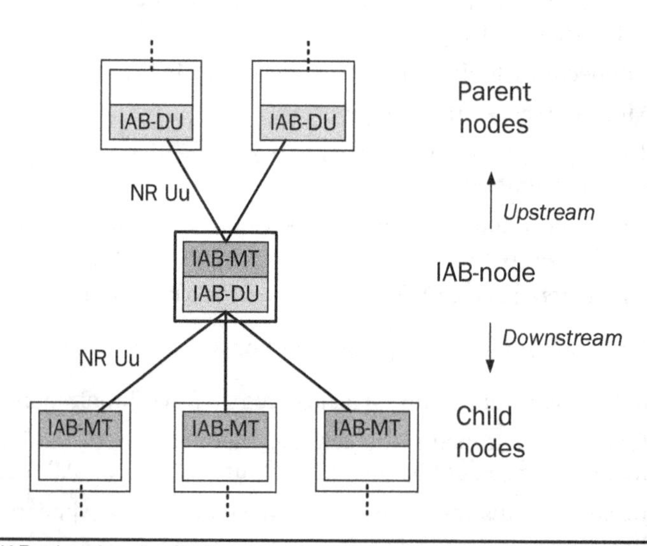

FIGURE 17.19 IAB-related nodes [14].

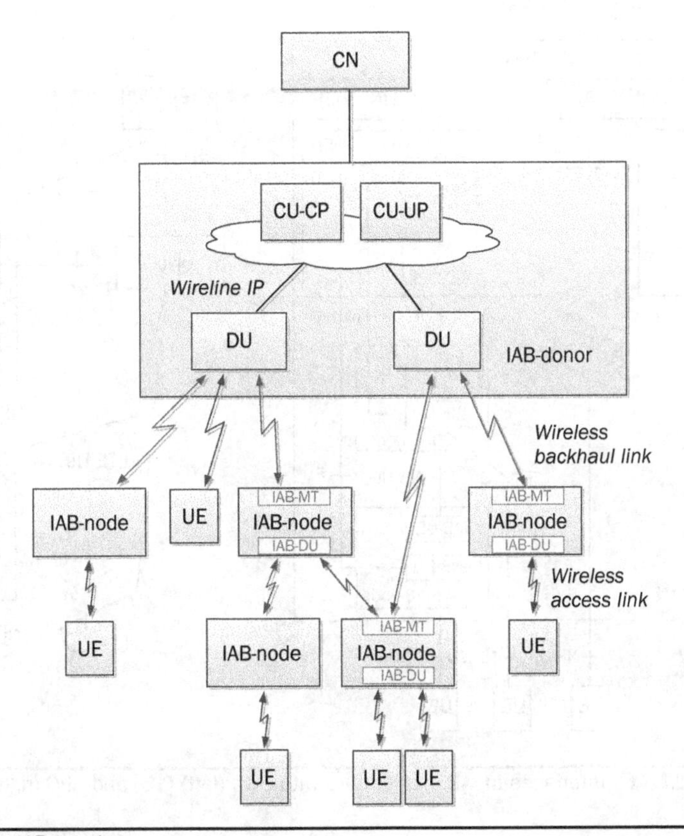

FIGURE 17.20 IAB topology.

The IAB-DU holds IP address for this IP layer, which is routable from IP layer on wireless fronthaul. The IAB-donor-DU implements an IP routing function. The IP-address management for this IP layer is performed within RAN.

The F1-U interface uses the same stack as for wireline deployment. It needs to be security-protected via IPsec using 3GPP network domain security framework.

The backhaul control plane (CP) protocols are depicted in Fig. 17.23. RLC channel and BAP layer are the same as for backhaul UP. IP layer is also the same as for backhaul UP.

F1-C also uses the same stack as for wireline deployment and needs to be security-protected via IPsec or DTLS using 3GPP network domain security framework.

The IAB node can connect to 5GC over NR ("SA mode") or connect to EPC ("NSA mode"). The IAB node function behaves as a UE toward the CN, and reuses UE procedures to connect to

- the gNB-DU on a parent IAB-node or IAB-donor for access and backhauling;
- the gNB-CU on the IAB-donor via RRC for control of the access and backhaul link;
- 5GC or EPC, e.g., AMF or MME, via NAS;
- OAM system via a PDU session or PDN connection (based on implementation).

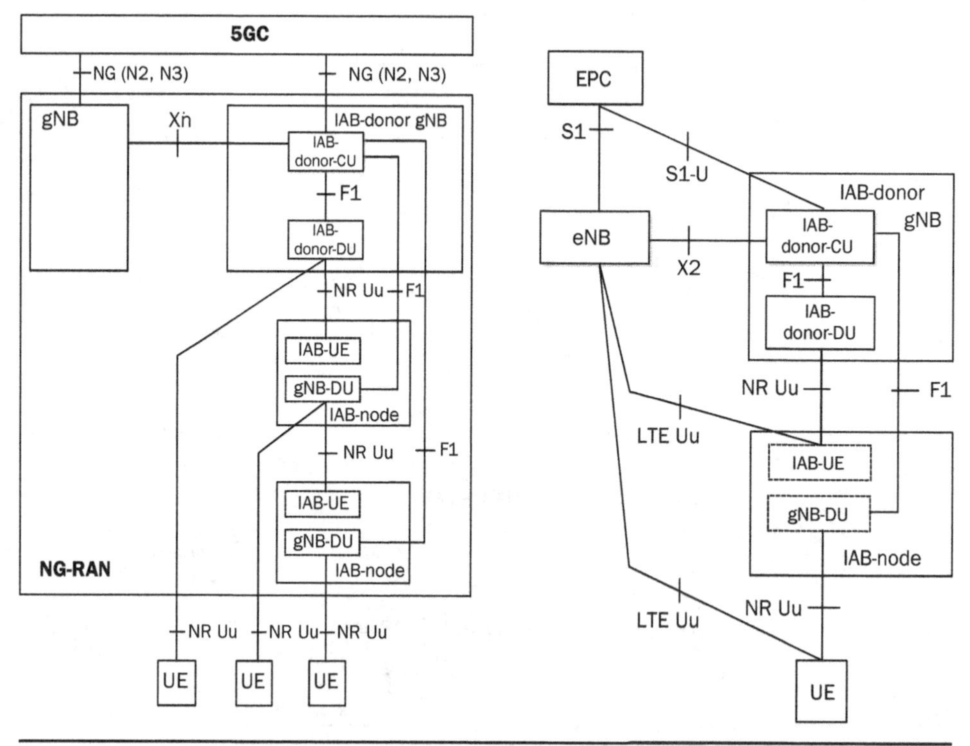

FIGURE 17.21 Interfaces in IAB architecture with 5GC (left) [15] and EPC (right) [16].

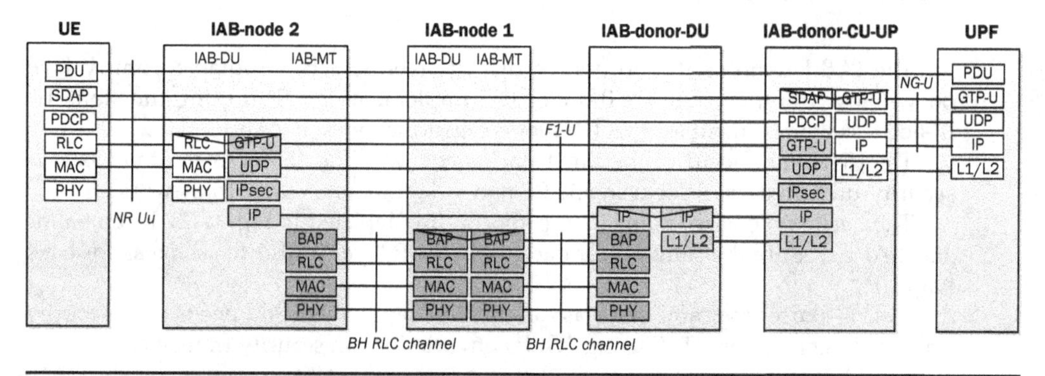

FIGURE 17.22 Backhaul U-plane protocols with 5GC.

In order to support IAB operation the registration procedure (in 5GC) or attach procedure (in EPC) is enhanced to indicate IAB-node's capability to the AMF or MME and a new UE subscription is required for IAB authorization. After registered to the 5GS or EPS, the IAB-node remains in connected state. More details for IAB system architecture and procedures can be found in Refs. [15] and [16] for 5GC and EPC, respectively.

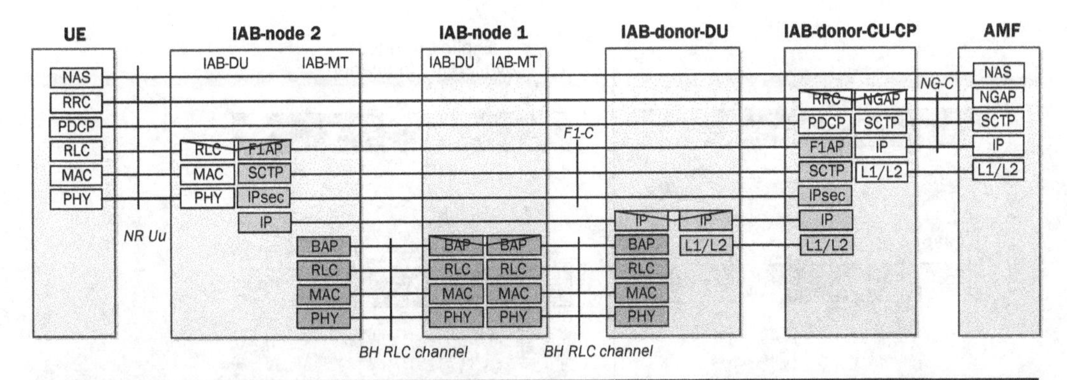

FIGURE 17.23 Backhaul C-plane protocols with 5GC.

17.8.2 IAB Physical Layer

One of the IAB characteristics that greatly modulates the design at the physical layer is the half-duplex constraint of the IAB node. In general, simultaneous operation of DU (communication with children nodes in downstream link) and MT (communication with parent node in upstream link) is not possible due to self-interference.

Release 16 assumed this half-duplex constraint in the design. As a result, the IAB-node operation is defined with TDM operation between DU and MT. There was a stretch goal to ensure forward compatibility with spatial and frequency multiplexing capabilities which would relax the TDM concern and the goal was achieved. Indeed, there is a Release 17 project for IAB [17] and we are expecting to look into solutions not resorting to the half-duplex constraint.

The parent DU controls the child MT resources via scheduling. The main issue is to coordinate resource for DUs in time domain.

DU resources can be classified to be

- Hard (**H**): The DU may use an H resource unconditionally, but it does not have to.
- Not available (**NA**): The DU can NOT use a NA resource, except if it matches an allocation for any of the following cell specific signals/channels: SSB transmission (both cell-defining SSB and non-cell-defining SSB), RACH reception, periodic CSI-RS transmission, SR reception. Note that the specifications use the term "Unavailable" instead of the one used here of "Not available" or "NA."
- Soft (**S**): The DU may use an S resource, if one of the following conditions is true:
  - Explicit indication: the parent node sends an indication to release the resource.
  - Implicit determination: the node determines that the use of the DU resource has no impact on what the MT is expected to do.
  - Same exception as for the NA case on cell-specific signals/channels.

Figure 17.24 depicts the different types of resources by way of two examples.

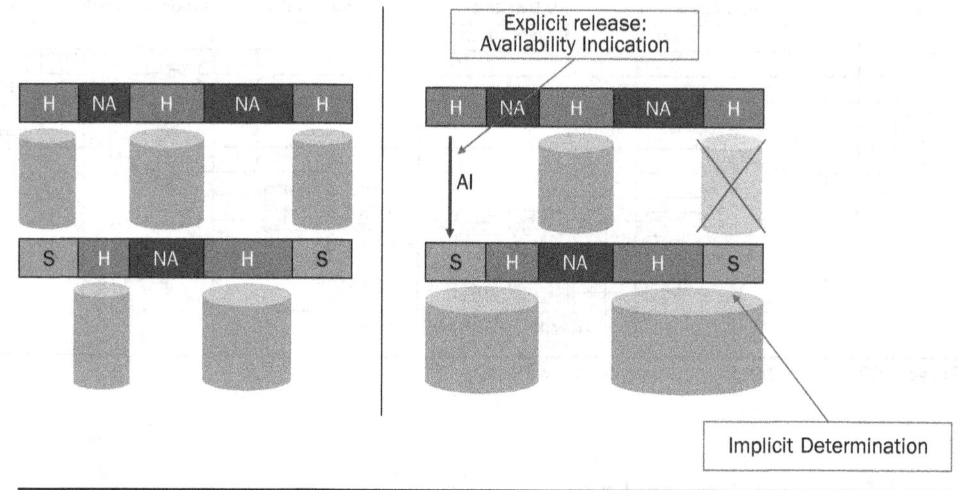

Figure 17.24 Illustration of resource designation.

17.8.3 Physical Layer Extensions for the Support of IAB

A number of physical layer extensions to support IAB operation have been identified. We present here some of them at very high level.

RACH enhancements:

- Increased RACH periodicity up to 640 ms
- Additional RACH occasions to enable time orthogonality among adjacent backhaul links and with Release 15 RACH occasions (obtained from scaling and offsetting existing Release 15 configurations)

SSB extensions for **inter-node discovery**:

- New SSBs orthogonal to cell-defining SSBs intended for Release 15 UEs.
- Half-duplex constraint forces time multiplexing of SSB transmission and reception.
    - To assure that IAB-nodes can discover/measure each other, their transmissions and reception (scan) windows should be coordinated. This coordination and pattern selection is left to the CU.
- Release 15 SMTC framework used as a baseline (see Chap. 6), some extensions introduced. Note that the Release 15 SSB mapping pattern within a half frame is reused for IAB node discovery and measurement.

Up to four independent SMTC windows can be configured per frequency per IAB node. Up to four independent SSB transmission configurations (STCs) can be configured per frequency per IAB node. The maximum SMTC periodicity is extended to 1280 ms.

OTA timing synchronization:

3GPP heavily debated multiple possible synchronization conventions. At the end, Release 16 will assume that the DL Tx slot timing is aligned at all DUs. The MT timing, in turn, follows Release 15 UE rules. Signaling of timing information from the parent node to the child node via MAC CE becomes necessary.

References

[1] 3GPP, "3rd Generation Partnership Project; Technical Specification Group Radio Access Network; Study on remote interference management for NR (Release 16)," 3GPP TR 38.866 V16.1.0, March 2019.

[2] 3GPP, "3rd Generation Partnership Project; Technical Specification Group Radio Access Network; Cross Link Interference (CLI) handling and Remote Interference Management (RIM) for NR; (Release 16)," 3GPP TR 38.828 V16.1.0, October 2019.

[3] 3GPP, "3rd Generation Partnership Project; Technical Specification Group Radio Access Network; NR; Study on User Equipment (UE) power saving in NR (Release 16)," 3GPP TR 38.840 V16.0.0, June 2019.

[4] 3GPP, "3rd Generation Partnership Project; Technical Specification Group Radio Access Network; Study on NR positioning support (Release 16)," 3GPP TR 38.855 V16.0.0, March 2019.

[5] 3GPP, "3rd Generation Partnership Project; Technical Specification Group Radio Access Network; Study on Non-Orthogonal Multiple Access (NOMA) for NR (Release 16)," 3GPP TR 38.812 V16.0.0, 3GPP TSG RAN#82, December 2018.

[6] 3GPP, "3rd Generation Partnership Project; Technical Specification Group Radio Access Network; NR; Study on Integrated Access and Backhaul; (Release 16)," 3GPP TR 38.874 V16.0.0, January 2019.

[7] 3GPP, "3rd Generation Partnership Project; Technical Specification Group Radio Access Network; NR; physical layer measurements (Release 16)," 3GPP TS 38.215 V16.0.0, January 2020.

[8] Ericsson, "Revised WID: Multi-RAT Dual-Connectivity and Carrier Aggregation enhancements," RP-192336, 3GPP TSG RAN#86, December 2019.

[9] CMCC, "Updated SID on remote interference management for NR," RP-181832, 3GPP TSG RAN#81, September 2018.

[10] Qualcomm, "Evaluation Methodology and Results for UE Power Saving," R1-1903015, 3GPP RAN1#96, April 2019.

[11] AT&T, Qualcomm, Samsung, "New SID Proposal: Study on Integrated Access and Backhaul for NR," RP-170831, 3GPP TSG RAN#75, March 2017.

[12] ITU, "Minimum requirements related to technical performance for IMT-2020 radio interface(s)," Report ITU-R M.2410-0, November 2017.

[13] Intel Corporation, "New WID: NR mobility enhancements," RP-192534, 3GPP TSG RAN#86, December 2019.

[14] Qualcomm, "New WID on Enhancements to Integrated Access and Backhaul," RP-193251, 3GPP TSG RAN#86, December 2019.

[15] 3GPP, "3rd Generation Partnership Project; Technical Specification Group System Architecture; System architecture for the 5G System (5GS) (Release 16)," 3GPP TS 23.501, V16.3.0, December 2019.

[16] 3GPP, "3rd Generation Partnership Project; Technical Specification Group System Architecture; General Packet Radio Service (GPRS) enhancements for Evolved Universal Terrestrial Radio Access Network (E-UTRAN) access (Release 16)," 3GPP TS 23.401, V16.5.0, December 2019.

[17] 3GPP, "3rd Generation Partnership Project; Technical Specification Group Radio Access Network; NR; NR and NG-RAN Overall Description; Stage 2 (Release 16)," 3GPP TS 38.300, V16.1.0, April 2020.

CHAPTER 18

A Look at Typical 5G Commercial Deployments

The early deployments of NR are varied. One obvious differentiator is NSA versus SA deployments [1]. Even with the same architecture option (e.g., NSA or SA), there are substantial differences in the spectrum type and topology among different deployments. In many cases, a gradual multistep approach can be used where the main goal of the initial phase is to provide capacity in hotspots in urban and dense-urban areas in the shortest time frame possible, while the later phases target extending coverage to less populated areas (e.g., suburban and rural). A possible strategy for the initial phase is to deploy NR either in FR1 mid- or high band or in FR2 in outdoor hotspots, and additionally in stadiums in the case of FR2. The early deployments in both cases rely on using NSA mode. The later phase is expected to reuse existing LTE cell sites and reuse the base station antennas and RF to extend coverage. This can be achieved with using low band LTE deployments and either refarm LTE channels or utilize DSS to simultaneously use LTE and NR in the same channel. This strategy can enable rapid deployment of NR and provide very good NR coverage in a short time frame. In the longer time frame, NR DC and NR CA is expected to be introduced to further increase data rate.

When considering the choice of refarmed LTE channel versus DSS, the trade-off evolves over time. Firstly, as it was mentioned earlier, DSS can be seen a transitionary phase, since once all UEs are NR UEs, there should be no motivation to maintain the LTE network. Even when the ratio of NR UEs to the total is not yet close to 100% but achieves, say, 50%, it is already reasonable to deploy the NR UEs on fully refarmed carriers without LTE UEs. However, when considering the scenario expected at the earlier deployment, the following may occur. Assume an operator has two 20 MHz carriers for which it needs to decide whether to designate as either LTE or NR. In the initial phase where there are, say, 100 LTE UEs and a single NR UE, it is obviously unappealing to designate one 20 MHz carrier as LTE for the 100 LTE UEs and one 20 MHz carrier as NR for the single NR UE. This is not only suboptimum but would also lead service quality degradation to the existing LTE users, who used to be distributed among two 20 MHz carriers but now they would have to all contend with obtaining resources in a single 20 MHz carrier. It is much more appealing, as a temporary solution, to keep using two 20 MHz carriers for LTE but start sharing one of the LTE carriers with the NR user. Once the user distribution is closer to 50 LTE UEs and 50 NR UEs, it will start to make more sense to designate one carrier as LTE and one carrier as NR.

As mentioned earlier, many of the initial NR deployments use high bands in FR1, such as 3.5 GHz, followed by 4.9 GHz, where wider spectrum is available. The higher frequency of these bands is both a challenge and opportunity. The challenge is represented by the fact that for the deployment new antenna and RF subsystem needs to be built, where the new antenna requires additional space and permitting. The opportunity is represented by the fact that the antenna element size at higher frequencies is smaller; therefore, more antenna elements can be enclosed in the same area/volume, which enables higher order beamforming and more dynamic beam adaptation, especially when coupled with reciprocity-based DL channel estimation.

Another set of challenges and opportunities is that at higher frequencies, the penetration losses increase; therefore, the coverage area that can be covered by a single cell site decreases. This requires deploying more cell sites, not necessarily co-sited with existing lower frequency deployments. On other hand, once the higher density cell sites are deployed, they provide cell-splitting gain, leading to even further system capacity gains over the existing deployments.

We note that the weaker propagation of radio waves at higher frequency compared to lower frequency is due to multiple factors but the most significant is penetration losses. The well-known f^2 factor in the free-space propagation loss formula would imply additional losses at higher frequency; however, this is compensated by the fact that in the same volume more antenna elements are used; therefore, the proportionally larger antenna gain offsets the larger free-space loss.

The fact that higher frequency cell sites and existing cell sites may not be colocated represents additional challenges. In particular, where the so-called coverage layer is in low band FR1 and the higher frequency capacity layer is in FR1 high band or in FR2, co-location in general is unlikely. In Figs. 18.1 through 18.10, various FR1-FR2 deployment scenarios are shown. Although all the examples illustrate FR1-FR2 deployments, most of them are equally applicable to the cases where FR1 is replaced with an FR1 low band and FR2 is replaced with an FR1 high band. Note that only NR carriers are shown, but the same basic scenarios can apply to NSA with the addition of LTE carriers (not shown).

Note that the deployment scenarios illustrated in Figs. 18.1 through 18.10 are not exhaustive, many other scenarios exist. Among the scenarios, Cases 1, 4, and 7, shown in Figs. 18.1, 18.4, 18.7, respectively, are less likely to deployed. On the other hand, Cases 3, 5, and 6, shown in Figs. 18.3, 18.5, and 18.6, respectively, are more likely.

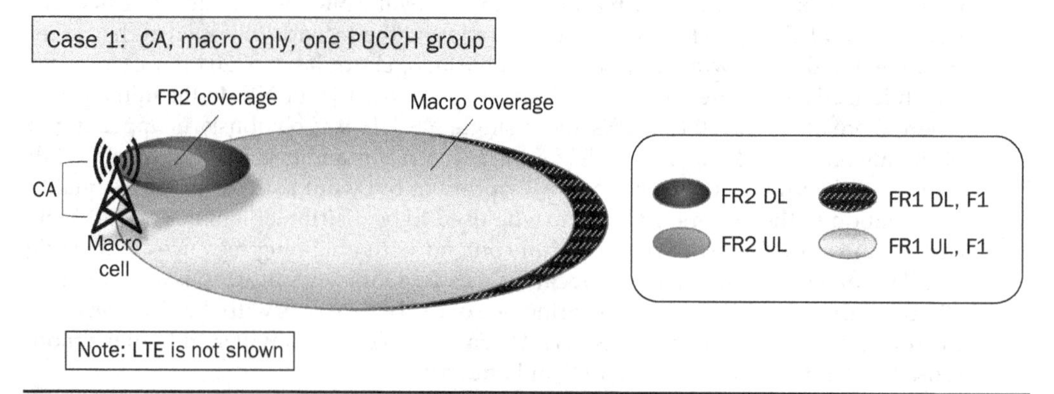

FIGURE 18.1 Release 15 FR1-FR2 deployment reusing existing sites only, CA, macro only, one PUCCH group.

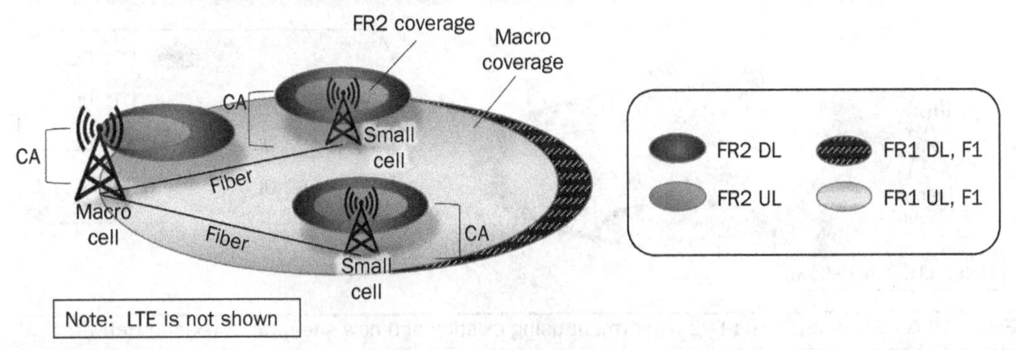

FIGURE 18.2 Release 15 FR1-FR2 deployment using existing and new sites, CA, FR2-only small cells, one PUCCH group.

Case 3: CA, FR2-only small cells, two PUCCH groups

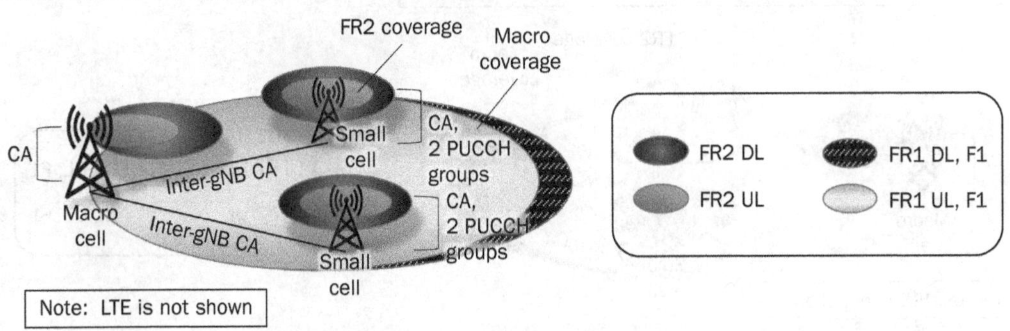

FIGURE 18.3 Release 15 FR1-FR2 deployment using existing and new sites, CA, FR2-only small cells, two PUCCH groups.

Case 4: CA + DC, FR2-only small cells

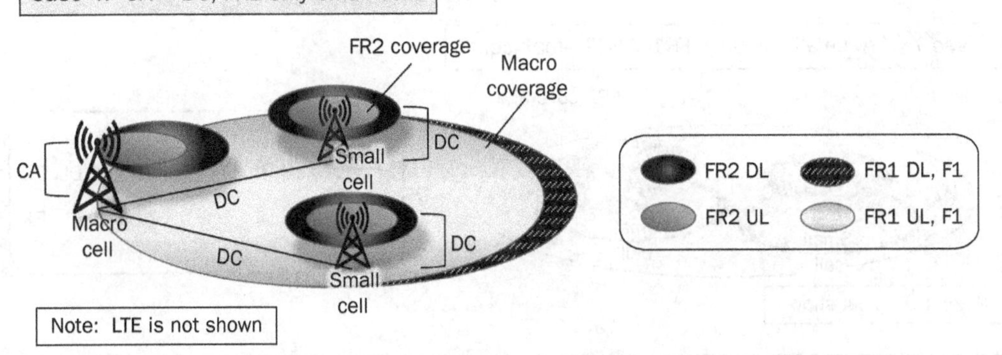

FIGURE 18.4 Release 15 FR1-FR2 deployment using existing and new sites, CA + DC, FR2-only small cells.

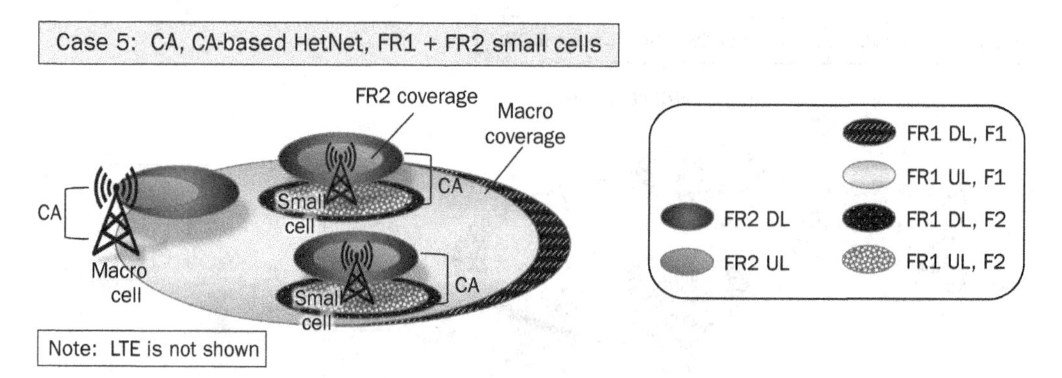

Figure 18.5 Release 15 FR1-FR2 deployment using existing and new sites, CA, CA-based HetNet, FR1 + FR2 small cells.

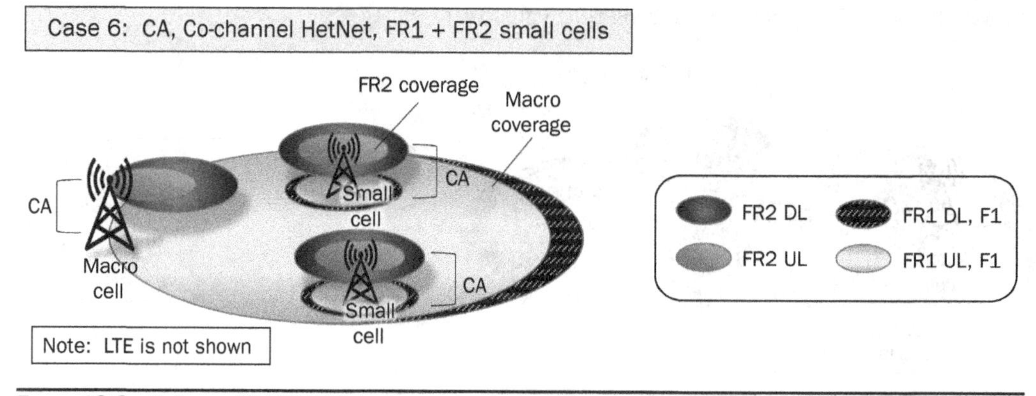

Figure 18.6 Release 15 FR1-FR2 deployment using existing and new sites, CA, Co-channel HetNet, FR1 + FR2 small cells.

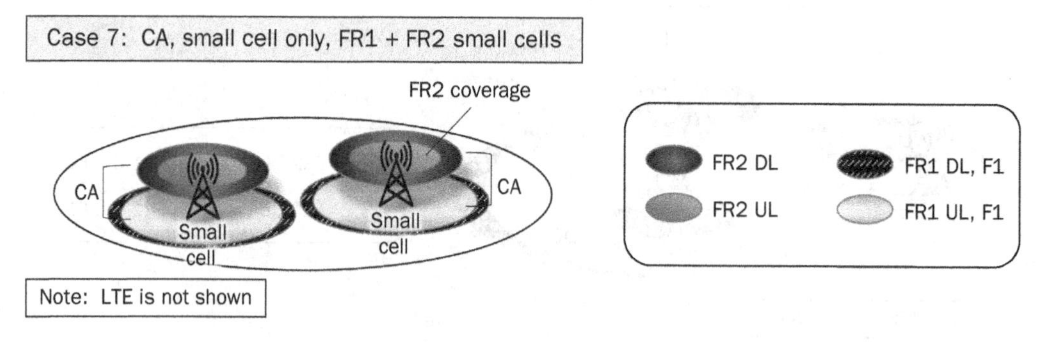

Figure 18.7 Release 15 FR1-FR2 deployment using new sites only, CA, small cell only, FR1 + FR2 small cells.

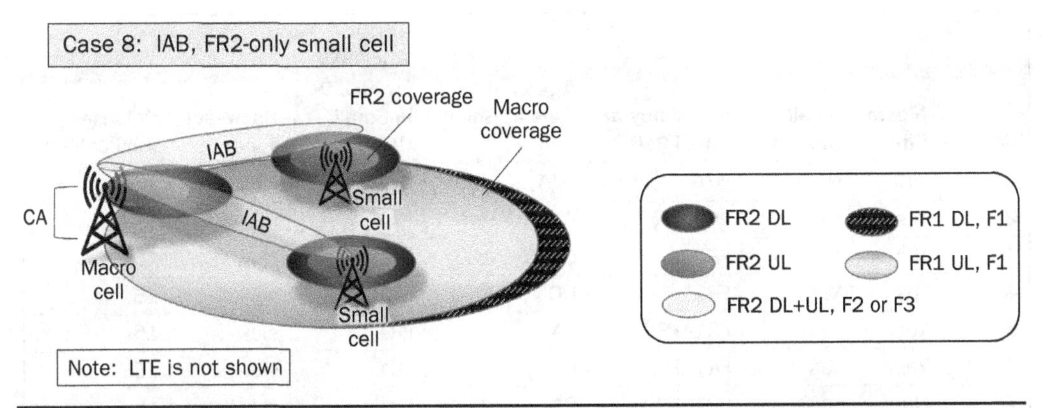

Figure 18.8 Release 16 FR1-FR2 deployment using existing and new sites, IAB backhaul, FR2-only small cells.

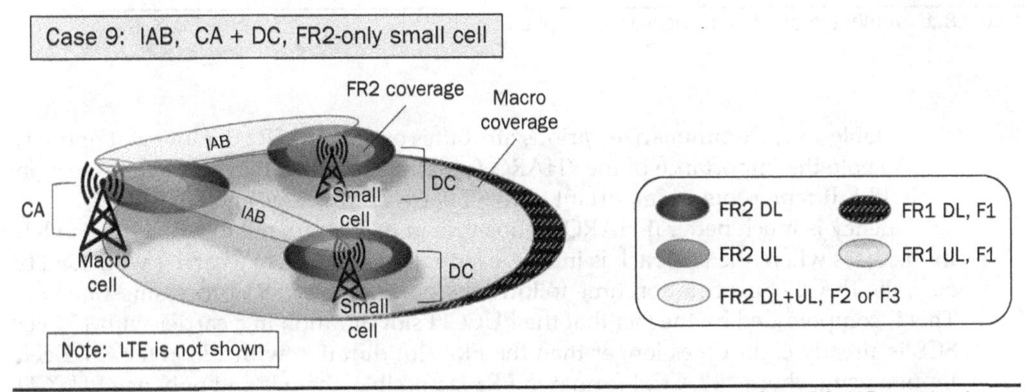

Figure 18.9 Release 16 FR1-FR2 deployment using existing and new sites, CA + DC, IAB backhaul, FR2-only small cells.

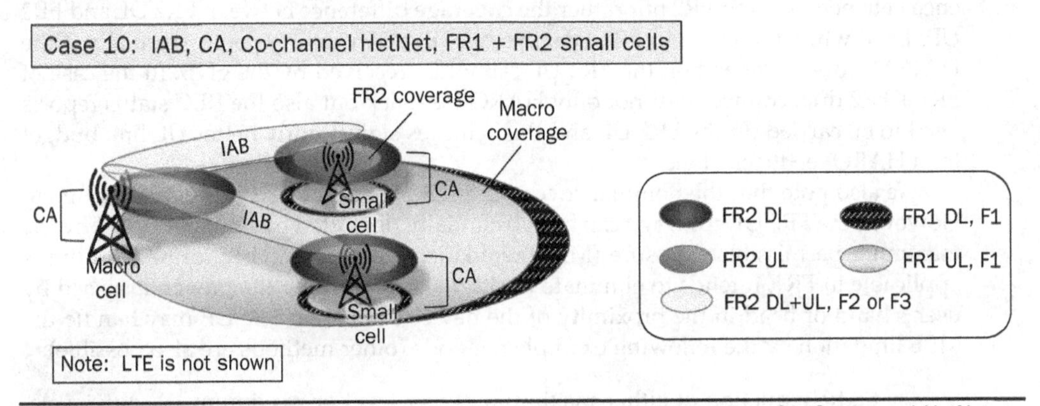

Figure 18.10 Release 16 FR1-FR2 deployment using existing and new sites, CA, Co-channel HetNet, IAB backhaul, FR1 + FR2 small cells.

Case	Macro Site	Small Cell Site	Frequency at Small Cell	FR1 and FR2 Aggregation Type at Small Cell	HARQ Feedback for FR2 DL in Small Cell	Backhaul Type	Release Applicability
1	Yes	No	N/A	N/A	N/A	N/A	15
2	Yes	Yes	FR2	CA	FR1	fast	15
3	Yes	Yes	FR2	CA	FR2	slow	15
4	Yes	Yes	FR2	DC	FR2	slow	15
5	Yes	Yes	FR1+FR2	CA	FR1	slow	15
6	Yes	Yes	FR1+FR2	CA	FR1	slow	15
7	No	Yes	FR1+FR2	CA	FR1	slow	15
8	Yes	Yes	FR2	N/A	FR2	IAB	16
9	Yes	Yes	FR2	DC	FR2	IAB	16
10	Yes	Yes	FR1+FR2	CA	FR1	IAB	16

Table 18.1 Attributes of FR1-FR2 Deployment Options

In Table 18.1, we summarize various attributes of the FR1-FR2 deployment options.

We note the importance of the "HARQ feedback for FR2 DL in small cell" entry in Table 18.1. It represents an important tradeoff between coverage and latency.

Latency is much better if HARQ feedback is in FR2 compared to when it is in FR1. In the cases where the feedback is in FR1, especially when it is in a carrier with 15 kHz SCS, the PUCCH preparation time follows the much slower FR1 processing timeline. This is compounded by the fact that the PUCCH slot duration in a carrier with 15 kHz SCS is already eight times longer than the FR2 slot duration with 120 kHz SCS. Even though using shorter PUCCH format in FR1 is possible, there is still only one PUCCH per slot allowed to carry HARQ ACK information in Release 15.

On the other hand, DL coverage can be much better if HARQ feedback is in FR1 compared to when it is in FR2. The reason for this is not so much the coverage difference between FR1 and FR2 but rather the coverage difference between FR2 DL and FR2 UL. Even when the UE is in FR2 DL coverage, the DL cannot operate normally if the HARQ feedback carried on the FR2 UL cannot be received by the gNB. In the case of FR1 + FR2 dual connectivity, not only HARQ feedback but also the RLC status reports need to be carried on the FR2 UL and this requires significantly better UL link budget than HARQ feedback alone.

We also note that the nominal coverage of FR2 DL and FR2 UL is actually similar. However, the FR2 UL coverage can be substantially degraded by limits imposed by the maximum permissible exposure (MPE) regulations. MPE, similarly to the SAR limits applicable in FR1, intends to eliminate health risks by limiting the power absorbed by user's hand or head in the proximity of the device's antenna. The UE may handle the MPE limitation by the following example methods (other methods are also possible):

- Static reduction of either maximum power level or maximum UL duty cycle, or both

- Dynamic reduction of either maximum power level or maximum UL duty cycle, or both, as a function of past transmission activity in a moving integration window
- Dynamic reduction of either maximum power level or maximum UL duty cycle, or both, based on proximity detection
- Dynamic reselection of UE antenna panel based on proximity detection

Further discussion of the various MPE techniques are omitted here, but we note that they can result in FR2 DL-UL coverage imbalance, and based on which technique is used, the UL coverage may change frequently.

As it was mentioned, the frequency choice for HARQ feedback has important impacts. We note that CA configurations already provide flexibility for the choice of the frequency used for HARQ feedback. FR1-FR2 CA can be operated in any of the modes showed in Fig. 18.11.

In Fig. 18.11, Cases 1, 1a, 2, and 3 are expected to be widely used, even though Case 3 may be replaced with dual connectivity. On the other hand, Cases 4 and 5 are using FR2 PCell and are more atypical. Supporting FR2 PCell is an optional UE capability.

To help the further discussion, we will refer to UE locations A, B, and C depicted in Fig. 18.12.

The UE locations are associated with the following UE coverage states

- A: In FR2 DL and in FR2 UL coverage
- B: In FR2 DL coverage, out of FR2 UL coverage; if FR1 exists, in FR1 DL and in FR1 UL coverage
- C: Out of FR2 DL coverage, out of FR2 UL coverage; if FR1 exists, in FR1 DL and in FR1 UL coverage

Since UEs are mobile, there needs to be additional consideration of how to handle the cases when the UE transitions between locations A and B or B and C. Transitioning

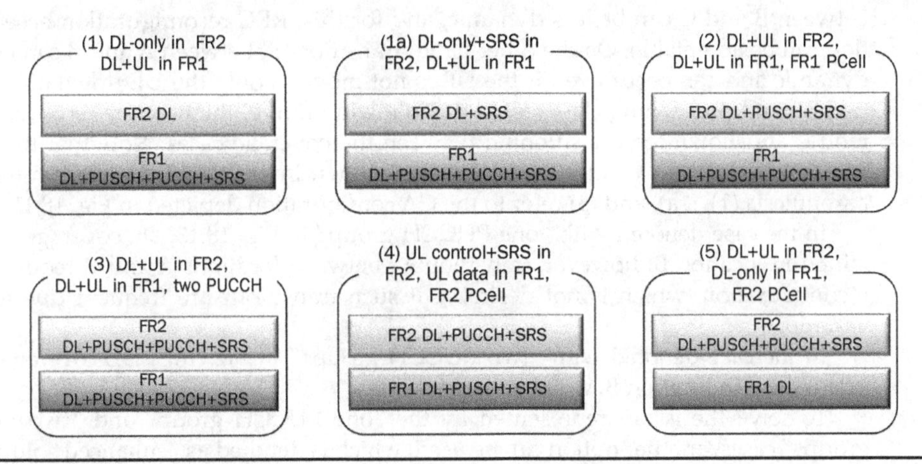

Figure 18.11 Release 15 CA channel configuration cases.

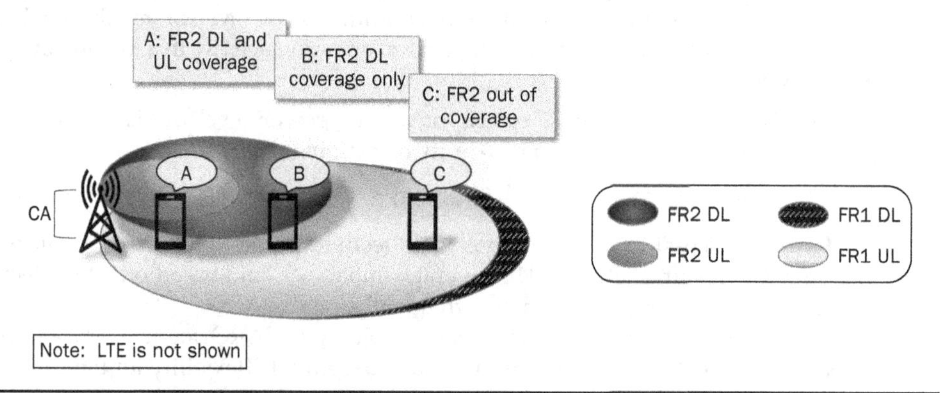

Figure 18.12 FR2 coverage cases.

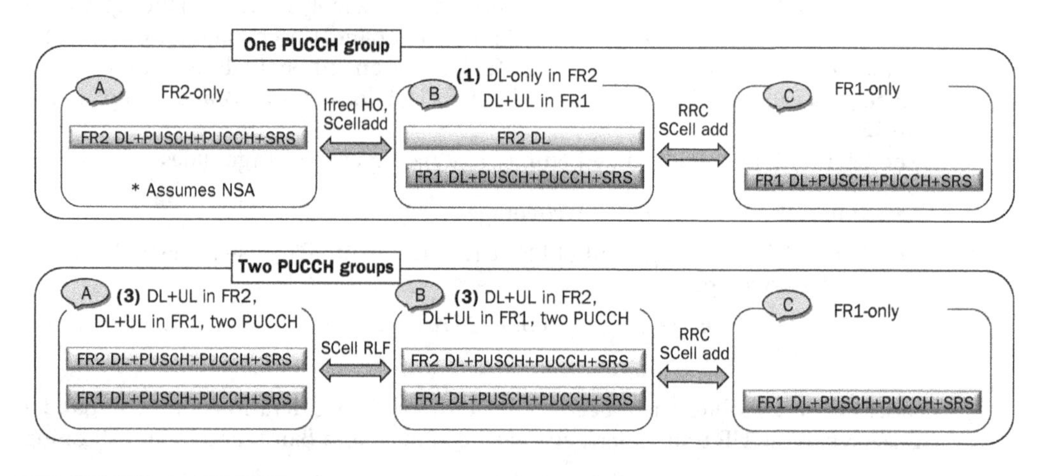

Figure 18.13 Baseline reconfiguration for the different coverage cases.

between B and C can be less dynamic, and for this, RRC reconfiguration-based solution can be acceptable. On the other hand, transitioning between A and B can be more dynamic and can occur even if the UE is not moving, only the UE orientation or the user's hand positioning changes. In the following figure, the baseline reconfiguration modes are shown for transitioning between the coverage cases. Note that the coverage cases A, B, and C refer to the location shown in Fig. 18.12. The case number in parenthesis (1), (1a), and (3) refer to the CA configuration depicted in Fig. 18.11.

In the case denoted with "one PUCCH group" in Fig. 18.13, DL coverage is maintained in location B; however, transitioning between locations A and B requires RRC reconfiguration, which is not desirable if such transitions are frequent due to MPE variations.

In the case denoted with "two PUCCH groups" in Fig. 18.3, DL coverage is not maintained in location B, which is a downside.

To solve the issues represented by the "one PUCCH group" and "two PUCCH groups" cases, another option can be used, which is denoted as "enhanced solution" in Fig. 18.14.

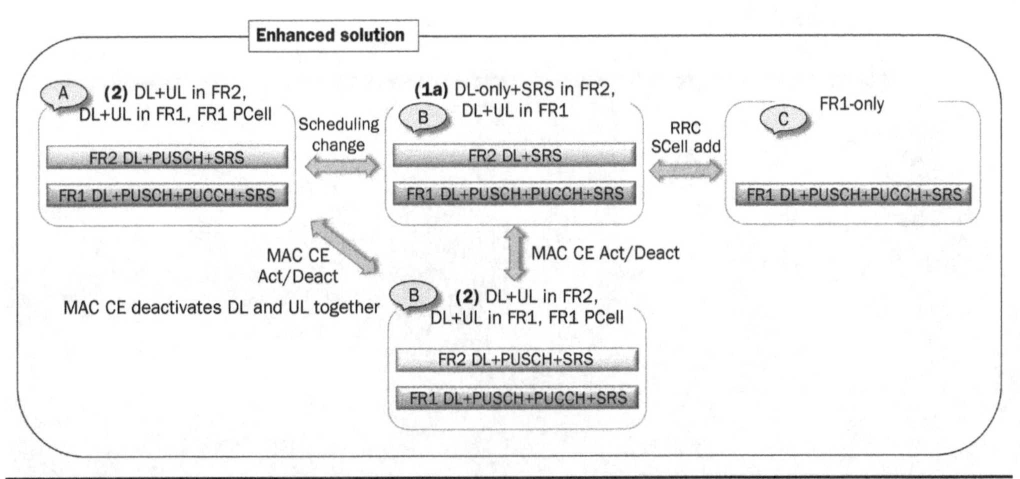

FIGURE 18.14 Enhanced reconfiguration for the different coverage cases.

The solution shown in Fig. 18.14 also assumes one PUCCH group. When transitioning between locations A and B, either the FR2 SCell is deactivated when there is no need for NR DL data, or simply the FR2 PUSCH scheduling is stopped. This transitioning can be very dynamic in response to MPE or FR2 power headroom changes.

Reference

[1] 3GPP, "3rd Generation Partnership Project; Technical Specification Group Radio Access Network; NR; NR and NG-RAN Overall Description; Stage 2 (Release 15)," 38.300, v15.8.0, December 2020.

CHAPTER 19

5G: What's Next?

19.1 Overview

At the time of the publishing of this book, the 3GPP package for Release17 had just been approved at the RAN Plenary [1] and SA Plenary [2] meetings of December 2019. This will constitute the third release of 5G NR and 5GC and was originally set for a duration of 15 months with the corresponding ASN.1 (Abstract Syntax Notation One) freeze scheduled for September 2021.

The cancelation of 3GPP face-to-face meetings throughout 2020 due to the COVID-19 outbreak has inevitably shifted the schedule of Release 17 as shown in Fig. 19.1 [3].

Similar to Release 16, the latest 3GPP release for 5G NR will continue making improvements for the smartphone industry (eMBB enhancements) and will continue expanding into new verticals. Some of the projects may benefit multiple use cases, e.g., sidelink enhancements will not only improve further NR-based V2X but will also enable NR-based sidelink for public safety and even commercial use cases of sidelink such as proximity discovery, direct communication between users and UE-to-network relay.

In this chapter, we provide an overview of the Release 17 projects and choose a small set to develop as they are expected to span multiple releases going forward. It is, however, difficult for us to predict what 3GPP will do beyond Release 17.

3GPP is expected to continue evolving the "5G platform" for a number of releases after which it will make sense to start looking into some new (future) requirements which would benefit from an entirely new platform, "6G."

Historically, each "G" or wireless generation spans around 10 years before the next one comes in, as discussed in the Chapter 0 "Introduction." For example, 3G in 3GPP spanned Release 99, Release 4, Release 5, Release 6, and Release 7, before LTE came in Release 8. Similarly, LTE spanned Release 8, Release 9, Release 10, Release 11, Release 12, Release 13, and Release 14, before NR came in Release 15. In both cases, there was one release where while the previous generation was being evolved, the new one was being studied. That happened in Release 7 timeframe for the 3G to 4G transition, and in Release 14 timeframe for the 4G to 5G transition. Taking the recent 4G to 5G transition as an example, we would expect 3GPP studies on 6G starting around Release 21 or 22.

The last section (Sec. 19.8) of this chapter will cover the authors' views on the future of wireless communications and what 6G could be addressing.

19.2 Radio Projects in Release 17

As discussed earlier, Release 17 continues with the fulfillment of the "5G vision." While Release 15 put forth the basis, Release 16 fanned it out into evolving the support of

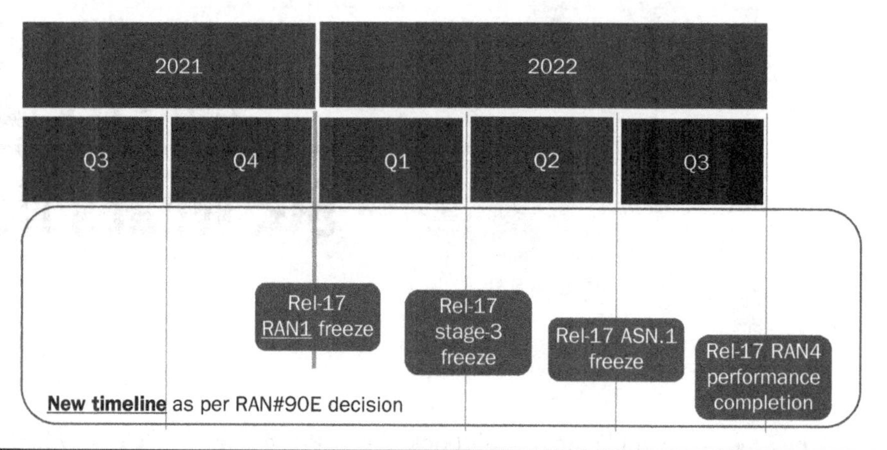

Figure 19.1 Overall timeline for Release 17 [3].

eMBB services, addressing new verticals and enabling more flexible deployments. We have covered the Release 16 scope either in dedicated chapters, including Chap. 17 on 5G miscellaneous, or embedded into other corresponding chapters.

In this section, we provide an overview of the radio Release 17 projects and give a high-level view of the objectives for each of them. This should give the reader a good sense of the problems that the different projects are trying to address and solve.

We divide the projects into the following categories: smartphone focus items, verticals, technology enablers, deployment flexibility, and other studies. The number of projects is fairly large which implies that the underlying scope of each individual item is rather focused.

Note that there is no specific item on unlicensed or shared spectrum operation. Instead, unlicensed or shared spectrum operation is being considered as a possible objective in every project and, by default, new functionalities apply to unlicensed operation, possibly without specific optimizations, unless explicitly sought. In addition, the project targeting supporting NR from 52.6 GHz to 71 GHz will provide specification support for NR in the (extended) 60 GHz unlicensed band.

Smartphone focus items:

- NR MIMO enhancements WI [4]
  - Multi-beam operation enhancements
  - Multi-TRP enhancements
  - Sounding enhancements
  - CSI measuring and reporting enhancements
- Dynamic spectrum sharing (DSS) enhancements WI [5]
  - Cross-carrier scheduling enhancements: Scell scheduling Pcell
  - Study and possibly specify a mechanism enabling a single DCI from one cell (Pcell, PScell, or Scell) to schedule PDSCH on multiple cells

- NR coverage enhancements SI [6]
  - Study coverage enhancement solutions for specific scenarios: Urban (outdoor gNB serving indoor UEs) scenario, and rural scenario (including extreme long-distance rural scenario) for FR1. Indoor scenario (indoor gNB serving indoor UEs), and urban/suburban scenario (including outdoor gNB serving outdoor UEs and outdoor gNB serving indoor UEs) for FR2. VoIP and eMBB services. These scenarios and services will be used for all the performance identifications and study of solutions in the rest of the study.
  - Identify baseline coverage performance for both DL and UL based on link-level simulations.
  - Identify the performance target for coverage enhancement and study potential coverage enhancements solutions.
- Multi-radio dual connectivity enhancements WI [7]
  - Efficient activation/de-activation mechanism for one SCG and Scells
  - Conditional PScell change/addition (scenarios not addressed in Release 16)
- Multi SIM WI [8] where UE SIMs may belong to same or different operator, and where USIM can be a physical SIM or eSIM
  - Enhancement(s) to address the collision due to reception of paging when the UE is in idle/inactive mode in both the networks associated with respective SIMs where Network A can be NR and Network B can be either LTE or NR (Applicable UE architecture: Single-Rx/Single-Tx)
  - Mechanism for UE to notify Network A of its switch from Network A (for MUSIM purpose). Network A is NR and Network B can be either LTE or NR (Applicable UE architecture: Single-Rx/Single-Tx, Dual-Rx/Single-Tx)
  - Mechanism for an incoming page to indicate to the UE whether the service is VoLTE/VoNR. Network A is either LTE or NR and Network B is either LTE or NR (Applicable UE architecture: Single-Rx/Dual-Rx/Single-Tx)
- UE power savings enhancements WI [9]
  - Power savings in RRC idle/inactive: Paging enhancements without impacting legacy UEs, possibility to use TRS/CSI-RS available in connected mode to assist idle/inactive mode UEs minimizing system overhead impact
- NR small data WI [10]
  - Enables UL small data transmission in RRC Inactive: RACH based and pre-configured PUSCH based
- NR SON/MDT enhancements WI [11]
  - Support of data collection for SON features, including capacity and coverage optimizations (CCO), inter-system inter-RAT energy saving, 2-step RACH optimization, mobility enhancement optimization, and leftovers of Release 16 SON/MDT WI (PCI selection, energy efficiency [OAM requirements], successful handovers reports, UE history information in EN-DC)

   - Support of data collection for MDT features for identified use cases, including 2-step RACH optimization, leftovers of Release 16 SON/MDT WI (MDT for MR-DC)
- NR quality of experience (QoE) SI [12]
   - Study the potential RAN side solution for supporting a generic framework for triggering, configuring, collecting measurements, and reporting for various 5G use cases
   - Study the potential interface impact and solutions (e.g., F1, NG, Xn interface) to support NR QoE functionality

Verticals:

- Sidelink enhancements WI [13]
   - Evaluation methodology update considering UE power savings [14] and LTE D2D [15] studies
   - Resource allocation enhancements for reduced power consumption and for enhanced reliability and reduced latency
   - Sidelink DRX for broadcast, multicast, and unicast
   - New sidelink frequency bands for operations (with a single carrier)
   - Mechanisms to confine sidelink operation to a predetermined geographic area for a given frequency range (non-ITS band)
- Reduced capability NR devices "NR-Light" SI [16]
   - Identify potential UE complexity reduction features
   - Power savings enhancements assuming low mobility and reduced complexity, and battery life extension assuming RRC idle and/or inactive
   - Coverage recovery mechanism from reduced complexity
   - Definition of the reduced capabilities (limited set of device types) and how to ensure those devices are only used for the intended use case
   - Capability to identify reduced capability terminals and possibility to restrict their access
- NB-IoT/eMTC enhancements WI [17]
   - NB-IoT: Add support of 16QAM for unicast in DL and UL; signaling for neighbor cell measurements and corresponding measurement triggering before RLF; NB-IoT carrier selection based on the coverage level, and associated carrier specific configuration
   - eMTC: Additional PDSCH scheduling delay for introduction of 14-HARQ processes in DL, for HD-FDD Cat M1 UEs; if found feasible, support power reduction for PRACH, PUCCH, and full-PRB PUSCH for UEs supporting sub-PRB PUSCH resource allocation
- NR for satellite access WI [18]
   - NR extensions for support of satellite communications
   - LEO and GEO with implicit compatibility to support high altitude platform station (HAPS) and air to ground (ATG) scenarios

- Assuming: FDD, UEs with GNSS capabilities (location/synchronization) and earth fixed tracking area (earth fixed and moving cells)

- NB-IoT/eMTC for satellite SI [19]
  - Identify applicable scenarios
  - Identify changes to support NB-IoT and eMTC over satellite reusing to the extent possible the conclusions of the studies performed for NR non-terrestrial network (NTN) [20]. GNSS capability in the UE is taken as a working assumption
- URLLC/IIoT enhancements WI [21]
  - Possible physical layer feedback enhancements (HARQ-ACK and CSI)
  - Identify extensions for unlicensed operation in controlled environments
  - Intra-UE multiplexing and prioritization of traffic with different priorities based on work done in Release 16
  - Enhancements for support of time synchronization (e.g., propagation delay compensation)
  - RAN enhancements based on new QoS-related parameters
- RAN slicing SI [22]
  - Study mechanisms to enable UE fast access to the cell supporting the intended slice (slice-based cell reselection, RACH configuration, access barring)
  - Study necessity and mechanisms to support service continuity (for intra-RAT handover)
- Nonpublic network enhancements for NR is subject to work in the system architecture working group (SA2). RAN impact, if any, will be assessed and would need to be carried out in a potential dedicated project (currently being discussed).

Technology enablers:

- NR up to 71 GHz SI [23] and WI [24]
  - Study item followed by corresponding work item with already agreed scope targeting extending NR operation up to 71 GHz leveraging FR2 to the extent possible
  - Required changes to NR using existing DL/UL NR waveform to support operation between 52.6 GHz and 71 GHz
  - Channel access mechanism for the 60 GHz (extended) unlicensed band
- Positioning enhancements SI [25]
  - Study solution enhancements targeting high accuracy positioning (horizontal and vertical) prioritizing enhancements to Release 16 techniques
  - Study solutions to support integrity and reliability of assistance data and position information
- NR multicast WI [26]
  - Broadcast/multicast for UEs in RRC connected: Group scheduling mechanism; dynamic service delivery between multicast and unicast with service

continuity; basic mobility with service continuity; reliability improvements, e.g., by UE feedback

  - Broadcast/multicast for UEs in RRC idle/inactive aiming at maximum commonality with RRC connected operation

Deployment flexibility:

- IAB enhancements WI [27]
  - Duplexing enhancements for resource multiplexing between child and parent link of an IAB node including simultaneous transmission/reception (with its corresponding specification impact on related procedures) and dual-connectivity scenarios for topology redundancy
  - Topology adaptation enhancements: inter-donor IAB migration with reduced service interruption, topology redundancy enhancements including CP/UP separation
  - Enhancements to improve topology-wide fairness, multi-hop latency and congestion mitigation
- NR sidelink relays SI [28]
  - Single-hop NR sidelink relay for UE-to-network and UE-to-UE relay.
  - Layer-2 and layer-3–based solutions are both in the scope of the study.
  - Specification impact for each of the solutions will be assessed for the following functions: relay selection/reselection criterion and procedure, relay and remote UE authorization, QoS for relaying functionality, service continuity, security of relayed connection, impact on UP protocol stack and CP procedure.
  - Note that Ref. [29] captures the outcome of a similar study done for LTE D2D. The recommended architecture for the layer-2 UE-to-network relay solution in Ref. [29] is taken as the starting point.
- LTE control/user plane split WI [30]
  - CP-UP separation and the interface between the CP and UP for eNB and ng-eNB (leftover from Release 16)

Other studies:

- XR evaluations over NR SI [31]
  - Confirm AR, VR, and cloud gaming applications of interest
  - Development of traffic models for those applications
  - KPIs and corresponding evaluation methodology
  - Performance characterization

In the sequel of this chapter, we choose a small set of areas to dive into some more details.

19.3 Systems Projects in Release 17

Similar to RAN in system area, 3GPP has agreed to work in a number of new projects either expanding the features provided by the 5GS in Release 16 (see Chap. 4) or even expanding and covering more use cases. The main projects approved as studies so far refer to the general architecture and will be addressed by SA WG2 of 3GPP, but is expected that further spin off projects will be approved later in time in order to cover other system aspects, e.g., related to security, multimedia services, and operation and management.

Regarding SA WG2 working projects, 3GPP approved a set of projects related also to new smartphone focus items, verticals, technology enablers, and other items.

Smartphone focus items:

- Multi SIM SI [32]
    - A mechanism for handling of mobile terminated service destined to USIM A while the UE is actively communicating with USIM B.
    - A mechanism allowing for coordinated leaving and resumption of an ongoing connection in the 3GPP system associated with USIM A, so that the UE can temporarily leave to the 3GPP system associated with USIM B, and then return to the 3GPP system associated with USIM A in a network-controlled manner.
    - A mechanism for enabling paging reception in a multi-USIM device. For this objective no E-UTRA radio interface impact is expected in RAN WGs.
    - Handling of service prioritization, i.e., the study shall determine whether the UE behavior upon reception of paging information is driven by USIM configuration or user preferences or both.

Verticals:

- Non-terrestrial networks architecture WI [33]
    - Mobility management with large coverage areas
    - Mobility management with moving coverage areas
    - Delay in satellite access
    - QoS with satellite access
    - QoS with satellite backhaul
    - RAN mobility with satellite access
    - Regulatory services with super-national satellite ground station.
- Enhancements to nonpublic networks SI [34]
    - Study enhancements to enable support for SNPN along with subscription / credentials owned by an entity separate from the SNPN
    - Study how to support UE onboarding and provisioning for nonpublic networks

  - Study enhancements to the 5GS for NPN to support service requirements for production of audio-visual content and services, e.g., for service continuity.
  - Study support for voice/IMS emergency services for SNPN
- Enhanced support of industrial IoT SI [35]
  - Enhanced support of integration with IEEE TSN. For example: provide support for Uplink synchronization via 5GS, support for multiple working clock domains connected to the UE (considering Uplink synchronization with UE as master), support for Time synchronization of UE(s) with the TSN GM attached to the UE side via 5G system
  - Enhanced support of deterministic applications including UE-to-UE time sensitive communication via the same UPF and exposure of network capability to support time sensitive communication, more specifically deterministic services and time synchronization
- Unmanned aerial vehicle (aka drone) identification and control SI [36]
  - Unmanned aerial vehicle (UAV) controller and UAV(s) identification and tracking, including studying the extent to which the 3GPP system is involved
  - UAV controller and UAV(s) authorization and authentication
  - Identifying the role of the 3GPP system, in authorization and/or authentication of UAV controller, UAV(s), UAV to controller to UAV(s) communications, and UAV to UAV communications
  - Identify the impacts on UAS operations of lack/revocation of authorization (e.g., lack of resources for use plane communications to carry UAV control messages, denied registration, etc.) while considering the need for the system to keep track of and control UAV(s)
  - Identify whether and what enhancements are needed to enable UAV(s) and a UAV controller to establish connectivity in the 3GPP system with the UAVs traffic management system for UAV operation
- Proximity services in 5G System SI [37]
  - Support for public safety and commercial-related proximity services, namely,
    - Direct discovery
    - Direct communication
    - UE-to-network relays
    - UE-to-UE relays
  - This study will consider using existing solutions as much as possible, e.g., PC5-based architecture and communications specified in R16 V2X as a basis.
- Further enhancements in architecture from advanced V2X services SI [38]
  - Enhanced support of V2X operation for pedestrian UEs (i.e., UEs for vulnerable road users), e.g., V2X communication with power efficiency.
- Enhancement of network slicing SI [39]
  - Identify the gaps in the currently defined 5GS procedures to support generic slice template (GST) parameters defined by GSMA and to study potential solutions that may address these gaps

Technology enablers:

- Architectural enhancements for 5G multicast-broadcast services SI [40]
  - Defines the framework, including the functional split between RAN and CN, to support multicast/broadcast services, e.g., ad-hoc multicast/broadcast streams, transparent IPv4/IPv6 multicast delivery, IPTV, software delivery over wireless, group communications and broadcast/multicast IoT applications, V2X applications, public safety
  - Support for different levels of services (e.g., transport only mode vs. full service mode)
  - Address whether and how relevant QoS and PCC rules are applicable to multicast/broadcast services
  - Support use cases and requirements (e.g., service continuity) for public safety, identified in SA WG1 and SA WG6 specifications
- Location services enhancements WI [41]
  - Support of service requirements for Industrial IoT for very low latency and very high accuracy positioning, including horizontal and vertical positioning service levels, 5G positioning service area
- Enhancements of support for edge computing in 5GC SI [42]
  - Forwarding some UE application traffic to the applications/contents deployed in edge computing environment
  - Improvements to 5GC support for seamless change of application server serving the UE
  - How to efficiently (with a low delay) provide local applications with information on, e.g., the expected QoS of the data path
- Enablers for network automation for 5G phase 2 SI [43]
  - UE driven analytics
  - How to ensure that a slice service level agreement (SLA) is guaranteed
  - Study of multiple NWDAF instances in one PLMN including hierarchies, roles, and inter-NWDAF instance cooperation
  - Interaction between NWDAF and AI Model & Training Service owned by the operator.

Other:

- Enhancement for advanced interactive services WI [44]
  - New standardized 5QI(s) corresponding to QoS requirements from SA WG1 for XR and AR services
    - Required latency for Uplink transmission from UE to UPF plus Downlink transmission from UPF to UE
    - Required reliability for Uplink sensor/pose data and Downlink pre-rendered/rendered audio/visual data
    - Required high data rate in Downlink direction related to SA1 agreed KPIs including FPS (frame-per-second) and resolution, etc.

- Enhancements to access traffic steering, switch, and splitting support in the 5G system architecture SI [45]
  - Investigate the following aspects for UEs that can connect to 5GC over both 3GPP and non-3GPP accesses:
    - Whether and how to support additional steering methods(s). Proposed solutions shall be based on IETF protocols or extension of such protocols (i.e., QUIC/MP-QUIC).
    - Whether and how to support multi-access PDU session with one access leg over EPC and the other access leg over non-3GPP access 5GS.

19.4 NR Expansion into Higher Frequencies

As discussed in earlier chapters, the first release of NR was defined for FR1 and FR2 spanning 410 MHz to 7.125 GHz and 24.25 to 52.6 GHz, respectively.

In Release 16, 3GPP approved study items for the range between 7.125 GHz and 24.25 GHz [46], which we call "FR3", and for 52.6 GHz to 114.25 GHz [47], which we call "FR4." The technical reports for those study items can be found at Refs. [48] and [49], respectively. Figure 19.2 shows a high-level illustration of the NR frequency ranges.

3GPP decided to take a phased approach where Release 17 will solely focus on extending NR operation up to 71 GHz without looking into new waveform choices and, hence, reusing FR2 procedures to the extent possible. This the reason why we denote the frequency range between 52.6 GHz and 71 GHz as FR2x in Fig. 19.2.

This extension will enable NR to offer operation in the extended 60 GHz band which is timely, mainly because regulations for this band are being revisited by regulators, e.g., ETSI.

In addition to the Release 16 study on FR4 mentioned above, 3GPP had already looked at the 60 GHz band in Release 14 with the outcome of the study captured in Ref. [50]. This technical report presents a nice survey of the band definition and the corresponding requirements at different regions.

Table 19.1 from Ref. [49] summarizes the spectrum allocation between 52.6 GHz and 71 GHz for the various geographical regions.

As we move to higher frequencies, propagation conditions worsen. It is expected that the decrease of the wavelengths, as we move to higher frequencies, will enable RFIC implementations with more antenna elements for a given size of an RFIC module to compensate the harsher propagation conditions. It is, however, important to have EIRP limits that would enable the increase in directionality.

The relatively low EIRP limit of the 60 GHz band (e.g., 40 dBm) makes it easy for typical RFIC implementations, already used for FR2 implementations, to reach it. As a result, over optimizing the underlying waveform was not deemed critical in comparison

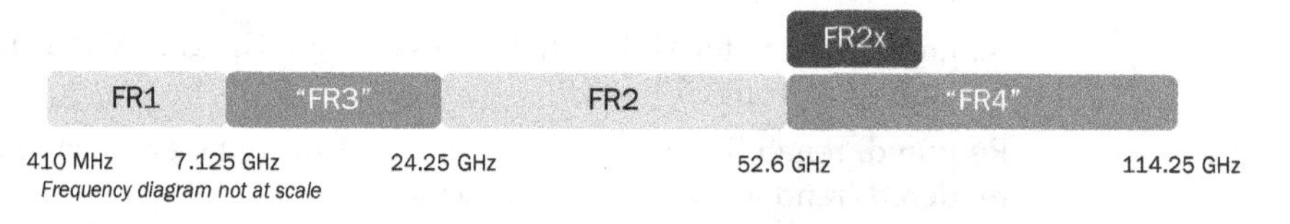

Figure 19.2 High-level illustration of NR frequency ranges.

<table>
<tr><th rowspan="2">Region</th><th rowspan="2">Country/Region</th><th colspan="11">Frequency (GHz)</th></tr>
<tr><th>52.6–54.25</th><th>54.25––55.78</th><th>55.78–56.9</th><th>56.9–57</th><th>57–58.2</th><th>58.2–59</th><th>59–59.3</th><th>59.3–64</th><th>64–65</th><th>65–66</th><th>66–71</th></tr>
<tr><td rowspan="3">ITU Region 1</td><td>Europe/CEPT</td><td></td><td></td><td></td><td></td><td colspan="7">U (Mobile)</td></tr>
<tr><td>Israel</td><td></td><td></td><td></td><td></td><td></td><td></td><td></td><td></td><td></td><td></td><td></td></tr>
<tr><td>South Africa</td><td></td><td></td><td></td><td></td><td colspan="6">U (Mobile)</td><td></td></tr>
<tr><td rowspan="4">ITU Region 2</td><td>USA</td><td></td><td></td><td></td><td></td><td colspan="7">U (Mobile)</td></tr>
<tr><td>Canada</td><td></td><td></td><td></td><td></td><td colspan="7">U (Mobile)</td></tr>
<tr><td>Brazil</td><td></td><td></td><td></td><td></td><td colspan="4">U (Mobile)</td><td></td><td></td><td></td></tr>
<tr><td>Mexico</td><td></td><td></td><td></td><td></td><td colspan="4">U (Mobile)</td><td></td><td></td><td></td></tr>
<tr><td rowspan="7">ITU Region 3</td><td>China</td><td></td><td></td><td></td><td></td><td></td><td></td><td colspan="2">U (Mobile)</td><td></td><td></td><td></td></tr>
<tr><td>Japan</td><td></td><td></td><td></td><td></td><td colspan="6">U (Mobile)</td><td></td></tr>
<tr><td>Korea</td><td></td><td></td><td></td><td></td><td colspan="6">U (Mobile)</td><td></td></tr>
<tr><td>India</td><td></td><td></td><td></td><td></td><td></td><td></td><td></td><td></td><td></td><td></td><td></td></tr>
<tr><td>Taiwan</td><td></td><td></td><td></td><td></td><td colspan="6">U (Mobile)</td><td></td></tr>
<tr><td>Singapore</td><td></td><td></td><td></td><td></td><td colspan="6">U (Mobile)</td><td></td></tr>
<tr><td>Australia</td><td></td><td></td><td></td><td></td><td colspan="7">U (Mobile)</td></tr>
</table>

NOTE:"U" stands for "Unlicensed."

TABLE 19.1 Spectrum Summary for 52.6 GHz to 71 GHz [49]

to having the ability to heavily leverage FR2 implementations which is what the FR2x concept is attempting.

As we move to higher frequencies, the phase noise of the corresponding oscillators will also worsen. OFDM waveform is inherently a block-based transmission (see Fig. 1.1) which enables one phase noise compensation per OFDM symbol. As a result, shorter OFDM symbols enable a better phase noise compensation than longer OFDM symbols. Shorter OFDM symbols correspond to higher subcarrier spacings (SCSs) and, hence, higher SCSs are better for the phase noise compensation capability viewpoint. As a result, it is expected that the FR2x study will determine the benefits of defining larger SCSs for improved resilience to phase noise at this higher frequency.

On the other hand, single-carrier waveforms enable a better tracking and correction of the phase noise as compensation within the time-duration corresponding to the equivalent OFDM block transmission is possible. Another advantage of single-carrier waveforms is its reduced peak to average power ratio (PAPR) compared to OFDM. Having a lower PAPR enables driving the power amplifiers (PAs) closer to their saturation point and, hence, improving the PA efficiency and possibly the link budget if we compare single-carrier and multi-carrier (e.g., OFDM) implementations with the same PA.

On the flip side, introducing single carrier in the DL of NR (as it is already supported for the UL) would require a major surgery of the specifications and the corresponding implementations, turning into a major drawback especially for a band where, as discussed earlier, the EIRP limit is relative low and can be easily achieved with envisioned RFIC implementations leveraging the FR2 support.

Note that channel access techniques for the 60 GHz band will need to be developed as part of the Release 17 project for operation in the 60 GHz band, where other unlicensed technologies could be deployed. Unlike a regular NR-U operation in the 5 GHz and 6 GHz band, coexistence in 60 GHz band is expected to be much easier mainly due to the narrow beam nature of the corresponding transmissions and receptions. Moreover, the relatively low transmit EIRP limit in this band also facilitates better coexistence of nodes. Nevertheless, mechanisms for spectrum sharing in the presence of narrow beam transmission and reception will have to be assessed and specified while complying with regulatory requirements. It is important for the cellular industry to play a role in the definition of the regulatory framework for coexistence in the 60 GHz band as it is being revisited. Thus, the Release 17 project came just in time for those discussions.

After Release 17, it is difficult to predict what will happen but there is a lot of spectrum between 71 GHz and 114.25 GHz with the already defined bands [47] or where we can expect regulators to identify new bands for mobile communications.

Since there is no perceived commercial urgency for deployments in frequency bands above 71 GHz, a 3GPP project assessing the advantages of changing the waveform for operation in these very high frequencies is not included in Release 17 but is expected to be in the near future.

19.5 Sidelink Beyond V2X

Chapter 15 has presented NR V2X in the context of evolving LTE V2X with increased functionality for new use cases for cellular V2X applications. As discussed in Sec. 19.2, there are two Release 17 projects related to NR sidelink operation, one for evolving sidelink operation [13] and the other to study sidelink relaying solutions [28].

As it may have become apparent from the corresponding objectives, the project on NR sidelink enhancements [13], in addition to further evolving NR V2X operation for increased reliability and reduced latency, will also look into introducing DRX in the sidelink to enable reduced power implementations. This will, in turn, enable taking the NR V2X module to handset devices to enable V2P support.

Once the V2X modules are in handset devices, a myriad of new use cases could be enabled for device-to-device (D2D) communications, not only for V2P applications, but also for other use cases, namely, public safety and commercial use cases.

Note that SA working groups (SA WG1 and SA WG2) have been looking at network controlled interactive services (NCIS) in Release 16 targeting those mentioned non-V2X sidelink use cases with more emphasis on higher data rates suited to NR radio, e.g., for VR and AR applications.

Interestingly, the story in LTE has not repeated in NR this time around. LTE started with the development of sidelink for D2D communications for proximity services (ProSe) and then migrated to sidelink for cellular V2X communications. On the other hand, NR has started its sidelink development targeting V2X applications and is leveraging it for non-V2X communications with the objective to reuse hardware implementations and to "ride" the same economy of scales.

We foresee 3GPP to spend more and more time on sidelink evolutions to augment the functionality of cellular V2X, as well as to make commercially available some of the "old ideas" of using D2D for public safety applications such as mission critical push to talk, and also to enable new use cases.

Putting this evolution path together with that of Sec. 19.2, one could easily see that sidelink operation for the 60 GHz band is becoming relevant in a not distant future in 3GPP.

19.6 Relaying Operation

Release 16 has introduced IAB, which is discussed in Chap. 17. As discussed in Sec. 19.2, there is one Release 17 project on sidelink relays looking into the pros and cons of layer 2 and layer 3 solutions. One can expect that the Release 16 IAB work will be heavily leveraged for layer 2 sidelink relays operation. On the other hand, little work would be necessary to enable a layer 3 sidelink relay operation. There could be a market for both types of solutions, which may be decided not just by 3GPP, but also by the market.

Once 3GPP defines sidelink relaying operation, applications using that new functionality could quickly emerge, providing commercial traction of the feature. Applications such as coverage extension, mesh networks, cooperative reception augmenting the MIMO channel to the aggregation of individual cooperating UEs' antennas would be possible.

Therefore, we expect this area to also span multiple releases beyond the work in Release 17 and will, for sure, bring interesting research opportunities and, possibly, subsequent new product offerings.

19.7 Edge Applications

The phenomenal increase of data rates with the advent of 5G networks brings about an interesting dichotomy: given the terminal communications and computation capabilities, where is it more efficient to carry on computations, in the terminal itself or at the edge of the network?

As explained in Chap. 3, 5GS already offers enablers for operators to deploy services closer to the edge in order to take advantage of the native low latency of NR but also potentially save in CAPEX in backhaul resources.

In Release 17, edge computing has been discussed in SA working groups in a more holistic manner:

- SA WG1: XR (and cloud gaming) use cases are outlined in the SA1 study item on NCIS [51].
- SA WG2: Work item on 5G system enhancement for advanced interactive services [44] proposes to introduce new 5QIs to identify the requirements on traffic from SA1 NCIS. In addition, the study item on architecture enhancements of support for edge computing [42] will study some additional enhancements to further optimize the use for edge computing services.
- SA WG4: XR use cases are discussed in detail in the SA4 study item XR in 5G [52].
- SA WG6: Edge computing is a network architecture to enable XR and cloud gaming and is under study in the SA6 study on application architecture for enabling edge applications [53].

RAN has approved a study in Release 17 to characterize XR and online gaming applications over NR [31].

As 5G networks get deployed and the adoption of these applications increases and becomes mainstream, we are expecting the learning from the deployments to come back to 3GPP to devise ways to improve the network ability to support these applications efficiently.

19.8 On the Path to 6G

Looking back at the original introduction of 5G in Release 15 and the first releases after that, we can see that the focus has been to complete the 5G vision to create a unified connectivity fabric for wireless communications. 5G will serve the smartphone industry and will feed verticals which may not have looked into wireless communications earlier.

The transformation of industries, which come with the 5G promises, has yet to be seen but the authors are confident that will happen big time. In that sense, 5G is expected to constitute the innovation platform for the next decade or so. During that time, we are expecting to witness incremental enhancements in technology and hardware maturity.

As happened in previous wireless generations, at some point, a new platform will be seen advantageous and 6G will then follow.

Despite the initial discussions and conferences on the 6G topics, it is too early to predict for which applications or use cases the 5G platform will no longer be adequate. It is also not clear at what point we decide to create a non-backwards compatible leap to support certain use cases or applications.

Two natural areas that come to mind as possibly disrupting are the application of machine learning (ML) and artificial intelligence (AI) to wireless communications, and communications approaching the THz range.

It is fair to say that, very likely, some RAN, CN, and UE implementations are already exploiting AI/ML techniques in their internal processing. As far as there is no need for messages to go over-the-air interface or the network interfaces, there is no role for 3GPP standardization to play. However, at the moment where inter-node collaboration or UE/network assistance is sought, 3GPP will have to get involved.

Given the complexity of the matters, one can expect a modest entrance of the AI/ML domain into 3GPP. Indeed, 3GPP is currently looking into starting a Release 17 project on AI/ML possibly involving inter-node communications and exchange of information with the addition of UE measurements and reports. How training, modeling, data sets, etc., are managed has not been subject of 3GPP discussions thus far. We can expect, however, this area to gain more interest as we traverse future releases of NR and as we get into 6G.

As discussed in Sec 19.2, 3GPP is looking into expanding NR operation into higher frequencies. While Release 17 will reach as high as 71 GHz, we can expect that, e.g., by Release 19, 3GPP may define NR operations up to 114.25 GHz. What happens after that is a big question mark and could redefine the rules of wireless communications.

Although the path to 6G is unknown at the time of writing this book, we are positive that 5G is a very powerful platform which will continue evolving on its own course for years to come. The knowledge and the lessons learned from the past five generations of wireless communications, together with the continued investment in new research and development in both industrial and academic fields, will jointly shape what 6G will turn to be.

References

[1] RAN Chairman, RAN1 Chairman, RAN2 Chairman, RAN3 Chairman, "Release 17 package for RAN," RP-193216, 3GPP RAN#86, December 2019.

[2] SA2 Chairman, "SA2 Rel-17 Prioritization," SP-191377, 3GPP SA#86, December 2019.

[3] TSG RAN Chairman, RAN1 Chairman, RAN2 Chairman, RAN3 Chairman, "Release 17 planning," RP-202868, 3GPP RAN#90e, December 2020.

[4] Samsung, "New WID: Further enhancements on MIMO for NR," RP-193133, 3GPP RAN#86, December 2019.

[5] Ericsson, "New WID on NR Dynamic Spectrum Sharing (DSS)," RP-193260, 3GPP RAN#86, December 2019.

[6] China Telecom, "New SID on NR coverage enhancement," RP-193240, 3GPP RAN#86, December 2019.

[7] Huawei, "New WID on further enhancements on Multi-Radio Dual-Connectivity," RP-193249, 3GPP RAN#86, December 2019.

[8] Vivo, China Telecom, China Unicom, "New WID: Support for Multi-SIM devices in Rel-17," RP-193263, 3GPP RAN#86, December 2019.

[9] MediaTek Inc., "New WID: UE Power Saving Enhancements," RP-193239, 3GPP RAN#86, December 2019.

[10] ZTE Corporation, "Work Item on NR small data transmissions in INACTIVE state," RP-193252, 3GPP RAN#86, December 2019.

[11] CMCC, "New WID on enhancement of data collection for SON/MDT in NR," in RP-193255, 3GPP RAN#86, December 2019.

[12] China Unicom, "New SID: Study on NR QoE management and optimizations for diverse services," RP-193256, 3GPP RAN#86, December 2019.
[13] LG Electronics, "New WID on NR sidelink enhancement," RP-193257, 3GPP RAN#86, December 2019.
[14] Ericsson, "New SID on support of reduced capability NR devices," RP-193238, 3GPP RAN#86, December 2019.
[15] Huawei, HiSilicon, "New WID on Rel-17 enhancements for NB-IoT and LTE-MTC," RP-193264, 3GPP RAN#86, December 2019.
[16] Thales, "Solutions for NR to support non-terrestrial networks (NTN)," RP-193234, 3GPP RAN#86, December 2019.
[17] MediaTek Inc., "New Study WID on NB-IoT/eTMC support for NTN," RP-193235, 3GPP RAN#86, December 2019.
[18] Nokia, Nokia Shanghai Bell, "New WID on enhanced Industrial Internet of Things (IoT) and URLLC support," RP-193233, 3GPP RAN#86, December 2019.
[19] CMCC, Verizon, "Study on enhancement of RAN Slicing," RP-193254, 3GPP RAN#86, December 2019.
[20] Intel Corporation, "New SID: Study on supporting NR from 52.6 GHz to 71 GHz," RP-193258, 3GPP RAN#86, December 2019.
[21] Qualcomm, "New WID on Extending current NR operation to 71 GHz," RP-193229, 3GPP RAN#86, December 2019.
[22] Qualcomm, "New SID on NR Positioning Enhancements," RP-193237, 3GPP RAN#86, December 2019.
[23] Huawei, "New Work Item on NR support of Multicast and Broadcast Services," RP-193248, 3GPP RAN#86, December 2019.
[24] Qualcomm, "New WID on Enhancements to Integrated Access and Backhaul," RP-193251, 3GPP RAN#86, December 2019.
[25] OPPO, "New SID: Study on NR sidelink relay," RP-193253, 3GPP RAN#86, December 2019.
[26] 3GPP, "3rd Generation Partnership Project; Technical Specification Group Radio Access Network; Study on further enhancements to LTE Device to Device (D2D), User Equipment (UE) to network relays for Internet of Things (IoT) and wearables (Release 15)," 3GPP TR 36.746 V15.1.1, April 2018.
[27] China Unicom, NTT Docomo, "New WID: Enhanced eNB(s) architecture evolution," RP-193181, 3GPP RAN#86, December 2019.
[28] Qualcomm, "New SID on XR Evaluations for NR," RP-193241, 3GPP RAN#86, December 2019.
[29] Huawei, Dish, "SID revision: Study on 7 - 24 GHz frequency range for NR," RP-192076, 3GPP RAN#85, September 2019.
[30] Intel Corporation, "Revised SID on Study on NR beyond 52.6 GHz," RP-182861, 3GPP RAN#82, December 2018.
[31] 3GPP, "3rd Generation Partnership Project; Technical Specification Group Radio Access Network; NR; 7 - 24 GHz frequency range (Release 16)," 3GPP TR 38.820 V2.0.0, March 2020.
[32] 3GPP, "3rd Generation Partnership Project; Technical Specification Group Radio Access Network; Study on requirements for NR beyond 52.6 GHz (Release 16)," 3GPP TR 38.807 V16.0.0, January 2020.
[33] 3GPP, "3rd Generation Partnership Project; Technical Specification Group Radio Access Network; NR; Study on UE Power Saving (Release 16)," 3GPP TR 38.840 V16.0.0, June 2019.

[34] 3GPP, "3rd Generation Partnership Project; Technical Specification Group Radio Access Network; Study on LTE Device to Device Proximity Services; Radio Aspects (Release 12)," 3GPP TR 36.843 V12.0.1, March 2014.
[35] 3GPP, "3rd Generation Partnership Project; Technical Specification Group Radio Access Network; Solutions for NR to support non-terrestrial networks (NTN) (Release 16)," 3GPP TR 38.821 V16.0.0, January 2020.
[36] 3GPP, "3rd Generation Partnership Project; Technical Specification Group Radio Access Network; Study on New Radio access technology; 60 GHz unlicensed spectrum (Release 14)," 3GPP TR 38.805 V14.0.0, March 2017.
[37] 3GPP, "3rd Generation Partnership Project; Technical Specification Group Services and System Aspects; Study on Network Controlled Interactive Services (Release 17)," 3GPP TR 22.842 V17.2.0, December 2019.
[38] 3GPP SA WG2, "New WID: 5G System Enhancement for Advanced Interactive Services," SP-190564, June 2019.
[39] 3GPP, "3rd Generation Partnership Project; Technical Specification Group Services and System Aspects; Extended Reality (XR) in 5G (Release 16)," 3GPP TR 26.928 V16.0.0, March 2020.
[40] 3GPP, "3rd Generation Partnership Project; Technical Specification Group Services and System Aspects; Study on application architecture for enabling Edge Applications; (Release 17)," 3GPP TR 23.758 V17.0.0, December 2019.
[41] Thales S. A., "Integration of satellite components systems in the 5G architecture," SP-191369, 3GPP SA#86, December 2019.
[42] Ericsson, "Study on enhanced support of Non-Public Networks," SP-200094, 3GPP SA#87-e, March 2020.
[43] Nokia, "Study on enhanced support of Industrial IoT," SP-200298, 3GPP SA#87-e, March 2020.
[44] Qualcomm, "Study on supporting Unmanned Aerial Systems Connectivity, Identification, and Tracking," SP-200097, 3GPP SA#87-e, March 2020.
[45] Intel, "Study on system enablers for multi-USIM devices," SP-200297, 3GPP SA#87-e, March 2020.
[46] Oppo, CATT, "Study on System enhancement for Proximity based Services in 5GS," SP-190443, 3GPP SA#84, June 2019.
[47] Huawei, "Study on architectural enhancements for 5G multicast-broadcast services," SP-200092, 3GPP SA#87-e, March 2020.
[48] LGE, "Study on architecture enhancements for 3GPP support of advanced V2X services – Phase 2," SP-190631, SA#85, September 2020.
[49] CATT, "Enhancement to the 5GC LoCation Services-Phase 2," SP-200082, SA#87-e, March 2020.
[50] Huawei, "Study on enhancement of support for Edge Computing in 5GC," SP-200093, SA#87-e, March 2020.
[51] China Mobile, Huawei, "Study on Enablers for Network Automation for 5G - phase 2," SP-200098, SA#87-e, March 2020.
[52] ZTE, " Feasibility Study on Enhancement of Network Slicing Phase 2," SP-190931, SA#85, September 2019.
[53] ZTE, "Study on Access Traffic Steering, Switch and Splitting support in the 5G system architecture Phase 2," SP-200095, SA#87-e, March 2020.

Index

Note: Page numbers followed by *f* indicate figures; by *t*, tables.

D

F

G

H

I

N

Q

R

S

T

V

W